GLENCOE
McGraw-Hill
New York, New York Columbus, Ohio Mission Hills, California Peoria, Illinois

Glencoe/McGraw-Hill

A Division of The ***McGraw-Hill*** *Companies*

Send all inquiries to:
Glencoe/McGraw-Hill
936 Eastwind Drive
Westerville, OH 43081-3329

ISBN: 0-02-825334-5

1 2 3 4 5 6 7 8 9 10 071/043 05 04 03 02 01 00 99 98 97

WHY IS ALGEBRA IMPORTANT?

Why do I need to study algebra? When am I ever going to have to use algebra in the real world?

Many people, not just algebra students, wonder why mathematics is important. ***Algebra 1*** is designed to answer those questions through **integration, applications,** and **connections.**

Geometry

Did you know that algebra and geometry are closely related? Topics from all branches of mathematics, like geometry and statistics, are integrated throughout the text.

You'll learn how to find the complement and supplement of an angle and how to find the measure of the third angle of a triangle. (Lesson 3–4*, pages 162–164)

Nutrition

I can't believe that a double cheeseburger has that much fat! Real-world uses of mathematics are presented.

The number of grams of fat in a double cheeseburger is determined by solving an open sentence. (Lesson 1–5*, page 32)

Biology

What does biology have to do with mathematics? Mathematical topics are connected to other subjects that you study.

Punnett squares, which are models that show the possible ways that genes combine, are connected to squaring a binomial. (Lesson 9–8*, page 542)

*Chapters 1–7 are contained in Volume One. Chapters 7–13 are contained in Volume Two.

Authors

WILLIAM COLLINS teaches mathematics at James Lick High School in San Jose, California. He has served as the mathematics department chairperson at James Lick and Andrew Hill High Schools. Mr. Collins received his B.A. in mathematics and philosophy from Herbert H. Lehman College in Bronx, New York, and his M.S. in mathematics education from California State University, Hayward. Mr. Collins is a member of the Association of Supervision and Curriculum Development and the National Council of Teachers of Mathematics, and is active in several professional mathematics organizations at the state level. He is also currently serving on the Teacher Advisory Panel of the *Mathematics Teacher.*

"In this era of educational reform and change, it is good to be part of a program that will set the pace for others to follow. This program integrates the ideas of the NCTM Standards with real tools for the classroom, so that algebra teachers and students can expect success every day."

GILBERT CUEVAS is a professor of mathematics education at the University of Miami in Miami, Florida. Dr. Cuevas received his B.A. in mathematics and M.Ed. and Ph.D., both in educational research, from the University of Miami. He also holds a M.A.T. in mathematics from Tulane University. Dr. Cuevas is a member of many mathematics, science, and research associations on the local, state, and national levels and has been an author and editor of several National Council of Teachers of Mathematics (NCTM) publications. He is also a frequent speaker at NCTM conferences, particularly on the topics of equity and mathematics for all students.

ALAN G. FOSTER is a former mathematics teacher and department chairperson at Addison Trail High School in Addison, Illinois. He obtained his B.S. from Illinois State University and his M.A. in mathematics from the University of Illinois. Mr. Foster is a past president of the Illinois Council of Teachers of Mathematics (ICTM) and was a recipient of the ICTM's T.E. Rine Award for Excellence in the Teaching of Mathematics. He also was a recipient of the 1987 Presidential Award for Excellence in the Teaching of Mathematics for Illinois. Mr. Foster was the chairperson of the MATHCOUNTS question writing committee in 1990 and 1991. He frequently speaks and conducts workshops on the topic of cooperative learning.

BERCHIE GORDON is the mathematics/science coordinator for the Northwest Local School District in Cincinnati, Ohio. Dr. Gordon has taught mathematics at every level from junior high school to college. She received her B.S. in mathematics from Emory University in Atlanta, Georgia, her M.A.T. in education from Northwestern University in Evanston, Illinois, and her Ph.D. in curriculum and instruction at the University of Cincinnati. Dr. Gordon has developed and conducted numerous inservice workshops in mathematics and computer applications. She has also served as a consultant for IBM, and has traveled throughout the country making presentations on graphing calculators to teacher groups.

"Using this textbook, you will learn to think mathematically for the 21st century, solve a variety of problems based on real-world applications, and learn the appropriate use of technological devices so you can use them as tools for problem solving."

Beatrice Moore-Harris is an educational specialist at the Region IV Education Service Center in Houston, Texas. She is also the Southwest Regional Director of the Benjamin Banneker Association. Ms. Moore-Harris received her B.A. from Prairie View A&M University in Prairie View, Texas. She has also done graduate work there, at Texas Southern University in Houston, Texas, and at Tarleton State University in Stephenville, Texas. Ms. Moore-Harris is a consultant for the National Council of Teachers of Mathematics (NCTM) and serves on the Editorial Board of the NCTM's *Mathematics Teaching in the Middle School.*

"This program will bring algebra to life by engaging you in motivating, challenging, and worthwhile mathematical tasks that mirror real-life situations. Opportunities to use technology, manipulatives, language, and a variety of other tools are an integral part of this program, which allows all students full access to the algebra curriculum."

James Rath has 30 years of classroom experience in teaching mathematics at every level of the high school curriculum. He is a former mathematics teacher and department chairperson at Darien High School in Darien, Connecticut. Mr. Rath earned his B.A. in philosophy from The Catholic University of America and his M.Ed. and M.A. in mathematics from Boston College. He has also been a Visiting Fellow in the mathematics department at Yale University in New Haven, Connecticut.

Dora Swart is a mathematics teacher at W.F. West High School in Chehalis, Washington. She received her B.A. in the mathematics education at Eastern Washington University in Cheney, Washington, and has done graduate work at Central Washington University in Ellensburg, Washington, and Seattle Pacific University in Seattle, Washington. Ms. Swart is a member of the National Council of Teachers of Mathematics, the Western Washington Mathematics Curriculum Leaders, and the Association of Supervision and Curriculum Development. She has developed and conducted numerous inservices and presentations to teachers in the Pacific Northwest.

"Glencoe's algebra series provides the best opportunity for you to learn algebra well. It explores mathematics through hands-on learning, technology, applications, and connections to the world around us. Mathematics can unlock the door to your success—this series is the key."

Leslie J. Winters is the former secondary mathematics specialist for the Los Angeles Unified School District and is currently supervising student teachers at California State University, Northridge. Mr. Winters received bachelor's degrees in mathematics and secondary education from Pepperdine University and the University of Dayton, and master's degrees from the University of Southern California and Boston College. He is a past president of the California Mathematics Council-Southern Section, and received the 1983 Presidential Award for Excellence in the Teaching of Mathematics and the 1988 George Polya Award for being the Outstanding Mathematics Teacher in the state of California.

Consultants, Writers, and Reviewers

Consultants

Robbie Bonneville
Consultant, 2-Volume Edition
Mathematics Coordinator
La Joya ISD
Alamo, Texas

Cindy J. Boyd
Mathematics Teacher
Abilene High School
Abilene, Texas

Gail Burrill
National Center for Research/Mathematical & Science Education
University of Wisconsin
Madison, Wisconsin

David Foster
Glencoe Author and Mathematics Consultant
Morgan Hill, California

Eva Gates
Independent Mathematics Consultant
Pearland, Texas

Joan Gell
Mathematics Department Chairman
Palos Verdes High School
Palos Verdes Estates, California

Bonnie Coulter Leech
Consultant, 2-Volume Edition
Mathematics Teacher
Marcus High School
Flower Mound, Texas

Carol Malloy
Consultant, 2-Volume Edition
Assistant Professor of Mathematics Education
University of North Carolina at Chapel Hill
Chapel Hill, North Carolina

Daniel Marks
Consultant, Real-World Applications
Associate Professor of Mathematics
Auburn University at Montgomery
Montgomery, Alabama

Melissa McClure
Consultant, Tech Prep, 2-Volume Edition
Mathematics Consultant
Teaching for Tomorrow
Fort Worth, Texas

Dr. Luis Ortiz-Franco
Consultant, Diversity
Associate Professor of Mathematics
Chapman University
Orange, California

Pamela Summers
Consultant, 2-Volume Edition
Secondary Mathematics and Science Coordinator
Lubbock ISD
Lubbock, Texas

Writers

David Foster
Writer, Investigations
Glencoe Author and Mathematics Consultant
Morgan Hill, California

Jeri Nichols-Riffle
Writer, Graphing Technology
Assistant Professor Teacher Education/Mathematics and Statistics
Wright State University
Dayton, Ohio

Reviewers

Susan J. Barr
Mathematics Department Chairperson
Dublin Coffman High School
Dublin, Ohio

William Biernbaum
Mathematics Teacher
Platteview High School
Springfield, Nebraska

John R. Blickenstaff
Mathematics Teacher
Martinsville High School
Martinsville, Indiana

Wayne Boggs
Mathematics Supervisor
Ephrata High School
Ephrata, Pennsylvania

Donald L. Boyd
Mathematics Teacher
South Charlotte Middle School
Charlotte, North Carolina

Judith B. Brigman
Mathematics Teacher
South Florence High School
Florence, South Carolina

William A. Brinkman
Mathematics Department Chairperson
Paynesville High School
Paynesville, Minnesota

Louis A. Bruno
Mathematics Department Chairman
Somerset Area Senior High School
Somerset, Pennsylvania

Luajean Nipper Bryan
Mathematics Teacher
McMinn County High School
Athens, Tennessee

Kenneth Burd, Jr.
Mathematics Teacher
Hershey Senior High School
Hershey, Pennsylvania

Todd W. Busse
Mathematics/Science Teacher
Wenatchee High School
Wenatchee, Washington

Esther Corn
Mathematics Department Chairperson
Renton High School
Renton, Washington

Kimberly C. Cox
Mathematics Teacher
Stonewall Jackson High School
Manassas, Virginia

Janis Frantzen
Mathematics Department Chairperson
McCullough High School
The Woodlands, Texas

Vicki Fugleberg
Mathematics Teacher
May-Port CG School
Mayville, North Dakota

Sabine Goetz
Mathematics Teacher
Hewitt-Trussville Junior High
Trussville, Alabama

Dee Dee Hays
Mathematics Teacher
Henry Clay High School
Lexington, Kentucky

Ralph Jacques
Mathematics Department Chairperson
Biddeford High School
Biddeford, Maine

Dixie T. Johnson
Mathematics Department Chairperson
Paul Blazer High School
Ashland, Kentucky

Jane Housman Jones
Mathematics/Technology Teacher
du Pont Manual High School
Louisville, Kentucky

Rebecca A. Luna
Mathematics Teacher
Austin Junior High School
San Juan, Texas

Sandra A. Nagy
Mathematics Resource Teacher
Mesa Schools
Mesa, Arizona

Linda Palmer
Mathematics Teacher
Clark High School
Plano, Texas

Cynthia H. Ray
Mathematics Teacher
Ravenswood High School
Ravenswood, West Virginia

Hazel Russell
Mathematics Teacher
Riverside Middle School
Fort Worth, Texas

Bernita Jones
Mathematics Department Chairperson
Turner High School
Kansas City, Kansas

Nancy W. Kinard
Mathematics Teacher
Palm Beach Gardens High School
Palm Beach Gardens, Florida

Frances J. MacDonald
Area Mathematics Coordinator, K–12
Pompton Lakes School District
Pompton Lakes, New Jersey

Roger O'Brien
Mathematics Supervisor
Polk County Schools
Bartow, Florida

Deborah Patonai Phillips
Mathematics Department Head
St. Vincent-St. Mary High School
Akron, Ohio

Carrol W. Rich
Mathematics Teacher
High Point Central
High Point, North Carolina

Judith Ryder
Mathematics Teacher
South Charlotte Middle School
Charlotte, North Carolina

Cheryl M. Jones
Mathematics Department Coordinator
Winnetonka High School
Kansas City, Missouri

Margaret M. Kirk
Mathematics Teacher
Welborn Middle School
High Point, North Carolina

George R. Murphy
Mathematics Department Chairperson
Pine-Richland High School
Gibsonia, Pennsylvania

Barbara S. Owens
Mathematics Teacher
Davidson Middle School
Davidson, North Carolina

Kristine Powell
Mathematics Teacher
Lehi Junior High School
Lehi, Utah

Mary Richards
Mathematics Teacher
Clyde A. Erwin High School
Asheville, North Carolina

Donna Schmitt
Reviewer, Technology
Mathematics and Computer Science Teacher
Dubuque Senior High School
Dubuque, Iowa

Mabel S. Sechrist
Mathematics Teacher
Mineral Springs Middle School
Winston-Salem, North Carolina

Diane Baish Stamy
Mathematics Chairperson
Big Spring School District
Newville, Pennsylvania

Robert L. Thompson
Mathematics Teacher
Raines High School
Jacksonville, Florida

John J. Turechek
Mathematics Department Chairperson
Bunnell High School
Stratford, Connecticut

Jan S. Wessell
Mathematics Supervisor
New Hanover County Schools
Wilmington, North Carolina

Kathy Zwanzig
Middle School Algebra Resource Teacher
Jefferson County Public Schools
Louisville, Kentucky

Richard K. Sherry
Mathematics Department Chairman
Roosevelt Junior High School
Altoona, Pennsylvania

Carolyn Stanley
Mathematics Department Chairperson
Tidewater Academy
Wakefield, Virginia

Linda K. Tinga
Mathematics Teacher
Laney High School
Wilmington, North Carolina

Tonya Dessert Urbatsch
Mathematics Curriculum Coordinator
Davenport Community Schools
Davenport, Iowa

Ted Wiecek
Mathematics Department Chairperson
Wells Community Academy
Chicago, Illinois

Catherine C. Sperando
Mathematics Teacher
Shades Valley High School
Birmingham, Alabama

Richard P. Strausz
Math/Computer Coordinator
Farmington Schools
Farmington, Michigan

Kimberly Troise
Mathematics Teacher
A. C. Reynolds High School
Asheville, North Carolina

Anthony J. Weisse
Mathematics Teacher
Central High School
LaCrosse, Wisconsin

Rosalyn Zeid
Mathematics Director
Union High School
Union, New Jersey

Table of Contents

Chapters 1–7 are contained in Volume One. Chapters 7–13 are contained in Volume Two.

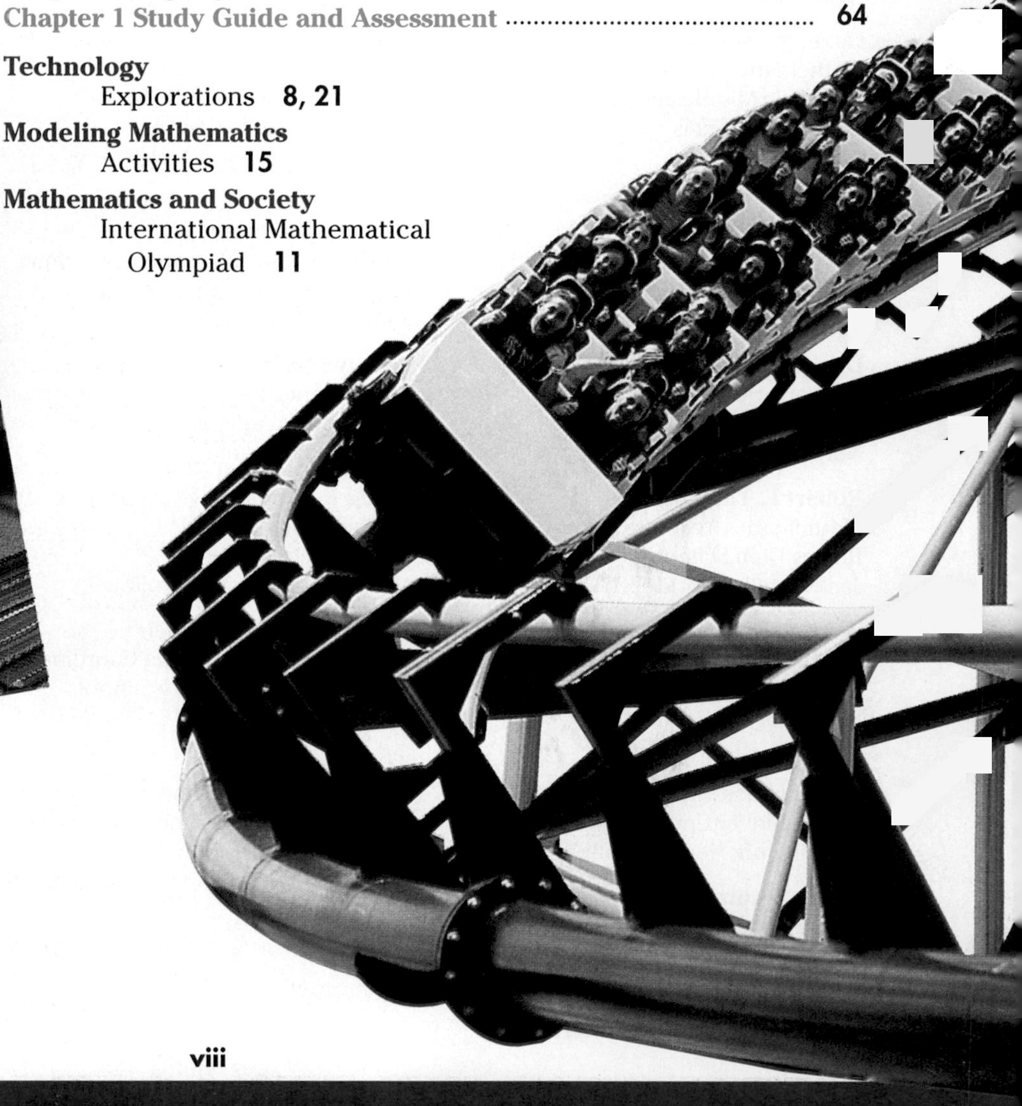

CHAPTER 2 Exploring Rational Numbers 70

CHAPTER 3 Solving Linear Equations 140

Content Integration

What does geometry have to do with algebra? Believe it or not, you can study most math topics from more than one point of view. Here are some examples.

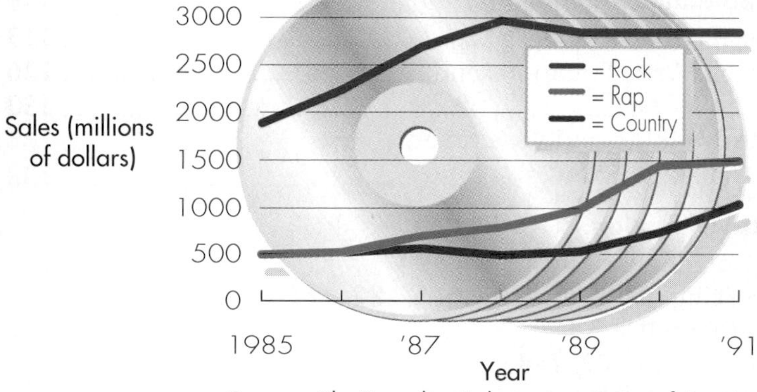

Source: The Recording Industry Association of America

◀ **Probability** You'll model the roll of two dice with relations and line plots. (Lesson 5–2*, Exercise 45)

▲ **Problem Solving** You'll interpret graphs and learn how to sketch graphs of real-world situations. (Lesson 1–9*, page 62)

LOOK BACK

You can refer to Lesson 1-7 for information on the distributive property.

Source: Lesson 9–6*, page 529

Look Back features refer you to skills and concepts that have been taught earlier in the book.

Discrete Mathematics ▶ You'll investigate patterns and sequences and use them to solve a variety of problems. (Lesson 1–2*, pages 12–18)

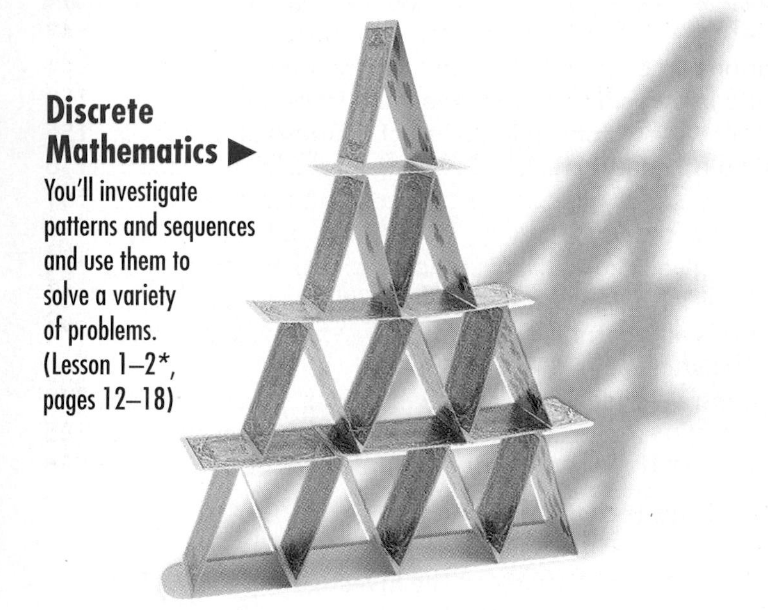

Statistics You'll learn how to display data about the speeds of the fastest animals on a line plot. (Lesson 2–2*, page 80) ▶

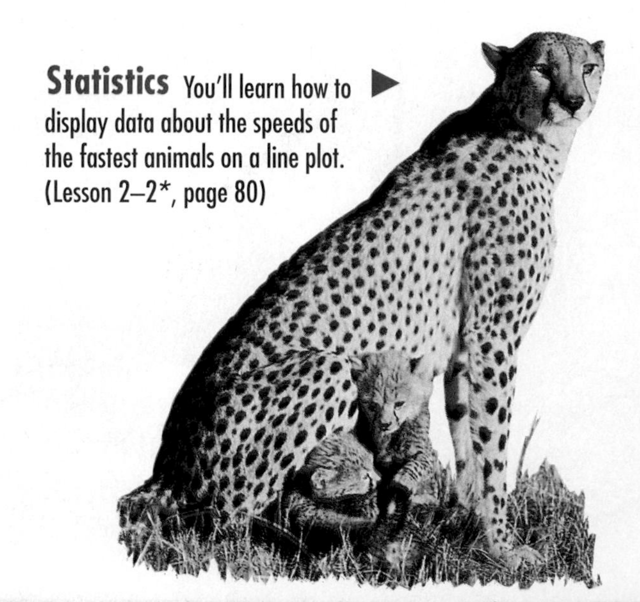

Geometry You'll use your algebra skills to find the missing measures of similar triangles. (Lesson 4–2*, pages 201–205)

*Chapters 1–7 are contained in Volume One.
Chapters 7–13 are contained in Volume Two.

CHAPTER 4 Using Proportional Reasoning 192

CHAPTER 5 Graphing Relations and Functions 252

CHAPTER 6 Analyzing Linear Equations 322

CHAPTER 7 Solving Linear Inequalities 382

Chapter B

APPLICATIONS

Real-Life Applications

Have you ever wondered if you'll ever actually use math? Every lesson in this book is designed to help show you where and when math is used in the real world. Since you'll explore many interesting topics, we believe you'll discover that math is relevant and exciting. Here are some examples.

Top Five List, FYI, and **Fabulous Firsts** contain interesting facts that enhance the applications.

Selling Prices of Paintings

1. *Portrait du Dr. Gachet* by van Gogh, $75,000,000
2. *Au Moulin de la Galette* by Renoir, $71,000,000
3. *Les Noces de Pierrette* by Picasso, $51,700,000
4. *Irises* by van Gogh, $49,000,000
5. *Yo Picasso* by Picasso, $43,500,000

Source: Lesson 9–4*, page 514

Football You'll see how punting a football is related to angle measure. (Lesson 3–4*, page 162)

Carpentry You'll study an industrial technology application that involves computing with rational numbers. (Lesson 2–7*, Exercise 47)

◀ Recreation Games from several world cultures that are similar to hopscotch will illustrate multiplying a polynomial by a monomial. (Lesson 9–6*, page 529)

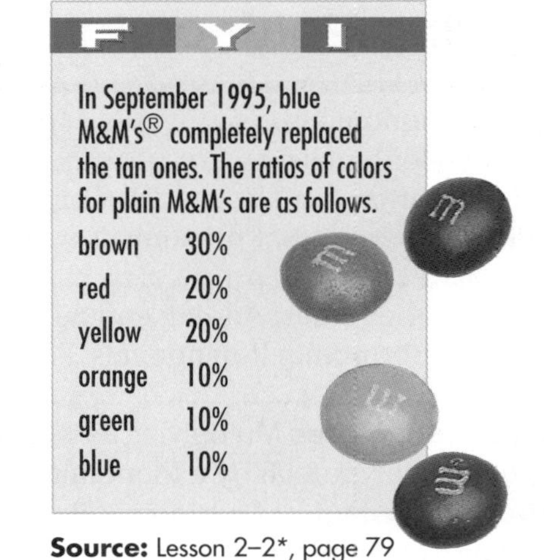

FYI

In September 1995, blue M&M's® completely replaced the tan ones. The ratios of colors for plain M&M's are as follows.

brown	30%
red	20%
yellow	20%
orange	10%
green	10%
blue	10%

Source: Lesson 2–2*, page 79

air shaft
King's chamber
Queen's chamber
escape shaft
unfinished chamber
grand gallery
ascending corridor
entrance
descending corridor

▲ World Cultures The Great Pyramid of Khufu is the setting for an application involving systems of linear equations. (Lesson 8–3*, Exercise 42)

fabulous FIRSTS

Elmer Simms Campbell (1906–1971)

Elmer Simms Campbell was the first African-American cartoonist to work for national publications. He contributed cartoons and other art work to *Esquire, Cosmopolitan,* and *Redbook,* as well as syndicated features in 145 newspapers.

Source: Lesson 8–3*, page 469

Space Science You'll discuss the extreme temperatures encountered during a walk in space as you learn about absolute value. (Lesson 2–3*, page 85)

Mathematics and SOCIETY

Teen Talk Barbies

What do Barbie dolls have to do with mathematics? Actual reprinted articles illustrate how mathematics is a part of our society. (Lesson 3–7*, page 183)

*Chapters 1–7 are contained in Volume One. Chapters 7–13 are contained in Volume Two.

CHAPTER 8 Solving Systems of Linear Equations and Inequalities 450

CHAPTER 9 Exploring Polynomials 494

CHAPTER 10 Using Factoring 556

CHAPTER 11 Exploring Quadratic and Exponential Functions 608

Interdisciplinary Connections

Did you realize that mathematics is used in biology? in history? in geography? Yes, it may be hard to believe, but mathematics is frequently connected to other subjects that you are studying.

GLOBAL CONNECTIONS

About 15,000 years ago, an important application of the lever appeared as hunters used an *atlatl,* the Aztec word for spear-thrower. This simple device was a handle that provided extra mechanical advantage by adding length to the hunter's arm. This enabled the hunter to use less motion to get greater results when hunting.

Source: Lesson 7–2*, page 392

◀ **Global Connections** features introduce you to a variety of world cultures.

Health You'll write inequalities to model target heart rates and exercise. (Lesson 7–8*, Exercise 47)

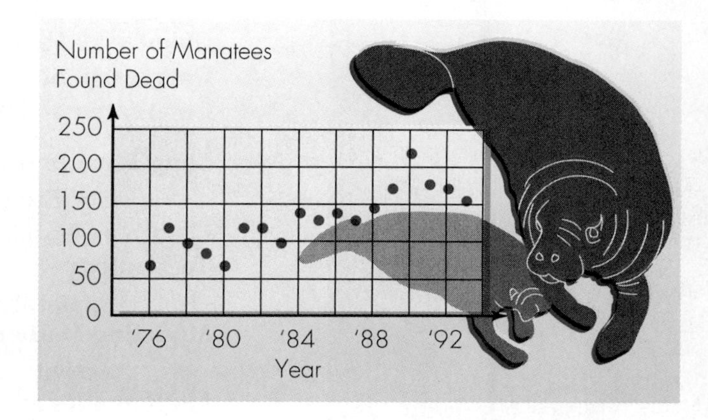

▲ **Biology** You'll study the mortality rates of Florida's manatees, an endangered species. (Lesson 5–2*, page 262)

CAREER CHOICES

A **zoologist** studies animals, their origin, life process, behavior, and diseases. They are usually identified by the animal group in which they specialize, for example herpetology (reptiles), ornithology (birds), and ichthyology (fish).

Zoology requires a minimum of a bachelor's degree. Many college positions in this field require a Ph.D. degree.

For more information, see:

Occupational Outlook Handbook, 1994–95, U.S. Department of Labor.

Source: Lesson 5–2*, page 264

◀ **Career Choices** features include information on interesting careers.

Geography In 1994, the population of New York was surpassed by Texas. This situation is modeled as a system of linear equations. (Lesson 8–2*, page 462)

▲ **Art** A Mondrian painting entitled *Composition with Red, Yellow, and Blue* is used to model polynomials. (Lesson 9–3*, page 514)

Math Journal exercises give you the opportunity to assess yourself and write about your understanding of key math concepts. (Lesson 7–1*, Exercise 6)

6. Sometimes statements we make can be translated into inequalities. For example, *In some states, you have to be at least 16 years old to have a driver's license* can be expressed as $a \geq 16$, and *Tomás cannot lift more than 72 pounds* can be translated into $w \leq 72$. Following these examples, write three statements that deal with your everyday life. Then translate each into a corresponding inequality.

*Chapters 1–7 are contained in Volume One.
Chapters 7–13 are contained in Volume Two.

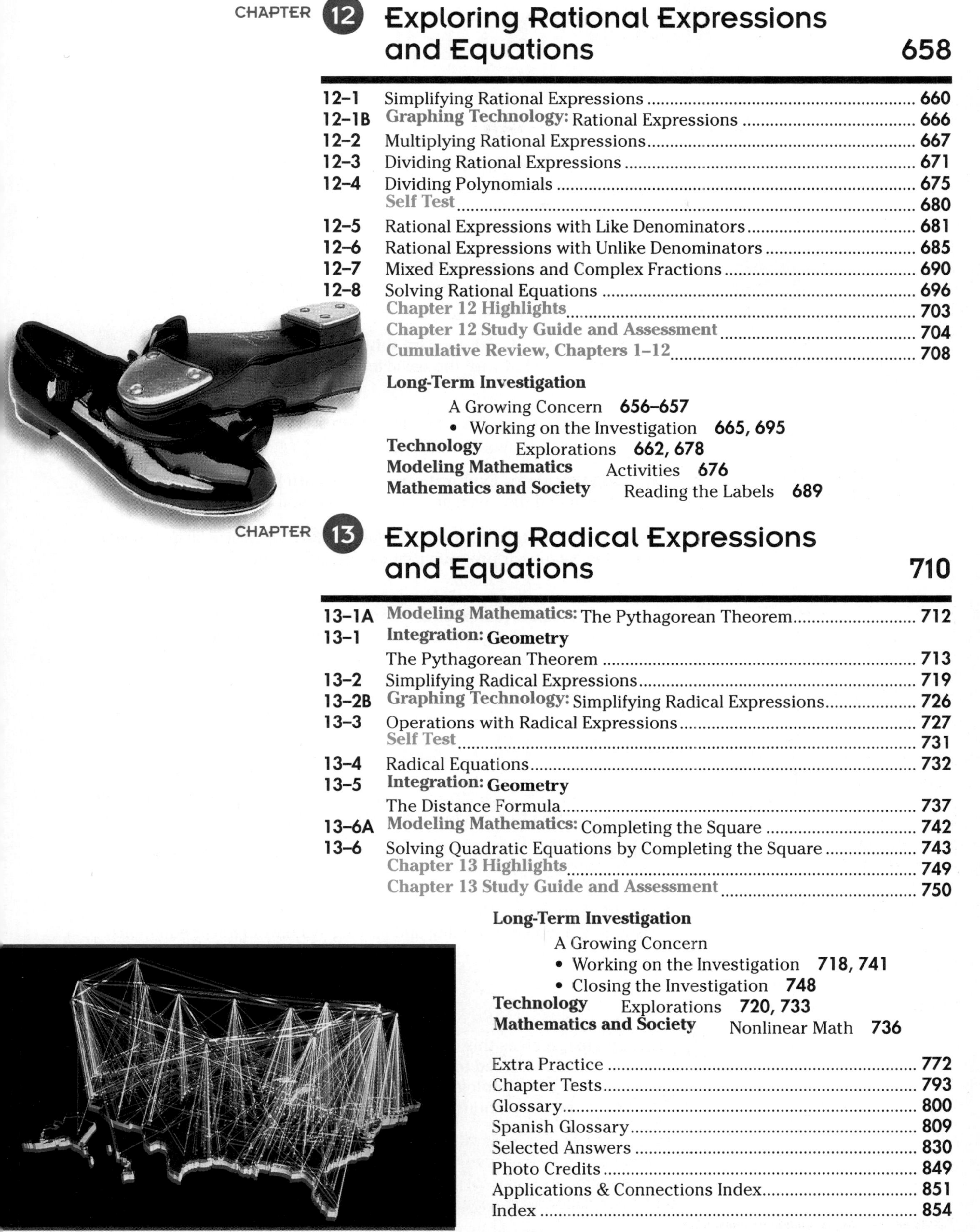

CHAPTER 12 Exploring Rational Expressions and Equations 658

CHAPTER 13 Exploring Radical Expressions and Equations 710

TECHNOLOGY

Do you know how to use computers and graphing calculators? If you do, you'll have a much better chance of being successful in today's high-tech society and workplace.

GRAPHING CALCULATORS

There are several ways in which graphing calculators are integrated.

- **Getting Acquainted with the Graphing Calculator** On pages 2–3, you'll get acquainted with the basic features and functions of a graphing calculator.
- **Graphing Technology Lessons** In Lesson 5–2A*, you'll learn how to plot points using a graphing calculator.
- **Graphing Calculator Explorations** You'll learn how to use a graphing calculator to find the mean and median in Lesson 3–7*.
- **Graphing Calculator Programs** In Lesson 3–5*, Exercise 39 includes a graphing calculator program that can be used to solve equations of the form $ax + b = cx + d$.
- **Graphing Calculator Exercises** Many exercises are designed to be solved using a graphing calculator. For example, see Exercises 40–42 in Lesson 7–8*.

COMPUTER SOFTWARE

- **Spreadsheets** On page 297 of Lesson 5–6*, a spreadsheet is used to help write an equation that models a relation.
- **BASIC Programs** The program on page 8 of Lesson 1–1* is designed to evaluate expressions.
- **Graphing Software** The graphing software Exploration on page 469 in Lesson 8–3* involves graphing and solving systems of linear equations.

Technology Tips, such as this one on page 216 of Lesson 4–4*, are designed to help you make more efficient use of technology through practical hints and suggestions.

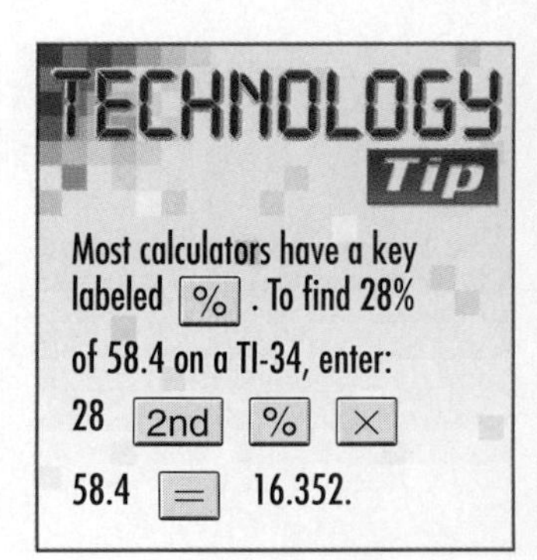

*Chapters 1–7 are contained in Volume One.
Chapters 7–13 are contained in Volume Two.

To The Student

Chapter B contains four sections: pretest, review lessons, chapter tests, and posttest. The pretest is a review of the concepts that you will need to succeed in the second half of Algebra 1. You should take the pretest to determine which concepts you need to review. The review lessons allow you to develop, and eventually master, the individual skills the pretest identified as needing to be reinforced. You should take the posttest to make sure you understand all of the concepts and to measure your progress.

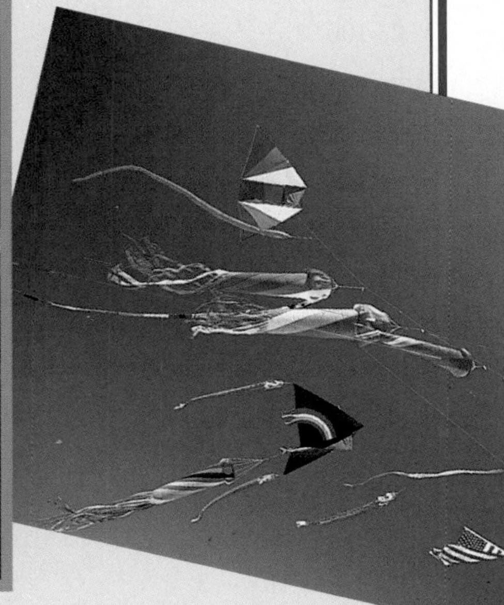

PRETEST

1. Write the expression $5 \cdot 5 \cdot m \cdot m \cdot m \cdot n \cdot n$ using exponents.

   A. $5m^3n^2$
   B. $5^2m^2n^2$
   C. $5^2m^3n^2$
   D. $5mn$

2. Name the property illustrated by $2(y + 5) = 2y + 10$.

   A. multiplicative identity property
   B. distributive property
   C. substitution property
   D. associative property of addition

3. Evaluate $3st^2 + 2s$ if $s = 5$ and $t = 3$.

   A. 145
   B. 55
   C. 100
   D. 141

4. Simplify $\frac{4^2 \div 2^3 + 11 \cdot 3}{10 - 3}$.

   A. 6
   B. $\frac{16}{287}$
   C. 5
   D. $\frac{39}{7}$

5. Simplify $-2x^2 + 3x + 5 + 4x^2 - 7x - 2$.

   A. $6x^2 - 4x + 3$
   B. $2x^2 - 10x + 3$
   C. $2x^2 + 4x - 3$
   D. $2x^2 - 4x + 3$

6. Simplify $\frac{5}{16} - \frac{2}{3}$.

   A. $-\frac{17}{48}$
   B. $\frac{17}{48}$
   C. $-\frac{3}{13}$
   D. $\frac{7}{24}$

7. Simplify $\frac{24x - 15y}{-3}$.

   A. $8x - 5y$
   B. $-8x + 5y$
   C. $-3xy$
   D. $-72x + 45y$

8. Evaluate $\sqrt{x^2 + y^2}$ for $x = 4$ and $y = 10$. Round to the nearest tenth if the result is not a whole number.

   A. 116
   B. 10.8
   C. 3.7
   D. 40

9. What is the solution of $45 = 39 - w$?

   A. 84
   B. 6
   C. -6
   D. 45

10. What is the solution of $\frac{1}{5}t = 35$?

    A. 7
    B. $\frac{1}{7}$
    C. 175
    D. 40

11. What is the solution of $5(x + 2) - 3 = 3x - 7$?

    A. -7
    B. -3
    C. 0
    D. $-\frac{7}{4}$

12. Solve $Q = \frac{13r}{5}$ for r.

    A. $r = \frac{5Q}{13}$
    B. $r = \frac{13Q}{5}$
    C. $r = \frac{Q - 5}{13}$
    D. $r = 65Q$

13. What is the solution of $\frac{5}{d} = \frac{25}{60}$?

A. 60

B. 5

C. 12

D. 275

14. What number is 16% of 160?

A. 10

B. 25.6

C. 1000

D. 185.6

15. A house built in 1982 originally cost $80,000. In 1994, the same house sold for $144,800. What was the percent of increase over the original price?

A. 45%

B. 80%

C. 81%

D. 90%

16. Jacqui Walker invested a portion of $20,000 at 5% simple interest and the balance at 8% simple interest. How much did she invest at each rate if her total annual income from both investments was $1510?

A. $5000 at 5% and $15,000 at 8%

B. $3000 at 5% and $17,000 at 8%

C. $8500 at 5% and $13,562.50 at 8%

D. $692.31 at 5% and $19,307.69 at 8%

Use the graph at right to answer Questions 17–19.

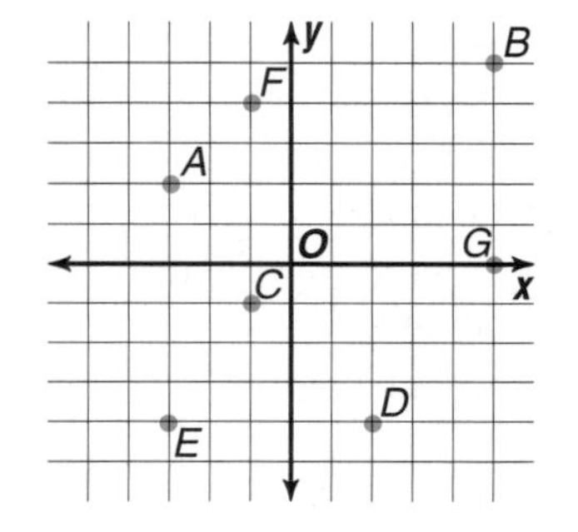

17. What is the ordered pair for point D?

A. $(2, -4)$

B. $(2, 4)$

C. $(-4, 2)$

D. $(4, -2)$

18. Name the quadrant in which point A is located.

A. I

B. II

C. III

D. IV

19. What is the range of the relation graphed?

A. $\{-3, -1, 2, 5\}$

B. $\{-4, -1, 0, 2, 4, 5\}$

C. {all the integers}

D. $\{-4, -1, 2, 4, 5\}$

20. What is the y-intercept of the graph of $2x + 3y = 15$?

A. 3

B. 5

C. 2

D. 15

21. Which equation has a graph that is a vertical line?

A. $x = 9$

B. $x + y = 0$

C. $y = -3$

D. $x - y = 0$

22. Find the slope of the line that passes through (3, 7) and (11, 19).

A. $\frac{3}{2}$

B. $-\frac{3}{2}$

C. $\frac{2}{3}$

D. $\frac{13}{7}$

23. Find an equation of the line through (0, 7) with slope 3.

A. $-3x + y = 7$

B. $3x + y = 7$

C. $-x + 3y = 7$

D. $x + 3y = -7$

24. A line that is parallel to the graph of $2x - 9y = 4$ has which of the following slopes?

A. 2

B. $\frac{2}{9}$

C. $-\frac{9}{2}$

D. $-\frac{2}{9}$

25. What are the coordinates of the midpoint of the segment whose endpoints are (7, 1) and (24, 13)?

A. (8.5, 6)

B. (15.5, 7)

C. (7, 15.5)

D. (10, 12.5)

Review 1–1

Variables and Expressions

Any letter used to represent an unspecified number is called a **variable**. You can use variables to translate verbal expressions into algebraic expressions.

Words	Symbols
4 more than a number	$y + 4$
a number decreased by 12	$b - 12$
the product of 3 and a number	$3t$
a number divided by 8	$h \div 8$ or $\frac{h}{8}$

The algebraic expression x^n represents a product in which each factor is the same. The small raised n is the exponent and it tells how many times the base, x, is used as a factor.

Example  **Evaluate 2^5.**

$2^5 = 2 \cdot 2 \cdot 2 \cdot 2 \cdot 2$ or 32

EXERCISES

Write a verbal expression for each algebraic expression.

1. $z - 1$ **2.** $\frac{2}{5}b^2$ **3.** $57 - 3q$

Write an algebraic expression for each verbal expression.

4. a number increased by 14
5. six times a number
6. 12 more than a number
7. a number divided by 4
8. the sum of a number and 17
9. 25 less than 7 times a number
10. three times the sum of a number and 13
11. 74 decreased by twice the cube of a number

Write each expression as an expression with exponents.

12. $9 \cdot 9 \cdot 9 \cdot 9$ **13.** $27 \cdot a \cdot a \cdot a$ **14.** $14(w)(w)$

Evaluate each expression.

15. 3^3 **16.** 5^2 **17.** 10^4

18. Environment According to the American Water Works Association, it takes an average of 20 gallons of water to wash dishes by hand and as much as 25 gallons of water to run a dishwasher.

a. Write an expression representing the amount of water used to wash dishes by hand for x washes.

b. Write an expression representing the amount of water needed to run a dishwasher for x washes.

c. Write an expression representing the difference between the amount of water needed to run a dishwasher and the amount of water used to wash dishes by hand for x washes.

Review 1–2

Patterns and Sequences

When solving certain problems, you must often look for a pattern. A **pattern** is a repeated design or arrangement.

Example 1 **Study the pattern at the right. Draw the next three figures in the pattern.**

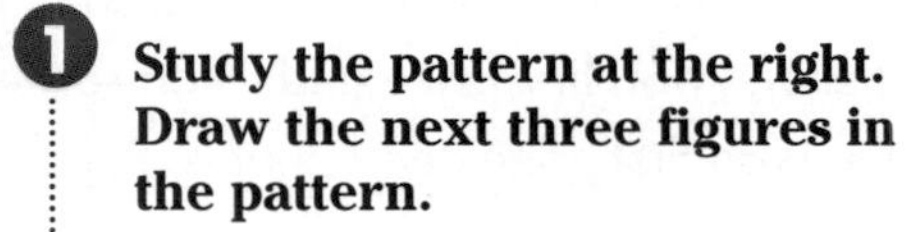

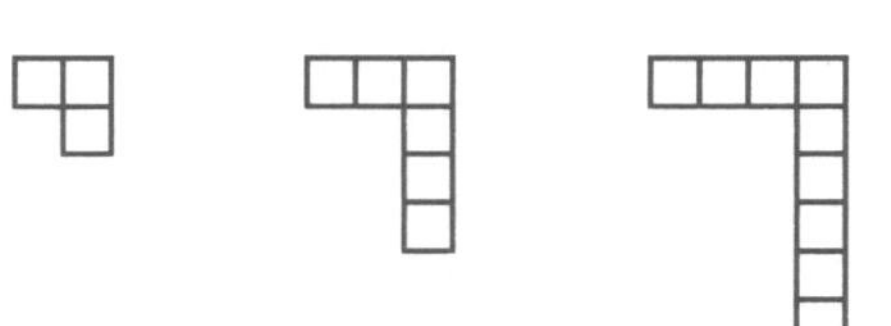

The pattern begins by adding one square to the left end of the horizontal piece and two squares to the bottom of the vertical piece, then continues in the same pattern. The next three figures are drawn below.

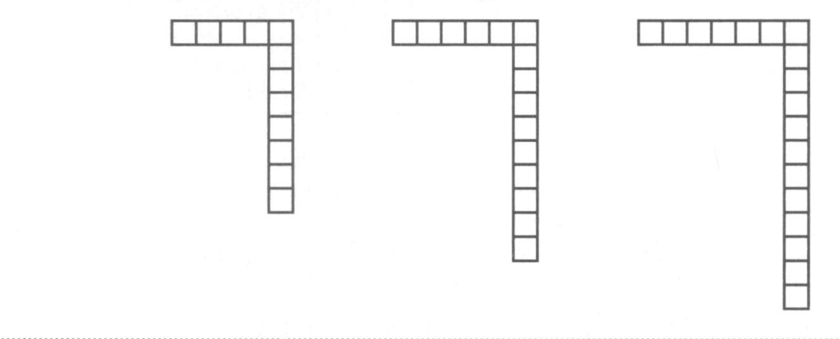

The numbers 1, 3, 5, 7, 9, and 11 form a **sequence**. A sequence is a set of numbers in a specific order. The numbers in a sequence are called **terms**.

Example 2 **Find the next three numbers in each sequence.**

a. 5, 10, 20, 40, . . . Study the pattern in the sequence. Each term is twice the term before it. The next three terms are 80, 160, and 320.

b. 729, 243, 81, . . . Study the pattern in the sequence. Each term is $\frac{1}{3}$ of the term before it. The next three terms are 27, 9, and 3.

EXERCISES

Give the next two items for each pattern.

1. 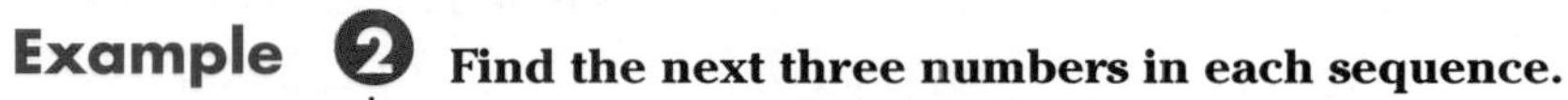

2. What does the 22nd figure in Exercise 1 look like? Explain your reasoning.
3. $w - 45, w - 40, w - 35, w - 30, \ldots$
4. 5, 7, 5, 9, 5, 11, 5, 13, 5, . . .
5. **Transportation** Keisha knows that trains arrive regularly at the train station. She has part of the afternoon schedule. When will the next train arrive at the station?

Train Schedule
Arrivals:
3:40 P.M.
4:02 P.M.
4:24 P.M.

Review 1–3

Order of Operations

Ian said Shelley won first prize and I won second prize. Without punctuation, this sentence has three possible meanings.

Ian said, "Shelley won first prize and I won second prize."
"Ian," said Shelley, "won first prize and I won second prize."
Ian said Shelley won first prize and I [that is, the speaker] won second prize.

In mathematics, in order to avoid confusion about meaning, an agreed-upon order of operations tells us whether a mathematical expression such as $45 - 15 \div 5$ means $(45 - 15) \div 5$ or $45 - (15 \div 5)$. That order is shown below.

Order of Operations
1. Simplify expressions inside grouping symbols. 2. Evaluate all powers. 3. Do all multiplications and divisions from left to right. 4. Do all additions and subtractions from left to right.

You can evaluate an algebraic expression when the value of each variable is known. Replace each variable with its value and then use the order of operations to perform the indicated operations. Remember to do all operations within grouping symbols first.

Example

a. Simplify $45 - 15 \div 5$.

$$\begin{aligned} 45 - 15 \div 5 &= 45 - 3 \\ &= 42 \end{aligned}$$

b. Evaluate $a^2 - 8(b + 7)$ if $a = 10$ and $b = 2$.

$$\begin{aligned} a^2 - 8(b + 7) &= 10^2 - 8(2 + 7) \\ &= 10^2 - 8(9) \\ &= 100 - 8(9) \\ &= 100 - 72 \text{ or } 28 \end{aligned}$$

EXERCISES

Evaluate each expression.

1. $18 - 2 \cdot 3$
2. $4^2 \div 8 - 3^3 \cdot \frac{1}{3} + 4 \cdot 6$
3. $5(36 - 31) + 2 \cdot 7$
4. $\frac{70 - 25}{7 + 8}$

Evaluate each expression when $a = 4$, $b = 2$, $x = \frac{1}{2}$, and $y = \frac{1}{3}$.

5. $8a - 2b$
6. $10x + 9y$
7. $48x^2 - (3a - 5b)$
8. $\frac{a^2 + b^2}{5y^2}$
9. **Accounting** Della and James are selling tickets for a dinner theater performance. Combination tickets for dinner and the performance cost $15. Performance only tickets cost $9.
   a. If Della and James sell both types of tickets, write an expression for the total amount of money they collect.
   b. Della sells 20 combination tickets and 15 performance tickets. James sells 18 combination tickets and 24 performance tickets. How much money have they collected?

Review 1–4

Integration: Statistics
Stem-and-Leaf Plots

Mrs. Cortez interviewed some students who were willing to help her with some typing. The number of words typed per minute by each student is listed below.

16 21 18 17 18 20 15 16

This data can be organized on a **stem-and-leaf plot**. The greatest common place value of each piece of data is used to form the **stem**. The next greatest place value is used to form the **leaves**.

Stem	Leaf
1	8 8 7 6 6 5
2	1 0

1|7 = 17 words per minute

Data with more than two digits may be rounded 13.5 ⇒ 14 or truncated 13.~~5~~ ⇒ 13. The rounded or truncated values of the data at the right can be compared in a back-to-back stem-and-leaf plot.

21.7	37.3	25.5
33.4	31.9	24.3
42.6	43.8	32.7
28.2	29.1	24.5

Rounded	Stem	Truncated
9 8 6 5 4 2	2	1 4 4 5 8 9
7 3 3 2	3	1 2 3 7
4 3	4	2 3

In a back-to-back stem-and-leaf plot, the same stem is used for the leaves of both plots.

EXERCISES

Solve each problem.

1. The stem-and-leaf plot at the right shows the height, in feet, of buildings in Boston that are at least 500 feet tall.
   a. How tall is the tallest building?
   b. What is the height of the shortest building represented in the plot?
   c. What building height occurs most frequently?

Stem	Leaf
5	5 3 2 1 0 0
6	1 0 0 0 0
7	5
8	0

5|2 = 520 feet

2. Danielle is a sales clerk at a bagel shop. Her sales for the first week were $58, $64, $26, $79, and $55. Her sales for the second week were $39, $72, $64, $52, and $78.
   a. Make a stem-and-leaf plot of Danielle's two weeks' sales.
   b. What is the greatest value of sales that Danielle made in one day during the two-week period?

Review 1–5

Open Sentences

Mathematical statements with one or more variables are called **open sentences**. Open sentences are **solved** by finding a replacement for the variable that results in a true sentence. The replacement is called a **solution**.

Example 1 **Replace y in $5y - 9 = 21$ with the value 6.**

$$5y - 9 = 21$$
$$5(6) - 9 = 21$$
$$30 - 9 = 21$$
$$21 = 21 \quad \text{true}$$

Since $y = 6$ makes the sentence $5y - 9 = 21$ true, 6 is a solution.

A set of numbers from which replacements for a variable may be chosen is called a **replacement set**. The set of all replacements for the variable in an open sentence that results in a true sentence is called the **solution set** for the sentence.

A sentence that contains an equals sign, =, is called an **equation** and sometimes may be solved by simply applying the order of operations. A sentence having the symbols $<$ or $>$ is called an **inequality**.

Example 2 **Solve $\frac{3(4 + 5)}{2 \cdot 4 - 3} = w$.**

$$\frac{3(4 + 5)}{2 \cdot 4 - 3} = w$$
$$\frac{3(9)}{8 - 3} = w$$
$$\frac{27}{5} = w$$

EXERCISES

State whether each equation is true or false for the value of the variable given.

1. $2q - 6 = 13, q = 10$
2. $3s^2 < 30, s = 4$
3. $5t + 2t^2 = 33, t = 3$
4. $\frac{4x + 5}{x^2} = 9, x = 1$

Solve each equation.

5. $d = 2(4 \cdot 5 - 3^2)$
6. $m = 5\frac{2}{5} - 1\frac{3}{10}$
7. $\frac{40 - 15}{5 + 6} = z$

8. **Business** Saundra must sell 4 t-shirts to make $10 profit.
   a. Write an equation that represents the number of t-shirts she must sell to make $60 profit.
   b. How many t-shirts would she have to sell?

Review 1–6

Identity and Equality Properties

The identity and equality properties in the chart below can help you solve algebraic equations and evaluate mathematical expressions.

Additive Identity Property	For any number a, $a + 0 = 0 + a = a$.
Multiplicative Identity Property	For any number a, $a \cdot 1 = 1 \cdot a = a$.
Multiplicative Property of Zero	For any number a, $a \cdot 0 = 0 \cdot a = 0$.
Substitution Property of Equality	For any numbers a and b, if $a = b$ then a may be replaced by b in any expression.
Reflexive Property of Equality	For any number a, $a = a$.
Symmetric Property of Equality	For any numbers a and b, if $a = b$, then $b = a$.
Transitive Property of Equality	For any numbers a, b, and c, if $a = b$ and $b = c$, then $a = c$.

Example 

Evaluate $36 \cdot 1 + 9 + 12(2 \cdot 3 - 6)$. Indicate the property used in each step.

$36 \cdot 1 + 9 + 12(2 \cdot 3 - 6) = 36 \cdot 1 + 9 + 12(6 - 6)$ *Substitution (=)*

$= 36 \cdot 1 + 9 + 12(0)$ *Substitution (=)*

$= 36 + 9 + 12(0)$ *Identity (×)*

$= 36 + 9 + 0$ *Mult. prop. of 0*

$= 45 + 0$ *Substitution (=)*

$= 45$ *Identity (+)*

EXERCISES

Solve each equation.

1. $w(34) = 0$
2. $7 \cdot x = 7$
3. $0 + r = 15$
4. $9(0) = a$

Name the property or properties illustrated by each statement.

5. $(0)3 = 0$
6. $10 \cdot 1 = 10$
7. $(1)52 = 52$
8. $12 + 5 = 12 + 5$
9. $17 + 0 = 17$
10. If $6 + 7 = 13$, then $13 = 6 + 7$.
11. $(60 - 20) - 10 = 40 - 10$
12. If $2^3 = 8$ and $8 = 10 - 2$, then $2^3 = 10 - 2$.

Evaluate each expression. Name the property used in each step.

13. $70 \div 10 + 3(6 - 3 \cdot 2) - 6 \div 2$
14. $8(7 - 6) + 48 \div 4^2$
15. **Telecommunications** A direct-dialed telephone call costs \$0.27 for the first minute and \$0.11 for each additional minute or fraction of a minute.
    a. Write an expression that represents the cost of a 7-minute call.
    b. Evaluate the expression. Name the property used in each step.
    c. How much will the 7-minute call cost?

Review

1–7

The Distributive Property

When you find the product of two integers, you find the sum of two partial products. For example, you can write

$$\begin{array}{r} 63 \\ \times\ 7 \\ \hline 441 \end{array} \quad \text{as} \quad \begin{array}{r} 60 + 3 \\ \times \quad\ \ 7 \\ \hline 420 + 21 \end{array} \leftarrow (60 \times 7) + (3 \times 7).$$

The statement $(60 + 3) \times 7 = (60 \times 7) + (3 \times 7)$ illustrates the **distributive property**. The multiplier 7 is distributed over the 60 and the 3.

Distributive Property
For any numbers a, b, and c, $a(b + c) = ab + ac$ and $(b + c)a = ba + ca$; $a(b - c) = ab - ac$ and $(b - c)a = ba - ca$.

You can use the distributive property to simplify algebraic expressions.

Example

Simplify $6(xy + z) + 9z$.

$6(xy + z) + 9z = 6xy + 6z + 9z$ *Distributive property*

$= 6xy + (6 + 9)z$ *Distributive property*

$= 6xy + 15z$ *Substitution (=)*

EXERCISES

Name the coefficient of each term. Then name the like terms in each list of terms.

1. $12r^3, 7r^2, 3r^3, 4r$

2. $3xy, 11x, 3y, 5x$

Use the distributive property to find each product.

3. $6 \cdot 27$

4. $4 \cdot 93$

Use the distributive property to rewrite each expression.

5. $7(5w + 3)$

6. $5a - 5b$

Simplify each expression, if possible. If not possible, write *in simplest form*.

7. $12c - 5c$

8. $6(4x + 9)$

9. $2t + 15t + 16y - 9y$

10. $26rs + 17st$

11. $2w - 3(4v + 3v)$

12. $0.9(0.3z - 2) + 0.7z$

13. Business Each month, a business owner pays $23 for telephone service and $9.95 for an on-line service.

a. Write an expression representing the owner's total costs for telephone and on-line services for an entire year.

b. What amount does the business owner pay for these services for a year?

Review 1–8

Commutative and Associative Properties

The commutative and associative properties can be used to simplify expressions.

Commutative Properties	For any numbers a and b, $a + b = b + a$ and $a \cdot b = b \cdot a$.
Associative Properties	For any numbers a, b, and c, $(a + b) + c = a + (b + c)$ and $(ab)c = a(bc)$.

Example **Simplify $4(3m + 5n) + 12m$.**

$4(3m + 5n) + 12m = (12m + 20n) + 12m$ *Distributive property*

$= (20n + 12m) + 12m$ *Commutative (+)*

$= 20n + (12m + 12m)$ *Associative (+)*

$= 20n + (12 + 12)m$ *Distributive property*

$= 20n + 24m$ *Substitution (=)*

EXERCISES

Simplify.

1. $5a + 6b + 12a$
2. $3mn + 14m + 9mn$
3. $5(3q + 2r) + 3(r + 5)$
4. $8x^2 + 9y^2 + 3x^2$
5. $6st + 9s^2t + 7st$
6. $3xy + 5x^2y + 10(3x^2y)$
7. $3(2q + r) - 4q + 6r$
8. $0.4(20m + 12n) + 3m$
9. $\frac{1}{5} + \frac{3}{10}(b + 2) + \frac{3}{5}$
10. $2(0.9w + 0.3v) + 4w$
11. $0.1(8s + 7t) + 4(s + t)$
12. $\frac{1}{2}c + 3d + \frac{1}{4}c + \frac{1}{2}d$

Name the property illustrated by each statement.

13. $13a + 4b = 4b + 13a$
14. $2(x + 7) = 2x + 2(7)$
15. $1 \cdot z^4 = z^4$
16. $(5w + 7) + 6z = 5w + (7 + 6z)$
17. **Automobiles** Consider the steps necessary to fill your car's gasoline tank. One step is to remove the gasoline tank cap and the other step is to pump the gasoline. Would you say that these steps are commutative?

Review 1–9

A Preview of Graphs and Functions

Thiel bought a used car with a mileage of 28,000 miles. He drives the car 1500 miles per month. Mathematically, the car's mileage can be defined by the open sentence mileage = 28,000 + 1,500t, where t is the length of ownership in months. The table at the right shows the car's mileage over a period of 4 months.

Length of Ownership	Car's Mileage
0 month	28,000
1 month	29,500
2 months	31,000
3 months	32,500
4 months	34,000

This information can also be represented in a graph. The graph shows the relationship between the car's mileage and the length of Thiel's ownership.

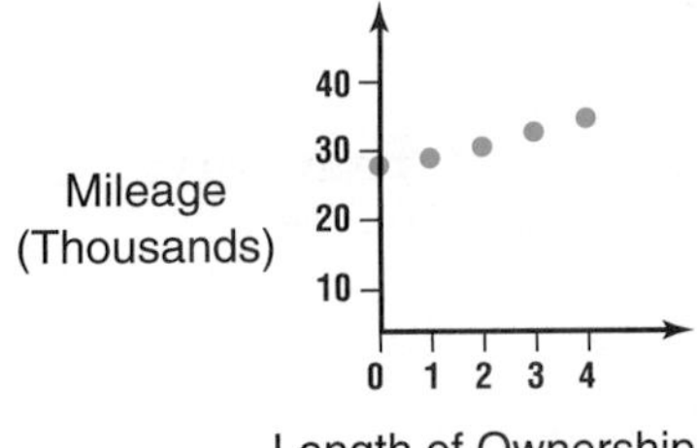

You can use a graph without scales on either axis to show the general shape of the graph that represents a situation.

Example

In a test of a car's brakes, the car is accelerated on a test track. Then the driver suddenly slams on the brakes to bring the car to a stop. Choose the graph that best represents this situation.

a.

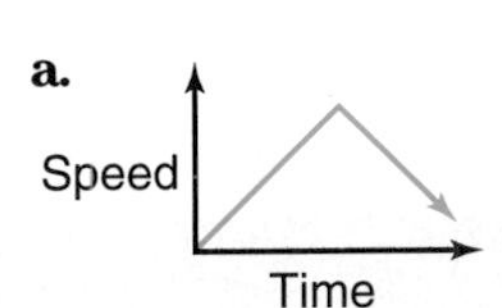

b.

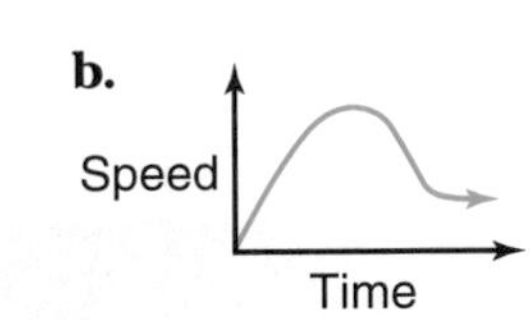

c.

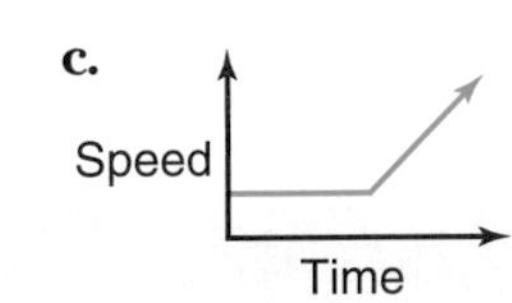

To select the graph that best represents this situation, find the graph that shows an increase in speed, followed by a decrease in speed. Graph a matches this description.

EXERCISES

1. Identify the graph that matches the following statement. Explain your answer. The outside temperature increases in the morning and falls in the afternoon.

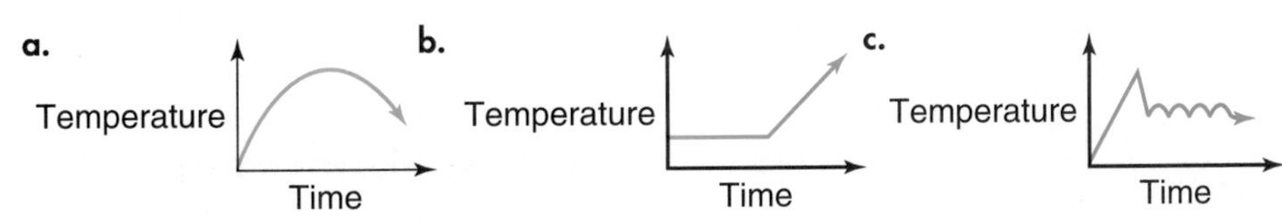

Sketch a reasonable graph for each situation.

2. Martien is in-line skating at a steady pace. Then she stops to drink some water. After her break, she continues her skating.
3. Tristan likes to trade Olympic pins. He started the week with lots of pins in his collection. Later in the week he lost and sold many of his pins. The next week he began building up his collection again.

1. Write an algebraic expression for the sum of 3 times a number and 7.
2. Rewrite $3 \cdot x \cdot x \cdot x \cdot x \cdot x$ using exponents.

Give the next two items for each pattern.

3. 1600, 800, 400, 200, . . .
4. 4, 9, 14, 19, 24, . . .

Evaluate each expression.

5. $105 - 45 \div 15 + 6$
6. $4^2[(20 - 4) \div 8]$

7. Write the members of the data set used to make the stem-and-leaf plot.

Stem	Leaf
7	0 4 6 7
8	1 2 9
9	3 3 5 8

8 | 1 = 81

8. Suppose the number 1816 is rounded to 1820 and plotted using stem 18 and leaf 20. Write the stem and leaf for 1738.

Solve each equation.

9. $q = 7.3 - 5.41$
10. $\frac{1}{5} + \frac{3}{10} = r$

State the property illustrated by each statement.

11. $5(13 + 2) = 5 \cdot 13 + 5 \cdot 2$
12. $0 + 17 = 17$
13. $(9 - 4) + b = 5 + b$
14. $2(st) = (2s)t$
15. $5 + t = t + 5$
16. If $w = y$ and $y = 4$, then $w = 4$.

17. Simplify $2m + 4(m + 3n)$.

18. Identify two ordered pairs on the graph at the right.

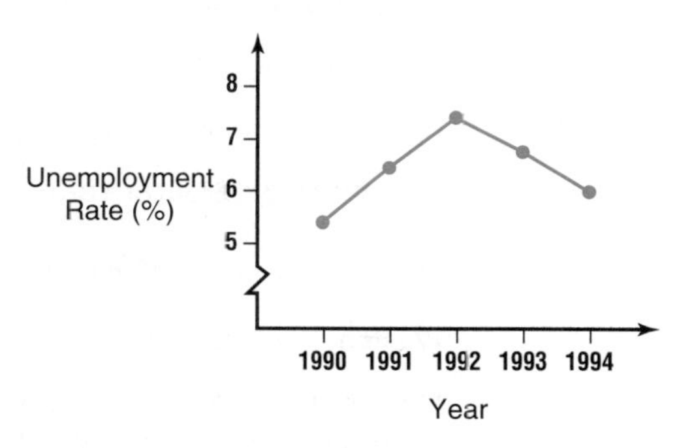

19. Sketch a reasonable graph for the following situation.
Miguel turns on the oven to preheat. He bakes a casserole for 1 hour at 350° F. Then he turns off the oven.

Review 2–1

Integers and the Number Line

The figure at the right is part of a number line. On a number line, the set of numbers that include 0 and numbers to the right of 0 are named by members of the set of **whole numbers**.

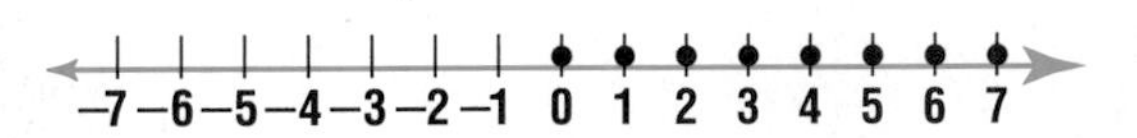

The set of numbers used to name the points marked on the number line at the right is called the set of **integers**.

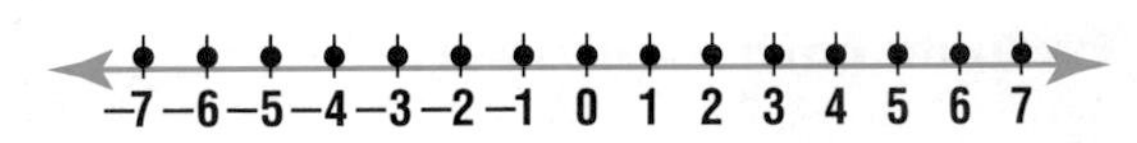

To graph a set of numbers means to locate the points named by those numbers on the number line. The number that corresponds to a point on the number line is called the **coordinate** of the point.

Name the coordinate of point K.

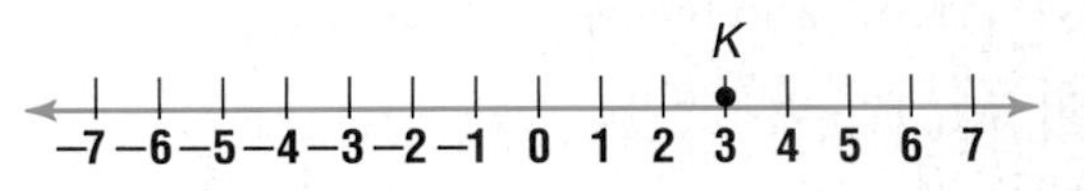

The coordinate of K is 3.

A number line is often used to show addition of integers. For example, to find the sum of 4 and -7, follow the steps at the right.

Step 1	Draw an arrow, starting at 0 and going to 4.
Step 2	Start at 4. Draw an arrow 7 units to the left.
Step 3	The second arrow points to the sum, -3.

EXERCISES

Name the coordinate of each point.

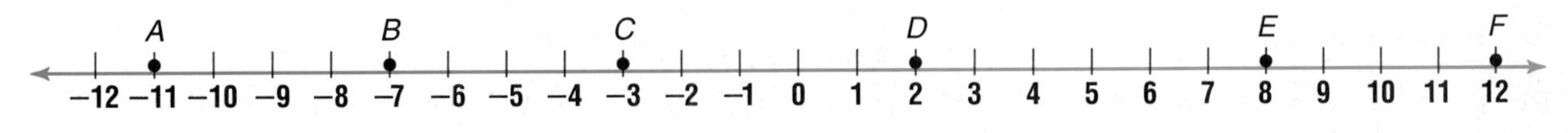

1. D **2.** B **3.** E **4.** C **5.** A **6.** F

Graph each set of numbers on a number line.

7. $\{-4, -2, 0, 3\}$ **8.** $\{-5, -1, 2, 5\}$ **9.** $\{-2, -1, 0, \ldots\}$

10. {integers less than 5 but greater than or equal to 0}

11. {integers greater than -4}

Find each sum. If necessary, use a number line.

12. $-3 + (-9)$ **13.** $-5 + 7$ **14.** $12 + (-14)$

15. $-3 + 3$ **16.** $8 + (-4)$ **17.** $-11 + 3$

18. Temperature The lowest temperature ever recorded in the state of Arizona was $-40°$. The highest temperature ever recorded in Arizona was 168° higher. What was the highest temperature ever recorded in Arizona?

Review 2–2

Integration: Statistics
Line Plots

The table below shows the amount Nicole earned each month for babysitting last year. The amounts are recorded in dollars.

Nicole's Earnings (dollars)		
36	45	60
55	39	57
45	42	58
33	45	64

Numerical information displayed on a number line is called a **line plot**. The line plot below is another way to show the data for Nicole's babysitting earnings.

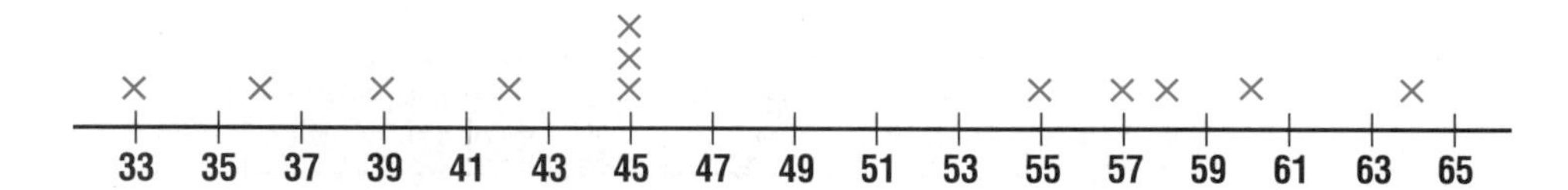

EXERCISES

1. The largest counties in the United States in 1994 are listed below.

County	Population
Los Angeles, CA	9,149,840
Cook, IL	5,141,375
Harris, TX	3,045,212
San Diego, CA	2,632,047
Orange, CA	2,543,124
Maricopa, AZ	2,346,610

   a. Make a line plot of the data.

   b. How many counties have at least 3 million residents?

   c. Write a sentence that compares the population of Los Angeles County with the populations of the three smallest counties in the table.

2. The table below lists the heights of marching band members at McKinley High School.

Heights of Marching Band Members (inches)							
63	67	63	64	62	65	66	64
72	70	65	69	65	66	68	

   a. Make a line plot of the data.

   b. How many band members are taller than 65 inches?

   c. Which height occurred most frequently?

Review 2–3

Adding and Subtracting Integers

Use the following definitions, rules, and properties when adding or subtracting integers.

Definition, Rule, or Property		Example
Definition of Absolute Value	For any real number a: if $a > 0$, then $\lvert a \rvert = a$, and if $a < 0$, then $\lvert a \rvert = -a$.	$\lvert 6 \rvert = 6$ $\lvert -6 \rvert = 6$
Adding Integers with the Same Sign	To add integers with the same Give sign, add their absolute values. the result the same sign as the integers.	$7 + 8 = 15$ $-7 + (-8) = -15$
Adding Integers with Different Signs	To add integers with different signs, subtract the lesser absolute value from the greater absolute value. Give the result the same sign as the integer with the greater absolute value.	$-10 + 3 = -7$ $6 + (-2) = 4$
Additive Inverse Property	For every number a, $a + (-a) = 0$.	$-5 + 5 = 0$
Subtracting Integers	To subtract a number, add its additive inverse. For any numbers a and b, $a - b = a + (-b)$.	$13 - (-8) = 13 + 8$ $= 21$

You can use the distributive property and the addition and subtraction rules for integers to simplify expressions with like terms.

Example  **Simplify $-3w - 2w + 10w$.**

$$\begin{aligned} -3w - 2w + 10w &= -3w + (-2w) + 10w \\ &= [-3 + (-2) + 10]w \\ &= (-5 + 10)w \\ &= 5w \end{aligned}$$

EXERCISES

Find each sum or difference.

1. $-13 + (-7)$
2. $96 + (-54)$
3. $39 + 62$
4. $18 - 32$
5. $26 - (-40)$
6. $-27 - (-14)$
7. $12a + (-17a) - 4a$
8. $-2t + 20t + (-7t)$
9. $-6x + 15x - (-20x)$

Evaluate each expression if $a = 2$, $b = -3$, and $c = -1$.

10. $212 + \lvert b \rvert$
11. $b + (-23) + \lvert c \rvert$
12. $-34 - \lvert a \rvert$
13. $16 - a - \lvert b \rvert$
14. **Elevators** The World Trade Center building in New York City has 110 stories. If an employee rode the elevator to her office on the 83rd floor, then came down 35 floors to deliver a report, which floor is she on now?

Review 2–4

Rational Numbers

Definition of a Rational Number	**A rational number is a number that can be expressed in the form $\frac{a}{b}$, where a and b are integers and b is not equal to 0.**

You can compare rational numbers by graphing them on a number line.

Comparing Numbers on the Number Line	**If a and b represent any numbers and the graph of a is to the left of the graph of b, then $a < b$. If the graph of a is to the right of the graph of b, then $a > b$.**
Comparison Property	**For any two numbers a and b, exactly one of the following sentences is true: $a < b$, $a = b$, or $a > b$.**

Example 1

a. $-5\frac{1}{3} < -2\frac{1}{3}$ The graph of $-5\frac{1}{3}$ is to the left of the graph of $-2\frac{1}{4}$.

b. $-1\frac{1}{8} > -4\frac{3}{8}$ The graph of $-1\frac{1}{8}$ is to the right of the graph of $-4\frac{4}{5}$.

Example 2

Replace __?__ with <, >, or = to make the sentence true.

-9 __?__ -11

$-9 > -11$ Since -9 is to the right of -11 on a number line, -9 is greater than -11.

You can use **cross products** to compare two fractions with different denominators.

Comparison Property for Rational Numbers	**For any rational numbers $\frac{a}{b}$ and $\frac{c}{d}$, with $b > 0$ and $d > 0$:** **1. if $\frac{a}{b} < \frac{c}{d}$, then $ad < bc$, and** **2. if $ad < bc$, then $\frac{a}{b} < \frac{c}{d}$.**

This property also holds if < is replaced by >, ≤, ≥, or =.

EXERCISES

Replace each __?__ with <, >, or = to make each sentence true.

1. -6 __?__ 3
2. -2 __?__ $-17 + 12$
3. $-3 - 4$ __?__ $-16 + 9$

Write the numbers in each set in order from least to greatest.

4. $-\frac{1}{5}$, 0.3, -2.5
5. $\frac{1}{8}$, -1, $-\frac{9}{7}$
6. Find a number between $\frac{1}{3}$ and $\frac{7}{15}$.
7. Which is the better buy? A 15-ounce box of cereal for \$1.99 or a 20-ounce box of cereal for \$2.49?

Review 2–5

Adding and Subtracting Rational Numbers

The rules for adding and subtracting integers also apply to adding and subtracting rational numbers.

Rational Number	Form $\frac{a}{b}$
5	$\frac{5}{1}$
$-1\frac{2}{3}$	$-\frac{5}{3}$
0.75	$\frac{3}{4}$

Example 1

a. Find $\left(-3\frac{1}{4}\right) + 6\frac{1}{2}$.

$$\left(-3\frac{1}{4}\right) + 6\frac{1}{2} = +\left(\left|6\frac{2}{4}\right| - \left|-3\frac{1}{4}\right|\right)$$
$$= +\left(6\frac{2}{4} - 3\frac{1}{4}\right)$$
$$= 3\frac{1}{4}$$

b. Find −6.14 − 9.87.

$$-6.14 - 9.87 = -6.14 + (-9.87)$$
$$= -16.01$$

To add three or more numbers, first group the numbers in pairs. Use the commutative and associative properties to rearrange the addends if necessary. Study the example below.

Example 2 **Find $-\frac{1}{4} + \frac{5}{6} + \left(-\frac{7}{4}\right)$.**

$$-\frac{1}{4} + \frac{5}{6} + \left(-\frac{7}{4}\right) = \left[-\frac{1}{4} + \left(-\frac{7}{4}\right)\right] + \frac{5}{6}$$
$$= -\frac{8}{4} + \frac{5}{6}$$
$$= -\frac{24}{12} + \frac{10}{12}$$
$$= -\frac{14}{12} \text{ or } -1\frac{1}{6}$$

EXERCISES

Find each sum or difference.

1. $-\frac{4}{13} + \left(-\frac{7}{13}\right)$
2. $\frac{3}{4} + \left(-\frac{7}{8}\right)$
3. $-0.019 + 0.062$
4. $\frac{4}{7} - \left(-\frac{2}{7}\right)$
5. $3.97 - 1.55$
6. $-\frac{2}{3} - \frac{4}{9}$

Evaluate each expression if $a = -0.25$ and $b = \frac{1}{3}$.

7. $a + 0.53$
8. $2 - b$
9. $-3 - a$
10. $b + \frac{2}{7}$

Find each sum.

11. $-44.1 + 62.7 + (-16.8)$
12. $-13q + (-41q) + 18q$
13. $\frac{1}{4} + \left(-\frac{3}{8}\right) + \frac{1}{2}$
14. **Fishing** The saltwater fish record for redeye bass is 8.75 pounds. The record for black sea bass is 0.75 pounds more than the redeye bass record. The record for barred sand bass is 3.75 pounds more than the black sea bass record. What is the record for barred sand bass?

Review 2–6

Multiplying Rational Numbers

You can use the rules below when multiplying rational numbers.

Rule or Property		Example
Multiplying Two Numbers with Different Signs	The product of two numbers that have different signs is negative.	$(-3m)9n = (-3)(9)mn$ $= -27mn$
Multiplying Two Numbers with the Same Sign	The product of two numbers that have the same sign is positive.	$(-11)(-4) = 44$
Multiplicative Property of −1	The product of any number and −1 is its additive inverse. $-1(a) = -a$ *and* $a(-1) = -a$	$(-2)(-7)(-1)(5) = 14(-1)(5)$ $= -14(5)$ $= -70$

To find the product of two or more numbers, first group the numbers in pairs.

Example  **Find (−3.1)(7.4)(− 0.1)(− 0.5).**

$(-3.1)(7.4)(-0.1)(-0.5) = [(-3.1)(7.4)]\,[(-0.1)(-0.5)]$ *Associative (×)*

$= (-22.94)(0.05)$ *Substitution (=)*

$= -1.147$

EXERCISES

Find each product.

1. $(-18)(-3)$
2. $(10.0)(-0.14)$
3. $(-1)(-5)(-3)$
4. $\left(\frac{1}{4}\right)(-16)(12)$
5. $(-4)(-55)\left(-\frac{3}{5}\right)$
6. $\left(\frac{5}{6}\right)(-2)(-1)(-3)$

Simplify.

7. $(-7)(2) + (3)(6)$
8. $\left(-\frac{1}{3}\right)\left(\frac{3}{5}\right) - \left(\frac{1}{2}\right)\left(\frac{1}{5}\right)$
9. $-3(5s - 2s) + 4(-st)$
10. $(-\ 9mn)4 - mn(7 - 5)$
11. $6.2(5x - 2y) - 1.5(2x + 8y)$
12. $(-8)(3) - (-4)(6)$
13. **Nutrition** The recommended daily allowance (RDA) of calcium for a teenaged female is 1200 mg. An eight-ounce serving of skim milk contains 302 mg of calcium. A one-ounce serving of cheddar cheese contains 204 mg of calcium. A medium orange contains 52 mg of calcium.
    a. How much calcium is in 3 eight-ounce servings of skim milk?
    b. Do 3 eight-ounce servings of skim milk satisfy the RDA of calcium for a teenaged female? Explain your reasoning.
    c. Would a serving of cheddar cheese and 2 oranges, in addition to 3 servings of skim milk, satisfy the RDA of calcium for a teenaged female? Explain your reasoning.

Review 2–7

Dividing Rational Numbers

Use the following rules to divide rational numbers.

Rule or Property		Example
Dividing Two Rational Numbers	The quotient of two numbers having the same sign is positive. The quotient of two numbers having different signs is negative.	$-40 \div (-8) = 5$ $-63 \div 7 = -9$
Multiplicative Inverse Property	For every nonzero number $\frac{a}{b}$, where $a, b \neq 0$, there is exactly one number $\frac{b}{a}$ such that $\frac{a}{b} \cdot \frac{b}{a} = 1$.	$\frac{3}{8} \div \frac{5}{4} = \frac{3}{8} \cdot \frac{4}{5}$ $= \frac{12}{40}$ or $\frac{3}{10}$
Division Rule	For all numbers a and b, with $b \neq 0$, $a \div b = \frac{a}{b} = a\left(\frac{1}{b}\right) = \frac{1}{b}(a)$.	$15 \div 3 = \frac{15}{3}$ $= 15\left(\frac{1}{3}\right)$ $= \frac{1}{3}(15)$

Since the fraction bar indicates division, you can use the division rules and the distributive property to simplify rational expressions.

Example **Simplify $\frac{-42p + 18}{6}$.**

$$\frac{-42p + 18}{6} = (-42p + 18)\left(\frac{1}{6}\right)$$
$$= (-42p)\left(\frac{1}{6}\right) + 18\left(\frac{1}{6}\right)$$
$$= -7p + 3$$

EXERCISES

Simplify.

1. $-\frac{54q}{9}$
2. $\frac{48t}{16}$
3. $\frac{120}{24}$
4. $\frac{45}{-15}$
5. $\frac{-78y}{3}$
6. $\frac{-96}{-37} \div \frac{96}{37}$
7. $\frac{52w}{13}$
8. $-\frac{1}{5} \div 5$
9. $\frac{24d - 16c}{-4}$
10. $\frac{18s}{2} \div \frac{1}{3s} + 3stu$
11. $\frac{44x - 28}{4}$
12. $\frac{-\frac{3}{7}}{3}$

13. **Cooking** Theo has a recipe for chocolate chip cookies that makes 6 dozen cookies. He only has enough flour to make 3 dozen cookies.
    a. If the original recipe calls for $2\frac{1}{4}$ cups margarine, how much margarine will he need to make 3 dozen cookies?
    b. If the original recipe calls for $3\frac{1}{2}$ teaspoons baking powder, how much baking powder will he need to make 3 dozen cookies?

Review

2–8

Square Roots and Real Numbers

Counting or Natural Numbers, N	{1, 2, 3, 4, . . . }
Whole Numbers, W	{0, 1, 2, 3, 4, . . . }
Integers, Z	{ . . ., −3, −2, −1, 0, 1, 2, 3, . . . }
Rational Numbers, Q	{all numbers that can be expressed in the form $\frac{a}{b}$, where a and b are integers and $b \neq 0$}
Irrational Numbers, I	{numbers that cannot be expressed in the form $\frac{a}{b}$, where a and b are integers and $b \neq 0$}
Real Numbers, R	{rational numbers and irrational numbers}

A **square root** is one of two equal factors of a number. For example, the square root of 49 is 7 and −7 since $7 \cdot 7$ is 49 and $(-7)(-7)$ is also 49. Since the square root of 49 is a rational number, it is a **perfect square**.

The symbol $\sqrt{}$ is called a **radical sign**. It indicates the nonnegative, or **principal**, square root of the expression under the radical sign.

Example

Find $\pm\sqrt{25}$. The symbol $\pm\sqrt{25}$ represents both square roots. Since $5^2 = 25$, we know that $\pm\sqrt{25} = \pm5$.

Numbers such as $\sqrt{2}$ and $\sqrt{3}$ are not perfect squares. Notice what happens when you find these square roots with your calculator. These numbers are not rational numbers since they are not repeating or terminating decimals. They are classified as **irrational numbers**.

EXERCISES

Find each square root. Use a calculator if necessary. Round to the nearest hundredth if the result is not a whole number.

1. $\sqrt{64}$ **2.** $\sqrt{0.00036}$ **3.** $-\sqrt{\frac{81}{49}}$ **4.** $-\sqrt{2500}$ **5.** $\pm\sqrt{\frac{144}{25}}$

Evaluate each expression. Use a calculator if necessary. Round to the nearest hundredth if the result is not a whole number.

6. $\sqrt{y}$, if $y = 76$ **7.** $\pm\sqrt{c + d}$, if $c = 14$ and $d = 45$

Name the set or sets of numbers to which each real number belongs. Use N for natural numbers, W for whole numbers, Z for integers, Q for rational numbers, and I for irrational numbers.

8. $\sqrt{24}$ **9.** 7.8 **10.** $\sqrt{100}$

11. Aviation The formula to determine the distance d in miles that an object can be seen on a clear day on the surface of the ocean is $d = 1.4\sqrt{h}$, where h is the height in feet the viewer's eyes are above the surface of the water. About how many miles can the pilot of an airplane see if the airplane is 2000 feet above the water?

Review

2–9

Problem Solving
Write Equations and Formulas

When solving a problem, you should read and explore the problem until you completely understand the relationships in the given information. Then you may translate the problem into an equation or formula. In an equation, you choose a variable to represent one of the unspecified numbers in the problem. This is called **defining the variable**. Then use the variable to write expressions for the other unspecified numbers in the problem. In a formula, an equation that states a rule for the relationship between certain quantities is formed.

Problem-Solving Plan
1. Explore the problem. 2. Plan the solution. 3. Solve the problem. 4. Examine the solution.

EXERCISES

Answer the related questions for the verbal problems below.

1. Will can mow a lawn in 3 hours. Tazeen can mow the same lawn in 2 hours.

a. How much of the lawn can Will mow in one hour?

b. How much of the lawn can Tazeen mow in one hour?

2. An airplane can hold a maximum of 120 people. The airplane is two-thirds full. One-fourth of the passengers remain on the airplane when it lands at the end of the first leg of its flight.

a. How many people are on the airplane on the first leg of its flight?

b. How many people remain on the airplane when it lands?

Translate each sentence into an equation, an inequality, or formula.

3. The area of a square is the length s of its side squared.

4. Two-thirds of the sum of m and n is 71.

5. The product of x and y is greater than 7 times the sum of x and y.

Define a variable, then write an equation for each problem. Do *not* try to solve.

6. One number is 65 less than a second number. The sum of the two numbers is 192. Find the numbers.

7. Andra has 90 books. If she has 10 more than one-third as many paperbacks as hardbacks, how many paperbacks does Andra have?

1. Graph $\{-3, -1, 0, 2, 5\}$ on a number line.

2. Name the coordinate of point P.

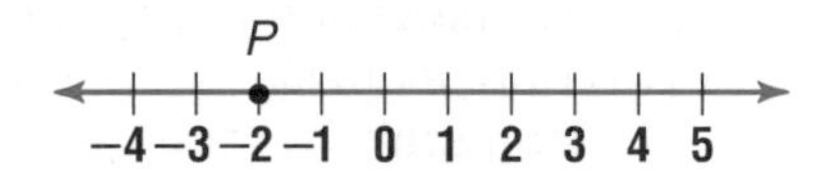

Use the table below for Exercises 3–4.

Biology Test Scores				
88	85	91	98	83
91	89	92	95	71
74	82	84	79	86

3. Make a line plot representing the scores on a biology test.

4. How many scores were 90 or above?

Find each sum or difference.

5. $-94 + 36$

6. $13.1 + (-5.6) + (-6.2) + 0.89$

7. $\frac{1}{8} + \left(-\frac{3}{16}\right)$

8. $4 - 13$

9. Find the value of $|y - x|$ if $x = 2$ and $y = -3$.

10. Write $-\frac{3}{4}$, 0.9, -1.4, 1.1 in order from least to greatest.

Simplify.

11. $(-9)(4) + (2)(11)$

12. $3.1(2x - 3y) - 4.5(8x + 2y)$

13. $\frac{-63q}{9}$

14. $\frac{-51}{-17} \div \frac{102}{34}$

15. $-\sqrt{\frac{64}{9}}$

16. $\sqrt{55 - 6}$

17. Translate the following sentence into an equation.
Twice the sum of a and b is 46.

18. Define a variable, then write an equation for the following. Do *not* try to solve.
One number is 61 more than a second number. The sum of the two numbers is 127. Find the numbers.

Review 3–1

Solving Equations with Addition and Subtraction

You can use the addition and subtraction properties of equality to solve equations. To check, substitute the solution for the variable in the original equation. If the resulting sentence is true, your solution is correct.

Addition Property of Equality	For any numbers a, b, and c, if $a = b$, then $a + c = b + c$.
Subtraction Property of Equality	For any numbers a, b, and c, if $a = b$, then $a - c = b - c$.

Example  **Solve $z - 15 = -3$.**

$$z - 15 = -3$$
$$z - 15 + 15 = -3 + 15$$
$$z = 12$$

Check:
$$z - 15 = -3$$
$$12 - 15 \stackrel{?}{=} -3$$
$$-3 = -3 \checkmark$$

Example  **Solve $q + 5 = -2$.**

$$q + 5 = -2$$
$$q + 5 - 5 = -2 - 5$$
$$q = -7$$

Check:
$$q + 5 = -2$$
$$-7 + 5 \stackrel{?}{=} -2$$
$$-2 = -2 \checkmark$$

Sometimes an equation can be solved more easily if it is rewritten first. Recall that subtracting a number is the same as adding its inverse. For example, the equation $b - (-3) = 11$ may be rewritten as $b + 3 = 11$.

EXERCISES

Solve each equation. Then check your solution.

1. $w - 7 = -34$
2. $y + 8 = 3$
3. $q + 5 = -20$
4. $-6 = m + 3$
5. $d + (-12) = 4$
6. $t - (-23) = 14$
7. $16 + a = -11$
8. $62 = 41 - s$
9. $-7.4 = c + (-1.6)$
10. $-32 - x = 22\frac{5}{8}$
11. $-\frac{1}{6} + b = \frac{2}{3}$
12. $\frac{4}{11} = -n + \frac{1}{3}$
13. **Clothing Design** A clothing designer is designing a blazer. The blazer will have three buttons on the front. The length of each button hole is to be $\frac{1}{8}$-inch longer than the sum of the diameter and the thickness of the button.
    a. Translate the description of the button hole length into a formula.
    b. Suppose the clothing designer decides that each button hole is to be $\frac{15}{16}$-inch long and each button is to be $\frac{5}{8}$-inch in diameter. What thickness should the button be?

Review

3–2

Solving Equations with Multiplication and Division

You can solve equations in which a variable has a coefficient by using the multiplication and division properties of equality.

Multiplication Property of Equality	For any numbers a, b, and c, if $a = b$, then $ac = bc$.
Division Property of Equality	For any numbers a, b, and c, with $c \neq 0$, if $a = b$, then $\frac{a}{c} = \frac{b}{c}$.

Example  **Solve $\frac{1}{7}p = 3$.**

$$\frac{1}{7}p = 3$$
$$7\left(\frac{1}{7}p\right) = 7(3)$$
$$p = 21$$

Check:
$$\frac{1}{7}p = 3$$
$$\frac{1}{7}(21) \stackrel{?}{=} 3$$
$$3 = 3 \checkmark$$

Example  **Solve $4x = 28$.**

$$4x = 28$$
$$\frac{4x}{4} = \frac{28}{4}$$
$$x = 7$$

Check:
$$4x = 28$$
$$4(7) \stackrel{?}{=} 28$$
$$28 = 28 \checkmark$$

EXERCISES

Solve each equation. Then check your solution.

1. $-6z = -72$
2. $-4q = 48$
3. $5r = -35$
4. $\frac{1}{7}a = -3$
5. $8s = \frac{4}{7}$
6. $1\frac{1}{3}t = -2\frac{2}{3}$

Define a variable, write an equation, and solve each problem. Then check your solution.

7. Seven times a number is 63. What is the number?
8. One third of a number is twelve. What is the number?
9. Negative three times a number is -93. What is the number?
10. Tyrone paid $82.50 for 3 ballet tickets. What is the cost of each ticket?

Complete.

11. If $5y = 80$, then $7y =$ ___.
12. If $-3r = 27$, then $4r =$ ___.
13. If $8b = -45$, then $-2b =$ ___.
14. If $3m - 2n = 33$, then $6m - 4n =$ ___.

Review 3–3

Solving Multi-Step Equations

When solving some equations you must perform more than one operation on both sides. First, determine what operations have been done to the variable. Then undo these operations in the reverse order.

Example  **How would you solve $\frac{1}{4} + 5 = 13$?**

$\frac{x}{4} + 5 = 13$ First, x was divided by 4. Then 5 was added.

To solve, first subtract 5 from each side. Then multiply each side by 4.

Procedure for Solving a Two-Step Equation	**1. Undo any indicated additions or subtractions.** **2. Undo any indicated multiplications or divisions involving the variable.**

Example  **$3w + 7 = 34$** *Addition of 7 is indicated.*

		Check:
$3w + 7 - 7 = 34 - 7$	*Therefore, subtract 7 from each side.*	$3w + 7 = 34$
$3w = 27$	*Multiplication by 3 is also indicated.*	$3(9) + 7 = 34$
$\frac{3w}{3} = \frac{27}{3}$	*Therefore, divide each side by 3.*	$27 + 7 = 34$
$w = 9$		$34 = 34$ ✔

EXERCISES

Solve each equation. Then check your solution.

1. $5r - 21 = 44$
2. $6y + 2 = 50$
3. $0.9a - 2.6 = 3.7$
4. $\frac{3m + 15}{2} = 22.5$
5. $21 = \frac{q + 35}{8}$
6. $14 + \frac{4z}{-11} = 66$
7. $\frac{3}{5}c - 6 = 9$
8. $\frac{t}{-2} + 17 = -9$
9. $-6 = \frac{5p - (-2)}{-7}$

Define a variable, write an equation, and solve each problem. Then check your solution.

10. Find three consecutive integers whose sum is 126.
11. Find two consecutive odd integers whose sum is 76.
12. Deborah, Travis, and Tito were each born in one of three consecutive years. The sum of their ages is 45. What are the three ages?

Review 3–4

Integration: Geometry Angles and Triangles

Supplementary Angles	Two angles are supplementary if the sum of their measures is 180°.
Complementary Angles	Two angles are complementary if the sum of their measures is 90°.
Sum of the Angles of a Triangle	The sum of the measures of the angles in any triangle is 180°.

Example 1 **The measure of an angle is four times the measure of its supplement. Find the measure of each angle.**

Let a = the lesser measure. Then $4a$ = the greater measure.

$a + 4a = 180$ *The sum of the measures is 180°.*

$5a = 180$ *Add a and 4a.*

$\frac{5a}{5} = \frac{180}{5}$ *Divide each side by 5.*

$a = 36$ The measures are 36° and $4 \cdot 36°$ or 144°.

Example 2 **The measures of two angles of a triangle are 42° and 83°. Find the measure of the third angle.**

Let x = the measure of the third angle.

$42 + 83 + x = 180$ *The sum of the measures of the angles is 180°.*

$125 + x = 180$ *Add 42 and 83.*

$125 - 125 + x = 180 - 125$ *Subtract 125 from each side.*

$x = 55$ The measure of the third angle is 55°.

EXERCISES

Find both the complement and the supplement of each measure.

1. 71°
2. 48°
3. 25°
4. $w°$
5. $(y + 14)°$
6. $(20 - q)°$

Find the measure of the third angle of each triangle in which the measures of two angles of the triangle are given.

7. 110°, 60°
8. 70°, 70°
9. $x°$, $2x°$

Write an equation and solve. Then check your solution.

10. The measure of an angle is 22° less than its complement. Find the measure of each angle.
11. One of two supplementary angles is 42° more than five times the other. Find the measure of each angle.
12. **Construction** A freeway exit ramp makes a 4° angle with the horizontal. What is the measure of the angle that the ramp makes with the vertical?

Review 3–5

Solving Equations with the Variable on Both Sides

When an equation contains parentheses or other grouping symbols, first use the distributive property to remove the grouping symbols. If the equation has variables on each side, use the addition or subtraction property of equality to write an equivalent equation that has all the variables on one side. Then solve the equation.

Example **Solve $6(3z - 4) = -5(z - 14) - 2$.**

$6(3z - 4) = -5(z - 14) - 2$

$18z - 24 = -5z + 70 - 2$ *Use the distributive property.*

$18z + 5z - 24 = -5z + 5z + 70 - 2$ *Add 5z to each side.*

$23z - 24 = 68$

$23z - 24 + 24 = 68 + 24$ *Add 24 to each side.*

$23z = 92$

$\frac{23z}{23} = \frac{92}{23}$ *Divide each side by 23.*

$z = 4$

Check: $6(3z - 4) = -5(z - 14) - 2$

$6(3 \cdot 4 - 4) \stackrel{?}{=} -5(4 - 14) - 2$

$6(12 - 4) \stackrel{?}{=} -5(-10) - 2$

$6(8) \stackrel{?}{=} 50 - 2$

$48 = 48$ ✓

Some equations may have *no solution,* and some equations may have *every number* in their solution set. An equation that is true for every value of the variable is called an **identity**.

EXERCISES

Solve each equation. Then check your solution.

1. $-7(b + 16) = 7(b - 4)$
2. $9 - z = 3z + 39$
3. $12c - 5c = 4c + 15$
4. $40 - 5s = -2(-1 + 3s)$
5. $2.8w + 5.3 = 3.3w - 0.7$
6. $4(m + 9) = 3(8 - m)$
7. $\frac{1}{8}k + 32 = \frac{1}{2}k - 1$
8. $\frac{3}{4}x - x = -\frac{1}{4}(2x + 10)$
9. $7(t + 1) = 5(t - 4) - 1$
10. $3d + 1.1 = 2.3 - d$
11. $-7(a + 2) + 15 = 1 - 7a$
12. $6(3n - 1) = 2(9n + 4)$
13. **Sales** Rundell owns a chain of coffee shops. His shops sell both doughnuts and bagels. Over the past few years he has found that the sales of doughnuts have been decreasing by 0.01 million dollars per year, and sales of bagels have been increasing by 0.06 million dollars per year. Last year, Rundell's shops sold 1.6 million dollars in doughnuts and 0.9 million dollars in bagels. If the sales trends continue, after how many years will sales of bagels and doughnuts be equal?

Review

3–6

Solving Equations and Formulas

If an equation that contains more than one variable is to be solved for a specific variable, use the properties of equality to isolate the specified variable on one side of the equation.

Example 1 **Solve $5s + t^2 = u - v$ for s.**

$$5s + t^2 = u - v$$

$$5s + t^2 - t^2 = u - v - t^2 \quad \text{Subtract } t^2 \text{ from each side.}$$

$$5s = u - v - t^2$$

$$\frac{5s}{5} = \frac{u - v - t^2}{5} \quad \text{Divide each side by 5.}$$

$$s = \frac{u - v - t^2}{5}$$

Example 2 **Solve $by = abx + cx - c$ for x.**

$$by = abx + cx - c$$

$$by + c = abx + cx - c + c \quad \text{Add } c \text{ to each side.}$$

$$by + c = abx + cx$$

$$by + c = x(ab + c) \quad \text{Distributive property}$$

$$\frac{by + c}{ab + c} = \frac{x(ab + c)}{ab + c} \quad \text{Divide each side by } (ab + c).$$

$$\frac{by + c}{ab + c} = x$$

EXERCISES

Solve for *y*.

1. $6 - 2y = z$
2. $y + 12q = 15$
3. $(y + b) + 14 = 20$
4. $ay + c = 4$
5. $y(3 + m) = 9n$
6. $2g + 5y = 8$
7. $4y + s = t$
8. $y(7 - d) = x$
9. $8y - 3v = 5u$
10. $\frac{b - y}{5} = a$

11. **Business** The formula $I = prt$ is the formula for computing simple interest, where I is the interest, p is the principal or the amount invested, r is the interest rate, and t is the time in years. Find the amount of interest earned if you were to invest $8000 at 4% interest (use 0.04) for 5 years.

Review 3–7

Integration: Statistics Measures of Central Tendency

In working with statistical data, it is often useful to have one value represent the complete set of data. For example, **measures of central tendency** represent centralized values of the data. Three measures of central tendency are the **mean**, **median**, and **mode**.

Definitions		Examples
Mean	The mean of a set of data is the sum of the numbers in the set divided by the number of numbers in the set.	Data: 14, 16, 17, 14, 22, 22 $\frac{14 + 16 + 17 + 14 + 22 + 22}{6} = 17.5$
Median	The median of a set of data is the middle number when the numbers in the set are arranged in numerical order. In an even number of elements, the median is halfway between the two middle numbers.	Data: 14, 14, 16, 17, 22, 22 $\frac{16 + 17}{2} = 16.5$
Mode	The mode of a set of data is the number that occurs most often in the set.	Data: 14, 16, 17, 14, 22, 22 There are two modes, 14 and 22.

EXERCISES

Find the mean, median, and mode(s) for each set of data.

1. 90, 88, 93, 87, 95

2.

```
                    ×
     ×    ×         ×
     × ×  ×         ×       × ×
|----+----+----+----+----+
0    1    2    3    4    5
```

3. 10, 18, 18, 18, 10, 10

4. 12, 7, 14, 30

5. $\frac{3}{4}, \frac{2}{3}, \frac{3}{5}, \frac{6}{8}$

6. 12.4, 6.8, 19.1, 30.4, 7.3

Find the median and mode(s) of the data shown in each stem-and-leaf plot.

7.

Stem	Leaf
5	3 8 8 9
6	1 2 2 2 4 6
7	7 7 7 8 9 9

6|2 = 62

8.

Stem	Leaf
11	1 3 4 5
12	2 2 4 4
13	1 3 3 3 4 5

13|3 = 133

9. Football The table at the right shows the teams that have played in the Super Bowl most often.

a. Find the mean, median, and mode of the data.

b. Which measure of central tendency best represents these data, and why?

Team	Times in Super Bowl
Dallas Cowboys	8
Miami Dolphins	5
Pittsburgh Steelers	5
San Francisco 49ers	5
Washington Redskins	5
Buffalo Bills	4
Denver Broncos	4
Minnesota Vikings	4

Solve each equation. Then check your solution.

1. $r - 4 = -48$
2. $h + (-9) = 13$
3. $-\frac{1}{3} + a = \frac{4}{5}$
4. $-3z = -39$
5. $7w = -42$
6. $\frac{1}{8}x = -5$
7. $11y + 9 = 64$
8. $34 = \frac{q + 17}{5}$
9. $\frac{3}{7}c + 3 = 9$
10. $-3(b + 8) = 3(b - 1)$
11. $1.9w + 7.9 = 0.9w - 3.8$
12. $\frac{1}{7}x - x = -\frac{1}{7}(2x + 84)$

13. Find the complement of an angle whose measure is 62°.

14. Find the supplement of an angle whose measure is 74°.

Solve for *d*.

15. $d + 17y = 21$
16. $d(7 - x) = 4v$
17. $7w + 3d = -12$

The list below gives the prices of the same pair of sneakers at different stores. Use this data to complete Exercises 18–20.

72, 63, 65, 72, 75, 70, 68, 72, 68

18. Find the mean of the sneaker prices.

19. Find the median of the sneaker prices.

20. Find the mode of the sneaker prices.

Review 4–1

Ratios and Proportions

In mathematics, a **ratio** compares two numbers by division. A ratio that compares a number x to a number y can be written in the following ways.

$$x \text{ to } y \qquad x{:}y \qquad \frac{x}{y}$$

When a ratio compares two quantities with different units of measure, that ratio is called a **rate**. For example, a 7°F rise in temperature per hour is a rate and can be expressed as $\frac{7 \text{ degrees}}{1 \text{ hour}}$, or 7 degrees per hour.

Proportions are often used to solve problems involving ratios. You can use the means-extremes property of proportions to solve equations that have the form of a proportion.

Definition of Proportion	**An equation of the form $\frac{a}{b} = \frac{c}{d}$ stating that two ratios are equal is called a proportion.**
Means-Extremes Property of Proportions	**In a proportion, the product of the extremes is equal to the product of the means. If $\frac{a}{b} = \frac{c}{d}$, then $ad = bc$.**

Example  **Solve $\frac{r}{8} = \frac{14}{3}$.**

$\frac{r}{8} = \frac{14}{3}$

$3r = 112$

$r = 37\frac{1}{3}$ The solution is $37\frac{1}{3}$.

EXERCISES

Solve each proportion.

1. $\frac{w-1}{9} = \frac{7}{9}$
2. $\frac{3}{11} = \frac{9}{z}$
3. $\frac{a}{7} = \frac{15}{35}$
4. $\frac{0.25}{4} = \frac{0.3}{y}$
5. $\frac{4}{x+2} = \frac{16}{28}$
6. $\frac{9-2t}{7+t} = \frac{15}{27}$
7. $\frac{2b}{13} = \frac{8}{20}$
8. $\frac{m+3}{-5} = \frac{20-m}{-4}$

Use a proportion to solve each problem.

9. To make a model of the Rio Grande River, Angelica used 1 inch of clay for 50 miles of the actual river's length. Her model river was 38 inches long. How long is the Rio Grande?

10. Jorge finished 28 math problems in one hour. At that rate, how many hours will it take him to complete 70 math problems?

Review 4–2

Integration: Geometry
Similar Triangles

Triangle *FGH* is similar to triangle *JKL*. The angles of the two triangles are congruent. They are called **corresponding angles**. The sides opposite corresponding angles are called **corresponding sides**. Proportions can be used to find the missing measures of similar triangles.

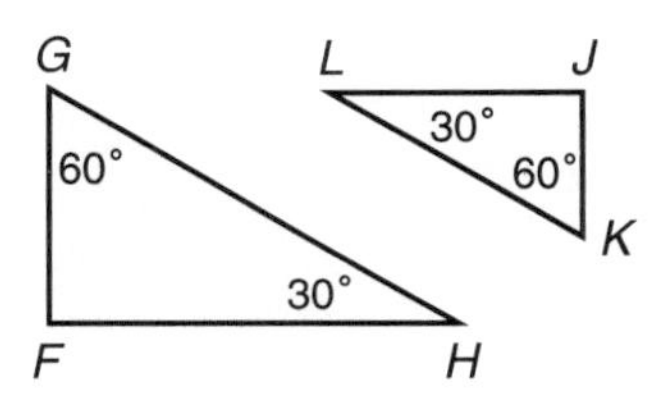

Similar Triangles	If two triangles are similar, the measures of their corresponding sides are proportional, and the measures of their corresponding angles are equal.

Example **Find the height of the apartment building.**

$\triangle ABC$ is similar to $\triangle AED$.

$$\frac{ED}{BC} = \frac{AD}{AC}$$

$$\frac{9}{x} = \frac{30}{390}$$

$$30x = 3510$$

$$x = 117$$

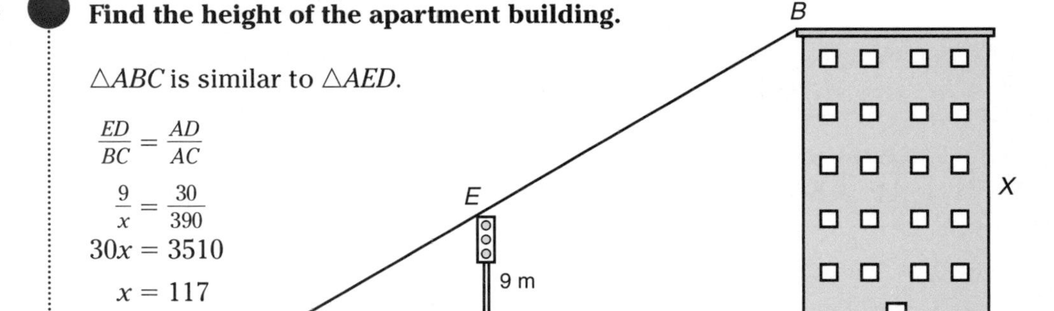

The apartment building is 117 meters high.

EXERCISES

Refer to the triangles below to answer the questions.

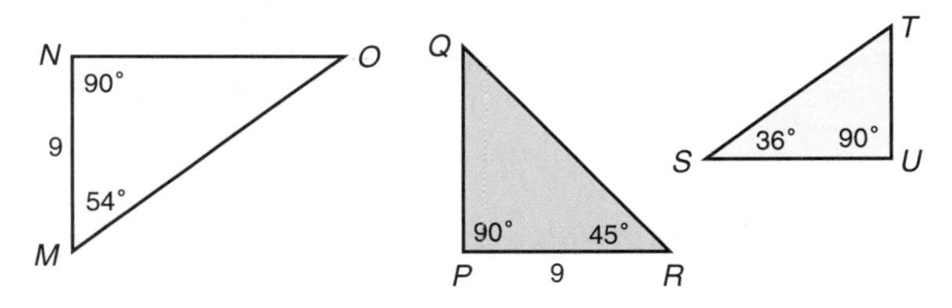

1. Which triangles are similar?
2. Name the corresponding angles of the similar triangles.
3. Name the corresponding sides of the similar triangles.
4. **Surveying** At a certain time of the day, the Empire State Building casts a shadow 140 feet long. At the same time of day, a post 10 feet tall casts a shadow 1.12 feet long. How tall is the Empire State Building?

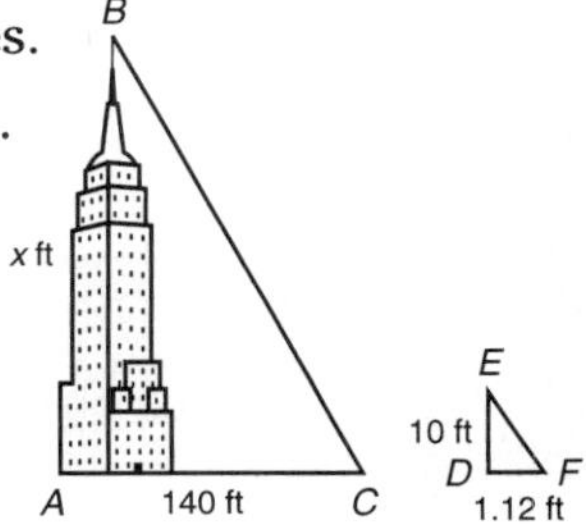

Review 4–3

Integration: Trigonometry
Trigonometric Ratios

For right triangles, **trigonometric ratios** can be defined.

For $\angle A$:

$\overline{BC}$ is *opposite* $\angle A$.

$\overline{AC}$ is *adjacent* to $\angle A$.

$\overline{AB}$ is the *hypotenuse* and is opposite the right angle C.

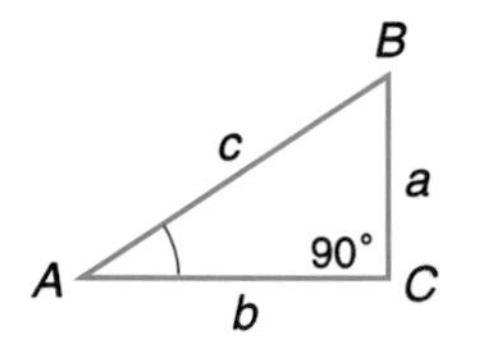

Definition of Trigonometric Ratios	
sine of $\angle A = \frac{\text{measure of leg opposite } \angle A}{\text{measure of hypotenuse}}$	$\sin A = \frac{a}{c}$
cosine of $\angle A = \frac{\text{measure of leg adjacent to } \angle A}{\text{measure of hypotenuse}}$	$\cos A = \frac{b}{c}$
tangent of $\angle A = \frac{\text{measure of leg opposite } \angle A}{\text{measure of leg adjacent to } \angle A}$	$\tan A = \frac{a}{b}$

Example 

a. Find cos *B* to the nearest thousandth.

$\cos B = \frac{a}{c} = \frac{6}{14}$ or 0.429

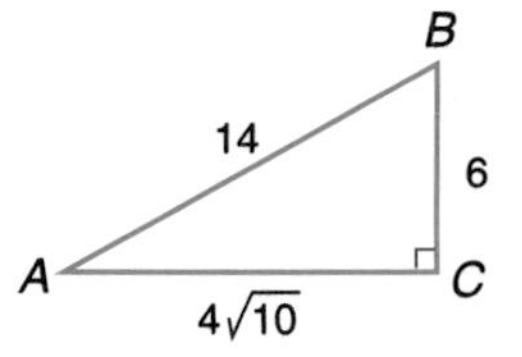

b. Use a scientific calculator to find the measure of $\angle B$.

Enter: .429 INV COS 64.59588564

The measure of $\angle B$ to the nearest degree is 65°.

EXERCISES

1. Find the sine, cosine, and tangent of each acute angle. Round your answers to the nearest thousandth.

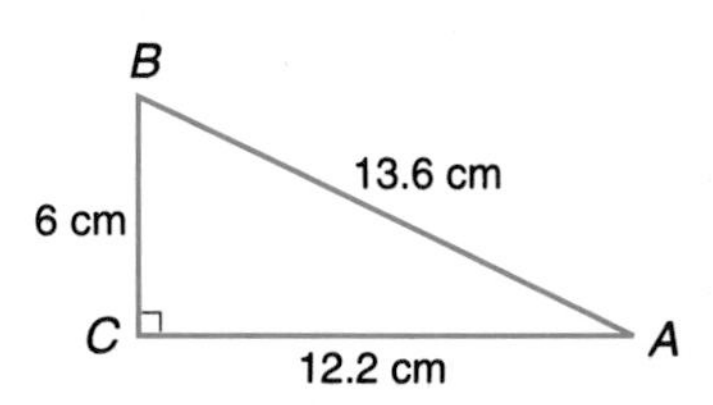

Find each value to the nearest ten thousandth.

2. sin 32°
3. cos 80°
4. tan 20°
5. sin 58°
6. cos 70°

Use a calculator to find the measure of each angle to the nearest degree.

7. $\sin Q = 0.4695$
8. $\tan D = 0.7002$
9. $\cos G = 0.500$

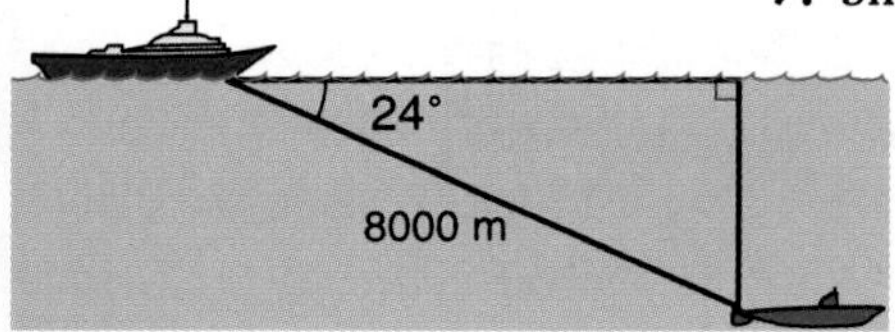

10. **Military** A navy ship has located a submarine using sonar. The sonar reading shows that the distance to the submarine is 8000 meters and the angle of depression to the submarine is 24°. What is the depth of the submarine to the nearest meter?

Review 4–4

Percents

A percent problem may be easier to solve if a proportion is used.

Percent Proportion
$\frac{\text{percentage}}{\text{base}} = \text{rate}$
or
$\frac{\text{percentage}}{\text{base}} = \frac{r}{100}$

Example

a. 48 is what percent of 60?

$$\frac{\textit{percentage}}{\textit{base}} \begin{matrix} \rightarrow \\ \rightarrow \end{matrix} \frac{48}{60} = \frac{r}{100} \leftarrow \textit{rate}$$

$$4800 = 60r$$

$$80 = r$$

48 is 80% of 60.

b. What number is 30% of 180?

$$\frac{\textit{percentage}}{\textit{base}} \begin{matrix} \rightarrow \\ \rightarrow \end{matrix} \frac{n}{180} = \frac{30}{100} \leftarrow \textit{rate}$$

$$n = \frac{30}{100}(180)$$

$$n = 54$$

54 is 30% of 180.

EXERCISES

Use a proportion to answer each question.

1. Forty-five is what percent of 80?
2. Twenty is what percent of 60?
3. What is 40% of 90?
4. Find 36% of 240.
5. Fifty-five is 44% of what number?
6. 8.4 is 12% of what number?
7. On Wednesday, Eric's Machine Shop received a shipment of 24 drill bits. Eric had ordered 40 drill bits. What percent of his order arrived on Wednesday?
8. Mackenzie received a commission of 6% on the sale of a house. If the amount of her commission was $8400, what was the selling price of the house?
9. According to the book *Are You Normal*, 33% of Americans are afraid of snakes, 8% are afraid of thunder and lightning, 4% by crowds and dogs, 2% of cats, 3% of driving a car or leaving the house, and $\frac{1}{3}$ are afraid of flying.
   a. In a group of 1025 people, about how many are afraid of snakes?
   b. About how many are afraid of crowds?
   c. About how many are afraid of flying?
10. In 1995, 351 million pairs of athletic shows were purchased. Women accounted for 42% of those sales, and men for about 33% of those sales. About how many pairs of athletic shoes were purchased by women and men?

Review 4–5

Percent of Change

Some percent problems involve finding a percent of increase or decrease.

Percent of Increase	Percent of Decrease
A basketball that cost \$20 last year costs \$22 this year. The price increased by \$2 since last year. *amount of increase* → / *original price* → $\frac{2}{20} = \frac{r}{100}$ $200 = 20r$ $10 = r$ or $r = 10$ The percent of increase is 10%.	A jacket that originally cost \$80 is now on sale for \$60. *amount of decrease* → / *original price* → $\frac{20}{80} = \frac{r}{100}$ $2000 = 80r$ $25 = r$ or $r = 25$ The percent of decrease is 25%.

The sales tax on a purchase is a percent of the purchase price. To find the total price, you must calculate the amount of sales tax and add it to the purchase price.

EXERCISES

Find the final price of each item. When there is a discount and sales tax, compute the discount first.

1. compact disc: \$12.00
   discount: 20%
2. two concert tickets: \$35.00
   student discount: 15%
3. airline ticket: \$348
   early booking discount: 30%
4. photo calendar: \$12.95
   sales tax: 6%
5. class ring: \$110
   group discount: 12%
   sales tax: 7%
6. multimedia software: \$49.95
   discount: 25%
   sales tax: 5%

Solve each problem. Round to the nearest tenth of a percent.

7. **Consumerism** According to the U.S. Bureau of Economic Analysis, Americans spent \$318.8 billion on recreation in 1992. In 1993, spending on recreation in the United States had grown to \$339.9 billion. What was the percent of increase in spending on recreation from 1992 to 1993?

8. **Agriculture** According to the U.S. Department of Agriculture, the number of farms in the United States in 1984 was 2,334,000. By 1994, this number had fallen to 2,065,000 farms. Find the percent of decrease in the number of farms from 1984 to 1994.

Review 4–6

Integration: Probability
Probability and Odds

The **probability** of an event is a ratio that tells how likely it is that the event will take place.

Definition of Probabilty
$P(\text{event}) = \frac{\text{number of favorable outcomes}}{\text{number of possible outcomes}}$

Example  **Ms. Michalski picks 8 of the 24 students in her class at random for a special project. What is the probability of being picked?**

$$P(\text{being picked}) = \frac{\text{number of students picked}}{\text{total number of students}}$$

The probability of being picked is $\frac{8}{24}$ or $\frac{1}{3}$.

The probability of any event has a value from 0 to 1. If the probability of an event is 0, it is impossible for the event to occur. An event that is certain to occur has a probability of 1. This can be expressed as $0 \leq P\,(\text{event}) \leq 1$.

The odds of an event occurring is the ratio of the number of ways an event can occur (successes) to the number of ways the event cannot occur (failures).

Definition of Odds
$\text{Odds} = \frac{\text{number of sucesses}}{\text{number of failures}}$

Example  **Find the odds that a member of Ms. Michalski's class will be picked for the special project.**

Number of successes: 8 Number of failures: 16

Odds of being picked = number of successes : number of failures

= 8:16 or 1:2

EXERCISES

Solve each problem.

1. There are 3 brown puppies, 4 white puppies, and 5 spotted puppies in a pen. What is the probability of pulling out a white puppy at random?
2. It will rain 5 times in March and snow 7 times. The other days it will be sunny. What is the probability of sun? What are the odds of sun?

There are 300 freshmen, 250 sophomores, 200 juniors, and 250 seniors at a high school. Solve each problem.

3. If one student is chosen at random, what is the probability that a freshman will be chosen?
4. What would be the odds of choosing a sophomore if all of the seniors were eliminated?
5. What are the odds that a junior will not be chosen?
6. What are the odds of choosing a senior at random?

Review 4–7

Weighted Averages

You can use charts to solve mixture problems.

Example  **Ho Lee invested a portion of $24,000 at 6% interest and the balance at 8% interest. How much did he invest at each rate if his total income from both investments was $1720 after one year?**

Amount Invested	Rate	Annual Income
x	0.06	$0.06x$
$24{,}000 - x$	0.08	$0.08(24{,}000 - x)$

$$\underset{\text{income from 6\% investment}}{0.06x} + \underset{\text{income from 8\% investment}}{0.08(24{,}000 - x)} = \underset{\text{total income}}{1720}$$

$$0.06x + 1920 - 0.08x = 1720$$

$$-0.02x = -200$$

$$x = 10{,}000 \text{ and } 24{,}000 - x = 14{,}000$$

Mr. Lee invested $10,000 at 6% and $14,000 at 8%.

When an object moves without changing its speed, it is said to be in **uniform motion**. The formula $d = rt$ is used to solve uniform motion problems.

Example  **Connie Cellucci left home driving at a speed of 62 miles per hour. How many hours did it take her to reach her destination 217 miles away?**

$d = rt$

$217 = 62t$

$3\frac{1}{2} = t$ It will take Ms. Cellucci $3\frac{1}{2}$ hours to drive 217 miles.

EXERCISES

1. How many grams of salt must be added to 90 grams of a 40% solution to obtain a 60% solution?
2. Ms. Sessa and Mr. Schroeder each drove home from a business meeting. Mr. Schroeder traveled north at 80 kilometers per hour and Ms. Sessa traveled south at 90 kilometers per hour. In how many hours were they 200 kilometers apart?
3. Coffee Heaven sells Basic Decaf coffee for $6 per pound and French Vanilla coffee for $9 per pound. How many pounds of Basic Decaf coffee must be added to 20 pounds of French Vanilla coffee to make a mixture that sells for $7 per pound?
4. Alfredo left home at 7:00 A.M., riding his bike at 5 miles per hour. His sister Anel left 1 hour later, riding her bike at 6 miles per hour. At what time will Anel catch up to Alfredo if Anel is delayed 30 minutes by a flat tire?

Review 4–8

Direct and Inverse Variation

If two variables x and y are related by the equation $y = kx$, where k is a nonzero constant, then the equation is called a **direct variation**, and k is called the **constant of variation**. If two variables x and y are related by the equation $xy = k$, where $k \neq 0$, then the equation is called an **inverse variation**.

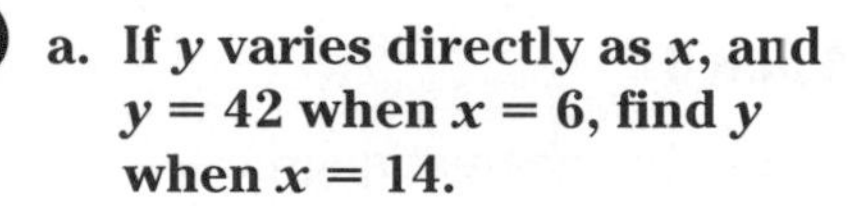

Example

a. If y varies directly as x, and $y = 42$ when $x = 6$, find y when $x = 14$.

$y = kx$	$y = kx$
$42 = 6k$	$y = 7(14)$
$\frac{42}{6} = \frac{6k}{6}$	$y = 98$
$7 = k$	

b. If y varies inversely as x, and $y = 8$ when $x = 9$, find x when $y = 24$.

$xy = k$	$xy = k$
$9(8) = k$	$24x = 72$
$72 = k$	$\frac{24x}{24} = \frac{72}{24}$
	$x = 3$

EXERCISES

Solve. Assume that y varies directly as x.

1. If $y = 15$ when $x = 3$, find y when $x = 8$.
2. If $y = -49$ when $x = 7$, find x when $y = 91$.
3. If $y = \frac{3}{8}$ when $x = 2$, find y when $x = \frac{1}{4}$.
4. If $y = \frac{4}{5}$ when $x = -\frac{3}{4}$, find x when $y = -\frac{7}{10}$.

Solve. Assume that y varies inversely as x.

5. If $y = 10$ when $x = 7.5$, find y when $x = 3$.
6. If $y = -6$ when $x = 14$, find y when $x = -5$.
7. If $y = 18.1$ when $x = 12.4$, find y when $x = 20$.
8. If $y = \frac{3}{8}$ and $x = \frac{1}{9}$, find y when $x = \frac{1}{6}$.
9. **Space** The weight of an object on Mars varies directly as its weight on Earth. An unmanned probe that weighs 500 pounds on Earth weighs 190 pounds on Mars. In the future, NASA hopes to send a manned mission to Mars. How much will an astronaut with gear weighing 220 pounds on Earth weigh on Mars?

CHAPTER 4 TEST

Solve each proportion.

1. $\frac{t-1}{7} = \frac{6}{7}$

2. $\frac{4}{19} = \frac{6}{z}$

3. $\frac{5}{x+2} = \frac{25}{30}$

4. At a certain time of the day, the NationsBank Tower in Atlanta, Georgia, casts a shadow 310 feet long. At the same time of day, a post 8 feet tall casts a shadow 2.42 feet long. How tall is the NationsBank Tower?

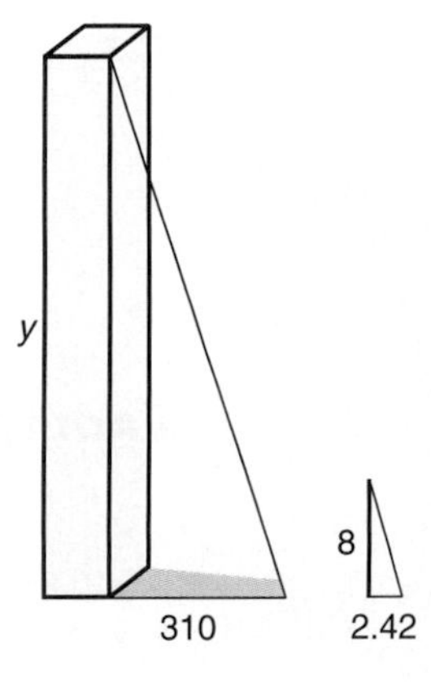

Use the triangle at the right for Exercises 5 and 6.

5. If $a = 9$ and $c = 16$, find the measure of $\angle A$ to the nearest degree.

6. If $a = 12$, $b = 15$, and $c = 3\sqrt{41}$, find tan B.

7. 54.6 is what percent of 130?

8. What number is 72% of 600?

9. In a survey, 29% of the respondents reported that they have returned rented videotapes to the store late. If 3200 people responded to the survey, how many have returned videotapes late?

10. The regular cost of a paperback book is $6.95. A discount store sells all of its paperbacks at a 25% discount. What is the cost of the paperback at the discount store?

11. The price of an oil change increased from $19.95 to $22.94. What was the percent of increase?

12. According to the NCAA, the annual number of Division I basketball games played was at an all-time high in 1992 at 8803 games. In 1995, only 8662 games were played. Find the percent of decrease in the number of Division I basketball games played from 1992 to 1995. Round to the nearest tenth of a percent.

13. If the probability in favor of an event is $\frac{8}{15}$, what are the odds of that event occurring?

14. A box of sports cards contains 9 baseball cards, 7 football cards, and 4 basketball cards. What is the probability that a football card is chosen at random?

15. A factory has an order for 3000 bricks of cheese. Machine A can process 120 bricks of cheese per hour while Machine B can process 110 bricks of cheese per hour. Machine A starts at 9:00 A.M. and Machine B starts at 11:00 A.M. At what time will the two machines complete the job?

16. A chemist has 40 liters of 20% saline solution. How many liters of 40% saline solution must be added to produce a 35% saline solution?

17. If y varies directly as x and $y = 12$ when $x = 14$, find y when $x = 50$.

18. If y varies inversely as x and $y = 3$ when $x = 60$, find y when $x = 100$.

Review

5–1

The Coordinate Plane

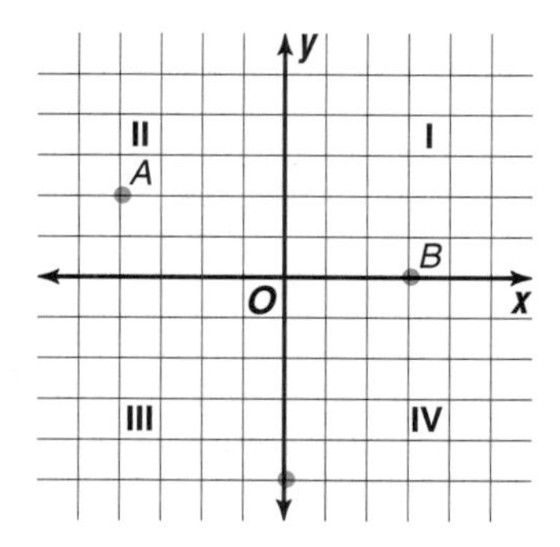

In the diagram at the right, the two perpendicular lines, called the x-axis and the y-axis, divide the coordinate plane into Quadrants I, II, III, and IV. The point where the two axes intersect is called the **origin**. The origin is represented by the ordered pair (0, 0).

Every other point in the coordinate plane is also represented by an ordered pair of numbers. The ordered pair for point A is $(-4, 2)$. We say that -4 is the x-coordinate of A and 2 is the y-coordinate of A.

Example  **Write the ordered pair for the point B on the diagram above.**

The x-coordinate is 3 and the y-coordinate is 0. Thus, the ordered pair is (3, 0).

To graph any ordered pair (x, y), begin at the origin. Move left or right x units. From there, move up or down y units. Draw a dot at that point.

EXERCISES

Graph each point on the same coordinate plane.

1. $C(-4, 0)$
2. $D(2, 2)$
3. $E(3, -5)$
2. $F(-2, 4)$
5. $G(-1, -3)$
6. $H(0, -3)$

Write the ordered pair for each point shown at the right. Name the quadrant in which the point is located.

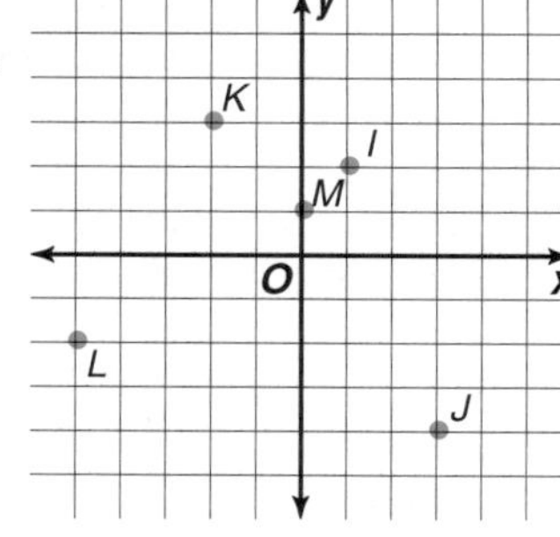

7. I
8. J
9. K
10. L
11. M

12. **Computers** Juan is a computer programmer. He is developing multimedia math software. Each screen in the program must be mapped out pixel by pixel. In the screen shown at the right, find the coordinates of each pixel making up the variable X.

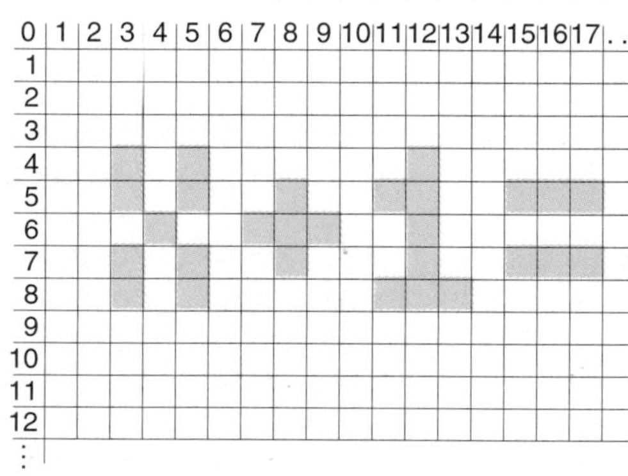

Review

5–2

Relations

A **relation** is a set of ordered pairs. The **domain** is the set of all first coordinates of the ordered pairs, and the **range** is the set of all second coordinates.

Example **State the domain and range of each relation.**

a. {(7, 1), (7, 3), (7, 5)} Domain = {7}; Range = {1, 3, 5}

b. {(2, 4), (3, 6), (4, 4)} Domain = {2, 3, 4}; Range = {4, 6}

Relations can be expressed as ordered pairs, tables, graphs, and mappings. The relation {(−3, 4), (0, 6), (2, −1)} can be expressed in each of the following ways.

Ordered Pairs

(−3, 4)
(0, 6)
(2, −1)

Table

x	y
−3	4
0	6
2	−1

Graph

Mapping

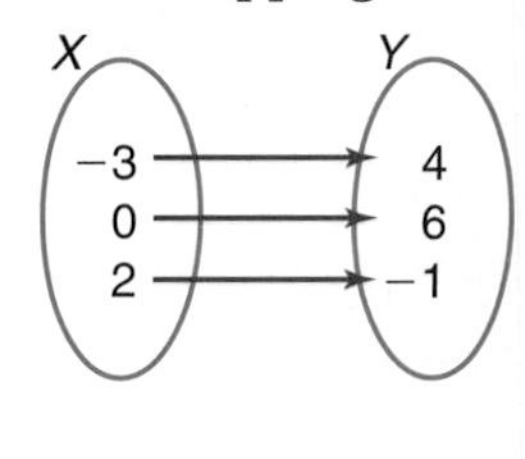

The **inverse** is obtained by switching the coordinates in each ordered pair.

EXERCISES

State the domain and range of each relation.

1. {(5, 4), (−2, 5), (−3, 0), (4, 5), (5, 0)}
2. {(1.25, −0.3), (−14, 12), (6, 1.25)}
3. $\left\{\left(\frac{1}{3}, \frac{3}{8}\right), \left(-\frac{7}{9}, 4\right), \left(2\frac{1}{5}, -\frac{1}{4}\right)\right\}$

Express each relation shown as a set of ordered pairs. Then state the domain, range, and inverse of the relation.

4.

x	y
3	−1
5	4
7	2

5. 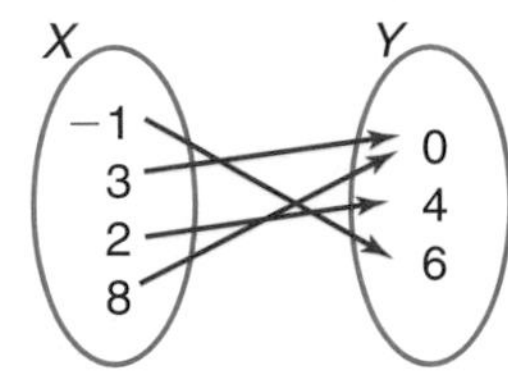

6. **Computers** The table gives the percentage of public schools in the United States that have computers capable of using CD-ROM software.

Percentage of Public Schools with CD-ROMs				
Year	1992	1993	1994	1995
Percentage	7	13	25	41

Source: Quality Education Data, Inc.

a. Determine the domain and range of the relation.

b. Graph the data.

c. What conclusions might you make from the graph of the data?

Review

5–3

Equations as Relations

An equation in two variables describes a relation. It is often easier to determine the solution of such an equation by solving for one of the variables.

Example **Solve $4x - 5y = 4$ if the domain is $\{-4, -\frac{3}{2}, 6\}$.**

First solve for y in terms of x.

$$4x - 5y = 4$$
$$-5y = 4 - 4x$$
$$y = \frac{4 - 4x}{-5}$$

Then substitute values of x.

x	$\frac{4-4x}{-5}$	y	(x, y)
-4	$\frac{4-4(-4)}{-5}$	-4	$(-4, -4)$
$-\frac{3}{2}$	$\frac{4-4\left(-\frac{3}{2}\right)}{-5}$	-2	$\left(-\frac{3}{2}, -2\right)$
6	$\frac{4-4(6)}{-5}$	4	$(6, 4)$

EXERCISES

Which ordered pairs are solutions of each equation?

1. $a + 3b = 10$ **a.** $(3, 1)$ **b.** $(1, 3)$ **c.** $(10, 0)$ **d.** $(-2, 4)$

2. $5x + 2y = 18$ **a.** $(4, -1)$ **b.** $(3, 1)$ **c.** $(2, 4)$ **d.** $(1, 7)$

Solve each equation if the domain is $\{-2, -1, 0, 1, 2\}$.

3. $x - y = 3$
4. $y = 2x + 6$
5. $2x - 3y = 7$
6. $x + 4y = 12$
7. $3x + 7y = 21$
8. $2x - 9 = y$
9. **Cooking** For mashed potatoes, some cooks recommend using one more potato than the number of people to be served.
   a. Write an equation describing the number of potatoes to use when making mashed potatoes. Let x be the number of people to be served. Let y be the number of potatoes to use.
   b. Make a table of the solution set if the domain is $\{3, 4, 5, 6\}$

Review 5–4

Graphing Linear Equations

An equation whose graph is a straight line is called a **linear equation**.

Definition of Linear Equation
A **linear equation** is an equation that can be written in the form $Ax + By = C$, where A, B, and C are any real numbers and A and B are not both zero.

Drawing the Graph of a Linear Equation

1. Solve the equation for one variable.
2. Set up a table of values for the variables.
3. Graph the ordered pairs and connect them with a line.

Example **Draw the graph of $y + 2x = 2$.**

$y + 2x = 2$

$y = 2 - 2x$

x	$2 - 2x$	y	(x, y)
−1	2 − 2(−1)	4	(−1, 4)
0	2 − 2(0)	2	(0, 2)
1	2 − 2(1)	0	(1, 0)

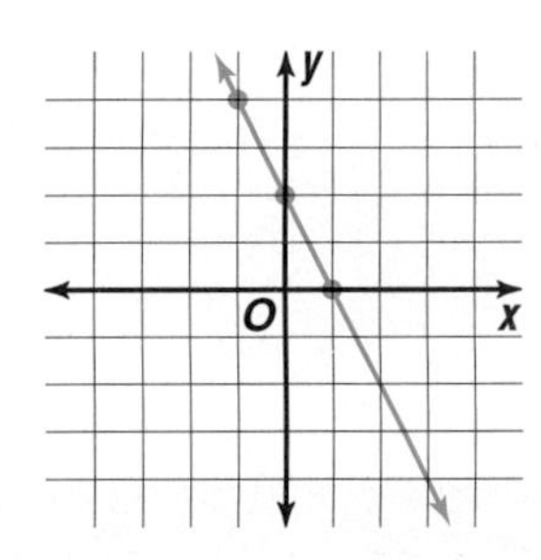

EXERCISES

Determine whether each equation is a linear equation. If an equation is linear, rewrite it in the form $Ax + By = C$.

1. $5x = 3y + 6$
2. $x^2 + 14 = 2y$
3. $\frac{y}{9} = 3$

Graph each equation.

4. $x + 3y = 7$
5. $2s - t = 4$
6. $a + b = -3$
7. $-2x + 4y = 6$
8. $m - 2n = -2$
9. $\frac{3}{4}x - \frac{1}{4}y = 8$
10. **Employment** Maria Tomasso works as a sales representative. She receives a salary of $2200 per month plus a 4% commission on monthly sales. She estimates that her sales in October, November, and December will be $2000, $3500, and $2800.
    a. Graph ordered pairs that represent her incomes y for the three monthly sales figures x.
    b. Will her total monthly income be more than $2300 in any of the three months? Explain.

Review

5–5

Functions

A special type of relation is called a **function**.

Definition of Function
A **function** is a relation in which each element of the domain is paired with *exactly* one element of the range.

Example  **Is {(4, 2), (7, 1), (5, −1), (3, 2)} a function? Is the inverse a function?**

Since each element of the domain is paired with exactly one element of the range, the relation is a function. The inverse is not a function because 2 is paired with more than one element of the range.

The equation $y = 3x - 1$ can be written as $f(x) = 3x - 1$. If $x = 4$, then $f(4) = 3(4) - 1$, or 11. Thus, $f(4)$, which is read "*f* of 4" is a way of referring to the value of y that corresponds to $x = 4$.

Example  **If $f(x) = 5x + 4$, find $f(2)$ and $f(-1)$.**

$$f(2) = 5(2) + 4 = 10 + 4 = 14$$

$$f(-1) = 5(-1) + 4 = -5 + 4 = -1$$

EXERCISES

Determine whether each relation is a function.

1. 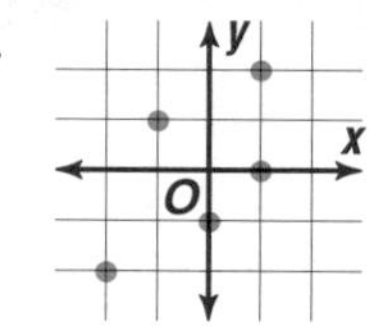

2. 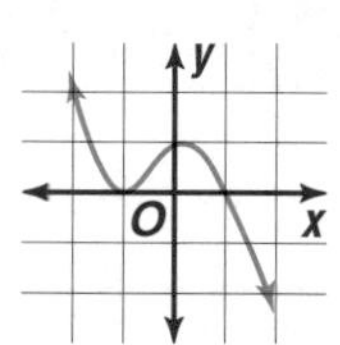

3. 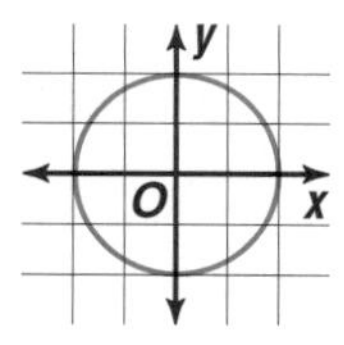

4. {(2, 1), (4, 4), (3, 1)}
5. {(7, 3), (6, 4), (7, 5)}
6. {(2, 1), (−2, 1), (3, 1)}
7. $7x + 7 = 5y$
8. $8 - y = 0$
9. $y = x^2 + 7$

Given $f(x) = 6x + 4$ and $g(x) = x^2 + 2x - 1$, find each value.

10. $f(2)$
11. $f(-3)$
12. $g(0)$
13. $g(2)$
14. $f(1)$
15. $g(-1)$
16. **Business** Viktor Alessandrovich owns a chain of ice cream parlors. He has noticed that his daily sales are dependent on the high temperature for the day. The formula for the relationship is $s = 2t^2 + 20t$, where t represents the daily high temperature in degrees Fahrenheit and s is the amount of daily sales.
    a. Suppose the domain for the function includes the set {60, 70, 75, 80, 85, 90}. Make a table using these values for t and graph the function.
    b. Describe the graph of the function. What trends do you see?

Review

5–6

Writing Equations from Patterns

You can find equations from relations. Suppose you purchased a number of packages of blank computer diskettes. If each package contained 10 diskettes, you could make a chart to show the relationship between the number of packages of diskettes and the total number of diskettes purchased. Use a for the number of packages and b for the number of diskettes.

a	1	2	3	4	5	6
b	10	20	30	40	50	60

This relationship can also be shown as an equation. Since b is always 10 times a, the equation is $b = 10a$. Another way to discover this relationship is to study the difference between successive values of a and b.

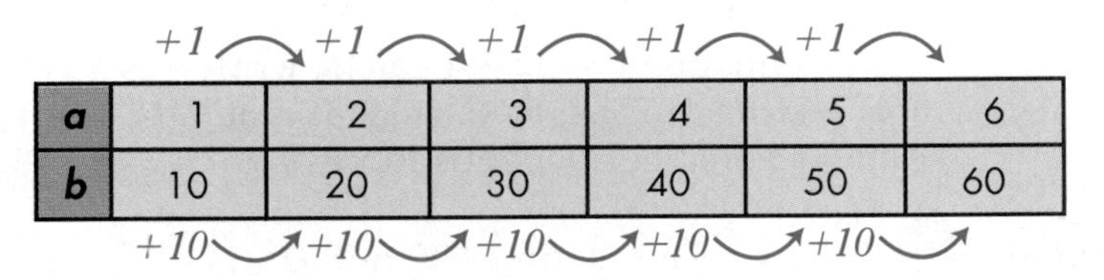

a	1	2	3	4	5	6
b	10	20	30	40	50	60

This suggests the relation $b = 10a$.

EXERCISES

Write an equation for each relation. Then complete each chart.

1.

a	0	1	2	3	4	5
b	0	$\frac{1}{4}$	$\frac{1}{2}$			

2.

x	−2	−1	0	1	2	3
y	0	3	6			

3.

x	−1	0	1	2	3	4
y	3	5	7			

4.

m	−3	−2	−1	0	1	2
n	1	0	−1			

5. **Sales** Kauffner's Sporting Goods, Inc., offers an annual bonus to the employees of its stores. The bonus is based on the amount of sales over the store's target sales level. The table below illustrates the relationship between sales and annual bonus.

Sales over target level	$10,000	$20,000	$25,000	$30,000	$50,000
Bonus	$200	$400	$500	$600	$1000

a. Write an equation in functional notation for the relation.

b. Estimate the amount of an employee's annual bonus if the store's sales are $17,000 over the target level.

Review 5–7

Integration: Statistics Measures of Variation

A *measure of variation* called the **range** describes the spread of numbers in a set of data. To find the range, determine the difference between the greatest and least values in the set.

Quartiles divide the data into four equal parts. The **upper quartile** divides the top half into two equal parts. The **lower quartile** divides the bottom half into two equal parts. Another measure of variation uses the upper and lower quartile values to determine the **interquartile range**. Study the data below.

lower quartile ↓ (between 20 and 33); median = 62 ↓ (between 51 and 73); upper quartile ↓ (between 90 and 90)

4 14 20 33 40 51 73 79 90 90 94 95

The lower quartile is the median of the lower half (26.5). The upper quartile is the median of the upper half (90). The range is $95 - 4 = 91$. The interquartile range is $90 - 26.5 = 63.5$.

EXERCISES

Find the range, median, upper quartile, lower quartile, and interquartile range for each set of data.

1.

Algebra Test Scores		
79	68	82
86	82	91
94	88	85
97	87	90

2.

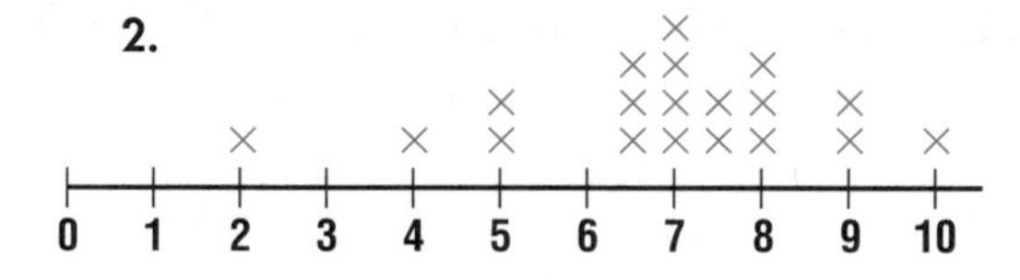

3. 20, 24, 36, 30, 28, 25, 35, 27, 33

4. 14, 19, 16, 13

5. 7, 2, 3, 5, 5, 10, 0, 6

6. 9.5, 12.5, 14, 13, 10.5,

7.

Stem	Leaf
5	3 8 8 9
6	1 2 2 2 4 6
7	7 7 7 8 9 9

6|2 = 62

8.

Stem	Leaf
11	1 3 4 5
12	2 2 4 4
13	1 3 3 3 4 5

13|3 = 133

9. The number of recreational visits to the most-visited sites in the U.S. National Park System in 1994 are given in the table at the right. Find the range, median, upper and lower quartiles, and interquartile range.

National Park Site	Number of visits (millions)
Blue Ridge Parkway	17
Golden Gate Recreational Area	15
Lake Mead Recreational Area	10
Great Smoky Mountains	9
George Washington Parkway	6
National Capital Parks	5
Natchez Trace	5
Cape Cod Seashore	5

Source: U.S. National Park Service

CHAPTER 5 TEST

Use the table to complete Exercises 1 and 2.

x	y
3	12
0	5
−4	−8

1. Write the relation as a set of ordered pairs.

2. State the domain, range, and inverse of the relation.

3. Graph these points on the same coordinate plane.

   $P(3, 0)$ $Q(-2, 6)$ $R(4, -1)$ $S(0, 5)$

4. Name the quadrant in which each point is located.

   $A(-5, -7)$ $B(2, 2)$ $C(9, -13)$

5. Draw a mapping for the relation $\{(0, 5), (6, 1), (3, 5), (4, -2)\}$.

6. Which ordered pairs are solutions of the equation $2x + 5y = -18$?

   a. $(-3, -2)$ b. $(10, -1)$ c. $(-9, 0)$ d. $(12, -2)$

Solve each equation if the domain is {−2, 0, 2, 4}.

7. $x - 2y = 8$

8. $y = 3x + 10$

Graph each equation.

9. $a - 2b = 9$

10. $3m + 2n = 5$

11. $x + y = -6$

12. Determine whether each relation is a function.

    a. $\{(3, 4), (5, 5), (6, 4)\}$

    b. $\{(1, 0), (2, 0), (5, 0)\}$

    c. $\{(1, 9), (-3, 6), (1, -1)\}$

13. Given $f(x) = 7x + 12$, find $f(3)$.

14. Given $g(x) = 2x^2 + 5$, find $g(2)$.

Write an equation for the relation in each chart.

15.

m	0	1	2	3	4	5
n	−2	1	4	7	10	13

16.

s	−2	−1	0	1	2
t	−19	−7	5	17	29

The circulations of the top public libraries in the United States in 1994–1995 are listed in the table at the right.

17. Find the range and median of the library circulation data.

18. Find the upper quartile, lower quartile, and interquartile range for the library circulation data.

Name of Public Library	Circulation (millions)
Los Angeles (CA)	18.3
Queens Borough (NY)	13.6
King County (Seattle, WA)	12.3
Los Angeles County (CA)	11.9
Cincinnati & Hamilton County (OH)	11.7
Baltimore County (MD)	11.0
Columbus (OH)	10.9
New York (NY)	10.2

Source: Public Library Association

Review 6–1

Slope

The ratio of *rise* to *run* is called **slope**. The slope of a line describes its steepness, or rate of change.

Definition of Slope

The slope m of a line is the ratio of the change in the y-coordinates to the corresponding change in the x-coordinates.

$$\text{Slope} = \frac{\text{change in } y}{\text{change in } x} \text{ or } m = \frac{\text{change in } y}{\text{change in } x}$$

On a coordinate plane, a line extending from lower left to upper right has a positive slope. A line extending from upper left to lower right has a negative slope. The slope of a horizontal line is zero. A vertical line has *no slope*.

The slope of a nonvertical line can be determined from the coordinates of any two points on the line.

Determining Slope Given Two Points

Given the coordinates of two points, (x_1, y_1) and the (x_2, y_2) on a line, the slope m can be found as follows:

$$m = \frac{y_2 - y_1}{x_2 - x_1}, \text{ where } x_1 \neq x_2.$$

Example **Determine the slope of the line that passes through (−3, 2) and (6, −4).**

$$m = \frac{y_2 - y_1}{x_2 - x_1}$$

$$= \frac{-4 - 2}{6 - (-3)}$$

$$= \frac{-6}{9} = -\frac{2}{3}$$

EXERCISES

Determine the slope of the line that passes through each pair of points.

1. (3, 7), (4, 5)
2. (10, −2), (6, 3)
3. (3, 0), (9, 1)
4. (0, 5), (8, −9)
5. (−2, −2), (−1, 7)
6. (4, −3), (−9, 0)

Determine the value of *r* so the line that passes through each pair of points has the given slope.

7. $(8, 6), (r, 2), m = \frac{4}{3}$
8. $(9, -7), (3, r), m = -1$
9. $(2, r), (0, -6), m = \frac{3}{2}$
10. $(r, 1), (6, -3), m = 2$
11. $(4, r), (-1, -5), m = \frac{8}{5}$
12. $(8, r), (r, 3), m = -\frac{1}{6}$
13. **Road Construction** A portion of the John Scott Memorial Highway in Steubenville, Ohio, has a grade of 12%. The length of this portion is approximately 1.1 miles. What is the change in elevation from the top of the grade to the bottom of the grade in feet?

Review

6–2

Writing Linear Equations in Point-Slope and Standard Forms

If you know the slope of a line and the coordinates of one point on the line, you can write an equation of the line by using the **point-slope form**. For a given point (x_1, y_1) on a nonvertical line with slope m, the point-slope form of a linear equation is $y - y_1 = m(x - x_1)$.

Standard Form

Any linear equation can be expressed in the form $Ax + By = C$, where A, B, and C are integers, $A \geq 0$, and A and B are not both zero. This is called the **standard form**.

Example 1

a. Write the point-slope form of an equation of the line that passes through (3, 8) and has a slope of $-\frac{2}{7}$.

$$y - y_1 = m(x - x_1)$$
$$y - 8 = -\frac{2}{7}(x - 3)$$

b. Write $y - 6 = 4(x + 3)$ in standard form.

$$y - 6 = 4(x + 3)$$
$$y - 6 = 4x + 12$$
$$-4x + y = 18$$
$$4x - y = -18$$

You can also find an equation of a line if you know the coordinates of two points on the line. First, find the slope of the line. Then write an equation of the line by using the point-slope form or the standard form.

EXERCISES

Write the standard form of an equation of the line that passes through the given point and has the given slope.

1. $(5, 6)$, $m = \frac{7}{10}$
2. $(-1, 0)$, $m = 3$
3. $(7, -8)$, $m = -3$
4. $(1, 4)$, $m = -\frac{8}{9}$
5. $(6, -5)$, $m = \frac{1}{6}$
6. $(9, 6)$, $m = 0$

Write the point-slope form of an equation of the line that passes through each pair of points.

7. $(7, 5)$, $(-3, 2)$
8. $(-6, 8)$, $(12, 4)$
9. $(4, 4)$, $(2, -8)$
10. $(-9, -4)$, $(4, -2)$
11. $(6, 7)$, $(7, 6)$
12. $(-4, -9)$, $(-6, -3)$

13. **Painting** Vanessa is painting a wall mural. To paint the perspective correctly, she must connect two points on the wall with a line. One point is two feet from the bottom of the painting, three feet from the left side of the painting, and has coordinates (3, 2). The other point is seven feet from the bottom of the painting and ten feet from the left of the painting.
    a. Find an equation for the line connecting these points in point-slope form.
    b. Vanessa has painted the line on her mural but is worried that the line isn't straight. The line she has painted goes through the point (6, 5). Has she painted the line correctly?

Review 6–3

Integration: Statistics Scatter Plots and Best-Fit Lines

A **scatter plot** is a graph that shows the relationship between paired data. The scatter plot may reveal a pattern, or association, between the paired data. This association can be negative or positive. The association is said to be positive when a line suggested by the points slants upward.

The scatter plot at the right represents the relationship between the amount of time Anita spends on her Spanish homework each week and her score on her weekly Spanish quiz. Since the points suggest a line that slants upward, there seems to be a positive relationship between the paired data. In general, the scatter plot seems to show that the more Anita studies, the better her quiz score.

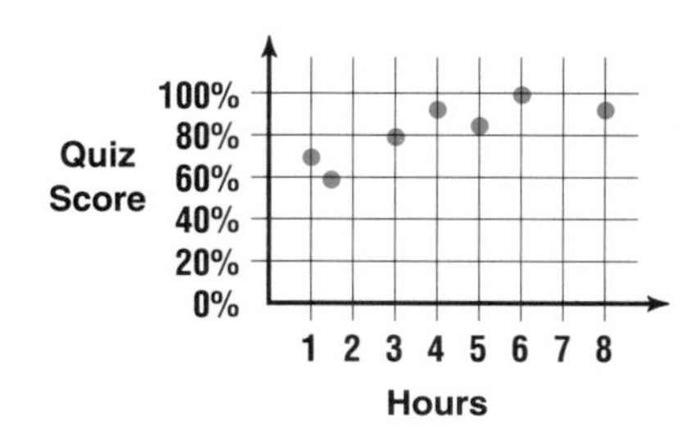

EXERCISES

Solve each problem.

1. The table at the right shows the gasoline/mileage record for a certain car. At each gasoline fill-up, the car's owner recorded the amount of gasoline used since the previous fill-up and the distance traveled on that amount of gasoline.
   a. Draw a scatter plot from the data in the table.
   b. What are the paired data?
   c. Is there a relationship between the gasoline used and the distance driven?

Gasoline/Mileage Record

Gallons of Gasoline	Miles Traveled
9.8	250
7.9	200
9.1	240
8.0	210
7.5	185
9.3	255
9.5	250

2. Alex's biking speed after 10 minutes was 25 mi/h; at 30 minutes, 22 mi/h; at 45 minutes, 20 mi/h; and at 60 minutes, 19 mi/h.
   a. Make a scatter plot pairing time biked with biking speed.
   b. How is the data related, positively, negatively, or not at all?
3. The table below shows the percent of children who are overweight ranked by the hours per day they spend watching television.
   a. Draw a scatter plot for these data.
   b. Write an equation for the best-fit line.
   c. What conclusion might you draw from these data?

Hours of TV Per Day	Percent Overweight
1	12
2	23
3	28
4	30
5 or more	33

Review 6–4

Writing Linear Equations in Slope-Intercept Form

The x-coordinate of the point where a line crosses the x-axis is called the x**-intercept**. Similarly, the y-coordinate of the point where the line crosses the y-axis is called the y**-intercept**.

Slope-Intercept Form of a Linear Equation
Given the slope m and the y-intercept b of a line, the slope-intercept form of an equation of the line is $y = mx + b$.

If an equation is given in standard form $Ax + By = C$ and B is not zero, the slope of the line is $-\frac{A}{B}$ and the y-intercept is $\frac{C}{B}$. The x-intercept is $\frac{C}{A}$, where $A \neq 0$.

Example **Find the x- and y-intercepts of the graph of $2x - 6y = -12$. Then write the equation in slope-intercept form.**

Since $A = 2$, $B = -6$, and $C = -12$,

$$\frac{C}{A} = \frac{-12}{2} = -6 \qquad \frac{C}{B} = \frac{-12}{-6} = 2 \qquad m = -\frac{A}{B} = \frac{1}{3}$$

Thus, the x-intercept is -6, and the y-intercept is 2. The equation of the line in slope-intercept form is $y = \frac{1}{3}x + 2$.

EXERCISES

Find the x- and y-intercepts of the graph of each equation.

1. $4x + 8y = 16$
2. $2x - 7y = -4$
3. $4x - 2y = 9$
4. $3x + 6y = -13$

Write an equation in slope-intercept form of a line with the given slope and y-intercept. Then write the equation in standard form.

5. $m = 0, b = 5$
6. $m = 3, b = 1$
7. $m = -9, b = 8$
8. $m = 2, b = -7$

Find the slope and y-intercept of the graph of each equation. Then write each equation in slope-intercept form.

9. $0.1x + 0.7y = 2.2$
10. $5x + 8y = 9$
11. $4x - y = 4$
12. $35x - 2y = 38$
13. **Computers** A popular on-line service offers a light-usage payment plan. With this plan, customers pay a $4.95 monthly charge, which includes three hours of use. Customers are charged $2.50 per hour for each additional hour of use beyond the first three hours.
    a. Write an equation to represent the line that shows the total monthly amount charged by the on-line service for a customer who spends more than three hours on-line per month.
    b. Suppose you spend an additional 7 hours on-line. Use the equation from part a to determine whether you should switch to the on-line service's unlimited usage plan for $19.95 per month.

Review 6–5

Graphing Linear Equations

There are three methods you can use for graphing equations. You can find two ordered pairs that satisfy the equation, the x- and y-intercepts, or the slope and y-intercept.

Example 1 **Graph $2x - 3y = 12$ by using the x- and y-intercepts.**

The equation is in standard form $Ax + By = C$.

The x-intercept is $\frac{C}{A}$, or 6.

The y-intercept is $\frac{C}{B}$, or -4.

Thus, the graph contains the points (6, 0) and (0, −4).

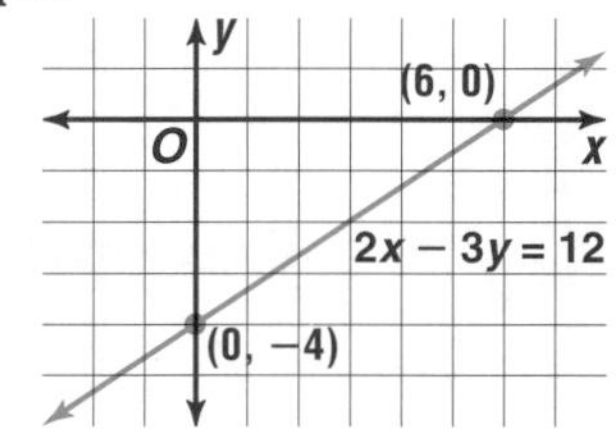

Example 2 **Graph $y = \frac{3}{4}x + 3$ by using the slope and y-intercept.**

The y-intercept is 3, and the slope is $\frac{3}{4}$.

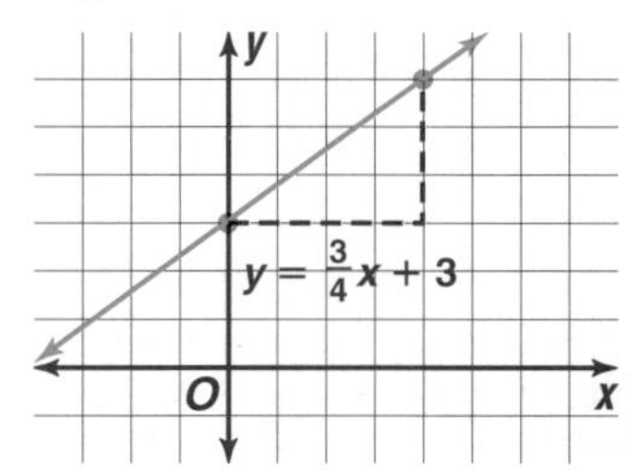

EXERCISES

Graph each equation by using the *x*- and *y*-intercepts.

1. $3x - 5y = 15$
2. $x + 7y = 14$

Graph each equation by using the slope and *y*-intercept.

3. $y = \frac{1}{5}x + 2$
4. $y = \frac{3}{4}x - 3$

5. **Biology** As the temperature increases, the number of times a cricket chirps per minute also increases. The temperature can be approximated by the equation $T = \frac{1}{4}c + 40$, where c represents the number of cricket chirps per minute and T is the temperature in degrees Fahrenheit.
   a. Graph the equation.
   b. On a summer evening you hear a cricket chirping outside your window. Suppose the cricket chirps 132 times per minute. What is the temperature outside?

Review

6–6

Integration: Geometry Parallel and Perpendicular Lines

When you graph two lines, you may encounter the two special types of graphs described at the right.

Parallel Lines and Perpendicular Lines
If two nonvertical lines have the same slope, then they are **parallel**. All vertical lines are parallel.
If the product of the slopes of two lines is -1, then the lines are **perpendicular**. In a plane, vertical lines and horizontal lines are perpendicular.

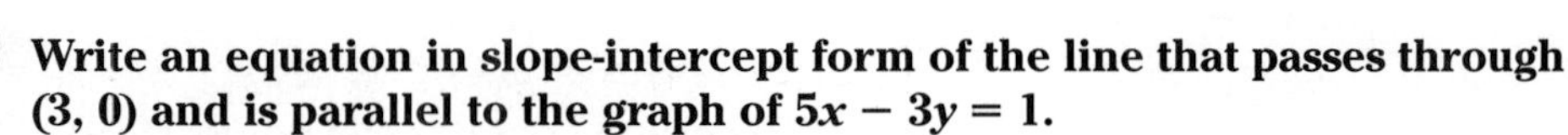

Example **Write an equation in slope-intercept form of the line that passes through (3, 0) and is parallel to the graph of $5x - 3y = 1$.**

The slope of the graph is $-\frac{A}{B} = -\frac{5}{-3} = \frac{5}{3}$.

The slope-intercept form of an equation whose graph is parallel to the original graph is $y = \frac{5}{3}x + b$. Substitute (3, 0) into the equation and solve for b.

$0 = \frac{5}{3}(3) + b$

$b = -5$ The y-intercept is -5.

The equation of the line is $y = \frac{5}{3}x - 5$.

Since $\frac{5}{3} \cdot \left(-\frac{3}{5}\right) = -1$, any line that is perpendicular to the line in the example has an equation of the form $y = -\frac{3}{5}x + b$. If the line includes the point (5, 7), then $7 = -\frac{3}{5}(5) + b$ and thus, $b = 10$. The equation of the line is $y = -\frac{3}{5}x + 10$.

EXERCISES

Write an equation in slope-intercept form of the line that passes through the given point and is parallel to the graph of each equation.

1. $y = 4x - 5$; (1, 6)
2. $y = -\frac{1}{3}x + 7$; (8, 0)
3. $4x - 9y = 18$, $(-3, 2)$

Write an equation in slope-intercept form of the line that passes through the given point and is perpendicular to the graph of each equation.

4. $x - 2y = 5$; (1, 4)
5. $y = 5x + 12$; $(-6, 2)$
6. $4x + y = 0$; $(3, -3)$

7. **Transportation** In an aerial photograph, one rail in a set of railroad tracks has the equation $y = 0.73x + 3.6$. Find the equation of the other rail.

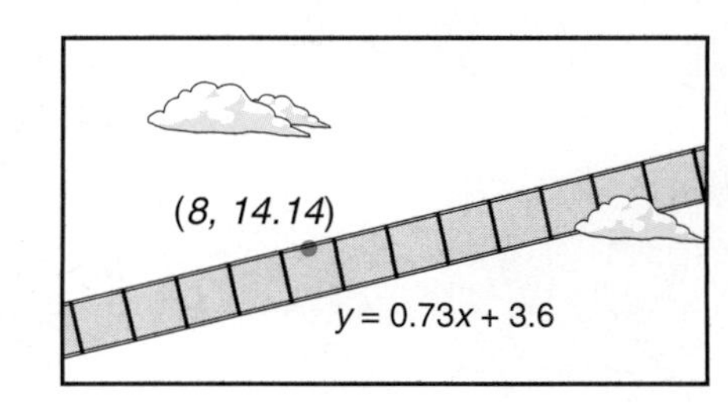

Review

6–7

Integration: Geometry Midpoint of a Line Segment

The **midpoint** of a line segment is the point that is halfway between the endpoints of the line segment.

Midpoint of a Line Segment
The coordinates of the midpoint of a line segment whose endpoints are at (x_1, y_1) and (x_2, y_2) are given by $\left(\frac{x_1 + x_2}{2}, \frac{y_1 + y_2}{2}\right)$.

Example

Find the coordinates of the midpoint of a segment whose endpoints are $A(-2, 3)$ and $B(8, 5)$.

$$(x, y) = \left(\frac{x_1 + x_2}{2}, \frac{y_1 + y_2}{2}\right)$$

$$= \left(\frac{-2 + 8}{2}, \frac{3 + 5}{2}\right)$$

$$= (3, 4)$$

The coordinates of the midpoint of $\overline{AB}$ are (3, 4).

EXERCISES

Find the coordinates of the midpoint of a segment with each pair of endpoints.

1. $E(7, 4), F(3, 2)$
2. $K(1, 2), L(7, 3)$
3. $A(0, -7), B(9, 7)$
4. $P\left(1, \frac{1}{3}\right), Q\left(3, \frac{1}{6}\right)$
5. $I(5, -5), J(3, -2)$
6. $S(2, 8), T(1, -5)$
7. $G(3, 7), H(9, 9)$
8. $M(64, 69), N(0, 10)$
9. $C(2a, 3b), D(4a, 3b)$

If *P* is the midpoint of the segment *MN*, find the coordinates of the missing point.

10. $M(7, 3), N(5, 4)$
11. $N(-9, 39), P(-1, 4)$
12. $N(-9, -2), P(-4, 0)$
13. $M(31, 50), P(40, 23)$
14. $M(10, -9), N(22, -15)$
15. $N(9, 3), P(-6, -2)$

16. **Employment** According to the U.S. Department of Labor, the average earnings of production workers in 1990 was \$10.01 per hour. In 1994 the average earnings of production workers was \$11.13 per hour.
    a. Use the data to form two ordered pairs.
    b. Graph the ordered pairs.
    c. Find the midpoint of the line segment joining the two pairs.
    d. What does this midpoint tell you?

CHAPTER 6 TEST

Determine the slope of the line that passes through each pair of points.

1. $(6, 9), (1, 10)$

2. $(8, -3), (2, 5)$

3. Determine the value of r so the line that passes through $(10, r)$ and $(8, 15)$ has a slope of $-\frac{7}{2}$.

Write the standard form of an equation of the line satisfying the given conditions.

4. passes through $(1, 3)$ and has a slope of $\frac{5}{8}$
5. passes through $(12, 3)$ and has a slope of 0
6. passes through $(4, 7)$ and $(-3, 2)$

The value of U.S. exports and imports (in billions of dollars) for selected years from 1970 to 1994 are listed in the table at the right. Use the data for Exercises 7 and 8.

Year	Exports	Imports
1970	43	40
1975	108	99
1980	221	245
1985	213	345
1990	394	495
1991	422	485
1992	448	533
1993	465	581
1994	513	663

7. Make a scatter plot of the data with the value of exports on the horizontal axis and the value of imports on the vertical axis.
8. Does the data have *positive, negative,* or *no* correlation?

9. Find the x- and y-intercepts of the graph of $9x - 3y = 5$.
10. Write an equation in slope-intercept form of a line that has a slope of -4 and y-intercept of 7.
11. Find the slope and y-intercept of the graph of $0.3x + 1.4y = 4$.
12. Graph $4x - y = 16$ by using the x- and y-intercepts.
13. Graph $y = \frac{3}{5}x - 4$ by using the slope and y-intercept.
14. Write an equation in slope-intercept form of the line that passes through $(-3, -5)$ and is parallel to the graph $y = -\frac{3}{4}x + 6$.

Write an equation in slope-intercept form of the line that passes through the given point and is perpendicular to the graph of each equation.

15. $5x - y = 5$; $(0, 9)$

16. $y = 3x + 17$; $(-7, 4)$

Find the coordinates of the midpoint of a segment with each pair of endpoints.

17. $A(3, -12), B(7, 9)$

18. $I(13, -4), J(7, -6)$

19. If P is the midpoint of the segment MN, find the coordinates of the missing point. $N(-7, -3), P(-14, 6)$

POSTTEST

1. Write the expression $7 \cdot a \cdot a \cdot b \cdot c \cdot c$ using exponents.
   - A. $7a^2c^2$
   - B. $7^2a^2bc^2$
   - C. $7a^2bc^2$
   - D. $7a^3bc^2$

2. Name the property illustrated by $3p(p + 3) = 3p^2 + 9p$.
   - A. associative property of addition
   - B. associative property of multiplication
   - C. distributive property
   - D. commutative property of multiplication

3. Evaluate $2x^2y^2 - 4xy$ if $x = 2$ and $y = 5$.
   - A. 160
   - B. -20
   - C. 40
   - D. 30

4. Simplify $\frac{6 \cdot 7 + 3}{3^2} - \frac{50 - 5^2}{5}$.
   - A. -400
   - B. $\frac{5}{3}$
   - C. 0
   - D. 5

5. Simplify $5x^2 - 9x + 18 - 2x^2 + 6x + 12$.
   - A. $7x^2 + 3x - 30$
   - B. $-7x^2 - 3x + 30$
   - C. $3x^2 + 3x - 30$
   - D. $3x^2 - 3x + 30$

6. Simplify $-1\frac{1}{2} + \frac{2}{3} - \frac{3}{4}$.
   - A. $-\frac{2}{13}$
   - B. $-\frac{1}{6}$
   - C. $-\frac{5}{12}$
   - D. $\frac{7}{24}$

7. Simplify $\frac{64m + 88n}{-4}$.
   - A. $-16m - 22n$
   - B. $16m - 22n$
   - C. $16m + 22n$
   - D. $-256m - 352n$

8. Evaluate $\sqrt{a^2 + b^2}$ for $a = 5$ and $b = 5$. Round to the nearest tenth if the result is not a whole number.
   - A. 25
   - B. 5
   - C. 0
   - D. 7.1

9. What is the solution of $77 = 103 - y$?
   - A. -26
   - B. 1.3
   - C. 33
   - D. 26

10. What is the solution of $\frac{1}{3}z = 48 - 29$?
    - A. 22
    - B. 19
    - C. $\frac{19}{3}$
    - D. 57

11. What is the solution of $6(a - 4) - 5 = 7a + 8$?
    - A. -37
    - B. $-\frac{37}{13}$
    - C. -17
    - D. 19

12. Solve $G = \frac{5h - 2}{3}$ for h.
    - A. $h = \frac{3G + 2}{5}$
    - B. $h = \frac{3G - 2}{5}$
    - C. $h = \frac{3}{5}(G + 2)$
    - D. $h = 5(3G + 2)$

13. What is the solution of $\frac{3}{c+3} = \frac{27}{63}$?

A. 63

B. 3

C. 7

D. 4

14. 34% of what number is 884?

A. 300.56

B. 918

C. 2600

D. 30,056

15. In 1985, a movie ticket cost \$4.50. In 1994, a movie ticket sold for \$7.65. What was the percent of increase over the original price?

A. 59%

B. 70%

C. 41%

D. 17%

16. A health food store owner has 20 pounds of granola that sells for \$4.60 per pound. How many pounds of raisins selling for \$2.50 per pound should she add to have a mixture that sells for \$3.40 per pound?

A. $26\frac{2}{3}$ pounds

B. 16 pounds

C. 80 pounds

D. 4 pounds

17. What is the domain of the relation $\{(3, 4), (6, 7), (5, 5), (3, 2), (-1, 7), (5, 0)\}$?

A. $\{-1, 0, 2, 3, 5, 6, 7\}$

B. $\{-1, 3, 5, 6\}$

C. $\{0, 2, 4, 5, 7\}$

D. $\{3, 5, 7\}$

18. Which of the following relations is a function?

A. $\{(-3, 4), (5, 7), (3, 4)\}$

B. $\{(5, 2), (0, -2), (4, 1), (0, 3)\}$

C. $2x^2 + y^2 = 10$

D. None of these relations are functions.

19. What is the x-intercept of the graph of $5x - y = -85$?

A. 17

B. -80

C. -17

D. -85

20. Which equation has a graph that is a horizontal line?

A. $3x + 2y = 2$

B. $6x + 7 = 9$

C. $xy = 1$

D. $4y - 10 = 3$

21. Find an equation of the line that passes through $(-4, 1)$ and $(5, -2)$.

A. $x + y = 3$

B. $3x + y = -11$

C. $x + 3y = -1$

D. $x + 3y = -7$

22. Which line at the right has a slope of $-\frac{1}{4}$ and contains $(4, 2)$?

A. k

B. l

C. m

D. n

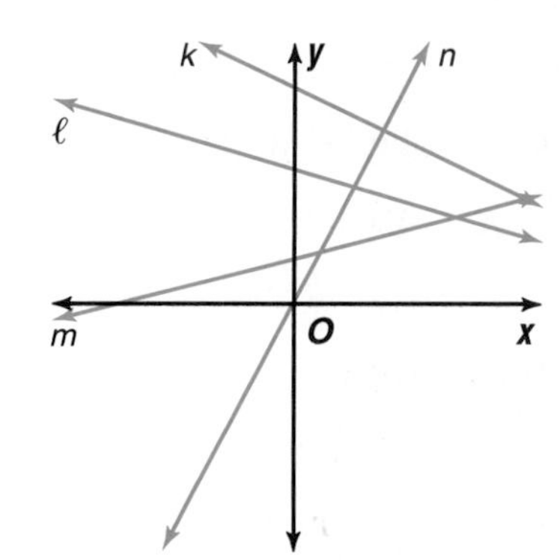

23. Which of the following equations has a graph that is parallel to the graph of $7x + 2y = 6$?

A. $y = \frac{2}{7}x + 3$

B. $y = -7x - 4$

C. $y = -\frac{7}{2}x + 5$

D. $y = \frac{7}{2}x + 3$

24. What are the coordinates of the midpoint of the segment whose endpoints are $(-3, -5)$ and $(-9, 5)$?

A. $(-6, -10)$

B. $(-6, 0)$

C. $(-3, 10)$

D. $(-3, 5)$

SYMBOLS AND MEASURES

Symbols

$=$	is equal to
$\neq$	is not equal to
$>$	is greater than
$<$	is less than
$\geq$	is greater than or equal to
$\leq$	is less than or equal to
$\approx$	is approximately equal to
$\sim$	is similar to
$\times$ or $\cdot$	times
$\div$	divided by
$-$	negative or minus
$+$	positive or plus
$\pm$	positive or negative
$-a$	opposite or additive inverse of a
$\lvert a \rvert$	absolute value of a
$a \stackrel{?}{=} b$	Does a equal b?
$a : b$	ratio of a to b
$\sqrt{a}$	square root of a
$P(A)$	probability of A
O	origin

π	pi
%	percent
$0.1\overline{2}$	decimal 0.12222...
$^\circ$	degree
$f(x)$	f of x, the value of f at x
(a, b)	ordered pair a, b
AB	line segment AB
$\widehat{AB}$	arc AB
$\overrightarrow{AB}$	ray AB
$\overleftrightarrow{AB}$	line AB
AB	measure of AB
$\angle$	angle
Δ	triangle
$\cos A$	cosine of A
$\sin A$	sine of A
$\tan A$	tangent of A
( )	parentheses; *also* ordered pairs
[ ]	brackets; *also* matrices
{ }	braces; *also* sets
$\varnothing$	empty set

Measures

mm	millimeter
cm	centimeter
m	meter
km	kilometer
g	gram
kg	kilogram
mL	milliliter
L	liter

in.	inch
ft	foot
yd	yard
mi	mile
in^2 or sq in.	square inch
s	second
min	minute
h	hour

GETTING ACQUAINTED WITH THE GRAPHING CALCULATOR

What is it?
What does it do?
How is it going to help me learn math?

These are just a few of the questions many students ask themselves when they first see a graphing calculator. Some students may think, "Oh, no! Do we *have* to use one?", while others may think, "All right! We get to use these neat calculators!" There are as many thoughts and feelings about graphing calculators as there are students, but one thing is for sure: a graphing calculator *can* help you learn mathematics.

So what is a graphing calculator? Very simply, it is a calculator that draws graphs. This means that it will do all of the things that a "regular" calculator will do, *plus* it will draw graphs of simple or very complex equations. In algebra, this capability is nice to have because the graphs of some complex equations take a lot of time to sketch by hand. Some are even considered impossible to draw by hand. This is where a graphing calculator can be very useful.

But a graphing calculator can do more than just calculate and draw graphs. You can program it, work with matrices, and make statistical graphs and computations, just to name a few things. If you need to generate random numbers, you can do that on the graphing calculator. If you need to find the absolute value of numbers, you can do that, too. It's really a very powerful tool—so powerful that it is often called a pocket computer. But don't let that intimidate you. A graphing calculator can save you time and make doing mathematics easier.

As you may have noticed, graphing calculators have some keys that other calculators do not. The Texas Instruments TI-82 will be used throughout this text. The keys located on the bottom half of the calculator are probably familiar to you as they are the keys found on basic scientific calculators. The keys located just below the screen are the graphing keys. You will also notice the up, down, left, and right arrow keys. These allow you to move the cursor around on the screen and to "trace" graphs that have been plotted. The other keys located on the top half of the calculator access the special features such as statistical and matrix computations.

There are some keystrokes that can save you time when using the graphing calculator. A few of them are listed below.

- Any light blue commands written above the calculator keys are accessed with the [2nd] key, which is also blue. Similarly, any gray characters above the keys are accessed with the [ALPHA] key, which is also gray.
- [2nd] [ENTRY] copies the previous calculation so you can edit and use it again.
- Pressing [ON] while the calculator is graphing stops the calculator from completing the graph.
- [2nd] [QUIT] will return you to the home (or text) screen.
- [2nd] [A-LOCK] locks the [ALPHA] key, which is like pressing "shift lock" or "caps locks" on a typewriter or computer. The result is that all caps will be typed and you do not have to hold the shift key down. (This is handy for programming.)
- [2nd] [OFF] turns the calculator off.

Some commonly used mathematical functions are shown in the table below. As with any scientific calculator, the graphing calculator observes the order of operations.

Mathematical Operation	Examples	Keys	Display
evaluate expressions	Find 2 + 5.	2 [+] 5 [ENTER]	2+5 7
exponents	Find 3^5.	3 [^] 5 [ENTER]	3^5 243
multiplication	Evaluate 3(9.1 + 0.8).	3 [×] [(] 9.1 [+] .8 [)] [ENTER]	3(9.1+.8) 29.7
roots	Find $\sqrt{14}$.	[2nd] [√] 14 [ENTER]	√ 14 3.741657387
opposites	Enter −3.	[(−)] 3	−3

Graphing on the TI–82

Before graphing, we must instruct the calculator how to set up the axes in the coordinate plane. To do this, we define a **viewing window.** The viewing window for a graph is the portion of the coordinate grid that is displayed on the **graphics screen** of the calculator. The viewing window is written as [left, right] by [bottom, top] or [Xmin, Xmax] by [Ymin, Ymax]. A viewing window of [−10, 10] by [−10, 10] is called the **standard viewing window** and is a good viewing window to start with to graph an equation. The standard viewing window can be easily obtained by pressing [ZOOM] 6. Try this. Move the arrow keys around and observe what happens. You are seeing a portion of the coordinate plane that includes the region from −10 to 10 on the x-axis and from −10 to 10 on the y-axis. Move the cursor, and you can see the coordinates of the points for the current position of the cursor.

Any viewing window can be set manually by pressing the [WINDOW] key. The window screen will appear and display the current settings for your viewing window. First press [ENTER]. Then, using the arrow and [ENTER] keys, move the cursor to edit the window settings. Xscl and Yscl refer to the x-scale and y-scale. This is the number of tick marks placed on the x- and y-axes. Xscl=1 means that there will be a tick mark for every unit of one along the x-axis. The standard viewing window would appear as follows.

Xmin = −10
Xmax = 10
Xscl = 1
Ymin = −10
Ymax = 10
Yscl = 1

Graphing equations is as simple as defining a viewing window, entering the equations in the Y= list, and pressing [GRAPH]. It is often important to view enough of a graph so you can see all of the important characteristics of the graph and understand its behavior. The term **complete graph** refers to a graph that shows all of the important characteristics such as intercepts or maximum and minimum values.

Example: Graph $y = x - 14$ in the standard viewing window.

Enter: 

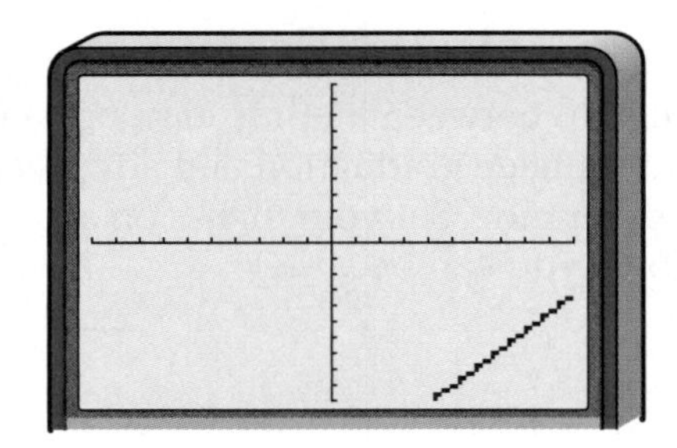

We see only a small portion of the graph. Why?

The graph of $y = x - 14$ is a line that crosses the x-axis at 14 and the y-axis at −14. The important features of the graph are plotted off the screen and so the graph is not complete. A better viewing window for this graph would be [−20, 20] by [−20, 20], which includes both intercepts. This is considered a complete graph.

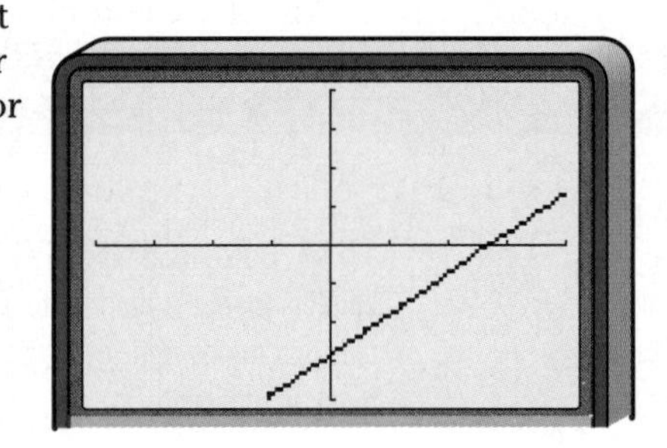

Programming on the TI–82

The TI–82 has programming features that allow us to write and execute a series of commands to perform tasks that may be too complex or cumbersome to perform otherwise. Each program is given a name. Commands begin with a colon (:), followed by an expression or an instruction. Most of the features of the calculator are accessible from program mode.

When you press [PRGM], you see three menus: EXEC, EDIT, and NEW. EXEC allows you to execute a stored program by selecting the name of the program from the menu. EDIT allows you to edit or change an existing program and NEW allows you to create a new program. To break during program execution, press [ON]. The following example illustrates how to create and execute a new program that stores an expression as Y and evaluates the expression for a designated value of X.

1. Enter [PRGM] [▶] [▶] [ENTER] to create a new program.
2. Type EVAL [ENTER] to name the program. (Make sure that the caps lock is on.) You are now in the program editor, which allows you to enter commands. The colon (:) in the first column of the line indicates that this is the beginning of the command line.
3. The first command lines will ask the user to designate a value for x. Enter [PRGM] [▶] 3 [2nd] [A-LOCK] ["] ENTER THE VALUE FOR X ["] [ALPHA] [ENTER] [PRGM] [▶] 1 [X,T,θ] [ENTER].
4. The expression to be evaluated for the value of x is $x - 7$. To store the expression as Y, enter [X,T,θ] [−] 7 [STO▶] [ALPHA] [Y] [ENTER].
5. Finally, we want to display the value for the expression. Enter [PRGM] [▶] 3 [ALPHA] [Y] [ENTER].
6. Now press [2nd] [QUIT] to return to the home screen.
7. To execute the program, press [PRGM]. Then press the down arrow to locate the program name and press [ENTER], or press the number or letter next to the progam name. The program asks for a value for x. You will input any value for which the expression is defined and press [ENTER]. To immediately re-execute the program, simply press [ENTER] when Done appears on the screen.

While a graphing calculator cannot do everything, it can make some things easier. To prepare for whatever lies ahead, you should try to learn as much as you can. The future will definitely involve technology, and using a graphing calculator is a good start toward becoming familiar with technology. Who knows? Maybe one day you will be designing the next satellite, building the next skyscraper, or helping students learn mathematics with the aid of a graphing calculator!

CHAPTER 7

Solving Linear Inequalities

Objectives

In this chapter, you will:

- solve inequalities,
- graph solutions of inequalities,
- graph solutions of open sentences that involve absolute value,
- solve problems by drawing a diagram, and
- use box-and-whisker plots to display and analyze data.

Source: *Chicago Tribune,* 1995

TIME *Line*

There is a direct relationship between lifetime earnings and educational attainment. College graduation and advanced degrees really do make a difference. Set your sights on a career and aim to be the best you can be.

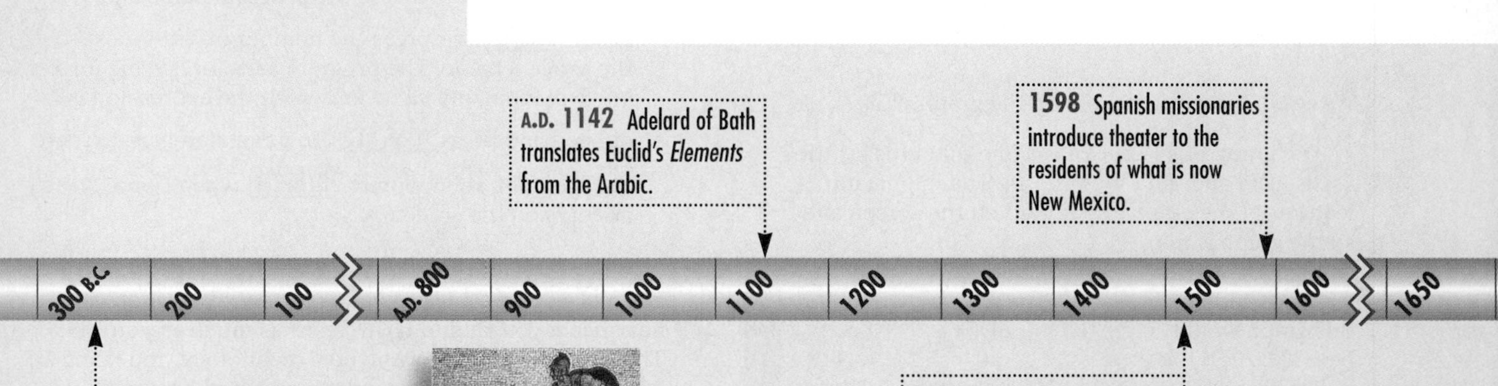

Pages 4–381 are contained in Volume One.

PEOPLE IN THE NEWS

Shakema Hodge is aiming for an advanced degree in microbiology. The young scientist from St. Thomas, Virgin Islands, earned her bachelor's degree in biology at the University of the Virgin Islands. Now she is enrolled in a 5-year Ph.D. program in molecular microbiology at the University of Rochester in New York. She would love to teach high school students, believing that "to capture a young person's mind, you have to expose them to the sciences at an early age."

Chapter Project

Kimana plans to study either botany, genetics, or microbiology at a nearby university. She hasn't ruled out teaching as a possible career choice, but is interested in finding other businesses in which she can apply the science she will be studying.

- Visit the public library or local college to investigate careers that relate to Kimana's interests.
- Which careers might offer Kimana an opportunity to earn an amount greater than that presented in the graph for a bachelor's degree?
- Which careers require a degree for which the years studied are greater than those for a bachelor's degree? Choose a career. Write an inequality that might express the time needed to study and train for that career.

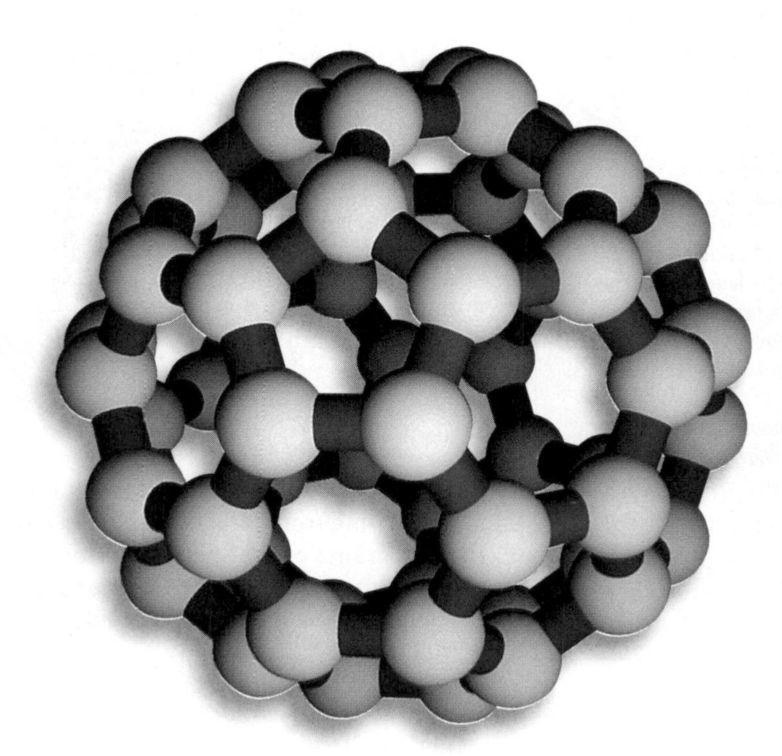

7–1 Solving Inequalities by Using Addition and Subtraction

What **YOU'LL LEARN**

- To solve inequalities by using addition and subtraction.

Why **IT'S IMPORTANT**

You can use inequalities to solve problems involving nutrition and personal finance.

CAREER CHOICES

Nutritionists plan dietary programs and supervise the preparation and serving of meals for hospitals, nursing homes, health maintenance organizations, and individuals. A bachelor's degree in dietetics, foods, or nutrition is required that includes studies in biology, physiology, mathematics, and chemistry.

For more information, contact:

The American Dietetic Association
216 West Jackson Blvd.
Chicago, IL 60606

Nutrition

In 1990, the U.S. Department of Agriculture issued new dietary guidelines. These guidelines recommend that people greatly reduce their fat intake. Your recommended calorie intake depends on your height, desired weight, and physical activity. The average 14-year-old is 5 feet 2 inches tall and weighs 107 pounds. Boys should consume about 2434 calories per day and girls 2208 calories per day to maintain this weight.

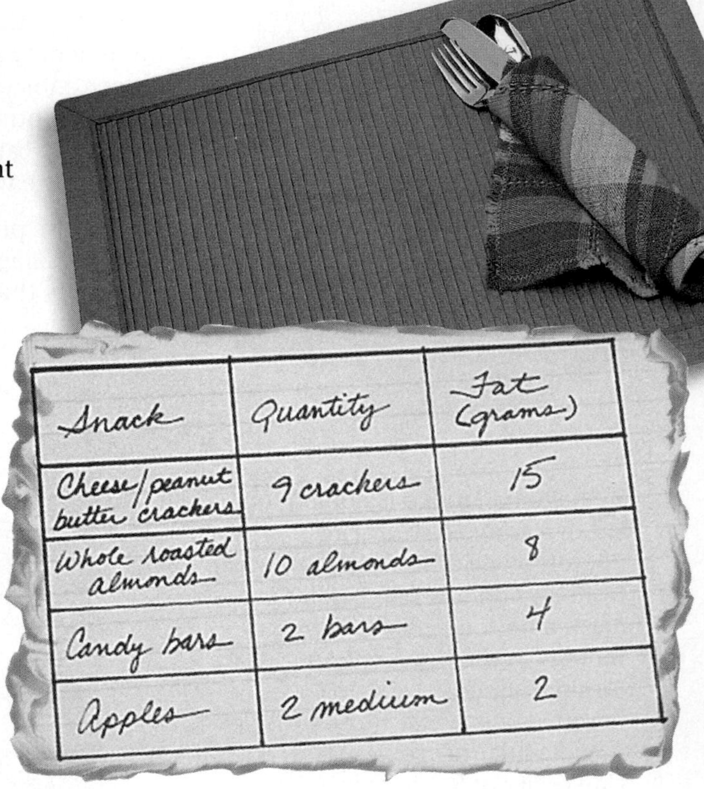

Snack	Quantity	Fat (grams)
Cheese/peanut butter crackers	9 crackers	15
Whole roasted almonds	10 almonds	8
Candy bars	2 bars	4
Apples	2 medium	2

Oliana learned in health class that no more than 30% of her calorie intake should come from fat. For her 2030-calorie-a-day diet, that means no more than 68 grams of fat. She keeps track of her snacks for one day and records their fat content, as shown in the table above. How many grams of fat can Oliana have in the other foods she eats that day and stay within the guidelines?

Let's write an inequality to represent the problem. Let g represent the remaining grams of fat that Oliana can eat that day.

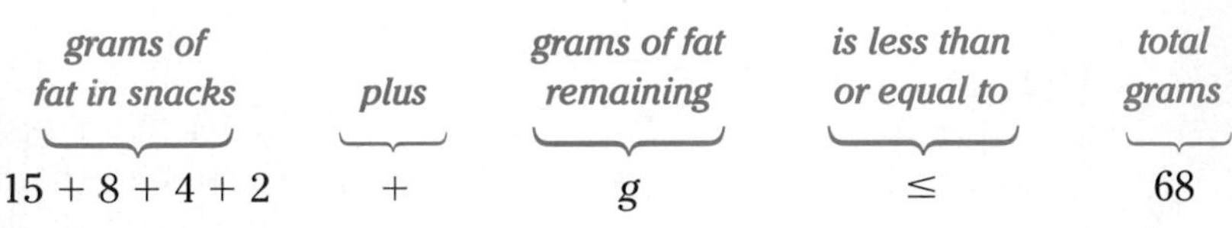

That is, $29 + g \leq 68$.

The symbol $\leq$ indicates *less than or equal to*. It is used in this situation because the total number of grams of fat in Oliana's daily diet should be no greater than 68. If this were an equation, we would subtract 29 from (or add -29 to) each side. Can the same procedure be used in an inequality? *This problem will be solved in Example 1.*

Let's explore what happens if inequalities are solved in the same manner as equations. We know that $7 > 2$. What happens when you add or subtract the same quantity to each side of the inequality? We can use number lines to model the situation.

Add 3 to each side.

$$7 > 2$$
$$7 + 3 \stackrel{?}{>} 2 + 3$$
$$10 > 5$$

Subtract 4 from each side.

$$7 > 2$$
$$7 - 4 \stackrel{?}{>} 2 - 4$$
$$3 > -2$$

In each case, the inequality holds true. These examples illustrate two properties of inequalities.

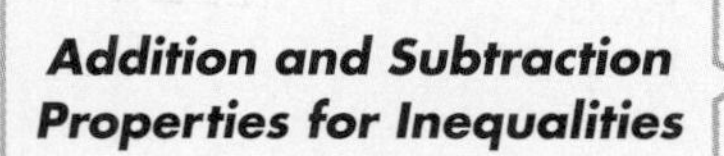

For all numbers a, b, and c, the following are true.
1. If $a > b$, then $a + c > b + c$ and $a - c > b - c$.
2. If $a < b$, then $a + c < b + c$ and $a - c < b - c$.

These properties are also true when $>$ and $<$ are replaced by $\geq$ and $\leq$. So, we can use these properties to obtain a solution to the application at the beginning of the lesson.

Example 

Refer to the application at the beginning of the lesson. Solve $29 + g \leq 68$.

$$29 + g \leq 68$$
$$29 - 29 + g \leq 68 - 29 \quad \textit{Subtract 29 from each side.}$$
$$g \leq 39 \quad \textit{This means all numbers less than or equal to 39.}$$

The solution set can be written as {all numbers less than or equal to 39}.

Check: To check this solution, substitute 39, a number less than 39, and a number greater than 39 into the inequality.

Let $g = 39$.	Let $g = 20$.	Let $g = 40$.
$29 + g \leq 68$	$29 + g \leq 68$	$29 + g \leq 68$
$29 + 39 \stackrel{?}{\leq} 68$	$29 + 20 \stackrel{?}{\leq} 68$	$29 + 40 \stackrel{?}{\leq} 68$
$68 \leq 68$ true	$49 \leq 68$ true	$69 \leq 68$ false

So, Oliana can have 39 or fewer grams of fat in other foods that day and stay within the dietary guidelines.

The solution to the inequality in Example 1 was expressed as a set. A more concise way of writing a solution set is to use **set-builder notation**. The solution in set-builder notation is $\{g \mid g \leq 39\}$. This is read *the set of all numbers g such that g is less than or equal to 39.*

You can refer to Lesson 1-5 for information on solution sets.

In Lesson 2–4, you learned that you can show the solution to an inequality on a graph. The solution to Example 1 is shown on the number line below.

28 29 30 31 32 33 34 35 36 37 38 39 40 41 42

The closed circle at 39 tells us that 39 is included in the inequality. The heavy arrow pointing to the left shows that it also includes all numbers less than 39. *If the inequality was $<$, the circle would be open.*

Example 2 **Solve $13 + 2z < 3z - 39$. Then graph the solution.**

$$13 + 2z < 3z - 39$$

$$13 + 2z - 2z < 3z - 2z - 39 \quad \textit{Subtract 2z from each side.}$$

$$13 < z - 39$$

$$13 + 39 < z - 39 + 39 \quad \textit{Add 39 to each side.}$$

$$52 < z$$

Since $52 < z$ is the same as $z > 52$, the solution set is $\{z \mid z > 52\}$.

The graph of the solution contains an open circle at 52 since 52 is not included in the solution, and the arrow points to the right.

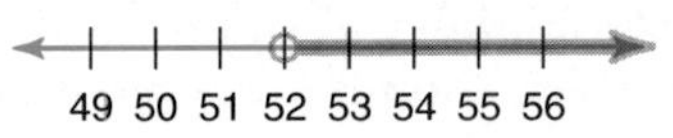

Verbal problems containing phrases like *greater than* or *less than* can often be solved by using inequalities. The following chart shows some other phrases that indicate inequalities.

Inequalities			
$<$	$>$	$\leq$	$\geq$
• less than • fewer than	• greater than • more than	• at most • no more than • less than or equal to	• at least • no less than • greater than or equal to

Example 3

Budgeting

Alvaro, Chip, and Solomon have earned \$500 to buy equipment for their band. They have already spent \$275 on a used guitar and a drum set. They are now considering buying a \$125 amplifier. What is the most they can spend on promotional materials and T-shirts for the band if they buy the amplifier?

Explore *At most* means that they cannot go over what is left of their budget of \$500. They have spent \$275, so they have \$225 left. Let $m =$ the amount of money for promotional materials and T-shirts.

Plan — *Total to spend* / *is at most* / *$225.*

$$125 + m \quad \leq \quad 225$$

Solve

$$125 + m \leq 225$$

$$125 - 125 + m \leq 225 - 125 \quad \text{Subtract 125 from each side.}$$

$$m \leq 100$$

The members of the band can spend $100 or less on promotional materials and T-shirts.

Examine Since $275 + $125 + $100 = $500, Alvaro, Chip, and Solomon can spend $100 or less on promotional materials and T-shirts.

When solving problems involving equations, it is often necessary to write an equation that represents the words in the problem. This is also true of inequalities.

Example 4

Write an inequality for the sentence below. Then solve the inequality and check the solution.

Three times a number is more than the difference of twice that number and three.

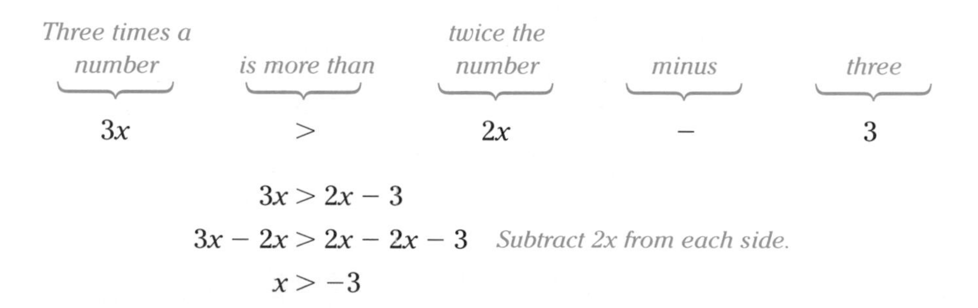

$$3x > 2x - 3$$

$$3x - 2x > 2x - 2x - 3 \quad \text{Subtract 2x from each side.}$$

$$x > -3$$

The solution set is $\{x \mid x > -3\}$.

CHECK FOR UNDERSTANDING

Communicating Mathematics

Study the lesson. Then complete the following.

1. **Write** three inequalities that are equivalent to $x < -10$.
2. **Explain** what $\{w \mid w > -3\}$ means.
3. **Explain** the difference between the solution sets for $x + 24 < 17$ and $x + 24 \leq 17$.
4. **Describe** how you would graph the solution to an inequality. Include examples and graphs in your explanation.
5. Is it possible for the solution set of an inequality to be the empty set? If so, give an example.

6. Sometimes statements we make can be translated into inequalities. For example, *In some states, you have to be at least 16 years old to have a driver's license* can be expressed as $a \geq 16$, and *Tomás cannot lift more than 72 pounds* can be translated into $w \leq 72$. Following these examples, write three statements that deal with your everyday life. Then translate each into a corresponding inequality.

Guided Practice

Match each inequality with the graph of its solution.

7. $b - 18 > -3$

8. $10 \geq -3 + x$

9. $x + 11 < 6$

10. $4c - 3 \leq 5c$

a. 10 11 12 13 14 15 16 17

b. −6 −5 −4 −3 −2 −1 0 1

c. 10 11 12 13 14 15 16 17

d. −7 −6 −5 −4 −3 −2 −1 0

Solve each inequality. Then check your solution.

11. $x + 7 > 2$

12. $10 \geq x + 8$

13. $y - 7 < -12$

14. $-81 + q > 16 + 2q$

Define a variable, write an inequality, and solve each problem. Then check your solution.

15. A number decreased by 17 is less than −13.

16. A number increased by 4 is at least 3.

EXERCISES

Practice

Solve each inequality. Then check your solution, and graph it on a number line.

17. $a - 12 < 6$

18. $m - 3 < -17$

19. $2x \leq x + 1$

20. $-9 + d > 9$

21. $x + \frac{1}{3} > 4$

22. $-0.11 \leq n - (-0.04)$

23. $2x + 3 > x + 5$

24. $7h - 1 \leq 6h$

Solve each inequality. Then check your solution.

25. $x + \frac{1}{8} < \frac{1}{2}$

26. $3x + \frac{4}{5} \leq 4x + \frac{3}{5}$

27. $3x - 9 \leq 2x + 6$

28. $6w + 4 \geq 5w + 4$

29. $-0.17x - 0.23 < 0.75 - 1.17x$

30. $0.8x + 5 \geq 6 - 0.2x$

31. $3(r - 2) < 2r + 4$

32. $-x - 11 \geq 23$

Define a variable, write an inequality, and solve each problem. Then check your solution.

33. A number decreased by -4 is at least 9.
34. The sum of a number and 5 is at least 17.
35. Three times a number is less than twice the number added to 8.
36. Twenty-one is no less than the sum of a number and -2.
37. The sum of two numbers is less than 53. One number is 20. What is the other number?
38. The sum of four times a number and 7 is less than 3 times that number.
39. Twice a number is more than the difference of that number and 6.
40. The sum of two numbers is 100. One number is at least 16 more than the other number. What are the two numbers?

If $3x \geq 2x + 5$, then complete each inequality.

41. $3x + 7 \geq 2x + \underline{\quad ? \quad}$
42. $3x - 10 \geq 2x - \underline{\quad ? \quad}$
43. $3x + \underline{\quad ? \quad} \geq 2x + 3$
44. $\underline{\quad ? \quad} \leq x$

Programming

45. **Geometry** For three line segments to form a triangle, the sum of the lengths of any two sides must exceed the length of the third side. Let the lengths of the possible sides be a, b, and c. Then these three inequalities must be true: $a + b > c$, $a + c > b$, and $b + c > a$. The graphing calculator program at the right uses these inequalities to determine if the three lengths can be measures of the sides of a triangle.

```
PROGRAM: TRIANGLE
: Disp "ENTER THREE LENGTHS"
: Prompt A, B, C
: If C≥A+B
: Then
: Goto 1
: End
: If B≥A+C
: Then
: Goto 1
: End
: If A≥B+C
: Then
: Goto 1
: End
: Disp "THIS IS A TRIANGLE."
: Stop
: Lbl 1
: Disp "NOT A TRIANGLE"
```

Use the program to determine whether segments with the given lengths can form a triangle.

Tip: To run the program again after trying one set of numbers, press ENTER.

a. 10 in., 12 in., 27 in.
b. 3 ft, 4 ft, 5 ft
c. 125 cm, 140 cm, 150 cm
d. 1.5 m, 2.0 m, 2.5 m

Critical Thinking

46. Using an example, show that even though $x > y$ and $t > w$, $x - t > y - w$ may be false.
47. What does the sentence $-2.4 < x < 3.6$ mean?

Applications and Problem Solving

Define a variable, write an inequality, and solve each problem.

48. Academics Josie must have at least 320 points in her math class to get a B. She needs a B or better to maintain her grade-point average so she can play on the basketball team. The grade is based on four 50-point tests, three 20-point quizzes, two 20-point projects, and a final exam worth 100 points. Josie's record of her grades are shown in the table below.

	Points	Total Points
Tests	40, 42, 41, 45	168
Quizzes	15, 12, 19	46
Projects	15, 18	33

a. Write an inequality to represent the range of scores on the final exam needed in order for Josie to make a B.

b. Solve the inequality and graph its solution.

49. Personal Finances Tanaka had \$75 to buy presents for his family. He bought a \$21.95 shirt for his dad, a \$23.42 necklace for his mother, and a \$16.75 CD for his sister. He still has to buy a present for his brother.

a. How much can he spend on his brother's present?

b. What factors not stated in the problem may affect how much money he can spend?

Mixed Review

50. Find the midpoint of the line segment whose endpoints are at $(-1, 9)$ and $(-5, 5)$. (Lesson 6–7)

Write an equation in slope-intercept form for a line that satisfies each condition. (Lesson 6–6)

51. perpendicular to $x + 7 = 3y$ that passes through $(1, 0)$

52. parallel to $\frac{1}{5}y - 3x = 2$ that passes through $(0, -3)$

53. Statistics Find the range, quartiles, and interquartile range for the set of data at the right. (Lesson 5–7)

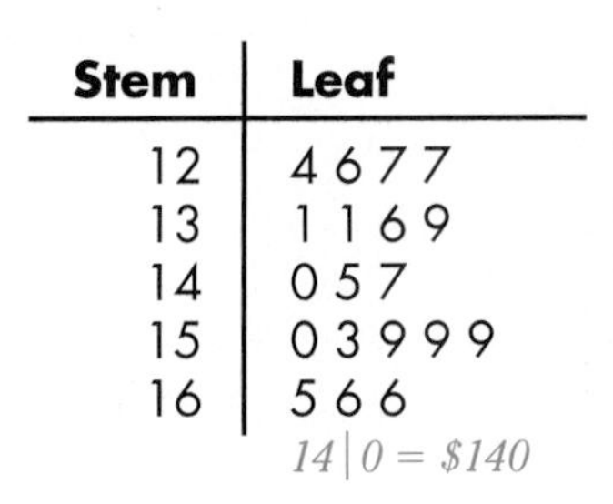

Stem	Leaf
12	4 6 7 7
13	1 1 6 9
14	0 5 7
15	0 3 9 9 9
16	5 6 6

14|0 = \$140

54. Solve $4x + 3y = 16$ if the domain is $\{-2, -1, 0, 2, 5\}$. (Lesson 5–3)

55. If y varies inversely as x and $y = 32$ when $x = 3$, find y when $x = 8$. (Lesson 4–8)

56. Solve $y - \frac{7}{16} = -\frac{5}{8}$ (Lesson 3–1)

57. Replace the __?__ with $<$, $>$, or $=$ to make $\frac{6}{13}$ __?__ $\frac{1}{2}$ true. (Lesson 2–4)

58. Look for a Pattern How many triangles are shown at the right? Count only the triangles pointing upward. (Lesson 1–2)

7–2A Solving Inequalities

Materials:  equation mat  cups and counters

 self-adhesive note

A Preview of Lesson 7–2

You can use an equation model to solve inequalities.

Activity **Model the solution for $-2x < 4$.**

Step 1 Use the note to cover the equals sign on the equation mat. Then write a $<$ symbol on the note. Label 2 cups with a negative sign, and place them on the left side. Place 4 positive counters on the right.

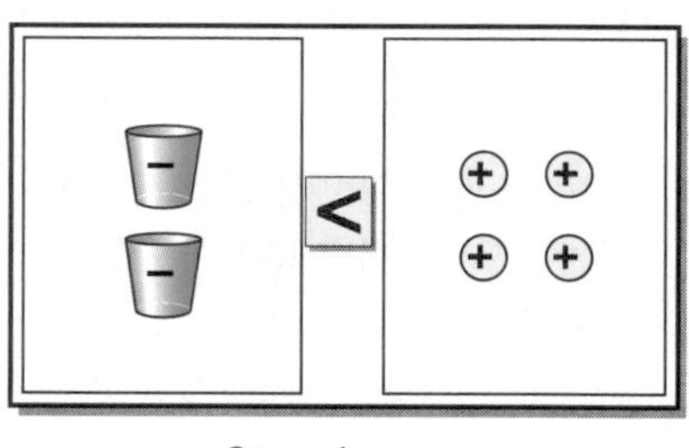

$-2x < 4$

Step 2 Since we cannot solve for a negative cup, we must eliminate the negative cups by adding 2 positive cups to each side. Remove the zero pairs.

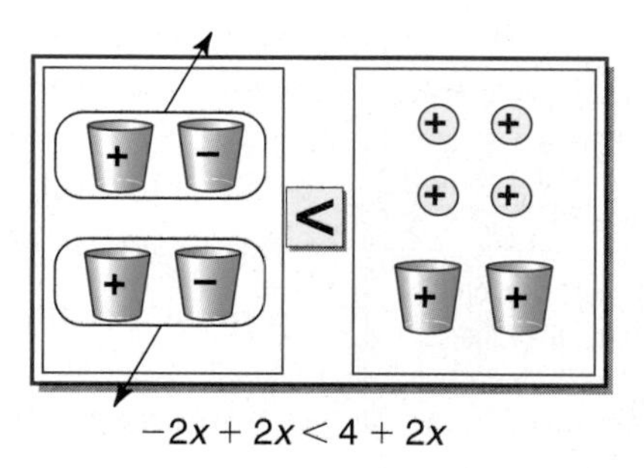

$-2x + 2x < 4 + 2x$

Step 3 Add 4 negative counters to each side to isolate the cups. Remove the zero pairs.

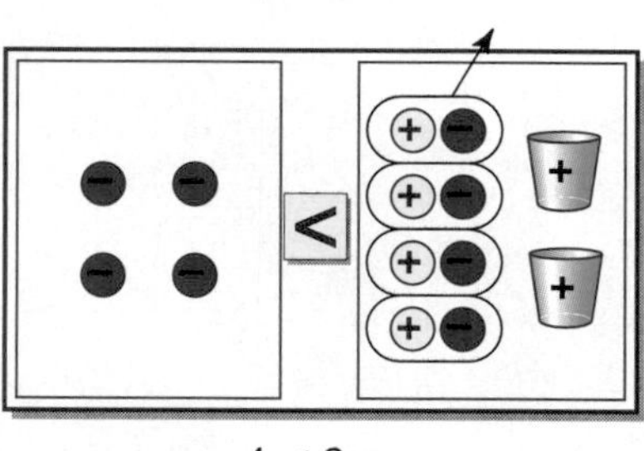

$-4 < 2x$

Step 4 Separate the counters into 2 groups.

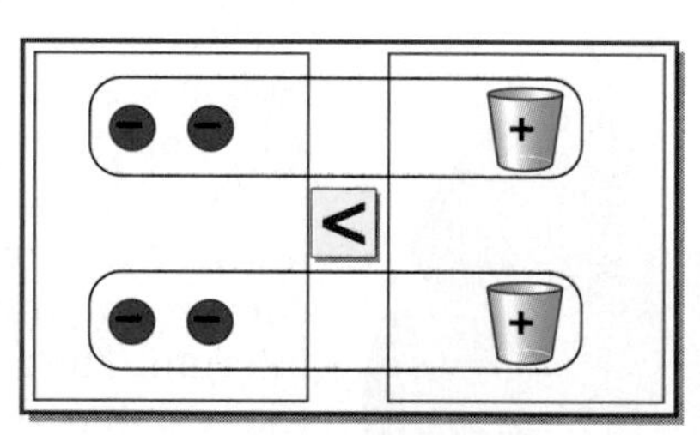

$-2 < x$ or $x > -2$

Model

1. Compare the symbol and location of the variable in the original problem with those in the solution. What do you find?
2. Model the solution for $3x > 12$. What do you find? How is this different from solving $-2x < 4$?

Write

3. Write a rule for solving inequalities involving multiplication.
4. Do you think the rule applies to inequalities involving division? *Remember that dividing by a number is the same as multiplying by its reciprocal.*

7–2 Solving Inequalities by Using Multiplication and Division

What YOU'LL LEARN

- To solve inequalities by using multiplication and division.

Why IT'S IMPORTANT

You can use inequalities to solve problems involving the physical and political sciences.

GLOBAL CONNECTIONS

About 15,000 years ago, an important application of the lever appeared as hunters used an *atlatl,* the Aztec word for spear-thrower. This simple device was a handle that provided extra mechanical advantage by adding length to the hunter's arm. This enabled the hunter to use less motion to get greater results when hunting.

Physical Science

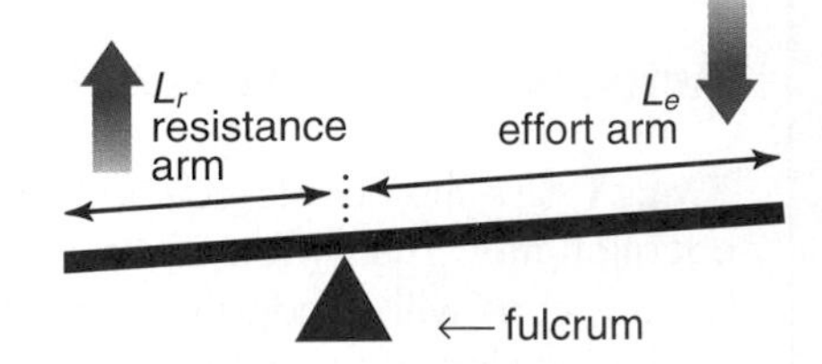

A lever can be used to multiply the effort force you exert when trying to move something. The fixed point or *fulcrum* of a lever separates the length of the lever into two sections—the *effort arm* on which the effort force is applied and the *resistance arm* that exerts the resistance force. The *mechanical advantage* of a lever is the number of times a lever multiplies that effort force.

The formula for determining the mechanical advantage MA of a lever can be expressed as $MA = \frac{L_e}{L_r}$, where L_e represents the length of the effort arm and L_r represents the length of the resistance arm.

Suppose a group of volunteers is clearing hiking trails at Yosemite National Park. They need to position a lever so that a mechanical advantage of at least 7 is achieved in order to remove a boulder blocking the trail. The volunteers place the lever on a rock so they can use the rock as a fulcrum. They will need the resistance arm to be 1.5 ft long so that it is long enough to get under the boulder. What should be the length of the lever in order to move the boulder?

We need to find the length of the effort arm to find the total length. Let L_e represent the length of the effort arm. We know that 1.5 feet is the length of the resistance arm L_r. Since the mechanical advantage must be at least 7, we can write an inequality using the formula.

$MA \geq 7$

$\frac{L_e}{L_r} \geq 7$ *Replace MA with* $\frac{L_e}{L_r}$.

$\frac{L_e}{1.5} \geq 7$ *Replace* L_r *with 1.5.* *You will solve this problem in Example 1.*

If you were solving the equation $\frac{L_e}{1.5} = 7$, you would multiply each side by 1.5. Will this method work when solving inequalities? Before answering this question, let's explore how multiplying (or dividing) an inequality by a positive or negative number affects the inequality. Consider the inequality $10 < 15$, which we know is true.

Multiply by 2.

$10 < 15$

$10(2) < 15(2)$

$20 < 30$ true

Multiply by −2.

$10 < 15$

$10(-2) < 15(-2)$ false

$-20 < -30$ false

$-20 > -30$ true

Divide by 5.

$10 < 15$

$\frac{10}{5} < \frac{15}{5}$

$2 < 3$ true

Divide by −5.

$10 < 15$

$\frac{10}{-5} < \frac{15}{-5}$ false

$-2 < -3$ false

$-2 > -3$ true

These results suggest the following.

- If each side of a true inequality is multiplied or divided by the same positive number, the resulting inequality is also true.
- If each side of a true inequality is multiplied or divided by the same negative number, the direction of the inequality symbol must be *reversed* so that the resulting inequality is also true.

Multiplication and Division Properties for Inequalities

For all numbers, *a*, *b*, and *c*, the following are true.

1. If *c* is positive and $a < b$, then $ac < bc$ and $\frac{a}{c} < \frac{b}{c}$, $c \neq 0$, and if *c* is positive and $a > b$, then $ac > bc$ and $\frac{a}{c} > \frac{b}{c}$, $c \neq 0$.

2. If *c* is negative and $a < b$, then $ac > bc$ and $\frac{a}{c} > \frac{b}{c}$, $c \neq 0$, and if *c* is negative and $a > b$, then $ac < bc$ and $\frac{a}{c} < \frac{b}{c}$, $c \neq 0$.

These properties also hold for inequalities involving ≤ and ≥.

Example 1

Physical Science

Refer to the connection at the beginning of the lesson. What should the minimum length of the lever be?

$\frac{L_e}{1.5} \geq 7$

$1.5 \cdot \frac{L_e}{1.5} \geq 1.5(7)$ *Multiply each side by 20.*

$L_e \geq 10.5$

The effort arm must be at least 10.5 feet long.

In order to find the length of the lever, add the lengths of the effort arm and the resistance arm. The lever should be at least 10.5 + 1.5 or 12 feet long.

Example 2 Solve $\frac{x}{12} \leq \frac{3}{2}$.

$$\frac{x}{12} \leq \frac{3}{2}$$

$$12 \cdot \frac{x}{12} \leq 12 \cdot \frac{3}{2}$$ *Multiply each side by 12.*

$$x \leq 18$$ *Since we multiplied by a positive number, the inequality symbol stays the same.*

The solution set is $\{x \mid x \leq 18\}$.

Since dividing is the same as multiplying by the reciprocal, there can be two methods to solve an inequality that involves multiplication.

Example 3 Solve $-3w > 27$.

Method 1

$$-3w > 27$$

$$\frac{-3w}{-3} < \frac{27}{-3}$$ *Divide each side by −3 and change > to <.*

$$w < -9$$

Method 2

$$-3w > 27$$

$$\left(-\frac{1}{3}\right)(-3w) < \left(-\frac{1}{3}\right)(27)$$ *Multiply each side by $-\frac{1}{3}$ and change > to <.*

$$w < -9$$

Check: Let w be any number less than −9.

$$-3w > 27$$

$$-3(-10) \stackrel{?}{>} 27$$ *Suppose we select −10.*

$$30 > 27 \quad \text{true}$$

Numbers less than −9 compose the solution set. The solution set is $\{w \mid w < -9\}$.

Example 4 **Angelica Moreno is a sales representative for an appliance distributor. She needs at least \$5000 in weekly sales of a particular TV model to qualify for a sales competition to win a trip to the Bahamas. If the TVs sell for \$250 each, how many TVs will Ms. Moreno have to sell to qualify?**

Business

Explore Let t represent the number of TVs to be sold. At least \$5000 means greater than or equal to \$5000.

Plan The price of one TV times the number of sets sold must be greater than or equal to the total amount of sales needed.

The price of one TV	*times*	*the number of TVs sold*	*is at least*	*\$5000.*
\$250	$\times$	t	$\geq$	\$5000

Solve

$$250t \geq 5000$$

$$\frac{250t}{250} \geq \frac{5000}{250}$$ *Divide each side by 250.*

$$t \geq 20$$

Examine Ms. Moreno must sell a minimum of 20 TVs to qualify for the contest. The amount of money from the sale of 20 TVs is \$250(20) or \$5000.

Example 5 **INTEGRATION Geometry**

Triangle XYZ is not an acute triangle. The greatest angle in the triangle has a measure of $(6d)°$. What are the possible values of d?

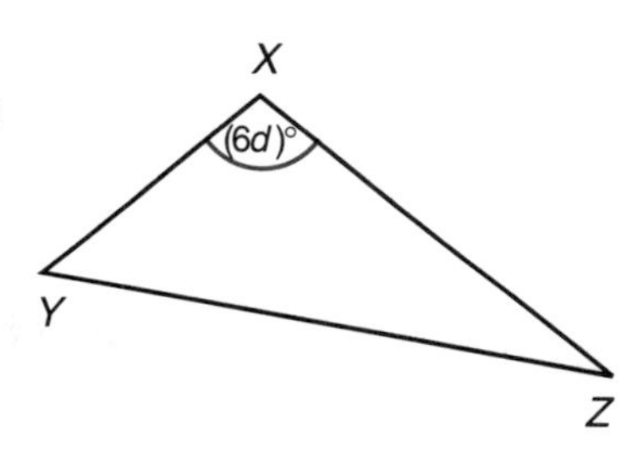

Since ΔXYZ is not acute, the measure of the greatest angle must be 90° or larger, but less than 180°. Thus, $6d \geq 90$ and $6d < 180$.

$6d \geq 90$

$\frac{6d}{6} \geq \frac{90}{6}$ *Divide each side by 6.*

$d \geq 15$

$6d < 180$

$\frac{6d}{6} < \frac{180}{6}$ *Divide each side by 6.*

$d < 30$

The value of d must be greater than or equal to 15 but less than 30.

CHECK FOR UNDERSTANDING

Communicating Mathematics

Study the lesson. Then complete the following.

1. **Classify** each statement as *true* or *false*. If false, explain how to change the inequality to make it true.
   a. If $x > 9$, then $-3x > -27$.
   b. If $x < 4$, then $3x < 12$.

2. **Complete** each statement.
   a. If each side of an inequality is multiplied by the same __?__ number, the direction of the inequality symbol must be reversed so that the resulting inequality is true.
   b. Multiplying by __?__ is the same as dividing by -6.
   c. An acute triangle has angles whose measures are all less than __?__.

3. **You Decide** Utina and Paige are discussing the rules for changes in the direction of the inequality symbol when solving inequalities. Paige says the rule is "Whenever you have a negative sign in the problem, the direction of the inequality symbol will change." Utina says that is not always true. Decide who is correct and give examples to support your answer.

MODELING MATHEMATICS

Use models to solve each inequality. Write the answer in set-builder notation.

4. $3x < 15$
5. $-6x < 18$
6. $2x + 6 > x - 7$
7. $-4x + 8 \geq 14$

8. Use models to determine the appropriate symbol (>, <, or =) to complete each comparison.
   a. $x + 5$ __?__ $x - 7$
   b. $x + 3 + (x - 6)$ __?__ $(2x + 7) - 4$
   c. $2x + 8$ __?__ $2x - 6 - x$

Guided Practice

State the number by which you multiply or divide to solve each inequality. Indicate whether the direction of the inequality symbol reverses. Then solve.

9. $-6y \geq -24$
10. $10x > 20$
11. $\frac{x}{4} < -5$
12. $-\frac{2}{7}z \geq -12$

Solve each inequality. Then check your solution.

13. $\frac{4}{5}x < 24$
14. $-\frac{v}{3} \geq 4$
15. $-0.1t \geq 3$
16. $5y > -25$

Define a variable, write an inequality, and solve each problem. Then check your solution.

17. One fifth of a number is at most 4.025.
18. The opposite of six times a number is less than 216.
19. **Geometry** Determine the value of s so that the area of the square is at least 144 square feet.

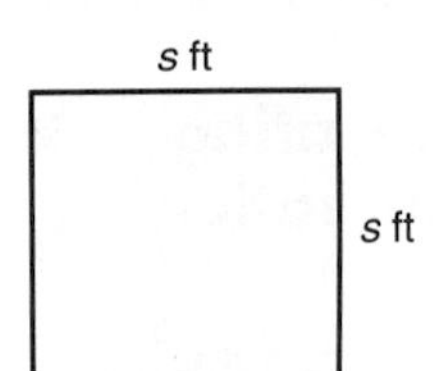

EXERCISES

Practice

Solve each inequality. Then check your solution.

20. $7a \leq 49$
21. $12b > -144$
22. $-5w > -125$
23. $-x \leq 44$
24. $4 < -x$
25. $-102 > 17r$
26. $\frac{b}{-12} \leq 3$
27. $\frac{t}{13} < 13$
28. $\frac{2}{3}w > -22$
29. $6 \leq 0.8g$
30. $-15b < -28$
31. $-0.049 \leq 0.07x$
32. $\frac{3}{7}h < \frac{3}{49}$
33. $\frac{12r}{-4} > \frac{3}{20}$
34. $\frac{3b}{4} \leq \frac{2}{3}$
35. $-\frac{1}{3}x > 9$
36. $\frac{y}{6} \geq \frac{1}{2}$
37. $\frac{-3m}{4} \leq 18$

Define a variable, write an inequality, and solve each problem. Then check your solution.

38. Four times a number is at most 36.
39. Thirty-six is at least one half of a number.
40. The opposite of three times a number is more than 48.
41. Three fourths of a number is at most -24.
42. Eighty percent of a number is less than 24.
43. The product of two numbers is no greater than 144. One of the numbers is -8. What is the other number?

44. **Geometry** Determine the value of x so that the area of the rectangle at the right is at least 918 square feet.

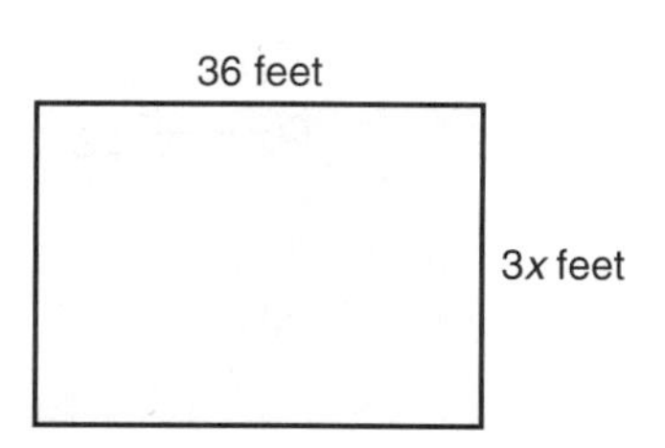

45. **Geometry** Determine the value of y so that the perimeter of the triangle at the right is less than 100 meters.

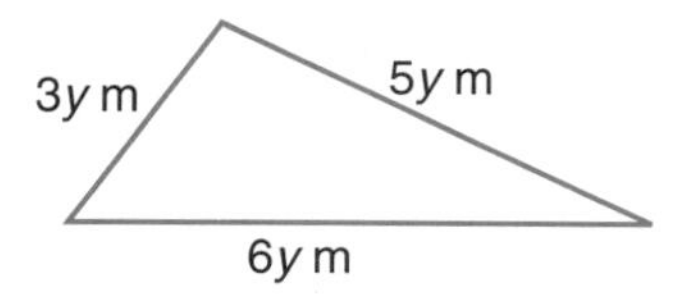

Complete.

46. If $24m \geq 16$, then _?_ ≥ 12.
47. If $-9 \leq 15b$, then $25b$ _?_ -15.
48. If $5y < -12$, then $20y <$ _?_.
49. If $-10a > 21$, then $30a$ _?_ -63.

Critical Thinking

50. Use an example to show that if $x > y$, then $x^2 > y^2$ is not necessarily true.

Applications and Problem Solving

Define a variable, write an inequality, and solve each problem.

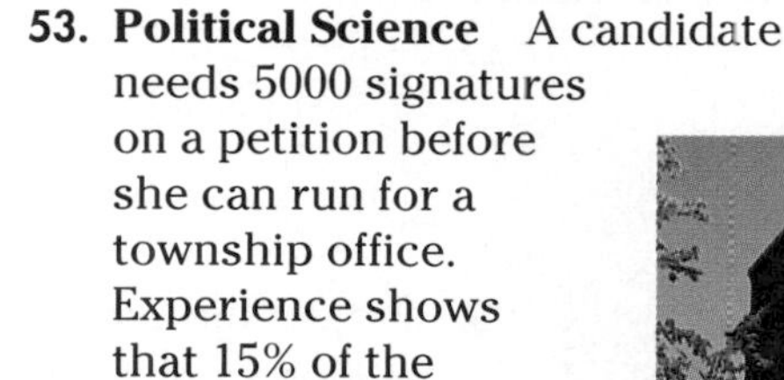

51. **Travel** The charge per mile for a compact rental car at 4-D Rentals is \$0.12. Mrs. Rodriguez is on a business trip and must rent a car to attend various meetings. She has a budget of \$50 per rental for mileage charges. What is the greatest number of miles Mrs. Rodriguez can travel without going over her budget?

52. **Physics** Refer to the application at the beginning of the lesson. A city worker needs to raise a utility access cover. She has an iron bar to use as a lever. When using a lever, the resistance force is equal to the effort force multiplied by the mechanical advantage, or $F_r = MA \cdot F_e$. The worker weighs 120 pounds, so she can supply that much effort force. All utility covers in the city weigh at least 360 pounds. What mechanical advantage does she need to lift the cover?

53. **Political Science** A candidate needs 5000 signatures on a petition before she can run for a township office. Experience shows that 15% of the signatures on petitions are not valid. What is the smallest number of signatures the candidate should get to end up with 5000 valid signatures?

Mixed Review

54. Define a variable, write an inequality, and solve the following problem. The difference of five times a number less four times that number plus seven is at most 34. (Lesson 7–1)

55. **Geometry** Three vertices of a square are at $(-5, -3)$, $(-5, 5)$, and $(3, -3)$. Suppose you were to inscribe a circle in this square. What would the coordinates of the center of the circle be? (Lesson 6–7)

56. **Geometry** Determine the slopes of the lines parallel and perpendicular to the graph of $3x - 6 = -y$. (Lesson 6–6)

57. Determine the slope of the line that passes through $(4, -9)$ and $(-2, 3)$. (Lesson 6–1)

58. Write an equation in functional notation for the relation at the right. (Lesson 5–6)

a	−2	0	2	4	6	8	10
b	−5	−3	−1	1	3	5	7

59. **Budgeting** JoAnne Paulsen's take-home pay is $1782 per month. She spends $325 on rent, $120 on groceries, and $40 on gas. She allows herself 12% of the remaining amount for entertainment. How much can she spend on entertainment each month? (Lesson 4–4)

60. What number is 47% of 27? (Lesson 4–4)

61. **Soccer** A soccer field is 75 yards shorter than 3 times its width. Its perimeter is 370 yards. Find its dimensions. (Lesson 3–5)

62. Evaluate $5y + 3$ if $y = 1.3$. (Lesson 1–3)

WORKING ON THE

In·ves·ti·ga·tion

Refer to the Investigation on pages 320–321.

Pages 320–321 are contained in Volume One.

Smoke Gets In Your Eyes

Currently, about 50 million Americans smoke, and each smoker averages about 30 cigarettes a day. Consider the amount of smoke produced by a single cigarette. Make some projections on the volume of cigarette smoke generated by all 50 million smoking Americans.

1 Find the volume of smoke produced by the average smoker.

2 Project the amount of smoke generated by 50 million smokers in a year.

3 Estimate the volume of air in your room at home. Write an inequality that would estimate the number of puffs p it would take to fill your room. Solve the inequality.

4 Suppose a smoker was locked in your room with several cartons of cigarettes. Would it be possible for the smoker to fill the room completely with smoke in the same concentration as the smoke exhaled? Explain your answer.

Add the results of your work to your Investigation Folder.

7-3

Solving Multi-Step Inequalities

What YOU'LL LEARN

- To solve linear inequalities involving more than one operation, and
- to find the solution set for a linear inequality when replacement values are given for the variables.

Why IT'S IMPORTANT

You can use inequalities to solve problems involving engineering and real estate.

APPLICATION

Engineering

Rosa Whitehair is a partner in an engineering consulting firm. Her fee for consulting on large construction projects is \$1000 plus 10% of the design fee that the company charges its clients. Ms. Whitehair is considering two construction projects: a 50-story office building and the design of an airport terminal. She is interested in both projects, but decides to choose the one for which her fee is higher. The company that is designing the office building has agreed to pay Ms. Whitehair a flat fee of \$5000. How much does the design company's fee need to be for her to choose the terminal?

Let x represent the design fee for the airport terminal.

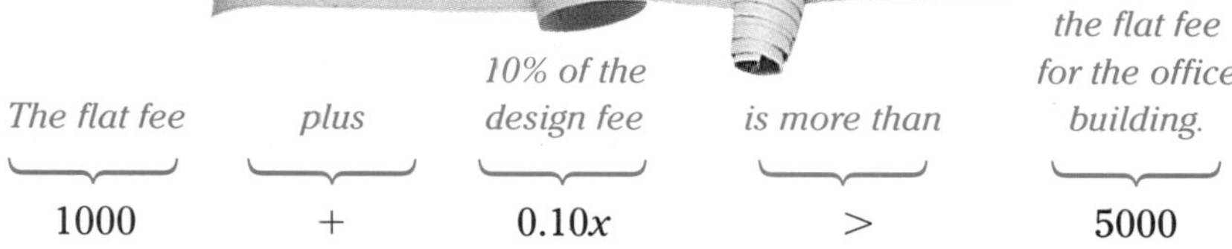

The flat fee	*plus*	*10% of the design fee*	*is more than*	*the flat fee for the office building.*
1000	+	$0.10x$	>	5000

LOOK BACK

You can review solving multi-step equations in Lesson 3-3.

This inequality involves more than one operation. It can be solved by undoing the operations in reverse of the order of operations in the same way you would solve an equation with more than one operation.

$$1000 + 0.10x > 5000$$

$$1000 - 1000 + 0.10x > 5000 - 1000 \quad \textit{Subtract 1000 from each side.}$$

$$0.10x > 4000$$

$$\frac{0.10x}{0.10} > \frac{4000}{0.10} \quad \textit{Divide each side by 0.10.}$$

$$x > 40{,}000$$

If the company charges design fees higher than \$40,000 for the terminal, Ms. Whitehair will choose them since her consulting fee is higher.

Example  Determine the value of x so that $\angle A$ is acute.

INTEGRATION
Geometry

Assume that x is positive.

For $\angle A$ to be acute, its measure must be less than 90°.

Thus, $3x - 15 < 90$.

$$3x - 15 < 90$$
$$3x - 15 + 15 < 90 + 15 \quad \textit{Add 15 to each side.}$$
$$3x < 105$$
$$\frac{3x}{3} < \frac{105}{3} \quad \textit{Divide each side by 3.}$$
$$x < 35$$

For $\angle A$ to be acute, x must be less than 35.

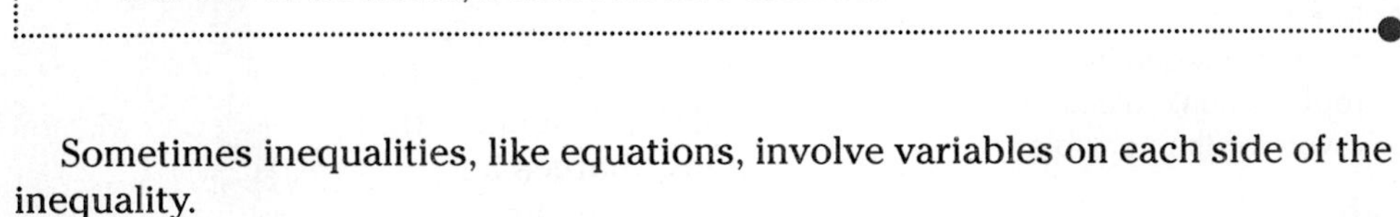

Sometimes inequalities, like equations, involve variables on each side of the inequality.

Example  Solve $-4w + 9 \leq w - 21$.

$$-4w + 9 \leq w - 21$$
$$-w - 4w + 9 \leq w - 21 - w \quad \textit{Subtract w from each side.}$$
$$-5w + 9 \leq -21$$
$$-5w + 9 - 9 \leq -21 - 9 \quad \textit{Subtract 9 from each side.}$$
$$-5w \leq -30$$
$$\frac{-5w}{-5} \geq \frac{-30}{-5} \quad \textit{Divide each side by } -5 \textit{ and change} \leq \textit{to} \geq.$$
$$w \geq 6$$

The solution set is $\{w \mid w \geq 6\}$.

When we solve an inequality, the solution set usually includes all numbers for a certain criteria, such as $\{x \mid x > 4\}$. Sometimes a replacement set is given from which the solution set can be chosen.

Example  Determine the solution set for $3x + 6 > 12$ if the replacement set for x is $\{-2, -1, 0, 1, 2, 3, 4, 5\}$.

Method 1

Substitute values into the inequality to find the values that satisfy the inequality. Try -2.

$$3x + 6 > 12$$
$$3(-2) + 6 > 12$$
$$0 > 12 \quad \text{false}$$

From this trial, we can estimate that the value of x must be much greater than -2 to make the inequality true. Try 2.

$$3x + 6 > 12$$
$$3(2) + 6 > 12$$
$$12 > 12 \quad \text{false}$$

From this trial, we see that values greater than 2 must be in the solution set. Try 3.

$$3x + 6 > 12$$
$$3(3) + 6 > 12$$
$$15 > 12 \quad \text{true}$$

The solution set is $\{3, 4, 5\}$.

Method 2

Solve the inequality for all values of x. Then determine which values from the replacement set belong to the solution set.

$$3x + 6 > 12$$
$$3x + 6 - 6 > 12 - 6$$
$$3x > 6$$
$$\frac{3x}{3} > \frac{6}{3}$$
$$x > 2$$

The solution set is those numbers from the replacement set that are greater than 2. Thus, the solution set is $\{3, 4, 5\}$.

EXPLORATION — GRAPHING CALCULATORS

You can use the inequality symbols in the TEST menu on the TI-82 graphing calculator to find the solution to an inequality in one variable.

Your Turn

a. Clear the [Y=] list. Enter $3x + 6 > 4x + 9$ as Y1. (The symbol $>$ is item 3 on the TEST menu.) Press [GRAPH]. Describe what you see.

b. Use the TRACE function to scan the values along the graph. What do you notice about the values of y on the graph?

c. Solve the inequality algebraically. How does your solution compare to the pattern you noticed in part **b**?

When solving some inequalities that contain grouping symbols, remember to first use the distributive property to remove the grouping symbols.

Example  **Solve $5(k + 4) - 2(k + 6) \geq 5(k + 1) - 1$. Then graph the solution.**

$$5(k + 4) - 2(k + 6) \geq 5(k + 1) - 1$$
$$5k + 20 - 2k - 12 \geq 5k + 5 - 1 \quad \textit{Distributive property.}$$
$$3k + 8 \geq 5k + 4 \quad \textit{Combine like terms.}$$
$$3k - 5k + 8 \geq 5k - 5k + 4 \quad \textit{Subtract 5k from each side.}$$
$$-2k + 8 \geq 4$$
$$-2k + 8 - 8 \geq 4 - 8 \quad \textit{Subtract 8 from each side.}$$
$$-2k \geq -4$$
$$\frac{-2k}{-2} \leq \frac{-4}{-2} \quad \textit{Divide each side by } -2 \textit{ and change } \geq \textit{ to } \leq.$$
$$k \leq 2$$

The solution set is $\{k \mid k \leq 2\}$.
The graph of $k \leq 2$ is shown at the right.

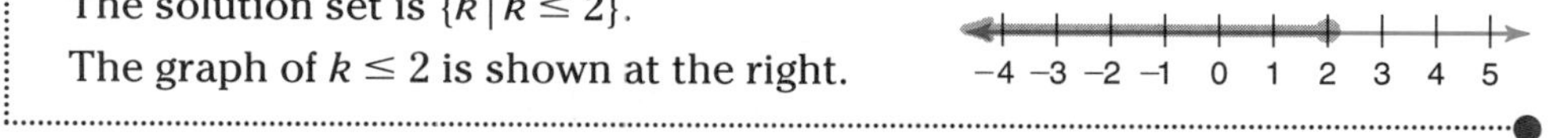

CHECK FOR UNDERSTANDING

Communicating Mathematics

Study the lesson. Then complete the following.

1. **Explain** each step in solving $-3x + 7 < 4x - 5$.
2. **Write** an inequality that expresses the fact that when you add 3 feet to the perimeter of a square of sides with length s, the resulting perimeter does not exceed 50 feet.
3. **Write** an inequality that has no solution.
4. **Describe** how you would solve $16 - 5w > 29$ without dividing by -5 or multiplying by $-\frac{1}{5}$.
5. Refer to Example 3. Which of the two methods seems more efficient in solving the inequality for the given replacement set? Explain your selection.

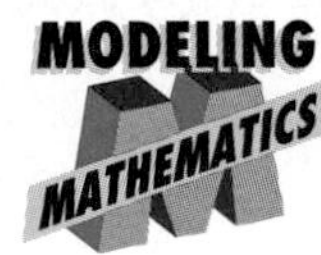

6. Use the method shown in Lesson 7–2A to model the solutions for each inequality.
   a. $3 - 4x \geq 15$ b. $6x - 1 < 5 + 3x$

Guided Practice

Choose the correct solution for each inequality.

7. Solve $2m + 5 \leq 4m - 1$.
   a. $m > -3$ b. $m < 3$ c. $m \geq 3$ d. $m \leq -3$
8. Solve $13r - 11 \geq 7r + 37$.
   a. $r < 8$ b. $r \geq 8$ c. $r \leq -8$ d. $r > -8$

Solve each inequality. Then check your solution.

9. $9x + 2 > 20$
10. $-4h + 7 > 15$
11. $-2 - \frac{d}{5} < 23$
12. $6a + 9 < -4a + 29$

Find the solution set of each inequality given the replacement set.

13. $3x - 1 > 4$, $\{-1, 0, 1, 2, 3\}$
14. $-7a + 6 \leq 48$, $\{-10, -9, -8, -7, -6, -5, -4, -3\}$
15. **Number Theory** Consider the sentence *The sum of two consecutive even integers is greater than 75.*
    a. Write an inequality for this statement.
    b. Solve the inequality.
    c. Name two consecutive even integers that meet the requirements of the statement.

EXERCISES

Practice

Find the solution set of each inequality if the replacement set for each variable is $\{-10, -9, -8, \ldots, 8, 9, 10\}$.

16. $n - 3 \geq \frac{n + 1}{2}$
17. $\frac{2(x + 2)}{3} < 4$
18. $1.3y - 12 < 0.9y + 4$
19. $-20 \geq 8 + 7k$

Solve each inequality. Then check your solution.

20. $2m + 7 > 17$
21. $-3 > -3t + 6$
22. $-2 - 3x \geq 2$
23. $\frac{2}{3}w - 3 \leq 7$
24. $7x - 1 < 29 - 2x$
25. $8n + 2 - 10n < 20$
26. $2x + 5 < 3x - 7$
27. $5 - 4m + 8 + 2m > -17$
28. $\frac{2x - 3}{5} < 7$
29. $x < \frac{2x - 15}{3}$
30. $9r + 15 \geq 24 + 10r$
31. $6p - 2 \leq 3p + 12$
32. $4y + 2 < 8y - (6y - 10)$
33. $3(x - 2) - 8x < 44$
34. $3.1q - 1.4 > 1.3q + 6.7$
35. $-5(k + 4) \geq 3(k - 4)$
36. $5(2h - 6) - 7(h + 7) > 4h$
37. $7 + 3y > 2(y + 3) - 2(-1 - y)$

Define a variable, write an inequality, and solve each problem. Then check your solution.

38. Two thirds of a number decreased by 27 is at least 9.
39. Three times the sum of a number and 7 is greater than 5 times the number less 13.

Number Theory

40. The sum of two consecutive odd integers is at most 123. Find the pair with the greatest sum.
41. Find all sets of two consecutive positive odd integers whose sum is no greater than 18.
42. Find all sets of three consecutive positive even integers whose sum is less than 40.

Solve each inequality. Then check your solution.

43. $3x + 4 > 2(x + 3) + x$
44. $3 - 3(y - 2) < 13 - 3(y - 6)$

Graphing Calculator

45. Use the methods presented in the Exploration to solve each inequality. Use the viewing window [−9.4, 9.4] by [−5, 5] with scale factors of 1.
    a. $-5 - 8x \geq 59$
    b. $13x - 11 > 7x + 37$
    c. $8x - (x - 5) > x + 17$
    d. $-5(x + 4) \geq 3(x - 4)$

Critical Thinking

46. We can write the expression *the numbers between −3 and 4* as $-3 < x < 4$. What does the algebraic sentence $-3 < x + 2 < 4$ represent?

Applications and Problem Solving

Define a variable, write an inequality, and solve each problem.

47. **Personal Finances** A couple does not want to have a charge of more than \$50 for dinner at a restaurant. A sales tax of 4% is added to the bill, and they plan to tip 15% after the tax has been added. What can the couple spend on the meal?
48. **Recreation** The admission fee to a video game arcade is \$1.25 per person, and it costs \$0.50 for each game played. Latoya and Donnetta have a total of \$10.00 to spend. What is the greatest number of games they will be able to play?

49. **Fund-raising** A university is running a drive to raise money. A corporation has promised to match 40% of whatever the university can raise from other sources. How much must the school raise from other sources to have a total of at least $800,000 after the corporation's donation?

50. **Real Estate** A homeowner is selling her house. She must pay 7% of the selling price to her real estate agent after the house is sold. To the nearest dollar, what must be the selling price of her house to have at least $90,000 after the agent is paid?

51. **Labor** A union worker is currently making $400 per week. His union is seeking a one-year contract. If there is a strike, the new contract will provide for a 6% raise; if there is no strike, the worker expects no increase in salary.
    a. Assuming the worker will not be paid during the strike, how many weeks could he go on strike and still make at least as much for the year as he would have made without a strike?
    b. How would your answer change if the worker was currently making $575 per week?
    c. How would your answer to part a change if the worker's union provided him with $120 per week during the strike?

Mixed Review

Solve each inequality.

52. $2r - 2.1 < -8.7 + r$ (Lesson 7–2)
53. $7 - 2y < -y - 3$ (Lesson 7–1)

54. Write the slope-intercept form of an equation for the line that passes through $(-12, 12)$ and $(-2, 7)$. (Lesson 6–4)

55. Write the standard form of an equation of the line that passes through $(2, 4)$ with slope $-\frac{3}{2}$. (Lesson 6–2)

56. **Business** The owner of No Spots City Car Wash found that if c cars were washed in a day, the average daily profit $P(c)$ was given by the formula $P(c) = -0.027c^2 + 8c - 280$. Find the values of $P(c)$ for various values of c to determine the least number of cars that must be washed each day for No Spots City to make a profit. (Lesson 5–5)

57. Find the domain for $3y - 2 = x$ if the range is $\left\{-1, 6, 0, -\frac{1}{3}, 2\right\}$. (Lesson 5–3)

58. **Advertising** For many years, the slogan of Crest® toothpaste has been "Four out of five dentists recommend that their patients use Crest." What are the odds that your dentist will *not* recommend you use Crest? (Lesson 4–6)

59. Jason scored the following points for his basketball team during the past ten games: 18, 32, 20, 21, 34, 9, 33, 37, 22, 25. Find the mean, median, and mode of his total points. (Lesson 3–7)

60. Find $18 - (-34)$. (Lesson 2–3)

7-4

Solving Compound Inequalities

What YOU'LL LEARN

- To solve problems by making a diagram,
- to solve compound inequalities and graph their solution sets, and
- to solve problems that involve compound inequalities.

Why IT'S IMPORTANT

You can use inequalities to solve problems involving chemistry and physics.

Chemistry

The largest fish that spends its whole life in fresh water is the rare Pla beuk, found in the Mekong River in China, Laos, Cambodia, and Thailand. The largest specimen was reportedly 9 feet $10\frac{1}{4}$ inches long and weighed 533.5 pounds.

Such rare fish are sometimes displayed in aquariums. Aquariums can house freshwater or marine life and must be closely monitored to maintain the correct temperature and pH for the animals to survive. pH is a measure of acidity. To determine pH, a scale with values from 0 to 14 is used. One such scale is shown in the diagram below.

FYI

The largest fish ever caught is a whale shark. Caught near Karachi, Pakistan in 1949, it was 41.5 feet long and weighed 16.5 tons. The smallest fish is a dwarf goby found in the Indian Ocean. Males of this species are only 0.34 inches long.

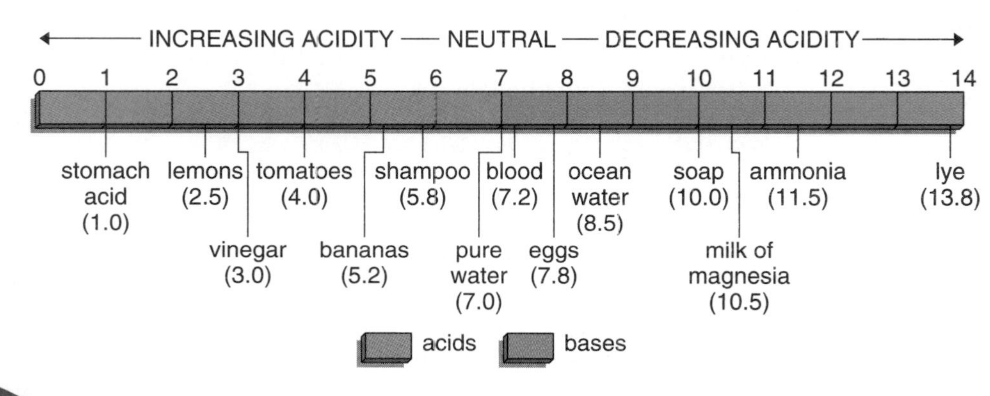

whale shark

dwarf goby

If we let p represent the value of the pH scale, we can express the different pH levels by using inequalities. For example, an acid solution will have a pH level of $p \geq 0$ and $p < 7$. When considered together, these two inequalities form a **compound inequality.** This compound inequality can also be written without using *and* in two ways.

$$0 \leq p < 7 \quad \text{or} \quad 7 > p \geq 0$$

The statement $0 \leq p < 7$ can be read *0 is less than or equal to p, which is less than 7.* The statement $7 > p \geq 0$ can be read *7 is greater than p, which is greater than or equal to 0.*

The pH levels of bases could be written as follows.

$$p > 7 \text{ and } p \leq 14 \quad \text{or} \quad 7 < p \leq 14 \quad \text{or} \quad 14 \geq p > 7$$

You will graph this inequality in Exercise 6.

You can **draw a diagram** to help solve many problems. Sometimes a picture will help you decide how to work the problem. Other times the picture will show you the answer to the problem.

Example 1 — **PROBLEM SOLVING** — **Draw a Diagram**

On May 6, 1994, President Francois Mitterrand of France and Queen Elizabeth II of England officially opened the Channel Tunnel connecting England and France. After the ceremonies, a group of 36 English and French government officials had dinner at a restaurant in Calais, France, to celebrate the occasion. Suppose the restaurant staff used small tables that seat four people each, placed end to end, to form one long table. How many tables were needed to seat everyone?

Draw a diagram to represent the tables placed end to end. Use Xs to indicate where the people are sitting. Let's start with a guess, say 10 tables.

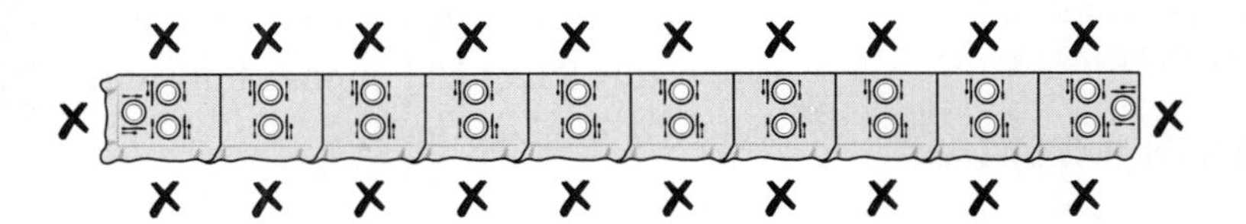

Ten tables will seat 22 people. If we use an extra table, we can seat 2 more people. Now, let's look for a pattern.

Number of tables	10	11	12	13	14	15	16	17
Number of people seated	22	24	26	28	30	32	34	36

This pattern shows that the restaurant needed 17 tables to seat all 36 officials.

The logic symbol for and is $\wedge$. You can write $x > 7$ and $x < 10$ as $(x > 7) \wedge (x < 10)$.

A compound inequality containing *and* is true only if *both* inequalities are true. Thus, the graphs of a compound inequality containing *and* is the **intersection** of the graphs of the two inequalities. The intersection can be found by graphing the two inequalities and then determining where these graphs overlap. In other words, draw a diagram to solve the inequality.

Example 2 **Graph the solution set of $x \geq -2$ and $x < 5$.**

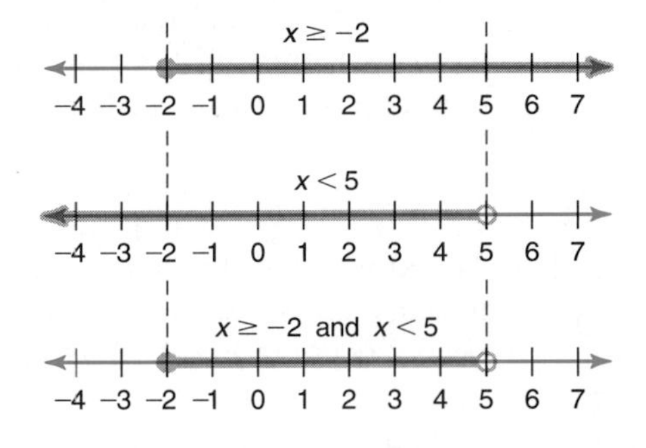

The solution set, shown in the bottom graph, is $\{x \mid -2 \leq x < 5\}$. Note that the graph of $x \geq -2$ includes the point -2. The graph of $x < 5$ does not include 5.

Example 3 **Solve $-1 < x + 3 < 5$. Then graph the solution set.**

First express $-1 < x + 3 < 5$ using *and.* Then solve each inequality.

$$-1 < x + 3 \quad \text{and} \quad x + 3 < 5$$
$$-1 - 3 < x + 3 - 3 \qquad x + 3 - 3 < 5 - 3$$
$$-4 < x \qquad x < 2$$

Now graph each solution and find the intersection of the solutions.

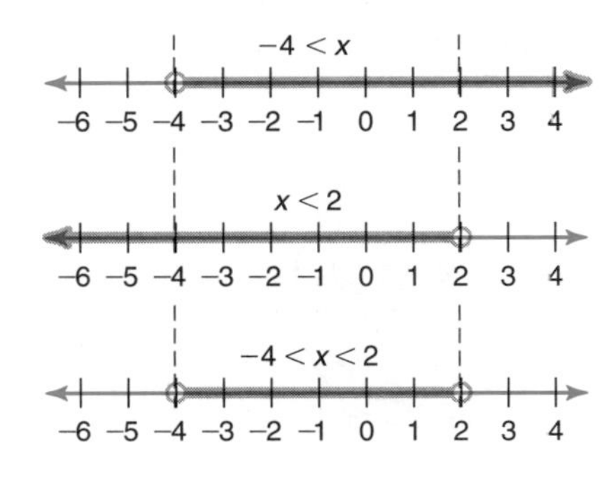

The solution set is $\{x \mid -4 < x < 2\}$.

The following example shows how you can solve a problem by using geometry, a diagram, and a compound inequality.

Example 4

PROBLEM SOLVING

Draw a Diagram

Mai and Luis hope someday to compete in the Olympics in pairs ice skating. Each day they travel from their homes to an ice rink to practice before going to school. Luis lives 17 miles from the rink, and Mai lives 20 miles from it. If this were all the information you were given, determine how far apart Mai and Luis live.

Explore Mai lives 20 miles from the rink and Luis lives 17 miles from the rink. We do not know the relative positions of Mai's and Luis's homes in relation to the rink.

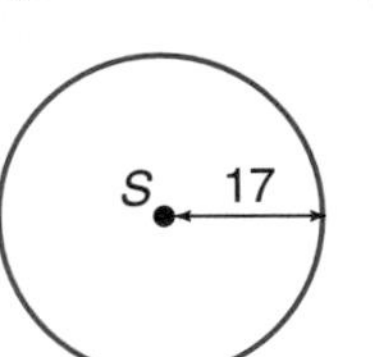

Plan Draw a diagram of the situation. Let S be the location of the skating rink. Since Luis lives 17 miles from the rink, we can draw a circle with a radius of 17. Luis will live somewhere on that circle. Likewise, we can draw another circle with radius of 20 for the location of Mai's home.

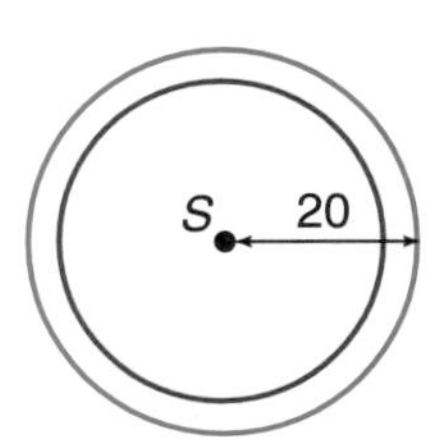

(continued on the next page)

Solve Let's examine three possibilities for the locations of Mai's and Luis's homes.

(1) Mai and Luis live along the same radius from the skating rink.
(2) Mai and Luis live on opposite radii from the skating rink.
(3) Mai and Luis live somewhere other than the locations described in (1) and (2).

Let L and M represent where Luis and Mai live, respectively.

LOOK BACK

You can find more information about the triangle inequality theorem in Exercise 45 (Programming) of Lesson 7–1.

(1) the same radius

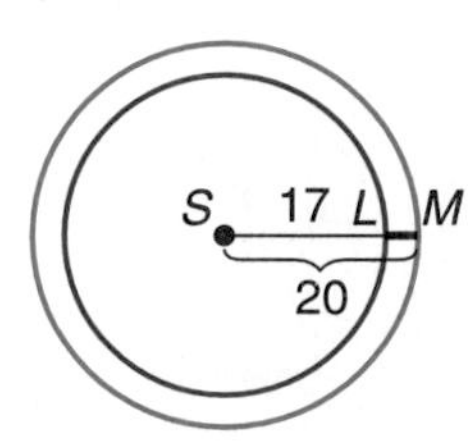

The distance is $20 - 17$ or 3 miles.

(2) opposite radii

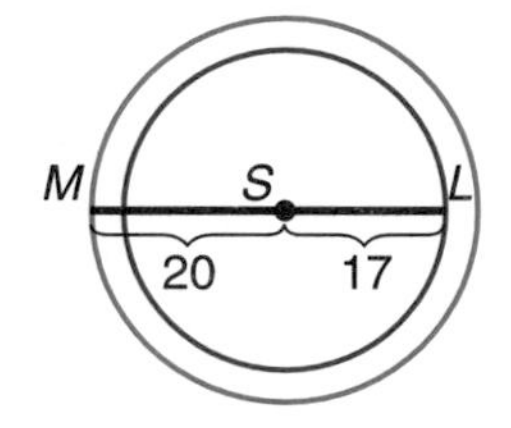

The distance is $20 + 17$ or 37 miles.

(3) somewhere other than (1) or (2)

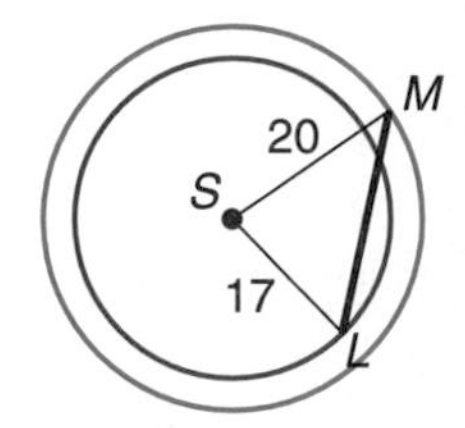

By the triangle inequality theorem, the distance must be less than 37 miles and greater than 3 miles.

The distance d can be described by the inequality $3 \leq d \leq 37$.

Examine Diagrams (1) and (2) show the least and greatest possiblities. To convince yourself that the statement with diagram (3) is always true, you may want to sketch a different triangle.

Another type of compound inequality contains the word *or* instead of *and.* A compound inequality containing *or* is true if one or more of the inequalities is true. The graph of a compound inequality containing *or* is the **union** of the graphs of the two inequalities.

Example  **Graph the solution set of $x \geq -1$ or $x < -4$.**

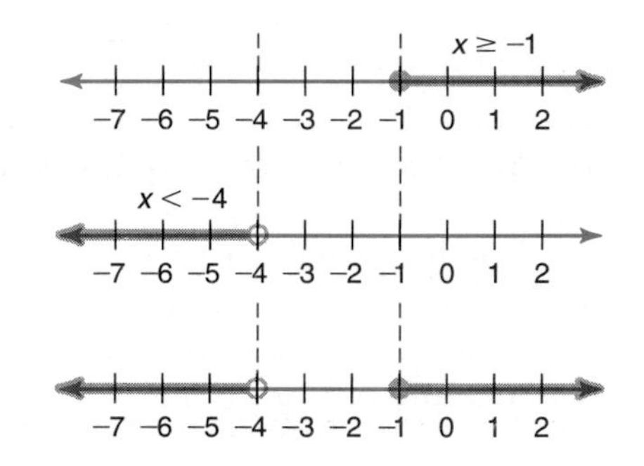

The last graph shows the solution set, $\{x \mid x \geq -1 \text{ or } x < -4\}$.

Example 6 **Solve $3w + 8 < 2$ or $w + 12 > 2 - w$. Graph the solution set.**

$$3w + 8 < 2 \qquad \text{or} \qquad w + 12 > 2 - w$$
$$3w + 8 - 8 < 2 - 8 \qquad\qquad w + w + 12 > 2 - w + w$$
$$3w < -6 \qquad\qquad 2w + 12 > 2$$
$$\frac{3w}{3} < \frac{-6}{3} \qquad\qquad 2w + 12 - 12 > 2 - 12$$
$$w < -2 \qquad\qquad 2w > -10$$
$$\frac{2w}{2} > \frac{-10}{2}$$
$$w > -5$$

Now graph each solution and find the union of the two.

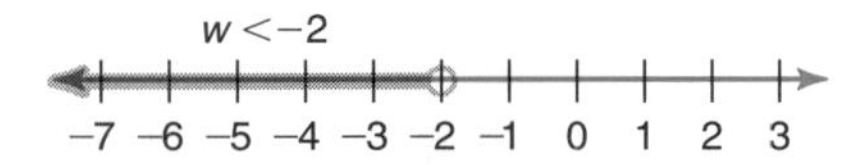

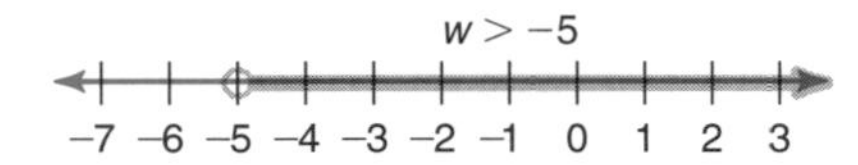

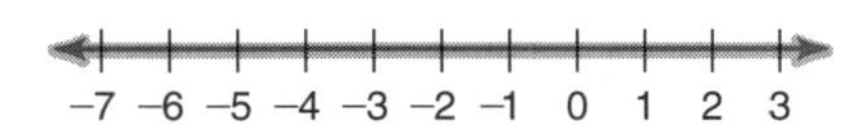

Find the union.

The last graph shows the solution set, $\{x \mid x \text{ is a real number}\}$.

CHECK FOR UNDERSTANDING

Communicating Mathematics

Study the lesson. Then complete the following.

1. **Write** a statement that could be represented by $\$7.50 < p \leq \$18.50$.
2. **Write** a compound inequality that describes the graph below.

   −6 −5 −4 −3 −2 −1 0 1 2 3

3. **List** two problem-solving strategies used in Example 1. Describe how each strategy helped in the solution of the problem.
4. **Describe** the difference between a compound inequality containing *and* and a compound inequality containing *or*.
5. **Name** two ways in which a picture or diagram can help you solve a problem more easily.
6. **Refer** to the connection at the beginning of the lesson. Graph the inequality that represents the pH levels of bases.

Guided Practice

Write each compound inequality without using *and*. Then graph the solution set.

7. $x < 9$ and $0 \leq x$
8. $x > -2$ and $x < 3$

Write a compound inequality for each solution set shown below.

9. −5 −4 −3 −2 −1 0 1 2 3 4
10. −5 −4 −3 −2 −1 0 1 2 3 4

11. Graph the solution set of $y > 5$ and $y < -3$. Describe what the solution set means.

Solve each compound inequality. Then graph the solution set.

12. $2 \le y + 6$ and $y + 6 < 8$
13. $4 + h \le -3$ or $4 + h \ge 5$
14. $b + 5 > 10$ or $b \ge 0$
15. $2 + w > 2w + 1 \ge -4 + w$

16. Write a compound inequality without using *and* for the following situation. Solve the inequality and then check the solution.
It costs the same to register mail containing articles with values from \$0 to \$100.

17. **Draw a Diagram** You can cut a pizza into seven pieces with only three straight cuts as shown at the right. Draw a diagram to show the greatest number of pieces you can make with five straight cuts.

EXERCISES

Practice

Graph the solution set of each compound inequality.

18. $m \ge -5$ and $m < 3$
19. $p < -8$ and $p > 4$
20. $s < 3$ or $s \ge 1$
21. $n \le -5$ or $n \ge -1$
22. $w > -3$ and $w < 1$
23. $x < -7$ or $x \ge 0$

Write a compound inequality for each solution set shown below.

24. −6 −5 −4 −3 −2 −1 0 1 2 3 4 5 6

25. −6 −5 −4 −3 −2 −1 0 1 2 3 4 5 6

26. −6 −5 −4 −3 −2 −1 0 1 2 3 4 5 6

27. −6 −5 −4 −3 −2 −1 0 1 2 3 4 5 6

Solve each compound inequality. Then graph the solution set.

28. $4m - 5 > 7$ or $4m - 5 < -9$
29. $x - 4 < 1$ and $x + 2 > 1$
30. $y + 6 > -1$ and $y - 2 < 4$
31. $x + 4 < 2$ or $x - 2 > 1$
32. $10 - 2p > 12$ and $7p < 4p + 9$
33. $6 - c > c$ or $3c - 1 < c + 13$
34. $4 < 2x - 2 < 10$
35. $14 < 3h + 2 < 2$
36. $8 > 5 - 3q$ and $5 - 3q > -13$
37. $-1 + x \le 3$ or $-x \le -4$
38. $3n + 11 \le 13$ or $2n \ge 5n - 12$
39. $3y + 1 > 10$ and $y \ne 6$
40. $4z + 8 \ge z + 6$ or $7z - 14 \ge 2z - 4$
41. $5x + 7 > 2x + 4$ or $3x + 3 < 24 - 4x$
42. $2 - 5(2y - 3) > 2$ or $3y < 2(y - 8)$
43. $5w > 4(2w - 3)$ and $5(w - 3) + 2 < 7$

Write a compound inequality for each solution set shown below.

44. −6 −5 −4 −3 −2 −1 0 1 2 3 4 5 6

45. 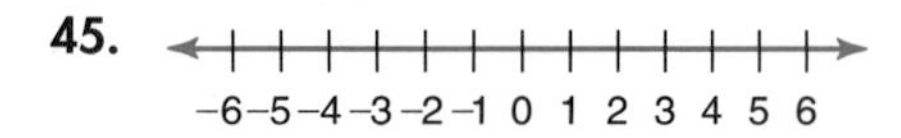

Define a variable, write a compound inequality, and solve each problem. Then check your solution.

46. When three times the distance to the finish line is increased by 5 km, the total kilometers covered in the race will be between 50 and 89 km.

47. The sum of a number and 2 is no more than 6 or no less than 10.

48. The sum of twice a number and 5 lies between 7 and 11.

49. Five less than 6 times a number is at most 37 and at least 31.

Solve each inequality. Then graph the solution set.

50. $3 + y > 2y > -3 - y$ **51.** $m > 2m - 1 > m - 5$ **52.** $\frac{5}{x} + 3 > 0$

Graphing Calculator

53. In Lesson 7–3, you learned how to use a graphing calculator to find graphically which values of x make a given inequality true. You can also use this method to test compound inequalities. The words *and* and *or* can be found in the LOGIC submenu of the TEST menu. Use this method to solve each of the following using your graphing calculator.

a. $3 + x < -4$ or $3 + x > 4$ **b.** $-2 \le x + 3$ and $x + 3 < 4$

Critical Thinking

54. For what values of a does the compound inequality $-a < x < a$ have no solution?

55. Write a compound inequality for the solution set shown below.

−6 −5 −4 −3 −2 −1 0 1 2 3 4 5 6

Applications and Problem Solving

56. Draw a Diagram There are eight houses on McArthur Street, all in a row. These houses are numbered from 1 to 8. Allison, whose house number is greater than 2, lives next door to her best friend, Adrienne. Belinda, whose house number is greater than 5, lives two doors away from her boyfriend, Benito. Cheri, whose house number is greater than Benito's, lives three doors away from her piano teacher, Mr. Crawford. Darryl, whose house number is less than 4, lives four doors away from his teammate, Don. Who lives in each house?

Define a variable, write a compound inequality, and solve each problem.

57. Physics According to Hooke's Law, the force F (in pounds) required to stretch a certain spring x inches beyond its natural length is given by $F = 4.5x$. If forces between 20 and 30 pounds, inclusive, are applied to the spring, what will be the range of the increased lengths of the stretched spring?

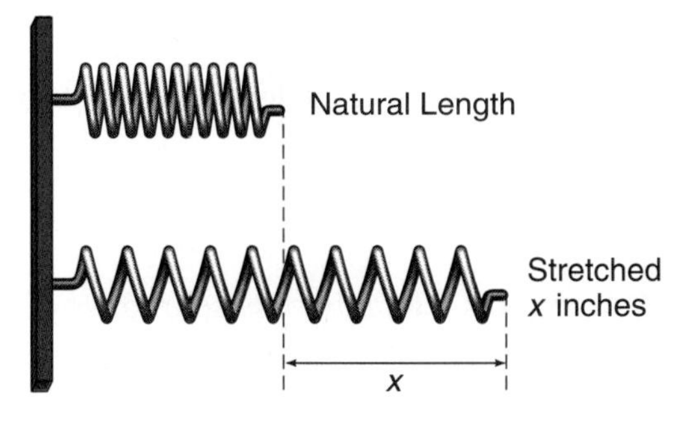

58. Statistics Clarissa must have an average of 92, 93, or 94 points to receive a grade of A– in social studies. She earned 92, 96, and 88 on the first three tests of the grading period. What range of scores on the fourth test will give her an A–?

Mixed Review

Define a variable, write an inequality, and solve each problem.

59. Three fourths of a number decreased by 8 is at least 3. (Lesson 7–3)

60. A number added to 23 is at most 5. (Lesson 7–1)

61. If $7x - 2 \leq 9x + 3$, then $2x \geq$ __?__ . (Lesson 7–1)

62. Geometry Line a is perpendicular to a line that is perpendicular to a line whose equation is $3x - 7y = -3$. If all lines are in the same plane, what is the slope of line a? (Lesson 6–6)

63. Graph $x = 2y - 4$ using the x- and y-intercepts. (Lesson 6–5)

64. Write an equation in functional notation for the relation graphed at the right. (Lesson 5–6)

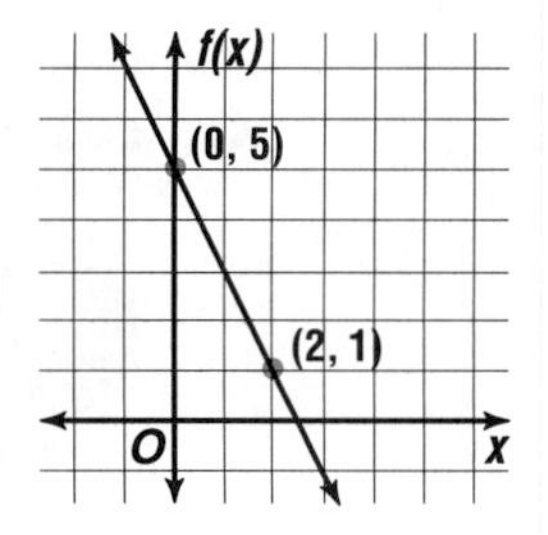

65. Sarah is visiting a friend about two miles away from her house. She sees a hot-air balloon directly above her house. Sara estimates that the angle of elevation formed by the hot-air balloon is about 15°. About how high is the hot-air balloon? (Lesson 4–3)

66. Solve $\frac{4m - 3}{-2} = 12$. (Lesson 3–3)

67. Simplify $-17px + 22bg + 35px + (-37bg)$. (Lesson 2–5)

68. Evaluate $3ab - c^2$ if $a = 6$, $b = 4$, and $c = 3$. (Lesson 1–3)

SELF TEST

Solve each inequality. Then check your solution. (Lessons 7–1, 7–2, and 7–3)

1. $y + 15 \geq -2$

2. $-102 > 17r$

3. $5 - 6n > -19$

4. $\frac{11 - 6w}{5} > 10$

5. $7(g + 8) < 3(g + 12)$

6. $0.1y - 2 \leq 0.3y - 5$

7. Choose an equivalent statement for $4 \geq x - 1 \geq -3$. (Lesson 7–4)

a. $5 \geq x \geq -4$ **b.** $3 \geq x \geq -4$ **c.** $5 \geq x \geq -2$ **d.** $3 \geq x \geq -2$

8. Solve $8 + 3t < 2$ or $-12 < 11t - 1$. Then graph the solution set. (Lesson 7–4)

9. Sports Jennifer has scored 18, 15, and 30 points in her last three starts on the junior varsity girls basketball team. How many points must she score in her next start so that her four-game average is greater than 20 points? (Lesson 7–3)

10. Draw a Diagram Suppose you roll a die two times. (Lesson 7–4)

a. Draw a diagram to show the possibilities of rolling an even number the first time and an odd number the second time.

b. How many of the possibilities have a sum of 7?

7-5 Integration: Probability Compound Events

What YOU'LL LEARN

- To find the probability of a compound event.

Why IT'S IMPORTANT

You can use tree diagrams to solve problems involving business, sports, and games.

APPLICATION

Archaeology

A group of college students are planning a trip to visit Virgin Islands National Park, where they will study the Carib Indian relics and the remnants of the forts built by the Danes. The Carib Indians were the original occupants of the islands, but had died or left by the early 1600s. The Danes formally claimed the islands in 1666, and they remained under Danish control until 1917.

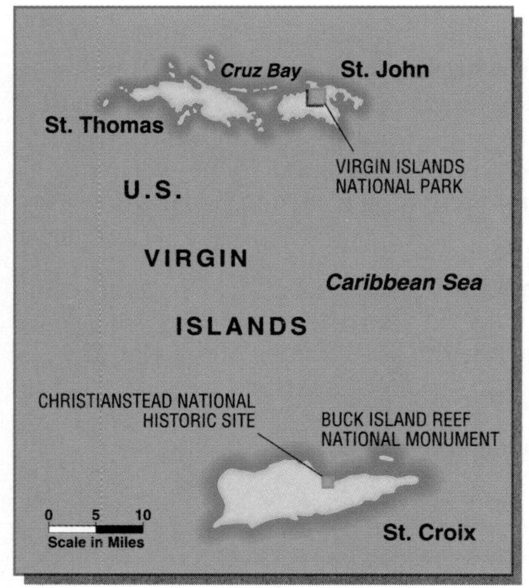

The group's advisor must plan how the group will get to the Virgin Islands. From their college, they will travel to Miami, Florida, by car, bus, train, or plane. Then to travel to St. Thomas in the Virgin Islands, they could take a plane or a ship. Suppose the advisor picks a mode of transportation at random. What is the probability that they will travel by car first and then fly?

To calculate this probability, you need to know all of the possible ways to get to St. Thomas. One method for finding this out is to draw a **tree diagram.** The tree diagram below shows how to get from the college to Miami and then to St. Thomas. The last column details all the possible combinations or **outcomes** of transportation.

Note that the number of outcomes is the product of the number of ways to Miami (4) and the number of ways to the Virgin Islands (2).

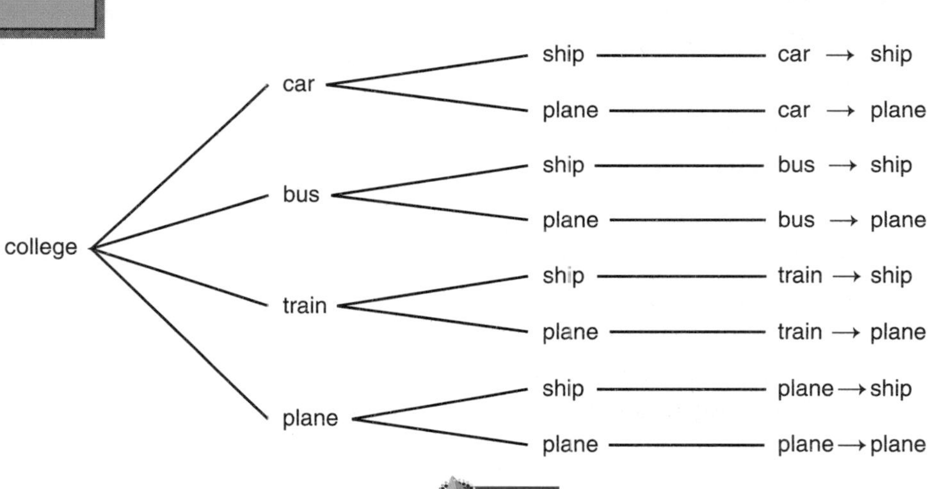

Since the mode of transportation is chosen at random, you can assume that all combinations of travel are equally likely. Since there are 8 outcomes, the probability of traveling by car first and then flying is $\frac{1}{8}$ or 0.125.

How is a compound event similar to a compound inequality?

This problem is an example of finding the probability of a **compound event.** A compound event consists of two or more **simple events.** Choosing a car, plane, train, or bus for the first part of the trip is a simple event. Then, selecting a plane or ship for the second stage of the trip is another simple event. The selection of a mode of transportation for each part of the trip is a compound event.

Example  **Use the tree diagram for the application at the beginning of the lesson to answer each question.**

a. What is the probability that the group will take a ship to get to St. Thomas?

b. What is the probability that the group will travel by plane for both parts of the trip?

LOOK BACK

You can find more information on probability in Lesson 4-6.

a. Since 4 of the 8 outcomes involve taking a ship, the probability is $\frac{4}{8}$ or 0.5.

b. Since only 1 outcome out of 8 involves taking a plane for both parts of the trip, the probability is $\frac{1}{8}$ or 0.125.

Refer to the application at the beginning of the lesson. Because one choice *does not affect* the others, we say that these are **independent events.** If the outcome of an event *does affect* the outcome of another event, we say that these are **dependent events.**

Example  **Booker T. Washington High School is having its annual Spring Carnival. The ninth grade class has decided to have a game booth. To win a small stuffed animal, a player will have to draw 2 marbles of the same color from a box containing 3 marbles—1 red, 1 white, and 1 yellow. First a marble is drawn, put back in the box, and then a second marble is drawn.**

a. What are the possible outcomes?

b. What is the probability of winning the game?

a. First, let's draw a tree diagram to see the possibilities of drawing 2 marbles of the same color.

Since the first marble is replaced, the outcome of the first selection does not affect the outcome of the second selection. Thus, they are independent events.

1st selection	2nd selection	Outcomes
red	red	red, red
	white	red, white
	yellow	red, yellow
white	red	white, red
	white	white, white
	yellow	white, yellow
yellow	red	yellow, red
	white	yellow, white
	yellow	yellow, yellow

There are 9 outcomes.

b. There are 3 ways out of 9 to draw the same color twice. The probability for each draw is equally likely, so the probability is $\frac{3}{9}$ or $0.\overline{3}$.

Probability can also play an important role in determining the possible outcomes of a sporting event.

Example 3

Basketball

The Houston Rockets and the New York Knicks are going to play a best two out of three exhibition game series.

a. What are the possible outcomes of the series?

b. Assuming that the teams are equally matched, what is the probability that the the series will end in two games?

a. The tree diagram below shows the possible outcomes of the first two games.

1st game winner	2nd game winner	Outcomes
Rockets	Rockets	Rockets win in two games.
	Knicks	3rd game is required.
Knicks	Rockets	3rd game is required.
	Knicks	Knicks win in two games.

There are 4 possible outcomes

b. The series can end in two games in two ways. The probability is $\frac{2}{4}$ or 0.5.

CHECK FOR UNDERSTANDING

Communicating Mathematics

Study the lesson. Then complete the following.

1. **Draw** a tree diagram to represent the outcomes of tossing a coin twice.
2. **Describe** a compound event that involves three stages. Draw a tree diagram for such an event.
3. **Explain** the difference between a simple event and a compound event.
4. **Compare and contrast** the tree diagram used in Example 2 and the one used in Example 3.

5. Write a paragraph describing how probabilities are used in real life. Give specific examples.

Guided Practice

6. **Games** In one turn in the game of Yahtzee®, there are 13 different categories you try to complete by rolling five dice. On the first try of your turn, you roll all five dice. Then you may pick up any or all of the dice and roll again on a second and third try. To get a Yahtzee, you must have the same number appearing on all five dice. What is the probability that you roll a Yahtzee on the first try of your turn?

7. **Dining** For lunch at the 66 Diner, you can select one item from each of the following categories for express service guaranteed in 15 minutes.

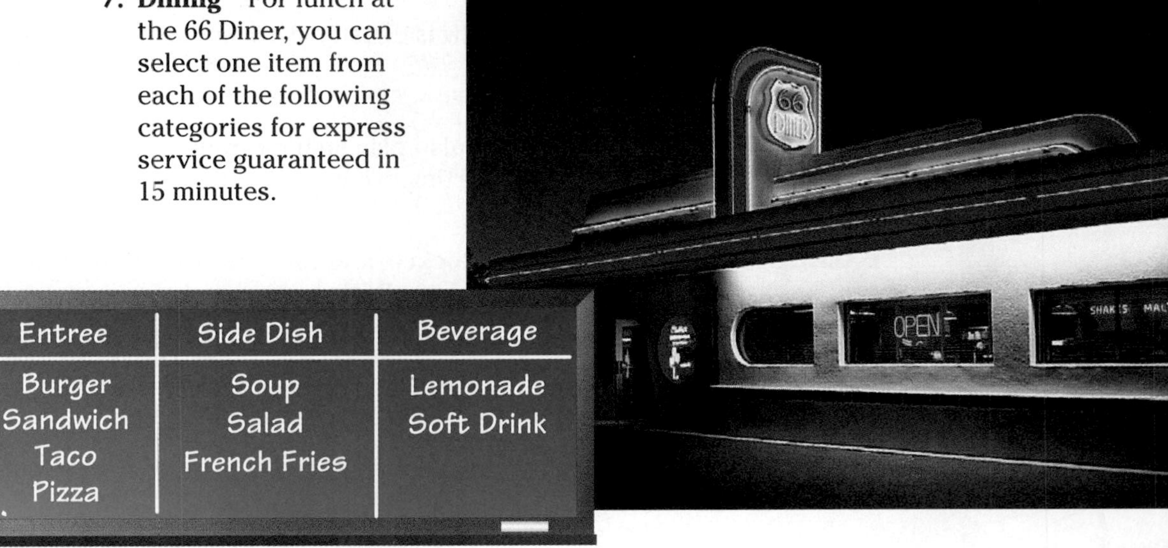

Entree	Side Dish	Beverage
Burger	Soup	Lemonade
Sandwich	Salad	Soft Drink
Taco	French Fries	
Pizza		

   a. Draw a tree diagram showing all of the meal combinations.
   b. What is the probability that a customer will have soup with the meal?
   c. What is the probability of selecting a burger with French fries?
   d. What is the probability of having pizza with a salad and a soft drink?

EXERCISES

Applications and Problem Solving

8. **Video Games** In a computer video game, you have a choice of five roads to collect an important clue to solve a mystery. At the end of each road, there are two doors. The clue can only be found behind one door.
   a. Draw a tree diagram to show the possibilities you have to find the clue.
   b. What is the probability that you will find the clue?

9. **Travel** Three different airlines fly from Bowling Green to Lexington. Those same three airlines and two other airlines fly from Lexington to Louisville. There are no direct flights from Bowling Green to Louisville.

   Louisville
   Lexington
   KENTUCKY
   Bowling Green

   a. How many ways can a traveler book flights from Bowling Green to Louisville?
   b. What is the probability that flights booked at random from Bowling Green to Louisville use the same airline?

10. **Business** On Wednesday, Ralph and Linda were talking about what they were going to wear on Friday since their company had declared it T-Shirt Day. Ralph said he only had three T-shirts he could wear to work. One was a solid color, one was an Earth Day shirt, and the other was a shirt with the company name and logo on it. He also said he would either wear tan pants or jeans. Linda is going to wear jeans and her company T-shirt. If Ralph selects a pair of pants and a T-shirt at random, what is the probability that he, too, will be wearing jeans and the company T-shirt?

11. **Food** Kita and Jason are working on the school newspaper's final layout for October. They decide that after they finish, they will go get pizza. They have a coupon for a large pizza with three toppings for $8.99. In discussing what the three toppings should be, they find they have four favorite toppings, and choosing three that both agreed upon was going to be difficult. After some discussion, they decided that pepperoni was the one topping they both agreed upon and they would put the names of the other three toppings in a bag and choose two at random. If the other three toppings are mushrooms, olives, and sausage, what is the probability they will get a pizza with mushrooms?

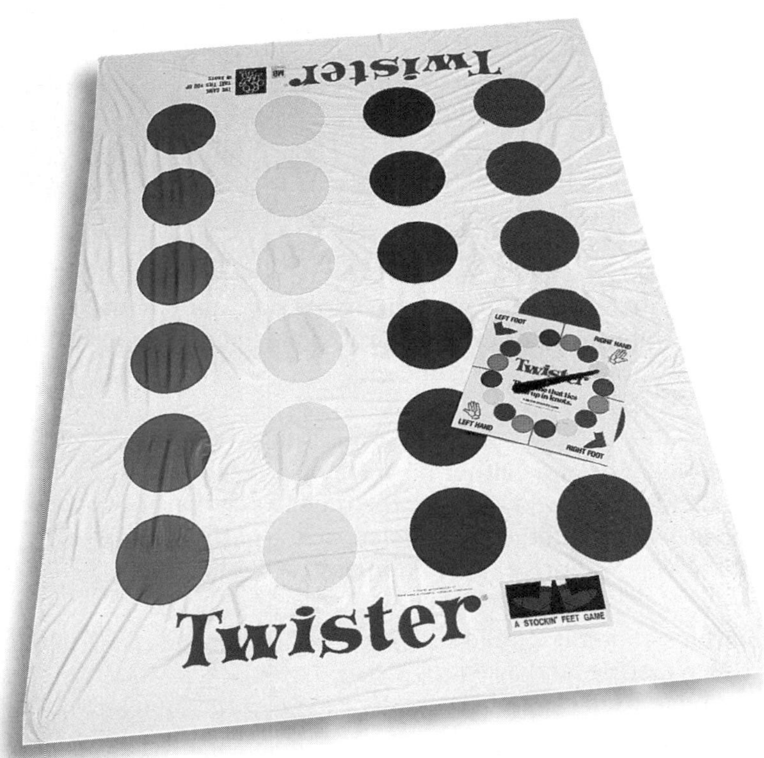

12. **Games** Twister® is a game composed of a mat with four rows, each having six circles of the same color. One person uses a spinner and calls out *hand* or *foot, right* or *left,* and a color (*blue, red, green,* or *yellow*). Each player must then place that particular body part on a circle of that color. The caller continues to call out body parts and colors with the players moving in the appropriate manner. If a player falls down in the process, he or she is disqualified. The winner is the last player still in position.
    a. Draw a tree diagram to show the possibilities for body parts and colors.
    b. What is the probability that the caller will spin and say *"right hand yellow?"*

13. **Playing Cards** The deck of cards for Rook® is composed of 57 cards. There are four suits (red, yellow, black, and green), each containing the numbers 1 to 14, and a rook card. Suppose there are two piles of five cards each with the cards turned face down. This first pile contains a red 3, a black 3, a red 5, a red 14, and a yellow 10. The second pile contains a green 5, a red 10, a black 10, a green 1, and a yellow 14. You are asked to draw a card from each pile.
    a. List all the possibilities of drawing two cards.
    b. What is the probability that both cards are red?
    c. What is the probability that both cards are 10s?
    d. What is the probability that both cards are green?
    e. What is the probability that the sum of the cards is at least 15?

14. **Testing** Marty is taking his final exam in algebra. He takes too long answering the calculation part of the exam to properly evaluate the last five questions, which are true-false. To keep from getting points off for not answering these questions at all, he guesses at them just as the bell rings.
    a. Draw a tree diagram to show the possible answer outcomes.
    b. What is the probability that he got at least two of the questions correct if the answers were T, F, F, T, F?
    c. Suppose Marty answered all the questions as true. Is this a good strategy to try to get the most correct without reading the question? Explain.

15. **Babies** According to the National Center for Health Statistics, each time a baby is born, the probability that it is a boy is always 51.3% and a girl is 48.7%, regardless of the sex of any previous births in the family. Suppose a couple has four children.
    a. What is the probability they are all girls?
    b. What is the probability there are one girl and three boys?
    c. How many children are there in your family? Find the probability of your family situation.

16. **Computers** The results of a survey in which students were asked how they used their computers are shown in the Venn diagram at the right. Based on these findings, what is the probability that a student, chosen at random, will use a computer only for word processing, and a second student will use it only for games?

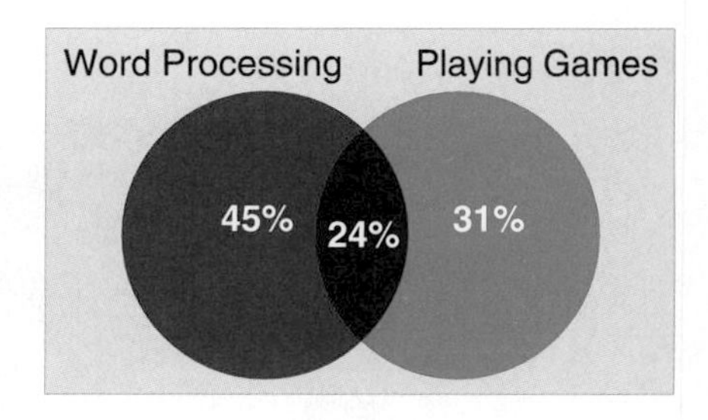

Critical Thinking

17. Every working day Adrianna either bicycles or drives to work. If she oversleeps, she drives to work 80% of the time. If she does not oversleep, she bicycles 80% of the time. She oversleeps 20% of the time. What is the probability that she will drive to work on any particular day?

18. A bag of marbles contains 2 yellow marbles, 1 blue marble, and 3 red marbles. Suppose you select one marble at random and *do not* put it back in the bag. Then you select another marble.
    a. Draw a tree diagram to model this situation. How many possible outcomes are there?
    b. What is the probability of selecting a red marble first and then a yellow one?

Mixed Review

19. **Statistics** Amy's scores on the first three of four 100-point biology tests were 88, 90, and 91. To get an A− in the class, her average must be between 88 and 92, inclusive, on all tests. What score must she receive on the fourth test to get an A− in biology? (Lesson 7–4)

20. Define a variable, write an inequality, and solve the following. Two thirds of a number is more than 99. (Lesson 7–2)

21. Determine the slope and y-intercept of the graph of $3x - y = 9$. (Lesson 6–4)

22. Solve $8r - 7t + 2 = 5(r + 2t) - 9$ for r. (Lesson 5–3)

23. **Optics** Peripheral vision is the ability to see around you when you keep your head and eyes pointed straight ahead. The average person with good peripheral vision can see about 180° when they are still. For a person in a moving vehicle, peripheral vision decreases as speed increases, as shown in the illustration below. (Lesson 3–4)

    a. If Seth Lytle is driving at 65 mph, approximately how large is his angle of vision?

    b. About how large an angle is he not seeing compared to his stationary angle of vision?

    c. If there were a deer standing 50° to the right of Seth's center view and he was driving at 45 mph, would he be able to see the deer?

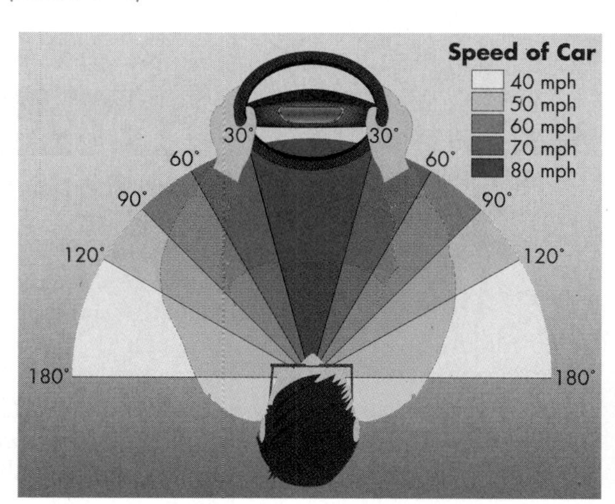

Computers and Chess

The excerpt below appeared in an article in the January 1995 issue of *Discover* magazine.

HOW SMART HAVE COMPUTERS BECOME? Very smart, at least by one measure: this past August, a computer defeated the human world champion at chess for the first time. Russian grand master Garry Kasparov was outwitted by Genius 2, a computer program designed by an English physicist, Richard Lang, who calls himself a mediocre chess player.... Chess programs like Lang's work by focusing on what computers do best: long and intricate calculations. Before deciding on a move, the program examines each of the roughly 36 possibilities in an average game situation. For each of these 36 "branches," it looks 16 moves ahead, determining its opponent's possible countermoves, its own possible responses to each countermove, and so on. Sixteen moves ahead, the number of possible board configurations is huge. The genius of Genius 2 lies in the way it prunes the possibility tree, ruling out bad moves from the start. But the software also had help from the hardware; Lang's computer uses a speedy Intel Pentium microprocessor that can execute 166 million instructions per second. "We wouldn't have won (the computer world championship) had we not had this processor," Lang concedes. ■

1. Many chess-playing programs are written by people who are not top-level chess players. How then can these programs defeat very good players?
2. If you want to improve your chess game by practicing against a computer, would you choose a program that is below your playing level, at your level, a little above your level, or far above your level? Why?
3. Do you think it's fair to match human chess players against computers? Why or why not?
4. How might a computer use probability in determining the proper move to make?

7-6 Solving Open Sentences Involving Absolute Value

What YOU'LL LEARN

- To solve open sentences involving absolute value and graph the solutions.

Why IT'S IMPORTANT

You can use open sentences to solve problems involving space exploration, law enforcement, and entertainment.

APPLICATION

Space Exploration

On Tuesday, February 7, 1995, the space shuttle *Discovery* maneuvered within 37 feet of the Russian space station *Mir,* 245 miles above Earth. To accomplish this feat, *Discovery* had to be launched within 2.5 minutes of a designated time (12:45 A.M.). This time period is known as the *launch window*. If t represents the time elapsed since the launch countdown began and there are 300 minutes scheduled from the beginning of the countdown to blast-off, you can write the following inequality to represent the launch window.

$|300 - t| \leq 2.5$ *The difference between 300 minutes and the actual time elapsed since the countdown began must be less than or equal to 2.5 minutes.*

We use absolute value because $300 - t$ cannot be negative.

fabulous FIRSTS

Sally Ride (1951–)

Sally Ride was the first American woman in space. She was a member of the *Challenger STS-7* crew that was launched June 18, 1983. She has a Ph.D. in physics from Stanford University.

There are three types of open sentences that can involve absolute value. They are as follows, when n is nonnegative.

$|x| = n$ $\quad\quad$ $|x| < n$ $\quad\quad$ $|x| > n$

First let's consider the case of $|x| = n$.

If $|x| = 4$, this means that the difference from 0 to x is 4 units.

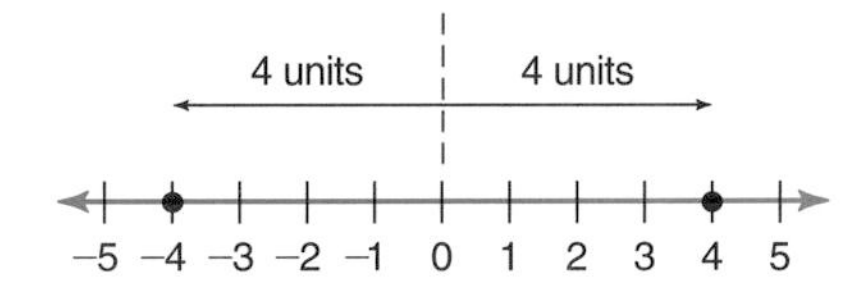

The solution set can also be written as $\{x \mid x = -4 \text{ or } x = 4\}$.

Therefore, if $|x| = 4$, then $x = -4$ or $x = 4$. The solution set is $\{-4, 4\}$. So, if $|x| = n$, then $x = -n$ or $x = n$.

Equations involving absolute value can be solved by graphing them on a number line or by writing them as a compound sentence and solving it.

Example  **Solve $|x - 3| = 5$.**

Method 1: Graphing

$|x - 3| = 5$ means the distance between x and 3 is 5 units. To find x on the number line, start at 3 and move 5 units in either direction.

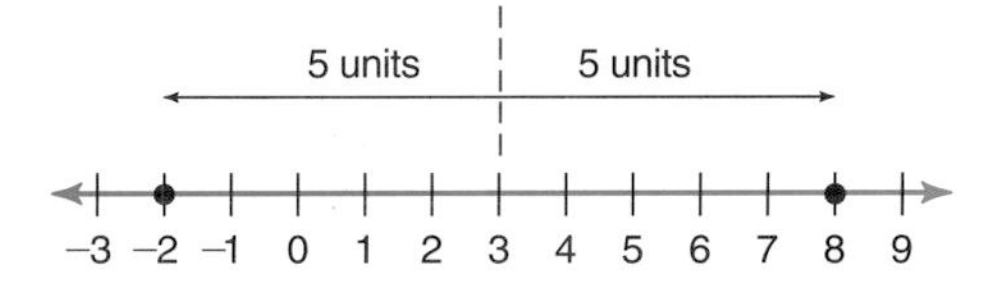

The solution set is $\{-2, 8\}$.

Method 2: Compound Sentence

$|x - 3| = 5$ also means $x - 3 = 5$ or $-(x - 3) = 5$.

$$x - 3 = 5 \quad \text{or} \quad -(x - 3) = 5$$
$$x - 3 + 3 = 5 + 3 \qquad x - 3 = -5 \quad \textit{Multiply each side by } -1.$$
$$x = 8 \qquad x - 3 + 3 = -5 + 3$$
$$x = -2$$

This verifies the solution set.

Now let's consider the case of $|x| < n$. Inequalities involving absolute value can also be represented on a number line or as compound inequalities. Let's examine $|x| < 4$.

$|x| < 4$ means that the distance from 0 to x is less than 4 units.

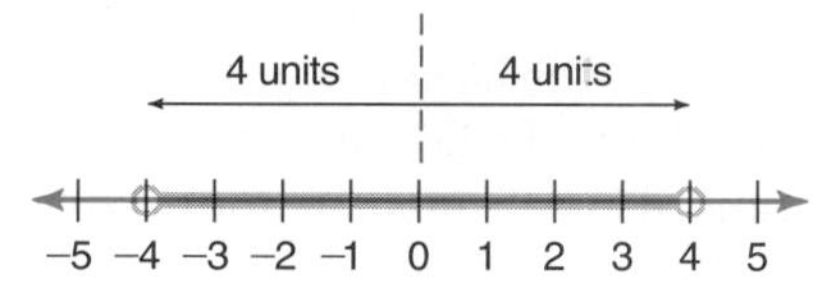

When the inequality is $<$ or $\leq$, the compound sentence uses and.

Therefore, $x > -4$ and $x < 4$. The solution set is $\{x \mid -4 < x < 4\}$. So, if $|x| < n$, then $x > -n$ and $x < n$.

Example 2 **Solve $|3 + 2x| < 11$ and graph the solution set.**

$|3 + 2x| < 11$ means $3 + 2x < 11$ and $3 + 2x > -11$.

$$3 + 2x < 11 \quad \text{and} \quad 3 + 2x > -11$$
$$3 - 3 + 2x < 11 - 3 \qquad 3 - 3 + 2x > -11 - 3$$
$$2x < 8 \qquad 2x > -14$$
$$\frac{2x}{2} < \frac{8}{2} \qquad \frac{2x}{2} > \frac{-14}{2}$$
$$x < 4 \qquad x > -7$$

The solution set is $\{x \mid x > -7 \text{ and } x < 4\}$, which can be written as $\{x \mid -7 < x < 4\}$.

Now graph the solution set.

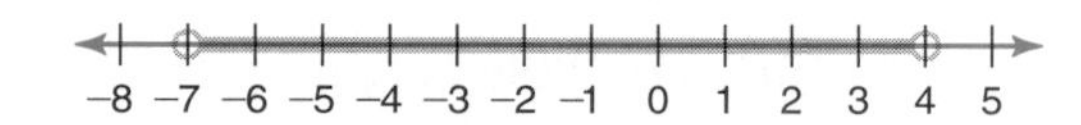

Finally let's examine $|x| > 4$. This means that the distance from 0 to x is greater than 4 units.

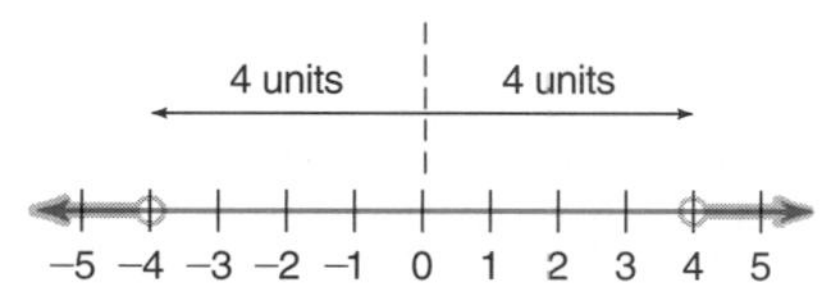

When the inequality is $>$ or $\geq$, the compound sentence uses or.

Therefore, $x < -4$ or $x > 4$. The solution set is $\{x \mid x < -4 \text{ or } x > 4\}$. So, if $|x| > n$, then $x < -n$ or $x > n$.

Example 3 **Solve $|5 + 2y| \geq 3$ and graph the solution set.**

$|5 + 2y| \geq 3$ means $5 + 2y \leq -3$ or $5 + 2y \geq 3$.

$$
\begin{array}{lcl}
5 + 2y \leq -3 & \text{or} & 5 + 2y \geq 3 \\
5 - 5 + 2y \leq -3 - 5 & & 5 - 5 + 2y \geq 3 - 5 \\
2y \leq -8 & & 2y \geq -2 \\
\frac{2y}{2} \leq \frac{-8}{2} & & \frac{2y}{2} \geq \frac{-2}{2} \\
y \leq -4 & & y \geq -1
\end{array}
$$

The solution set is $\{y \mid y \leq -4 \text{ or } y \geq -1\}$.

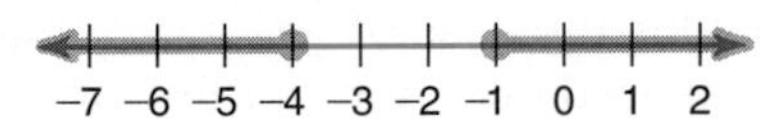

Organizations such as OSHA set standards for buildings to meet the needs of those using the building. Building code standards are often written as maximums or minimums that must be met. These standards can be written as inequalities.

Example 4

Construction

There are a number of specifications in the building industry that address the needs of physically-challenged persons. For example, hallways in hospitals must have handrails. The handrails must be placed within a range of 2 inches from a height of 36 inches.

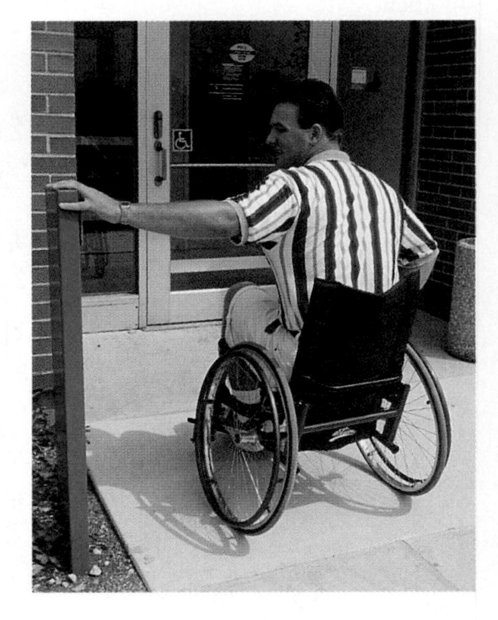

a. Write an open sentence that involves absolute value to represent the range of acceptable heights for hallway handrails.

b. Find and graph the corresponding compound sentence.

a. Let h be an acceptable height for the handrail. Then h can differ from 36 inches by no more than 2 inches. Write an open sentence that represents the range of acceptable heights.

h differs from 36	*by less than or equal to*	*2.*
$\lvert h - 36 \rvert$	$\leq$	2

Now, solve $|h - 36| \leq 2$ to find the compound sentence.

$$
\begin{array}{lcl}
h - 36 \geq -2 & \text{and} & h - 36 \leq 2 \\
h - 36 + 36 \geq -2 + 36 & & h - 36 + 36 \leq 2 + 36 \\
h \geq 34 & & h \leq 38
\end{array}
$$

The compound sentence is $34 \leq h \leq 38$.

b. The graph of this sentence is shown below.

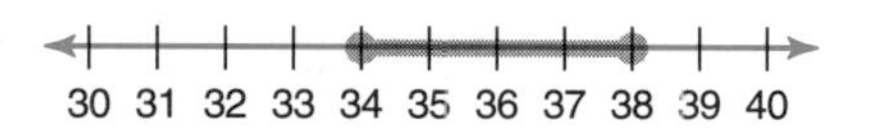

So, the handrail must be placed between 34 and 38, inclusive, inches from the floor.

CHECK FOR UNDERSTANDING

Communicating Mathematics

Study the lesson. Then complete the following.

1. **Describe** two ways you can solve an open sentence involving an absolute value.
2. **Explain** the difference in the solution for $|x + 7| > 4$ and the solution for $|x + 7| < 4$.
3. If $x < 0$ and $|x| = n$, describe n in terms of x.
4. **You Decide** Jamila's teacher said they should work out each problem and test points to determine whether their solutions are correct. Jamila thinks that she knows the solution for problems like $|w + 5| < 0$, where the absolute value quantity is always less than 0, without testing points. Is she correct? Why or why not?

Guided Practice

Choose the replacement set that makes each open sentence true.

5. $|x - 7| = 2$
   a. $\{9, -5\}$ **b.** $\{5, -9\}$ **c.** $\{5, 9\}$ **d.** $\varnothing$
6. $|x - 2| > 4$
   a. $\{x \mid -2 < x < 6\}$ **b.** $\{x \mid x = 6 \text{ or } x = 2\}$
   c. $\{x \mid x > 6 \text{ or } x > -2\}$ **d.** $\{x \mid x < -2 \text{ or } x > 6\}$

State which graph below matches each open sentence in Exercises 7–10.

7. $|y| = 3$
8. $|y| > 3$
9. $|y| < 3$
10. $|y| \leq 3$

a. –5 –4 –3 –2 –1 0 1 2 3 4 5

b.

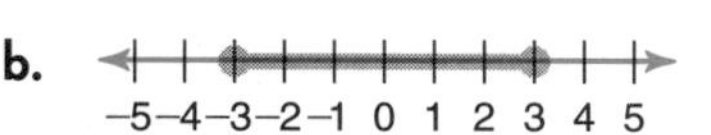

c. –5 –4 –3 –2 –1 0 1 2 3 4 5

d. –5 –4 –3 –2 –1 0 1 2 3 4 5

Solve each open sentence. Then graph the solution set.

11. $|m| \geq 5$
12. $|n| < 6$
13. $|r + 3| < 6$
14. $|8 - t| \geq 3$

For each graph, write an open sentence involving absolute value.

15. –6 –5 –4 –3 –2 –1 0 1 2 3 4 5 6

16.

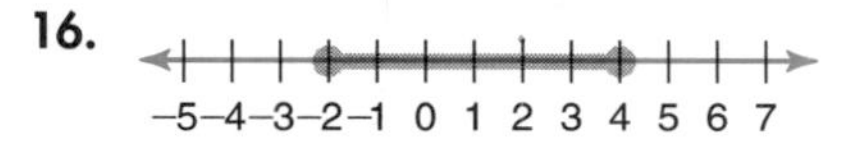

EXERCISES

Practice

Solve each open sentence. Then graph the solution set.

17. $|y - 2| = 4$
18. $|3 - 3x| = 0$
19. $|7x + 2| = -2$
20. $|w + 8| \geq 1$
21. $|2 - y| \leq 1$
22. $|t + 4| \geq 3$
23. $|4y - 8| < 0$
24. $|2x + 5| < 4$
25. $|3e - 7| < 2$
26. $|3x + 4| < 8$
27. $|1 - 3y| > -2$
28. $3 + |x| > 3$
29. $|8 - (w - 1)| \leq 9$
30. $|6 - (11 - b)| = -3$
31. $|2.2y - 1.1| = 5.5$
32. $|3r - 0.5| \geq 5.5$
33. $\left|\frac{2 - 3x}{5}\right| \geq 2$
34. $\left|\frac{1}{2} - 3p\right| < \frac{7}{2}$

Express each statement in terms of an inequality involving absolute value. Do *not* try to solve.

35. The diameter of the lead in a pencil p must be within 0.01 millimeters of 1 millimeter.
36. The cruise control of a car set at 55 mph should keep the speed s within 3 mph of 55 mph.
37. A liquid at 50°C will change to a gas or liquid if the temperature t increases or decreases more than 50°C.

For each graph, write an open sentence involving absolute value.

38. 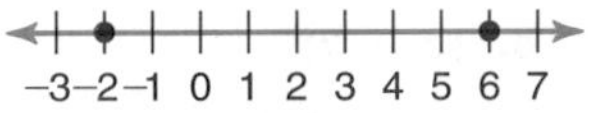 −3 −2 −1 0 1 2 3 4 5 6 7
39. 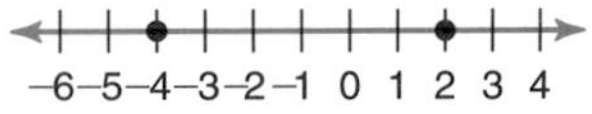 −6 −5 −4 −3 −2 −1 0 1 2 3 4
40. 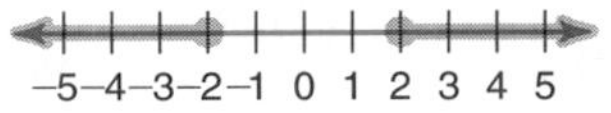 −5 −4 −3 −2 −1 0 1 2 3 4 5
41. 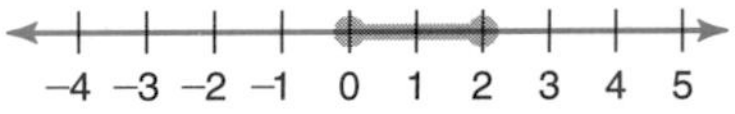 −4 −3 −2 −1 0 1 2 3 4 5
42. 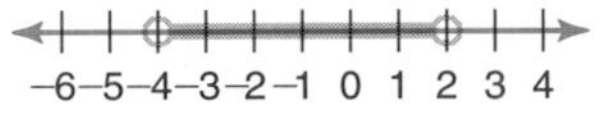 −6 −5 −4 −3 −2 −1 0 1 2 3 4
43. 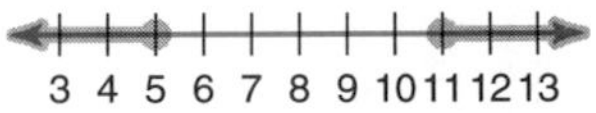 3 4 5 6 7 8 9 10 11 12 13

44. Find all integer solutions of $|x| < 4$.
45. Find all integer solutions of $|x| \leq 2$.
46. If $a > 0$, how many integer solutions exist for $|x| < a$?
47. If $a > 0$, how many integer solutions exist for $|x| \leq a$?

Critical Thinking

48. Solve $|y - 3| = |2 + y|$.
49. Under what conditions is $-|a|$ negative? positive?
50. Suppose $8 \leq x \leq 12$. Write an absolute value inequality that is equivalent to this compound inequality.

Applications and Problem Solving

51. **Probability** Suppose $|x| \leq 6$ and x is an integer. Find the probability of $|x|$ being a factor of 18.
52. **Chemistry** For hydrogen to be a liquid, its temperature must be within 2°C of −257°C. What is the range of temperatures for this substance to remain a liquid?

53. **Law Enforcement** A radar gun used to determine the speed of passing cars must be within 7 mph of the actual speed of a selected car. If a highway patrol officer reads a speed of 59 mph for a car, does he have irrefutable evidence that the car was speeding in a 55 mph zone? Explain your reasoning and include a graph.

54. **Space Exploration** Refer to the application at the beginning of the lesson. Find how much time can elapse from the beginning of the countdown to remain within the launch window.

55. **Entertainment** Luis Gomez is a contestant on the *Price is Right.* He must guess within $1500 of the actual price of a Jeep Cherokee without going over in order to win the vehicle. The actual price of the Jeep is $18,000. What is the range of guesses in which Luis can win the vehicle?

56. **Spending** The graph below shows the spending power of kids aged 3 to 17.

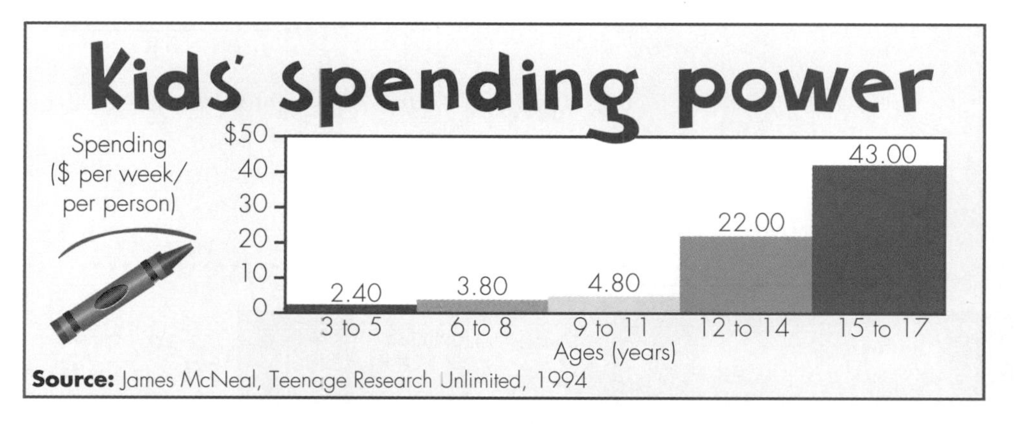

a. Write an inequality that represents the spending power of 3- to 17-year olds. Then write an absolute value inequality that describes their spending.

b. Write an inequality that represents the spending power of 12- to 17-year olds. Then write an absolute value inequality that describes their spending.

c. Keep a record of how much money you spend in a week. How does your spending compare with the data in this graph?

Mixed Review

57. **Draw a Diagram** The Sanchez family acts as a host family for foreign exchange students during each quarter of the year. Suppose it is equally likely that they get a boy or a girl each quarter. (Lesson 7–5)

a. Draw a tree diagram to represent the possible orders of boys (B) and girls (G) during the four quarters of the year. List the possible outcomes.

b. From the tree diagram, what is the probability that all of the students will be girls? boys?

c. From the tree diagram, what is the probability that they will host two boys and two girls?

58. Peter wants to buy Crystal an engagement ring. He wants to spend between \$1700 and \$2200. If he goes shopping during a 12%-off sale to celebrate the store's 12th anniversary, what would his price range be? (Lesson 7–4)

59. Solve $10x - 2 \geq 4(x - 2)$. (Lesson 7–3)

Solve each inequality.

60. $396 > -11t$ (Lesson 7–2)

61. $-11 \leq k - (-4)$ (Lesson 7–1)

62. Find the coordinates of the midpoint of the line segment whose endpoints are at $(-4, 1)$ and $(10, 3)$. (Lesson 6–7)

63. Graph $2x - 9 = 2y$. (Lesson 5–4)

64. **Architecture** Julie is making a model of a house for her drafting class. One part of the house has a triangle like the one below. She wants the base of the triangle to measure 2 inches. What will be the measures of the other two sides? (Lesson 4–2)

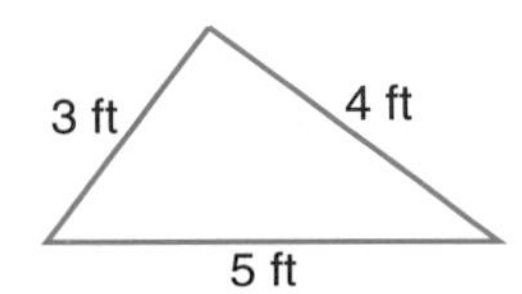

65. Solve $2m + \frac{3}{4}n = \frac{1}{2}m - 9$ for m. (Lesson 3–6)

66. Simplify $\frac{-42r + 18}{3}$. (Lesson 2–7)

WORKING ON THE

In·ves·ti·ga·tion

Refer to the Investigation on pages 320–321.

Smoke Gets In Your Eyes

Smoking increases the chance of health problems. Constant exposure to secondhand smoke also puts the nonsmoker at risk. According to a statement from a tobacco company, a nonsmoker working among smoking coworkers inhales the smoke of 1.25 cigarettes per month. A restaurant worker in the smoking section breathes just 2 cigarettes per month. The Environmental Protection Agency (EPA) disagrees with these data.

1 Write an equation to express the equivalent number of cigarettes one would breathe if he/she worked in the smoking section of a restaurant for one year.

2 A lawsuit involving the tobacco industry and the EPA states that the EPA did not use the standard significance level of 5% in evaluating their data from various studies. A 5% level means that their numerical conclusions have a 5% margin of error. How does the margin of error relate to absolute value?

3 In January, 1995, the average price of a pack of cigarettes was \$2.06, of which 56¢ is federal and state excise tax. The price of cigarettes is on the rise. Write an inequality to represent how much is spent by the average smoker in a year. Solve the inequality.

4 How much tax revenue is generated by the sale of cigarettes to the estimated number of smokers in the United States?

Add the results of your work to your Investigation Folder.

7-7

Integration: Statistics Box-and-Whisker Plots

What **YOU'LL LEARN**

- To display and interpret data on box-and-whisker plots.

Why **IT'S IMPORTANT**

Box-and-whisker plots are a useful way to display data. They allow you to see important characteristics of the data at a glance.

APPLICATION

Olympics

In 1992, the site of the Summer Games of the XXV Olympiad was Barcelona, Spain. More than 14,000 athletes from 172 nations competed for medals in 257 events. The table at the right shows the number of gold medals won by the top 16 medal-winning teams.

1992 SUMMER OLYMPICS Top 16 Medal Winners

Team	Number of Medals Won: Gold	Number of Medals Won: Total
Unified Team (formerly USSR)	45	112
USA	37	108
Germany	33	82
China	16	54
Cuba	14	31
Hungary	11	30
South Korea	12	29
France	8	29
Australia	7	27
Spain	13	22
Japan	3	22
Britain	5	20
Italy	6	19
Poland	3	19
Canada	6	18
Romania	4	18

Source: *The World Almanac, 1995*

We can describe these data using the mean, median, and mode. We can also use the median, along with the quartiles and interquartile range, to obtain a graphic representation of the data. A type of diagram, or graph, that shows quartiles and extreme values of data is called a **box-and-whisker plot.**

Box-and-whisker plots are sometimes called box plots.

Suppose we wanted to make a box-and-whisker plot of the numbers of gold medals won by each of the nations in the table. First, arrange the data in numerical order. Next, compute the median and quartiles. Also, identify the extreme values.

3 3 4 5 6 6 7 (8 ↑ 11) 12 13 14 16 33 37 45

median (Q2)

LOOK BACK

You can refer to Lesson 3-7 for information on measures of central tendency and Lesson 5-7 for measures of variation.

The median for this set of data is the average of the eighth and ninth values.

$$\text{median} = \frac{8 + 11}{2} \text{ or } 9.5$$

Recall that the *lower quartile* (Q1) is the median of the lower half of the distribution of values. The *upper quartile* (Q3) is the median of the upper half of the data.

The median is not included in either half of the data.

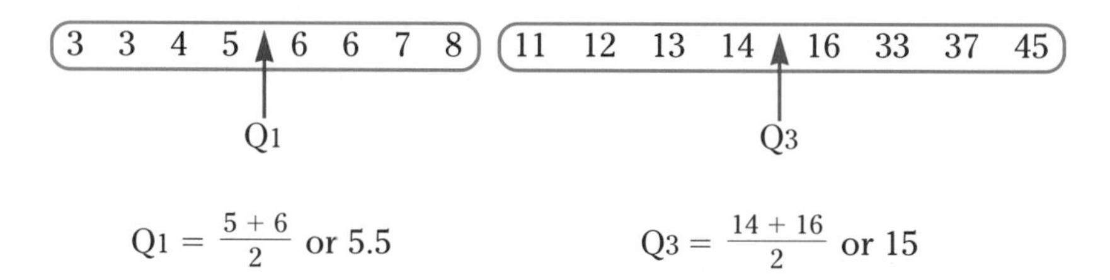

$$Q1 = \frac{5 + 6}{2} \text{ or } 5.5 \qquad Q3 = \frac{14 + 16}{2} \text{ or } 15$$

The **extreme values** are the least value (LV), 3, and the greatest value (GV), 45.

Now we have the information we need to draw a box-and-whisker plot.

Step 1 Draw a number line. Assign a scale to the number line that includes the extreme values. Plot dots to represent the extreme values (LV and GV), the upper and lower quartile points (Q_3 and Q_1), and the median (Q_2).

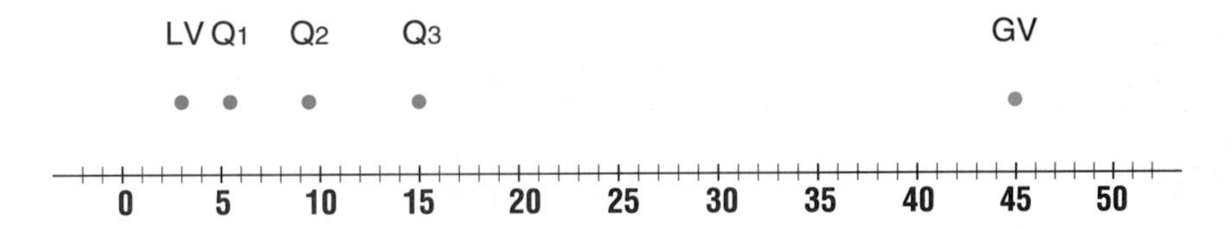

The median line will not always divide the box into equal parts.

Step 2 Draw a box to designate the data falling between the upper and lower quartiles. Draw a vertical line through the point representing the median. Draw a segment from the lower quartile to the least value and one from the upper quartile to the greatest value. These segments are the **whiskers** of the plot.

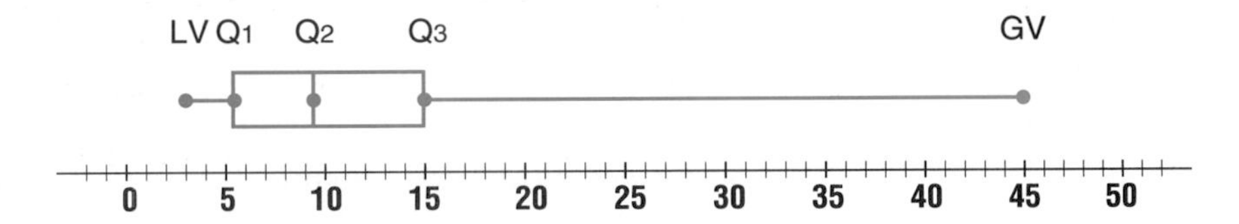

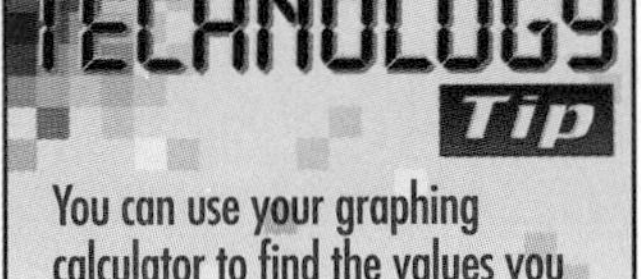

You can use your graphing calculator to find the values you need to make a box-and-whisker plot. Refer to Lesson 5–7A for instructions on finding these values.

Even though the whiskers are different lengths, each whisker contains at least one fourth of the data while the box contains one half of the data. Compound inequalities can be used to describe the data in each fourth. Assume that the replacement set for x is the set of data.

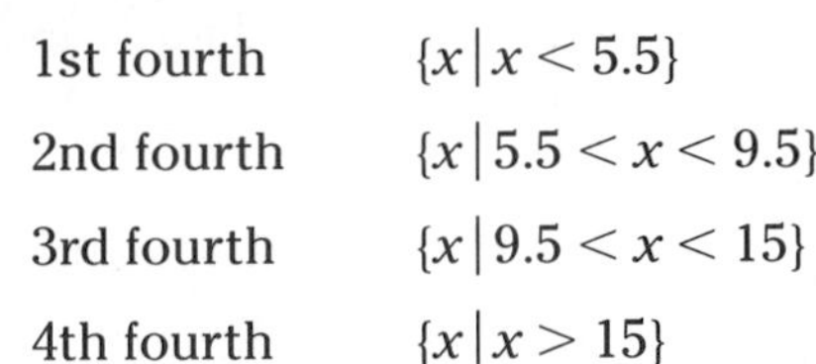

1st fourth	$\{x \mid x < 5.5\}$
2nd fourth	$\{x \mid 5.5 < x < 9.5\}$
3rd fourth	$\{x \mid 9.5 < x < 15\}$
4th fourth	$\{x \mid x > 15\}$

Step 3 Before finishing the box-and-whisker plot, check for outliers. In Lesson 5–7, you learned that an outlier is any element of the set of data that is at least 1.5 interquartile ranges above the upper quartile or below the lower quartile. Recall that the *interquartile range* (IQR) is the difference between the upper and lower quartiles, or in this case, $15 - 5.5$ or 9.5.

$x \geq Q_3 + 1.5(\text{IQR})$	or	$x \leq Q_1 - 1.5(\text{IQR})$
$x \geq 15 + 1.5(9.5)$		$x \leq 5.5 - 1.5(9.5)$
$x \geq 15 + 14.25$		$x \leq 5.5 - 14.25$
$x \geq 29.25$		$x \leq -8.75$

Step 4 If x is an outlier in this set of data, then the outliers can be described as $\{x \mid x \leq -8.75 \text{ or } x \geq 29.5\}$. In this case, there are no data less than -8.75. However, 45, 37, and 33 are greater than 29.25, so they are outliers. We now need to revise the box-and-whisker plot. Outliers are plotted as isolated points, and the right whisker is shortened to stop at 16.

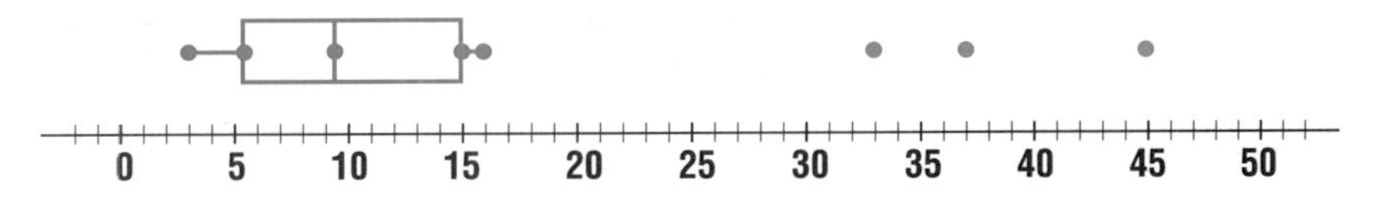

Example

Refer to the application at the beginning of the lesson. Use the box-and-whisker plot for the gold medals to answer each question.

a. What percent of the teams won between 6 and 15 gold medals?

b. What does the box-and-whisker plot tell us about the upper half of the data compared to the lower half?

a. The box in the plot indicates 50% of the values in the distribution. Since the box goes from 5.5 to 15, we know that 50% of the teams won between 6 and 15 gold medals.

b. The upper half of the data is spread out while the lower half is fairly clustered together.

CHECK FOR UNDERSTANDING

Communicating Mathematics

Study the lesson. Then complete the following.

1. **Explain** how to determine the scale of the number line in a box-and-whisker plot.
2. **Describe** which two points the two whiskers of a box-and-whisker plot connect.
3. What does Q2 represent?
4. Refer to the box-and-whisker plot at the right. Assume that LV, Q1, Q2, Q3, and GV are whole numbers.

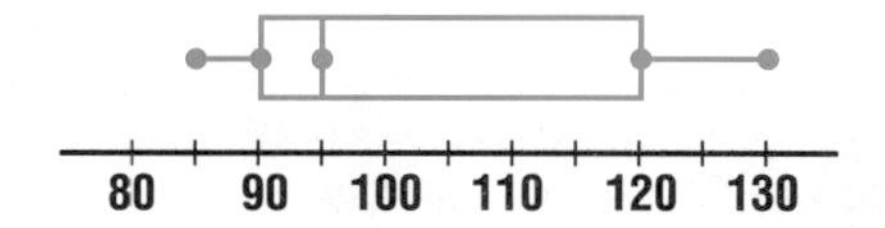

   a. What percent of the data is between 85 and 90?
   b. Between what two values does the middle 50% of the data lie?
   c. What outliers are represented in the box-and-whisker plot?
5. **Describe** what characteristics of a set of data you can gather from its box-and-whisker plot. What are some things you cannot gather from a box-and-whisker plot?

Guided Practice

6. The table at the right shows the birthrates for 15 selected countries.
   a. Find the median, upper quartile, lower quartile, and interquartile range for each year's data.
   b. Are there any outliers? If so, name them.
   c. Draw a box-and-whisker plot for each set of data on the same number line.
   d. Which set of data seems to be more clustered? Why?

Birth Rates for Selected Countries (per 1000 population)

Country	1985	1992
Australia	15.7	15.1
Cuba	18.0	14.5
Denmark	10.6	13.1
France	13.9	12.9
Hong Kong	14.0	11.9
Israel	23.5	21.5
Italy	10.1	9.9
Japan	11.9	9.7
Netherlands	12.3	13.0
Panama	26.6	23.3
Poland	18.2	13.4
Portugal	12.8	11.4
Singapore	16.6	17.7
Switzerland	11.6	12.6
United States	15.7	15.7

Source: United Nations, *Monthly Bulletin of Statistics*, May 1994

7. Refer to box-and-whisker plots A and B.

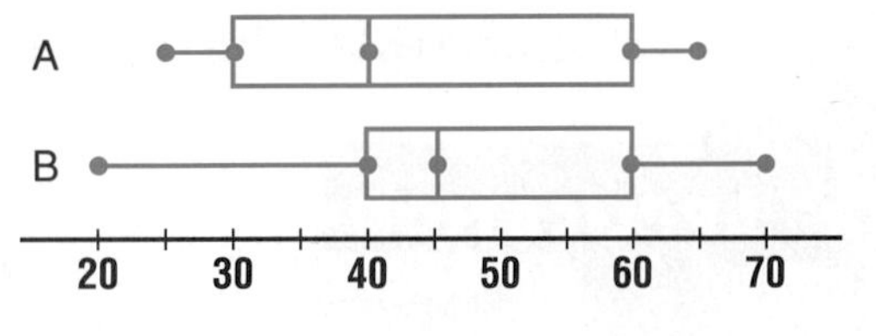

   a. Estimate the least value, greatest value, lower quartile, upper quartile, and median for each plot. Assume that these values are whole numbers.
   b. Which set of data contains the least value?
   c. Which plot has the greatest interquartile range?
   d. Which plot has the greatest range?

EXERCISES

Applications and Problem Solving

8. **Manufacturing** The box-and-whisker plots at the right show the results of testing the useful life of 10 light bulbs each from two manufacturers.

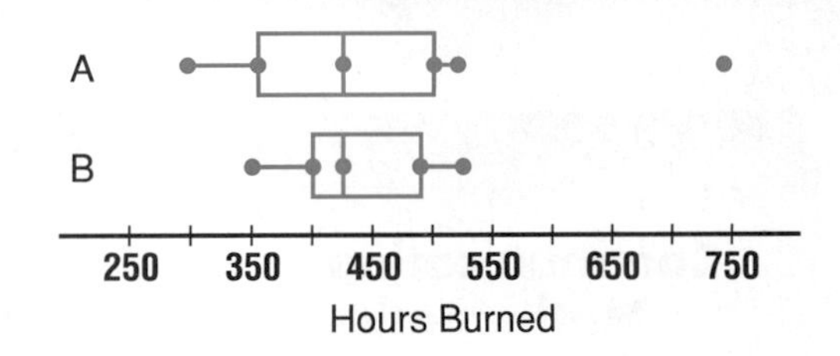

   a. Which test had the most varied results?
   b. Were there any outliers? If so, for which brand?
   c. How would you compare the medians of the tests?
   d. Based on this plot, from which manufacturer would you buy your light bulbs? Why?

9. **Meteorology** Meteorologists keep track of temperatures for four 90-day periods during the year to predict the trends in weather for future years. The following low temperatures were recorded during a 2-week cold snap in Indianapolis during 1993.

$$30°, 20°, 2°, 12°, 5°, 4°, 17°, 7°, 6°, 16°, 5°, 0°, 5°, 16°$$

   a. Find the median, upper quartile, lower quartile, and interquartile range.
   b. Are there any outliers? If so, name them.
   c. Draw a box-and-whisker plot of the data.

10. **Football** The table below shows the American Football Conference's leading quarterbacks in touchdowns for the 1993 season. Make a box-and-whisker plot for the data.

1993 AFC Individual Leaders in Passing

Player	Team	Touchdowns
Steve DeBerg	Miami	7
John Elway	Denver	25
Boomer Esiason	N.Y. Jets	16
John Friesz	San Diego	6
Jeff George	Indianapolis	8
Jeff Hostetler	L.A. Raiders	14
Jim Kelly	Buffalo	18
Scott Mitchell	Miami	12
Joe Montana	Kansas City	13
Warren Moon	Houston	21
Neil O'Donnell	Pittsburgh	14
Vinny Testaverde	Cleveland	14

Source: *The World Almanac, 1995*

11. Demographics According to the 1990 Census, the American Indian population in the United States is 1.959 million. Many American Indian people live on reservations or trust lands. The stem-and-leaf plot shows the number of reservations in the 34 states that have them.

Stem	Leaf
0	1 1 1 1 1 1 1 1 1 1 1 1 1 2
•	3 3 3 3 3 4 4 4 4 7 7 8 8 9
1	1 4 9
2	3 5 7
9	6 9|6 = 96

a. Make a box-and-whisker plot of these data.
b. Describe the distribution of the data.
c. Why do you think there are so many outliers?
d. The mean of these data is about 8.8. How does this compare with the median?

12. Environment The table below shows the number of hazardous waste sites in 25 states.

Hazardous Waste Sites in the United States (selected states)									
State	No.	State	No.	State	No.	State	No.	State	No.
AL	13	FL	57	LA	13	OH	38	TN	17
AR	12	GA	13	MD	13	OK	11	TX	30
CA	96	IL	37	MI	77	OR	12	VA	25
CO	18	IN	33	NY	85	PA	101	WA	56
CT	16	KY	20	NC	22	SC	24	WV	6

Source: Environmental Protection Agency, May 1994

a. Make a box-and-whisker plot of the data.
b. What is the median number of waste sites for the states listed?
c. Which states, if any, are outliers?
d. Which half of the data is more widely dispersed?

13. Education The graph below shows the average American College Testing (ACT) Program mathematics scores for students from 1985–1993.

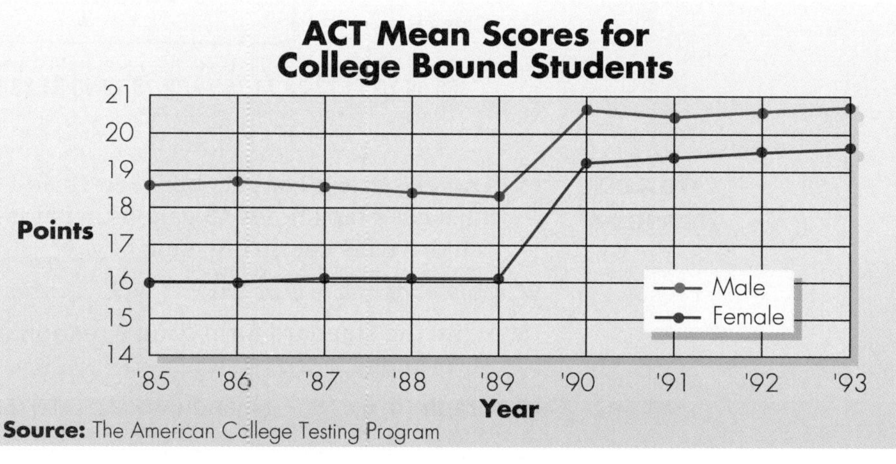

(continued on the next page)

a. Make a box-and-whisker plot of the scores for the male students and another for the female students using the same scale. Compare the plots.

b. In a particular year, an entirely new ACT Assessment was given that emphasized rhetorical skills, advanced mathematics items, and a new reading test. From the data on the graph, in which year do you think this occurred? Why?

14. History Did you know that there have been more U.S. vice-presidents than presidents? As of 1995, there have been 41 presidents and 45 vice-presidents. Some presidents had more than one vice-president, and some had none. The line plot below shows the ages of vice-presidents on their inauguration days.

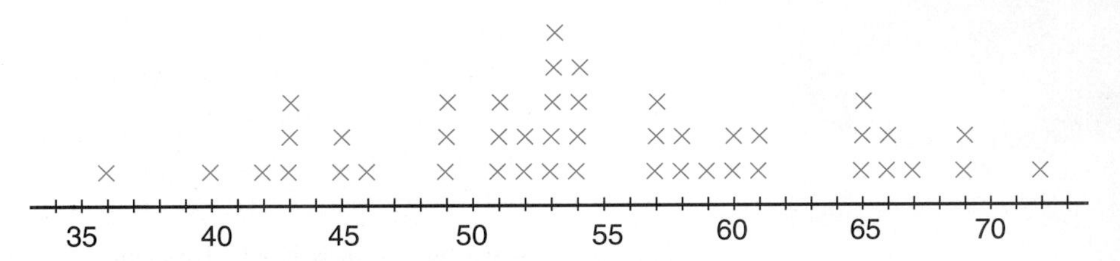

Source: *The World Almanac*, 1995

a. Make a box-and-whisker plot of these data.

b. The ages for presidents on their inauguration days have the following statistics: LV = 42, Q1 = 51, Q2 = 55, Q3 = 58, GV = 68, and 69 is an outlier. Make a box-and-whisker plot to represent these ages.

c. Which set of data is more clustered? Explain your answer.

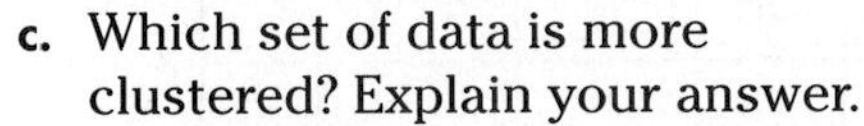

d. Fourteen vice-presidents went on to become presidents. Which ages of vice-presidents were definitely not those who went on to become president?

Critical Thinking

15. The box-and-whisker plots shown below picture the distribution of test scores in two algebra classes taught by two different teachers. If you could select one of the classes to be in, which one would it be, and why?

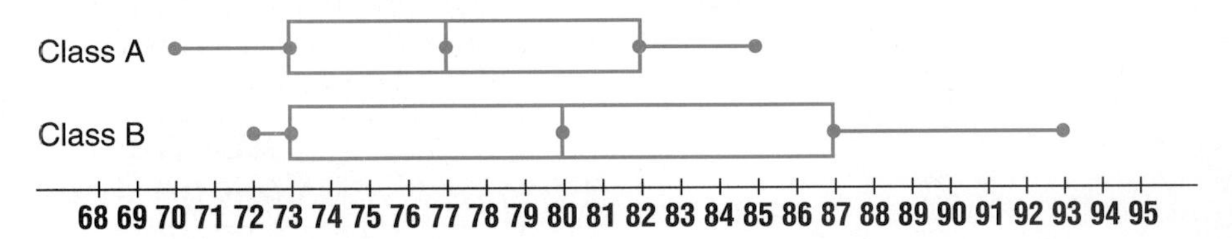

Mixed Review

16. **Travel** Greg's car gets between 18 and 21 miles per gallon of gasoline. If his car's tank holds 15 gallons, what is the range of distances that Greg can drive his car on one tank of gas? (Lesson 7–6)

17. Solve $2m - 3 > 7$ or $2m + 7 > 9$. (Lesson 7–4)

18. Write the standard form of an equation of the line that passes through (4, 7) and (1, −2). (Lesson 6–2)

19. Graph (0, 6), (8, −1), and (−3, 2). (Lesson 5–1)

20. Solve $\frac{6}{x-3} = \frac{3}{4}$. (Lesson 4–1)

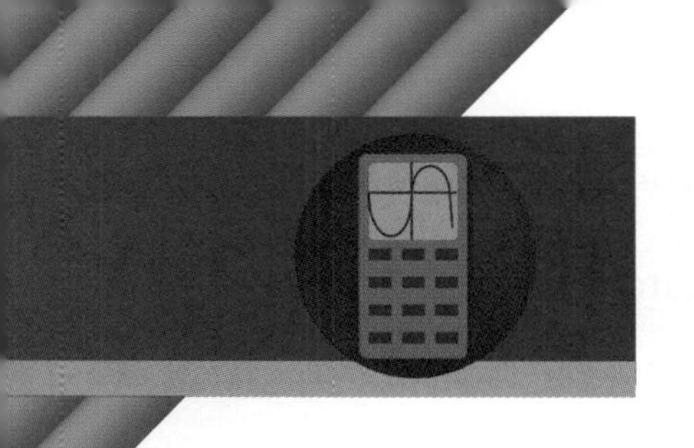

7–7B Graphing Technology Box-and-Whisker Plots

An Extension of Lesson 7–7

You can use a graphing calculator to compare two sets of data by using a double box-and-whisker plot. The plots that the calculator draws, however, do not account for outliers. If you want to use a graphing calculator to help you sketch a plot, it will be necessary for you to check for outliers and adjust the graph as needed.

In an experiment, the ability of boys and girls to identify objects held in their left hands versus those held in their right hands was tested. The left side of the body is controlled by the right side of the brain and vice versa. The results of the experiment found that the boys did not identify objects with their right hand as well as with their left. The girls could identify objects equally as well with either hand.

Texas Instruments decided to test this premise by conducting the same tests with 12 male and 10 female employees, chosen at random. Thirty small objects were selected and separated into two groups—one for the right hand and one for the left hand. Blindfolded employees felt each of the objects with the prescribed hand and tried to identify them. The results are presented in the chart below.

LOOK BACK

For more information on entering data into lists on the graphing calculator, see Lesson 5-7A.

Each employee had a score for left and right hands.

Correct Responses			
Female Left Hand	**Female Right Hand**	**Male Left Hand**	**Male Right Hand**
8	4	7	12
9	3	8	6
12	7	7	12
11	12	5	12
10	11	7	7
8	11	8	11
12	13	11	12
7	12	4	8
9	11	10	12
11	12	14	11
		13	9
		5	9

Source: *TI-82 Graphics Calculator Guidebook*

Make a double box-and-whisker plot of the data to compare the results of Female Left Hand with Female Right Hand using a graphing calculator.

Step 1 Clear lists L1, L2, L3, and L4. Enter the data from each column of the table into lists L1, L2, L3, and L4, respectively.

Step 2 Select the box-and-whisker plot and define which list will be used.

Enter: 2nd STAT PLOT 1 ENTER *Turns plot on.*

▼ ▶ ▶ ENTER *Selects box-and-whisker plot.*

If L1 is not highlighted in the Xlist, use the down arrow and ENTER to highlight it and make sure the frequency is set for 1.

Repeat the process to assign Plot2 as a box-and-whisker plot using L2.

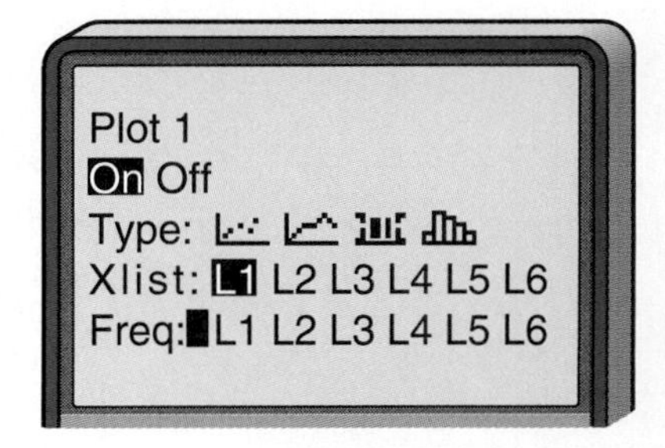

Step 3 Clear the Y= list. Set the WINDOW settings for Xscl = 1, Ymin = 0, and Yscl = 0. Ignore the other settings. Press ZOOM 9 to select ZoomStat. This sets the other settings and displays the box-and-whisker plots. It will only display those plots that you have turned on.

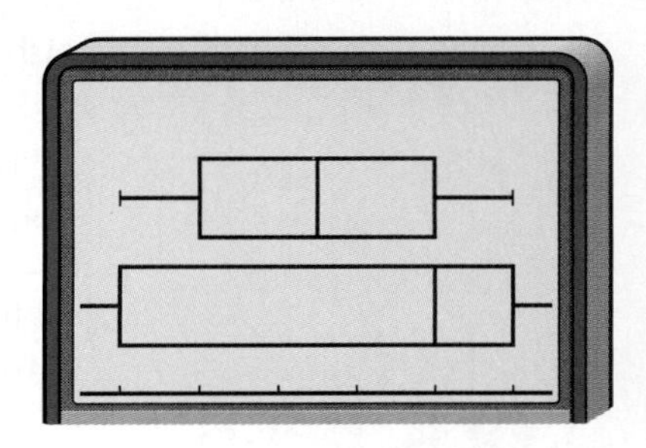

Step 4 Use TRACE to examine the minX (least value), Q1 (lower quartile), Med (median), Q3 (upper quartile), and maxX (greatest value).

EXERCISES

1. Which set of data does the upper plot represent?
2. Observe the two graphs. Does it appear that the females guessed correctly more often with the left hand or the right? How do you know?
3. Reset your calculator to define Plot1 as L3 and Plot 2 as L4 to examine the males' data. What do you observe in these plots?
4. Reset the calculator to compare the left-hand results of males and females. Were the males or females better at guessing with their left hands?
5. Reset the calculator to compare the right-hand results of males and females. Which group seemed more adept at identifying objects with their right hands?
6. How do the results of this experiment compare with the study of boys and girls mentioned at the beginning of this lesson? What reasons may account for any discrepancies?

7–8A Graphing Technology Graphing Inequalities

A Preview of Lesson 7–8

Inequalities in two variables can be graphed on a graphing calculator using the "Shade(" command, which is option 7 on the DRAW menu. You must enter *two* functions to activate the shading since the calculator always shades between two specified functions. The first function entered defines the lower boundary of the region to be shaded. The second function defines the upper boundary of the region. The calculator graphs both functions and shades between the two.

Example  **Graph $y \geq 2x - 3$ in the standard viewing window.**

Before using the "Shade(" option, be sure to clear any equations stored in the Y= *list, and press* ZOOM *6 for the standard viewing window.*

The inequality refers to points at which y is *greater than or equal to* $2x - 3$. This means we want to shade above the graph of $y = 2x - 3$. Since the calculator screen shows only part of the coordinate plane, we can use the top of the screen, Ymax or 10, as the upper boundary and $2x - 3$ as the lower boundary.

Enter:  2nd DRAW 7 2 X,T,θ  — 3 , 10) ENTER

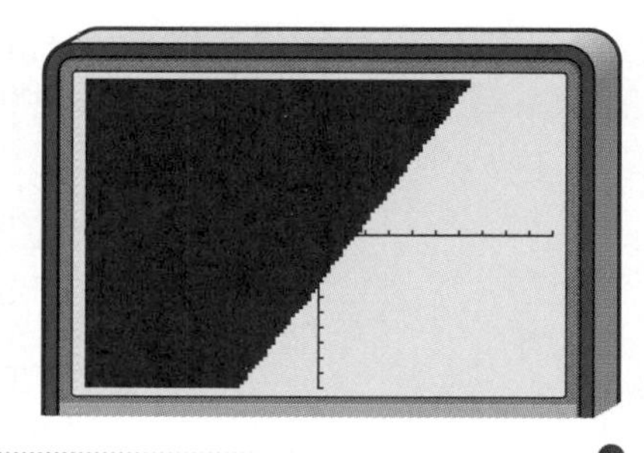

Since both the x- and y-intercepts of the graph and the origin are within the current viewing window, the graph of the inequality is complete.

When finished, press 2nd DRAW *1 to clear the screen.*

Example  **Graph $y - x \leq 1$ in the standard viewing window.**

First solve the inequality for y: $y \leq x + 1$. This inequality refers to points where y is *less than or equal to* $x + 1$. This means we want to shade below the graph of $y = x + 1$. We can use the bottom of the screen, Ymin or -10, as the lower boundary and $x + 1$ as the upper boundary.

Enter: 2nd DRAW 7  (–) 10 ,  X,T,θ + 1) ENTER

Don't forget to clear the screen when finished.

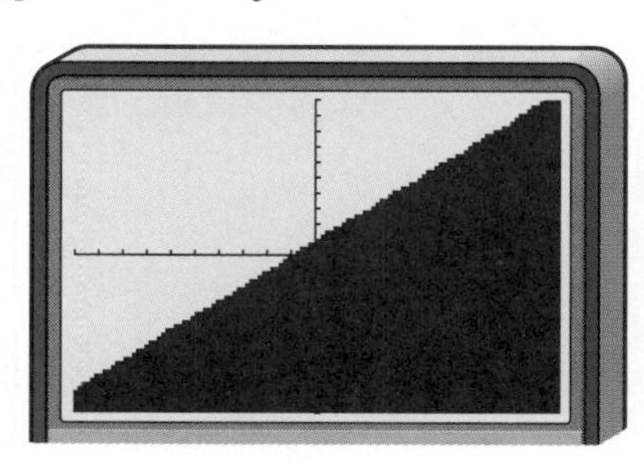

EXERCISES

Use a graphing calculator to graph each inequality. Sketch each graph on a sheet of paper.

1. $y \geq x + 2$
2. $y \leq -2x - 4$
3. $y + 1 \leq 0.5x$
4. $y \geq 4x$
5. $x + y \leq 0$
6. $2y + x \geq 4$
7. $3x + y \leq 18$
8. $y \geq 3$
9. $0.2x + 0.1y \leq 1$

7-8 Graphing Inequalities in Two Variables

What YOU'LL LEARN

- To graph inequalities in the coordinate plane.

Why IT'S IMPORTANT

You can graph inequalities to solve problems involving manufacturing and health.

Manufacturing

Rapid Cycle, Inc. is a manufacturer and distributor of racing bicycles. It takes 3 hours to assemble a bicycle and 1 hour to road test a bicycle. Each technician in the company works no more than 45 hours a week. How many racing bikes can one technician assemble, and how many can he or she road test in one week?

Let x represent the number of bikes that are assembled in a week, and let y represent the number of bikes that are road-tested. Then the following inequality can be used to represent the solution.

Total time to assemble x bikes	*plus*	*Total time to road test y bikes*	*is no more than*	*45 hours.*
$3x$	$+$	y	$\leq$	45

There are an infinite number of ordered pairs that are solutions to this inequality. The easiest way to show all of these solutions is to draw a *graph* of the inequality. Before doing this, let's consider some simpler inequalities.
This problem will be solved in Example 3.

Example 1 **From the set {(3, 4), (0, 1), (1, 4), (1, 1)}, which ordered pairs are part of the solution set for $4x + 2y < 8$?**

Let's use a table to substitute the x and y values of each ordered pair into the inequality.

x	y	$4x + 2y < 8$	True or False?
3	4	$4(3) + 2(4) < 8$ $20 < 8$	false
0	1	$4(0) + 2(1) < 8$ $2 < 8$	true
1	4	$4(1) + 2(4) < 8$ $12 < 8$	false
1	1	$4(1) + 2(1) < 8$ $6 < 8$	true

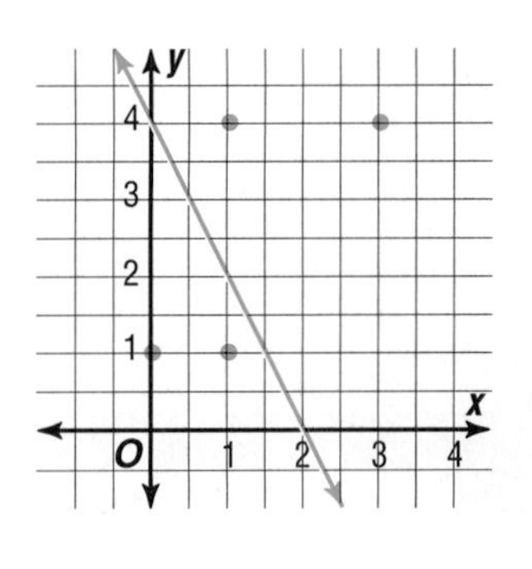

The ordered pairs {(0, 1), (1, 1)} are part of the solution set of $4x + 2y < 8$. The graph above shows the four ordered pairs of the replacement set and the equation $4x + 2y = 8$. Notice the location of the two ordered pairs that are solutions for $4x + 2y < 8$ in relation to the graph of the line.

EXPLORATION — PROGRAMMING

You can use the following graphing calculator program to find out if a given ordered pair (x, y) is a solution for the inequality $5x - 3y \geq 15$.

```
PROGRAM: XYTEST
: Disp "IS (X, Y) A ","SOLUTION?"
: Prompt X,Y
: If 5X−3Y ≥ 15
: Then
: Disp "YES"
: Else
: Disp "NO"
```

To run the program for other ordered pairs, simply press ENTER and the program will begin again.

Your Turn

a. Try the program for ten ordered pairs (x, y). Keep a list of which ordered pairs you tried and which ones were solutions.

b. How do you think you could change this program to test the inequality $2x + y \geq 2y$?

c. Use your changed program to find the solution set if $x = \{-1, 0, 1\}$ and $y = \{-2, -1, 0, 1\}$.

The solution set for an inequality contains many ordered pairs when the domain and range are the set of real numbers. The graphs of all of these ordered pairs fill an area on the coordinate plane called a **half-plane.** An equation defines the **boundary** or edge for each half-plane. For example, suppose you wanted to graph the inequality $y > 5$ on the coordinate plane.

First determine the boundary by graphing $y = 5$.

Since the inequality involves only $>$, the line should be dashed. The boundary divides the coordinate plane into two half-planes.

If the inequality contains $\geq$ or $\leq$, the graph of the boundary equation would be drawn as a solid line.

To determine which half-plane contains the solution, choose a point from each half-plane and test it in the inequality.

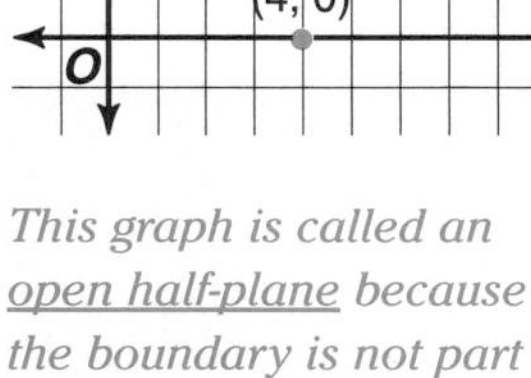

Try (7, 10).		Try (4, 0).	
$y > 5$	*$y = 10$*	$y > 5$	*$y = 0$*
$10 > 5$	true	$0 > 5$	false

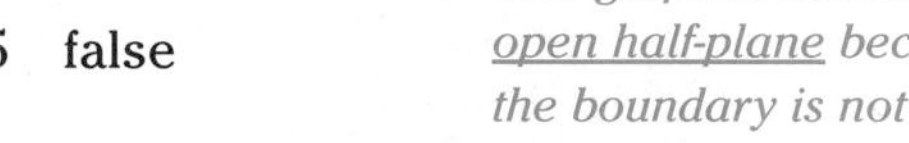

This graph is called an open half-plane because the boundary is not part of the graph.

The half-plane that contains (7, 10) contains the solution. Shade that half-plane.

Example  **Graph $y + 2x \leq 3$.**

First solve for y in terms of x.

$$y + 2x \leq 3$$
$$y + 2x - 2x \leq 3 - 2x \quad \textit{Subtract 2x from each side.}$$
$$y \leq 3 - 2x$$

(continued on the next page)

Graph $y = 3 - 2x$. Since $y \leq 3 - 2x$ means $y < 3 - 2x$ or $y = 3 - 2x$, the boundary is included in the graph and should be drawn as a solid line.

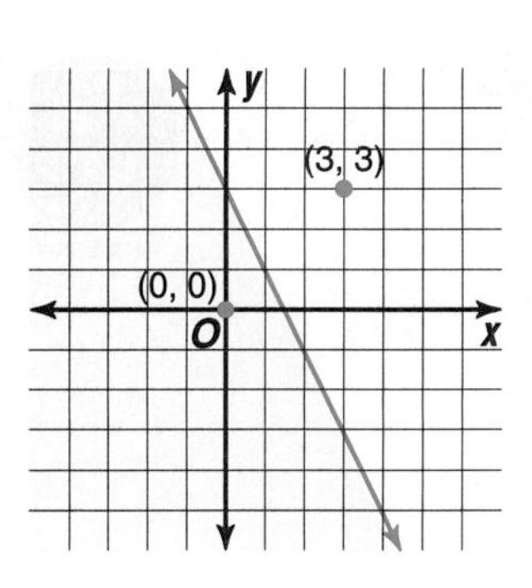

This graph is called a closed half-plane because the boundary line is included.

Select a point in one of the half-planes and test it. For example, use the origin (0, 0).

The origin is often used as a test point because the values are easy to substitute into the inequality.

The half-plane that contains the origin should be shaded.

Check: Test a point in the other half-plane, for example (3, 3).

Since the statement is false, the half-plane containing (3, 3) is not part of the solution.

When solving real-life inequalities, the domain and range of the inequality are often restricted to nonnegative numbers or whole numbers.

Example 3 **Refer to the application at the beginning of the lesson. How many racing bikes will the technician be able to assemble and road test?**

APPLICATION
Manufacturing

FYI

The world speed record for bicycles over a 200-meter course using a flying start is 65.484 mph by Fred Markham at Mono Lake, California, set in 1986.

First, solve for y in terms of x.

$3x + y \leq 45$

$3x - 3x + y \leq 45 - 3x$

$y \leq 45 - 3x$

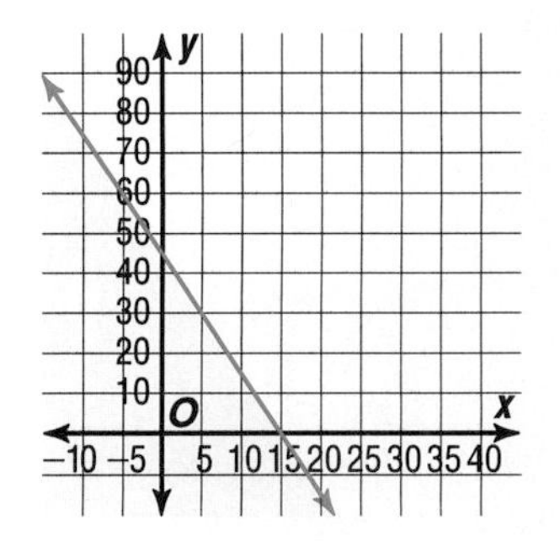

Since the open sentence includes the equation, graph $y = 45 - 3x$ as a solid line. Test a point in one of the half-planes, for example (0, 0). The half-plane containing (0, 0) represents the solution since $3(0) + 0 \leq 45$ is true.

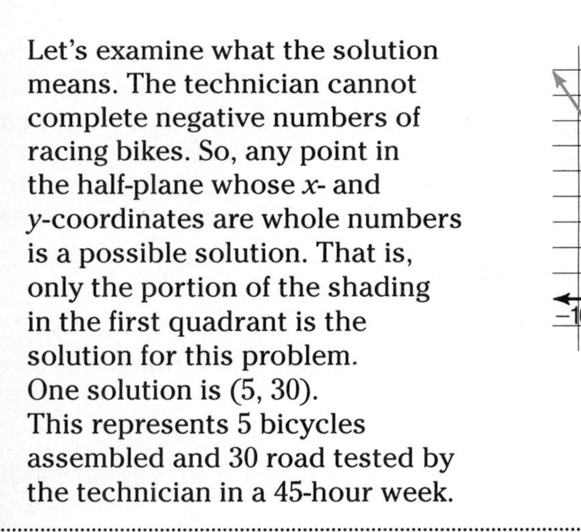

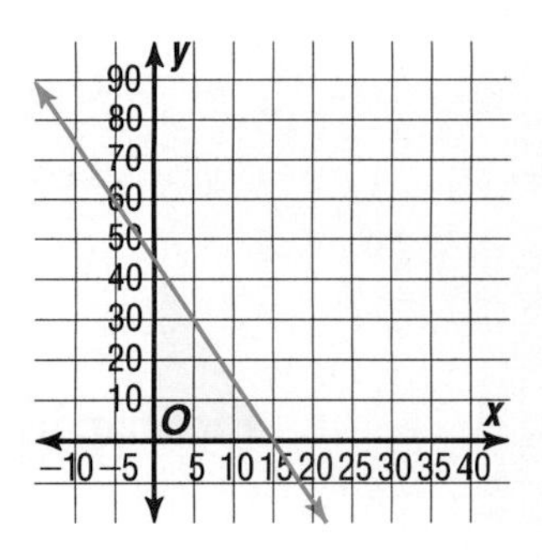

Let's examine what the solution means. The technician cannot complete negative numbers of racing bikes. So, any point in the half-plane whose x- and y-coordinates are whole numbers is a possible solution. That is, only the portion of the shading in the first quadrant is the solution for this problem.
One solution is (5, 30).
This represents 5 bicycles assembled and 30 road tested by the technician in a 45-hour week.

CHECK FOR UNDERSTANDING

Communicating Mathematics

Study the lesson. Then complete the following.

1. **a. Graph** $y \geq x + 1$.
   b. Identify the boundary and indicate whether it is included or not.
   c. Identify the half-plane that is part of the graph.
   d. Write the coordinates of a point not on the boundary that satisfy the inequality.
2. **Explain** how you would check whether a point is part of the graph of an inequality.

3. **Assess Yourself** What do you think was the most challenging concept you learned in this chapter? Give an example of that concept and tell why you thought it was challenging.

Guided Practice

Match each inequality with its graph.

4. $y \geq \frac{1}{2}x - 2$
5. $y \leq 0.5x - 2$
6. $y \geq \frac{2}{3}x + 2$
7. $y \leq \frac{2}{3}x + 2$

a.

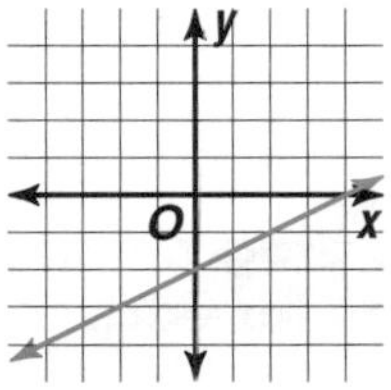

b.

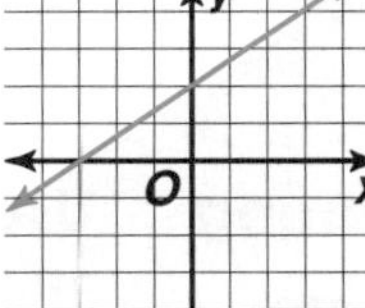

c.
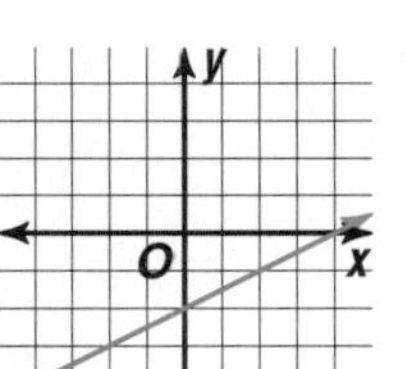

d.
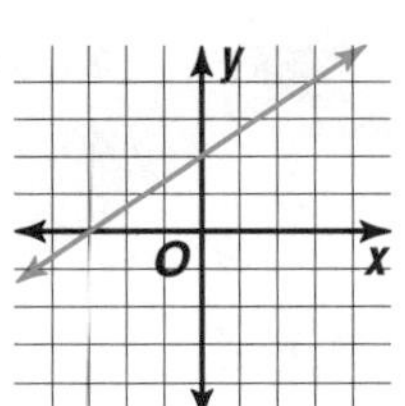

Determine which ordered pairs are solutions to the inequality. State whether the boundary is included in the graph.

8. $y \leq x$ — **a.** $(-3, 2)$ **b.** $(1, -2)$ **c.** $(0, -1)$
9. $y > x - 1$ — **a.** $(0, 0)$ **b.** $(2, 0)$ **c.** $(1, 3)$
10. Find which ordered pairs from the set $\{(-2, 2), (-2, 3), (2, 2), (2, 3)\}$ are part of the solution set for $a + b < 1$.

Graph each inequality.

11. $y > 3$
12. $x + y > 1$
13. $2x + 3y \geq -2$
14. $-x < -y$

EXERCISES

Practice

Copy each graph. Shade the appropriate half-plane to complete the graph of the inequality.

15. $x > 4$

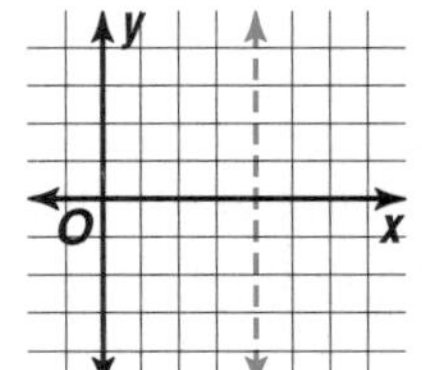

16. $3y < x$

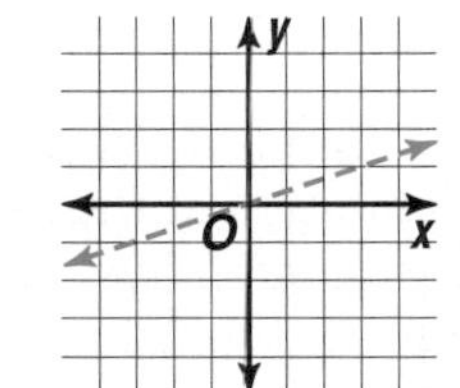

17. $3x + y > 4$

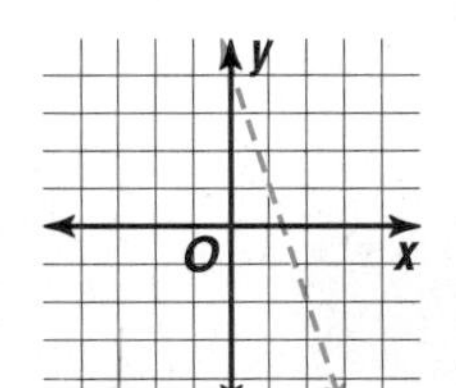

18. $2x - y \leq -2$

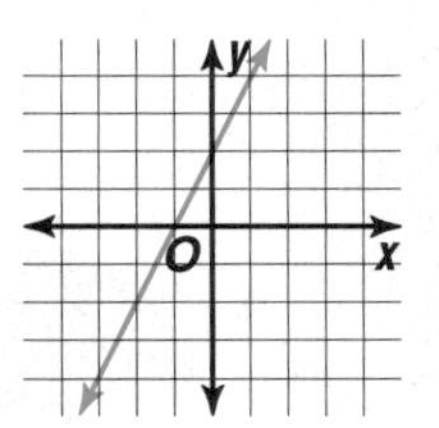

Find which ordered pairs from the given set are part of the solution set for each inequality.

19. $y < 3x$, $\{(-3, 1), (-3, 2), (1, 1), (1, 2)\}$
20. $y - x > 0$, $\{(1, 1), (1, 2), (4, 1), (4, 2)\}$
21. $2y + x \geq 4$, $\{(-1, -3), (-1, 0), (-2, -3), (-2, 0)\}$

Graph each inequality.

22. $x > -5$
23. $y < -3$
24. $3y + 6 > 0$
25. $4x + 8 < 0$
26. $y \leq x + 1$
27. $x + y > 2$
28. $x + y < -4$
29. $3x - 1 \geq y$
30. $3x + y < 1$
31. $x - y \geq -1$
32. $x < y$
33. $-y > x$
34. $2x - 5y \leq -10$
35. $8y + 3x < 16$
36. $|y| \geq 2$
37. $y > |x + 2|$
38. $y > 2$ and $x < 3$
39. $y \leq -x$ and $x \geq -3$

Graphing Calculator

Use a graphing calculator to graph each inequality. Make a sketch of the graph.

40. $y > x - 1$
41. $4y + x < 16$
42. $x - 2y < 4$

Programming

Use the program in the Exploration to determine which pairs are solutions for each inequality.

43. $x + 2y \geq 3$ **a.** $(-2, 2)$ **b.** $(4, -1)$ **c.** $(3, 1)$ **d.** $(0, 0)$
44. $2x - 3y \leq 1$ **a.** $(2, 1)$ **b.** $(5, -1)$ **c.** $(1, 1)$ **d.** $(0, 0)$
45. $-2x < 8 - y$ **a.** $(5, 10)$ **b.** $(3, 6)$ **c.** $(-4, 0)$ **d.** $(0, 0)$

Critical Thinking

46. What compound inequality is described by the graph at the right? Find a simple inequality that also describes this graph.

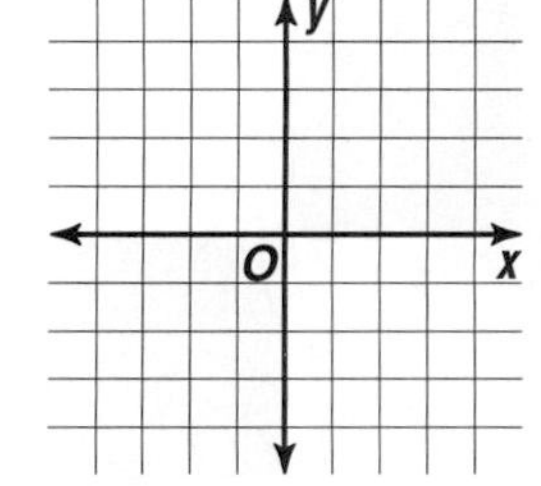

Applications and Problem Solving

47. **Health** The graph below shows the effective heart rate ranges for each type of exercise goal.

Workout Goals for Exercise

Goals
Boost performance as a competitive athlete
Improve cardiovascular conditioning
Lose weight
Improve overall health and reduce risk of heart attack

40% 50% 60% 70% 80% 90% 100%
Target Heart Rate Range

Source: *Vitality*, May 1994

Arrio is 35 years old and just beginning a bench-stepping aerobics class. In the orientation at the beginning of the first class, he learned that during exercise an effective minimum heart rate (beats/minute) should be 70% of the difference of 220 and his age. A maximum heart rate should be 80% of the difference of 220 and his age.

a. Write a compound inequality that expresses the effective rate zone for a person a years of age.

b. In class, the participants take a break and count their heart beats for 15 seconds. What should be Arrio's effective heart rate zone for that 15-second count?

c. According to the graph and the heart rate range given, what is the goal of the bench-stepping class?

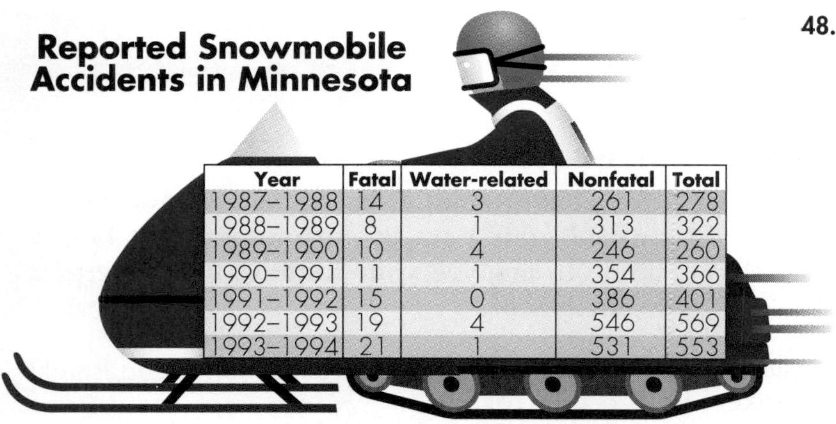

Reported Snowmobile Accidents in Minnesota

Year	Fatal	Water-related	Nonfatal	Total
1987–1988	14	3	261	278
1988–1989	8	1	313	322
1989–1990	10	4	246	260
1990–1991	11	1	354	366
1991–1992	15	0	386	401
1992–1993	19	4	546	569
1993–1994	21	1	531	553

Source: Minnesota Department of Natural Resources

48. **Snowmobiling** Although snowmobiling is an exhilarating sport, it can be very dangerous. The chart below shows the reported accidents in Minnesota involving snowmobiles.

a. A study suggests that the average number of nonfatal snowmobile accidents per year nationwide is about 350. Suppose x represents the total number of fatal snowmobile accidents in Minnesota and y represents the total number of snowmobile accidents in Minnesota. In what years is $x + 350 > y$?

b. Graph $x + 350 > y$.

c. The snowmobiling season in Minnesota goes from November to March. The lack of snow during the 1994–1995 season caused many to resort to riding their snowmobiles on icy lakes instead of on land. As of December 1, there had been 7 fatal accidents, 1 water-related fatal accident, and 81 nonfatal accidents. Do you think that the 1994–1995 season's accident count was greater or less than the 1993–1994 season? Explain your answer.

Mixed Review

49. **Statistics** Make a box-and-whisker plot of the total number of snowmobiling accidents shown in the chart for Exercise 48. (Lesson 7–7)

50. Solve $5 - |2x - 7| > 2$. (Lesson 7–6)

51. Write an equation in slope-intercept of the line that is parallel to the graph of $8x - 2y = 7$ and whose y-intercept is the same as the line whose equation is $2x - 9y = 18$. (Lesson 6–4)

52. Determine the value of r so that the line passing through $(r, 4)$ and $(-4, r)$ has a slope of 4. (Lesson 6–1)

53. Graph $-y + \frac{2}{7}x = 1$. (Lesson 5–4)

54. State the inverse of the relation $\{(4, -1), (3, 2), (-4, 0), (17, 9)\}$. (Lesson 5–2)

55. What is 98.5% of \$140.32? (Lesson 4–4)

56. Find three consecutive integers whose sum is 87. (Lesson 3–3)

CLOSING THE

In·ves·ti·ga·tion

Smoke Gets In Your Eyes

Refer to the Investigation on pages 320–321. Add the results of your work below to your Investigation Folder.

The EPA's most recent long-term study on the effects of secondhand smoke shows that non-smokers married to smokers have a 19% increased risk of having lung cancer. Lung cancer is not the only danger of secondhand smoke. Twelve studies show that heart disease is another danger. Nonsmokers who are exposed to their spouses' smoke have a 30% increased chance of death from heart disease than do other nonsmokers. After reviewing a number of studies, the EPA's risk analysis has also concluded that secondhand smoke causes an extra 150,000 to 300,000 respiratory infections a year among the nations 5.5 million children under the age of 18 months.

Katharine Hammond, an environmental-health expert at the University of California, Berkeley, has also conducted a study on the *carcinogenic* components of secondhand smoke. The carcinogenic components are the parts of smoke that are known to cause cancer in humans. She found that "in the same room, at the same time, the nonsmoker is getting as much benzene (a chemical that is known to cause cancer in humans) as a smoker gets smoking six cigarettes."

James Repace and Alfred Lowery, two statistical researchers who study the effects of secondhand smoke, have concluded that a lifetime increase in lung-cancer risk of 1 in 1000 could be caused by long-term exposure to air containing more than 6.8 micrograms of nicotine per cubic meter of air.

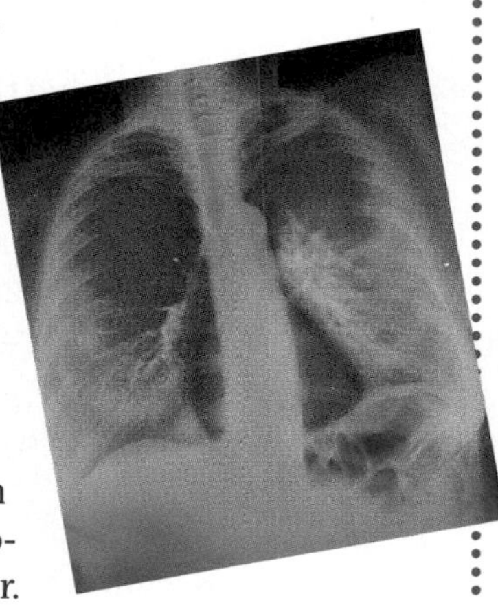

Analyze

You have conducted experiments and organized your data in various ways. It is now time to analyze your findings and state your conclusions.

PORTFOLIO ASSESSMENT

You may want to keep your work on this Investigation in your portfolio.

1. True secondhand smoke consists mostly of sidestream smoke. This is the smoke that comes from the smoldering cigarette. This smoke is much more toxic than inhaled smoke. How does this information affect your conclusions about the amount of smoke inhaled by nonsmokers in a room?
2. If you are a nonsmoker and live with a smoker, what are the cost factors involved? Explain your calculations.
3. Describe your personal experience with secondhand smoke.

Write

You want to inform people of the effects of secondhand smoke. You decide to write a letter to the editor of a local paper describing your investigation on the effects of secondhand smoke.

4. Use the information above and the results of your experiments and explorations to write a paper regarding the health risks of secondhand smoke.
5. You may want to do further research. The American Cancer Society and other agencies have information regarding the health risks of smoking and secondhand smoke.
6. Use data, charts, and graphs to justify your position. Use mathematics to help convince your readers of the conclusions you drew from this Investigation.

CHAPTER 7 HIGHLIGHTS

VOCABULARY

After completing this chapter, you should be able to define each term, property, or phrase and give an example or two of each.

Algebra

addition property for inequality (p. 385)
boundary (p. 437)
compound inequality (p. 405)
division property for inequality (p. 393)
half-plane (p. 437)
intersection (p. 406)
multiplication property for inequality (p. 393)
set-builder notation (p. 385)
subtraction property for inequality (p. 385)
union (p. 408)

Statistics

box-and-whisker plot (p. 427)
extreme values (p. 427)
whiskers (p. 428)

Probability

compound event (p. 414)
outcomes (p. 413)
simple events (p. 414)
tree diagram (p. 413)

Problem Solving

draw a diagram (p. 406)

UNDERSTANDING AND USING THE VOCABULARY

Choose the letter of the term that best matches each statement, algebraic expression, or algebraic sentence.

1. If $\frac{1}{2}x \leq -5$, then $x \leq -10$.
2. If $8 > 4$, then $8 + 5 > 4 + 5$.
3. $\{h \mid h > 43\}$
4. $x \geq -3$ or $x < -10$
5. $x \geq -4$ and $x < 2$
6. If $4x - 1 < 7$, then $4x - 4 < 4$.
7. If $-3x < 9$, then $x > -3$.
8. $>$
9. $<$
10. $7 > x > 1$
11. $|x + 6| > 12$ means $x + 6 > 12$ or $-(x + 6) > 12$.

a. absolute value inequality
b. addition property for inequality
c. compound inequality
d. division property for inequality
e. greater than
f. intersection
g. less than
h. multiplication property for inequality
i. set builder notation
j. subtraction property for inequality
k. union

STUDY GUIDE AND ASSESSMENT

SKILLS AND CONCEPTS

OBJECTIVES AND EXAMPLES

Upon completing this chapter, you should be able to:

- solve inequalities by using addition and subtraction (Lesson 7–1)

$56 > m + 16$

$56 - 16 > m + 16 - 16$

$40 > m$

$\{m \mid m < 40\}$

REVIEW EXERCISES

Use these exercises to review and prepare for the chapter test.

Solve each inequality. Then check your solution.

12. $r + 7 > -5$
13. $-35 + 6n < 7n$
14. $2t - 0.3 \leq 5.7 + t$
15. $-14 + p \geq 4 - (-2p)$

Define a variable, write an inequality, and solve each problem. Then check your solution.

16. The difference of a number and 3 is at least 2.
17. Three times a number is greater than four times the number less eight.

- solve inequalities by using multiplication and division (Lesson 7–2)

$\frac{-5}{6}m > 25$

$\frac{-6}{5}\left(\frac{-5}{6}m\right) > \frac{-6}{5}(25)$

$m < -30$

$\{m \mid m < -30\}$

Solve each inequality. Then check your solution.

18. $7x \geq -56$
19. $90 \leq -6w$
20. $\frac{2}{3}k \geq \frac{2}{15}$
21. $9.6 < 0.3x$

Define a variable, write an inequality, and solve each problem. Then check your solution.

22. Six times a number is at most 32.4.
23. Negative three fourths of a number is no more than 30.

- solve linear inequalities involving more than one operation (Lesson 7–3)

$15b - 12 > 7b + 60$

$15b - 7b - 12 > 7b - 7b + 60$

$8b - 12 > 60$

$8b - 12 + 12 > 60 + 12$

$8b > 72$

$\frac{8b}{8} > \frac{72}{8}$

$b > 9$

$\{b \mid b > 9\}$

Find the solution set of each inequality if the replacement set for each variable is {−5, −4, −3, . . . 3, 4, 5}.

24. $\frac{x-5}{3} > -3$
25. $3 \leq -4x + 7$

Solve each inequality. Then check your solution.

26. $2r - 3.1 > 0.5$
27. $4y - 11 \geq 8y + 7$
28. $-3(m - 2) > 12$
29. $-5x + 3 < 3x + 23$
30. $4(n - 1) < 7n + 8$
31. $0.3(z - 4) \leq 0.8(0.2z + 2)$

OBJECTIVES AND EXAMPLES

- solve compound inequalities and graph their solution sets (Lesson 7–4)

$$2a > a - 3 \quad \text{and} \quad 3a < a + 6$$
$$2a - a > a - a - 3 \qquad 3a - a < a - a + 6$$
$$a > -3 \qquad 2a < 6$$
$$\frac{2a}{2} < \frac{6}{2}$$
$$a < 3$$
$$\{a \mid -3 < a < 3\}$$

−4 −3 −2 −1 0 1 2 3 4

REVIEW EXERCISES

Solve each compound inequality. Then graph the solution set.

32. $x - 5 < -2$ and $x - 5 > 2$

33. $2a + 5 \leq 7$ or $2a \geq a - 3$

34. $4r \geq 3r + 7$ and $3r + 7 < r + 29$

35. $-2b - 4 \geq 7$ or $-5 + 3b \leq 10$

36. $a \neq 6$ and $3a + 1 > 10$

- find the probability of a compound event (Lesson 7–5)

Draw a tree diagram to show the possibilities for boys and girls in a family of 3 children. Assume that the probabilities for girls and boys being born are the same.

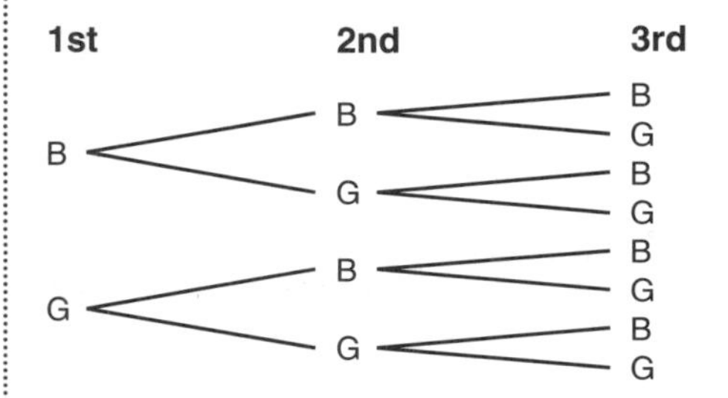

The probability that the family has exactly 3 girls is $\frac{1}{8}$ or 0.125, because there is 1 way out of 8 for this to happen. The probability that the family has exactly 2 boys and 1 girl is $\frac{3}{8}$ or 0.375, because there are 3 ways out of 8 for this to happen.

37. With each shrimp, salmon, or crab dinner at the Seafood Palace, you may have soup or salad. With shrimp, you may have broccoli or a baked potato. With salmon, you may have rice or broccoli. With crab, you may have rice, broccoli, or a potato. If all combinations are equally likely, find the probability of an order containing each item.

a. salmon **b.** soup
c. rice **d.** shrimp and rice
e. salad and broccoli
f. crab, soup, and rice

38. Matthew has 2 brown and 4 black socks in his dresser. While dressing one morning, he pulled out 2 socks without looking. What is the probability that he chose a matching pair?

- solve open sentences involving absolute value and graph the solutions (Lesson 7–6)

$$|2x + 1| > 1$$
$$2x + 1 > 1 \quad \text{or} \quad 2x + 1 < -1$$
$$2x + 1 - 1 > 1 - 1 \qquad 2x + 1 - 1 < -1 - 1$$
$$\frac{2x}{2} > \frac{0}{2} \qquad \frac{2x}{2} < \frac{-2}{2}$$
$$x > 0 \qquad x < -1$$
$$\{x \mid x > 0 \text{ or } x < -1\}$$

−4 −3 −2 −1 0 1 2 3

Solve each open sentence. Then graph the solution set.

39. $|y + 5| > 0$

40. $|1 - n| \leq 5$

41. $|4k + 2| \leq 14$

42. $|3x - 12| < 12$

43. $|13 - 5y| \geq 8$

44. $\left|2p - \frac{1}{2}\right| > \frac{9}{2}$

OBJECTIVES AND EXAMPLES

- display and interpret data on box-and-whisker plots (Lesson 7–7)

The following high temperatures were recorded during a two-week cold spell in St. Louis. Make a box-and-whisker plot of the temperatures.

20° 2° 12° 5° 4° 16° 17°

7° 6° 16° 5° 0° 5° 30°

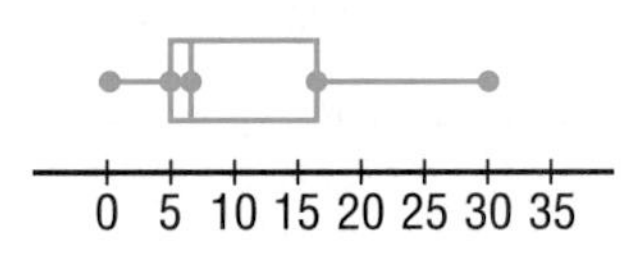

REVIEW EXERCISES

The number of calories in a serving of french fries at 13 restaurants are 250, 240, 220, 348, 199, 200, 125, 230, 274, 239, 212, 240, and 327.

45. Make a box-and-whisker plot of these data.

46. Are there any outliers? If so, name them.

- graph inequalities in the coordinate plane (Lesson 7–8)

Graph $2x + 3y < 9$.

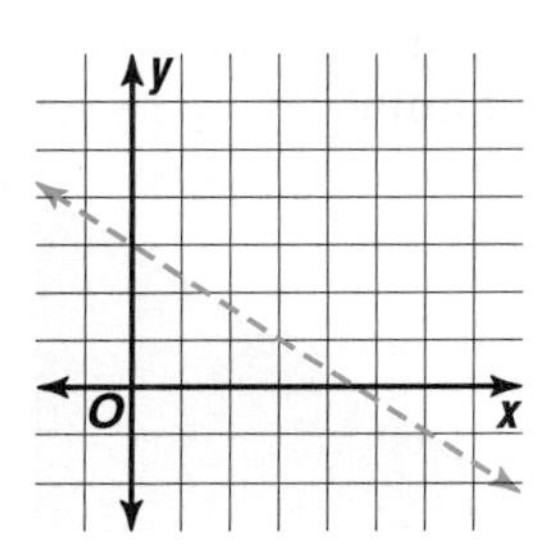

Find which ordered pairs from the given set are part of the solution set for each inequality.

47. $3x + 4y < 7$, $\{(1, 1), (2, -1), (-1, 1), (-2, 4)\}$

48. $4y - 8 \geq 0$, $\{(5, -1), (0, 2), (2, 5), (-2, 0)\}$

49. $-2x < 8 - y$, $\{(5, 10), (3, 6), (-4, 0), (-3, 6)\}$

Graph each inequality.

50. $x + 2y > 5$

51. $4x - y \leq 8$

52. $\frac{1}{2}y \geq x + 4$

53. $3x - 2y < 6$

APPLICATIONS AND PROBLEM SOLVING

54. Number Theory The sum of three consecutive integers is less than 100. Find the three integers with the greatest sum. (Lesson 7–3)

55. Shipping An empty book crate weighs 30 pounds. The weight of a book is 1.5 pounds. For shipping, the crate must weigh at least 55 pounds and no more than 60 pounds. What is the acceptable number of books that can be packed in the crate? (Lesson 7–4)

56. Automobiles An automobile dealer has cars available painted red or blue, with 4-cylinder or 6-cylinder engines, and with manual or automatic transmissions. (Lesson 7–5)

a. What is the probability of selecting a car with manual transmission?

b. What is the probability of selecting a car with a 4-cylinder engine and a manual transmission?

c. What is the probability of selecting a blue car with a 6-cylinder engine and an automatic transmission?

A practice test for Chapter 7 is provided on page 793.

ALTERNATIVE ASSESSMENT

COOPERATIVE LEARNING PROJECT

Statistics In this chapter, you learned how to make and interpret a box-and-whisker plot. Making a box-and-whisker plot and interpreting the data from a box-and-whisker plot are two different skills, however. One can go through the routine of drawing the plot but not be able to use the data portrayed by the plot to answer appropriate questions.

For this project, suppose there are two companies that manufacture window glass. They have each submitted a bid to a contractor who is building a library. Since glass that varies in thickness can cause distortions, the contractor has decided to measure the thickness of panes of glass from each factory, at several locations on each pane. The table below shows measurements for the two panes, one from each manufacturer.

Glass Thickness (mm)	
Company A	**Company B**
10.2	9.4
12.0	13.0
11.6	8.2
10.1	14.9
11.2	12.6
9.7	7.7
10.7	13.2
11.6	12.2
10.4	10.2
9.8	9.5
10.6	9.9
10.3	9.7
8.5	11.5
10.2	11.5
9.7	10.5
9.2	10.6
8.6	6.4
11.3	13.5

Prepare a stem-and-leaf plot and a box-and-whisker plot to organize and compare the two sets of data.

Follow these steps to organize your data.

- Determine how to set up the stem-and-leaf plot using decimals.
- Determine if using a common stem would be useful in comparing the two sets of data.
- Compare the ranges of the two sets of data.
- For which company is there "bunching" or "spreading out evenly" of the data?
- Compare the stem-and-leaf plot shape with the box-and-whisker plot shape.
- Compare the middle half of the data for each company.
- Write a comparative description of the sets of data and determine, with support, which company's glass should have less distortion.

THINKING CRITICALLY

- Why are multiplication and division the only two out of the four operations for which it is necessary to distinguish between positive and negative numbers when solving linear inequalities?
- Under what conditions will the compound sentence $x < a$ and $-a < x$ have no solutions?

PORTFOLIO

Select one of the assignments from this chapter for which you felt organization of the problem and reevaluation of the answer were important in order to get an accurate answer. Revise your work as necessary and place it in your portfolio. Explain why organization and reevaluation were important.

SELF EVALUATION

Do you look beyond the obvious in your math answers? Many times math students will work through a math problem rather routinely and not evaluate or check their answer. An answer must make sense and be accurate.

Assess yourself. Do you take the obvious solution as the whole answer or do you evaluate your answers for accuracy and rationalness? List two problems in mathematics and/or your daily life whereby the obvious answer was incorrect, so you needed to evaluate your solution for accuracy.

LONG-TERM PROJECT

In·ves·ti·ga·tion

Ready, Set, Drop!

MATERIALS NEEDED

- construction paper
- metric ruler

- paper clips
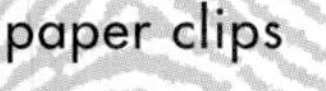
- scissors
- stopwatch
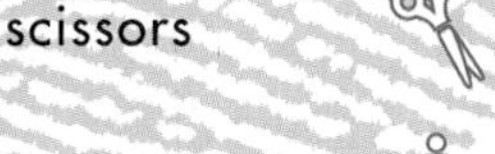
- tape

- tissue paper
- washers

- wire

Hang gliding became popular in the United States in the early 1970s. In most states, a hang gliding certification is required before you are allowed to participate in the sport. The U.S. Hang Gliding Association is located in Los Angeles and certifies instructors and safety officers to train would-be hang gliders.

A hang glider looks like a manned kite. It consists of a triangular sail of synthetic fabric attached to an aluminum frame. The pilot hangs from a harness and steers the glider with a control bar that adjusts as the pilot shifts his or her body weight.

Hang gliders can be launched in several ways. The pilot can hold the glider and run down a hill until the glider is airborne. In areas with high cliffs, the pilot can run and jump from the cliff's edge, using the air currents below to fly. In flatter landscapes, the glider is often launched by towing it with a rope from a truck or boat and releasing it at an altitude of 400–500 feet.

Imagine that you are an engineer for an aeronautical engineering firm. A group of people who are interested in hang gliding have asked your firm to design a hang glider that can be used for recreational purposes. Your task is to design a hang glider that is as compact as possible, yet is safe for flight and landings. You have no previous experience designing hang gliders. You don't know what size hang glider is needed or whether or not the size of a hang glider depends on the size of its load. (The people range in size.)

With so many unknowns, you decide to conduct some tests to understand the principles involved. In this Investigation, you will use mathematics to examine the relationship between the speed of descent and the size of a hang glider. As part of a three-member research team, you will use tissue-paper triangles to study hang gliders.

Make an Investigation Folder in which you can store all of your work on this Investigation for future use.

TRIANGLE TEST				
Test	5 cm	10 cm	20 cm	35 cm
perimeter				
surface area				
1				
2				
3				
4				
5				

THE EXPERIMENT

1. Begin by copying the chart above.
2. Cut out four equilateral triangles from tissue paper. The sides of the triangles should be 5, 10, 20, and 35 centimeters long, respectively. These triangles will serve as models of hang gliders. Find the perimeter and surface area of each of these triangles and record them in your chart.
3. Measure a height of five feet on a wall. Mark this height with a piece of masking tape.
4. Hold the smallest triangle parallel to the ground at a height of 5 feet. Have a second person ready to use a stopwatch to time how long it takes the triangle to reach the floor. A third person should give the verbal command, "Ready, set, drop." At the drop command, the person holding the tissue paper glider should let go of the paper. The timer starts the stopwatch at the verbal command and stops it when the glider hits the ground. Repeat this process until five drops have been made, recording your data after each drop.
5. Repeat Step 4 for the other three gliders, recording the data for each drop.
6. Review the data that you collected. What observations can you make? Are there any relationships that you can see from the data? Do the perimeter and surface area have a relationship with the time of the drop? Explain.

You will continue working on this Investigation throughout Chapters 8 and 9.

Be sure to keep your triangle models, charts, and other materials in your Investigation Folder.

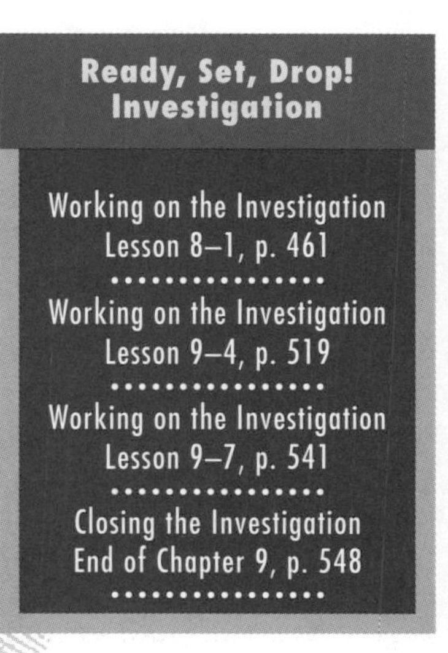

CHAPTER

8

Solving Systems of Linear Equations and Inequalities

Objectives

In this chapter, you will:

- graph systems of equations,
- solve systems of equations using various methods,
- organize data to solve problems, and
- solve systems of inequalities by graphing.

Games People Play

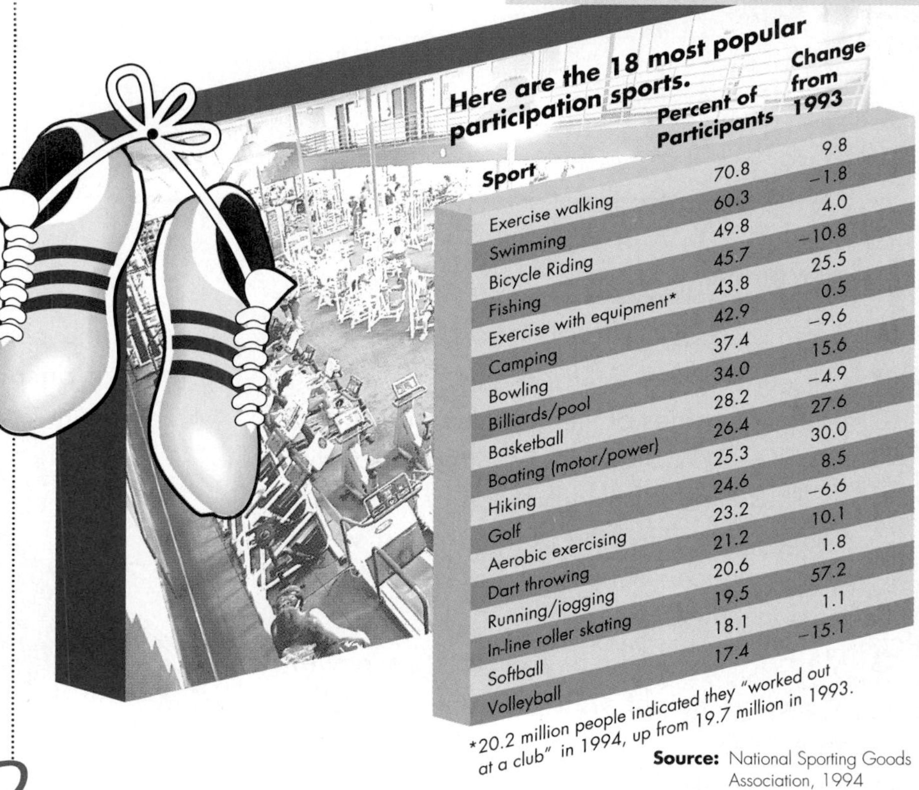

Here are the 18 most popular participation sports.

Sport	Percent of Participants	Change from 1993
Exercise walking	70.8	9.8
Swimming	60.3	−1.8
Bicycle Riding	49.8	4.0
Fishing	45.7	−10.8
Exercise with equipment*	43.8	25.5
Camping	42.9	0.5
Bowling	37.4	−9.6
Billiards/pool	34.0	15.6
Basketball	28.2	−4.9
Boating (motor/power)	26.4	27.6
Hiking	25.3	30.0
Golf	24.6	8.5
Aerobic exercising	23.2	−6.6
Dart throwing	21.2	10.1
Running/jogging	20.6	1.8
In-line roller skating	19.5	57.2
Softball	18.1	1.1
Volleyball	17.4	−15.1

*20.2 million people indicated they "worked out at a club" in 1994, up from 19.7 million in 1993.

Source: National Sporting Goods Association, 1994

TIME*Line*

Do you dream of starring in the NBA or playing for the New York Yankees? Do you devote all your leisure time to mainly one sport? Maybe you should try something new in the world of sports. Lacrosse, in-line skating, or judo, anyone?

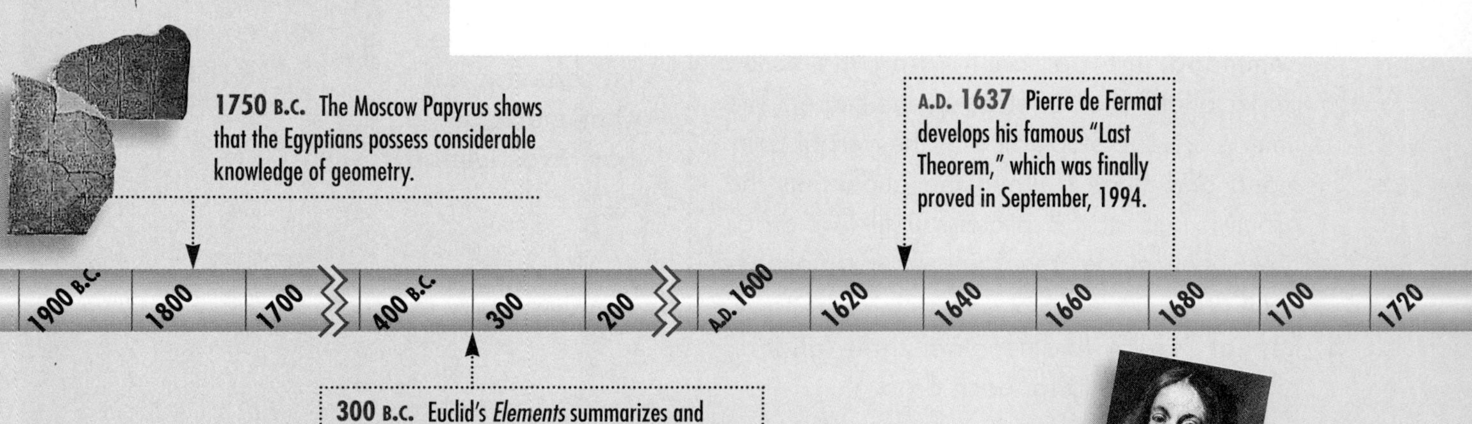

1750 B.C. The Moscow Papyrus shows that the Egyptians possess considerable knowledge of geometry.

300 B.C. Euclid's *Elements* summarizes and organizes the mathematical knowledge developed in Greece in the three preceding centuries.

A.D. 1637 Pierre de Fermat develops his famous "Last Theorem," which was finally proved in September, 1994.

PEOPLE IN THE NEWS

Short-track speed skaters **Julie Goskowicz**, 15, and **Tony Goskowicz**, 18, are a brother-and-sister team aiming for the 1998 Olympics. They started skating eight years ago in their hometown of New Berlin, Wisconsin, when their father gave them each a pair of skates. Although both finished last in their first race, they enjoyed the sport and continued to train. Hard work has earned them a place at the U.S. Olympic Education Center in Marquette, Michigan, where they study and train while participating in the racing circuit.

Chapter Project

In 1994, five-time Boston Marathon winner Jim Knaub tested his wheelchair's aerodynamics in the same wind tunnel that the Chrysler Corporation uses to test its car and truck designs. Knaub gained invaluable information concerning racing posture as well as helmet, wheel, and seat design. Earlier in the year, Knaub's Boston-Marathon-winning streak ended when he had to pull over twice during the race to make repairs. The winner was Heinz Frei of Switzerland.

- Suppose during a race, Frei's speed is 45 mph and Knaub is 264 feet ahead of him, racing at 36 mph.
- Write a system of equations to represent this situation. (*Hint:* Convert units from miles per hour to feet per second.)
- If their speeds remained constant, when would Frei catch up with Knaub? Explain how you know using graphing.

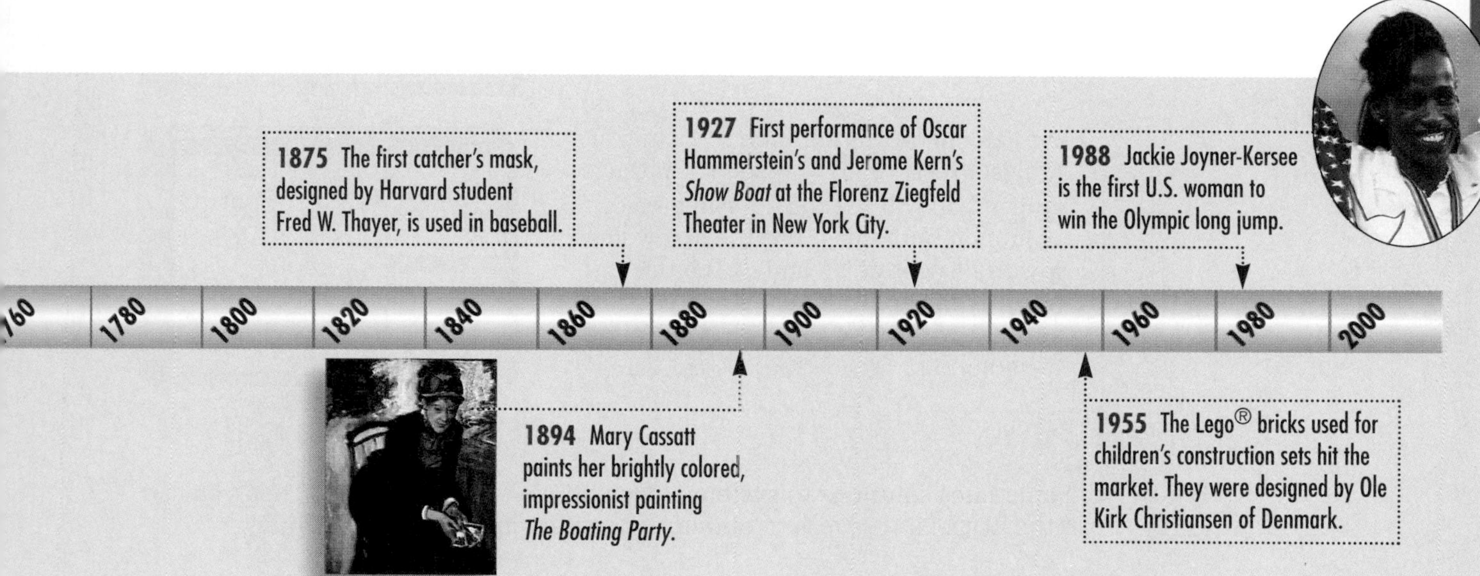

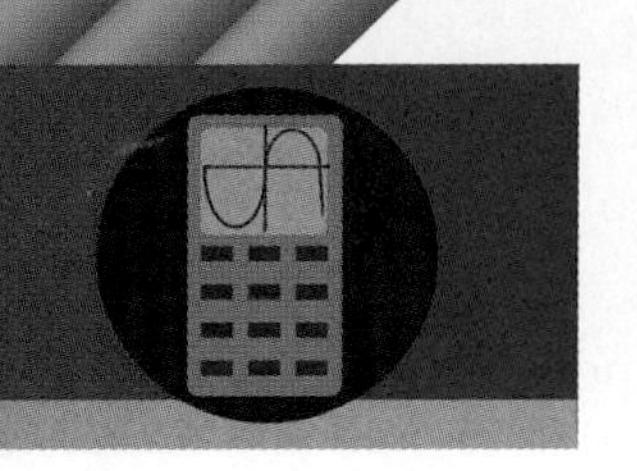

8–1A Graphing Technology Systems of Equations

A Preview of Lesson 8–1

When solving systems of linear equations graphically, each equation is graphed on the same coordinate plane. The coordinates of the point at which the graphs intersect is the solution of the system. The graphing calculator permits us to graph several equations on the same coordinate plane and approximate the coordinates of the intersection point.

Example  **Use a graphing calculator to solve the system of equations.**

$$x + y = 9$$
$$2x - y = 15$$

Begin by rewriting each equation in an equivalent form by solving for y.

$x + y = 9$	$2x - y = 15$
$y = -x + 9$	$2x - 15 = y$

Graph each equation in the integer window $[-47, 47]$ by $[-31, 31]$. Recall that the integer window can be obtained by entering ZOOM 6 ZOOM 8 ENTER.

Enter:  Y= (−)  X,T,θ  + 9 ENTER 2 X,T,θ − 15 ZOOM 6 ZOOM 8 ENTER

The graphs intersect in one point. The coordinates of this point are the solution to the system of equations. Press the TRACE key and use the arrow keys to move the cursor to the point of intersection. The coordinates of the point are (8, 1). Thus, the solution is (8, 1).

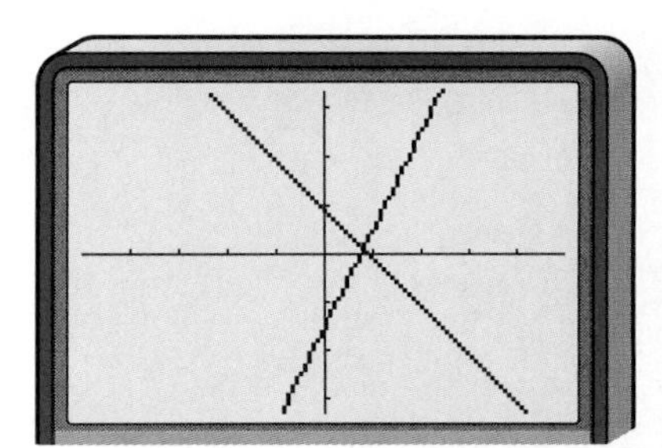

We can check this solution by using tables. Press 2nd TABLE. On the screen you will see the coordinates of points on both lines. Use the arrow keys to scroll up or down and watch the trend of the coordinates. When you find a row at which $Y_1 = Y_2$, you have found the solution. *The solution checks.*

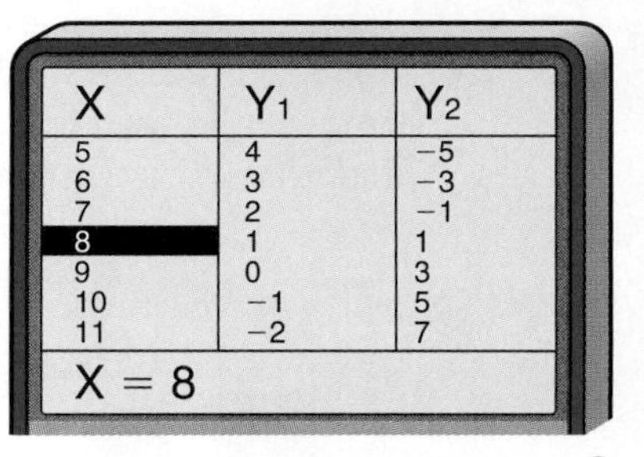

X	Y_1	Y_2
5	4	−5
6	3	−3
7	2	−1
8	1	1
9	0	3
10	−1	5
11	−2	7

X = 8

Sometimes solutions to systems of equations are not integers. Then you can use the ZOOM IN process to obtain an accurate approximate solution.

Example 2 **Use a graphing calculator to solve this system of equations to the nearest hundredth.**

$\mathbf{y = 0.35x - 1.12}$
$\mathbf{y = -2.25x - 4.05}$

Begin by graphing the equations in the standard viewing window.

Enter:  [Y=] .35 [X,T,θ] [−] 1.12

[ENTER] [(−)] 2.25 [X,T,θ]

[−] 4.05 [ZOOM] 6

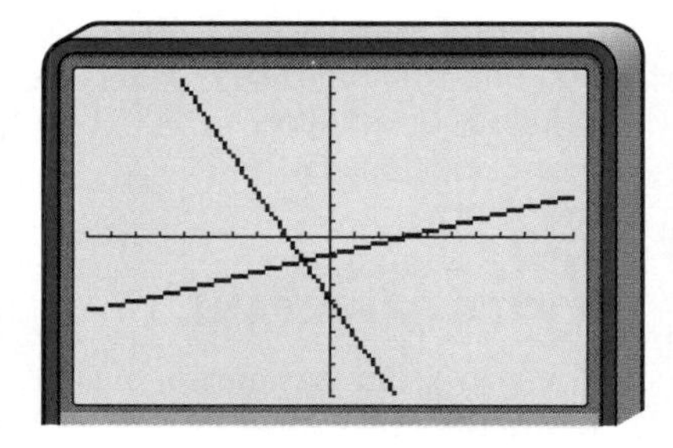

The graphs intersect at a point in the third quadrant. Use the TRACE function and the arrow keys to determine an approximation for the coordinates of the point of intersection. The ZOOM IN feature of the calculator is very useful for determining the coordinates of the intersection point with greater accuracy. Begin by placing the cursor on the intersection point and observing the coordinates, then press [ZOOM] 2 [ENTER]. Repeat this process as many times as necessary to get a more accurate answer.

You may want to use the INTERSECT feature to find the coordinates of the point of intersection. Press [2nd]

[CALC] 5 [ENTER] [ENTER] [ENTER].

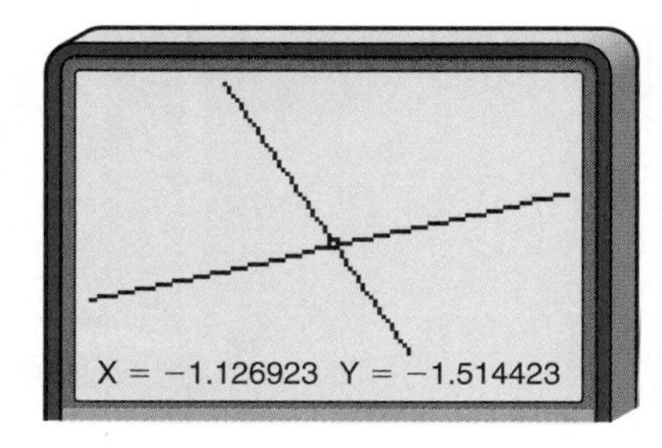

The solution is $(-1.13, -1.51)$.

EXERCISES

Use a graphing calculator to solve each system of equations. State each decimal solution to the nearest hundredth.

1. $y = x + 7$
   $y = -x + 9$
2. $x + y = 27$
   $3x - y = 41$
3. $y = 3x - 4$
   $y = -0.5x + 6$
4. $x - y = 6$
   $y = 9$
5. $x + y = 5.35$
   $3x - y = 3.75$
6. $5x - 4y = 26$
   $4x + 2y = 53.3$
7. $2x + 3y = 11$
   $4x + y = -6$
8. $2.93x + y = 6.08$
   $8.32x - y = 4.11$
9. $125x - 200y = 800$
   $65x - 20y = 140$
10. $0.22x + 0.15y = 0.30$
    $-0.33x + y = 6.22$

8–1 Graphing Systems of Equations

What YOU'LL LEARN

- To solve systems of equations by graphing, and
- to determine whether a system of equations has one solution, no solution, or infinitely many solutions by graphing.

Why IT'S IMPORTANT

You can graph systems of equations to solve problems involving business and geometry.

World Records

Cape Verde is a group of islands located off the westernmost point of Africa. In December, 1994, Frenchman Guy Delage set off from these islands for a 2400-mile swim across the Atlantic Ocean. He arrived at Barbados in the West Indies eight weeks later. Every day he would swim a while and then rest while floating with the current on a huge raft equipped with a fax machine, a computer, and a two-way radio.

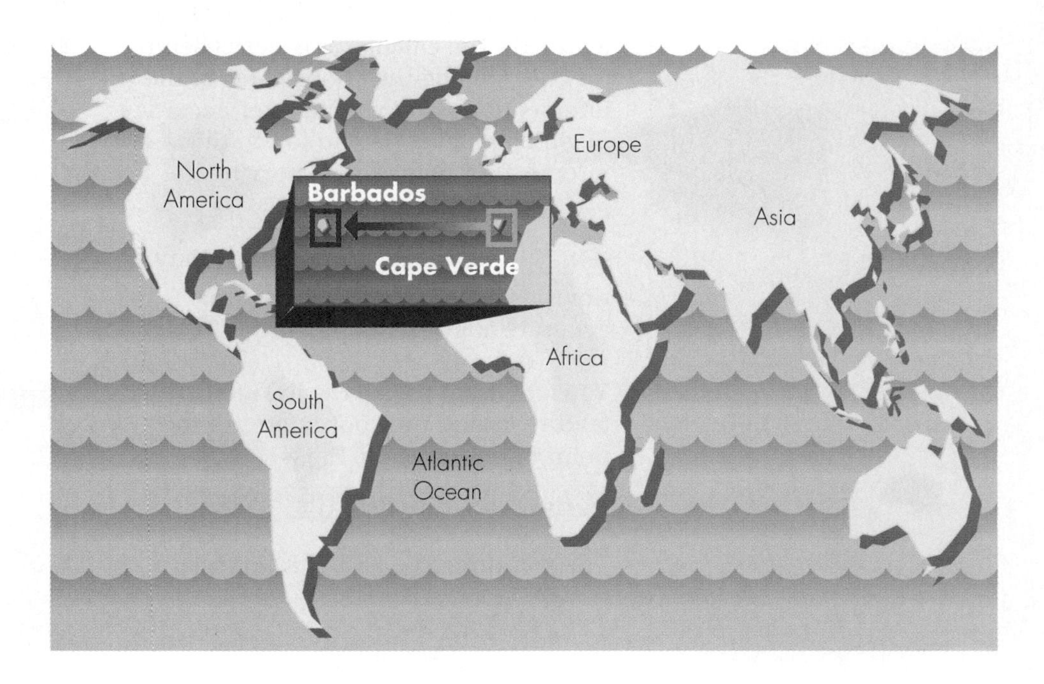

People are already considering trying to break Guy's record, but before a challenger makes the attempt, he or she should know what is required. Guy traveled approximately 44 miles per day. A good swimmer like Guy can swim about 3 miles per hour for an extended period, and the Atlantic currents will float a raft about 1 mile per hour. To match Guy's record, how many hours per day would one have to swim? How many hours would one be able to spend floating on the raft?

Fastest swimmers to cross the English Channel (hr:min)

1. Penny Lee Dean, 7:40
2. Philip Rush, 7:55
3. Richard Davey, 8:05
4. Irene van der Laan, 8:06
5. Paul Asmuth, 8:12

To solve this problem, let s represent the number of hours Guy swam, and let f represent the number of hours he floated. Then $3s$ represents the number of miles he traveled while swimming and $1f$ represents the number of miles he traveled while floating. You can write two equations to represent this situation.

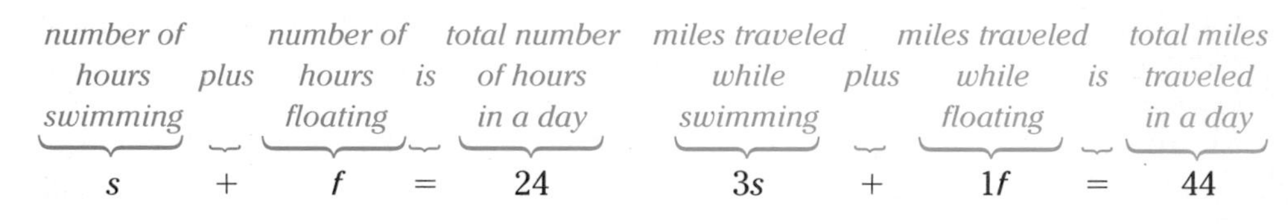

$$s + f = 24 \qquad 3s + 1f = 44$$

The equations $s + f = 24$ and $3s + f = 44$ together are called a **system of equations**. The solution to this problem is the ordered pair of numbers that satisfies both of these equations.

One method for solving a system of equations is to carefully graph the equations on the same coordinate plane. The coordinates of the point at which the graphs intersect is the solution of the system.

Guy Delage's raft

With most graphs of systems of equations, we can only estimate the solution. In this case, the graphs of $s + f = 24$ and $3s + f = 44$ appear to intersect at the point with coordinates (10, 14).

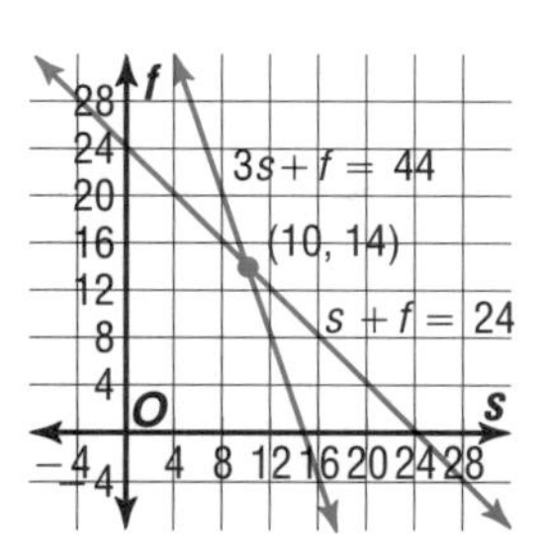

Check: In each equation, replace s with 10 and f with 14.

$$s + f = 24 \qquad 3s + f = 44$$

$$10 + 14 \stackrel{?}{=} 24 \qquad 3(10) + 14 \stackrel{?}{=} 44$$

$$24 = 24 \ ✔ \qquad 44 = 44 \ ✔$$

The solution of the system of equations $s + f = 24$ and $3s + f = 44$ is (10, 14). The ordered pair (10, 14) means that a person trying to match Guy Delage's record would have to spend approximately 10 hours a day swimming and 14 hours floating.

Example 1

Graph the system of equations to find the solution.

$$x + 2y = 1$$
$$2x + y = 5$$

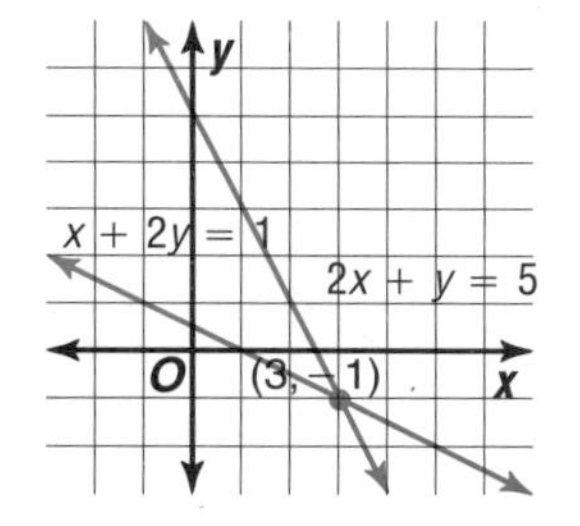

The graphs appear to intersect at the point with coordinates (3, −1). Check this estimate by replacing x with 3 and y with −1 in each equation.

Check: $x + 2y = 1 \qquad 2x + y = 5$

$$3 + 2(-1) \stackrel{?}{=} 1 \qquad 2(3) + (-1) \stackrel{?}{=} 5$$

$$1 = 1 \ ✔ \qquad 5 = 5 \ ✔$$

The solution is (3, −1).

CAREER CHOICES

Graphing systems of equations often arises in the study of populations of groups, and is used by people with careers in biological science, such as **ecologists**. An ecologist uses graphs of systems to study populations of organisms and how they relate to their environment.

A career as an ecologist usually requires a Ph.D. in a biological science and several years of laboratory work.

For more information, contact:

Ecological Society of America
2010 Massachusetts Ave.
Suite 400
Washington, D.C. 20036

A system of two linear equations has exactly one ordered pair as its solution when the graphs of the equations intersect at exactly one point. If the graphs coincide, they are the same line and have infinitely many points in common. In either case, the system of equations is said to be **consistent**. That is, it has *at least* one ordered pair that satisfies both equations.

It is also possible for the two graphs to be *parallel*. In this case, the system of equations is **inconsistent** because there is *no* ordered pair that satisfies both equations.

Another way to classify a system is by the number of solutions it has.

- If a system has exactly one solution, it is **independent**.
- If a system has an infinite number of solutions, it is **dependent**.

Thus, the system in Example 1 is said to be *consistent and independent*.

The chart below summarizes the possible solutions to systems of linear equations.

Graphs of Equations	Number of Solutions	Terminology
intersecting lines	exactly one	consistent and independent
same line	infinitely many	consistent and dependent
parallel lines	none	inconsistent

Example 2 Graph each system of equations to determine the number of solutions.

a. $x + y = 4$
$x + y = 1$

The graphs of the equations are parallel lines. Since they do not intersect, there is no solution to this system of equations. Notice that the two lines have the same slope but different y-intercepts.

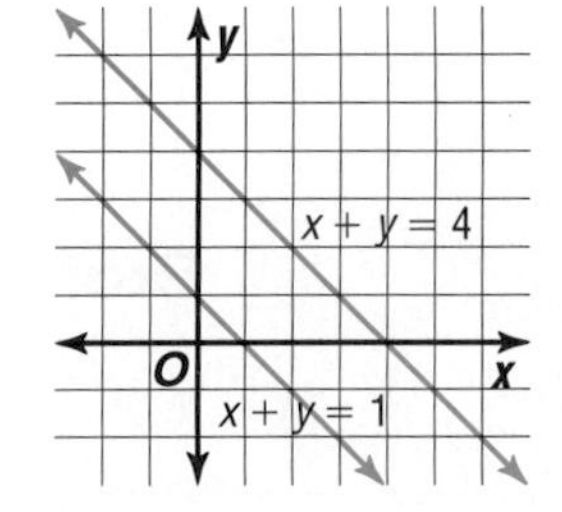

Recall that a system of equations that has no solution is said to be *inconsistent*.

b. $x - y = 3$
$2x - 2y = 6$

Each equation has the same graph. Any ordered pair on the graph will satisfy both equations. Therefore, there are infinitely many solutions of this system of equations. Notice that the graphs have the same slope and intercepts.

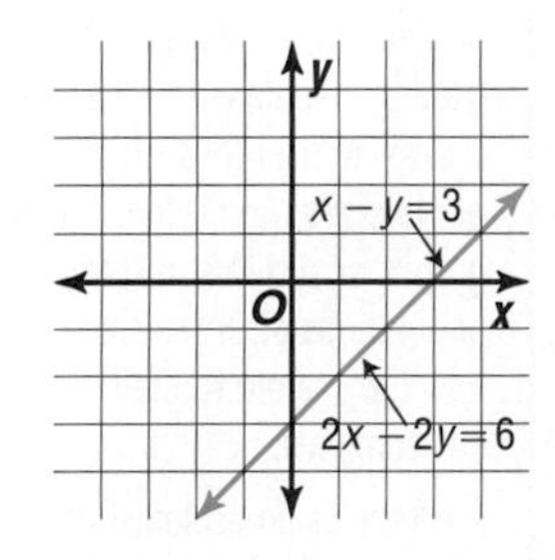

Recall that a system of equations that has infinitely many solutions is said to be *consistent and dependent*.

Check: Verify that the point at (4, 1) lies on both lines.

$x - y = 3$ $\qquad$ $2x - 2y = 6$

$4 - 1 \stackrel{?}{=} 3$ $\qquad$ $2(4) - 2(1) \stackrel{?}{=} 6$

$3 = 3$ ✓ $\qquad$ $6 = 6$ ✓

The methods you use to solve algebra problems are often useful in solving problems involving geometry.

Example 3 The points $A(-1, 6)$, $B(4, 8)$, $C(8, 3)$ and $D(-2, -1)$ are vertices of a quadrilateral.

Geometry

a. Use a graph to determine the point of intersection of the diagonals of quadrilateral *ABCD*.

b. Find the equations of the lines containing the diagonals to verify the solution.

a. Draw quadrilateral *ABCD* with diagonals $\overline{AC}$ and $\overline{BD}$. The diagonals appear to intersect at the point (2, 5).

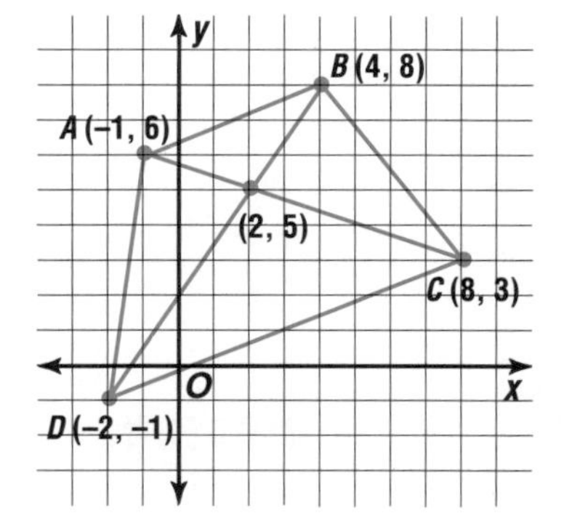

b. To check the solution, find the equations of lines *AC* and *BD* and then verify that (2, 5) is a solution of both equations. First, find the slope of each line using

$m = \frac{y_2 - y_1}{x_2 - x_1}$.

Slope of $\overleftrightarrow{AC}$

$m = \frac{3 - 6}{8 - (-1)}$

$= \frac{-3}{9}$ or $-\frac{1}{3}$

Slope of $\overleftrightarrow{BD}$

$m = \frac{-1 - 8}{-2 - 4}$

$= \frac{-9}{-6}$ or $\frac{3}{2}$

Then use the slope-intercept form, $y = mx + b$, to determine the equations.

Equation for $\overleftrightarrow{AC}$

$y = mx + b$

$6 = -\frac{1}{3}(-1) + b$ *Replace m with $-\frac{1}{3}$ and (x, y) with (−1, 6).*

$\frac{17}{3} = b$ The equation for $\overleftrightarrow{AC}$ is $y = -\frac{1}{3}x + \frac{17}{3}$.

Equation for $\overleftrightarrow{BD}$

$y = mx + b$

$8 = \frac{3}{2}(4) + b$ *Replace m with $\frac{3}{2}$ and (x, y) with (4, 8).*

$2 = b$ The equation for $\overleftrightarrow{BD}$ is $y = \frac{3}{2}x + 2$.

Check that (2, 5) is a solution to both equations.

$y = -\frac{1}{3}x + \frac{17}{3}$

$5 \stackrel{?}{=} -\frac{1}{3}(2) + \frac{17}{3}$ *(x, y) = (2, 5)*

$5 = 5$ ✓

$y = \frac{3}{2}x + 2$

$5 \stackrel{?}{=} \frac{3}{2}(2) + 2$ *(x, y) = (2, 5)*

$5 = 5$ ✓

The solution checks.

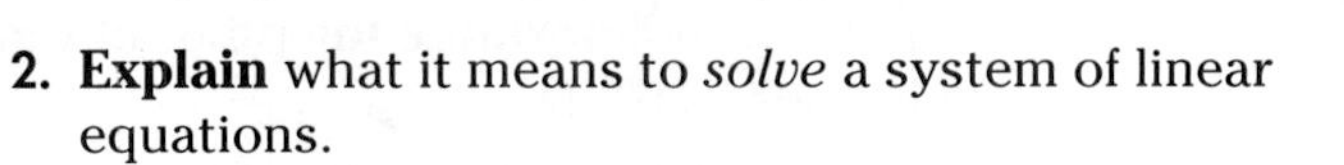

CHECK FOR UNDERSTANDING

Communicating Mathematics

Study the lesson. Then complete the following.

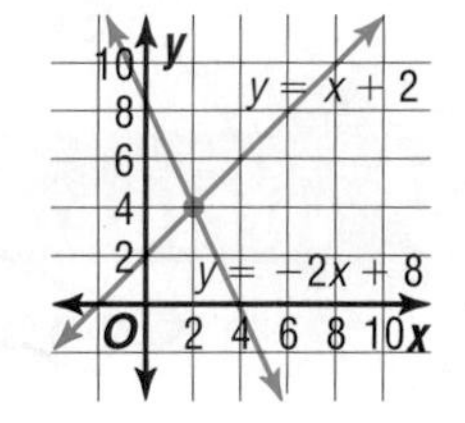

1. **State** the solution of the system of equations shown in the graph at the right. Justify your answer.
2. **Explain** what it means to *solve* a system of linear equations.
3. **Describe** the graph of a linear system that has infinitely many solutions.
4. **Name** two of the solutions for the system of equations in Example 2b. Verify your answers algebraically.
5. **Write** a system of linear equations that has $(-3, 5)$ as its only solution.
6. **Sketch** the graph of a linear system that has *no* solution.

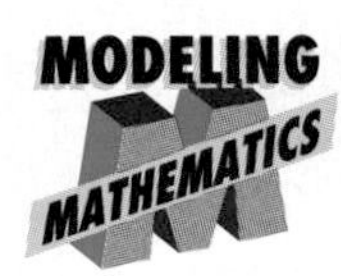

7. Use a geoboard and rubber bands to model a system of two equations that has the solution (3, 2). Let the lower left point on the geoboard represent the origin.

Guided Practice

Use the graphs at the right to determine whether each system has *one* solution, *no* solution, or *infinitely many* solutions. If the system has one solution, name it.

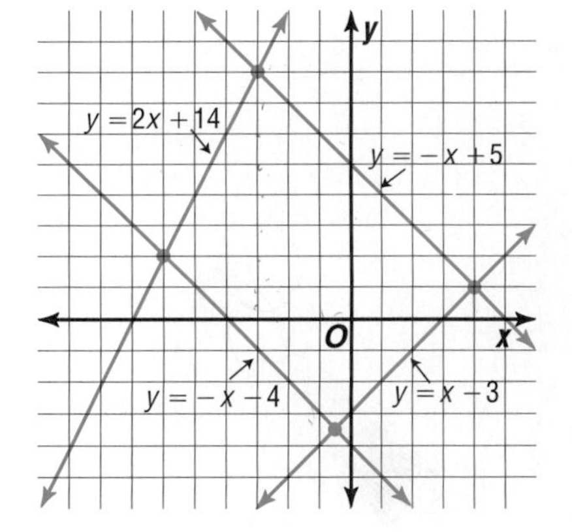

8. $y = -x + 5$
   $y = x - 3$
9. $y = -x - 4$
   $y = -x + 5$
10. $y = 2x + 14$
    $y = -x + 5$
11. $y = -x - 4$
    $y = 2x + 14$

State whether the given ordered pair is a solution to each system. Write *yes* or *no*.

12. $x - y = 6$
    $2x + y = 0$ $(-2, -4)$
13. $2x - y = 4$
    $3x + y = 1$ $(1, -2)$

Graph each system of equations. Then determine whether the system has *one* solution, *no* solution, or *infinitely many* solutions. If the system has one solution, name it.

14. $y = 3x - 4$
    $y = -3x - 4$
15. $y = -x + 8$
    $y = 4x - 7$
16. $x + 2y = 5$
    $2x + 4y = 2$
17. $y = -6$
    $4x + y = 2$
18. $2x + 3y = 4$
    $-4x - 6y = -8$
19. $2x + y = -4$
    $5x + 3y = -6$
20. a. Graph the line $y - x = 6$.
    b. Slide the entire line four units to the right and down one unit. Draw the new line.
    c. Describe this system of equations.

EXERCISES

Practice

Use the graphs below to determine whether each system has *one* solution, *no* solution, or *infinitely many* solutions. If the system has one solution, name it.

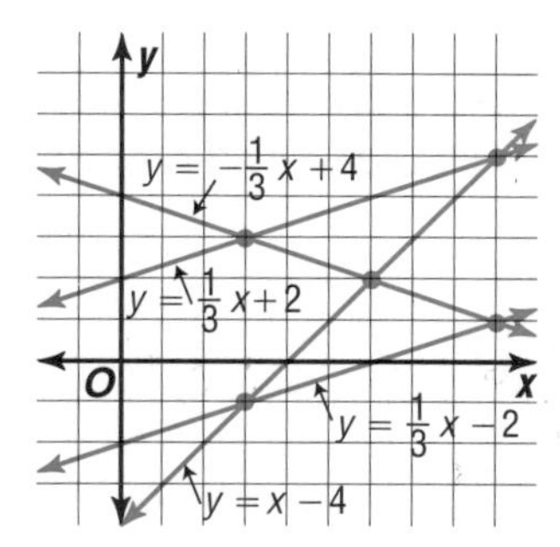

21. $y = x - 4$
$y = \frac{1}{3}x - 2$

22. $y = x - 4$
$y = -\frac{1}{3}x + 4$

23. $y = \frac{1}{3}x + 2$
$y = \frac{1}{3}x - 2$

24. $y = x - 4$
$y = \frac{1}{3}x + 2$

25. $y = -\frac{1}{3}x + 4$
$y = \frac{1}{3}x + 2$

26. $y = \frac{1}{3}x - 2$
$y = -\frac{1}{3}x + 4$

Graph each system of equations. Then determine whether the system has *one* solution, *no* solution, or *infinitely many* solutions. If the system has one solution, name it.

27. $y = -x$
$y = 2x - 6$

28. $y = 2x + 6$
$y = -x - 3$

29. $x + y = 2$
$y = 4x + 7$

30. $2x + y = 10$
$y = \frac{1}{2}x$

31. $x + y = 2$
$2y - x = 10$

32. $3x + 2y = 12$
$3x + 2y = 6$

33. $x - 2y = 2$
$3x + y = 6$

34. $x - y = 2$
$3y + 2x = 9$

35. $3x + y = 3$
$2y = -6x + 6$

36. $2x + 3y = -17$
$y = x - 4$

37. $y = \frac{2}{3}x - 5$
$3y = 2x$

38. $4x + 3y = 24$
$5x - 8y = -17$

39. $\frac{1}{2}x + \frac{1}{3}y = 6$
$y = \frac{1}{2}x + 2$

40. $6 - \frac{3}{8}y = x$
$\frac{2}{3}x + \frac{1}{4}y = 4$

41. $2x + 4y = 2$
$3x + 6y = 3$

Geometry

42. The graphs of the equations $-x + 2y = 6$, $7x + y = 3$, and $2x + y = 8$ contain the sides of a triangle. Find the coordinates of the vertices of the triangle.

43. Graph the system of equations below. Then find the area of the geometric figure.
$2x - 4 = 0$
$y = 8$
$x = 5$
$3y - 9 = 0$

Graphing Calculator

Use a graphing calculator to solve each system of equations. Approximate the coordinates of the point of intersection to the nearest hundredth.

44. $y = x + 2$
$y = -x - 1$

45. $y = \frac{1}{4}x - 3$
$y = -\frac{1}{3}x - 2$

46. $6x + y = 5$
$y = 9 + 3x$

47. $3 + y = x$
$2 + y = 5x$

Critical Thinking

48. If (0, 0) and (2, 2) are known to be solutions of a system of two linear equations, does the system have any other solutions? Justify your answers.

49. The solution to the system of equations $Ax + y = 5$ and $Ax + By = 7$ is $(-1, 2)$. What are the values of A and B?

Applications and Problem Solving

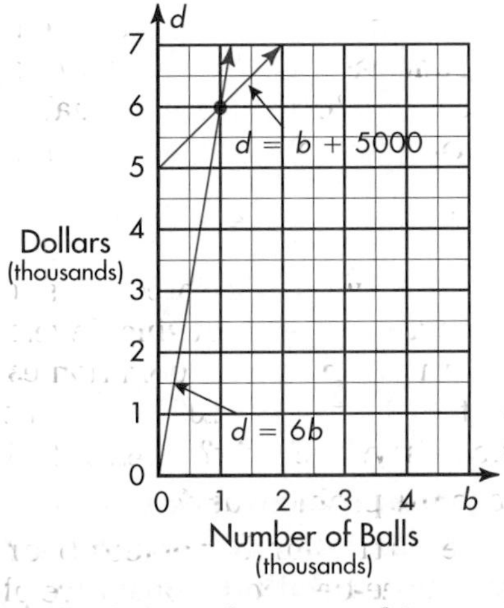

50. **Business** Mary Rodas is an 18-year-old toy specialist who tests and evaluates products at Catco, Inc., a company in New York City. She also helps design new toys, such as the Balzac Balloon Balls. Suppose the income from Balzac Balloon Balls is represented by the equation $d = 6b$ and the expenses are represented by the equation $d = b + 5000$. In both equations, b is the number of balls, and d is the number of dollars. Use the graph at the left to answer the following questions.

   a. Find the solution to this system of equations. This solution is called the *break-even point*. What does this point represent?

   b. A profit is made if income is greater than expenses. When is a profit made from the toys? How can you tell this from the graph?

   c. Money is lost if expenses are greater than income. When is money lost from the Balzac Balloon Balls? How can you tell this from the graph?

FYI

Gupta doctors developed plastic surgery. Metallurgists made iron columns that are still free from rust after more than 1500 years. Gupta mathematicians developed a system of numbers that was later adopted by the Arabs.

51. **World Cultures** The Golden Age of India was during the expansion of the Gupta Empire, beginning in A.D. 320. India became a center of art, medicine, science, and mathematics. Suppose $P = \frac{1}{2}t + 22$ represents the percent of Indian people in the Gupta Empire, at time t. Let $P = -\frac{1}{2}t + 78$ represent the percent of Indian people that were not Guptas. Graph the system of equations and estimate the year in which the percent of Guptas equaled the percent of Indians that were not Gupta. (*Hint:* Let $t = 0$ correspond to A.D. 320.)

Mixed Review

52. Graph $y - 7 > 3x$. (Lesson 7–8)

53. Solve $|2m + 15| = 12$. (Lesson 7–6)

54. Solve $10p - 14 < 8p - 17$. (Lesson 7–3)

55. Write an equation for the line that passes through the point at $(2, -2)$ and is parallel to $y = -2x + 21$. (Lesson 6–6)

56. Statistics Find the range, median, upper and lower quartiles, and interquartile range of the data in the stem-and-leaf plot at the right. (Lesson 5–7)

Stem	Leaf
43	3 5 6 6 9
44	1 4 4 4 9 9
45	0 2 7 7 8
46	5 7 $44 \mid 9 = 449$

57. Finance Patricia invested \$5000 for one year. Martin also invested \$5000 for one year. Martin's account earned interest at a rate of 10% per year. At the end of the year, Martin's account had earned \$125 more than Patricia's account. What was the annual interest rate on Patricia's account? (Lesson 4–4)

58. Solve $\frac{a-x}{-3} = \frac{-2}{b}$ for x. (Lesson 3–6)

59. Architecture Answer the related questions for the verbal problem below. A developer is designing a housing development. She proposes to have four times as many three-bedroom homes as four-bedroom homes. If the development is planned for 100 homes, how many three- and four-bedroom homes will be built? (Lesson 2–9)

a. What does the problem ask?

b. If h represents the number of four-bedroom homes that are planned, how many three-bedroom homes are planned?

c. If 20 four-bedroom homes are planned, how many three-bedroom homes should be built?

60. Evaluate $\frac{6ab}{3x + 2y}$ if $a = 6$, $b = 4$, $x = 0.2$, and $y = 1.3$. (Lesson 1–3)

WORKING ON THE In·ves·ti·ga·tion

Refer to the Investigation on pages 448–449.

Ready, Set, Drop!

Your research team determines that hang gliders are not just dropped from a point as you did with your tissue paper triangles. They are always launched into forward motion before gliding. The team decides that scale models are needed in order to get a feel for the launching and landing aspects of a real hang glider.

1 Each team in your class will construct a hang glider model using tissue paper and wire. A table top will act as the top of the cliff from which the glider is to be launched.

2 Each team should discuss different types of methods for launching their hang glider models from the table top. They should present their ideas to the class, and the class should agree upon which method they prefer to use. Then each team tests their glider using the method that the class has chosen.

3 Launch the glider 10 times. For each trial, measure the horizontal distance (along the floor) from the table to the spot at which the glider lands. Record this measurement and the height of the launch site.

4 Use these data to write a linear equation that describes the path of your glider. What is the slope of the path for your glider?

5 Using the linear equations from each of the other teams' data, would your glider collide with any of the other teams' gliders if they were launched at the same time from cliffs that are opposite each other? Write a detailed report on your conclusions.

Add the results of your work to your Investigation Folder.

8-2 Substitution

What YOU'LL LEARN

- To solve systems of equations by using the substitution method, and
- to organize data to solve problems.

Why IT'S IMPORTANT

You can use systems of equations to solve problems involving geography and accounting.

FYI

California has 12% of the entire U.S. population.

Alaska, the largest state in the U.S., has the second smallest population.

Geography

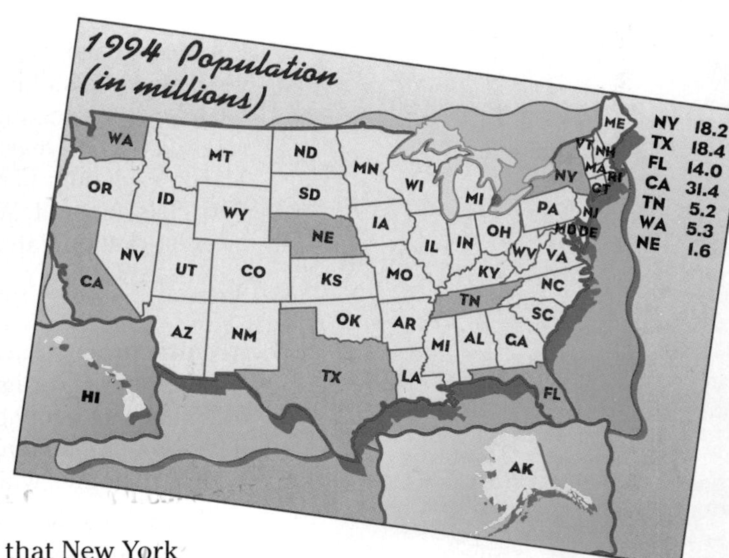

A recent article in *USA Today* reported that New York lost its position as the second most populous state when Texas slipped into the No. 2 spot at the end of 1994. Census Bureau projections show that New York will likely be pushed even further down the population ladder when Florida catches up early in the 21st century.

If New York's population grows at a constant rate of 0.02 million people per year and Florida's population grows at a constant rate of 0.26 million people per year, when would Florida catch up to New York in population? What would their populations be then?

Let P represent the population in millions, and let t represent the amount of time in years. The information above can be described by the following system of equations.

$$P = 18.2 + 0.02t$$
$$P = 14 + 0.26t$$

You could try to solve this system of equations by graphing, as shown at the right. Notice that the *exact* coordinates of the point where the lines intersect cannot be easily determined from this graph. An estimate is (18, 18).

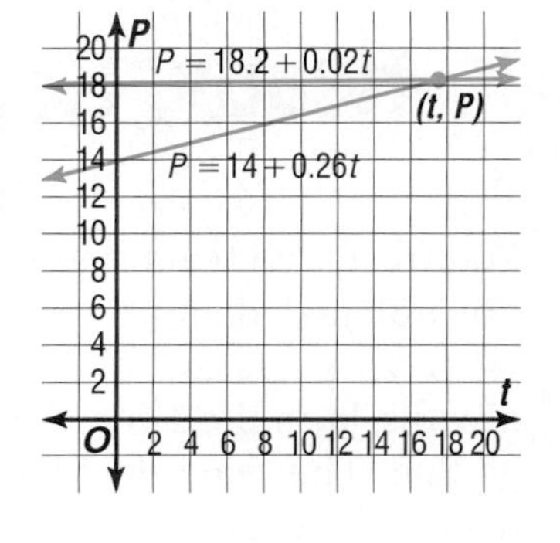

The exact solution of this system of equations can be found by using algebraic methods. One such method is called **substitution**.

From the first equation in the system, $P = 18.2 + 0.02t$, you know that P is equal to $18.2 + 0.02t$. Since P must have the same value in *both* equations, you can substitute $18.2 + 0.02t$ for P in the second equation $P = 14 + 0.26t$.

$$
\begin{aligned}
P &= 14 + 0.26t \\
18.2 + 0.02t &= 14 + 0.26t \\
0.02t &= -4.2 + 0.26t \\
-0.24t &= -4.2 \\
t &= 17.5
\end{aligned}
$$

Substitute 18.2 + 0.02t for P so the equation will have only one variable.

Now find the value of P by substituting 17.5 for t in either equation.

$$P = 18.2 + 0.02t$$
$$= 18.2 + 0.02(17.5)$$ *You could also substitute 17.5 for t in P = 14 + 0.26t.*
$$= 18.55$$

Check: In each equation, replace t with 17.5 and P with 18.55.

$$P = 18.2 + 0.02t \qquad P = 14 + 0.26t$$
$$18.55 \stackrel{?}{=} 18.2 + 0.02(17.5) \qquad 18.55 \stackrel{?}{=} 14 + 0.26(17.5)$$
$$18.55 = 18.55 \; \checkmark \qquad 18.55 = 18.55 \; \checkmark$$

The solution of the system of equations is (17.5, 18.55). Therefore after 17.5 years, or in 2011, the populations of New York and Florida would both be 18.55 million. *Compare this result to the estimate we obtained from the graph.*

You can use substitution to solve systems of equations even when the equations are more complex.

Example 1 **Use substitution to solve each system of equations.**

a. $\mathbf{x + 4y = 1}$
$\mathbf{2x - 3y = -9}$

Solve the first equation for x since the coefficient of x is 1.

$$x + 4y = 1$$
$$x = 1 - 4y$$

Next, find the value of y by substituting $1 - 4y$ for x in the second equation.

$$2x - 3y = -9$$
$$2(1 - 4y) - 3y = -9$$
$$2 - 8y - 3y = -9$$
$$-11y = -11$$
$$y = 1$$

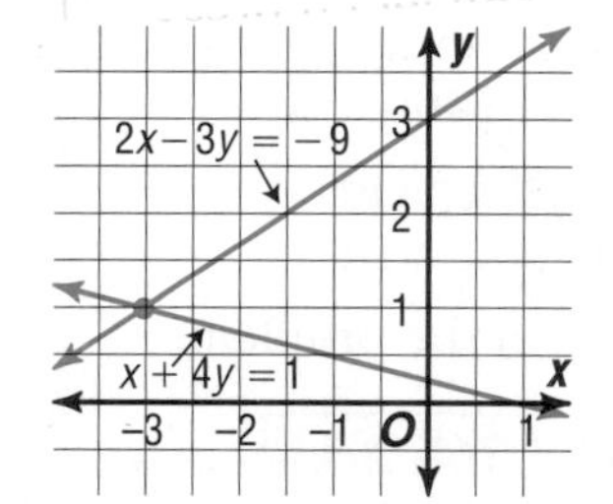

Then substitute 1 for y in either of the original equations and find the value of x. *Choose the equation that is easier for you to solve.*

$$x + 4y = 1$$
$$x + 4(1) = 1$$
$$x + 4 = 1$$
$$x = -3$$

The solution of this system is $(-3, 1)$. *Use the graph at the left to verify this result.*

b. $\mathbf{\frac{5}{2}x + y = 4}$

$\mathbf{5x + 2y = 8}$

Solve the first equation for y since the coefficient of y is 1.

$$\frac{5}{2}x + y = 4$$
$$y = 4 - \frac{5}{2}x$$

Next, find the value of x by substituting $4 - \frac{5}{2}x$ for y in the second equation.

$$5x + 2y = 8$$
$$5x + 2\left(4 - \frac{5}{2}x\right) = 8$$
$$5x + 8 - 5x = 8$$
$$8 = 8$$

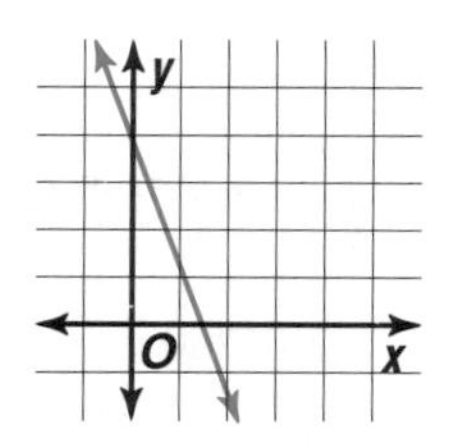

The statement $8 = 8$ is true. This means that there are infinitely many solutions to the system of equations. This is true because the slope intercept form of both equations is $y = 4 - \frac{5}{2}x$. That is, the equations are equivalent, and both have the same graph.

In general, if you solve a system of linear equations and the result is a true statement (an identity such as $8 = 8$), the system has an infinite number of solutions; if the result is a false statement (for example, $8 = 12$), the system has no solution.

MODELING MATHEMATICS — Systems of Equations

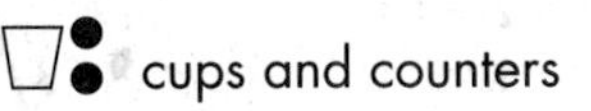

Materials: cups and counters, equation mat

Use a model to solve the system of equations.

$4x + 3y = 8$

$y = x - 2$

Your Turn

a. Let a cup represent the unknown value x. If $y = x - 2$, how can you represent y?

b. Represent $4x + 3y = 8$ on the equation mat. On one side of the mat, place four cups to represent $4x$ and three representations of y from step a. On the other side of the mat, place eight positive counters.

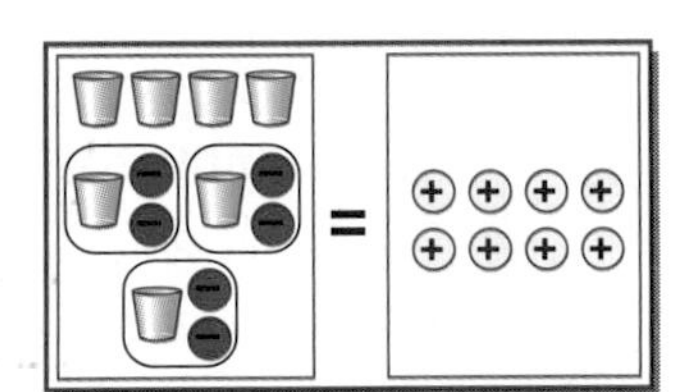

c. Use what you know about equation mats and zero pairs to solve the equation. What value of x is the solution of the system of equations?

d. Use the value of x from step c and the equation $y = x - 2$ to find the value of y.

e. What is the solution of the system of equations?

Sometimes it is helpful to **organize data** before solving a problem. Some ways to organize data are to use tables, charts, different types of graphs, or diagrams.

Example 2 **PROBLEM SOLVING** **Organize Data**

EJH Labs needs to make 1000 gallons of a 34% acid solution. The only solutions available are 25% acid and 50% acid. How many gallons of each solution should be mixed to make the 34% solution?

Explore Let a represent the number of gallons of 25% acid. Let b represent the number of gallons of 50% acid.

Make a table to organize the information in the problem.

	25% Acid	50% Acid	34% Acid
Total Gallons	a	b	1000
Gallons of Acid	$0.25a$	$0.50b$	$0.34(1000)$

Plan The system of equations is $a + b = 1000$ and $0.25a + 0.50b = 0.34(1000)$. Use substitution to solve this system.

Solve Since $a + b = 1000$, $a = 1000 - b$.

$$0.25a + 0.50b = 0.34(1000)$$

$$0.25(1000 - b) + 0.50b = 340 \quad \textit{Substitute } 1000 - b \textit{ for } a.$$

$$250 - 0.25b + 0.50b = 340 \quad \textit{Solve for } b.$$

$$0.25b = 90$$

$$b = 360$$

$$a + b = 1000$$

$$a + 360 = 1000 \quad \textit{Substitute 360 for } b.$$

$$a = 640 \quad \textit{Solve for } a.$$

Thus, 640 gallons of the 25% acid solution and 360 gallons of the 50% acid solution should be used.

Examine The 34% acid solution contains $0.25(640) + 0.50(360) = 160 + 180$ or 340 gallons of acid. Since $0.34(1000) = 340$, the answer checks.

Systems of equations can be useful in representing real-life situations and solving real-life problems.

Example 3 **The Williams family is going to the Johnstown Summer Carnival. They have two ticket options, as shown in the table below.**

Ticket Option	Admission Price	Price Per Ride
A	\$5	30¢
B	\$3	80¢

a. Write an equation that represents the cost per person for each option.

b. Graph the equations and estimate a solution. Explain what the solution means.

c. Solve the system using substitution.

d. Write a short paragraph advising the Williams family which option to choose.

a. Let r represent the number of rides. The total cost C for each person will be the cost of admission plus the cost of the rides.

Option A: $C = 5 + 0.30r$ *The cost of the rides is the price per*

Option B: $C = 3 + 0.80r$ *ride × number of rides, r.*

b. We can estimate from the graph that the solution is about (4, 6). This means that when the number of rides equals 4, both ticket options cost about \$6 per person.

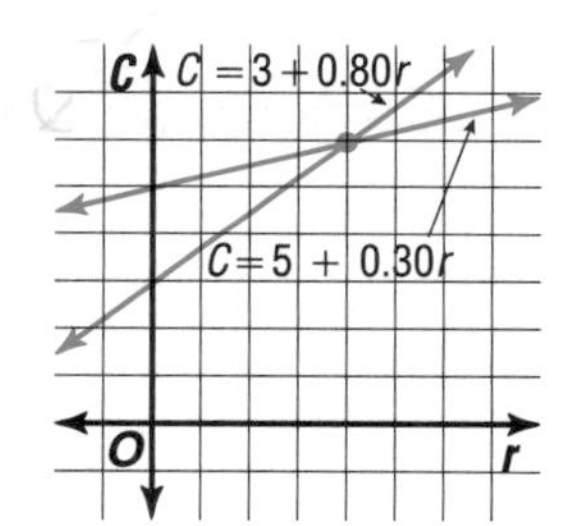

c. Use substitution to solve this system.

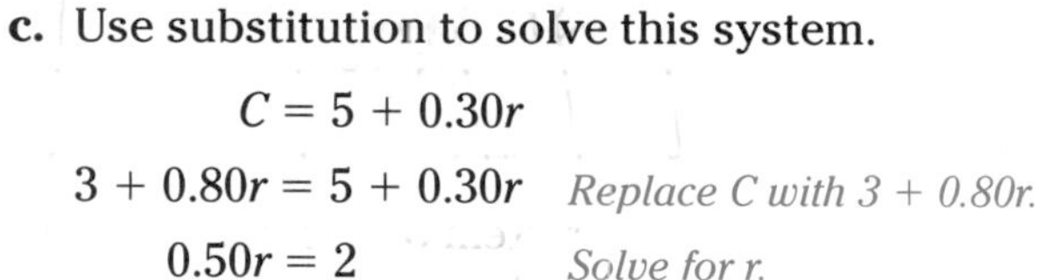

$$C = 5 + 0.30r$$

$$3 + 0.80r = 5 + 0.30r \quad \text{Replace } C \text{ with } 3 + 0.80r.$$

$$0.50r = 2 \quad \text{Solve for } r.$$

$$r = 4$$

$$C = 5 + 0.30r$$

$$= 5 + 0.30(4) \quad \text{Replace } r \text{ with 4.}$$

$$= 6.2 \quad \text{Solve for } C.$$

The solution is (4, 6.2). This means that if a person rides 4 rides, both options cost the same, \$6.20. From the graph, you can see that Option A tickets will cost less if a person rides more than 4 rides. Option B tickets will cost less if a person rides less than 4 rides.

d. You should advise the Williams family to purchase Option A tickets for those who plan to ride more than 4 rides and purchase Option B tickets for the rest of the family.

CHECK FOR UNDERSTANDING

Communicating Mathematics

Study the lesson. Then complete the following.

1. **Explain** why, when solving the system $y = 2x - 4$ and $4x - 2y = 0$, you can substitute $2x - 4$ for y in the second equation.
2. **State** what you would conclude if the solution to a system of linear equations yields the equation $8 = 0$.
3. **Describe** how you can tell just by looking at the equations $y = 9x + 2$ and $y = 9x - 5$ whether or not the system has a solution.
4. **Explain** why graphing a system of equations may not give you an exact solution.

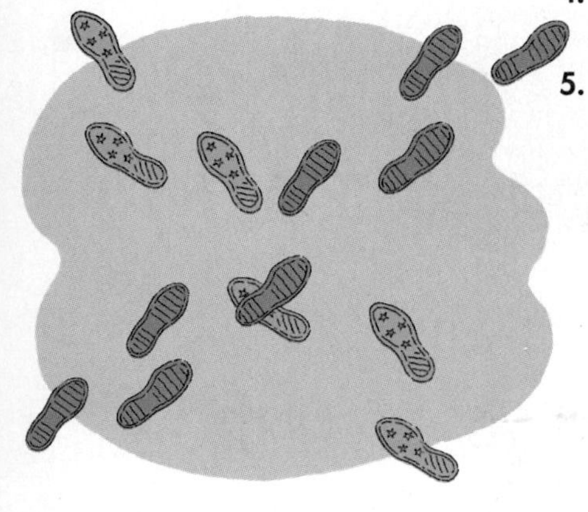

5. Yolanda is walking across campus when she sees Adele walking about 30 feet ahead of her. In each graph, t represents time in seconds and d represents distance in feet. Describe what happens in each case and how it relates to the solution.

a.

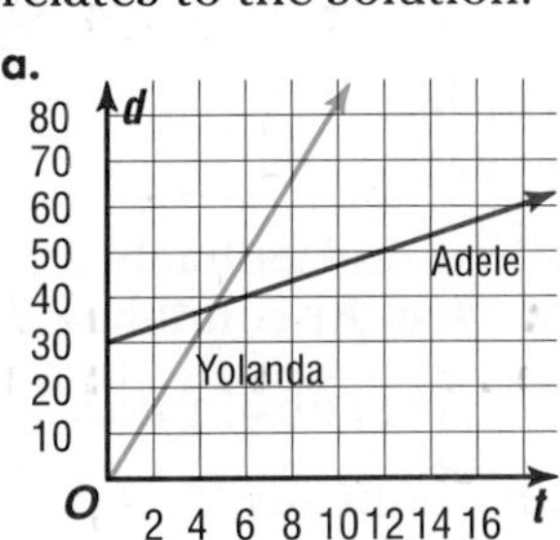

b.

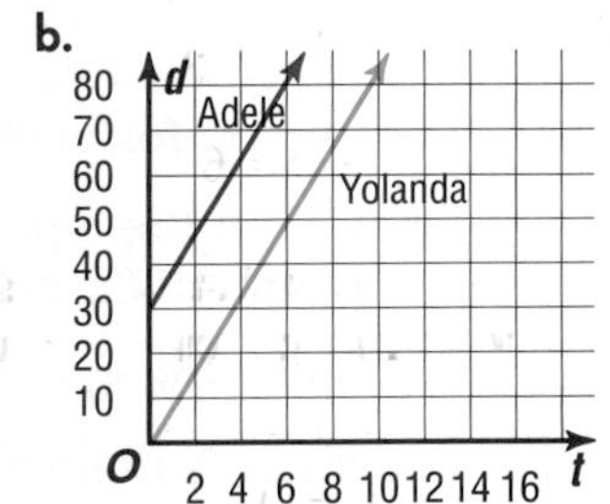

c.

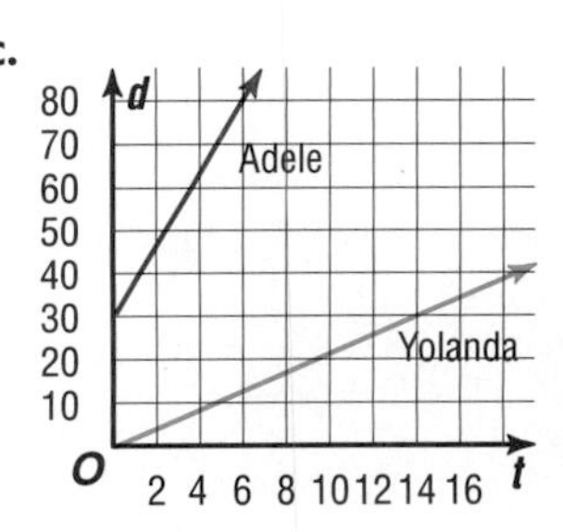

MODELING MATHEMATICS

6. Use cups and counters to model and solve the system of equations.

$y = 2x - 6$

$3x + 2y = 9$

Guided Practice

Solve each equation for *x*. Then, solve each equation for *y*.

7. $x + 4y = 8$
8. $3x - 5y = 12$
9. $0.8x + 6 = -0.75y$

Use substitution to solve each system of equations. If the system *does not* have exactly one solution, state whether it has *no* solution or *infinitely many* solutions.

10. $y = 3x$
    $x + 2y = -21$
11. $x = 2y$
    $4x + 2y = 15$
12. $x + 5y = -3$
    $3x - 2y = 8$
13. $8x + 2y = 13$
    $4x + y = 11$
14. $2x - y = -4$
    $-3x + y = -9$
15. $6x - 2y = -4$
    $y = 3x + 2$

16. **Sales** Maria spent a long day working the cash register at Musicville during a sale on CDs. For this sale, all CDs in the store were marked either \$12 or \$10. Just when she thought she could go home, the store manager gave Maria the job of figuring out how many CDs they had sold at each price, so they could write the total in the store records. Maria doesn't want to sort through hundreds of sales slips, so she decided on an easier way. The counter at the exit of the store says that 500 people left with CDs (limit one per customer) during the sale, and the cash register contains \$5750 from the day's sales. Maria wrote a system of equations for the number of \$10 CDs and the number of \$12 CDs.
    a. What was the system of equations?
    b. How many CDs were sold at each price?

EXERCISES

Practice

Use substitution to solve each system of equations. If the system *does not* have exactly one solution, state whether it has *no* solution or *infinitely many* solutions.

17. $y = 3x - 8$
$y = 4 - x$

18. $2x + 7y = 3$
$x = 1 - 4y$

19. $x + y = 0$
$3x + y = -8$

20. $4c = 3d + 3$
$c = d - 1$

21. $4x + 5y = 11$
$y = 3x - 13$

22. $3x - 5y = 11$
$x - 3y = 1$

23. $c - 5d = 2$
$2c + d = 4$

24. $3x - 2y = 12$
$x + 2y = 6$

25. $x + 3y = 12$
$x - y = 8$

26. $x - 3y = 0$
$3x + y = 7$

27. $5r - s = 5$
$-4r + 5s = 17$

28. $2x + 3y = 1$
$-3x + y = 15$

29. $8x + 6y = 44$
$x - 8y = -12$

30. $0.5x - 2y = 17$
$2x + y = 104$

31. $-0.3x + y = 0.5$
$0.5x - 0.3y = 1.9$

32. $x = \frac{1}{2}y + 3$
$2x - y = 6$

33. $y = \frac{1}{2}x + 3$
$y = 2x - 1$

34. $y = \frac{3}{5}x$
$3x - 5y = 15$

Use substitution to solve each system of equations. Write each solution as an ordered triple of the form (*x*, *y*, *z*).

35. $x + y + z = -54$
$x = -6y$
$z = 14y$

36. $2x + 3y - z = 17$
$y = -3z - 7$
$2x = z + 2$

37. $12x - y + 7z = 99$
$x + 2z = 2$
$y + 3z = 9$

Critical Thinking

38. **Number Theory** If 36 is subtracted from certain two-digit positive integers, their digits are reversed. Find all integers for which this is true.

Applications and Problem Solving

39. **Entertainment** American songwriter Cole Porter completed his first professional score in 1916 at age 23. At Harding High, this year's spring musical is *Anything Goes,* which Porter completed in 1934. The production is going to be part of a dinner theater; each ticket includes dinner and the show. The total cost of producing the show (stage, costumes, and so on) is \$1000, and each dinner costs \$5 to prepare. The drama club is going to sell tickets for \$13 each.

    a. Write a system of equations to represent the cost of and the income from the production.

    b. How many tickets do they need to sell to break even?

40. **Humor** Refer to the cartoon below. Solve the problem that is sending Peppermint Patty into a frenzy. Find how much cream and milk must be mixed together to obtain 50 gallons of cream containing $12\frac{1}{2}\%$ butterfat.

Peanuts®

41. **Athletes** According to *Health* magazine, top women athletes are narrowing the gap between their performances and those of their male counterparts. Speed skater Bonnie Blair's fastest time in the 500-meter would have won an Olympic gold medal in every men's 500-meter competition through 1976. The women's record time for the 500-meter in speed skating is 39.1 seconds, and the men's is 36.45 seconds. Suppose the women's record time decreases at an average rate of 0.20 second per year and the men's record time decreases at an average rate of 0.10 second per year.

    a. When would the women's record time equal the men's?

    b. What would the time be?

    c. Do you think this could actually happen? Why or why not?

42. **Accounting** Sometimes accountants must figure out how many stock shares to transfer from one person to another to reach a certain proportion of ownership. Suppose Rebeca Avila owns $3000 worth of stock in a new company that has no other stockholders. For tax purposes, the company is going to issue new stock to Muriel Eppick so that Ms. Avila owns 80%, rather than 100% of the total stock. Let S represent the new total value of company stock and let x represent the value of stock that Ms. Eppick is to receive. Use the equations below to find the value of stock to be issued to Ms. Eppick.

    $S = 3000 + x$ *New total stock = Ms. Avila's share + Ms. Eppick's share.*

    $3000 = 0.80S$ *Ms. Avila's share is 80% of new total stock.*

    $x = 0.20S$ *Ms. Eppick's share is 20% of new total stock.*

43. **Organize Data** For thousands of years, gold has been considered one of Earth's most precious metals. When archaeologist Howard Carter discovered King Tutankhamun's tomb in 1922, he exclaimed that the tomb was filled with "strange animals, statues, and gold—everywhere the glint of gold." One hundred percent pure gold is 24-carat gold. If 18-carat gold is 75% gold and 12-carat gold is 50% gold, how much of each would be used to make a 14-carat gold bracelet weighing 300 grams? (*Hint:* 14-carat gold is about 58% gold.)

    a. Make a table to organize the data.

    b. Write a system of equations that represents this problem.

    c. How much 18-carat gold and 12-carat gold would it take to make a 14-carat gold bracelet weighing 300 grams?

Mixed Review

44. Graph the system of equations below. Determine whether the system has *one* solution, *no* solutions, or *infinitely many* solutions. If the system has one solution, name it. (Lesson 8–1)

    $y = 2x + 1$

    $7y = 14x + 7$

45. **Finance** Michael uses at most 60% of his annual FlynnCo stock dividend to purchase more shares of FlynnCo stock. If his dividend last year was $885 and FlynnCo stock is selling for $14 per share, what is the greatest number of shares that he can purchase? (Lesson 7–2)

46. Graph $y = \frac{1}{5}x - 3$ using the slope and y-intercept. (Lesson 6–5)

47. Solve $3a - 4 = b$ if the domain is $\{-1, 4, 7, 13\}$. (Lesson 5–3)

48. What is 25% less than 94? (Lesson 4–5)

49. Solve $-8 - 12x = 28$. (Lesson 3–3)

50. Graph the solution set of $n \leq -2$ on a number line. (Lesson 2–8)

51. Write an algebraic expression for *twelve less than m*. (Lesson 1–1)

8-3 Elimination Using Addition and Subtraction

What **YOU'LL LEARN**

- To solve systems of equations by using the elimination method with addition or subtraction.

Why **IT'S IMPORTANT**

You can use systems of equations to solve problems involving entertainment and testing.

Entertainment

Disney cartoons are animated using an expensive computer process that makes the action flow smoothly and seem lifelike. In 1994, Disney's animated feature *The Lion King* was the top-grossing film of the year, making an estimated $300.4 million at the box office.

On a Saturday afternoon, the Johnson and Olivera families decided to go see *The Lion King* together. The Johnson family, two adults and four children, can afford to spend $30 from their entertainment budget this weekend for the movie tickets, while the Olivera family, two adults and two children, can afford to spend $21.50. Different theaters around town charge different amounts for adult and child tickets. What price can the Johnsons and Oliveras afford to pay for each adult and each child?

Let a represent the ticket price for one adult, and let c represent the ticket price for one child. Then the information in this problem can be represented by the following system of equations.

$$2a + 4c = 30$$
$$2a + 2c = 21.5$$

From the graph at the right, an estimated solution is ($7, $4). To get an exact solution, solve algebraically. You could solve this system by first solving either of the equations for a or c and then using substitution.

However, a simpler method of solution is to subtract one equation from the other since the coefficients of the variable a are the same. This method is called **elimination** because the subtraction eliminates one of the variables. First, write the equations in column form and subtract.

Recall that subtraction is the same as adding the opposite.

$$\begin{array}{r} 2a + 4c = 30 \\ \underline{(-)\ 2a + 2c = 21.5} \end{array} \quad \text{Multiply by } -1. \rightarrow \quad \begin{array}{r} 2a + 4c = 30 \\ \underline{(+)\ -2a - 2c = -21.5} \\ 2c = 8.5 \\ c = 4.25 \end{array}$$

Then, substitute 4.25 for c in either equation and find the value of a.

$$\begin{aligned} 2a + 2c &= 21.5 \\ 2a + 2(4.25) &= 21.5 \quad \textit{Substitute 4.25 for c.} \\ 2a + 8.5 &= 21.5 \\ 2a &= 13 \\ a &= 6.5 \end{aligned}$$

Is (6.5, 4.25) a solution of the system?

***fabulous* FIRSTS**

Elmer Simms Campbell (1906–1971)

Elmer Simms Campbell was the first African-American cartoonist to work for national publications. He contributed cartoons and other art work to *Esquire, Cosmopolitan,* and *Redbook,* as well as syndicated features in 145 newspapers.

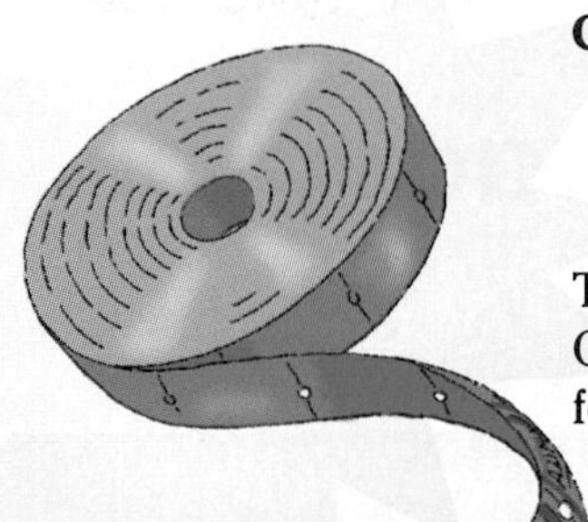

Check:

$$2a + 4c = 30$$
$$2(6.5) + 4(4.25) \stackrel{?}{=} 30$$
$$30 = 30 \checkmark$$

$$2a + 2c = 21.5$$
$$2(6.5) + 2(4.25) \stackrel{?}{=} 21.5$$
$$21.5 = 21.5 \checkmark$$

The solution of this system of equations is (6.5, 4.25). Thus, the Johnsons and Oliveras should look for a theater that charges \$6.50 for each adult and \$4.25 for each child.

In some systems of equations, the coefficients of terms containing the same variable are additive inverses. For these systems, the elimination method can be applied by adding the equations.

Example 1 **Use elimination to solve the system of equations.**

$\mathbf{3x - 2y = 4}$

$\mathbf{4x + 2y = 10}$

Since the coefficients of the y-terms, -2 and 2, are additive inverses, you can solve the system by adding the equations.

$$\begin{array}{rl} 3x - 2y = 4 & \textit{Write the equations in column form and add.} \\ (+)\ 4x + 2y = 10 & \textit{Notice that the variable y is eliminated.} \\ \hline 7x \qquad = 14 & \\ x = 2 & \end{array}$$

Now substitute 2 for x in either equation to find the value of y.

$$\begin{aligned} 3x - 2y &= 4 \\ 3(2) - 2y &= 4 \\ 6 - 2y &= 4 \\ -2y &= -2 \\ y &= 1 \end{aligned}$$

The solution of this system is (2, 1). *Check this result.*

Use subtraction to solve a system of two linear equations whenever one of the variables has the same coefficient in both equations.

Example 2 **The sum of two numbers is 18. The sum of the greater number and twice the smaller number is 25. Find the numbers.**

Number Theory

Let x = the greater number and let y = the lesser number. Since the sum of the numbers is 18, one equation is $x + y = 18$. Since the sum of the greater number and twice the smaller number is 25, the other equation is $x + 2y = 25$. Use elimination to solve this system.

$$\begin{array}{rl} x + \ y = 18 & \\ (-)\ x + 2y = 25 & \textit{Since the coefficients of the x terms are} \\ \hline -y = -7 & \textit{the same, use elimination by subtraction.} \\ y = 7 & \end{array}$$

Find x by substituting 7 for y in one of the equations.

$x + y = 18$

$x + 7 = 18$ *Substitute 7 for y.*

$x = 11$ *Solve for x.*

The solution is (11, 7), which means that the numbers are 7 and 11. *Check this result.*

Several software packages can be used to help you solve systems of equations.

EXPLORATION GRAPHING SOFTWARE

The *Mathematics Exploration Toolkit (MET)* can be used to graph and solve systems of equations. Use the following CALC commands.

CLEAR F (clr f)	Removes previous graphs from the graphing window.
GRAPH (gra)	Graphs the most recent equation in the expression window.
SCALE (sca)	Sets limits on the x- and y-axes.

To set up the graphing window, enter clr f. Then enter sca 10. This sets limits on the axes at -10 to 10 for x and y. Since only two points are needed to graph a line, use the command gra 2.

Your Turn

a. Check the solutions of the examples using *MET*.

b. Graph the system $3x + 2y = 7$ and $5x + 2y = 17$ and find the solution using *MET*.

c. Describe the difference between solving a system of equations using graphing software and using a graphing calculator. Which do you prefer, and why?

Example 3

Testing

Lina is preparing to take the Scholastic Assessment Test (SAT). She has been taking practice tests for a year, and her scores are steadily improving. She always scores about 150 points higher on the math test than on the verbal test. She needs a combined score of 1270 to get into the college she has chosen. If she assumes she will still have that 150-point difference between the two tests, how high does she need to score on each part?

Let m represent Lina's math score. Let v represent Lina's verbal score. Since the sum of her scores is 1270, one equation is $m + v = 1270$. Since the difference of her scores is 150, another equation is $m - v = 150$. Use elimination to solve this system.

$$\begin{array}{r} m + v = 1270 \\ (+)\ m - v = 150 \\ \hline 2m \quad = 1420 \\ m = 710 \end{array}$$

Since the coefficients of the v term are additive inverses, use elimination by addition.

Find v by substituting 710 for m in one of the equations.

$$\begin{aligned} m + v &= 1270 \\ 710 + v &= 1270 \quad \textit{Substitute 710 for m.} \\ v &= 560 \end{aligned}$$

The solution is (710, 560), which means that Lina must score 710 on the math portion and 560 on the verbal portion of the SAT.

CHECK FOR UNDERSTANDING

Communicating Mathematics

Study the lesson. Then complete the following.

1. **Explain** when it is easier to solve a system of equations in each way.
   a. by elimination using subtraction
   b. by elimination using addition

2. **a. State** the result when you add $3x - 8y = 29$ and $-3x + 8y = 16$. What does this result tell you about the system of equations?
   b. What does this result tell you about the graph of the system?

3. **You Decide** Maribela says that a system of equations has no solution if both variables are eliminated by addition or subtraction. Devin argues that there may be an infinite number of solutions. Who is correct? Explain your answer.

Guided Practice

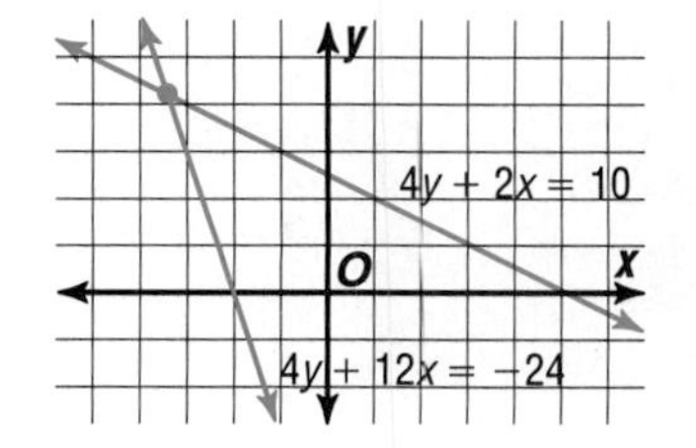

4. Refer to the graph at the right.
   a. Estimate the solution of the system.
   b. Use elimination to find the exact solution.

State whether addition, subtraction, or substitution would be most convenient to solve each system of equations. Then solve the system.

5. $3x - 5y = 3$ $4x + 5y = 4$	**6.** $3x + 2y = 7$ $y = 4x - 2$	**7.** $-4m + 2n = 6$ $-4m + n = 8$
8. $8a + b = 1$ $8a - 3b = 3$	**9.** $3x + y = 7$ $2x + 5y = 22$	**10.** $2b + 4c = 8$ $c - 2 = b$

11. **Statistics** The mean of two numbers is 28. Find the numbers if three times one of the numbers equals half the other number.

EXERCISES

Practice

For Exercises 12–14,
a. estimate the solution of each system of linear equations, and
b. use elimination to find the exact solution of each system.

12.

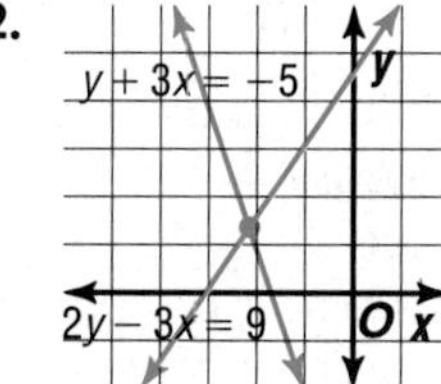

13.
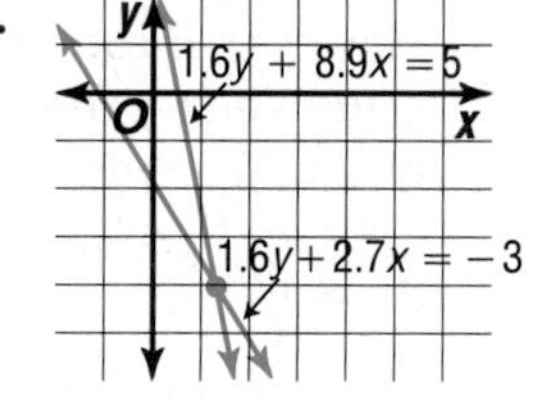

14.
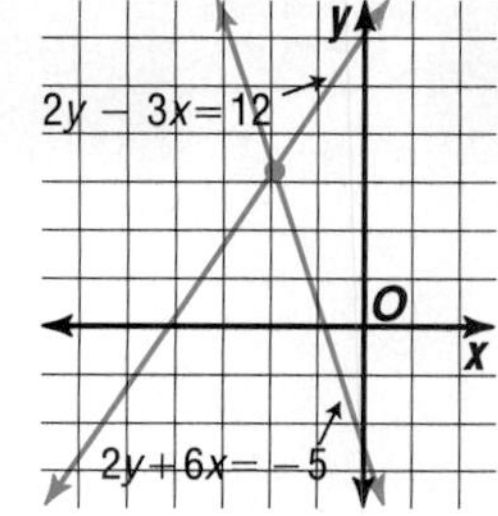

State whether addition, subtraction, or substitution would be most convenient to solve each system of equations. Then solve the system.

15. $x + y = 8$
    $x - y = 4$
16. $2r + s = 5$
    $r - s = 1$
17. $x - 3y = 7$
    $x + 2y = 2$
18. $3x + y = 5$
    $2x + y = 10$
19. $5s + 2t = 6$
    $9s + 2t = 22$
20. $4x - 3y = 12$
    $4x + 3y = 24$
21. $2x + 3y = 13$
    $x - 3y = 2$
22. $2m - 5n = -6$
    $2m - 7n = -14$
23. $x - 2y = 7$
    $-3x + 6y = -21$
24. $3r - 5s = -35$
    $2r - 5s = -30$
25. $13a + 5b = -11$
    $13a + 11b = 7$
26. $a - 2b - 5 = 0$
    $3a - 2b - 9 = 0$
27. $4x = 7 - 5y$
    $8x = 9 - 5y$
28. $\frac{2}{3}x + y = 7$
    $\frac{10}{3}x + 5y = 11$
29. $\frac{3}{5}c - \frac{1}{5}d = 9$
    $\frac{7}{5}c + \frac{1}{5}d = 11$
30. $0.6m - 0.2n = 0.9$
    $0.3m = 0.45 - 0.1n$
31. $1.44x - 3.24y = -5.58$
    $1.08x + 3.24y = 9.99$
32. $7.2m + 4.5n = 129.06$
    $7.2m + 6.7n = 136.54$

Number Theory

Use a system of equations and elimination to solve each problem.

33. Find two numbers whose sum is 64 and whose difference is 42.
34. Find two numbers whose sum is 18 and whose difference is 22.
35. Twice one number added to another number is 18. Four times the first number minus the other number is 12. Find the numbers.
36. If $x + y = 11$ and $x - y = 5$, what does xy equal?

Use elimination twice to solve each system of equations. Write the solution as an ordered triple of the form (x, y, z).

37. $x + y = 5$
    $y + z = 10$
    $x + z = 9$
38. $2x + y + z = 13$
    $x - y + 2z = 8$
    $4x - 3z = 7$
39. $x + 2z = 2$
    $y + 3z = 9$
    $12x - y + 7z = 99$

Critical Thinking

40. The graphs of $Ax + By = 7$ and $Ax - By = 9$ intersect at $(4, -1)$. Find A and B.

Applications and Problem Solving

41. **On-Line Entertainment** On June 27, 1994, Aerosmith became the first major rock band to release a song distributed exclusively in the U.S. through a computer on-line service. Users of the commercial service, CompuServe, were able to download the Aerosmith song *Head First* for free. However, it took a long time to download the song, which itself lasted only 3 minutes, 14 seconds, because of the high-fidelity sound. José and Ling share a personal computer, and one evening they each downloaded the song without realizing that the other had done it. Ling also wasted 18 minutes because he typed the wrong word and had to start over. At the end of the month, the bill from CompuServe said they had used a total of 2.6 hours of time that evening. How long did it take to download the song each time? (*Hint:* 18 minutes = 0.3 hour.)

FYI

The Great Pyramid of King Khufu, built around 2600 B.C., has only one entrance that is nearly impossible to find. Centuries later, however, robbers stole everything from the pyramid, including the mummy of the king.

42. **World Cultures** The ancient Egyptians believed that the pharaohs lived forever after death in their houses of eternity, the pyramids. Suppose the side of the pyramid containing the entrance is represented by the line $13x + 10y = 9600$, the opposite side of the pyramid, by the line $13x - 10y = 0$, and the descending corridor leading to the entrance, by the line $3x - 10y = 1500$, where x is the distance in feet and y is the height in feet.

    a. Find the coordinates of the entrance.

    b. Find the height of the pyramid.

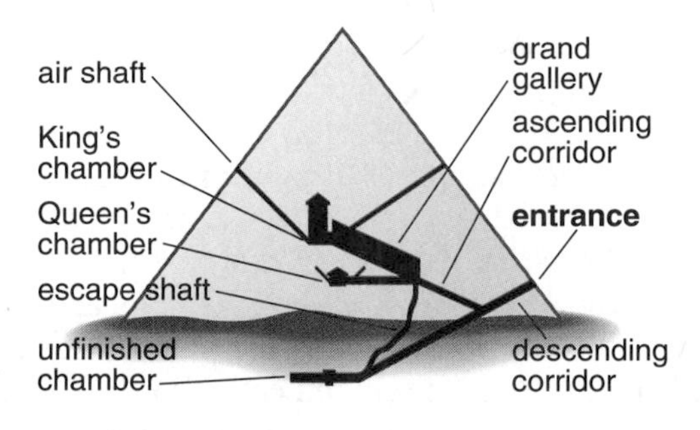

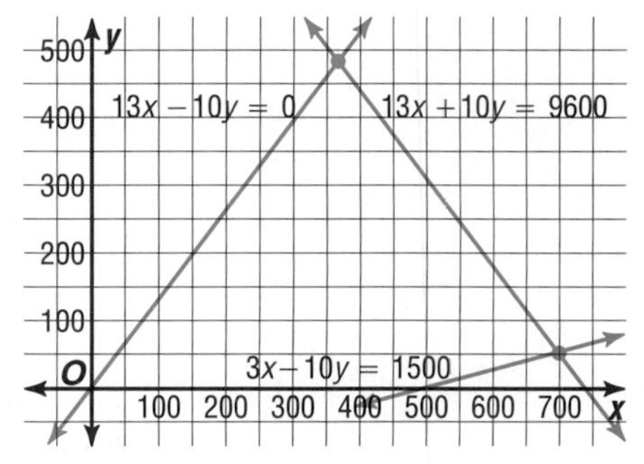

Mixed Review

43. **Chemistry** MX Labs needs to make 500 gallons of a 34% acid solution. The only solutions available are 25% acid and 50% acid. How many gallons of each solution should be mixed to make the 34% solution? Write and solve a system of equations by using substitution. (Lesson 8–2)

44. Solve $5 - 8h \leq 9$. (Lesson 7–1)

45. Determine the slope of the line that passes through the points at $(2, -9)$ and $(-1, 0)$. (Lesson 6–1)

46. Graph $6x - \frac{1}{2}y = -10$. (Lesson 5–4)

47. Solve $-2(3t + 1) = 5$. (Lesson 3–3)

48. Find an approximation to the nearest hundredth for $\sqrt{15}$. (Lesson 2–8)

49. Name the property illustrated by the following statement. (Lesson 1–6)

    If $6 = 2a$ *and* $a = 3$, *then* $6 = 2 \cdot 3$.

SELF TEST

Graph each system of equations. Then determine if the system has *one* solution, *no* solution, or *infinitely many* solutions. If the system has one solution, name it. (Lesson 8–1)

1. $x - y = 3$
   $3x + y = 1$

2. $2x - 3y = 7$
   $3y = 7 + 2x$

3. $4x + y = 12$
   $x = 3 - \frac{1}{4}y$

Use substitution to solve each system of equations. (Lesson 8–2)

4. $y = 5x$
   $x + 2y = 22$

5. $2y - x = -5$
   $y - 3x = 20$

6. $3x + 2y = 18$
   $x + \frac{8}{3}y = 12$

Use elimination to solve each system of equations. (Lesson 8–3)

7. $x - y = -5$
   $x + y = 25$

8. $3x + 5y = 14$
   $2x - 5y = 1$

9. $5x + 4y = 12$
   $3x + 4y = 4$

10. **Recreation** At a recreation and sports facility, 3 members and 3 nonmembers pay a total of \$180 to take an aerobics class. A group of 5 members and 3 nonmembers pay \$210 to take the same class. How much does it cost members and nonmembers to take an aerobics class? (Lesson 8–3)

8–4

Elimination Using Multiplication

What YOU'LL LEARN

- To solve systems of equations by using the elimination method with multiplication and addition, and
- to determine the best method for solving systems of equations.

Why IT'S IMPORTANT

You can use systems of equations to solve problems involving telecommunications and geography.

FYI

On January 14, 1876, Alexander Graham Bell beat Elisha Gray by only a few hours when filing a patent application for his telephone. Both men had invented workable prototypes simultaneously.

APPLICATION

Telecommunications

GBT Mobilnet provides monthly plans for cellular phone customers. Carla Ramos and Robert Johnson both selected Plan B for which monthly charges are based on per-minute rates of calls during peak and nonpeak hours. In one month, Carla made 75 minutes of peak calls and 30 minutes of nonpeak calls. Her bill was \$40.05. During the same period, Robert made 50 minutes of peak calls and 60 minutes of nonpeak calls. His bill was \$35.10. What is GBT Mobilnet's charge per minute for peak and nonpeak calls on Plan B?

Let p represent the rate per minute for peak calls, and let n represent the rate per minute for nonpeak calls. Then the information in this problem can be represented by the following system of equations.

$$75p + 30n = 40.05$$
$$50p + 60n = 35.10$$

So far, you have learned four methods for solving a system of two linear equations.

Method	The Best Time to Use
Graphing	if you want to estimate the solution, since graphing usually does not give an exact solution
Substitution	if one of the variables in either equation has a coefficient of 1 or −1
Addition	if one of the variables has opposite coefficients in the two equations
Subtraction	if one of the variables has the same coefficient in the two equations

The system above is not easily solved using any of these methods. However, there is an extension of the elimination method that can be used. Multiply one of the equations by some number so that adding or subtracting eliminates one of the variables.

For this system, multiply the first equation by −2 and add. Then the coefficient of n in both equations will be 60 or −60.

$75p + 30n = 40.05$ **Multiply by −2.** $\rightarrow$ $-150p - 60n = -80.10$

$50p + 60n = 35.10$ $\qquad\qquad (+)\ 50p + 60n = 35.10$

$$-100p = -45$$
$$p = 0.45$$

Now, solve for n by replacing p with 0.45.

$$75p + 30n = 40.05$$
$$75(0.45) + 30n = 40.05 \quad \textit{Substitute 0.45 for p.}$$
$$33.75 + 30n = 40.05 \quad \textit{Solve for n.}$$
$$30n = 6.3$$
$$n = 0.21 \quad \text{Is (0.45, 0.21) a solution?}$$

Check:

$$75p + 30n = 40.05 \qquad\qquad 50p + 60n = 35.10$$
$$75(0.45) + 30(0.21) \stackrel{?}{=} 40.05 \qquad\qquad 50(0.45) + 60(0.21) \stackrel{?}{=} 35.10$$
$$40.05 = 40.05 \checkmark \qquad\qquad 35.10 = 35.10 \checkmark$$

The solution of this system is (0.45, 0.21). Thus, the per-minute rate for peak-hour calls is 45¢, and the per-minute rate for nonpeak calls is 21¢ on this plan.

For some systems of equations, it is necessary to multiply *each* equation by a different number in order to solve the system by elimination. You can choose to eliminate either variable.

Example 

Use elimination to solve the system of equations in two different ways.

$\mathbf{2x + 3y = 5}$
$\mathbf{5x + 4y = 16}$

Method 1

You can eliminate the variable x by multiplying the first equation by 5 and the second equation by -2 and then adding the resulting equations.

$2x + 3y = 5$ → **Multiply by 5.** → $10x + 15y = 25$

$5x + 4y = 16$ → **Multiply by −2.** → $(+)\ -10x - 8y = -32$

$$7y = -7$$
$$y = -1$$

Now find x using one of the original equations.

$$2x + 3y = 5$$
$$2x + 3(-1) = 5 \quad \textit{Substitute } -1 \textit{ for } y.$$
$$2x - 3 = 5 \quad \textit{Solve for } x.$$
$$2x = 8$$
$$x = 4$$

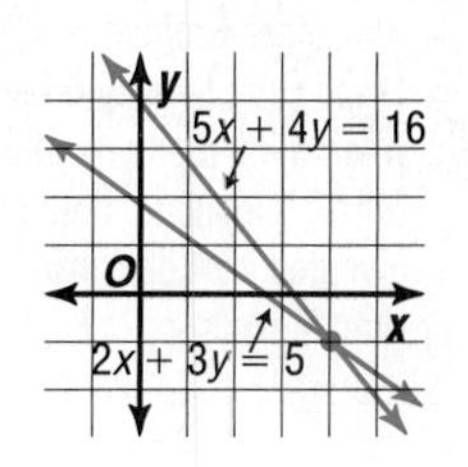

The solution of the system is $(4, -1)$.

Method 2

You can also solve this system by eliminating the variable y. Multiply the first equation by -4 and the second equation by 3. Then add.

$2x + 3y = 5$ → **Multiply by −4.** → $-8x - 12y = -20$

$5x + 4y = 16$ → **Multiply by 3.** → $(+)\ 15x + 12y = 48$

$$7x = 28$$
$$x = 4$$

Now find y.

$$2x + 3y = 5$$
$$2(4) + 3y = 5 \quad \textit{Substitute 4 for } x.$$
$$8 + 3y = 5 \quad \textit{Solve for } y.$$
$$3y = -3$$
$$y = -1$$

The solution is $(4, -1)$, which matches the result obtained with Method 1.

Example 2

Testing

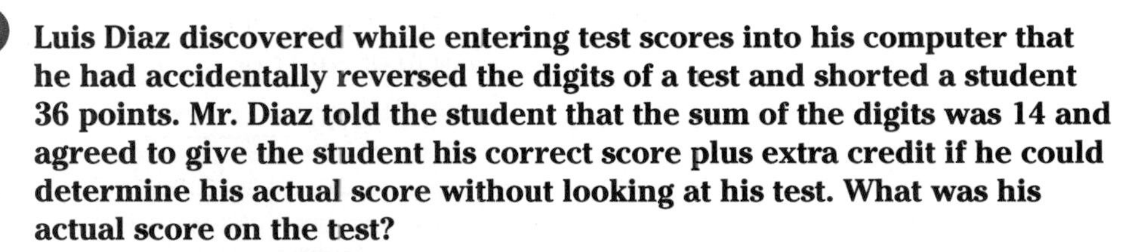

Luis Diaz discovered while entering test scores into his computer that he had accidentally reversed the digits of a test and shorted a student 36 points. Mr. Diaz told the student that the sum of the digits was 14 and agreed to give the student his correct score plus extra credit if he could determine his actual score without looking at his test. What was his actual score on the test?

Explore Let t represent the tens digit of the score.
Let u represent the units digit.

The actual score on the test can be represented by $10t + u$. The amount entered in the computer can be represented by $10u + t$. *Why?*

Plan Since the sum of the digits is 14, one equation is $t + u = 14$. Since the teacher accidentally shorted the student by 36 points, another equation is $(10t + u) - (10u + t) = 36$ or $9t - 9u = 36$.

Solve

$t + u = 14$ → Multiply by 9. → $9t + 9u = 126$

$9t - 9u = 36$ → $(+)\ 9t - 9u = 36$

$$18t = 162$$
$$t = 9$$

Now find u using one of the original equations.

$t + u = 14$

$9 + u = 14$ *Substitute 9 for t.*

$u = 5$ *Solve for u.*

The solution is (9, 5), which means that the student's actual test score was $10(9) + 5$, or 95 points.

Examine The sum of the digits, $9 + 5$, is 14 and $95 - 59$ is 36.

You can use systems of equations to solve problems involving the distance formula, $rt = d$.

Example 3

Uniform Motion

A riverboat on the Mississippi River travels 48 miles upstream in 4 hours. The return trip takes the riverboat only 3 hours. Find the rate of the current.

Explore Let r represent the rate of the riverboat in still water. Let c represent the rate of the current.

Then $r + c$ represents the rate of the riverboat traveling downstream *with* the current and $r - c$ represents the rate of the riverboat traveling upstream *against* the current.

FYI

The world's largest riverboat is the 418-ft *American Queen.* Passengers paid up tp $9400 to take its first cruise in June, 1995.

(continued on the next page)

Plan Use the formula rate × time = distance, or $rt = d$, to write a system of equations. Then solve the system to find the value of c.

	r	t	d	$rt = d$
Downstream	$r + c$	3	48	$3r + 3c = 48$
Upstream	$r - c$	4	48	$4r - 4c = 48$

Solve $3r + 3c = 48$ **Multiply by 4.** $12r + 12c = 192$

$4r - 4c = 48$ **Multiply by −3.** $(+)\ -12r + 12c = -144$

$$24c = 48$$
$$c = 2$$

The rate of the current is 2 miles per hour.

Examine Find the value of r for this system and then check the solution.

CHECK FOR UNDERSTANDING

Communicating Mathematics

Study the lesson. Then complete the following.

1. **Write** a problem about a real-life situation in which only an estimate of the solution is needed rather than the exact solution. The problem should involve a system of equations.
2. **Explain** why you might need to multiply each equation by a different number when using elimination to solve a system of equations.
3. **Write** a system of equations that could best be solved by using multiplication and then elimination using addition or subtraction.

4. **Assess Yourself** Describe the method you like to use best when solving systems of linear equations. Explain your reasons.

Guided Practice

Explain the steps you would follow to eliminate the variable x in each system of equations. Then solve the system.

5. $x + 5y = 4$
   $3x - 7y = -10$
6. $2x - y = 6$
   $3x + 4y = -2$
7. $-5x + 3y = 6$
   $x - y = 4$

Explain the steps you would follow to eliminate the variable y in each system of equations. Then solve the system.

8. $4x + 7y = 6$
   $6x + 5y = 20$
9. $3x - 8y = 13$
   $4x - 5y = 6$
10. $2x - 3y = 2$
    $5x + 4y = 28$

Match each system of equations with the method that could be most efficiently used to solve it. Then solve the system.

11. $3x - 7y = 6$
    $2x + 7y = 4$
12. $y = 4x + 11$
    $3x - 2y = -7$
13. $4x + 3y = 19$
    $3x - 4y = 8$

a. substitution
b. elimination using addition or subtraction
c. elimination using multiplication

14. Uniform Motion A riverboat travels 36 miles downstream in 2 hours. The return trip takes 3 hours.

a. Find the rate of the riverboat in still water.

b. Find the rate of the current.

EXERCISES

Practice

Use elimination to solve each system of equations.

15. $2x + y = 5$
$3x - 2y = 4$

16. $4x - 3y = 12$
$x + 2y = 14$

17. $3x - 2y = 19$
$5x + 4y = 17$

18. $9x = 5y - 2$
$3x = 2y - 2$

19. $7x + 3y = -1$
$4x + y = 3$

20. $6x - 5y = 27$
$3x + 10y = -24$

21. $8x - 3y = -11$
$2x - 5y = 27$

22. $11x - 5y = 80$
$9x - 15y = 120$

23. $4x - 7y = 10$
$3x + 2y = -7$

24. $3x - \frac{1}{2}y = 10$
$5x + \frac{1}{4}y = 8$

25. $2x + \frac{2}{3}y = 4$
$x - \frac{1}{2}y = 7$

26. $\frac{2x + y}{3} = 15$
$\frac{3x - y}{5} = 1$

27. $7x + 2y = 3(x + 16)$
$x + 16 = 5y + 3x$

28. $0.4x + 0.5y = 2.5$
$1.2x - 3.5y = 2.5$

29. $1.8x - 0.3y = 14.4$
$x - 0.6y = 2.8$

Number Theory

Use a system of equations and elimination to solve each problem.

30. The sum of the digits of a two-digit number is 14. If the digits are reversed, the new number is 18 less than the original number. Find the original number.

31. Three times one number equals twice a second number. Twice the first number is 3 more than the second number. Find the numbers.

32. The ratio of the tens digit to the units digit of a two-digit number is 1:4. If the digits are reversed, the sum of the new number and the original number is 110. Find the original number.

Determine the best method to solve each system of equations. Then solve the system.

33. $9x - 8y = 17$
$4x + 8y = 9$

34. $3x - 4y = -10$
$5x + 8y = -2$

35. $x + 2y = -1$
$2x + 4y = -2$

36. $5x + 3y = 12$
$4x - 5y = 17$

37. $\frac{2}{3}x - \frac{1}{2}y = 14$
$\frac{5}{6}x - \frac{1}{2}y = 18$

38. $\frac{1}{2}x - \frac{2}{3}y = \frac{7}{3}$
$\frac{3}{2}x + 2y = -25$

Use elimination to solve each system of equations.

39. $\frac{1}{x - 5} - \frac{3}{y + 6} = 0$
$\frac{2}{x + 7} - \frac{1}{y - 3} = 0$

40. $\frac{2}{x} + \frac{3}{y} = 16$
$\frac{1}{x} + \frac{1}{y} = 7$

41. $\frac{1}{x - y} = \frac{1}{y}$
$\frac{1}{x + y} = 2$

Programming

42. The graphing calculator program at the right finds the solution of two linear equations written in standard form.

$ax + by = c$
$dx + ey = f$

The formulas for the solution of this system are as follows.

$$x = \frac{ce - bf}{ae - bd}, \quad y = \frac{af - cd}{ae - bd}$$

```
PROGRAM:SOLVE
: Disp "ENTER COEFFICIENTS"
: Prompt A, B, C, D, E, F
: If AE-BD = 0
: Then
: Goto 1
: End
: (CE-BF)/(AE-BD) → X
: (AF-CD)/(AE-BD) → Y
: Disp "THE SOLUTION IS"
: Disp "X= ", X
: Disp "Y= ", Y
: Stop
: Lbl 1
: If CE-BF=0 or AF-CD=0
: Then
: Disp "INFINITELY", "MANY"
: Else
: Disp "NO SOLUTION"
```

Use the program to solve each system.

a. $8x + 2y = 0$
   $12x + 3y = 0$

b. $x - 2y = 5$
   $3x - 5y = 8$

c. $5x + 5y = 16$
   $2x + 2y = 5$

d. $7x - 3y = 5$
   $14x - 6y = 10$

Critical Thinking

43. The graphs of the equations $5x + 4y = 18$, $2x + 9y = 59$, and $3x - 5y = -4$ contain the sides of a triangle. Determine the coordinates of the vertices of the triangle.

Applications and Problem Solving

44. **Geography** Benjamin Banneker, a self-taught mathematician and astronomer, was the first African-American to publish an almanac. He is most noted for being the assistant surveyor on the team that designed the ten-mile square of Washington, D.C. The White House is located in the center of the square, at the intersection of Pennsylvania Avenue and New York Avenue. Let $-5x + 7y = 0$ represent New York Avenue and let $3x + 8y = 305$ represent Pennsylvania Avenue. Find the coordinates for the White House.

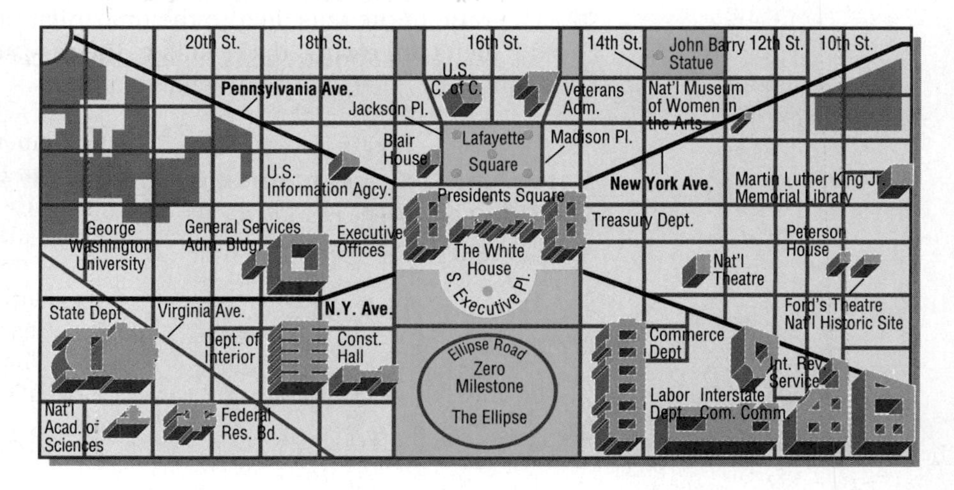

45. **Organize Data** At the new Cozy Inn Restaurant, which is still under construction, the owners have hired enough waiters and waitresses to handle 17 tables of customers. The fire marshall has looked at the plans for the restaurant and says he will approve it for a limit of 56 customers. The restaurant owners are now deciding how many two-seat tables and how many four-seat tables to buy for the restaurant. How many of each kind should they buy?

46. Information Highway The Mercury Center provides a reference and research service for on-line computer users based on peak and nonpeak usage. Miriam and Lesharo are both subscribers. The chart below displays the number of peak and nonpeak minutes each of them spent on-line in one month and how much it cost. Use the information to find the Mercury Center's rate per minute for its peak and nonpeak on-line research service.

User	Number of Peak Minutes	Number of Nonpeak Minutes	Cost
Lesharo	45	50	$27.75
Miriam	70	30	$36

Mixed Review

47. Use elimination to solve the system of equations. (Lesson 8–3)

$2x - y = 10$
$5x + 3y = 3$

48. Statistics What is the outlier in the box-and-whisker plot? (Lesson 7–7)

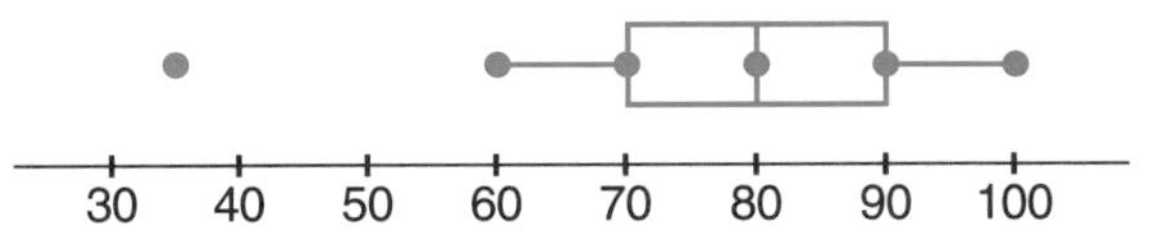

49. Probability If Bill, Raul, and Kenyatta each have an equal chance of winning a bicycle race, find the probability that Raul finishes last. (Lesson 7–5)

50. Find the coordinates of the midpoint of the line segment whose endpoints are at (1, 6) and (−3, 4). (Lesson 6–7)

51. Track Alfonso runs a 440-yard race in 55 seconds, and Marcus runs it in 88 seconds. To have Alfonso and Marcus finish at the same time, how much of a head start should Alfonso give Marcus? (Lesson 4–7)

52. If $12m = 4$, then $3m =$ __?__. (Lesson 3–2)

53. Cooking If there are four sticks in a pound of butter and each stick is $\frac{1}{2}$ cup, how many cups of butter are in a pound of butter? (Lesson 2–6)

54. Evaluate $288 \div [3(9 + 3)]$. (Lesson 1–3)

Mathematics and SOCIETY

High-Tech Checkout Lanes

The article below appeared in *Progressive Grocer* in February, 1994.

KMART...HAS INSTALLED A NEW technology in 48 stores that helps ensure that enough checklanes are open to serve customers in a store. The system counts the number of adults and children who are in a store at any given moment. The system, called ShopperTrak...uses infrared technology on door-mounted units to give a continuous count of shoppers entering and exiting a store...The data is channeled to software in a PC called FastLane, which uses it to calculate how many checklanes should be open during the next 20 minutes so that no more than two to three people are in line at each lane. Managers read the data at monitors stationed at the checkout area. ■

1. How do you think the ShopperTrak system can count the number of children entering and leaving the store as well as the number of adults? Why might separate counts of children and adults be useful?
2. Do you think the average shopping times would differ between men and women, boys and girls, or senior citizens and young people? Why do you think the ShopperTrak system doesn't consider these factors?

8–5 Graphing Systems of Inequalities

What YOU'LL LEARN

- To solve systems of inequalities by graphing.

Why IT'S IMPORTANT

You can use systems of inequalities to solve problems involving travel and nutrition.

Unita likes her job as a baby-sitter, but it pays only \$3 per hour. She has been offered a job as a tutor that pays \$6 per hour. Because of school, her parents only allow her to work a maximum of 15 hours per week. How many hours can Unita tutor *and* baby-sit and still make at least \$65 per week?

Let x represent the number of hours Unita can baby-sit each week. Let y represent the number of hours she can tutor each week. Since both x and y represent a number of hours, neither can be a negative number. Thus, $x \geq 0$ and $y \geq 0$. Then the following **system of inequalities** can be used to represent the conditions of this problem.

$x \geq 0$

$y \geq 0$

$3x + 6y \geq 65$ *She wants to earn at least \$65.*

$x + y \leq 15$ *She can work up to 15 hours.*

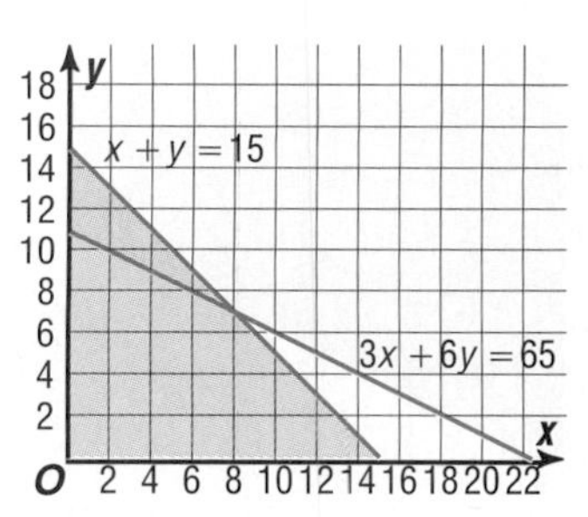

The solution of this system is the set of all ordered pairs that satisfies both inequalities and lies in the first quadrant. The solution can be determined by graphing each inequality on the same coordinate plane.

Recall that the graph of each inequality is called a *half-plane*. The intersection of the two half-planes represents the solution to the system of inequalities. This solution is a region that contains the graphs of an infinite number of ordered pairs. The boundary line of the half-plane is solid and is included in the graph if the inequality is $\leq$ or $\geq$. The boundary line of the half-plane is dashed and is not included in the graph if the inequality is $<$ or $>$.

LOOK BACK

You can refer to Lesson 7-8 for information on graphing inequalities in two variables.

The graphs of $3x + 6y = 65$ and $x + y = 15$ are the boundaries of the region and are included in the graph of this system. This region is shown in green above. Only the portion in the first quadrant is shaded since $x \geq 0$ and $y \geq 0$. Every point in this region is a possible solution to the system. For example, since the graph of (5, 9) is a point in the region, Unita could baby-sit for 5 hours and tutor for 9 hours. In this case, she would make \$3(5) + \$6(9) or \$69. *Does this meet her requirements of time and earnings?*

Example 1 **Solve each system of inequalities by graphing.**

a. $y < 2x + 1$
$y \geq -x + 3$

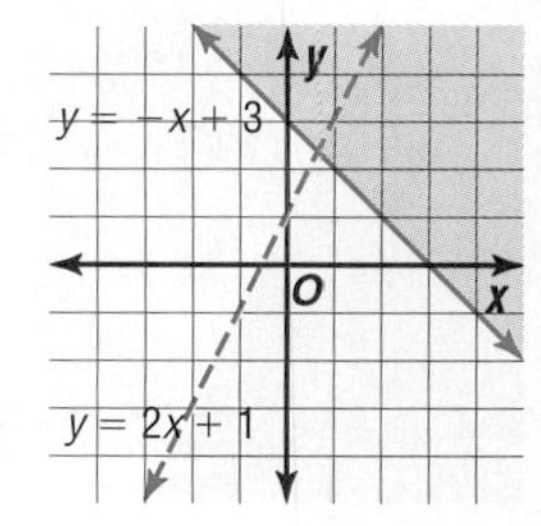

The solution includes the ordered pairs in the intersection of the graphs of $y < 2x + 1$ and $y \geq -x + 3$. This region is shaded in green at the right. The graphs of $y = 2x + 1$ and $y = -x + 3$ are the boundaries of this region. The graph of $y = 2x + 1$ is dashed and is *not* included in the graph of $y < 2x + 1$. The graph of $y = -x + 3$ is included in the graph of $y \geq -x + 3$.

b. $2x + y \geq 4$
$y \leq -2x - 1$

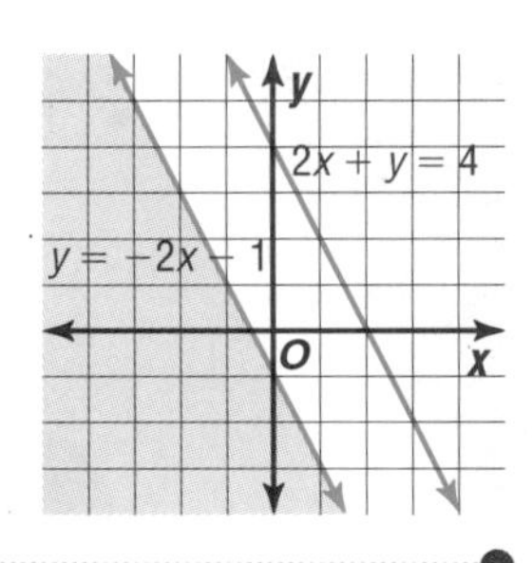

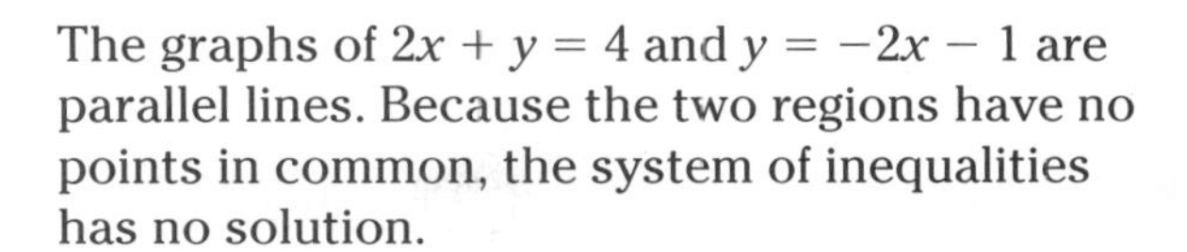

The graphs of $2x + y = 4$ and $y = -2x - 1$ are parallel lines. Because the two regions have no points in common, the system of inequalities has no solution.

Sometimes in real-life problems involving systems, only whole-number solutions make sense.

Example 2

Vacations

Elena Ayala wants to spend no more than \$700 for hotels while vacationing in Hawaii. She wants to stay at the Hyatt Resort at least one night and at the Coral Reef Hotel for the remainder of her stay. The Hyatt Resort costs \$130 per night, and the Coral Reef Hotel costs \$85 per night.

a. If she wants to stay in Hawaii at least 6 nights, how many nights could she spend at each hotel and still stay within her budget?

b. What advice might you give Elena concerning her options?

a. Let c represent the number of nights she will stay at the Coral Reef Hotel. Let h represent the number of nights she will stay at the Hyatt Resort.

Then the following system of inequalities can be used to represent the conditions of this problem.

$h + c \geq 6$	*Elena wants to stay <u>at least</u> 6 nights.*
$h \geq 1$	*She wants to stay <u>at least</u> 1 night at the Hyatt Resort.*
$130h + 85c \leq 700$	*She wants to spend <u>no more</u> than \$700.*

The solution is the set of all ordered pairs whose graphs are in the intersection of the graphs of these inequalities. This region is shown in brown at the right.

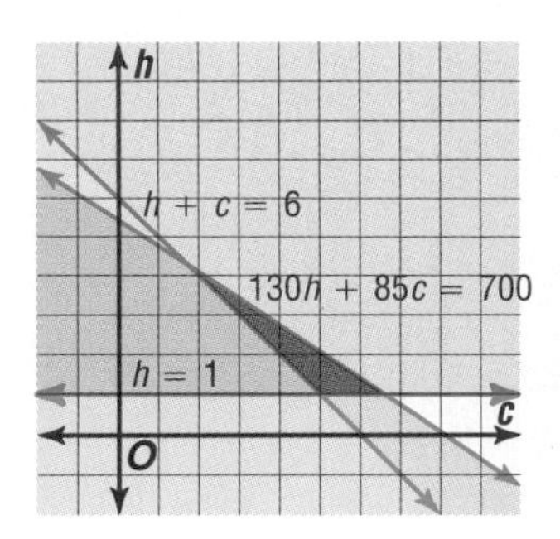

Any point in this region is a possible solution; however, only whole-number solutions make sense in this problem. *Why?* For example, since (3, 3) is a point in the region, Elena could stay 3 nights at each hotel. In this case, she would spend 3(\$130) or \$390 at the Hyatt Resort and 3(\$85) or \$255 at the Coral Reef Hotel for a total of \$645. The other solutions are (5, 1), (6, 1), (4, 2), (5, 2) and (2, 4). *Check this result.*

b. You could advise Elena that she could stay in Hawaii a maximum of 7 nights if she stayed at the Hyatt Resort only 1 or 2 nights and stayed at the Coral Reef Hotel for the remainder of her vacation.

GLOBAL CONNECTIONS

Polynesians from the Marquesas Islands settled in Hawaii in about A.D. 400. A second wave of immigration arrived from Tahiti approximately 400 to 500 years later.

A graphing calculator is a useful tool for graphing systems of inequalities. It is important to enter the functions in the correct order, since this determines the shading.

EXPLORATION — GRAPHING CALCULATORS

You can use a graphing calculator to solve systems of inequalities. The TI-82 graphs functions and shades above the first function entered and below the second function entered. Select 7 on the DRAW menu to choose the SHADE feature. First, enter the function that is the lower boundary of the region to be shaded. (Note that inequalities that have $>$ or $\geq$ are lower boundaries and inequalities that have $<$ or $\leq$ are upper boundaries.) Press [,]. Then enter the function that is the upper boundary of the region. Press [ )] [ENTER].

Your Turn

a. Use a graphing calculator to graph the system of inequalities.

$y \geq 4x - 3$

$y \leq -2x + 9$

b. Use a graphing calculator to work through the examples in this lesson. List and explain any disadvantages that you discovered when using the graphing calculator to graph systems of inequalities.

c. Describe the process of using a graphing calculator to solve systems of linear inequalities in your own words.

CHECK FOR UNDERSTANDING

Communicating Mathematics

Study the lesson. Then complete the following.

1. **Explain** how to determine whether boundary lines should be included in the graph of a system of inequalities.

2. **You Decide** Joshua says that the intersection point of the boundary lines is always a solution of a system of inequalities. Rolanda says the point of intersection may not be part of the solution set. Explain who is correct and give an example to support your answer.

3. **Write** a system of inequalities that has no solutions. Describe the graph of your system.

4. **State** which points are solutions to the system of inequalities graphed at the right. Explain how you know.

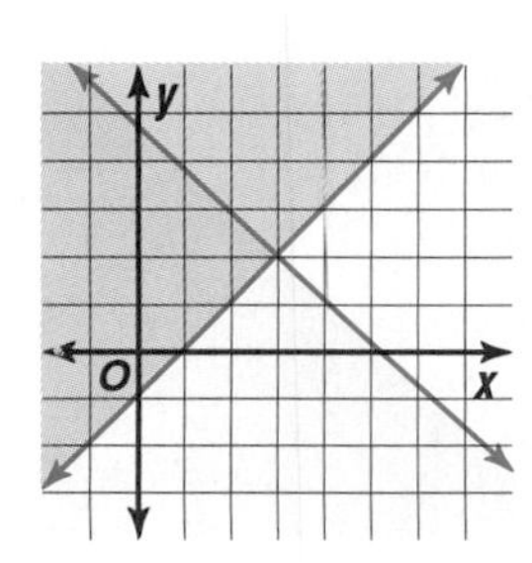

a. (0, 0) **b.** (−1, 4)

c. (2, 5) **d.** (0.5, −1.7)

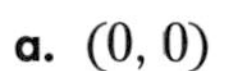

5. Describe a real-life situation that you can model using a system of linear inequalities.

Guided Practice

Solve each system of inequalities by graphing.

6. $x < 1$
   $x > -4$
7. $y \geq -2$
   $y - x < 1$
8. $y \geq 2x + 1$
   $y \leq -x + 1$
9. $y \geq 3x$
   $3y < 5x$
10. $y - x < 1$
    $y - x > 3$
11. $2x + y \leq 4$
    $3x - y \geq 6$

Write a system of inequalities for each graph.

12.

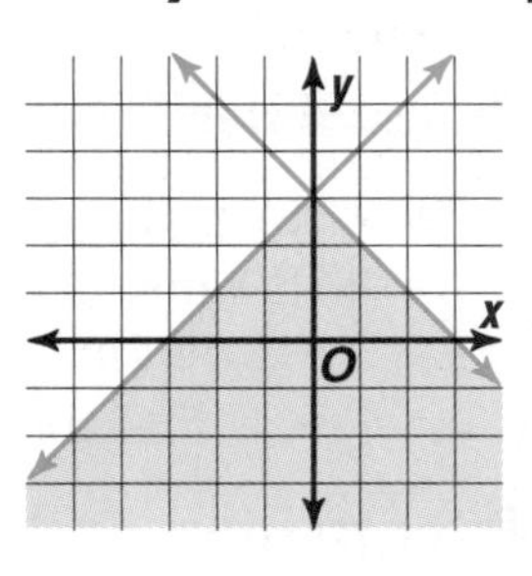

13.

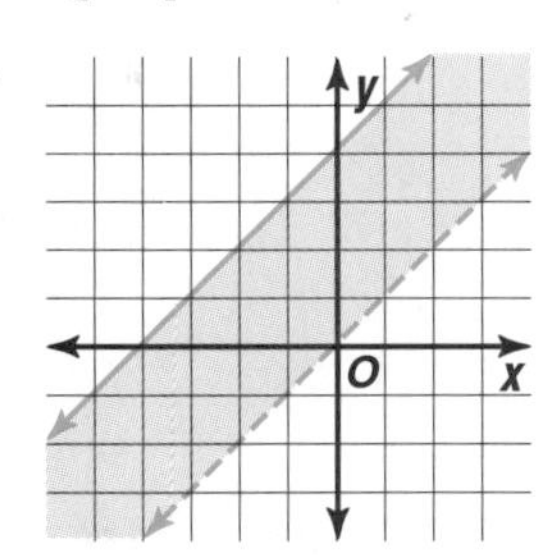

14. **Sales** Ms. Johnson's homeroom class can order up to \$90 of free pizzas from Angelino's Pizza as a reward for selling the most magazines during the magazine drive. They need to order at least 6 large pizzas in order to serve the entire class. If a pepperoni pizza costs \$9.95 and a supreme pizza costs \$12.95, how many of each type can they order? List three possible solutions.

EXERCISES

Practice

Solve each system of inequalities by graphing.

15. $x > 5$
    $y \leq 4$
16. $y < 0$
    $x \geq 0$
17. $y > 3$
    $y > -x + 4$
18. $x \leq 2$
    $y - 4 \geq 5$
19. $x \geq 2$
    $y + x \leq 5$
20. $y < -3$
    $x - y > 1$
21. $y \leq 2x + 3$
    $y < -x + 1$
22. $y - x < 3$
    $y - x \geq 2$
23. $y \geq 3x$
    $7y < 2x$
24. $x - y < -1$
    $x - y > 3$
25. $2y + x < 6$
    $3x - y > 4$
26. $3x - 4y < 1$
    $x + 2y \leq 7$
27. $y - 4 > x$
    $y + x < 4$
28. $5y \geq 3x + 10$
    $2y \leq 4x - 10$
29. $y + 2 \leq x$
    $2y - 3 > 2x$
30. $2x + y \geq -4$
    $-5x + 2y < 1$
31. $x + y > 4$
    $-2x + 3y < -12$
32. $-4x + 5y \leq 41$
    $x + y > -1$

Write a system of inequalities for each graph.

33.

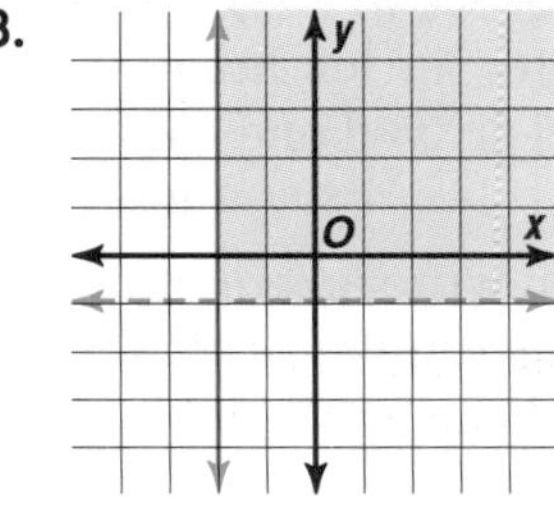

34.

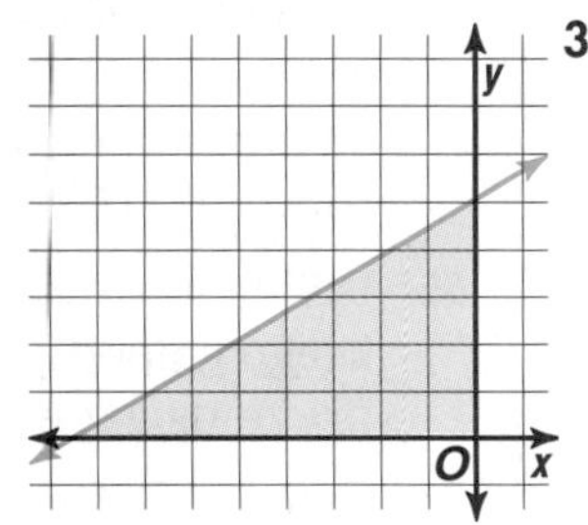

35.

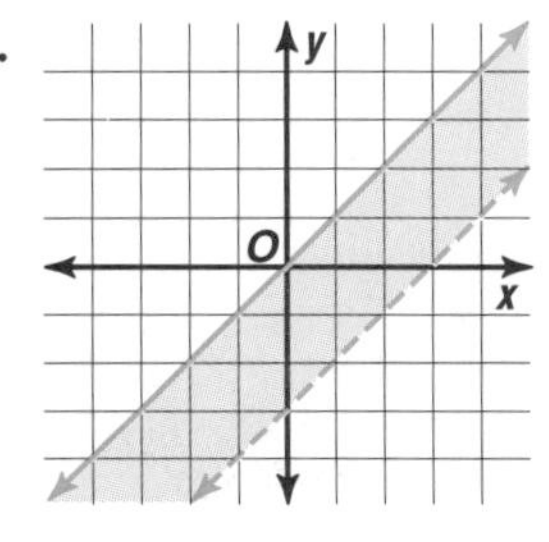

Write a system of inequalities for each graph.

36.

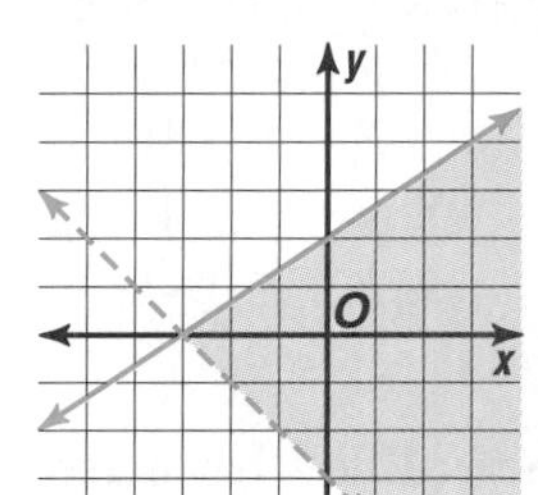

37.

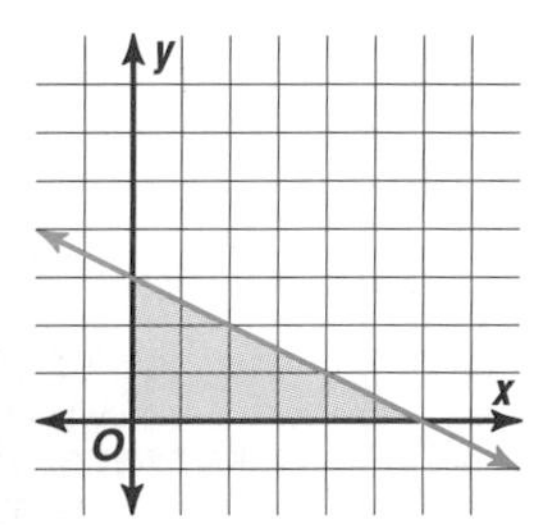

38.

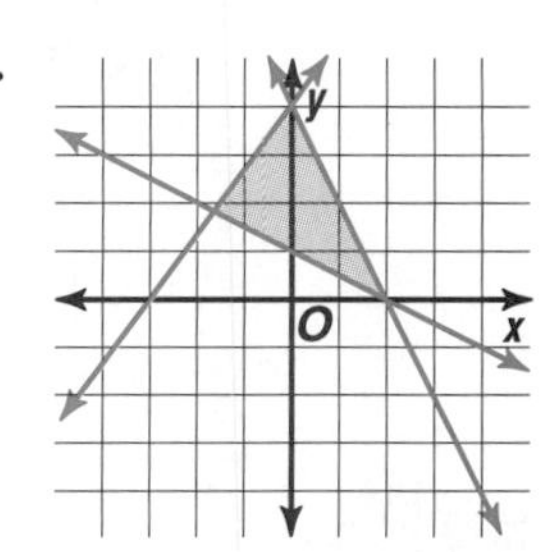

Solve each system of inequalities by graphing.

39. $x - 2y \leq 2$
$3x + 4y \leq 12$
$x \geq 0$

40. $x - y \leq 5$
$5x + 3y \geq -6$
$y \leq 3$

41. $x < 2$
$4y > x$
$2x - y < -9$
$x + 3y < 9$

Graphing Calculator

Use a graphing calculator to solve each system of inequalities.

42. $y \geq 3x - 6$
$y \leq x + 1$

43. $y \leq x + 9$
$y > -x - 4$

44. $y < 2x + 10$
$y \geq 7x + 15$

Critical Thinking

45. Solve the inequality $|y| \leq 3$ by graphing. (*Hint:* Graph as a system of inequalities.)

Applications and Problem Solving

Graph a system of inequalities to solve each problem.

46. **Nutrition** Young people between the ages of 11 and 18 should get at least 1200 milligrams of calcium each day. One ounce of mozzarella cheese has 147 milligrams of calcium, and one ounce of Swiss cheese has 219 milligrams. If you wanted to eat no more than 8 ounces of cheese, how much of each type could you eat and still get your daily requirement of calcium? List three possible solutions.

47. **Organize Data** Kenny Choung likes to exercise every day by walking and jogging at least 3 miles. Kenny walks at a rate of 4 mph and jogs at a rate of 8 mph. If he has only a half hour to exercise, how much time can he spend walking and jogging and cover at least 3 miles? List 3 possible solutions.

Mixed Review

48. **Number Theory** If the digits of a two-digit positive integer are reversed, the result is 6 less than twice the original number. Find all such integers for which this is true. (Lesson 8–4)

49. **Organize Data** When Roberta cashed her check for \$180, the bank teller gave her 12 bills, each one worth either \$5 or \$20. How many of each bill did she receive? (Lesson 8–2)

50. Solve $4 > 4a + 12 > 24$ and graph the solution set. (Lesson 7–4)

51. Write an equation in slope-intercept form of a line that passes through the points at (3, 3) and (−1, 5). (Lesson 6–2)

52. Solve $y = -\frac{1}{2}x + 3$ if the domain is {2, 4, 6}. (Lesson 5–3)

53. What number increased by 40% equals 14? (Lesson 4–5)

54. **Travel** Paloma Rey drove to work on Wednesday at 40 miles per hour and arrived one minute late. She left home at the same time on Thursday, drove 45 miles per hour, and arrived one minute early. How far does Ms. Rey drive to work? (*Hint:* Convert hours to minutes.) (Lesson 3–5)

55. Define a variable, then write an equation for the following problem. Diego gained 134 yards running. This was 17 yards more than in the previous game. How many yards did he gain in both games? (Lesson 2–9)

56. Name the property illustrated by $(3 \cdot x) \cdot y = 3 \cdot (x \cdot y)$. (Lesson 1–8)

CHAPTER 8 HIGHLIGHTS

VOCABULARY

After completing this chapter, you should be able to define each term, property, or phrase and give an example or two of each.

Algebra

consistent (p. 455)
dependent (p. 456)
elimination (p. 469)
independent (p. 456)
inconsistent (p. 456)
substitution (p. 462)
system of equations (p. 455)
system of inequalities (p. 482)

Problem Solving

organize data (p. 464)

UNDERSTANDING AND USING THE VOCABULARY

Choose the correct term to complete each statement.

1. The method used in solving the following system of equations is (*elimination*, *substitution*).

$$\left.\begin{aligned} x &= 4y + 1 \\ x + y &= 6 \end{aligned}\right\} \rightarrow \quad \begin{aligned} (4y + 1) + y &= 6 \\ 5y + 1 &= 6 \\ 5y &= 5 \\ y &= 1 \end{aligned} \qquad \begin{aligned} x &= 4(1) + 1 \\ x &= 4 + 1 \\ x &= 5 \\ \text{solution: } & (5, 1) \end{aligned}$$

2. If a system of equations has exactly one solution, it is (*dependent, independent*).
3. If the graph of a system of equations is parallel lines, the system of equations is said to be (*consistent, inconsistent*).
4. A system of equations that has infinitely many solutions is (*dependent, independent*).
5. The method used in solving the following system of equations is (*elimination, substitution*).

$$\left.\begin{aligned} -2c + b &= 3 \\ -b - c &= -6 \end{aligned}\right\} \rightarrow \quad \begin{aligned} b - 2c &= 3 \\ (+)\ -b - c &= -6 \\ \hline -3c &= -3 \\ c &= 1 \end{aligned} \qquad \begin{aligned} -b - (1) &= -6 \\ -b - 1 &= -6 \\ -b &= -5 \\ b &= 5 \end{aligned} \quad \text{solution: } (5, 1)$$

6. If a system of equations has the same slope and different intercepts, the graph of the system is (*intersecting lines, parallel lines*).
7. If a system of equations has the same slope and intercepts, the system has (*exactly one, infinitely many*) solutions.
8. The solution to a system of equations is (3, −5); therefore, this system is (*consistent, inconsistent*).
9. The graph of a system of equations is shown at the right. This system has (*infinitely many, no*) solution.
10. The solution to a system of inequalities is the (*intersection, union*) of two half-planes.
11. A system of inequalities that includes $x < 0$ and $y > 0$ is in the (*second, fourth*) quadrant.

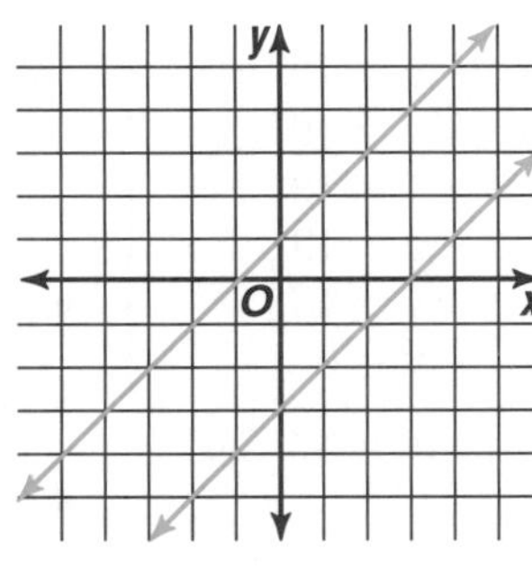

SKILLS AND CONCEPTS

OBJECTIVES AND EXAMPLES

Upon completing this chapter, you should be able to:

- solve systems of equations by graphing (Lesson 8–1)

Graph $x + y = 6$ and $x - y = 2$. Then find the solution.

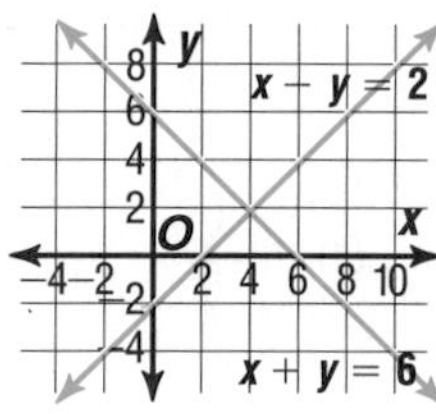

The solution is (4, 2).

REVIEW EXERCISES

Use these exercises to review and prepare for the chapter test.

Graph each system of equations to find the solution.

12. $y = 2x - 7$
$x + y = 11$

13. $x + 2y = 6$
$2y - 8 = -x$

14. $3x + y = -8$
$x + 6y = 3$

15. $5x - 3y = 11$
$2x + 3y = -25$

- determine whether a system of equations has one solution, no solution, or infinitely many solutions by graphing (Lesson 8–1)

Graph $3x + y = -4$ and $6x + 2y = -8$. Then determine the number of solutions.

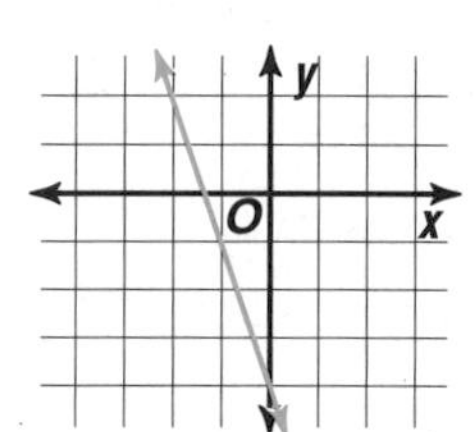

There are infinitely many solutions.

Graph each system of equations. Then determine whether the system of equations has *one* solution, *no* solution, or *infinitely many* solutions. If the system has one solution, name it.

16. $x - y = 9$
$x + y = 11$

17. $9x + 2 = 3y$
$y - 3x = 8$

18. $2x - 3y = 4$
$6y = 4x - 8$

19. $3x - y = 8$
$3x = 4 - y$

- solve systems of equations by using the substitution method (Lesson 8–2)

Use substitution to solve the system of equations.

$y = x - 1$
$4x - y = 19$

$$\begin{aligned} 4x - y &= 19 \\ 4x - (x - 1) &= 19 \\ 4x - x + 1 &= 19 \\ 3x + 1 &= 19 \\ 3x &= 18 \\ x &= 6 \end{aligned} \qquad \begin{aligned} y &= x - 1 \\ y &= 6 - 1 \\ y &= 5 \end{aligned}$$

The solution is (6, 5).

Use substitution to solve each system of equations. If the system *does not* have exactly one solution, state whether it has *no* solution or *infinitely many* solutions.

20. $2m + n = 1$
$m - n = 8$

21. $3a - 2b = -4$
$3a + b = 2$

22. $x = 3 - 2y$
$2x + 4y = 6$

23. $3x - y = 1$
$2x + 4y = 3$

OBJECTIVES AND EXAMPLES

- solve systems of equations by using the elimination method with addition or subtraction (Lesson 8–3)

Use elimination to solve the system of equations.

$2m - n = 4$
$m + n = 2$

$2m - n = 4$	$m + n = 2$
$(+)\ m + n = 2$	$2 + n = 2$
$3m = 6$	$n = 0$
$m = 2$	

The solution is (2, 0).

REVIEW EXERCISES

Use elimination to solve each system of equations.

24. $x + 2y = 6$
$x - 3y = -4$

25. $2m - n = 5$
$2m + n = 3$

26. $3x - y = 11$
$x + y = 5$

27. $3s + 6r = 33$
$6r - 9s = 21$

28. $3x + 1 = -7y$
$6x + 7y = 0$

29. $12x - 9y = 114$
$7y + 12x = 82$

- solve systems of equations by using the elimination method with multiplication and addition (Lesson 8–4)

Use elimination to solve the system of equations.

$3x - 4y = 7$
$2x + y = 1$

$3x - 4y = 7 \qquad\qquad\qquad 3x - 4y = 7$

$2x + y = 1$ **Multiply by 4.** → $(+)\ 8x + 4y = 4$

$11x = 11$

$x = 1$

$2x + y = 1$
$2(1) + y = 1$
$y = -1$

The solution is (1, −1).

Use elimination to solve each system of equations.

30. $x - 5y = 0$
$2x - 3y = 7$

31. $x - 2y = 5$
$3x - 5y = 8$

32. $2x + 3y = 8$
$x - y = 2$

33. $-5x + 8y = 21$
$10x + 3y = 15$

34. $5m + 2n = -8$
$4m + 3n = 2$

35. $6x + 7y = 5$
$2x - 3y = 7$

- determine the best method for solving systems of equations (Lesson 8–4)

Use the best method to solve the system of equations.

$x + 2y = 8$
$3x + 2y = 6$

$3x + 2y = 6$	$x + 2y = 8$
$(-)\ x + 2y = 8$	$-1 + 2y = 8$
$2x = -2$	$2y = 9$
$x = -1$	$y = \frac{9}{2}$

The solution is $\left(-1, \frac{9}{2}\right)$.

Determine the best method to solve each system of equations. Then solve the system.

36. $y = 2x$
$x + 2y = 8$

37. $9x + 8y = 7$
$18x - 15y = 14$

38. $2x - y = 36$
$3x - 0.5y = 26$

39. $3x + 5y = 2x$
$x + 3y = y$

40. $5x - 2y = 23$
$5x + 2y = 17$

41. $2x + y = 3x - 15$
$x + 5 = 4y + 2x$

OBJECTIVES AND EXAMPLES	REVIEW EXERCISES

- solve systems of inequalities by graphing (Lesson 8–5)

Solve the system of inequalities.

$x \geq -3$
$y \leq x + 2$

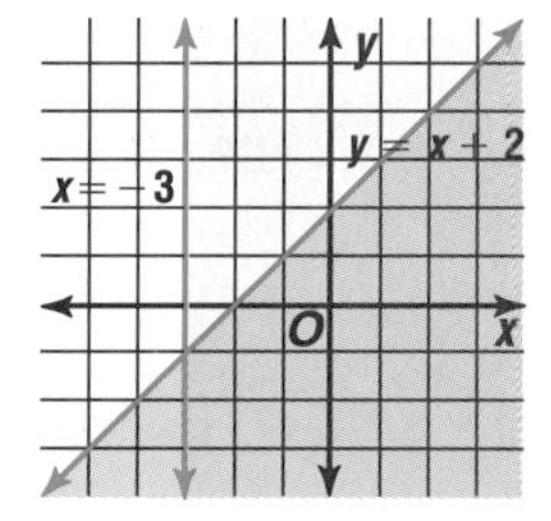

Solve each system of inequalities by graphing.

42. $y < 3x$
$x + 2y \geq -21$

43. $y > -x - 1$
$y \leq 2x + 1$

44. $2x + y < 9$
$x + 11y < -6$

45. $x \geq 1$
$y + x \leq 3$

46. $y \geq x - 3$
$y \geq -x - 1$

47. $x - 2y \leq -4$
$4y < 2x - 4$

APPLICATIONS AND PROBLEM SOLVING

48. Ballooning A hot-air balloon is 10 meters above the ground rising at a rate of 15 meters per minute. Another balloon is 150 meters above the ground descending at a rate of 20 meters per minute. (Lesson 8–1)

a. After how long will the balloons be at the same height?

b. What is that height?

49. Number Theory A two-digit number is 7 times its units digit. If 18 is added to the number, its digits are reversed. Find the original number. (Lesson 8–2)

50. Travel While driving to Fullerton, Mrs. Sumner travels at an average speed of 40 mph. On the return trip, she travels at an average speed of 56 mph and saves two hours of travel time. How far does Mrs. Sumner live from Fullerton? (Lesson 8–4)

51. Sales The Beach Resort is offering two weekend specials. One includes a 2-night stay with 3 meals and costs \$195. The other includes a 3-night stay with 5 meals and costs \$300. (Lesson 8–4)

a. What is the cost of a 1-night stay?

b. What is the cost per meal?

52. Organize Data Abby plans to spend at most \$24 to buy cashews and peanuts for her Fourth of July party. The Nut Shoppe sells peanuts for \$3 a pound and cashews for \$5 a pound. If Abby needs to have at least 5 pounds of nuts for the party, how many of each type can she buy? List three possible solutions. (Lesson 8–5)

A practice test for Chapter 8 is provided on page 794.

ALTERNATIVE ASSESSMENT

COOPERATIVE LEARNING PROJECT

Landscaping Gavin Royse needs a layout of his yard. He is doing some landscaping and needs to have a grid of his yard for the landscape workers to use when putting in his fence and the two new trees that he bought. His house sits on an angle on the lot, which measures 80 feet by 80 feet. The boundaries of his house are described by the four equations below.

$3x - 4y = 55$
$3x + 2y = 85$
$3x - 4y = -125$
$3x + 2y = 175$

The house faces southwest.

The fence will run from the farthest north corner of the house to (5, 80) and then run along the north lot line to the corner of the lot (80, 80). It will then head south on the lot line to (80, 30). From there it will angle toward the house to (55, 5) and then head to the farthest east corner of the house.

A shade tree will be placed at the intersection of $3x - 4y = 55$ and $x + 3y = 170$, while an ornamental tree will be placed 15 feet south of the west corner of the house.

Prepare a graph for the landscape workers to use as a guide. Follow these steps to organize your data.

- Determine how to set up the graph and its scale.
- Construct a colorful and detailed graph that Gavin could give to the landscape workers.
- Create the system of inequalities that describes Gavin's fenced area.
- Determine the vertices of the house.
- Determine the vertices of the two trees.

Write a detailed description for the workers of the work that needs to be done and the area where the work is to be done.

THINKING CRITICALLY

- How does the elimination or substitution method show that a system of equations is inconsistent or that a system of equations is consistent and dependent?
- Create a system of equations for which there are no solutions. What were your criteria for creating this system?

PORTFOLIO

Select one of the systems of equations from this chapter that could be solved by various methods. Use this system of equations and solve it using each of the methods introduced in this chapter: graphing, substitution, elimination using addition and subtraction, and elimination using multiplication and addition. Write an explanation involving the pros and cons of using each of these methods for this problem.

SELF EVALUATION

Are you a team player? Do you pull your end of the load? Do you pull too much of the load? Since you will often need to work in cooperative groups, you need to be responsible for your actions and understand how they affect the group. A person who is too dominant can stifle the learning process of other students. On the other end of the spectrum, a person who is too passive can get left out and not experience the whole learning process.

Assess yourself. What kind of a group worker are you? Do you take the initiative to enhance the whole group or do you look out only for your own learning? Think about a group that you were recently a part of, either in mathematics or your daily life. List two positive actions that you observed in the group that helped the group. Then list two negative actions that you observed in the group that hindered the group.

CUMULATIVE REVIEW

CHAPTERS 1–8

SECTION ONE: MULTIPLE CHOICE

There are eight multiple-choice questions in this section. After working each problem, write the letter of the correct answer on your paper.

1. Choose the statement that is true for a system of two linear equations.

A. There are no solutions when the graphs of the equations are perpendicular lines.

B. There is exactly one solution when the graphs of the equations are one line.

C. A system can only be solved by graphing the equations.

D. There are infinitely many solutions when the graphs of the equations have the same slope and intercepts.

2. The scale on a map is 2 centimeters to 5 kilometers. Doe Creek and Kent are 15.75 kilometers apart. How far apart are they on the map?

A. 7.88 cm

B. 39.38 cm

C. 6.3 cm

D. 31.5 cm

3. Choose the graph that represents $2x - y < 6$.

A.

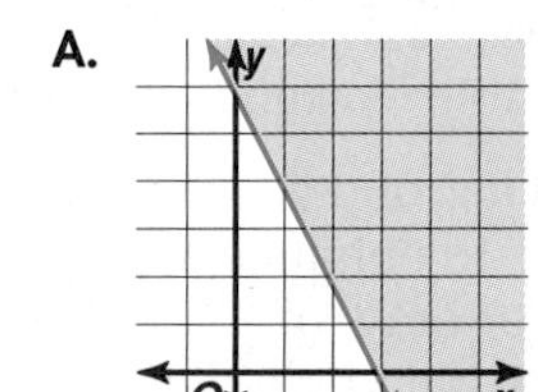

B.

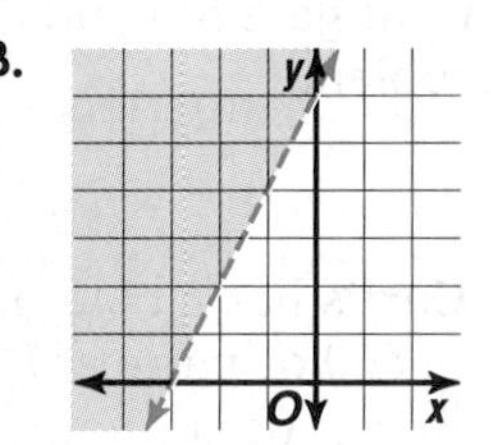

C.

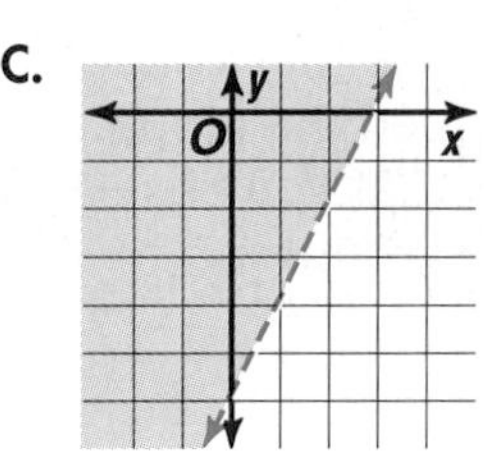

D.

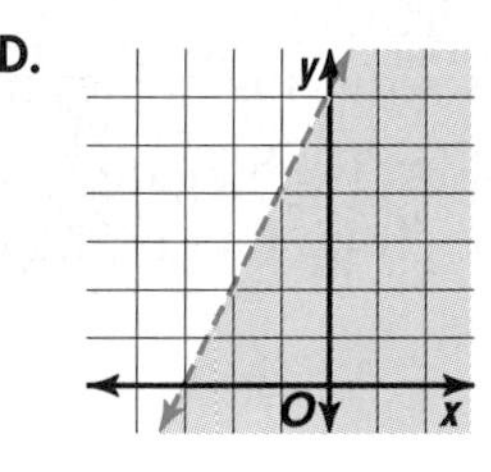

4. The units digit of a two-digit number exceeds twice the tens digit by 1. Find the number if the sum of its digits is 7.

A. 25 **B.** 16

C. 34 **D.** 61

5. State which region in the graph shown below is the solution of the system.

$y \geq 2x + 2$

$y \leq -x - 1$

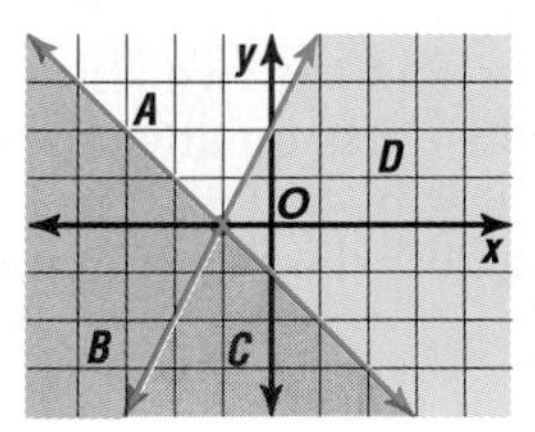

A. Region A

B. Region B

C. Region C

D. Region D

6. Abeytu's scores on the first four of five 100-point tests were 85, 89, 90, and 81. What score must she receive on the fifth test to have an average of at least 87 points for all of the tests?

A. at least 86 points

B. at least 90 points

C. at least 69 points

D. at least 87 points

7. The frequency of a vibrating violin string is inversely proportional to its length. If a 10-inch violin string vibrates at a frequency of 512 cycles per second, find the frequency of an 8-inch violin string.

A. 284.4 cycles per second

B. 409.6 cycles per second

C. 514 cycles per second

D. 640 cycles per second

8. Find the mean of $3\frac{1}{2}$, 5, $4\frac{1}{8}$, $7\frac{3}{4}$, 4, and $6\frac{5}{8}$.

A. $5\frac{1}{6}$

B. 6

C. $6\frac{1}{5}$

D. 31

SECTION TWO: SHORT ANSWER

This section contains seven questions for which you will provide short answers. Write your answer on your paper.

9. Eric is preparing to run in the Bay Marathon. One day, he ran and walked a total of 16 miles. He ran the first mile, and after that walked one mile for every two miles he ran. How many miles did he run, and how many did he walk?

10. The number of Calories in a serving of French fries at 13 restaurants are 250, 240, 220, 348, 199, 200, 125, 230, 274, 239, 212, 240, and 327. Make a box-and-whisker plot of these data.

11. Find three consecutive odd integers whose sum is 81.

12. The concession stand sells hot dogs and soda during Beck High School football games. John bought 6 hot dogs and 4 sodas and paid $6.70. Jessica bought 4 hot dogs and 3 sodas and paid $4.65. At what prices are the hot dogs and sodas sold?

13. Draw a tree diagram to show the possible meals that could be created from the following choices.

Meat: chicken, steak

Vegetable: broccoli, baked potato, tossed salad, carrots

Drink: milk, cola, juice

14. Patricia is going to purchase graduation gifts for her friends, Sarah and Isabel. She wants to spend at least $5 more on Isabel's gift than on Sarah's. She can afford to spend at most a total of $56. Draw a graph showing the possible amounts she can spend on each gift.

15. The cost of a one-day car rental from Rossi Rentals is given by the formula $C(m) = 31 + 0.13m$, where m is the number of miles that the car is driven, $0.13 is the cost per mile driven, and $C(m)$ is the total cost. If Sheila drove a distance of 110 miles and back in one day, what is the cost of the car rental?

SECTION THREE: OPEN-ENDED

This section contains two open-ended problems. Demonstrate your knowledge by giving a clear, concise solution to each problem. Your score on these problems will depend on how well you do the following.

- Explain your reasoning.
- Show your understanding of the mathematics in an organized manner.
- Use charts, graphs, and diagrams in your explanation.
- Show the solution in more than one way or relate it to other situations.
- Investigate beyond the requirements of the problem.

16. Curtis has a coupon for 33% off any purchase of $50 or more at First Place Sports Shop. His grandmother has given him $75 for his birthday. Write a problem about how Curtis spends his money if he wants to buy several items, including a baseball hat for $18.75, a pair of cleats for $53.95, and a $26.95 sweatshirt discounted 15%. Then solve the problem.

17. The graphs of the equations $y = x + 3$, $2x - 7y = 4$, and $2y + 3x = 6$ contain the sides of a triangle. Write a problem about these graphs that requires using systems of equations to solve. Then solve the problem.

CHAPTER

9 Exploring Polynomials

Objectives

In this chapter, you will:

- solve problems by looking for a pattern,
- multiply and divide monomials,
- express numbers in scientific notation, and
- add, subtract, and multiply polynomials.

Keeping Up With Technology

You've probably heard lots of excuses for not having homework done, but this one is probably a new one. Do you use a computer to do your homework? What are the advantages and disadvantages of doing your homework this way?

239 B.C. Leap year introduced into the Egyptian calendar.

A.D. 600 Zu Chong-zhi and his son Zu Geng-shi of China calculate π to be between 3.1415926 and 3.1415927.

1149 Hildegard von Bingen, a female medical practitioner, is named a supervisor at St. Ruperts on the Rhine.

1364 Aztecs build their capital Tenochtitlan at the site of present-day Mexico City.

400 B.C. | 300 | 200 | A.D. 500 | 600 | 700 | 800 | 900 | 1000 | 1100 | 1200 | 1300 | 1400

PEOPLE IN THE NEWS

Jenny Slabaugh and **Amy Gusfa** help produce a weekly video show created by students at Dearborn High School in Michigan. When they were only 15 years old, their expertise in electronics and video won each a $1000 scholarship to the Sony Institute of Technology in Hollywood. They were the first females to receive a scholarship from the Sony Institute, and they spent a week there studying theoretical and applied electronics.

Their interest in video production came from Dearborn High's nationally honored video program. They are combining their interests in science, math, electronics, music, and art to create video programs. They both think that more girls should investigate video technology, and they see themselves contributing to this field in the future.

Chapter Project

A byte is a single unit of information such as a number or a letter processed by a computer. A megabyte is 1.048576×10^6 bytes, which is just over 1 million bytes.

- Research three personal computers and three laptop computers. Find out the amount of RAM and hard drive memory in megabytes for each computer. Write each number in standard notation and in scientific notation.
- Use scientific notation to write the amount of memory in bytes for each computer.

- Find the mean of the memory in megabytes for the three personal computers. Find the mean of the memory in megabytes for the three laptop computers. Write the answers in scientific notation.
- Write the ratio of the mean memory for the personal computers to the mean memory for the laptop computers. Write the ratio as a decimal. Discuss the significance of this ratio.
- Investigate the size of the memory of computers that are being developed for the future. How many bytes are in a gigabyte? How many times larger is a gigabyte than a megabyte?

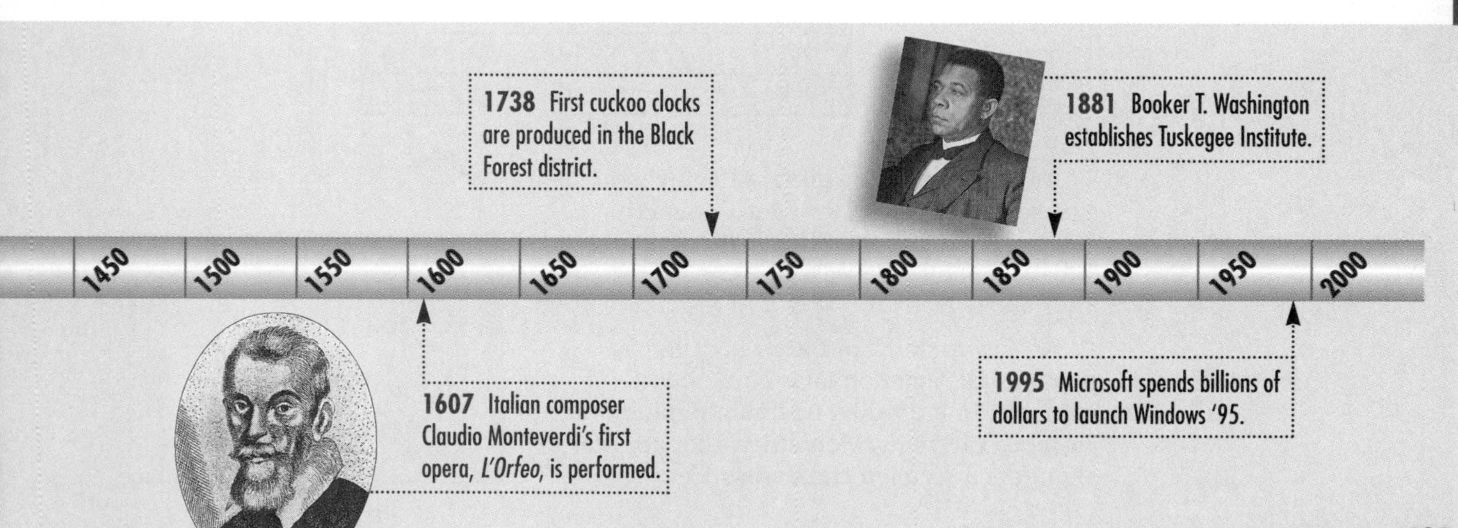

Multiplying Monomials

What YOU'LL LEARN

- To multiply monomials,
- to simplify expressions involving powers of monomials, and
- to solve problems by looking for a pattern.

What YOU'LL LEARN

You can use monomials to solve problems involving finance and geometry.

fabulous FIRSTS

Muriel Siebert (1932–   )

The first woman to hold a seat on the New York Stock Exchange was Muriel ("Mickey") Siebert. She paid $445,000 and was admitted as a full member on December 28, 1967.

Finance

Since 1983, the Beardstown Business and Professional Women's Investment Club ("Beardstown Ladies") has been able to earn enough of a profit in the stock market to make any market expert envious. The club was started when each of 16 women from a small town in Illinois contributed $100 to start an investment fund. Dividends and monthly dues of $25 are also invested. The Beardstown Ladies have earned an average annual profit of 23% on their investments. How has each woman's initial investment of $100 increased over the years?

At the end of the first year (1984), each initial investment would be worth $100(1 + 0.23), or $123. By the end of the second year, the initial investment would have grown to $100(1 + 0.23)(1 + 0.23), which is the same as $100(1 + 0.23)^2$ or $151.29. The table below shows the value of an initial investment of $100 for each of the first 13 years.

Year	Yearly Calculation	Value
1984	$100(1 + 0.23)$	$123.00
1985	$100(1 + 0.23)^2$	$151.29
1986	$100(1 + 0.23)^3$	$186.09
1987	$100(1 + 0.23)^4$	$228.89
1988	$100(1 + 0.23)^5$	$281.53
1989	$100(1 + 0.23)^6$	$346.28
1990	$100(1 + 0.23)^7$	$425.93
1991	$100(1 + 0.23)^8$	$523.89
1992	$100(1 + 0.23)^9$	$644.39
1993	$100(1 + 0.23)^{10}$	$792.59
1994	$100(1 + 0.23)^{11}$	$974.89
1995	$100(1 + 0.23)^{12}$	$1199.12
1996	$100(1 + 0.23)^{13}$	$1474.91

After 13 years, the initial $100 investment was worth $1474.91! If we let x equal the factor $(1 + 0.23)$, then the value of the initial investment after 13 years can be represented by $100x^{13}$.

An expression like $100x^{13}$ is called a **monomial.** A monomial is a number, a variable, or a product of a number and one or more variables. Monomials that are real numbers are called **constants.**

Monomials	Not Monomials
12	$a + b$
q	$\frac{a}{b}$
$4x^3$	$5 - 7d$
$11ab$	$\frac{5}{a^2}$
$\frac{1}{3}xyz^{12}$	$\frac{5a}{7b}$

Recall that an expression of the form x^n is a *power*. The base is x, and the exponent is n. A table of powers of 2 is shown below.

2^1	2^2	2^3	2^4	2^5	2^6	2^7	2^8	2^9	2^{10}
2	4	8	16	32	64	128	256	512	1024

In the following products, each number can be expressed as a power of 2. Study the pattern of the exponents.

Number	$8(32) = 256$	$8(64) = 512$	$4(16) = 64$	$16(32) = 512$
Power	$2^3(2^5) = 2^8$	$2^3(2^6) = 2^9$	$2^2(2^4) = 2^6$	$2^4(2^5) = 2^9$
Pattern of Exponents	$3 + 5 = 8$	$3 + 6 = 9$	$2 + 4 = 6$	$4 + 5 = 9$

These examples suggest that you can multiply powers with the same base by adding exponents.

Product of Powers	**For any number a, and all integers m and n, $a^m \cdot a^n = a^{m+n}$.**

Example 1 **Simplify each expression.**

a. $(3a^6)(a^8)$

$$(3a^6)(a^8) = 3a^{6+8}$$
$$= 3a^{14}$$

b. $(8y^3)(-3x^2y^2)\left(\frac{3}{8}xy^4\right)$

$$(8y^3)(-3x^2y^2)\left(\frac{3}{8}xy^4\right)$$
$$= \left(8 \cdot (-3) \cdot \frac{3}{8}\right)(x^2 \cdot x)(y^3 \cdot y^2 \cdot y^4)$$
$$= -9x^{2+1}y^{3+2+4}$$
$$= -9x^3y^9$$

We looked for a pattern to discover the product of powers property. **Look for a pattern** is an important strategy in problem solving.

Example 2 **Solve by extending the pattern.**

PROBLEM SOLVING
Look for a Pattern

$$4 \times 6 = 24$$
$$14 \times 16 = 224$$
$$24 \times 26 = 624$$
$$34 \times 36 = 1224$$
$$124 \times 126 = ?$$

Explore Look at the problem. You need to find a pattern to determine the product of 124 and 126.

Plan The last two digits of the product are always 24. To find the first digit(s) of the product, look at the tens place of each pair of factors. Notice that $0 \times 1 = 0$, $1 \times 2 = 2$, $2 \times 3 = 6$, and $3 \times 4 = 12$. Extend this pattern to find the product.

Solve $12 \times 13 = 156$

Therefore, $124 \times 126 = 15{,}624$

Examine Use a calculator to verify that the product is 15,624. The pattern remains true and the product is correct.

Study the examples below.

$$(8^3)^5 = (8^3)(8^3)(8^3)(8^3)(8^3) \qquad (y^7)^3 = (y^7)(y^7)(y^7)$$
$$= 8^{3+3+3+3+3} \longleftarrow \textit{Product of powers} \longrightarrow = y^{7+7+7}$$
$$= 8^{15} \qquad = y^{21}$$

Therefore, $(8^3)^5 = 8^{15}$ and $(y^7)^3 = y^{21}$. These examples suggest that you can find the power of a power by multiplying the exponents.

Power of a Power	**For any number a, and all integers m and n, $(a^m)^n = a^{mn}$.**

Look for a pattern in the examples below.

$$(ab)^4 = (ab)(ab)(ab)(ab)$$
$$= (a \cdot a \cdot a \cdot a)(b \cdot b \cdot b \cdot b)$$
$$= a^4b^4$$
$$(5pq)^5 = (5pq)(5pq)(5pq)(5pq)(5pq)$$
$$= (5 \cdot 5 \cdot 5 \cdot 5 \cdot 5)(p \cdot p \cdot p \cdot p \cdot p)(q \cdot q \cdot q \cdot q \cdot q)$$
$$= 5^5p^5q^5 \text{ or } 3125p^5q^5$$

LOOK BACK

You can refer to Lesson 1-1 for information on using a calculator to find a power of a number.

These examples suggest that the power of a product is the product of the powers.

Power of a Product	**For all numbers a and b, and any integer m, $(ab)^m = a^mb^m$.**

The power of a power property and the power of a product property can be combined into the following property.

Power of a Monomial	**For all numbers a and b, and all integers m, n, and p, $(a^mb^n)^p = a^{mp}b^{np}$.**

Example 3 **Simplify $(2a^4b)^3[(-2b)^3]^2$.**

$(2a^4b)^3[(-2b)^3]^2 = (2a^4b)^3(-2b)^6$	*Power of a power property*
$= 2^3(a^4)^3b^3(-2)^6b^6$	*Power of a product property*
$= 8a^{12}b^3(64)b^6$	*Power of a power property*
$= 512a^{12}b^9$	*Product of powers property*

To simplify an expression involving monomials, write an equivalent expression in which:

- there are no powers of powers,
- each base appears exactly once, and
- all fractions are in simplest form.

CHECK FOR UNDERSTANDING

Communicating Mathematics

Study the lesson. Then complete the following.

1. **Write** in your own words.
   a. the product of powers property
   b. the power of a power property
   c. the power of product property
2. **Explain** why the product of powers property does not apply when the bases are different.
3. **You Decide** Luisa says $10^4 \times 10^5 = 100^9$, but Taryn says that $10^4 \times 10^5 = 10^9$. Who is correct? Explain your answer.

4. **Write** 64 in six different ways using exponents; for example, $64 = (2^2)^3$.

Guided Practice

Determine whether each pair of monomials is equivalent. Write *yes* or *no*.

5. $2d^3$ and $(2d)^3$
6. $(xy)^2$ and x^2y^2
7. $-x^2$ and $(-x)^2$
8. $5(y^2)^2$ and $25y^4$

Simplify.

9. $a^4(a^7)(a)$
10. $(xy^4)(x^2y^3)$
11. $[(3^2)^4]^2$
12. $(2a^2b)^2$
13. $(-27ay^3)\left(-\frac{1}{3}ay^3\right)$
14. $(2x^2)^2\left(\frac{1}{2}y^2\right)^2$
15. **Geometry** Find the measure of the area of the rectangle at the right.

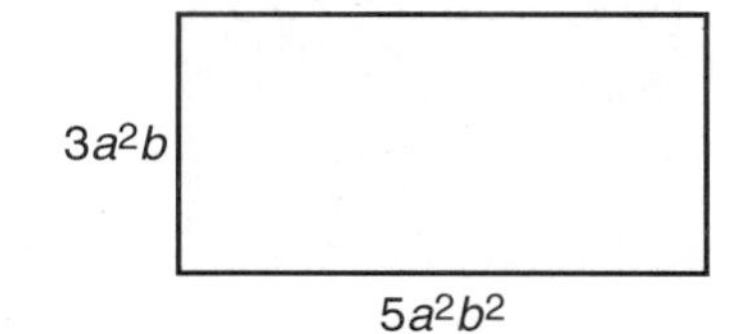

EXERCISES

Practice

Simplify.

16. $b^3(b)(b^5)$
17. $(m^3n)(mn^2)$
18. $(a^2b)(a^5b^4)$
19. $[(2^3)^2]^2$
20. $(3x^4y^3)(4x^4y)$
21. $(a^3x^2)^4$
22. $m^7(m^3b^2)$
23. $(3x^2y^2z)(2x^2y^2z^3)$
24. $(0.6d)^3$
25. $(ab)(ac)(bc)$
26. $-\frac{5}{6}c(12a^3)$
27. $\left(\frac{2}{5}d\right)^2$
28. $-3(ax^3y)^2$
29. $(0.3x^3y^2)^2$
30. $(-3ab)^3(2b^3)$
31. $\left(\frac{3}{10}y^2\right)^2(10y^2)^3$
32. $(3x^2)^2\left(\frac{1}{3}y^2\right)^2$
33. $\left(\frac{2}{5}a\right)^2(25a)(13b)\left(\frac{1}{13}b^4\right)$
34. $(3a^2)^3 + 2(a^3)^2$
35. $(-2x^3)^3 - (2x)^9$

Critical Thinking

36. Explain why $(x + y)^z$ does not equal $x^z + y^z$.
37. Explain why -2^4 does not equal $(-2)^4$.

Applications and Problem Solving

38. **Investments** Refer to the application at the beginning of the lesson. Each of the Beardstown Ladies added $25 to the investments each month. This amounts to $300 a year. Assume that each member invested the $300 at the beginning of each year starting in 1984. You can use the formula $T = p\left[\frac{(1 + r)^t - 1}{r}\right]$ to determine how each member's money grew. T represents the total amount, p represents the regular payment, r represents the annual interest rate, and t represents the time in years.
   a. How much money did each member make from their additional investments from 1984 to 1996?
   b. What was the total value of each member's investment in 1996?

39. **Look for a Pattern** The symbol used for a U.S. dollar is a capital S with a vertical line through it. The line separates the S into 4 parts, as shown at the right. How many parts would there be if the S had 100 vertical lines through it?

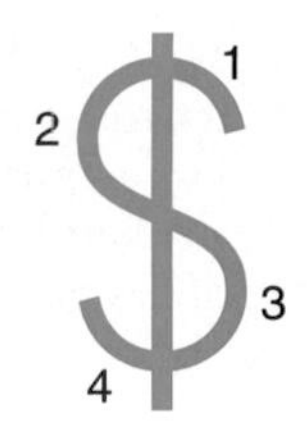

Mixed Review

40. Write a system of inequalities for the graph at the right. (Lesson 8–5)

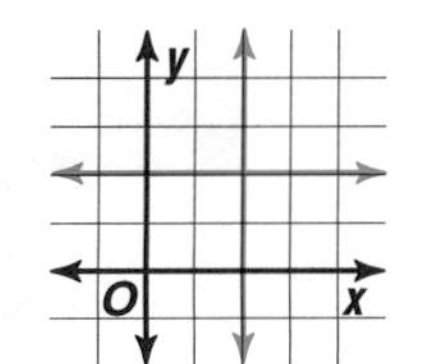

41. Graph the system of equations. Determine whether the system has *one* solution, *no* solution, or *infinitely many* solutions. If the system has one solution, name it. (Lesson 8–1)
$x + 2y = 0$
$y + 3 = -x$

42. **Business** Jorge Martinez has budgeted $150 to have business cards printed. A card printer charges $11 to set up each job and an additional $6 per box of 100 cards printed. What is the greatest number of cards Mr. Martinez can have printed? (Lesson 7–3)

43. Write an equation from the relation shown in the chart below. Then copy and complete the chart. (Lesson 5–6)

m	−3	−2	−1	0	1
n	−5	−3	−1		

44. **Travel** Tiffany wants to reach Dallas at 10 A.M. If she drives at 36 miles per hour, she would reach Dallas at 11 A.M. But if she drives 54 miles per hour, she would arrive at 9 A.M. At what average speed should she drive to reach Dallas exactly at 10 A.M.? (Lesson 4–8)

45. **Geometry** Find the supplement of 44°. (Lesson 3–4)

46. Simplify $-16 \div 8$. (Lesson 2–7)

47. Simplify $0.3(0.2 + 3y) + 0.21y$. (Lesson 1–8)

9-2

Dividing by Monomials

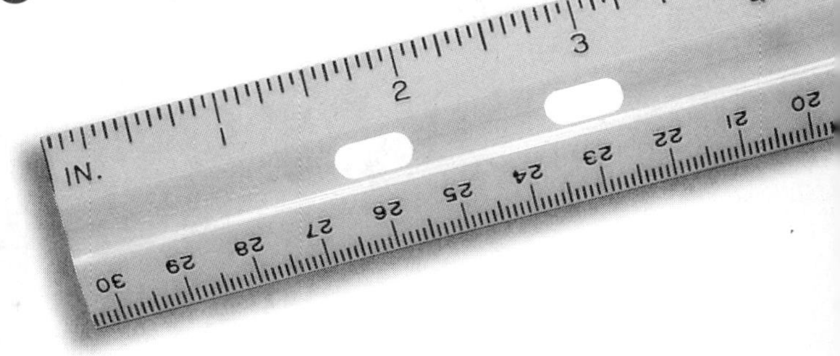

What YOU'LL LEARN

- To simplify expressions involving quotients of monomials, and
- to simplify expressions containing negative exponents.

Why IT'S IMPORTANT

You can use monomials to solve problems involving finance and geometry.

Geometry

The volume of a cube with each side s units long is s^3 cubic units. So, the ratio of the measure of the volume of a cube to the measure of the length of each side is $\frac{s^3}{s}$. How can you express this ratio in simplest form?

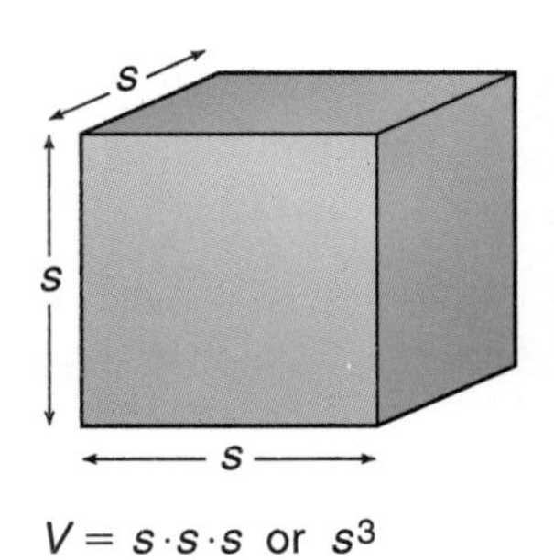

$V = s \cdot s \cdot s$ or s^3

Just as we used a pattern to discover the product of powers property, we can use a pattern to discover a property for a quotient of powers such as $\frac{s^3}{s}$. In the following quotients, each number can be expressed as a power of 2. Study the pattern of exponents.

Number	$\frac{64}{32} = 2$	$\frac{32}{8} = 4$	$\frac{64}{8} = 8$	$\frac{32}{2} = 16$
Power	$\frac{2^6}{2^5} = 2^1$	$\frac{2^5}{2^3} = 2^2$	$\frac{2^6}{2^3} = 2^3$	$\frac{2^5}{2^1} = 2^4$
Pattern of Exponents	$6 - 5 = 1$	$5 - 3 = 2$	$6 - 3 = 3$	$5 - 1 = 4$

These examples suggest that you can divide powers with the same base by subtracting exponents.

Quotient of Powers	**For all integers m and n, and any nonzero number a, $\frac{a^m}{a^n} = a^{m-n}$.**

To write $\frac{s^3}{s}$ in simplest form, subtract the exponents.

$\frac{s^3}{s} = s^{3-1}$ *Recall that $s = s^1$, and apply the quotient of powers property.*

$= s^2$

In simplest form, $\frac{s^3}{s} = s^2$.

Example 1 **Simplify $\frac{y^4z^3}{y^2z^2}$.**

$\frac{y^4z^3}{y^2z^2} = \left(\frac{y^4}{y^2}\right)\left(\frac{z^3}{z^2}\right)$ *Group the powers with the same base, y^4 with y^2 and z^3 with z^2.*

$= (y^{4-2})(z^{3-2})$ *Quotient of powers property.*

$= y^2z$

A calculator can be used to find the value of expressions with 0 as an exponent as well as expressions with negative exponents.

EXPLORATION — SCIENTIFIC CALCULATORS

You can use the $\boxed{y^x}$ key to find the value of expressions with exponents.

Your Turn

a. Copy the following table. Use a scientific calculator to complete the table.

Exponential Expression	2^4	2^3	2^2	2^1	2^0	2^{-1}	2^{-2}	2^{-3}	2^{-4}
Decimal Form									
Fraction Form									

b. According to your calculator display, what is the value of 2^0?

c. Compare the values of 2^2 and 2^{-2}.

d. Compare the values of 2^4 and 2^{-4}.

e. What happens when you evaluate 0^0?

Study the following methods used to simplify $\frac{b^4}{b^4}$ where $b \neq 0$.

How does this value for b^0 compare with the value displayed on your calculator for 2^0?

Method 1
Definition of Powers or Expanded Form

$$\frac{b^4}{b^4} = \frac{\overset{1}{(\not b)}\overset{1}{(\not b)}\overset{1}{(\not b)}\overset{1}{(\not b)}}{\underset{1}{(\not b)}\underset{1}{(\not b)}\underset{1}{(\not b)}\underset{1}{(\not b)}}$$
$$= 1$$

Method 2
Quotient of Powers

$$\frac{b^4}{b^4} = b^{4-4}$$
$$= b^0$$

Since $\frac{b^4}{b^4}$ cannot have two different values, we can conclude that $b^0 = 1$. This example and the Exploration suggest the following definition.

Zero Exponent	**For any nonzero number a, $a^0 = 1$.**

We can also simplify $\frac{r^3}{r^7}$ in two ways.

Did the display on your calculator for 2^{-4} equal $\frac{1}{2^4}$?

Method 1
Definition of Powers or Expanded Form

$$\frac{r^3}{r^7} = \frac{\overset{1}{(\not r)}\overset{1}{(\not r)}\overset{1}{(\not r)}}{\underset{1}{(\not r)}\underset{1}{(\not r)}\underset{1}{(\not r)}(r)(r)(r)(r)}$$
$$= \frac{1}{r^4}$$

Method 2
Quotient of Powers

$$\frac{r^4}{r^7} = r^{3-7} \quad \textit{Quotient of powers}$$
$$= r^{-4}$$

Since $\frac{r^3}{r^7}$ cannot have two values, we conclude that $\frac{1}{r^4} = r^{-4}$. This example and the Exploration suggest the following definition.

Negative Exponents	**For any nonzero number a and any integer n, $a^{-n} = \frac{1}{a^n}$.**

Example 2 Simplify each expression.

a. $\frac{-9m^3n^5}{27m^{-2}n^5y^{-4}}$

$$\frac{-9m^3n^5}{27m^{-2}n^5y^{-4}} = \left(\frac{-9}{27}\right)\left(\frac{m^3}{m^{-2}}\right)\left(\frac{n^5}{n^5}\right)\left(\frac{1}{y^{-4}}\right)$$

$$= \frac{-1}{3}m^{3-(-2)}n^{5-5}y^4 \quad \textit{$\frac{1}{y^{-4}} = y^4$}$$

$$= -\frac{1}{3}m^5n^0y^4 \quad \textit{Subtract the exponents.}$$

$$= -\frac{m^5y^4}{3} \quad \textit{$n^0 = 1$}$$

b. $\frac{(5p^{-2})^{-2}}{(2p^3)^2}$

$$\frac{(5p^{-2})^{-2}}{(2p^3)^2} = \frac{5^{-2}p^4}{2^2p^6} \quad \textit{Power of a monomial property}$$

$$= \left(\frac{1}{2^2}\right)\left(\frac{1}{5^2}\right)\left(\frac{p^4}{p^6}\right)$$

$$= \left(\frac{1}{4}\right)\left(\frac{1}{25}\right)p^{4-6} \quad \textit{Quotient of powers property}$$

$$= \frac{1}{100}p^{-2}$$

$$= \frac{1}{100p^2} \quad \textit{Definition of negative exponents}$$

CHECK FOR UNDERSTANDING

Communicating Mathematics

Study the lesson. Then complete the following.

1. **Explain** why 0^0 is not defined. (*Hint:* Think of computing $\frac{0^m}{0^m}$.)
2. **Explain** why a cannot equal zero in the negative exponents property.
3. **Write** a convincing argument to show that $3^0 = 1$ using the following pattern.

$$3^5 = 243, 3^4 = 81, 3^3 = 27, 3^2 = 9, \ldots$$

4. **Study** the pattern below.

5^5	5^4	5^3	5^2	5^1	5^0	5^{-1}	5^{-2}	5^{-3}	5^{-4}
3125	625	125	25	5	1				

Copy the table and complete the pattern. Write an explanation of the pattern you used.

5. **You Decide** Taigi and Isabel each simplified $\left(\frac{x^{-2}y^3}{x}\right)^{-2}$ correctly as shown below.

Taigi

$$\left(\frac{x^{-2}y^3}{x}\right)^{-2} = \frac{x^4y^{-6}}{x^{-2}}$$
$$= x^{4-(-2)}y^{-6}$$
$$= x^6y^{-6}$$
$$= \frac{x^6}{y^6}$$

Isabel

$$\left(\frac{x^{-2}y^3}{x}\right)^{-2} = (x^{-2-1}y^3)^{-2}$$
$$= (x^{-3}y^3)^{-2}$$
$$= x^6y^{-6}$$
$$= \frac{x^6}{y^6}$$

Whose method do you prefer? Why?

6. **Assess Yourself** Which properties of monomials do you find easy to understand? Which properties do you need to study more?

Guided Practice

Simplify. Assume that no denominator is equal to zero.

7. 11^{-2}
8. $(6^{-2})^2$
9. $\left(\frac{1}{4} \cdot \frac{2}{3}\right)^{-2}$
10. $a^4(a^{-7})(a^0)$
11. $\frac{6r^3}{r^7}$
12. $\frac{(a^7b^2)^2}{(a^{-2}b)^{-2}}$
13. **Geometry** Write the ratio of the area of the circle to the area of the square in simplest form.

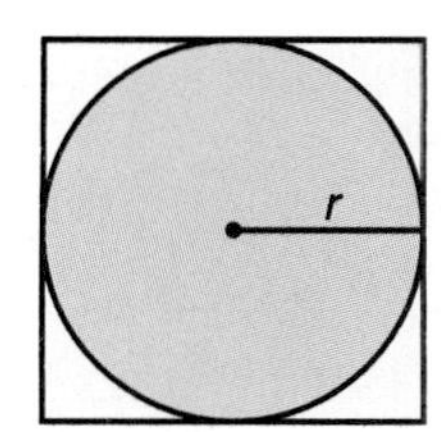

EXERCISES

Practice

Simplify. Assume that no denominator is equal to zero.

14. $a^0b^{-2}c^{-1}$
15. $\frac{a^0}{a^{-2}}$
16. $\frac{5n^5}{n^8}$
17. $\frac{m^2}{m^{-4}}$
18. $\frac{b^5d^2}{b^3d^8}$
19. $\frac{10m^4}{30m}$
20. $\frac{(-y)^5m^8}{y^3m^{-7}}$
21. $\frac{b^6c^5}{b^{14}c^2}$
22. $\frac{22a^2b^5c^7}{-11abc^2}$
23. $\frac{(a^{-2}b^3)^2}{(a^2b)^{-2}}$
24. $\frac{7x^3z^5}{4z^{15}}$
25. $\frac{(-r)^5s^8}{r^5s^2}$
26. $\frac{(r^{-4}k^2)^2}{(5k^2)^2}$
27. $\frac{16b^4}{-4bc^3}$
28. $\frac{27a^4b^6c^9}{15a^3c^{15}}$
29. $\frac{(4a^{-1})^{-2}}{(2a^4)^2}$
30. $\left(\frac{3m^2n^2}{6m^{-1}k}\right)^0$
31. $\frac{r^{-5}s^{-2}}{(r^2s^5)^{-1}}$
32. $\left(\frac{7m^{-1}n^3}{n^2r^{-1}}\right)^{-1}$
33. $\frac{(-b^{-1}c)^0}{4a^{-1}c^2}$
34. $\left(\frac{3xy^{-2}z}{4x^{-2}y}\right)^{-2}$

Critical Thinking

Simplify. Assume that no denominator is equal to zero.

35. $m^3(m^n)$
36. $y^{2c}(y^{5c})$
37. $(3^{2x+1})(3^{2x-7})$
38. $\dfrac{r^{y-2}}{r^{y+3}}$
39. $\dfrac{(q^{y-7})^2}{(q^{y+2})^2}$
40. $\dfrac{y^x}{y^{a-x}}$

Applications and Problem Solving

41. **Finance** You can use the formula $P = A\left[\dfrac{i}{1-(1+i)^{-n}}\right]$ to determine the monthly payment on a home. P represents the monthly payment, A represents the price of the home less the down payment, i represents the *monthly* interest rate (annual rate ÷ 12), and n is the total number of monthly payments. Find the monthly payment on a \$180,000 home with 10% down and an *annual* interest rate of 8.6% over 30 years.

42. **Finance** You can use the formula $B = P\left[\dfrac{1-(1+i)^{k-n}}{i}\right]$ to calculate the balance due on a car loan after a certain number of payments have been made. B represents the balance due (or payoff), P represents the current monthly payment, i represents the *monthly* interest rate (annual rate ÷ 12), k represents the total number of monthly payments already made, and n is the total number of monthly payments. Find the balance due on a 48-month loan for \$10,562 after 20 monthly payments of \$265.86 have been made at an *annual* interest rate of 9.6%.

Mixed Review

43. Simplify $(2a^3)(7ab^2)^2$. (Lesson 9–1)

44. **Aviation** Flying with the wind, a plane travels 300 miles in 40 minutes. Flying against the wind, it travels 300 miles in 45 minutes. Find the air speed of the plane. (Lesson 8–4)

45. Write an open sentence involving the absolute value for the graph at the right. (Lesson 7–6)

–5 –4 –3 –2 –1 0 1 2 3

46. Solve $-\frac{2}{5} > \frac{4z}{7}$. (Lesson 7–2)

47. **Geometry** Find the coordinates of the midpoint of the line segment whose endpoints are $(5, -3)$ and $(1, -7)$. (Lesson 6–7)

48. **Sales** Latoya bought a new dress for \$32.86. This included 6% sales tax. What was the cost of the dress before tax? (Lesson 4–5)

49. Solve $\frac{4-x}{3+x} = \frac{16}{25}$. (Lesson 4–1)

50. Simplify $41y - (-41y)$. (Lesson 2–3)

9-3 Scientific Notation

***What* YOU'LL LEARN**

- To express numbers in scientific and standard notation, and
- to find products and quotients of numbers expressed in scientific notation.

***Why* IT'S IMPORTANT**

You can use scientific notation to express the solutions to many problems.

Transportation

The numbers of passengers arriving at and departing from some major U. S. airports for 1993 are listed in the table below.

Airport	City	Number of Passengers Arriving and Departing (nearest million)
O'Hare International	Chicago	65,000,000
Dallas/Ft. Worth International	Dallas/Ft. Worth	50,000,000
Los Angeles International	Los Angeles	48,000,000
Hartsfield Atlanta International	Atlanta	48,000,000
San Francisco International	San Francisco	32,000,000
Miami International	Miami	29,000,000
J. F. Kennedy International	New York	27,000,000
Newark International	Newark	26,000,000
Detroit Metropolitan Wayne County	Detroit	24,000,000
Logan International	Boston	24,000,000

Source: Air Transport Association of America

When dealing with very large numbers, keeping track of place value can be difficult. For this reason, it is not always desirable to express numbers in standard notation as shown in the chart. Large numbers such as these may be expressed in **scientific notation.**

Definition of Scientific Notation	**A number is expressed in scientific notation when it is in the form $a \times 10^n$, where $1 \leq a < 10$ and n is an integer.**

For example, the number of passengers arriving at and departing from Miami International was about 29,000,000. To write this number in scientific notation, express it as a product of a number greater than or equal to 1, but less than 10, and a power of 10.

$$29{,}000{,}000 = 2.9 \times 10{,}000{,}000$$
$$= 2.9 \times 10^7$$

Scientific notation is also used to express very small numbers. When numbers between zero and one are written in scientific notation, the exponent of 10 is negative.

Example 1

Express each number in scientific notation.

a. 98,700,000,000

$$98{,}700{,}000{,}000 = 9.87 \times 10{,}000{,}000{,}000 = 9.87 \times 10^{10}$$

b. 0.0000056

$$0.0000056 = 5.6 \times 0.000001 = 5.6 \times \frac{1}{1{,}000{,}000} = 5.6 \times \frac{1}{10^6} = 5.6 \times 10^{-6}$$

Example 2

Express each number in standard notation.

a. 3.45×10^5

$$3.45 \times 10^5 = 3.45 \times 100{,}000 = 345{,}000$$

b. 9.72×10^{-4}

$$9.72 \times 10^{-4} = 9.72 \times \frac{1}{10^4} = 9.72 \times \frac{1}{10{,}000} = 9.72 \times 0.0001 = 0.000972$$

CAREER CHOICES

A **pilot** is a highly trained professional who flies planes and helicopters for the transportation of people and cargo, law enforcement, or emergency rescues.

Pilots are required to have a license, which is earned through extensive flight training and written examinations. A college degree is recommended for pilot candidates.

For more information, contact:

Future Aviation Professionals of America
4291J Memorial Dr.
Atlanta, GA 30032

You can use scientific notation to simplify computation with very large numbers and/or very small numbers.

EXPLORATION — GRAPHING CALCULATORS

You can use a graphing calculator to solve problems using scientific notation. First, put your calculator in scientific mode. To enter 3.5×10^9, enter 3.5 [X] 10 [^] 9.

Your Turn

a. Use your calculator to find $(3.5 \times 10^9)(2.36 \times 10^{-3})$.

b. Explain how the calculator calculated the product in part a.

c. Write the product for part a in standard notation.

d. Use your calculator to find $(5.544 \times 10^3) \div (1.54 \times 10^7)$.

e. Explain how the calculator calculated the quotient in part d.

f. Write the quotient for part d in standard notation.

Example 3

Use scientific notation to evaluate each expression.

a. (610)(2,500,000,000) *Estimate: 2.5 billion × 600 = 1.5 trillion*

$$(610)(2{,}500{,}000{,}000) = (6.1 \times 10^2)(2.5 \times 10^9)$$
$$= (6.1 \times 2.5)(10^2 \times 10^9) \quad \textit{Associative property}$$
$$= 15.25 \times 10^{11}$$
$$= 1.525 \times 10^{12} \text{ or } 1{,}525{,}000{,}000{,}000$$

b. (0.000009)(3700) *Estimate: 0.00001 × 3700 = 0.037*

$$(0.000009)(3700) = (9 \times 10^{-6})(3.7 \times 10^3)$$
$$= (9 \times 3.7)(10^{-6} \times 10^3) \quad \textit{Associative property}$$
$$= 33.3 \times 10^{-3}$$
$$= 3.33 \times 10^{-2} \text{ or } 0.0333$$

c. $\dfrac{2.0286 \times 10^8}{3.15 \times 10^3}$ *Estimate:* $\dfrac{210{,}000{,}000}{3000} = 70{,}000$

$$\frac{2.0286 \times 10^8}{3.15 \times 10^3} = \left(\frac{2.0286}{3.15}\right)\left(\frac{10^8}{10^3}\right)$$
$$= 0.644 \times 10^5$$
$$= 6.44 \times 10^4 \text{ or } 64{,}400$$

Scientific notation is extensively used by scientists in fields such as physics and astronomy.

Example 4

Astronomy

A black hole is a region in space where matter seems to disappear. A star becomes a black hole when the radius of the star reaches a certain critical value called the *Schwarzschild radius*. The value is given by the equation $R_s = \frac{2GM}{c^2}$, where R_s is the Schwarzschild radius in meters, G is the gravitational constant (6.7×10^{-11}), M is the mass in kilograms, and c is the speed of light $(3 \times 10^8$ meters per second).

a. The mass of the sun is 2×10^{30} kilograms. Find the Schwarzschild radius of the sun.

b. The actual radius of the sun is 700,000 kilometers. Is it in danger of becoming a black hole in the near future?

FYI

Stephen W. Hawking is a theoretical physicist from Great Britain. Dr. Hawking has made many contributions to the field of science including extensive work dealing with black holes.

a. Use your knowledge of exponents and scientific notation to evaluate the expression for the sun's Schwarzschild radius.

$$R_s = \frac{2GM}{c^2}$$
$$= \frac{2(6.7 \times 10^{-11})(2 \times 10^{30})}{(3 \times 10^8)^2}$$
$$= \frac{2(6.7 \times 10^{-11})(2 \times 10^{30})}{3^2 \times 10^{16}} \quad \textit{Power of a monomial property}$$
$$= \left(\frac{2(6.7)(2)}{3^2}\right)\left(\frac{10^{-11}(10^{30})}{10^{16}}\right)$$
$$= \left(\frac{26.8}{9}\right)10^{-11+30-16} \quad \textit{Product and quotient of powers}$$
$$\approx 2.98 \times 10^3 \text{ or } 2980$$

The Schwarzschild radius for the sun is about 2980 meters.

b. Since the actual radius of the sun is 700,000 kilometers, it does not seem to be in danger of becoming a black hole in the near future.

CHECK FOR UNDERSTANDING

Communicating Mathematics

Study the lesson. Then complete the following.

1. When do you use positive exponents in scientific notation?
2. When do you use negative exponents in scientific notation?
3. **Explain** how you can find the product of (1.2×10^5) and (4×10^8) without using pencil and paper or a calculator.
4. **Explain** how you can find the quotient of (4.4×10^4) and (4×10^7) without using pencil and paper or a calculator.

Guided Practice

Express each number in the second column in standard notation. Express each number in the third column in scientific notation.

	Planet	Maximum Distance from Sun (miles)	Radius (miles)
5.	Mercury	4.34×10^7	1515
6.	Earth	9.46×10^7	3963
7.	Jupiter	5.07×10^8	44,419
8.	Uranus	1.8597×10^9	15,881
9.	Pluto	4.5514×10^9	714

fabulous **FIRSTS**

Maria Goeppert-Mayer (1906–1972)

Maria Goeppert-Mayer was the first American woman to win the Nobel Prize for Physics. She won in 1963 for her work on the theory about the stability of atomic nuclei.

Express each number in scientific notation.

10. **Chemistry** The wavelength of cadmium's green line is 0.0000509 centimeters.
11. **Physics** The mass of a proton is 0.000000000000000000001672 milligrams.
12. **Health** The length of the AIDS virus is 0.00011 millimeters.
13. **Biology** The diameter of an organism called the *Mycoplasma laidlawii* is 0.000004 inch.

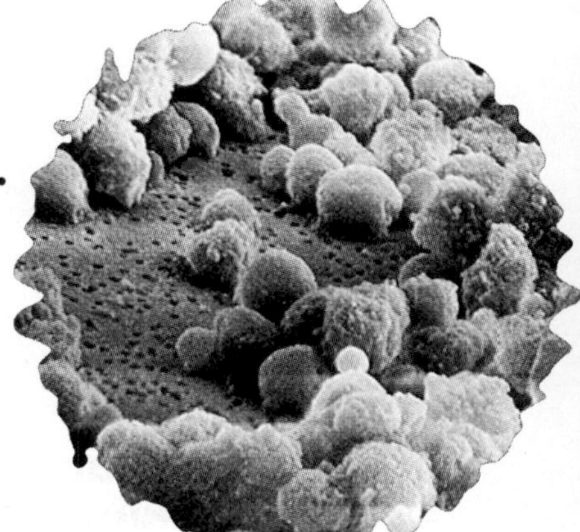

AIDS virus

Evaluate. Express each result in scientific and standard notation.

14. $(3.24 \times 10^3)(6.7 \times 10^4)$
15. $(0.2 \times 10^{-3})(31 \times 10^{-4})$
16. $\dfrac{8.1 \times 10^2}{2.7 \times 10^{-3}}$
17. $\dfrac{52{,}440{,}000{,}000}{(2.3 \times 10^6)(38 \times 10^{-5})}$

EXERCISES

Practice

Express each number in scientific notation.

18. 9500
19. 0.0095
20. 56.9
21. 87,600,000,000
22. 0.000000000761
23. 312,720,000
24. 0.00000008
25. 0.090909
26. 355×10^7
27. 78.6×10^3
28. 112×10^{-8}
29. 0.007×10^{-7}
30. 7830×10^{-2}
31. 0.99×10^{-5}

Evaluate. Express each result in scientific and standard notation.

32. $(6.4 \times 10^3)(7 \times 10^2)$

33. $(4 \times 10^2)(15 \times 10^{-6})$

34. $360(5.8 \times 10^7)$

35. $(5.62 \times 10^{-3})(16 \times 10^{-5})$

36. $\dfrac{6.4 \times 10^9}{1.6 \times 10^2}$

37. $\dfrac{9.2 \times 10^3}{2.3 \times 10^5}$

38. $\dfrac{1.035 \times 10^{-3}}{4.5 \times 10^2}$

39. $\dfrac{2.795 \times 10^{-7}}{4.3 \times 10^{-2}}$

40. $\dfrac{3.6 \times 10^2}{1.2 \times 10^7}$

41. $\dfrac{5.412 \times 10^{-2}}{8.2 \times 10^3}$

42. $\dfrac{(35{,}921{,}000)(62 \times 10^3)}{3.1 \times 10^5}$

43. $\dfrac{1.6464 \times 10^5}{(98{,}000)(14 \times 10^3)}$

Graphing Calculator

Use a graphing calculator to evaluate each expression. Express each answer in scientific notation.

44. $(4.8 \times 10^6)(5.73 \times 10^2)$

45. $(5.07 \times 10^{-4})(4.8 \times 10^2)$

46. $(9.1 \times 10^6) \div (2.6 \times 10^{10})$

47. $(9.66 \times 10^3) \div (3.45 \times 10^{-2})$

Programming

48. The graphing calculator program at the right evaluates the expression $(2ab^2)^3$ for the values you input. It expresses the result in scientific notation. If $a = 9$ and $b = 10$, the result is 5.832E9.

```
PROGRAM:SCINOT
: Sci
: Prompt A, B, C
: Disp "(2AB^2)^3 =",
  (2AB^2)^3
```

Edit the program to evaluate each expression. Then evaluate for $a = 4$, $b = 6$, and $c = 8$.

a. $a^2b^3c^4$ b. $(-2a)^2(4b)^3$ c. $(4a^2b^4)^3$ d. $(ac)^3 + (3b)^2$

Critical Thinking

49. Use a calculator to multiply 3.7×10^{112} and 5.6×10^{10}.
   a. Describe what happens when you multiply these values.
   b. Describe how you could find the product.
   c. Write the product in scientific notation.

Applications and Problem Solving

50. **Biology** Seeds come in all sizes. The largest seed is the seed from the double coconut tree, which can have a mass as great as 23 kilograms. In contrast, the mass of the seed of an orchid is about 3.5×10^{-6} grams. Use a calculator to find how many times greater the mass of the seed of the double coconut tree is than the mass of the seed of the orchid. Express your answer in scientific notation.

Germination of wheat seedling

51. **Movies** In the movie *I.Q.*, Albert Einstein, played by Walter Matthau, tries to start a romance between his niece Catherine Boyd, played by Meg Ryan, and an auto mechanic named Ed Walters, played by Tim Robbins. When Ed asks Catherine to estimate the number of stars in the sky, she answers "$10^{12} + 1$." Write this number in standard notation.

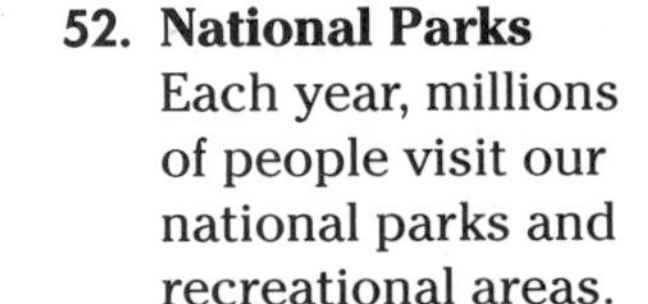

52. **National Parks** Each year, millions of people visit our national parks and recreational areas. The five most popular locations are listed at the right. Write the number of visitors to each location in scientific notation.

Location	State	Visitors in 1994 (nearest thousand)
Golden Gate National Recreation Area	California	15,309,000
Great Smokey Mountains National Park	Tennessee	9,227,000
Lake Mead National Recreational Area	Nevada	9,022,000
Gulf Islands National Seashore	Florida	5,460,000
Cape Cod National Seashore	Massachusetts	5,153,000

Source: *Good Housekeeping*, Sept., 1994

53. **Biology** There are an average of 25 billion red blood cells in the human body and about 270 million hemoglobin molecules in each red blood cell. Use a calculator to find the average number of hemoglobin molecules in the human body. Write your answer in scientific notation.

54. **Health** Laboratory technicians look at bacteria through microscopes. A microscope set on 1000× makes an organism appear to be 1000 times larger than its actual size. Most bacteria are between 3×10^{-4} and 2×10^{-3} millimeters in diameter.
   a. How large would bacteria appear under a microscope set on 1000×?
   b. Do you think a microscope set on 1000× would allow the technician to see all the bacteria? Explain your answer.

55. **Economics** Suppose you try to feed all of the people on Earth using the 4.325×10^{11} kilograms of food produced each year in the United States and Canada. You need to divide this food among the population of the world so that each person receives the same amount of food each day. The population of Earth is about 4.8×10^9.
   a. How much food will each person have each day?
   b. Do you think the amount in part a is enough to live on? Justify your answer.

Mixed Review

56. Simplify $\frac{24x^2y^7z^3}{-6x^2y^3z}$. (Lesson 9–2)

57. Graph the system of inequalities. (Lesson 8–5)

$y \leq -x$
$x \geq -3$

58. Use substitution to solve the system of equations. (Lesson 8–2)

$x = 2y - 12$
$x - 3y = 8$

59. **Statistics** What percent of the data represented at the right is between 15 and 25? (Lesson 7–7)

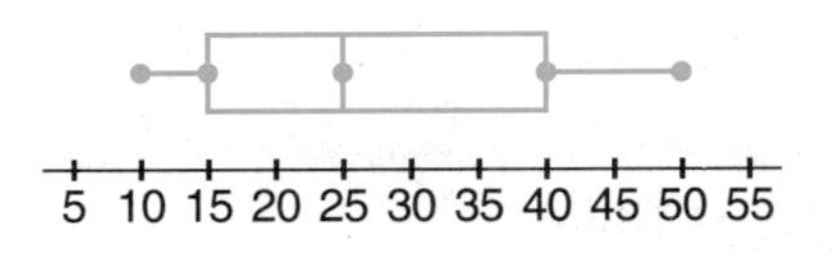

60. Graph the solution set of $x < -3$ or $x \geq 1$. (Lesson 7–4)

61. **Geometry** Write an equation of the line that is perpendicular to the line $5x + 5y = 35$ and passes through the point at $(-3, 2)$. (Lesson 6–6)

62. **Aviation** An airplane passing over Sacramento at an elevation of 37,000 feet begins its descent to land at Reno, 140 miles away. If the elevation of Reno is 4500 feet, what should be the approximate slope of descent? (Lesson 6–1)

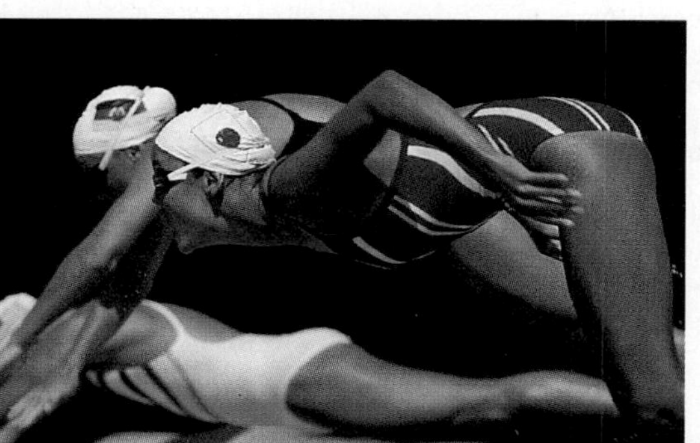

63. Is $\{(3, 4), (5, 4), (7, 5), (9, 5)\}$ a function? (Lesson 5–5)

64. State the domain of $\{(4, 4), (0, 1), (21, 5), (13, 0), (3, 9)\}$. (Lesson 5–2)

65. **Geometry** Suppose $\sin K = 0.4563$. Find the measure of $\angle K$ to the nearest degree. (Lesson 4–3)

66. **Swimming** Rosalinda swims the 50-yard freestyle for the Wachung High School swim team. Her times in the last six meets were 26.89 seconds, 26.27 seconds, 25.18 seconds, 25.63 seconds, 27.16 seconds, and 27.18 seconds. Find the mean and median of her times. (Lesson 3–7)

67. Solve $x - 44 = -207$. (Lesson 3–1)

Measurements Great and Small

The excerpt below appeared in an article in the *New York Times* on September 12, 1993.

Once upon a time, when the world was simpler, a foot was really as long as someone's foot, and a cubit the distance from someone's elbow to the end of the middle finger. Now we measure things that are much smaller than shoes and larger than arms, but there is still a desire to put them into a human context. Take the micron, for example, a unit so small—one millionth of a meter—that it seems hard to cast in a human context. But that doesn't stop newspapers from trying; they inevitably hitch it to something else: a human hair. Earlier this year, describing the one-micron width of a computer circuit, one newspaper article said a hair, by comparison, is 100 microns across. But another article, on cancer-causing soot particles smaller than 10 microns, said a human hair was 75 microns in diameter. Last year, in an article on fiber-optic beams, a human hair was 70 microns; in 1982, in one on coatings for cutting tools, the hair was down to 25 microns. . . . It can be hard to grope with the big, too. *Strategically Speaking,* a marketing newsletter, said recently that 1.8 billion slices of frozen pizza are sold each year—enough to cover 511,366 square miles. How big is that? The newsletter had the sense not to give the answer in microns: enough to cover New York, California, Texas, Maine, Delaware, and Rhode Island combined. ■

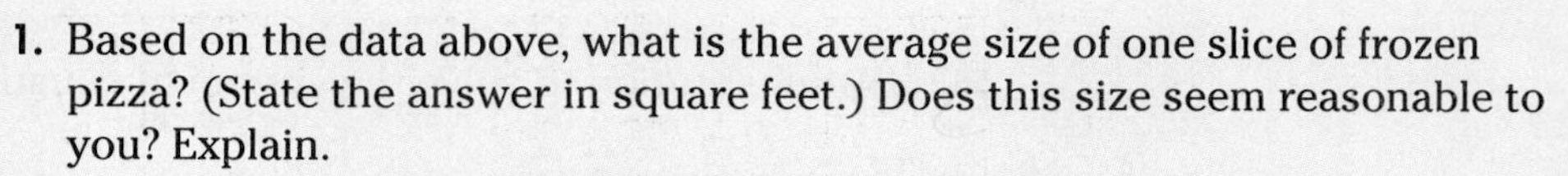

1. Based on the data above, what is the average size of one slice of frozen pizza? (State the answer in square feet.) Does this size seem reasonable to you? Explain.
2. Did you have any difficulty performing the calculations for Exercise 1? How is scientific notation useful in calculations with very large numbers?
3. When you can, do you check the numbers you see presented in newspapers or magazines to see if they are reasonable? Why or why not?

A Preview of Lesson 9–4

9–4A Polynomials

Materials: algebra tiles

Algebra tiles can be used as a model for polynomials. A **polynomial** is a monomial or the sum of monomials. The diagram below shows the models.

Polynomial Models	
Polynomials are modeled using three types of tiles.	1 x x^2
Each tile has an opposite.	−1 $-x$ $-x^2$

Activity Use algebra tiles to model each polynomial.

a. $2x^2$

To model this monomial, you will need 2 blue x^2-tiles.

b. $x^2 - 3x$

To model this polynomial, you will need 1 blue x^2-tile and 3 red x-tiles.

c. $x^2 + 2x - 3$

To model this polynomial, you will need 1 blue x^2-tile, 2 green x-tiles, and 3 red 1-tiles.

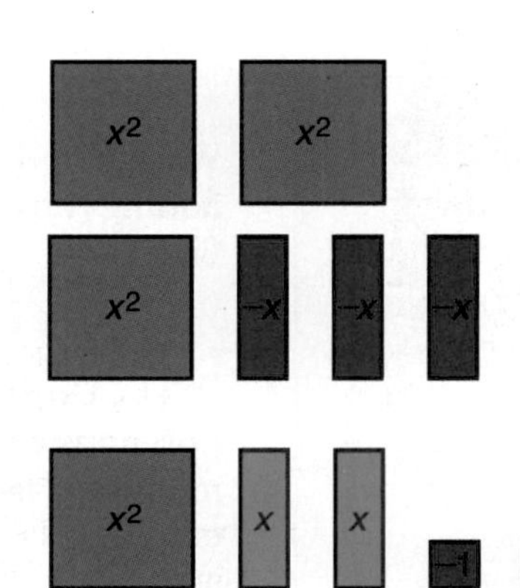

Model Use algebra tiles to model each monomial or polynomial. Then draw a diagram of your model.

1. $-3x^2$
2. $2x^2 - 3x + 5$
3. $2x^2 - 7$
4. $6x - 4$

Write Write each model as an algebraic expression.

5. x^2 $-x$ $-x$ $-x$ 1   1

6. $-x^2$ $-x^2$ x x x x

7. 8.

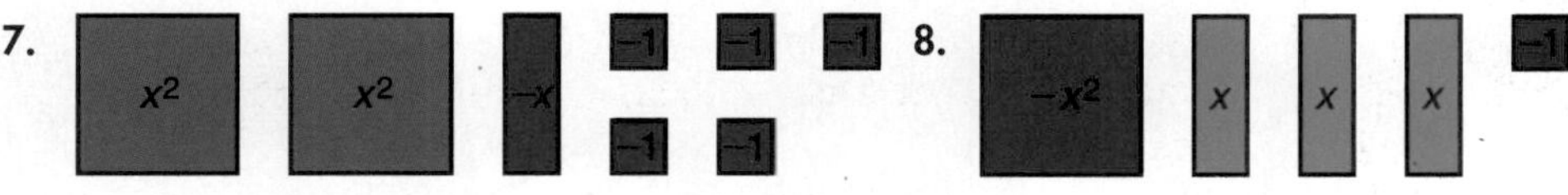

9. Write a few sentences giving reasons why algebra tiles are sometimes called *area tiles.*

9–4 Polynomials

What YOU'LL LEARN

- To find the degree of a polynomial, and
- to arrange the terms of a polynomial so that the powers of a variable are in ascending or descending order.

Why IT'S IMPORTANT

You can use polynomials to solve problems involving agriculture and biology.

Art

The picture at the right shows the painting *Composition with Red, Yellow, and Blue* by the Dutch painter Piet Mondrian. Mondrian preferred abstraction and simplification in his art. He liked to limit his palette to the primary colors and to use straight lines and right angles. The style of painting that Mondrian developed is called *neoplasticism,* and he is considered one of the most influential painters of the 20th century.

Consider a portion of the painting *Composition with Red, Yellow, and Blue.* The length of the sides and the area of each section are given.

	x	y	z
r	rx	ry	rz
s	sx	sy	sz
t	tx	ty	tz

We can find the area of this portion of the painting by adding the areas of each section.

$$rx + ry + rz + sx + sy + sz + tx + ty + tz$$

The expression representing the area of this portion of the painting is called a **polynomial**. A polynomial is a monomial or a sum of monomials. Recall that a monomial is a number, a variable, or a product of numbers and variables. The exponents of the variables of a monomial must be positive. A **binomial** is the sum of two monomials. A **trinomial** is the sum of three monomials. Here are some examples of each. *Polynomials with more than three terms have no special names.*

Monomial	Binomial	Trinomial
$3y^2$	$4x - 7$	$a + 2b + 4c$
$2abc^2$	$2x + 9y$	$x^2 + 8x + 9$
-9	$3x^2 - 11xy$	$x^2 + 2xy + y^2$
$14m$	$2 + 13x$	$3a - 7b^2 - 4c$

Selling Prices of Paintings

1. *Portrait du Dr. Gachet* by van Gogh, $75,000,000
2. *Au Moulin de la Galette* by Renoir, $71,000,000
3. *Les Noces de Pierrette* by Picasso, $51,700,000
4. *Irises* by van Gogh, $49,000,000
5. *Yo Picasso* by Picasso, $43,500,000

***Irises* by van Gogh**

***Au Moulin de la Galette* by Renoir**

Example 1 **State whether each expression is a polynomial. If it is a polynomial, identify it as either a *monomial, binomial,* or *trinomial.***

a. $3a - 7bc$

The expression $3a - 7bc$ can be written as $3a + (-7bc)$. Therefore, it is a polynomial. Since $3a - 7bc$ can be written as a sum of two monomials, $3a$ and $-7bc$, it is a binomial.

LOOK BACK

You can refer to Lesson 1-7 for information on simplifying expressions.

b. $3x^2 + 7a - 2 + a$

The expression $3x^2 + 7a - 2 + a$ can be written as $3x^2 + 8a + (-2)$. Therefore, it is a polynomial. Since it can be written as the sum of three monomials, $3x^2$, $8a$, and -2, it is a trinomial.

c. $\frac{7}{2r^2} + 6$

The expression $\frac{7}{2r^2} + 6$ is not a polynomial because $\frac{7}{2r^2}$ is not a monomial.

Polynomials can be used to represent savings accumulated over several years.

Example 2

Finance

Each of the three summers before Li Chiang attends college, she plans to work as a lifeguard and save \$2000 towards her college expenses. She plans to invest the money in a savings account at her local bank. The value of each year's savings can be found by the expression px^t where p represents the amount invested, x represents the sum of 1 and the annual interest rate, and t represents the time in years.

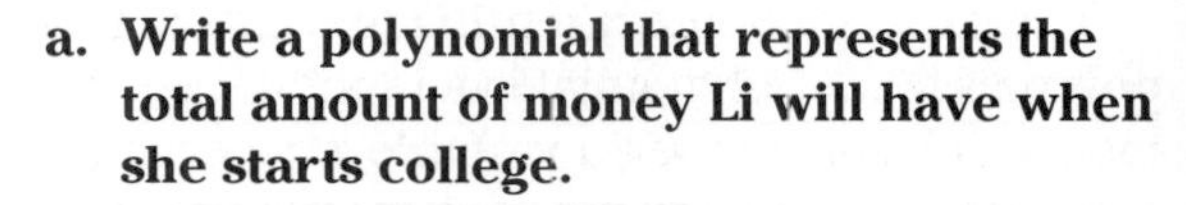

a. Write a polynomial that represents the total amount of money Li will have when she starts college.

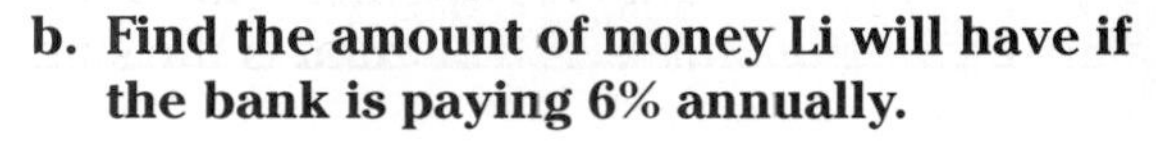

b. Find the amount of money Li will have if the bank is paying 6% annually.

a. By the time Li goes to college, the money she makes the last summer will be worth \$2000. The money she makes the second summer will be worth $2000x$, and money she makes the first summer will be worth $2000x^2$. Altogether, she will have $2000 + 2000x + 2000x^2$.

b. Replace x with 1.06 and calculate the total amount of money Li will have.

$$2000 + 2000(1.06) + 2000(1.06)^2 = 6367.20 \quad x = 1 + 6\% \text{ or } 1.06$$

Li will have \$6367.20 for college.

The **degree** of a monomial is the sum of the exponents of its variables.

Monomial	Degree
$8y^3$	3
$4y^2ab$	$2 + 1 + 1 = 4$
-14	0
$42abc$	$1 + 1 + 1 = 3$

Remember that $a = a^1$ and $b = b^1$.

Remember that $x^0 = 1$ and $-14 = -14x^0$.

To find the degree of a polynomial, you must find the degree of each term. The greatest degree of any term is the degree of the polynomial.

Polynomial	Terms	Degree of the Terms	Degree of the Polynomial
$3x^2 + 8a^2b - 4$	$3x^2, 8a^2b, -4$	2, 3, 0	3
$7x^4 - 9x^2y^7 + 4x$	$7x^4, -9x^2y^7, 4x$	4, 9, 1	9

Example 3 **Find the degree of each polynomial.**

a. $\mathbf{9xy + 2}$

The degree of $9xy$ is 2.

The degree of 2 is 0.

Thus, the degree of $9xy + 2$ is 2.

b. $\mathbf{18x^2 + 21xy^2 + 13x - 2abc}$

The degree of $18x^2$ is 2.

The degree of $21xy^2$ is 3.

The degree of $13x$ is 1.

The degree of $2abc$ is 3.

Thus, the degree of $18x^2 + 21xy^2 + 13x - 2abc$ is 3.

The terms of a polynomial are usually arranged so that the powers of one variable are in ascending or descending order. Later in this chapter, you will learn to add, subtract, and multiply polynomials. These operations are easier to perform if the polynomials are arranged in one of these orders.

Ascending Order	Descending Order
$4 + 5a - 6a^2 + 2a^3$	$2a^3 - 6a^2 + 5a + 4$
$-5 - 2x + 4x^5$	$4x^5 - 2x - 5$
(in x) $8xy - 3x^2y + x^5 - 2x^7y$	(in x) $-2x^7y + x^5 - 3x^2y + 8xy$
(in y) $2x^4 - 3x^3y + 2x^2y^2 - y^{12}$	(in y) $-y^{12} + 2x^2y^2 - 3x^3y + 2x^4$

CHECK FOR UNDERSTANDING

Communicating Mathematics

Study the lesson. Then complete the following.

1. **Explain** why the degree of an integer like -27 is 0.
2. **Explain** why $m + \frac{34}{n}$ is not a binomial.
3. In this lesson, you were introduced to the words *polynomial*, *monomial*, *binomial*, and *trinomial*. These words begin with the prefixes poly-, mono-, bi-, and tri- respectively. Find the meaning of each prefix. List two other words that begin with each prefix, and define each of these words.

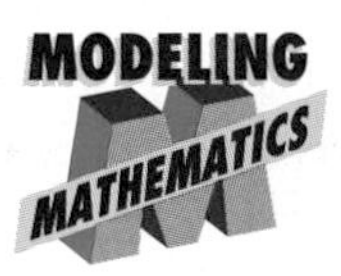

4. The model below represents a polynomial.

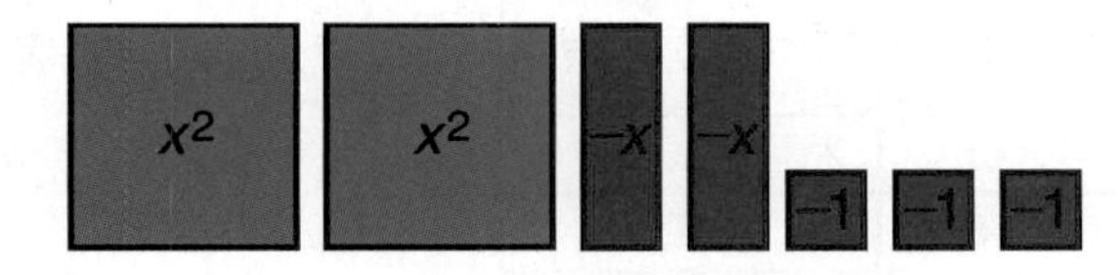

Write this polynomial in simplest form.

Guided Practice

State whether each expression is a polynomial. If the expression is a polynomial, identify it as a *monomial*, a *binomial*, or a *trinomial*.

5. $4x^3 - 11ab + 6$
6. $x^3 - \frac{7}{4}x + \frac{y}{x^2}$
7. $4c + ab - c$

Find the degree of each polynomial.

8. $11d$
9. 10
10. $42x^{12}y^3 - 23x^8y^6$

11. Arrange the terms of $-11x + 5x^3 - 12x^6 + x^8$ so that the powers of x are in descending order.

12. Arrange the terms of $y^4x + y^5x^3 - x^2 + yx^5$ so that the powers of x are in ascending order.

13. **Geometry** The area of a rectangle equals the length times the width. The area of a square equals the square of the length of a side. The area of a circle equals the number pi (π) times the square of the radius. The figure at the right consists of a rectangle, a circle, and a square.

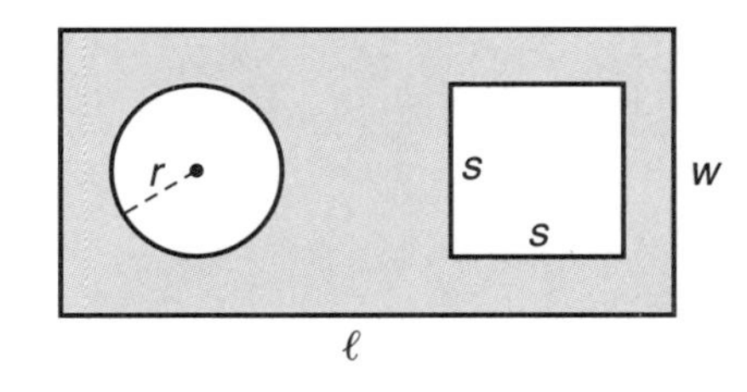

   a. Write a polynomial expression that represents the area of the shaded region at the right.
   b. Find the area of the shaded region if $w = 8$, $\ell = 15$, $r = 2$, and $s = 4$.

EXERCISES

Practice

State whether each expression is a polynomial. If the expression is a polynomial, identify it as a *monomial*, a *binomial*, or a *trinomial*.

14. $\frac{r^5}{26}$
15. $x^2 - \frac{1}{3}x + \frac{y}{234}$
16. $\frac{y^3}{12x}$
17. $5a - 6b - 3a$
18. $\frac{7}{t} + t^2$
19. $9ag^2 + 1.5g^2 - 0.7ag$

Find the degree of each polynomial.

20. $6a^2$
21. $15t^3y^2$
22. 24
23. $m^2 + n^3$
24. $x^2y^3z - 4x^3z$
25. $3x^2y^3z^4 - 18a^5f^3$
26. $8r - 7y + 5d - 6h$
27. $9 + t^2 - s^2t^2 + rs^2t$
28. $-4yzw^4 + 10x^4z^2w$

Arrange the terms of each polynomial so that the powers of x are in descending order.

29. $5 + x^5 + 3x^3$
30. $8x - 9x^2y + 5 - 2x^5$
31. $abx^2 - bcx + 34 - x^7$
32. $7a^3x + 9ax^2 - 14x^7 + \frac{12}{19}x^{12}$

Arrange the terms of each polynomial so that the powers of x are in ascending order.

33. $1 + x^3 + x^5 + x^2$
34. $4x^3y + 3xy^4 - x^2y^3 + y^4$
35. $7a^3x - 8a^3x^3 + \frac{1}{5}x^5 + \frac{2}{3}x^2$
36. $\frac{3}{4}x^3y - x^2 + 4 + \frac{2}{3}x$

Geometry

Write a polynomial to represent the area of each shaded region. Then find the area of each region if $a = 20$, $b = 6$, $c = 2$, $r = 5$, and $x = 1$.

37.

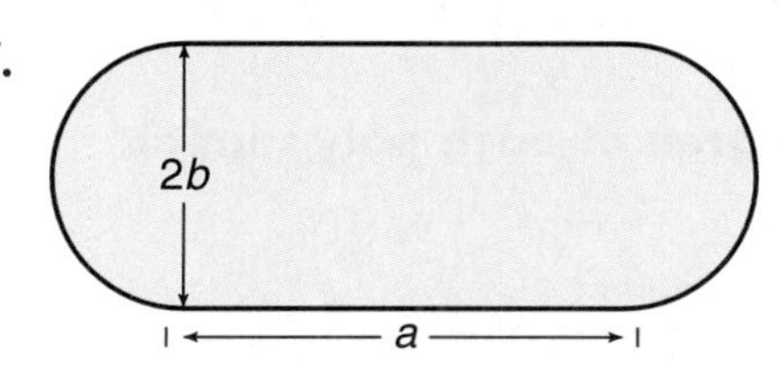

38.

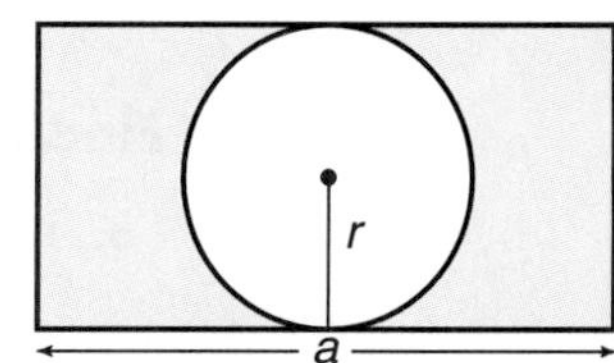

39.

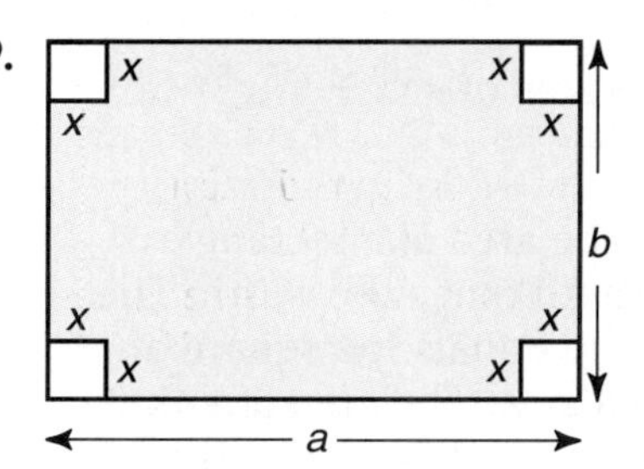

40.

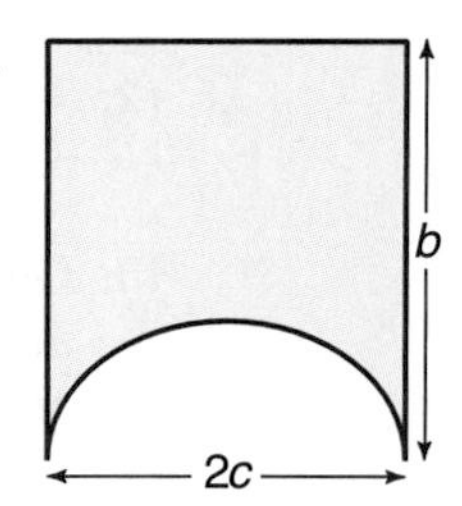

Critical Thinking

41. You can write a numeral in base 10 in polynomial form. For example, $3892 = 3(10)^3 + 8(10)^2 + 9(10)^1 + 2(10)^0$.

a. Write the year of your birth in polynomial form.

b. Suppose 89435 is a numeral in base a. Write 89435 in polynomial form.

Applications and Problem Solving

42. Banking Tawana Hodges inherited \$25,000. She invested the money at an annual interest rate of 7.5%. Each year she added \$2000 of her own money. Will her original investment of \$25,000 double in 7 years? If not, how many years will it take to double the \$25,000?

43. Biology The width of the abdomen of a certain type of female moth is useful in estimating the number of eggs that she can carry. The average number of eggs can be estimated by $14x^3 - 17x^2 - 16x + 34$, where x represents the width of the abdomen in millimeters. About how many eggs would you expect this type of moth to produce if her abdomen measures 2.75 millimeters?

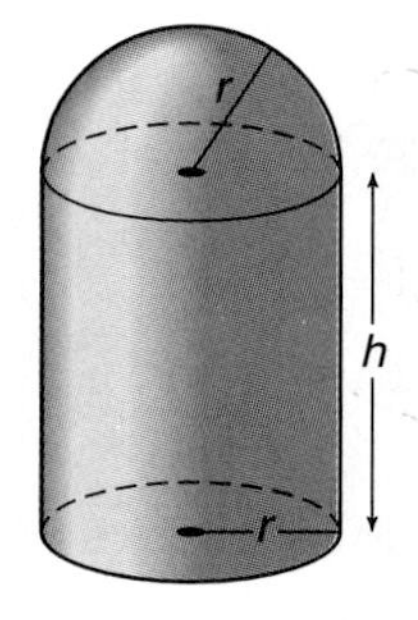

44. Agriculture A diagram of a silo is shown at the left. The volume of a cylinder is the product of π, the square of the radius, and the height. The volume of a sphere is the product of $\frac{4}{3}$, π, and the cube of the radius.

a. Write a polynomial that represents the volume of the silo.

b. If the height is 40 feet and the radius is 8 feet, find the volume of the silo.

Mixed Review

45. Express 42,350 in scientific notation. (Lesson 9–3)

46. Use elimination to solve the system of equations. (Lesson 8–3)

$$\frac{3}{2}x + \frac{1}{5}y = 5$$

$$\frac{3}{4}x - \frac{1}{5}y = -5$$

47. Number Theory If 6.5 times an integer is increased by 11, the result is between 55 and 75. What is the integer? List all possible answers. (Lesson 7–4)

48. **Manufacturing** During one month, Tanisha's Sporting Equipment manufactured a total of 3250 tennis racket covers. Assuming that the planned production of tennis racket covers can be represented by a straight line, determine how many covers will be manufactured by the end of the year. (Lesson 6–4)

49. Solve $4x + 3y = 12$ if the domain is $\{-3, 1, 3, 9\}$. (Lesson 5–3)

50. **Probability** Suppose you meet someone whose birthday is in March. What is the probability that the day of the person's birthday has a 3 in the numeral? (Lesson 4–6)

51. **Insurance** Insurance claims are usually paid based on the depreciated value of the item in question. An item stolen or destroyed with half its useful life remaining would bring a payment equal to half of its original cost. If a battery was stolen when it was 5 months old, find the depreciated value if the battery cost \$45 new and was guaranteed for 36 months. (Lesson 4–1)

52. Solve $ax - by = 2cz$ for y. (Lesson 3–6)

WORKING ON THE Investigation

Refer to the Investigation on pages 448–449.

You can determine an average drop time for each triangle and then use these data to predict glide times for triangles of different sizes.

1. Using either the mean, median, or mode, find an average glide time from the data for each of the four triangles. For each average, explain which measure you used and why you selected that measure to give you the best average or most typical time for the experiment. Are the average glide times best represented by scientific notation? Explain.
2. Describe a hypothetical case for which the mode would be the most appropriate average time, one for which the median would be the most appropriate average time, and one for which the mean would be most appropriate.
3. At this point, you have four average times—one for the typical glide time of each of the four triangles. Draw a scatter plot of the data. Let the independent variable be the perimeter, and let the dependent variable be the average glide time.
4. Describe the graph. Describe any mathematical relationships that you see. What is the relationship of the perimeter of a triangular piece of tissue paper to the speed of the glide?
5. Draw a best-fit line for your data. Write a linear equation to model this best fit line.
6. Draw a second scatter plot in which the horizontal axis represents the area of the triangles and the vertical axis represents the average glide time.
7. Use your scatter plots to predict the glide time for a triangle with a perimeter of 36 centimeters. Use that glide time and the second scatter plot to predict the surface area of that triangle. How does your prediction compare with the actual area of an equilateral triangle whose perimeter is 36 centimeters? Explain your results.

Add the results of your work to your Investigation Folder.

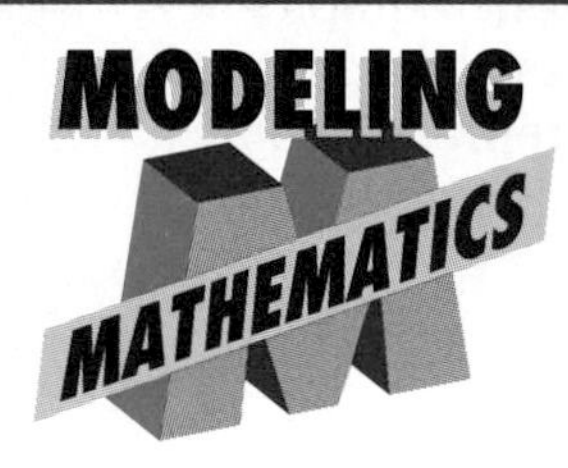

A Preview of Lesson 9–5

9–5A Adding and Subtracting Polynomials

Materials: algebra tiles

Monomials such as $4x^2$ and $-7x^2$ are called *like terms* because they have the same variable to the same power. When you use algebra tiles, you can recognize like terms because the individual tiles have the same size and shape.

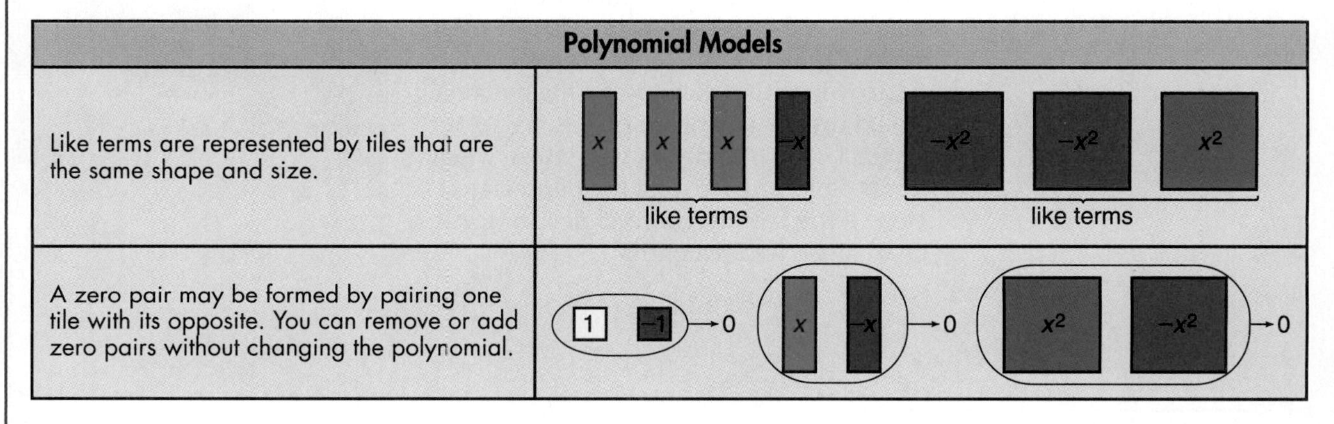

Polynomial Models	
Like terms are represented by tiles that are the same shape and size.	x x x −x (like terms) −x² −x² x² (like terms)
A zero pair may be formed by pairing one tile with its opposite. You can remove or add zero pairs without changing the polynomial.	1 −1 → 0; x −x → 0; x² −x² → 0

Activity 1 Use algebra tiles to find $(2x^2 + 3x + 2) + (x^2 - 5x - 5)$.

Step 1 Model each polynomial. You may want to arrange like terms in columns for convenience.

$2x^2 + 3x + 2 \rightarrow$ $2x^2$, $+3x$, $+2$

$x^2 - 5x - 5 \rightarrow$ x^2, $-5x$, -5

Step 2 Combine like terms and remove all zero pairs.

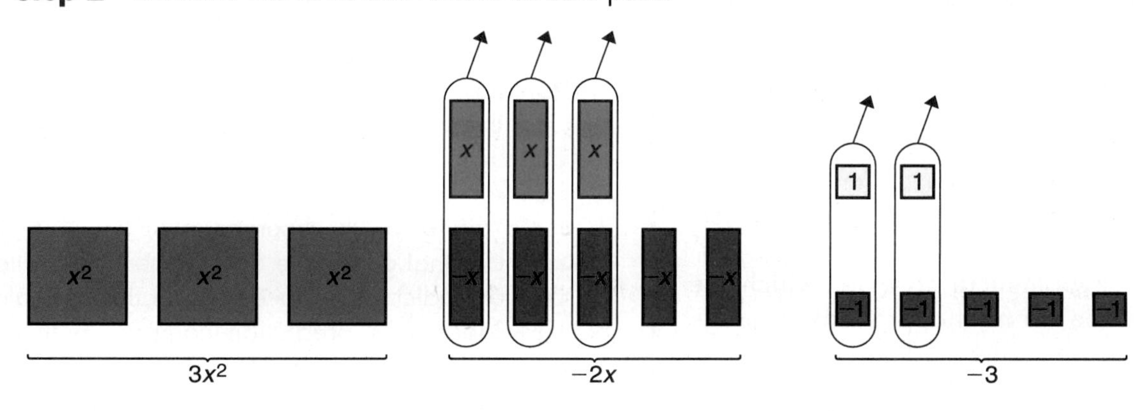

Step 3 Write the polynomial for the tiles that remain.
$(2x^2 + 3x + 2) + (x^2 - 5x - 5) = 3x^2 - 2x - 3$

Activity 2 Use algebra tiles to find $(2x + 5) - (-3x + 2)$.

Step 1 Model the polynomial $2x + 5$.

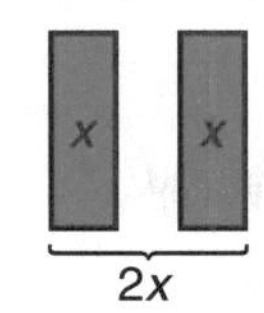

Step 2 To subtract $-3x + 2$, you must remove 3 red x-tiles and 2 yellow 1-tiles. You can remove the yellow 1-tiles, but there are no red x-tiles. Add 3 zero pairs of x-tiles. Then remove the 3 red x-tiles.

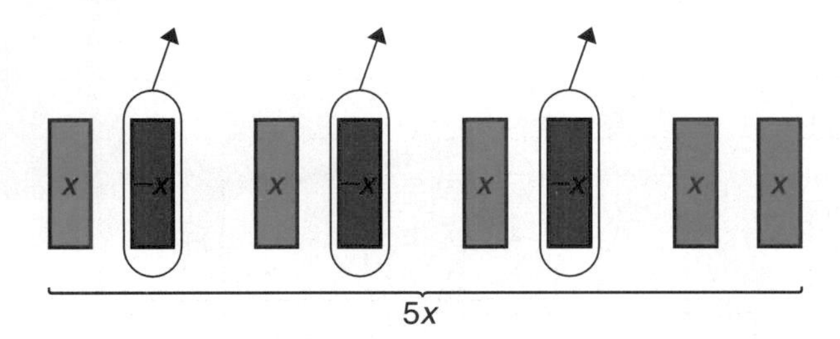

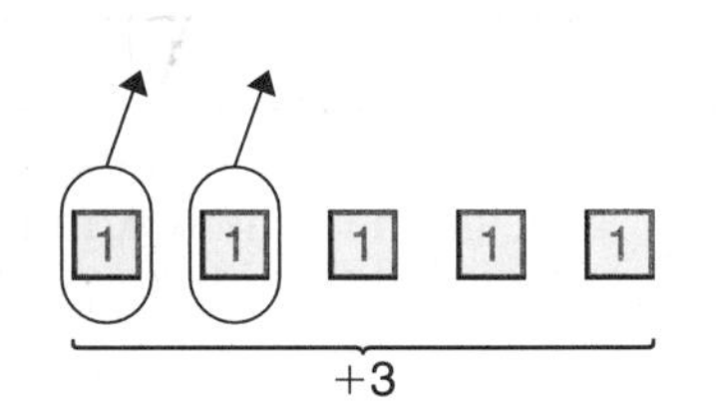

Remember that the value of a zero pair is 0.

Step 3 Write the polynomial for the tiles that remain.
$(2x + 5) - (-3x + 2) = 5x + 3$

Recall that you can subtract a number by adding its additive inverse or opposite. Similarly, you can subtract a polynomial by adding its opposite.

Activity 3 Use algebra tiles and the additive inverse, or opposite, to find $(2x + 5) - (-3x + 2)$.

Step 1 To find the difference of $2x + 5$ and $-3x + 2$, add $2x + 5$ and the opposite of $-3x + 2$.

$2x + 5 \rightarrow$

The opposite of $-3x + 2$ is $3x - 2$. $\rightarrow$

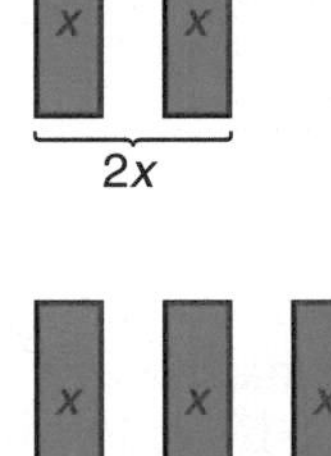

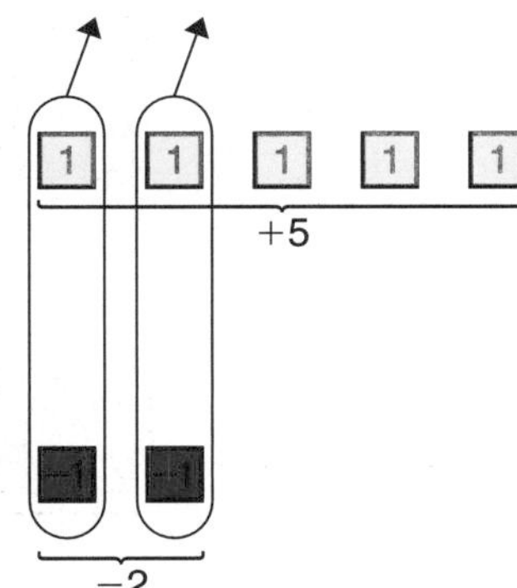

Step 2 Write the polynomial for the tiles that remain.
$(2x + 5) - (-3x + 2) = 5x + 3$ *This is the same answer as in Activity 2.*

Model **Use algebra tiles to find each sum or difference.**

1. $(2x^2 - 7x + 6) + (-3x^2 + 7x)$
2. $(-2x^2 + 3x) + (-7x - 2)$
3. $(x^2 - 4x) - (3x^2 + 2x)$
4. $(3x^2 - 5x - 2) - (x^2 - x + 1)$
5. $(x^2 + 2x) + (2x^2 - 3x + 4)$
6. $(2x^2 + 3x - 4) - (3x^2 - 4x + 1)$

Draw **Is each statement *true* or *false*? Justify your answer with a drawing.**

7. $(3x^2 + 2x - 4) + (-x^2 + 2x - 3) = 2x^2 + 4x - 7$
8. $(x^2 - 2x) - (-3x^2 + 4x - 3) = -2x^2 - 6x - 3$

Write

9. Find $(x^2 - 2x + 4) - (4x + 3)$ using each method from Activity 2 and Activity 3. Illustrate with drawings and explain in writing how zero pairs are used in each case.

9–5 Adding and Subtracting Polynomials

What YOU'LL LEARN

- To add and subtract polynomials.

Why IT'S IMPORTANT

You can use polynomials to solve problems involving architecture and geometry.

Postal Service

The U.S. Postal Service has restrictions on the sizes of boxes that may be shipped by parcel post. The length plus the girth of the box must not exceed 108 inches. Girth is the shortest distance around the package. It is defined as follows.

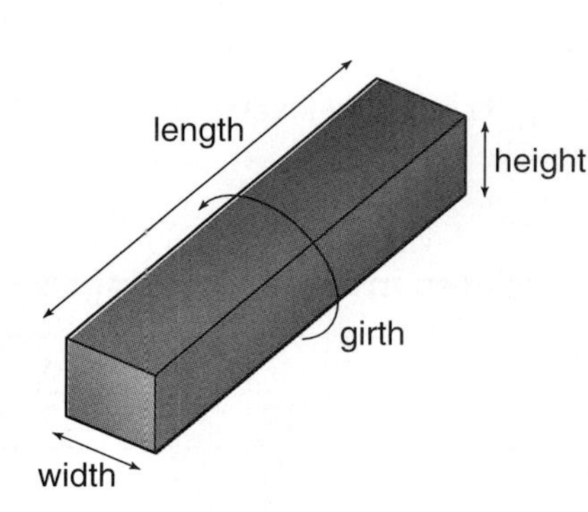

girth = twice the width + twice the height or $w + w + h + h$

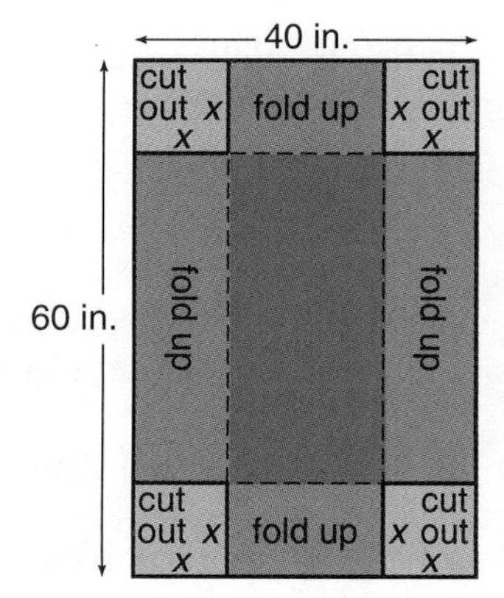

Mrs. Diaz wants to send a package to her daughter, who is a student at Auburn University. She can't find any cardboard boxes at home; all she has is a 60-by-40 inch rectangle of cardboard. She decides that she can make a box out of it by cutting squares out of each corner and folding up the flaps. Then she will use another rectangle of cardboard for the top of the box. She doesn't know how big the squares should be, though. For now, she is calling the side of each square x.

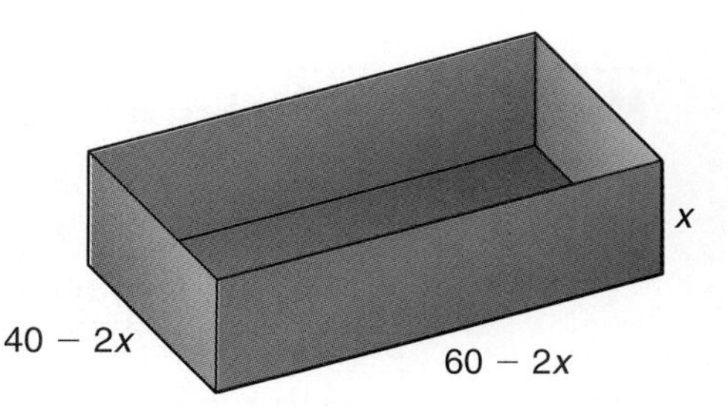

$$\begin{aligned}\text{Girth} &= w + w + h + h \\ &= (40 - 2x) + (40 - 2x) + x + x\end{aligned}$$

In order to mail this box, the length plus the girth must be at most 108 inches. That is, $(60 - 2x) + (40 - 2x) + (40 - 2x) + x + x \leq 108$. *You will solve this problem in Exercise 42.*

To add polynomials, you can group like terms and then find the sum, or you can write them in column form and then add.

Example 1 **Find $(4a^2 + 7a - 12) + (-9a^2 - 6 + 2a)$.**

Method 1

Group the like terms together.

$$\begin{aligned}(4a^2 + 7a - 12) &+ (-9a^2 - 6 + 2a)\\ &= [4a^2 + (-9a^2)] + (7a + 2a) + [-12 + (-6)]\\ &= [4 + (-9)]a^2 + (7 + 2)a + (-18) \quad \textit{Distributive property}\\ &= -5a^2 + 9a - 18\end{aligned}$$

Method 2

Arrange the like terms in column form and add.

$$\begin{array}{rrrr} & 4a^2 & + 7a & - 12 \\ (+) & -9a^2 & + 2a & - 6 \\ \hline & -5a^2 & + 9a & - 18 \end{array}$$

Notice that terms are in descending order and like terms are aligned.

Recall that you can subtract a rational number by adding its opposite or additive inverse. Similarly, you can subtract a polynomial by adding its additive inverse. To find the additive inverse of a polynomial, replace each term with its additive inverse or opposite.

Polynomial	Additive Inverse
$2a - 3b$	$-2a + 3b$
$4x^2 + 7x - 18$	$-4x^2 - 7x + 18$
$-9y + 4x - 2z$	$9y - 4x + 2z$
$7x^3 + 12x^2 + 21$	$-7x^3 - 12x^2 - 21$

Example 2 **Find $(6a^2 - 8a + 12b^3) - (-11a^2 + 6b^3)$.**

Method 1

Find the additive inverse of $-11a^2 + 6b^3$. Then group the like terms and add.

The additive inverse of $-11a^2 + 6b^3$ is $11a^2 - 6b^3$.

$$\begin{aligned}(6a^2 - 8a + 12b^3) &- (-11a^2 + 6b^3)\\ &= (6a^2 - 8a + 12b^3) + (11a^2 - 6b^3)\\ &= (6a^2 + 11a^2) + (-8a) + [12b^3 + (-6b^3)]\\ &= (6 + 11)a^2 - 8a + [12 + (-6)]b^3\\ &= 17a^2 - 8a + 6b^3\end{aligned}$$

Method 2

Arrange like terms in column form and then subtract by adding the additive inverse.

$$\begin{array}{rrrr} & 6a^2 & - 8a & + 12b^3 \\ (-) & -11a^2 & & + 6b^3 \\ \hline \end{array} \quad \rightarrow \quad \begin{array}{rrrr} & 6a^2 & - 8a & + 12b^3 \\ (+) & 11a^2 & & - 6b^3 \\ \hline & 17a^2 & - 8a & + 6b^3 \end{array}$$

Polynomials can be used to represent measures of geometric figures.

Example 3 **The measure of the perimeter of the triangle at the right is represented by $11x^2 - 29x + 10$. Find the polynomial that represents the measure of the third side of the triangle.**

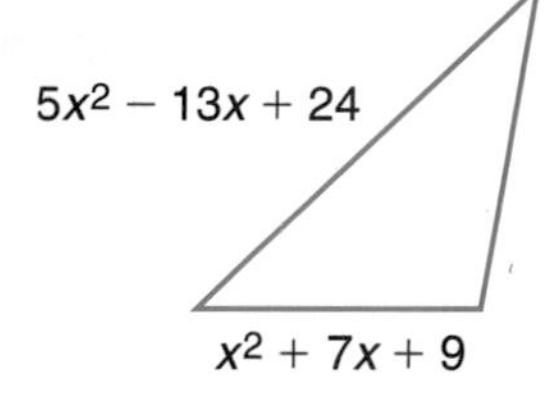

Explore Look at the problem. You know the perimeter of the triangle and the measures of two sides. You need to find the measure of the third side.

Plan The perimeter of a triangle is the sum of the measures of the three sides. To find the measure of the missing side, subtract the two given measures from the perimeter.

Solve

$$\begin{aligned}
&(11x^2 - 29x + 10) - [(5x^2 - 13x + 24) + (x^2 + 7x + 9)]\\
&= (11x^2 - 29x + 10) - (5x^2 - 13x + 24) - (x^2 + 7x + 9)\\
&= (11x^2 - 29x + 10) + (-5x^2 + 13x - 24) + (-x^2 - 7x - 9)\\
&= [11x^2 + (-5x^2) + (-x^2)] + [-29x + 13x + (-7x)] +\\
&\quad [10 + (-24) + (-9)]\\
&= 5x^2 - 23x - 23
\end{aligned}$$

The measure of the third side of the triangle is $5x^2 - 23x - 23$.

Examine The sum of the measures of the three sides should equal the perimeter.

$$\begin{array}{r}
5x^2 - 13x + 24\\
x^2 + 7x + 9\\
(+)\ 5x^2 - 23x - 23\\
\hline
11x^2 - 29x + 10
\end{array}$$

The sum of the measures of the three sides of the triangle equals the measure of the perimeter. The answer is correct.

CHECK FOR UNDERSTANDING

Communicating Mathematics

Study the lesson. Then complete the following.

1. **Describe** the first step you take when you add or subtract polynomials in column form.
2. **Write** three like terms containing powers of a and b. Find the sum of your three terms.
3. **Explain** how to check your answer when you subtract two polynomials.
4. **Write** a paragraph explaining how to subtract a polynomial.

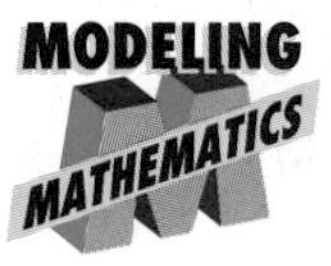

5. Use algebra tiles to find each sum or difference.
   a. $(3x^2 + 2x - 7) + (-2x^2 + 15)$ **b.** $(4x + 1) - (x^2 - 2x + 3)$

Guided Practice

Find the additive inverse of each polynomial.

6. $5y - 7z$
7. $-6a^2 + 3$
8. $7y^2 - 3x^2 + 2$
9. $-4x^2 - 3y^2 + 8y + 7x$

Name the like terms in each group.

10. $3m, 8n, 4mn, 5n, 6m$
11. $-8y^2, 2x, 3y^2, 4x, 2z$
12. $2x^3, 5xy, -x^2y, 14xy, 12xy$
13. $3p^3q, -2p, 10p^3q, 15pq, -p$

Find each sum or difference.

14. $\begin{array}{r} 5ax^2 + 3a^2x \quad\quad - 5x \\ (+)\ 2ax^2 \quad\quad - 5ax + 7x \\ \hline \end{array}$

15. $\begin{array}{r} 11m^2n^2 + 2mn - 11 \\ (-)\ 5m^2n^2 - 6mn + 17 \\ \hline \end{array}$

16. $(4x^2 + 5x) + (-7x^2 + x)$
17. $(3y^2 + 5y - 6) - (7y^2 - 9)$
18. $(5b - 7ab + 8a) - (5ab - 4a)$
19. $(6p^3 + 3p^2 - 7) + (p^3 - 6p^2 - 2p)$
20. **Geometry** The sum of the degree measures of the angles of a triangle is 180.
    a. Write a polynomial to represent the measure of the third angle of the triangle at the right.
    b. If $x = 15$, find the measures of the three angles of the triangle.

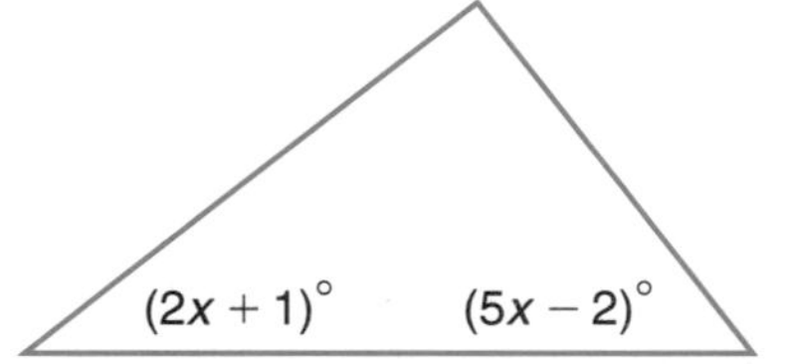

EXERCISES

Practice **Find each sum or difference.**

21. $\begin{array}{r} 4x^2 + 5xy - 3y^2 \\ (+)\ 6x^2 + 8xy + 3y^2 \\ \hline \end{array}$

22. $\begin{array}{r} 6x^2y^2 - 3xy - 7 \\ (-)\ 5x^2y^2 + 2xy + 3 \\ \hline \end{array}$

23. $\begin{array}{r} a^3 \quad\quad\quad\quad\quad - b^3 \\ (+)\ 3a^3 + 2a^2b - b^2 + 2b^3 \\ \hline \end{array}$

24. $\begin{array}{r} 3a^2 \quad\quad - 8 \\ (-)\ 5a^2 + 2a + 7 \\ \hline \end{array}$

25. $\begin{array}{r} 3a + 2b - 7c \\ -4a + 6b + 9c \\ (+)\ -3a - 2b - 7c \\ \hline \end{array}$

26. $\begin{array}{r} 2x^2 - 5x + 7 \\ 5x^2 \quad\quad - 3 \\ (+)\ x^2 - x + 11 \\ \hline \end{array}$

27. $(5a - 6m) - (2a + 5m)$
28. $(3 + 2a + a^2) + (5 - 8a + a^2)$
29. $(n^2 + 5n + 13) + (-3n^2 + 2n - 8)$
30. $(5x^2 - 4) - (3x^2 + 8x + 4)$
31. $(13x + 9y) - 11y$
32. $(5ax^2 + 3ax) - (2ax^2 - 8ax + 4)$
33. $(3y^3 + 4y - 7) + (-4y^3 - y + 10)$
34. $(7p^2 - p - 7) - (p^2 + 11)$
35. $(4z^3 + 5z) + (-2z^2 - 4z)$
36. $(x^3 - 7x + 4x^2 - 2) - (2x^2 - 9x + 4)$

Geometry

The measures of two sides of a triangle are given. *P* represents the measure of the perimeter. Find the measure of the third side.

37. $P = 5x + 2y$

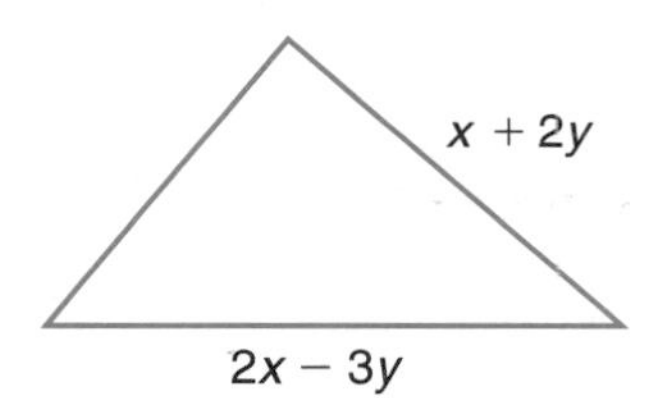

38. $P = 13x^2 - 14x + 12$

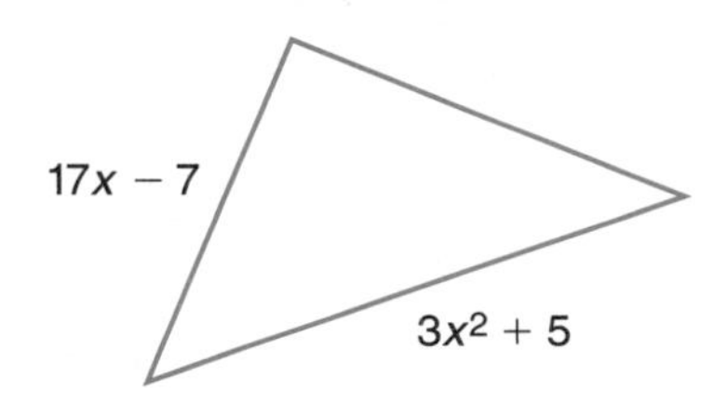

The sum of the degree measures of the angles of a quadrilateral is 360. Find the measure of the fourth angle of each quadrilateral given the degree measures of the other three angles.

39. $4x + 12, 8x - 10, 6x + 5$

40. $x^2 - 2, x^2 + 5x - 9, -8x - 42$

Critical Thinking

41. Your teacher gives you two polynomials. When the first polynomial is subtracted from the second polynomial, the difference is $2n^2 + n - 4$. What is the difference when the second polynomial is subtracted from the first?

Applications and Problem Solving

42. Postal Service Refer to the application at the beginning of the lesson.

a. Solve the inequality to find the possible values of x Mrs. Diaz could use in designing her package.

b. For reasons other than Postal Service regulations, what is the greatest integral value x could have?

c. Use a calculator to find the volume of the box when x is the minimum value and when x is the maximum value.

d. What can you conclude about the volume of this box at various heights?

43. Architecture The Sears Tower in Chicago is one of the tallest structures in the world. It is actually a building of varying heights as shown in the photo at the left. The diagram below indicates the height of each section in stories.

x	50 stories	89 stories	66 stories
x	110 stories	110 stories	89 stories
x	66 stories	89 stories	50 stories
	x	x	x

Use stories as a unit of measure. Assume that each section is x stories long and x stories wide. Write an expression for the volume of the Sears Tower.

44. Look for a Pattern At City Center Mall, there are 25 lockers numbered 1 through 25. Suppose a shopper opens every locker. Then a second shopper closes every second locker. Next a third shopper changes the state of every third locker. (If it's open, the shopper closes it. If it's closed, the shopper opens it.) Suppose this process continues until the 25th shopper changes the state of the 25th locker.

a. Which lockers will still be open?

b. Describe the numbers in your answer for part a.

c. Give the next three numbers for the pattern you found in part a.

d. If there were n lockers in this pattern, which lockers will be open?

Mixed Review

45. Arrange the terms of $-3x + 4x^5 - 2x^3$ so that the powers of x are in ascending order. (Lesson 9–4)

46. Finance The current balance on a car loan can be found by evaluating the expression $P\left[\frac{1-(1+r)^{k-n}}{r}\right]$, where P is the monthly payment, r is the monthly interest rate, k is the number of payments already made, and n is the total number of monthly payments. Find the current balance if $P = \$256$, $r = 0.01$, $k = 20$, and $n = 60$. (Lesson 9–2)

47. Solve the system of equations by graphing. Then state the solution of the system of equations. (Lesson 8–1)

$5x - 3y = 12$
$2x - 5y = 1$

48. **Statistics** What is the upper quartile of the data represented by the box-and-whisker plot at the right? (Lesson 7–7)

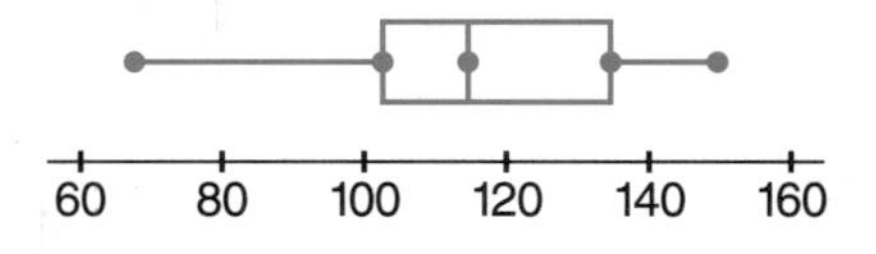

49. **Business** Janet's Garage charges \$83 for a two-hour repair job and \$185 for a five-hour repair job. Define the variables and write a linear equation that Janet can use to bill customers for repair jobs of any length of time. (Lesson 6–4)

50. **Finance** The selling price of \$145,000 for a home included a 6.5% commission for a real estate agent. How much money did the owners receive from the sale? (Lesson 4–5)

51. Solve $\frac{-3n - (-4)}{-6} = -9$. (Lesson 3–3)

52. **Tennis** The diameter of a circle is the distance across the circle. If the diameter of a tennis ball is $2\frac{1}{2}$ inches, how many tennis balls will fit in a can 12 inches high? (Lesson 2–7)

SELF TEST

Simplify. Assume that no denominator is equal to zero. (Lessons 9–1 and 9–2)

1. $(-2n^4y^3)(3ny^4)$
2. $(-3a^2b^5)^2$
3. $\frac{24a^3b^6}{-2a^2b^2}$
4. $\frac{(5r^{-1}s)^3}{(s^2)^3}$

Express each number in scientific notation. (Lesson 9–3)

5. 5,670,000
6. 0.86×10^{-4}
7. **Space Exploration** A space probe that is 2.85×10^9 miles away from Earth sends radio signals back to NASA. If the radio signals travel at the speed of light (186,000 miles per second), how long will it take the signals to reach NASA? (Lesson 9–3)
8. Find the degree of the polynomial $11x^2 + 7ax^3 - 3x + 2a$. Then write the polynomial so that the powers of x are in ascending order. (Lesson 9–4)

Find each sum or difference. (Lesson 9–5)

9. $(x^2 + 3x - 5) + (4x^2 - 7x - 9)$
10. $(2a - 7) - (2a^2 + 8a - 11)$

MODELING MATHEMATICS

9–6A Multiplying a Polynomial by a Monomial

A Preview of Lesson 9–6

Materials: algebra tiles product mat

You have used rectangles to model multiplication. In this activity, you will use algebra tiles to model the product of simple polynomials. The width and length of a rectangle will represent a monomial and a polynomial, respectively. The area of the rectangle will represent the product of the monomial and the polynomial.

Activity 1 Use algebra tiles to find $x(x - 4)$.

The rectangle will have a width of x units and a length of $(x - 4)$ units. Use your algebra tiles to mark off the dimensions on a product mat. Then make the rectangle with algebra tiles.

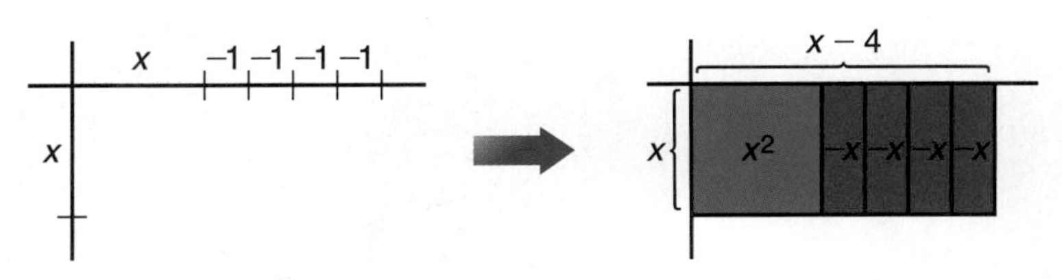

The rectangle consists of 1 blue x^2-tile and 4 red x-tiles.
The area of the rectangle is $x^2 - 4x$. Therefore, $x(x - 4) = x^2 - 4x$.

Activity 2 Use algebra tiles to find $2x(x + 2)$.

The rectangle will have a width of $2x$ units and a length of $(x + 2)$ units. Make the rectangle with algebra tiles.

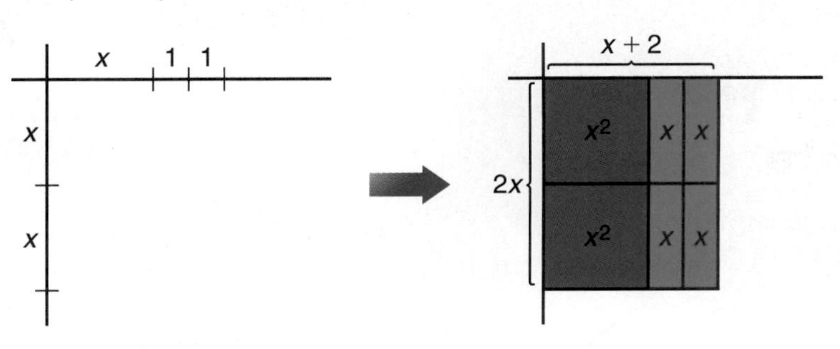

The rectangle consists of 2 blue x^2-tiles and 4 green x-tiles.
The area of the rectangle is $2x^2 + 4x$. Therefore, $2x(x + 2) = 2x^2 + 4x$.

Model **Use algebra tiles to find each product.**

1. $x(x + 2)$
2. $x(x - 3)$
3. $2x(x + 1)$
4. $2x(x - 3)$
5. $x(2x + 1)$
6. $3x(2x - 1)$

Draw **Is each statement *true* or *false*? Justify your answer with a drawing.**

7. $x(2x + 4) = 2x^2 + 4x$
8. $2x(3x - 4) = 6x^2 - 8$

Write

9. Suppose you have a square storage building that measures x feet on a side. You triple the length of the building and increase the width by 15 feet.
   a. What will be the dimensions of the new building?
   b. What is the area of the new building? Write your solution in paragraph form, complete with drawings.

9-6 Multiplying a Polynomial by a Monomial

What YOU'LL LEARN

- To multiply a polynomial by a monomial, and
- to simplify expressions involving polynomials.

Why IT'S IMPORTANT

You can use monomials and polynomials to solve problems involving travel and recreation.

APPLICATION

Recreation

Have you ever played the game of hopscotch? Children from all over the world like to play some form of this game. The following diagrams show versions from different countries.

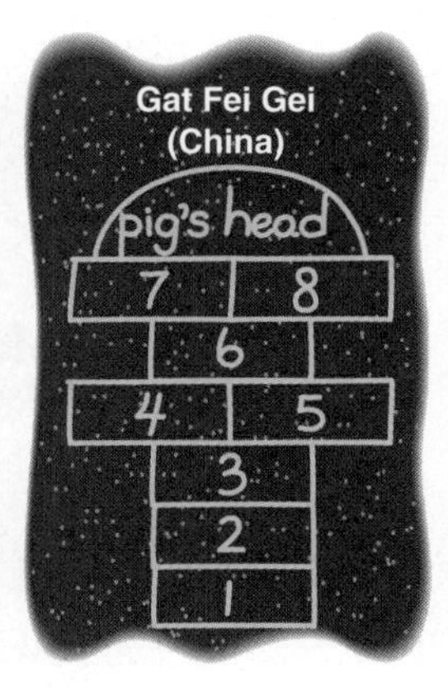

FYI

Tourists can still see the hopscotch pattern carved into the floor of the ancient Forum in Rome.

On the Caribbean island of Trinidad, the children play Jumby. The pattern for this game is shown at the right. Suppose the dimensions of each rectangle are x and $x + 14$. To find the area of each rectangle, you must multiply its length by its width.

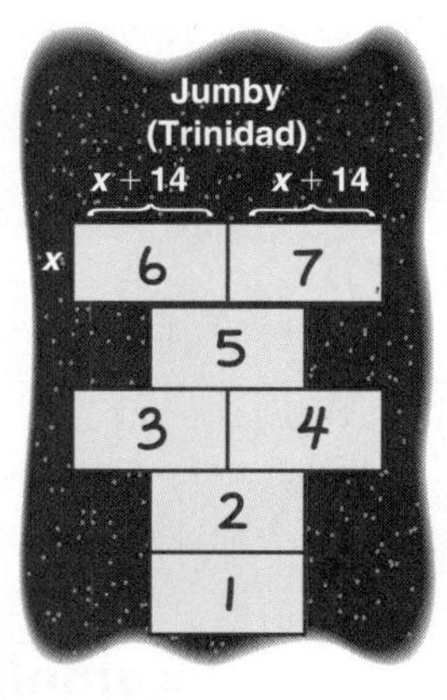

This diagram of one of the rectangles shows that the area is $x(x + 14)$.

	$x + 14$
x	$x(x + 14)$

This diagram of the same rectangle shows that the area is $x^2 + 14x$.

	x	14	
x	x^2	$14x$	x

Since the areas are equal, $x(x + 14) = x^2 + 14x$.

The application above shows how the distributive property can be used to multiply a polynomial by a monomial.

Example 1 **Find each product.**

a. $7b(4b^2 - 18)$

You can multiply horizontally or vertically.

Method 1: Horizontal

$$7b(4b^2 - 18) = 7b(4b^2) - 7b(18)$$
$$= 28b^3 - 126b$$

Method 2: Vertical

$$\begin{array}{r} 4b^2 - 18 \\ (\times) \quad 7b \\ \hline 28b^3 - 126b \end{array}$$

b. $-3y^2(6y^2 - 8y + 12)$

$$-3y^2(6y^2 - 8y + 12) = -3y^2(6y^2) - (-3y^2)(8y) + (-3y^2)(12)$$
$$= -18y^4 + 24y^3 - 36y^2$$

Some expressions may contain like terms. In these cases, you will need to simplify by combining like terms.

Example 2 **Find $-3pq(p^2q + 2p - 3p^2q)$.**

Method 1

Multiply first and then simplify by combining like terms.

$$-3pq(p^2q + 2p - 3p^2q) = -3pq(p^2q) + (-3pq)(2p) - (-3pq)(3p^2q)$$
$$= -3p^3q^2 - 6p^2q + 9p^3q^2$$
$$= 6p^3q^2 - 6p^2q$$

Method 2

Simplify by combining like terms and then multiply.

$$-3pq(p^2q + 2p - 3p^2q) = -3pq(-2p^2q + 2p)$$
$$= -3pq(-2p^2q) + (-3pq)(2p)$$
$$= 6p^3q^2 - 6p^2q$$

p^2q and $-3p^2q$ are like terms.

Example 3

APPLICATION

Track

The runners in a 200-meter dash race around the curved part of a track. If the runners start and finish at the same line, the runner on the outside lane would run farther than the other runners. To compensate for this situation, the starting points of the runners are staggered. If the radius of the inside lane is x and each lane is 2.5 feet wide, how far apart should the officials start the runners in the two inside lanes?

The formula for the circumference C of a circle is $C = 2\pi r$, where r is the radius of the circle. The distance around half of a circle is πr. Use this information to find the distance around the curve for the two inside lanes and subtract the quantities to find the stagger distance.

Finish

x + 2.5

x

Start

$$\underbrace{\pi(x + 2.5)}_{\text{outside semicircle}} - \underbrace{\pi x}_{\text{inside semicircle}} = \pi x + 2.5\pi - \pi x \quad \textit{Distributive property}$$

$$= 2.5\pi \quad \textit{Combine like terms.}$$

The two runners should start 2.5π, or about 7.9, feet apart.

Many equations contain polynomials that must be added, subtracted, or multiplied before the equation can be solved. The distributive property is frequently used as at least one of the steps in solving equations.

Example 4 **Solve $x(x + 3) + 7x - 5 = x(8 + x) - 9x + 14$.**

$$x(x + 3) + 7x - 5 = x(8 + x) - 9x + 14$$

$$x^2 + 3x + 7x - 5 = 8x + x^2 - 9x + 14 \quad \textit{Distributive property}$$

$$x^2 + 10x - 5 = x^2 - x + 14 \quad \textit{Combine like terms.}$$

$$10x - 5 = -x + 14 \quad \textit{Subtract } x^2 \textit{ from each side.}$$

$$11x - 5 = 14 \quad \textit{Add } x \textit{ to each side.}$$

$$11x = 19 \quad \textit{Add 5 to each side.}$$

$$x = \frac{19}{11} \quad \textit{Divide each side by 11.}$$

The solution is $\frac{19}{11}$.

CHECK FOR UNDERSTANDING

Communicating Mathematics

Study the lesson. Then complete the following.

1. **Name** the property used to simplify $3a(5a^2 + 2b - 3c^2)$.
2. Refer to the application at the beginning of the lesson.
   a. **Describe** how you could find the area of the entire pattern used to play Jumby.
   b. **Write** an expression in simplest form for the area of this pattern.
   c. If x represents 8 inches, find the area of the pattern.
3. Refer to Example 4.
   a. **Explain** how you could check the solution to the equation.
   b. **Check** the solution.

4. A rectangular garden is $2x + 3$ units long and $3x$ units wide.
   a. Draw a model of the garden.
   b. Find the area of the garden.

Guided Practice

Find each product.

5. $-7b(9b^3c + 1)$
6. $4a^2(-8a^3c + c - 11)$
7. $\begin{array}{r} 5y - 13 \\ \underline{(\times)\ 2y} \end{array}$
8. $\begin{array}{r} 2ab - 5a \\ \underline{(\times)\ 11ab} \end{array}$

Simplify.

9. $w(3w - 5) + 3w$

10. $4y(2y^3 - 8y^2 + 2y + 9) - 3(y^2 + 8y)$

Solve each equation.

11. $12(b + 14) - 20b = 11b + 65$

12. $x(x - 4) + 2x = x(x + 12) - 7$

13. **Number Theory** Suppose a is an even integer.
    a. Write the product, in simplest form, of a and the next integer after it.
    b. Write the product, in simplest form, of a and the next even integer after it.

EXERCISES

Practice

Find each product.

14. $-7(2x + 9)$

15. $\frac{1}{3}x(x - 27)$

16. $3st(5s^2 + 2st)$

17. $-4m^3(5m^2 + 2m)$

18. $3d(4d^2 - 8d - 15)$

19. $5m^3(6m^2 - 8mn + 12n^3)$

20. $7x^2y(5x^2 - 3xy + y)$

21. $-4d(7d^2 - 4d + 3)$

22. $2m^2(5m^2 - 7m + 8)$

23. $-8rs(4rs + 7r - 14s^2)$

24. $-\frac{3}{4}ab^2\left(\frac{1}{3}abc + \frac{4}{9}a - 6\right)$

25. $\frac{4}{5}x^2(9xy + \frac{5}{4}x - 30y)$

Simplify.

26. $b(4b - 1) + 10b$

27. $3t(2t - 4) + 6(5t^2 + 2t - 7)$

28. $8m(-9m^2 + 2m - 6) + 11(2m^3 - 4m + 12)$

29. $8y(11y^2 - 2y + 13) - 9(3y^3 - 7y + 2)$

30. $\frac{3}{4}t(8t^3 + 12t - 4) + \frac{3}{2}(8t^2 - 9t)$

31. $6a^2(3a - 4) + 5a(7a^2 - 6a + 5) - 3(a^2 + 6a)$

Solve each equation.

32. $2(5w - 12) = 6(-2w + 3) + 2$

33. $7(x - 12) = 13 + 5(3x - 4)$

34. $\frac{1}{2}(2d - 34) = \frac{2}{3}(6d - 27)$

35. $p(p + 2) + 3p = p(p - 3)$

36. $y(y + 12) - 8y = 14 + y(y - 4)$

37. $x(x - 3) - x(x + 4) = 17x - 23$

38. $a(a + 8) - a(a + 3) - 23 = 3a + 11$

39. $t(t - 12) + t(t + 2) + 25 = 2t(t + 5) - 15$

Geometry

Find the measure of the area of each shaded region in simplest terms.

40.

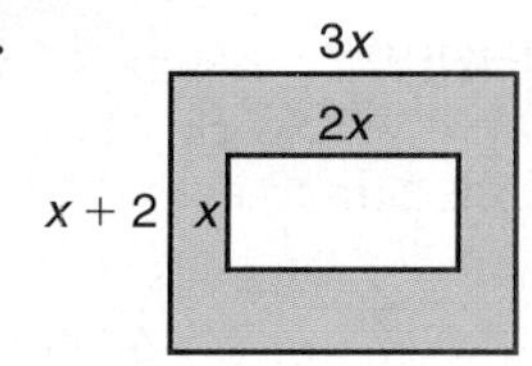

41.

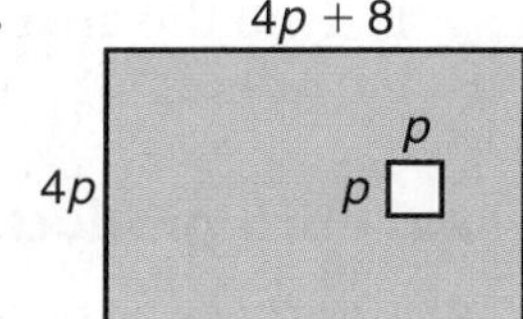

42.

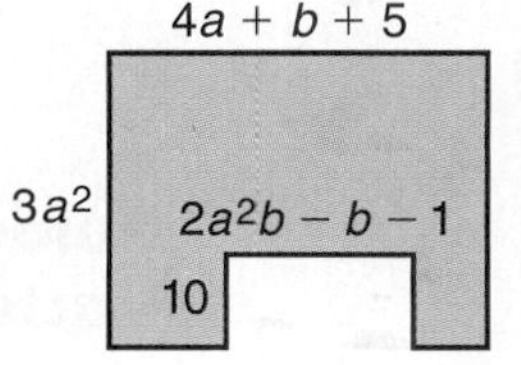

Critical Thinking

43. Write eight multiplication problems whose product is $8a^2b + 18ab$.

Applications and Problem Solving

44. Recreation In Honduras, children play a form of hopscotch called La Rayuela. The pattern for this game is shown at the right. Suppose that each rectangle is $2y + 1$ units long and y units wide.

a. Write an expression in simplest form for the area of the pattern.

b. If y represents 9 inches, find the area of the pattern.

45. Travel The Drama Club of Lincoln High School is visiting New York City. They plan to take taxis from the World Trade Center to the Metropolitan Museum of Art. The fare for a taxi is \$2.75 for the first mile and \$1.25 for each additional mile. Suppose the distance between the two locations is m miles and t taxis are needed to transport the entire group. Write an expression in simplest form for the cost to transport the group to the Metropolitan Museum of Art excluding the tip.

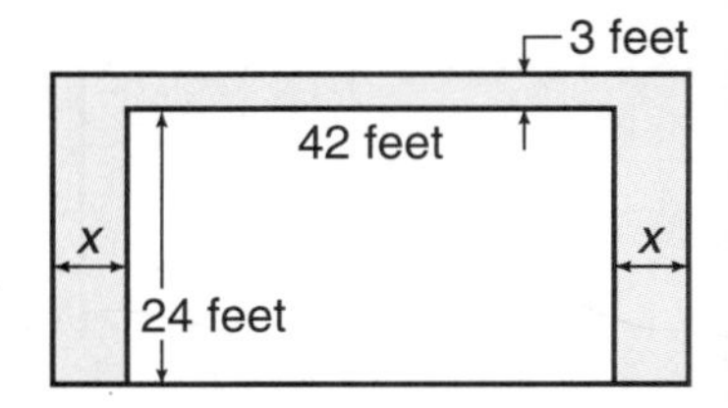

46. Construction A landscaper is designing a rectangular garden for an office complex. There will be a concrete walkway on three sides of the garden, as shown at the right. The width of the garden will be 24 feet, and the length will be 42 feet. The width of the longer portion of the walkway will be 3 feet. The concrete will cost \$20 per square yard, and the builders have told the landscaper that she can spend \$820 on the concrete. How wide should the two remaining sides be?

47. Geometry The number of diagonals that can be drawn for a polygon with n sides is represented by the expression $\frac{1}{2}n(n - 3)$.

a. Draw polygons with 3, 4, 5, and 6 sides. Show that this expression is true for these polygons.

b. Find the product of this expression.

c. How many diagonals can be drawn for a polygon with 15 sides?

Mixed Review

48. Find $(3a - 4ab + 7b) - (7a - 3b)$. (Lesson 9–5)

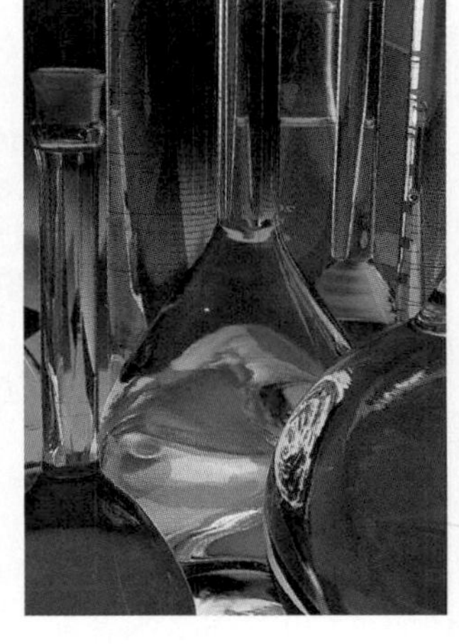

49. Chemistry One solution is 50% glycol, and another is 30% glycol. How much of each solution should be mixed to make a 100-gallon solution that is 45% glycol? (Lesson 8–2)

50. Determine the slope of the line that passes through $A(1, 5)$ and $B(-3, 0)$. (Lesson 6–1)

51. If $g(x) = x^2 + 2x$, find $g(a - 1)$. (Lesson 5–5)

52. Draw the graph of $2x - y = 8$. (Lesson 5–4)

53. Finance Antonio earned \$340 in 4 days by mowing lawns and doing yard work. At this rate, how long will it take him to earn \$935? (Lesson 4–8)

54. Geometry Find the measure of an angle that is 44° less than its complement. (Lesson 3–4)

55. Evaluate $(15x)^3 - y$ if $x = 0.2$ and $y = 1.3$. (Lesson 1–3)

A Preview of Lesson 9–7

9–7A Multiplying Polynomials

Materials: algebra tiles product mat

You can find the product of binomials by using algebra tiles.

Activity 1 **Use algebra tiles to find $(x + 2)(x + 3)$.**

The rectangle will have a width of $x + 2$ and a length of $x + 3$. Use your algebra tiles to mark off the dimensions on a product mat. Then make the rectangle with algebra tiles.

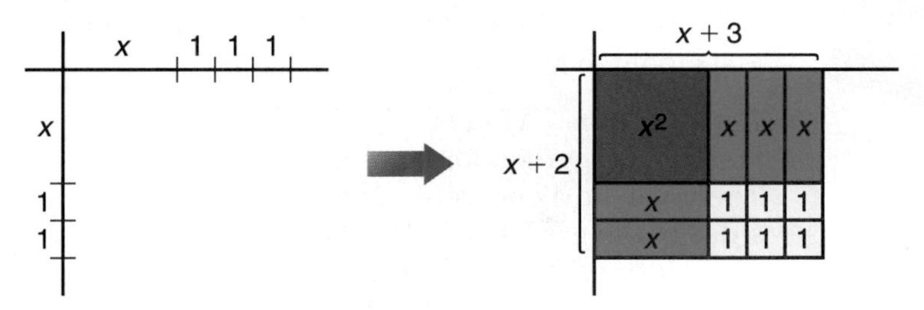

The rectangle consists of 1 blue x^2-tile, 5 green x-tiles, and 6 yellow 1-tiles.
The area of the rectangle is $x^2 + 5x + 6$. Therefore, $(x + 2)(x + 3) = x^2 + 5x + 6$.

Activity 2 **Use algebra tiles to find $(x - 1)(x - 3)$.**

Step 1 The rectangle will have a width of $(x - 1)$ units and a length of $(x - 3)$ units. Use your algebra tiles to mark off the dimensions on a product mat. Then begin to make the rectangle with algebra tiles.

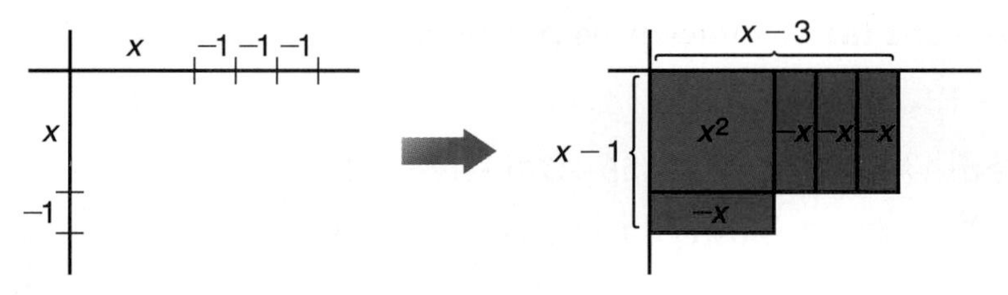

Step 2 Determine whether to use 3 yellow 1-tiles or 3 red 1-tiles to complete the rectangle. Remember that the numbers at the top and side give the dimensions of the tile needed. The area of each tile is the product of -1 and -1. This is represented by a yellow 1-tile. Fill in the space with 3 yellow 1-tiles to complete the rectangle.

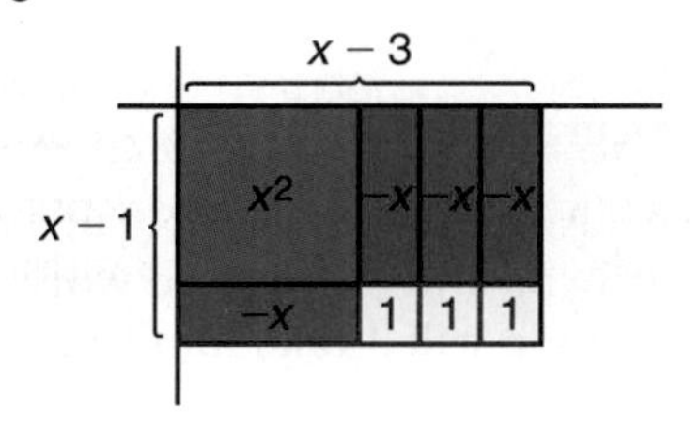

The rectangle consists of 1 blue x^2-tile, 4 red x-tiles, and 3 yellow 1-tiles.
The area of the rectangle is $x^2 - 4x + 3$. Therefore, $(x - 1)(x - 3) = x^2 - 4x + 3$.

Activity 3 Use algebra tiles to find $(x + 1)(2x - 1)$.

Step 1 The rectangle will have a width of $(x + 1)$ units and a length of $(2x - 1)$. Use your algebra tiles to mark off the dimensions on a product mat. Then begin to make the rectangle with algebra tiles.

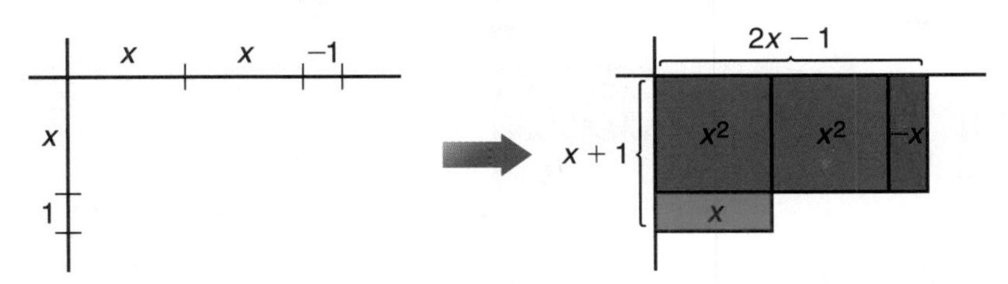

Step 2 Determine what color x-tile and what color 1-tile to use to complete the rectangle. The area of the x-tile is the product of x and 1. This is represented by a green x-tile. The area of the 1-tile is represented by the product of -1 and 1. This is represented by a red 1-tile.

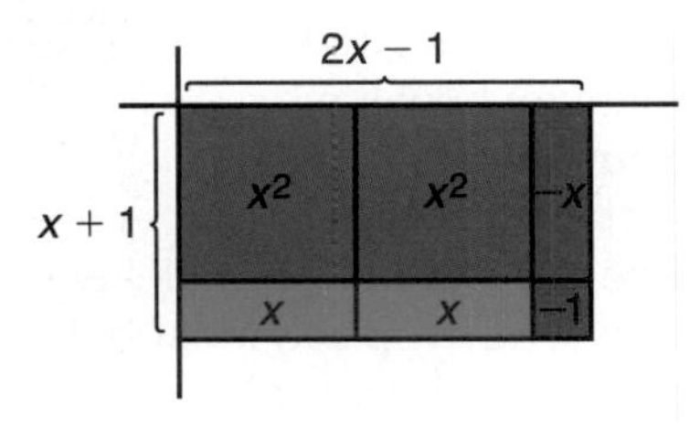

Step 3 Rearrange the tiles to simplify the polynomial you have formed. Notice that a zero pair is formed by the x-tiles.

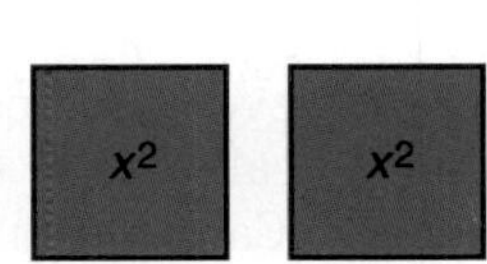

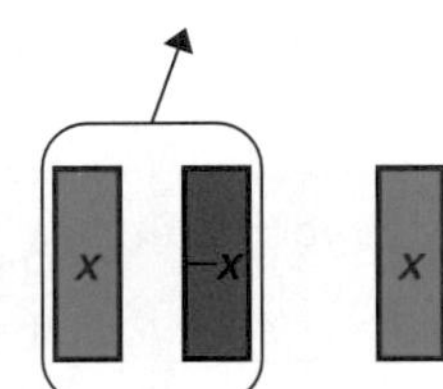

There are 2 blue x^2-tiles, 1 green x-tile, and 1 red 1-tile left. In simplest form, $(x + 1)(2x - 1) = 2x^2 + x - 1$.

Model

Use algebra tiles to find each product.

1. $(x + 1)(x + 2)$
2. $(x + 1)(x - 3)$
3. $(x - 2)(x - 4)$
4. $(x + 1)(2x + 2)$
5. $(x - 1)(2x + 2)$
6. $(x - 3)(2x - 1)$

Draw

Is each statement *true* or *false*? Justify your answer with a drawing.

7. $(x + 4)(x + 6) = x^2 + 24$
8. $(x + 3)(x - 2) = x^2 + x - 6$
9. $(x - 1)(x + 5) = x^2 - 4x - 5$
10. $(x - 2)(x - 3) = x^2 - 5x + 6$

Write

11. You can also use the distributive property to find the product of two binomials. The figure at the right shows the model for $(x + 3)(x + 2)$ separated into four parts. Write a paragraph explaining how this model shows the use of the distributive property.

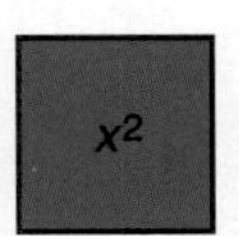

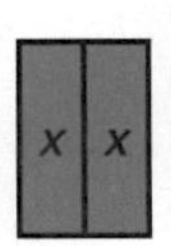

9–7 Multiplying Polynomials

What YOU'LL LEARN
- To use the FOIL method to multiply two binomials, and
- to multiply any two polynomials by using the distributive property.

Why IT'S IMPORTANT

You can use polynomials to solve problems involving art and business.

Art

Have you ever flown over farm land and looked down? You probably saw various fields that gave the appearance of a patchwork quilt. However, if you had flown over a field designed by Stan Herd, you may have seen some sunflowers in a vase or a picture of Will Rogers. Since 1981, Stan Herd has been combining his interests in art and agriculture to form crop art. Most of Herd's work is harvested, and therefore is only visible for a short time.

In 1991, however, Herd created the picture above using native perennials. This picture is called *Little Girl in the Wind*. It depicts a Kickapoo Indian girl by the name of Carole Cadue. If you fly near Salina, Kansas, you may see this work of art.

GLOBAL CONNECTIONS

There is an historic work of art in Ohio that is best experienced from a bird's eye view. It is a mound built by the Adena Indians perhaps 30 centuries ago. The mound is in the shape of a snake and is about one-quarter mile long.

Suppose the measure of the length of the field used for *Little Girl in the Wind* can be represented by the polynomial $7x + 2$ units and the width can be represented by $5x + 1$. You know that the area of a rectangle is the product of its length and width. You can multiply $7x + 2$ and $5x + 1$ to find the area of the rectangle.

$$\begin{aligned}(7x + 2)(5x + 1) &= 7x(5x + 1) + 2(5x + 1) && \textit{Distributive property}\\ &= 7x(5x) + 7x(1) + 2(5x) + 2(1) && \textit{Distributive property}\\ &= 35x^2 + 7x + 10x + 2 && \textit{Substitution property}\\ &= 35x^2 + 17x + 2 && \textit{Combine like terms.}\end{aligned}$$

The area can also be determined by finding the sum of the areas of four smaller rectangles.

	$5x$	1
$7x$	$7x \cdot 5x$	$7x \cdot 1$
2	$2 \cdot 5x$	$2 \cdot 1$

$$\begin{aligned}(7x + 2)(5x + 1) &= 7x \cdot 5x + 7x \cdot 1 + 2 \cdot 5x + 2 \cdot 1 && \textit{Find the sum of the four areas.}\\ &= 35x^2 + 7x + 10x + 2 && \textit{Substitution property}\\ &= 35x^2 + 17x + 2 && \textit{Combine like terms.}\end{aligned}$$

This example illustrates a shortcut of the distributive property called the **FOIL method**. You can use the FOIL method to multiply two binomials.

$$(7x + 2)(5x + 1) = \underset{F}{(7x)(5x)} + \underset{O}{(7x)(1)} + \underset{I}{(2)(5x)} + \underset{L}{(2)(1)}$$
$$= 35x^2 + 7x + 10x + 2$$
$$= 35x^2 + 17x + 2$$

FOIL Method for Multiplying Two Binomials	**To multiply two binomials, find the sum of the products of** **F** the ***First*** **terms,** **O** the ***Outer*** **terms,** **I** the ***Inner*** **terms, and** **L** the ***Last*** **terms.**

Example  **Find each product.**

a. $\mathbf{(x - 4)(x + 9)}$

$$(x - 4)(x + 9) = \overset{F}{(x)(x)} + \overset{O}{(x)(9)} + \overset{I}{(-4)(x)} + \overset{L}{(-4)(9)}$$
$$= x^2 + 9x - 4x - 36$$
$$= x^2 + 5x - 36 \quad \textit{Combine like terms.}$$

b. $\mathbf{(4x + 7)(3x - 8)}$

$$(4x + 7)(3x - 8) = \overset{F}{(4x)(3x)} + \overset{O}{(4x)(-8)} + \overset{I}{(7)(3x)} + \overset{L}{(7)(-8)}$$
$$= 12x^2 - 32x + 21x - 56$$
$$= 12x^2 - 11x - 56 \quad \textit{Combine like terms.}$$

The distributive property can be used to multiply any two polynomials.

Example  **Find each product.**

a. $\mathbf{(2y + 5)(3y^2 - 8y + 7)}$

$(2y + 5)(3y^2 - 8y + 7)$

$= 2y(3y^2 - 8y + 7) + 5(3y^2 - 8y + 7)$ *Distributive property*

$= (6y^3 - 16y^2 + 14y) + (15y^2 - 40y + 35)$ *Distributive property*

$= 6y^3 - 16y^2 + 14y + 15y^2 - 40y + 35$

$= 6y^3 - y^2 - 26y + 35$ *Combine like terms.*

b. $\mathbf{(x^2 + 4x - 5)(3x^2 - 7x + 2)}$

$(x^2 + 4x - 5)(3x^2 - 7x + 2)$

$= x^2(3x^2 - 7x + 2) + 4x(3x^2 - 7x + 2) - 5(3x^2 - 7x + 2)$

$= (3x^4 - 7x^3 + 2x^2) + (12x^3 - 28x^2 + 8x) - (15x^2 - 35x + 10)$

$= 3x^4 - 7x^3 + 2x^2 + 12x^3 - 28x^2 + 8x - 15x^2 + 35x - 10$

$= 3x^4 + 5x^3 - 41x^2 + 43x - 10$ *Combine like terms.*

Polynomials can also be multiplied in column form. Be careful to align the like terms.

Example 3 **Find $(x^3 - 8x^2 + 9)(3x + 4)$ using column form.**

Since there is no x term in $x^3 - 8x^2 + 9$, $0x$ is used as a placeholder.

$$\begin{array}{rrl} & x^3 - 8x^2 + 0x + 9 & \\ (\times) & 3x + 4 & \\ \hline & 4x^3 - 32x^2 + \ \ 0x + 36 & \leftarrow \textit{product of } x^3 - 8x^2 + 0x + 9 \textit{ and } 4 \\ 3x^4 - 24x^3 + \ \ 0x^2 + 27x & & \leftarrow \textit{product of } x^3 - 8x^2 + 0x + 9 \textit{ and } 3x \\ \hline 3x^4 - 20x^3 - 32x^2 + 27x + 36 & & \leftarrow \textit{sum of the partial products} \end{array}$$

Example 4

The volume V of a prism equals the area of the base B times the height h.

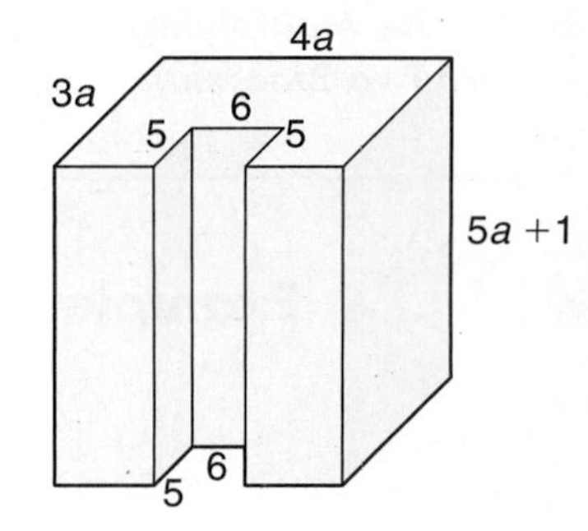

a. Write a polynomial expression that represents the volume of the prism shown at the right.

b. Find the volume if $a = 5$.

a. A diagram of the base is shown below.

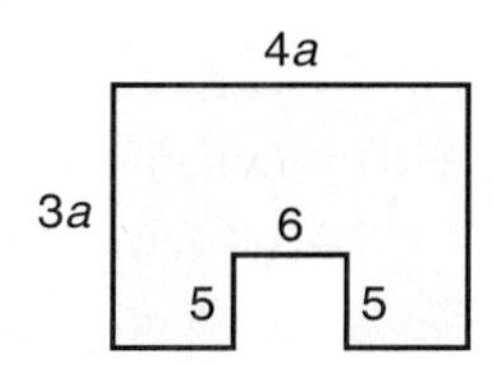

To find the area of the base, first find the area of a rectangle that is $3a$ by $4a$. Then subtract the area of a rectangle that is 6 by 5.

$B = 3a(4a) - 6(5)$

$\quad = 12a^2 - 30$

The volume of this prism equals the product of the base and the height, $12a^2 - 30$ and $5a + 1$, respectively. Use FOIL to find the product.

$$(12a^2 - 30)(5a + 1) = \overset{F}{(12a^2)(5a)} + \overset{O}{(12a^2)(1)} + \overset{I}{(-30)(5a)} + \overset{L}{(-30)(1)}$$
$$= 60a^3 + 12a^2 - 150a - 30$$

The volume of the prism is $(60a^3 + 12a^2 - 150a - 30)$ cubic units.

b. Substitute 5 for a and evaluate the expression.

$60(5^3) + 12(5^2) - 150(5) - 30 = 7500 + 300 - 750 - 30$
$= 7020$

If $a = 5$, the volume of the prism is 7020 cubic units.

CHECK FOR UNDERSTANDING

Communicating Mathematics

Study the lesson. Then complete the following.

1. Use the FOIL method to evaluate each product.
   a. 42(27) (*Hint:* Rewrite as (40 + 2) (20 + 7) or (40 + 2)(30 − 3).)
   b. $4\frac{1}{2} \cdot 6\frac{3}{4}$

2. **You Decide** Adita and Delbert used the following methods to find the product of $(t^3 - t^2 + 5t)$ and $(6t^2 + 8t - 7)$.

Adita:

$$\begin{aligned}(t^3 - t^2 + 5t)(6t^2 + 8t - 7) \\ &= t^3(6t^2 + 8t - 7) - t^2(6t^2 + 8t - 7) + 5t(6t^2 + 8t - 7) \\ &= 6t^5 + 8t^4 - 7t^3 - 6t^4 - 8t^3 + 7t^2 + 30t^3 + 40t^2 - 35t \\ &= 6t^5 + 2t^4 + 15t^3 + 47t^2 - 35t\end{aligned}$$

Delbert:

$$\begin{array}{r} t^3 - t^2 + 5t \\ (\times)\ 6t^2 + 8t - 7 \\ \hline -7t^3 + 7t^2 - 35t \\ 8t^4 - 8t^3 + 40t^2 \quad\quad \\ 6t^5 - 6t^4 + 30t^3 \quad\quad\quad\quad \\ \hline 6t^5 + 2t^4 + 15t^3 + 47t^2 - 35t \end{array}$$

Which method do you prefer? Why?

3. **Draw** a diagram to show how you would use algebra tiles to find the product of $(2x - 3)$ and $(x + 2)$.

4. **Write** two binomials whose product is represented at the right.

$2ax$	$3a$
$2x^2$	$3x$

Guided Practice

Find each product.

5. $(d + 2)(d + 8)$
6. $(r - 5)(r - 11)$
7. $(y + 3)(y - 7)$
8. $(3p - 5)(5p + 2)$
9. $(2x - 1)(x + 5)$
10. $(2m + 5)(3m - 8)$
11. $(2a + 3b)(5a - 2b)$
12. $(2x - 5)(3x^2 - 5x + 4)$

13. a. **Number Theory** Find the product of three consecutive integers if the least integer is a.
   b. Choose an integer as the first of three consecutive integers. Find their product.
   c. Evaluate the polynomial in part a for these integers. Describe the result.

EXERCISES

Practice

Find each product.

14. $(y + 5)(y + 7)$
15. $(c - 3)(c - 7)$
16. $(x + 4)(x - 8)$
17. $(w + 3)(w - 9)$
18. $(2a - 1)(a + 8)$
19. $(5b - 3)(2b + 1)$
20. $(11y + 9)(12y + 6)$
21. $(13x - 3)(13x + 3)$
22. $(8x + 9y)(3x + 7y)$
23. $(0.3v - 7)(0.5v + 2)$
24. $\left(3x + \frac{1}{3}\right)\left(2x - \frac{1}{9}\right)$
25. $\left(a - \frac{2}{3}b\right)\left(\frac{2}{3}a + \frac{1}{2}b\right)$
26. $(2r + 0.1)(5r - 0.3)$
27. $(0.7p + 2q)(0.9p + 3q)$
28. $(x + 7)(x^2 + 5x - 9)$
29. $(3x - 5)(2x^2 + 7x - 11)$
30. $\begin{array}{r} a^2 - 3a + 11 \\ (\times)\ 5a + 2 \\ \hline \end{array}$
31. $\begin{array}{r} 3x^2 - 7x + 2 \\ (\times)\ 3x - 8 \\ \hline \end{array}$
32. $\begin{array}{r} 5x^2 + 8x - 11 \\ (\times)\ x^2 - 2x - 1 \\ \hline \end{array}$
33. $\begin{array}{r} 5d^2 - 6d + 9 \\ (\times)\ 4d^2 + 3d + 11 \\ \hline \end{array}$

Find each product.

34. $(x^2 - 8x - 1)(2x^2 - 4x + 9)$
35. $(5x^2 - x - 4)(2x^2 + x + 12)$
36. $(-7b^3 + 2b - 3)(5b^2 - 2b + 4)$
37. $(a^2 + 2a + 5)(a^2 - 3a - 7)$

Geometry

Find the measure of the volume of each prism.

38. 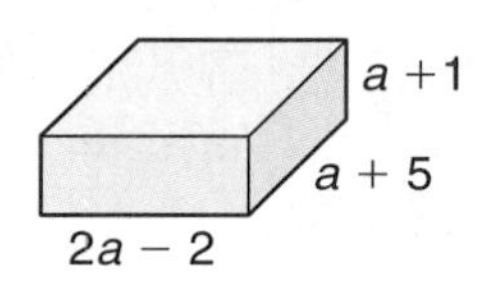

39. 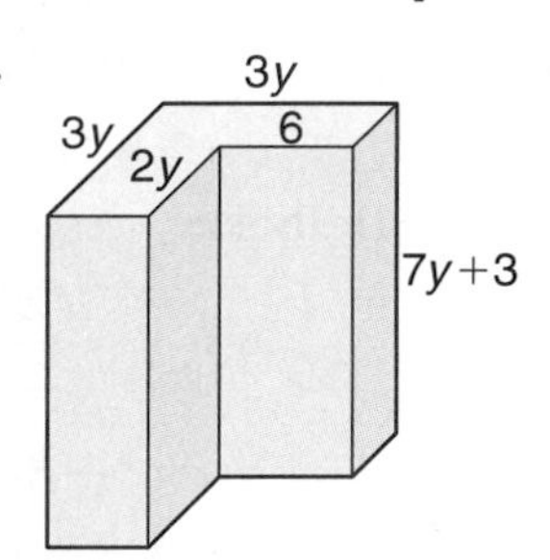

40. 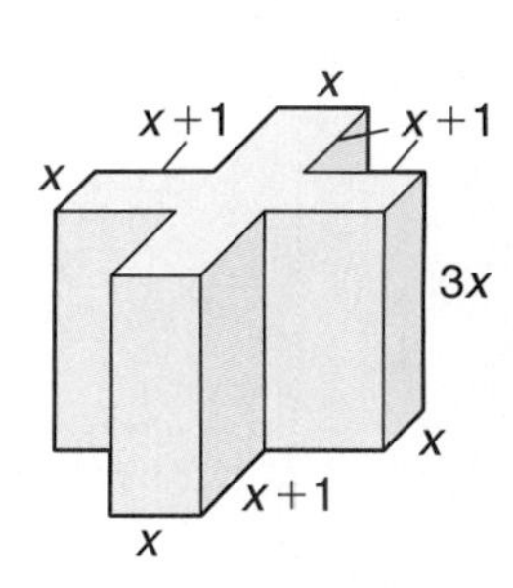

41. **Geometry** Refer to the prism in Exercise 38. Suppose a represents 15 centimeters.
   a. Find the length, width, and height of the prism.
   b. Use the values in part a to find the volume of the prism.
   c. Evaluate your answer for Exercise 38 if $a = 15$.
   d. How do your answers for parts b and c compare?

Critical Thinking

If $A = 3x + 4$, $B = x^2 + 2$, and $C = x^2 + 3x - 2$, find each of the following.

42. $AC + B$
43. $2B(3A - 4C)$
44. ABC
45. $(A + B)(B - C)$

Applications and Problem Solving

46. **Construction** A homeowner is considering installing a swimming pool in his backyard. He wants its length to be 5 yards longer than its width, to make room for a diving area at one end. Then he wants to surround it with a concrete walkway 4 yards wide. After finding out the price of concrete, he decides that he can afford 424 square yards of it for the walkway. What should the dimensions of the pool be?

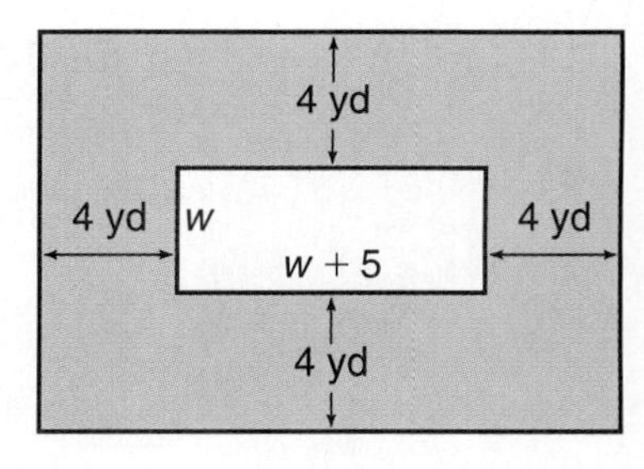

47. **Business** Raul Agosto works for a company that has modular offices. His office space is presently a square. A new floor plan calls for his office to become 2 feet shorter in one direction and 3 feet longer in the other.
   a. Write expressions that represent the new dimensions of Mr. Agosto's office.
   b. Find the area of his new office.
   c. Suppose his office is presently 8 feet by 8 feet. Will his new office be bigger or smaller than this office? by how much?

Mixed Review

48. Find $\frac{3}{4}a(6a + 12)$. (Lesson 9–6)

49. Solve $6 - 9y < -10y$. (Lesson 7–3)

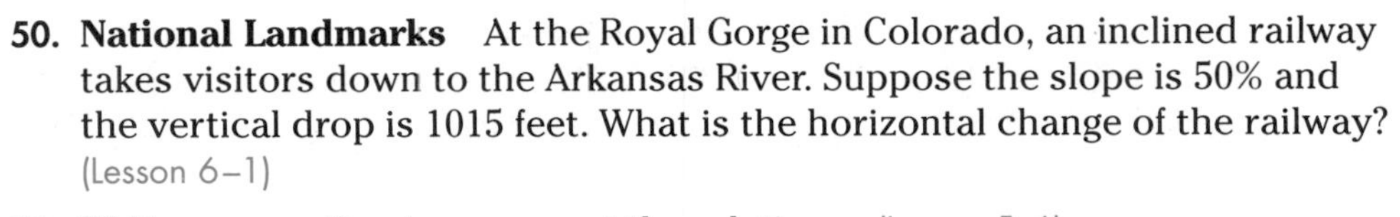

50. **National Landmarks** At the Royal Gorge in Colorado, an inclined railway takes visitors down to the Arkansas River. Suppose the slope is 50% and the vertical drop is 1015 feet. What is the horizontal change of the railway? (Lesson 6–1)

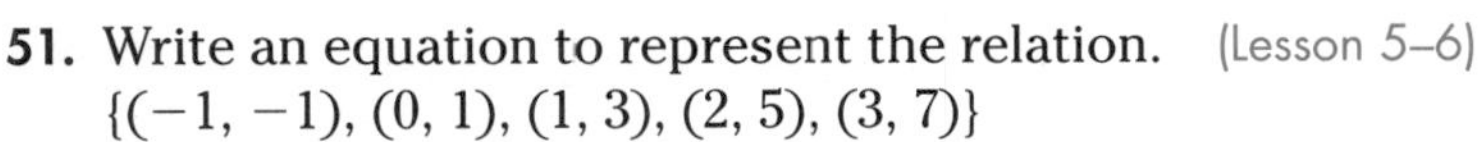

51. Write an equation to represent the relation. (Lesson 5–6)
$\{(-1, -1), (0, 1), (1, 3), (2, 5), (3, 7)\}$

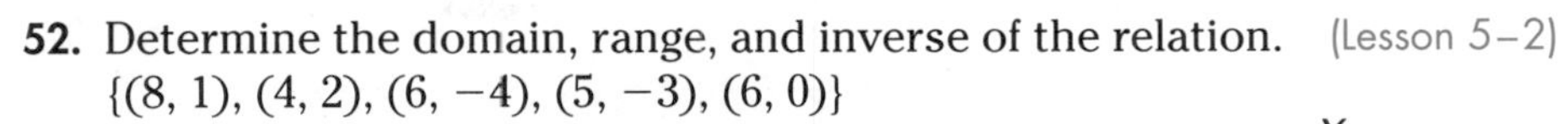

52. Determine the domain, range, and inverse of the relation. (Lesson 5–2)
$\{(8, 1), (4, 2), (6, -4), (5, -3), (6, 0)\}$

53. **Geometry** ΔABC and ΔXYZ are similar. Find the values of a and y. (Lesson 4–2)

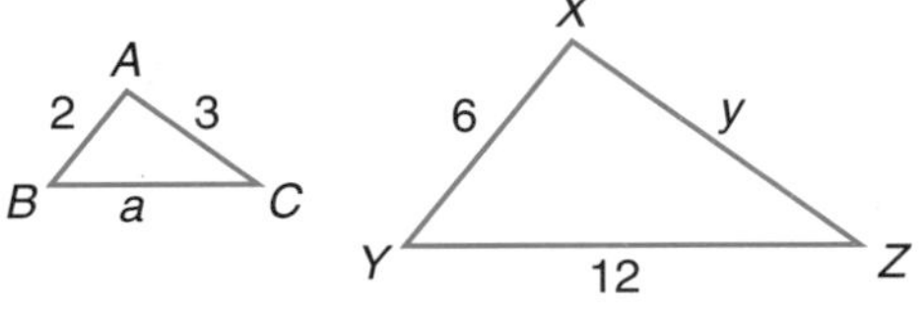

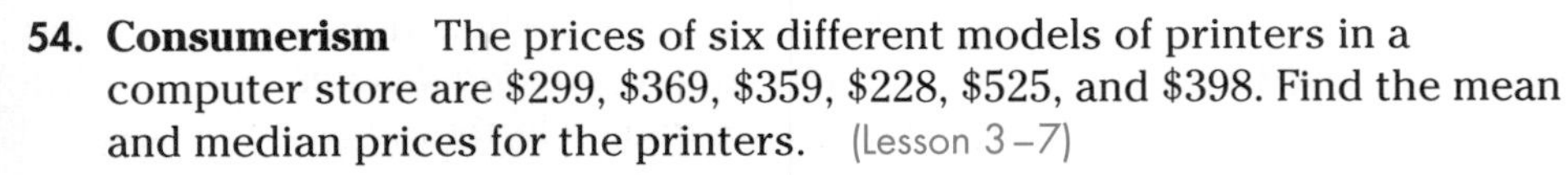

54. **Consumerism** The prices of six different models of printers in a computer store are \$299, \$369, \$359, \$228, \$525, and \$398. Find the mean and median prices for the printers. (Lesson 3–7)

55. **Temperature** The formula for finding the Celsius temperature C when you know the Fahrenheit temperature F is $C = \frac{5}{9}(F - 32)$. Find the Celsius temperature when the Fahrenheit temperature is 59°. (Lesson 2–9)

56. Replace the variable to make the sentence $\frac{3}{4}s = 6$ true. (Lesson 1–5)

$359

WORKING ON THE

In·ves·ti·ga·tion

Refer to the Investigation on pages 448–449.

You have experimented with various sizes of gliders to explore their flying abilities. You now need to investigate how the size of the glider and the weight of the load are related.

1. Cut out two equilateral triangles from construction paper so that the side of one is 6 centimeters long, and the side of the other is 12 centimeters long. What is the surface area and perimeter of each triangle?
2. Straighten two paper clips and bend them into the shape shown at the right. Punch the bent end of the paper clip through the center of the triangle, and tape it into place so that the hook-end hangs down under the other side of the triangle.

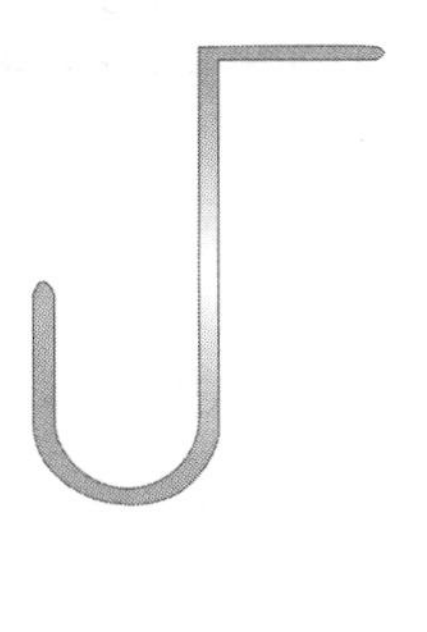

3. One at a time, drop each hang glider from the top of the bleachers or out of a second-floor window. (In order to get more accurate data, a greater height is needed than those used for previous experiments.) Record the glide time of each hang glider.
4. Add one washer onto the hook of each glider. Repeat the dropping procedure and record the times. How do those times compare with the first drop? Continue to add washers, one at a time, to each glider and record the glide times in a chart that compares the glide times with the number of washers carried by each glider.
5. Graph the relationship between the weight and the glide times. Let the independent variable be the weight, and let the dependent variable be the glide time. Analyze your findings.
6. Look at the weight, glide time, surface area, and perimeter in your data. Write a polynomial expression to relate some, if not all, of these measures.

Add the results of your work to your Investigation Folder.

9–8 Special Products

What YOU'LL LEARN

- To use patterns to find $(a + b)^2$, $(a - b)^2$, and $(a + b)(a - b)$.

Why IT'S IMPORTANT

You can use polynomials to solve problems involving biology and history.

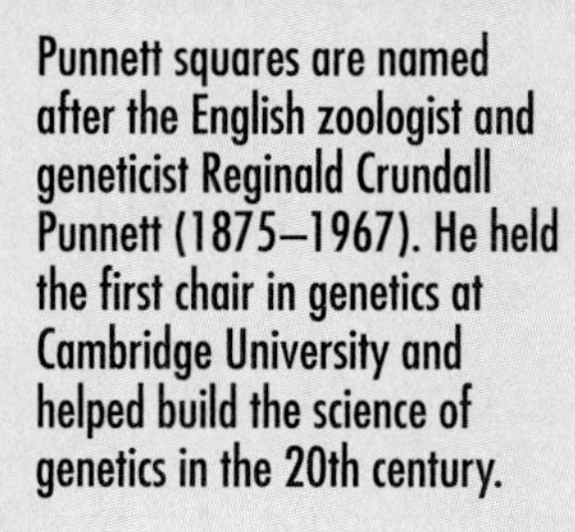

FYI

Punnett squares are named after the English zoologist and geneticist Reginald Crundall Punnett (1875–1967). He held the first chair in genetics at Cambridge University and helped build the science of genetics in the 20th century.

CONNECTION

Biology

Punnett squares are diagrams that are used to show the possible ways that genes can combine at fertilization. In a Punnett square, *dominant* genes are shown with capital letters. Recessive genes are shown with lowercase letters. Letters representing the parents' genes are placed on two of the outer sides of the Punnett square. Letters inside the boxes of the square show the possible gene combinations of their offspring.

The Punnett square below represents a cross between tall pea plants and short pea plants. Let T represent the dominant gene for tallness. Let t represent the recessive gene for shortness. The parents are called *hybrids,* since they have one of each kind of gene.

Hybrid tall × Hybrid tall

Tall = $\boldsymbol{T}$

Short = $\boldsymbol{t}$

Offspring

$\frac{1}{4}$ or 25% pure tall (TT)

$\frac{2}{4}$ or 50% hybrid tall (Tt)

$\frac{1}{4}$ or 25% pure short (tt)

hybrid tall	T	t
T	TT	Tt
t	Tt	tt

Because the parent plants have both a dominant tall gene and a recessive short gene, biologists know that their offspring can be predicted by squaring the binomial $(0.5T + 0.5t)^2$. Therefore, the following must be true.

$$\begin{aligned}(0.5T + 0.5t)^2 &= (0.5T + 0.5t)(0.5T + 0.5t)\\ &= 0.5T(0.5T) + 0.5T(0.5t) + 0.5t(0.5T) + 0.5t(0.5t)\\ &= 0.25T^2 + 0.25Tt + 0.25Tt + 0.25t^2\\ &= 0.25T^2 + 0.50Tt + 0.25t^2\end{aligned}$$

T^2 and t^2 represent TT and tt, respectively.

You can use the diagram below to derive a general form for the expression $(a + b)^2$.

$$(a + b)^2 = a^2 + ab + ab + b^2$$
$$= a^2 + 2ab + b^2$$ *Check this result by using FOIL.*

In general, the square of a binomial that is a sum can be found by using the following rule.

Square of a Sum

$$(a + b)^2 = (a + b)(a + b) = a^2 + 2ab + b^2$$

Example 1 **Find each product.**

a. $(y + 7)^2$

Method 1

Use the square of a sum rule.

$(a + b)^2 = a^2 + 2ab + b^2$

$(y + 7)^2 = y^2 + 2(y)(7) + 7^2$

$= y^2 + 14y + 49$

Method 2

Use FOIL.

$(y + 7)^2 = (y + 7)(y + 7)$

$= y^2 + 7y + 7y + 49$

$= y^2 + 14y + 49$

b. $(6p + 11q)^2$

$(a + b)^2 = a^2 + 2ab + b^2$

$(6p + 11q)^2 = (6p)^2 + 2(6p)(11q) + (11q)^2$ *$a = 6p$ and $b = 11q$*

$= 36p^2 + 132pq + 121q^2$

The square of a sum rule can be used with other rules to simplify products of polynomials.

Example 2

History

Tourists to the southern part of England can visit the historic Gwennap Pit. In the 16th century, the pit of a tin mine was converted into an amphitheater. During the 18th century, John Wesley spoke to overflow crowds in this amphitheater. Gwennap Pit consists of a circular stage surrounded by circular levels used for seating. Each seating level is 1 meter wide. Suppose the radius of the stage is s meters. Find the area of the third seating level.

(continued on the next page)

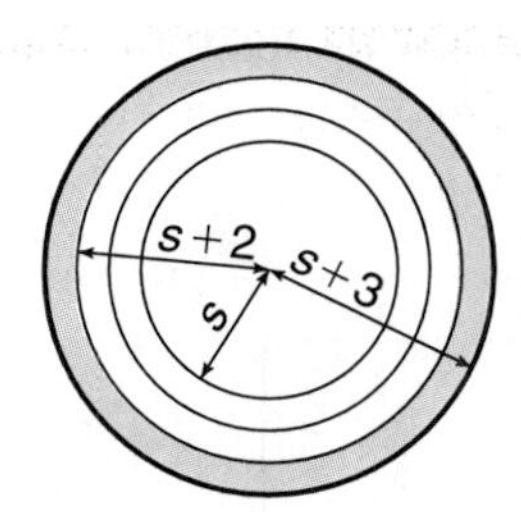

The area of a circle equals πr^2. The radius of the second seating level is $s + 2$ meters, and the radius of the third seating level is $s + 3$. The area of the third seating level can be found by subtracting the areas of two circles.

$$A = \underbrace{\pi(s+3)^2}_{\text{area of third level}} - \underbrace{\pi(s+2)^2}_{\text{area of second level}}$$

$$= \pi(s^2 + 6s + 9) - \pi(s^2 + 4s + 4) \quad \textit{Square of a sum rule}$$

$$= (\pi s^2 + 6\pi s + 9\pi) - (\pi s^2 + 4\pi s + 4\pi) \quad \textit{Distributive property}$$

$$= \pi s^2 + 6\pi s + 9\pi - \pi s^2 - 4\pi s - 4\pi$$

$$= 2\pi s + 5\pi \quad \textit{Combine like terms.}$$

The area of the third seating level is $2\pi s + 5\pi$, or about $6.3s + 15.7$ square meters.

To find $(a - b)^2$, write $(a - b)$ as $[a + (-b)]$ and square it.

$$(a - b)^2 = [a + (-b)]^2$$

$$= a^2 + 2(a)(-b) + (-b)^2$$

$$= a^2 - 2ab + b^2$$

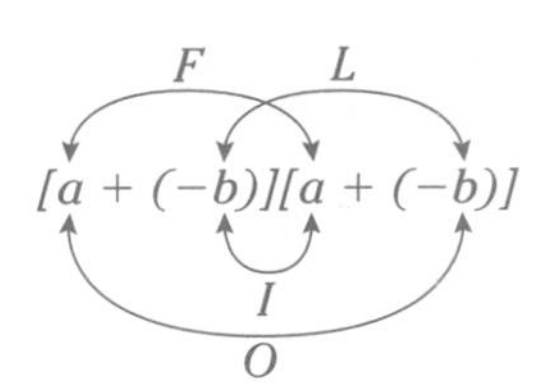

In general, the square of a binomial that is a difference can be found by using the following rule.

Square of a Difference	$(a - b)^2 = (a - b)(a - b)$ $= a^2 - 2ab + b^2$

Example  Find each product.

a. $(r - 6)^2$

Method 1

Use the square of a difference rule.

$$(a - b)^2 = a^2 - 2ab + b^2$$

$$(r - 6)^2 = r^2 - 2(r)(6) + 6^2$$

$$= r^2 - 12r + 36$$

Method 2

Use FOIL.

$$(r - 6)^2 = (r - 6)(r - 6)$$

$$= r^2 - 6r - 6r + 36$$

$$= r^2 - 12r + 36$$

b. $(4x^2 - 7t)^2$

$$(a - b)^2 = a^2 - 2ab + b^2$$

$$(4x^2 - 7t)^2 = (4x^2)^2 - 2(4x^2)(7t) + (7t)^2 \quad a = 4x^2 \text{ and } b = 7t$$

$$= 16x^4 - 56x^2t + 49t^2$$

MODELING MATHEMATICS

Product of a Sum and a Difference

Materials:  algebra tiles product mat

You have learned how to use algebra tiles to find the product of two binomials. In this activity, you will use algebra tiles to study a special situation.

Your Turn

a. Use algebra tiles to find each product.

$(x + 3)(x - 3)$ $(x + 5)(x - 5)$

$(x + 1)(x - 1)$ $(x + 2)(x - 2)$

$(x + 6)(x - 6)$ $(x + 4)(x - 4)$

b. What do you notice about the binomials used as factors in part a?

c. What pattern do the products in part a have?

You can use the FOIL method to find the product of a sum and a difference of the same two numbers.

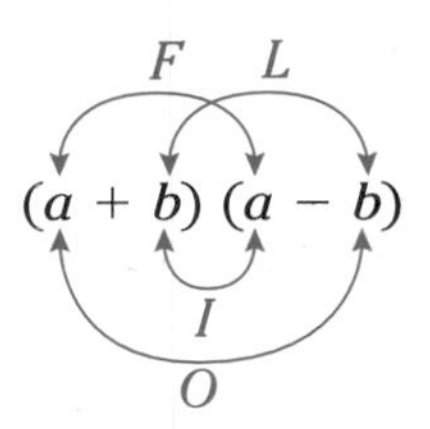

$$\begin{aligned}(a + b)(a - b) &= a(a) + a(-b) + b(a) + b(-b)\\ &= a^2 - ab + ab - b^2\\ &= a^2 - b^2\end{aligned}$$

The resulting product, $a^2 - b^2$, has a special name. It is called a **difference of squares.**

Difference of Squares	$(a + b)(a - b) = (a - b)(a + b) = a^2 - b^2$

Example  **Find each product.**

a. $\mathbf{(m - 2n)(m + 2n)}$

$$\begin{aligned}(a - b)(a + b) &= a^2 - b^2\\ (m - 2n)(m + 2n) &= m^2 - (2n)^2 \quad a = m \text{ and } b = 2n\\ &= m^2 - 4n^2\end{aligned}$$

b. $\mathbf{(0.3t + 0.25w^2)(0.3t - 0.25w^2)}$

$$\begin{aligned}(a + b)(a - b) &= a^2 - b^2\\ (0.3t + 0.25w^2)(0.3t - 0.25w^2) &= (0.3t)^2 - (0.25w^2)^2 \quad a = 0.3t \text{ and } b = 0.25w^2\\ &= 0.09t^2 - 0.0625w^4\end{aligned}$$

CHECK FOR UNDERSTANDING

Communicating Mathematics

Study the lesson. Then complete the following.

1. **Explain** how the square of a difference and the square of a sum are different.
2. **Compare and contrast** the square of a difference and the difference of two squares.
3. **Explain** how you could mentally multiply 29×31. (*Hint:* $29 = 30 - 1$ and $31 = 30 + 1$)

4. Draw a diagram to represent each of the following.
   a. $(x + y)^2$
   b. $(x - y)^2$
5. What does the diagram at the right represent if the shading represents regions to be removed or subtracted?

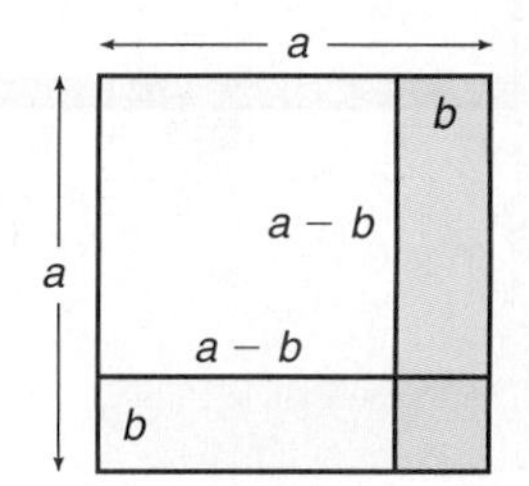

Guided Practice

Find each product.

6. $(2x + 3y)^2$
7. $(m - 3n)^2$
8. $(2a + 3)(2a - 3)$
9. $(m^2 + 4n)^2$
10. $(4y + 2z)(4y - 2z)$
11. $(5 - x)^2$
12. **Recreation** In India, children play a form of hopscotch called Chilly. One of the three possible patterns for this game is shown at the right. Suppose each side of the small squares is $2x + 5$ units long. Find the area of this Chilly pattern.

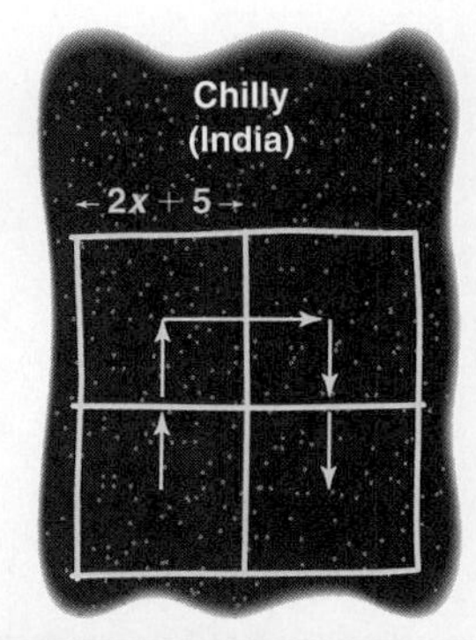

EXERCISES

Practice

Find each product.

13. $(x + 4y)^2$
14. $(m - 2n)^2$
15. $(3b - a)^2$
16. $(3x + 5)(3x - 5)$
17. $(9p - 2q)(9p + 2q)$
18. $(5s + 6t)^2$
19. $(5b - 12a)^2$
20. $(2a + 0.5y)^2$
21. $(x^3 + a^2)^2$
22. $\left(\frac{1}{2}b^2 - a^2\right)^2$
23. $(8x^2 - 3y)(8x^2 + 3y)$
24. $(7c^2 + d^3)(7c^2 - d^3)$
25. $(1.1g + h^5)^2$
26. $(9 - z^9)(9 + z^9)$
27. $\left(\frac{4}{3}x^2 - y\right)\left(\frac{4}{3}x^2 + y\right)$
28. $\left(\frac{1}{3}v^2 - \frac{1}{2}w^3\right)^2$
29. $(3x + 1)(3x - 1)(x - 5)$
30. $(x - 2)(x + 5)(x + 2)(x - 5)$
31. $(a + 3b)^3$
32. $(2m - n)^4$

Critical Thinking

33. Find $(x + y + z)^2$. Draw a diagram to show each term of the polynomial.

Applications and Problem Solving

34. **Biology** Refer to the application at the beginning of the lesson.
    a. Make a Punnett square for pea plants if one parent is pure short (tt) and the other parent is hybrid tall (Tt).
    b. What percent of the offspring will be pure short?
    c. What percent of the offspring will be hybrid tall?
    d. What percent of the offspring will be pure tall?

35. **History** Refer to Example 2.
    a. Write an expression for the area of the fourth seating level in the Gwennap Pit.
    b. The radius of the stage level of the Gwennap Pit is 3 meters. Find the area of the stage.
    c. Find the area of the fourth seating level in the Gwennap Pit.

36. **Photography** Lenora cut off a 0.75-inch strip all around a square photograph so it would fit in an envelope she was mailing to her aunt. She decided to have the photo lab make a copy of the photo from the negative, but she forgot to measure how large the original photo was. All she had was the strips she cut off, whose area was 33.75 square inches. What were the original dimensions of the photograph?

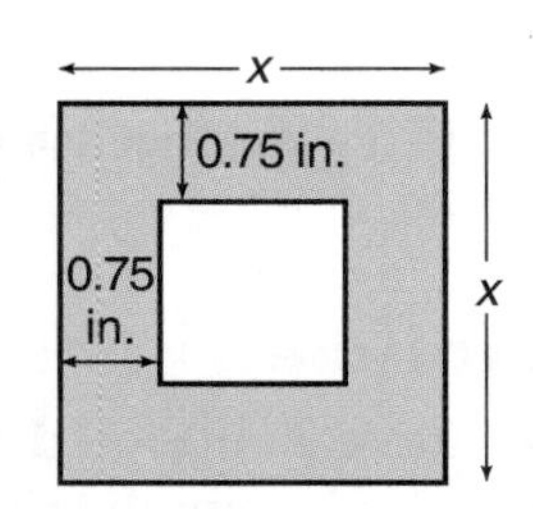

Mixed Review

37. Find $(3t - 3)(2t + 1)$. (Lesson 9–7)

38. Solve $-13z > -1.04$. (Lesson 7–2)

39. **Statistics** The table below shows the heights and weights of each of 12 players on a pro basketball team. (Lesson 6–3)

Height (in.)	75	82	75	74	80	80	75	79	80	78	76	81
Weight (lb)	180	235	184	185	230	205	185	230	221	195	205	215

    a. Make a scatter plot of these data.
    b. Describe the correlation between height and weight.

40. Write the standard form of the line that passes through (3, 1) and has a slope of $\frac{2}{7}$. (Lesson 6–2)

41. Graph the points $A(4, 2)$, $B(-3, 1)$, and $C(-2, -3)$. (Lesson 5–1)

42. **Electricity** The resistance R of a power circuit is 4.5 ohms. How much current I, in amperes, can the circuit generate if it can produce at most 1500 watts of power? Use $I^2R = P$. (Lesson 2–8)

43. Evaluate $5(9 \div 3^2)$. (Lesson 1–6)

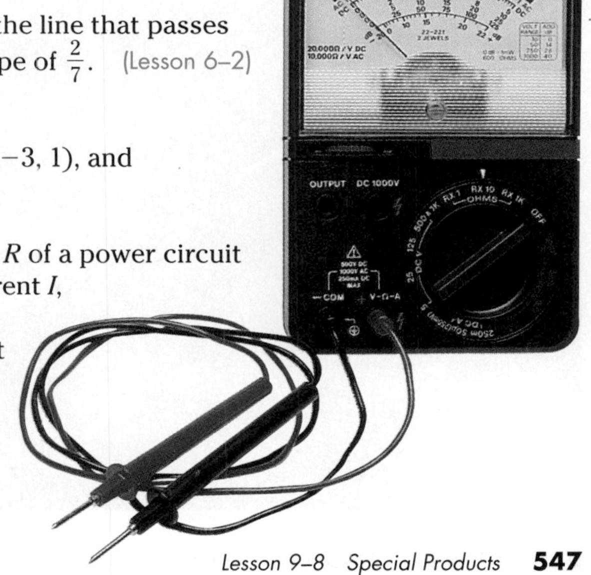

CLOSING THE

In·ves·ti·ga·tion

Refer to the Investigation on pages 448–449.

Analyze

You have conducted several experiments and organized your data in various ways. It is now time to analyze your findings and state your conclusions.

PORTFOLIO ASSESSMENT

You may want to keep your work on this Investigation in your portfolio.

1 Look over your data and organize it in such a way that the various relationships are obvious.

2 Describe the relationships in the data. What does weight have to do with glide time? Does perimeter or surface area have an effect on glide time? What other factors need to be considered?

Write

The report to the people interested in hang gliding should explain your process for investigating these hang glider models and what you found from your investigations.

3 Begin the report by stating the process you used to investigate the matter. Explain all the experiments conducted. State the purpose and findings for each.

4 Show the data you collected in tables, charts, and graphs. Explain your analysis of the data and conclusions you found.

5 Make a recommendation to the group about the size of hang glider(s) that is most suited for them. Include the weight of the object that the hang glider(s) should carry to be most efficient.

6 While you were conducting experiments, one of your team members found that the frames for most hang gliders are 32 feet wide. Explain how this information affects your generalizations regarding the weight and size of hang gliders.

7 Summarize your findings in a concluding statement to the group.

CHAPTER 9 HIGHLIGHTS

VOCABULARY

After completing this chapter, you should be able to define each term, property, or phrase and give an example or two of each.

Algebra

binomial (p. 514)
constants (p. 496)
degree of monomial (p. 515)
degree of polynomial (p. 516)
difference of squares (p. 545)
FOIL method (p. 537)
monomial (p. 496)
negative exponent (p. 503)
polynomial (pp. 513, 514)
power of a monomial (p. 498)
power of a power (p. 498)
power of a product (p. 498)
product of powers (p. 497)
quotient of powers (p. 501)
scientific notation (p. 506)
square of a difference (p. 544)
square of a sum (p. 543)
trinomial (p. 514)
zero exponent (p. 502)

Problem Solving

look for a pattern (p. 497)

UNDERSTANDING AND USING THE VOCABULARY

Choose the letter of the term that best matches each example.

1. $4^{-3} = \frac{1}{4^3}$ or $\frac{1}{64}$
2. $(x + 2y)(x - 2y) = x^2 - 4y^2$
3. $\frac{4x^2y}{8xy^3} = \frac{x}{2y^2}$
4. $4x^2$
5. $x^2 - 3x + 1$
6. $2^0 = 1$
7. $x^4 - 3x^3 + 2x^2 - 1$
8. $(x + 3)(x - 4) = x^2 - 4x + 3x - 12$
9. $x^2 + 2$
10. $(a^3b)(2ab^2) = 2a^4b^3$

a. binomial
b. difference of squares
c. FOIL method
d. monomial
e. negative exponent
f. polynomial
g. product of powers
h. quotient of powers
i. trinomial
j. zero exponent

CHAPTER 9 STUDY GUIDE AND ASSESSMENT

SKILLS AND CONCEPTS

OBJECTIVES AND EXAMPLES

Upon completing this chapter, you should be able to:

- multiply monomials and simplify expressions involving powers of monomials (Lesson 9–1)

$(2ab^2)(3a^2b^3) = (2 \cdot 3)(a \cdot a^2)(b^2 \cdot b^3)$
$= 6a^3b^5$

$(2x^2y^3)^3 = 2^3(x^2)^3(y^3)^3$
$= 8x^6y^9$

REVIEW EXERCISES

Use these exercises to review and prepare for the chapter test.

Simplify.

11. $y^3 \cdot y^3 \cdot y$
12. $(3ab)(-4a^2b^3)$
13. $(-4a^2x)(-5a^3x^4)$
14. $(4a^2b)^3$
15. $(-3xy)^2(4x)^3$
16. $(-2c^2d)^4(-3c^2)^3$
17. $-\frac{1}{2}(m^2n^4)^2$
18. $(5a^2)^3 + 7(a^6)$

- simplify expressions involving quotients of monomials and negative exponents (Lesson 9–2)

$\frac{2x^6y}{8x^2y^2} = \frac{2}{8} \cdot \frac{x^6}{x^2} \cdot \frac{y}{y^2}$
$= \frac{x^4}{4y}$

$\frac{3a^{-2}}{4a^6} = \frac{3}{4}(a^{-2-6})$
$= \frac{3}{4}(a^{-8}) \text{ or } \frac{3}{4a^8}$

Simplify. Assume that no denominator is equal to zero.

19. $\frac{y^{10}}{y^6}$
20. $\frac{(3y)^0}{6a}$
21. $\frac{42b^7}{14b^4}$
22. $\frac{27b^{-2}}{14b^{-3}}$
23. $\frac{(3a^3bc^2)^2}{18a^2b^3c^4}$
24. $\frac{-16a^3b^2x^4y}{-48a^4bxy^3}$

- express numbers in scientific and decimal notation (Lesson 9–3)

$3{,}600{,}000 = 3.6 \times 1{,}000{,}000$
$= 3.6 \times 10^6$

$0.0021 = 2.1 \times 0.001$
$= 2.1 \times 10^{-3}$

Express each number in scientific notation.

25. 240,000
26. 0.000314
27. 4,880,000,000
28. 0.00000187
29. 796×10^3
30. 0.03434×10^{-2}

- find products and quotients of numbers expressed in scientific notation (Lesson 9–3)

$(2 \times 10^2)(5.2 \times 10^6) = (2 \times 5.2)(10^2 \times 10^6)$
$= 10.4 \times 10^8$
$= 1.04 \times 10^9$

$\frac{1.2 \times 10^{-2}}{0.6 \times 10^3} = \frac{1.2}{0.6} \times \frac{10^{-2}}{10^3}$
$= 2 \times 10^{-5}$

Evaluate. Express each result in scientific notation.

31. $(2 \times 10^5)(3 \times 10^6)$
32. $(3 \times 10^3)(1.5 \times 10^6)$
33. $\frac{5.4 \times 10^3}{0.9 \times 10^4}$
34. $\frac{8.4 \times 10^{-6}}{1.4 \times 10^{-9}}$
35. $(3 \times 10^2)(5.6 \times 10^{-4})$
36. $34(4.7 \times 10^5)$

OBJECTIVES AND EXAMPLES

- find the degree of a polynomial (Lesson 9–4)

Find the degree of $2xy^3 + x^2y$.

degree of $2xy^3$: $1 + 3$ or 4

degree of x^2y: $2 + 1$ or 3

degree of $2xy^3 + x^2y$: 4

REVIEW EXERCISES

Find the degree of each polynomial.

37. $n - 2p^2$
38. $29n^2 + 17n^2t^2$
39. $4xy + 9x^3z^2 + 17rs^3$
40. $-6x^5y - 2y^4 + 4 - 8y^2$
41. $3ab^3 - 5a^2b^2 + 4ab$
42. $19m^3n^4 + 21m^5n^2$

- arrange the terms of a polynomial so that the powers of a variable are in ascending or descending order (Lesson 9–4)

Arrange the terms of $4x^2 + 9x^3 - 2 - x$ in descending order.

$$9x^3 + 4x^2 - x - 2$$

Arrange the terms of each polynomial so that the powers of *x* are in descending order.

43. $3x^4 - x + x^2 - 5$
44. $-2x^2y^3 - 27 - 4x^4 + xy + 5x^3y^2$

- add and subtract polynomials (Lesson 9–5)

$$\begin{array}{r} 4x^2 - 3x + 7 \\ (+)\ 2x^2 + 4x \phantom{{}+7} \\ \hline 6x^2 + x + 7 \end{array}$$

$$\begin{aligned}(7r^2 + 9r) - (12r^2 - 4) &= 7r^2 + 9r - 12r^2 + 4 \\ &= (7r^2 - 12r^2) + 9r + 4 \\ &= -5r^2 + 9r + 4\end{aligned}$$

Find each sum or difference.

45. $(2x^2 - 5x + 7) - (3x^3 + x^2 + 2)$
46. $(x^2 - 6xy + 7y^2) + (3x^2 + xy - y^2)$

47. $\begin{array}{r} 11m^2n^2 + 4mn - 6 \\ (+)\ 5m^2n^2 - 6mn + 17 \\ \hline \end{array}$

48. $\begin{array}{r} 7z^2 \phantom{{}+2z} + 4 \\ (-)\ 3z^2 + 2z - 6 \\ \hline \end{array}$

49. $\begin{array}{r} 13m^4 - 7m - 10 \\ (+)\ 8m^4 - 3m + 9 \\ \hline \end{array}$

50. $\begin{array}{r} -5p^2 + 3p + 49 \\ (-)\ 2p^2 + 5p + 24 \\ \hline \end{array}$

- multiply a polynomial by a monomial (Lesson 9–6)

$ab(-3a^2 + 4ab - 7b^3) = -3a^3b + 4a^2b^2 - 7ab^4$

Find each product.

51. $4ab(3a^2 - 7b^2)$
52. $7xy(x^2 + 4xy - 8y^2)$
53. $4x^2y(2x^3 - 3x^2y^2 + y^4)$
54. $5x^3(x^4 - 8x^2 + 16)$

- simplify expressions involving polynomials (Lesson 9–6)

$$\begin{aligned}x^2(x + 2) + 3(x^3 + 4x^2) &= x^3 + 2x^2 + 3x^3 + 12x^2 \\ &= 4x^3 + 14x^2\end{aligned}$$

Simplify.

55. $2x(x - y^2 + 5) - 5y^2(3x - 2)$

56. $x(3x - 5) + 7(x^2 - 2x + 9)$

OBJECTIVES AND EXAMPLES

- use the FOIL method to multiply two binomials and multiply any two polynomials by using the distributive property (Lesson 9–7)

$$(3x+2)(x-2) = \overset{F}{(3x)(x)} + \overset{O}{(3x)(-2)} + \overset{I}{(2)(x)} + \overset{L}{(2)(-2)}$$
$$= 3x^2 - 6x + 2x - 4$$
$$= 3x^2 - 4x - 4$$

$$(4x-3)(3x^2 - x + 2)$$
$$= 4x(3x^2 - x + 2) - 3(3x^2 - x + 2)$$
$$= (12x^3 - 4x^2 + 8x) - (9x^2 - 3x + 6)$$
$$= 12x^3 - 4x^2 + 8x - 9x^2 + 3x - 6$$
$$= 12x^3 - 13x^2 + 11x - 6$$

REVIEW EXERCISES

Find each product.

57. $(r - 3)(r + 7)$
58. $(x + 5)(3x - 2)$
59. $(4x - 3)(x + 4)$
60. $(2x + 5y)(3x - y)$
61. $(3x + 0.25)(6x - 0.5)$
62. $(5r - 7s)(4r + 3s)$
63. $x^2 + 7x - 9$ $(\times)\ 2x + 1$
64. $a^2 - 17ab - 3b^2$ $(\times)\ 2a + b$

OBJECTIVES AND EXAMPLES

- use patterns to find $(a + b)^2$, $(a - b)^2$, and $(a + b)(a - b)$ (Lesson 9–8)

$$(x + 4)^2 = x^2 + 2(4x) + 4^2$$
$$= x^2 + 8x + 16$$
$$(r - 5)^2 = r^2 - 2(5r) + 5^2$$
$$= r^2 - 10r + 25$$
$$(b + 9)(b - 9) = b^2 - 9^2$$
$$= b^2 - 81$$

REVIEW EXERCISES

Find each product.

65. $(x - 6)(x + 6)$
66. $(7 - 2x)(7 + 2x)$
67. $(4x + 7)^2$
68. $(8x - 5)^2$
69. $(5x - 3y)(5x + 3y)$
70. $(a^2 + b)^2$
71. $(6a - 5b)^2$
72. $(3m + 4n)^2$

APPLICATIONS AND PROBLEM SOLVING

73. **Finance** Find the current monthly payment on a 36-month car loan for \$18,543. Twenty-five monthly payments have already been made at an annual interest rate of 8.7%. There is a balance due of \$3216.27 at this time. Use the formula $B = P\left[\frac{1 - (1 + i)^{k-n}}{i}\right]$, where B represents the balance, P represents the current monthly payment, i represents the *monthly* interest rate (annual rate ÷ 12), k represents the total number of monthly payments already made, and n is the total number of monthly payments. (Lesson 9–2)

74. **Health** A radio station advertised the Columbus Marathon by saying that about 19,500,000 Calories would be burned in one day. If there were 6500 runners, about how many Calories did each runner burn? (Use scientific notation to solve.) (Lesson 9–3)

75. **Finance** Upon his graduation from college, Mark Price received \$10,000 in a trust fund from his grandparents. If he invests this money in an account with an annual interest rate of 6% and adds \$1000 of his own money to the account at the end of each year, will his money have doubled after 5 years? If not, when? (Lesson 9–4)

A practice test for Chapter 9 is provided on page 795.

ALTERNATIVE ASSESSMENT

COOPERATIVE LEARNING PROJECT

Saving for College In this chapter, you developed the concept of polynomials. You performed operations on polynomials, simplified polynomials, and solved polynomial equations. They were helpful in setting up a general formula to be used for inputting various data.

In this project, you will forecast a friend's finances. Jane has received $75 from her grandparents on every birthday since she was one year old. She has been saving the money in an account that pays 5% interest. She is saving her money to help pay for her college education, which she will start this fall after her 18th birthday. She also has been receiving birthday checks from her other relatives, but these didn't start until she was 12 years old. The amounts of these checks from her 12th birthday until her 18th birthday are $45, $45, $55, $50, $55, $60, and $65.

How much money will she have saved just from her birthdays by the time she starts college? Is this a reasonable amount to pay for a used car during her junior year in college? If she had invested her money in a different account that had earned 7% interest, how much more money would she have saved?

Follow these steps to accomplish your task.

- Construct a pattern for this situation.
- Develop a polynomial model to describe the amount of money she has each year.
- Determine the amount of money she received on birthdays 12 through 18.
- Determine what needs to be changed in your model when changing the interest rate.
- Write a paragraph describing the problem and your solution.

THINKING CRITICALLY

- Can $(-b)^2$ ever equal $-b^2$? Explain and give an example to support your answer.
- For all numbers a and b and any integer m, is $(a + b)^m = a^m + b^m$ a true sentence? Explain and give examples.

PORTFOLIO

Error analysis shows common mistakes that happen when performing an operation. Here is an example of an error when multiplying like bases.

$$4^3 \cdot 4^4 = 16^7$$

Actually, $4^3 \cdot 4^4 = 4^7$. The error of multiplying the bases while adding the exponents was incorrect. The base should stay the same while adding the exponent.

From the material in this chapter, find a problem that occurs often and write an error analysis for it. Describe the situation, give an example of the incorrect method, give the correct method for that example, and write a paragraph about it. Place this in your portfolio.

SELF EVALUATION

In this chapter, there are several words that have prefixes or suffixes that can be analyzed to determine what the word means. Do you break down words to find their meanings or do you just skip over those words and look for the meaning in the context of the sentence or paragraph? Maybe you go straight to the dictionary to get the meaning.

Assess yourself. How do you best learn new vocabulary words? After learning the meaning of the new word, do you then try to use that new word in your speaking and/or writing? Describe the plan that you use when learning a new word and how you accomplish it. Give an example of a new math-related word and a new word used in your daily life and explain how you found the meaning of each of these words.

LONG-TERM PROJECT

In·ves·ti·ga·tion

the BRICKYARD

MATERIALS NEEDED

construction paper

scissors

ruler

You work for a construction company that specializes in brick patios. Your job is to create custom-designed patios. The company manufactures square and rectangular bricks. Recently, your manager sent your department the memo shown below.

In this Investigation, you must design brick patios that fit the specifications given in the memo. The design plans must be explicit and detailed, so that the construction crew can build them accordingly. Your design team consists of three people.

Make an Investigation Folder in which you can store all of your work on this Investigation for future use.

MEMO

To: Custom Design Department
From: Joanna Brown, Manager *JB*

We have a problem, and I need your help in solving it. We have an excess inventory of three types of bricks:

- small square bricks,
- large square bricks, and
- rectangular bricks that are as long as the large square brick and as wide as the small square brick.

We need to move this inventory, so I am asking you to investigate the possible patio design patterns using these three types of bricks.

I don't know if this is helpful, but the length of the larger brick is the same length as the diagonal of the smaller square brick.

Our other custom designs include using triangles, rectangles, and hexagons exclusively to create repeating patterns. However, for these surplus bricks, we need to concentrate on repeating rectangular patterns.

Please create several patio designs that will utilize these bricks. Submit at least three different plans, explaining the materials required for each patio. I am anxious to see the different ways in which these bricks can be arranged to form rectangular patios. Is there a general formula or pattern we can use to design these in the future? I look forward to your report on helping us solve our inventory problem.

PATIO SKETCH #____	DIMENSIONS	MATERIALS USED		
	small square: ___ × ___	bricks	number	total area
	large square: ___ × ___	sm. squares		
	rectangle: ___ × ___	lg. squares		
	design size: ___ × ___	rectangles		
		TOTAL		

CREATE MODELS

1 Copy the table above. You will use it, along with other tables, to record the data as you explore the designs that are possible.

2 Using the measurement requirements from Ms. Brown's memo, create a model of each of the three sizes of bricks.

3 After determining that your three models comply with the given specifications, use construction paper to make several copies of each model.

ANALYZE THE MODELS

4 Share the dimensions of your models with the other design teams in your class. Obviously the models of each design team will not be the same size, but each set of models must comply with the requirements Ms. Brown wrote in her memo.

5 Explain how the three sizes relate to each other. Make a chart listing the dimensions of each of the models from the other design teams. Do each set of dimensions relate in the same way that your dimensions do? Should they? Explain.

6 How do the lengths of the two square bricks relate? If you were to make one row of small squares and below it make one row of large squares, how many small squares would it take to match the exact length of the row of large squares? Explain your answer mathematically.

You will continue working on this Investigation throughout Chapters 10 and 11.

Be sure to keep your individual brick models, chart, and other materials in your Investigation Folder.

The Brickyard Investigation

Working on the Investigation
Lesson 10–4, p. 586

Working on the Investigation
Lesson 10–6, p. 600

Working on the Investigation
Lesson 11–1, p. 617

Working on the Investigation
Lesson 11–2, p. 627

Closing the Investigation
End of Chapter 11, p. 650

CHAPTER

10 Using Factoring

The Impact of the Media

Objectives

In this chapter, you will:

- find the prime factorization of integers,
- find the greatest common factors (GCF) for sets of monomials,
- factor polynomials,
- solve problems by using guess and check, and
- use the zero product property to solve equations.

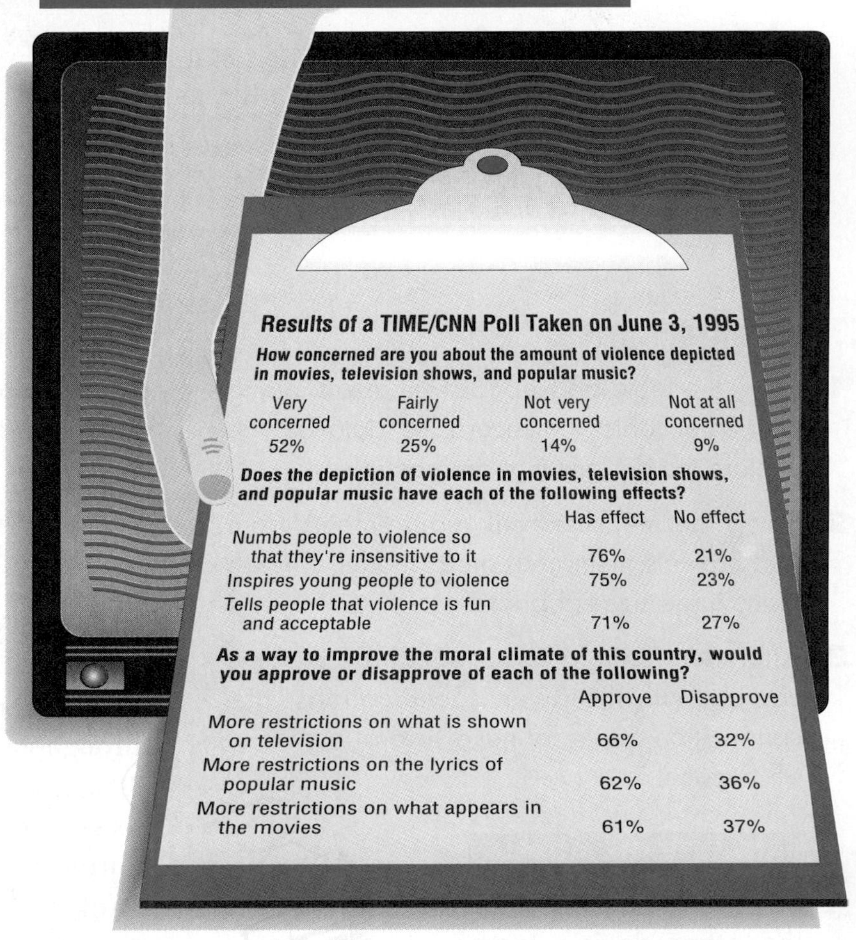

Results of a TIME/CNN Poll Taken on June 3, 1995

How concerned are you about the amount of violence depicted in movies, television shows, and popular music?

Very concerned	Fairly concerned	Not very concerned	Not at all concerned
52%	25%	14%	9%

Does the depiction of violence in movies, television shows, and popular music have each of the following effects?

	Has effect	No effect
Numbs people to violence so that they're insensitive to it	76%	21%
Inspires young people to violence	75%	23%
Tells people that violence is fun and acceptable	71%	27%

As a way to improve the moral climate of this country, would you approve or disapprove of each of the following?

	Approve	Disapprove
More restrictions on what is shown on television	66%	32%
More restrictions on the lyrics of popular music	62%	36%
More restrictions on what appears in the movies	61%	37%

Source: *Time Magazine,* June 12, 1995

TIME Line

Is American culture too violent? Do movies, television, magazines, and music reflect a true picture of America or do they contribute to the violence in our culture? What impact does the media have on American youth?

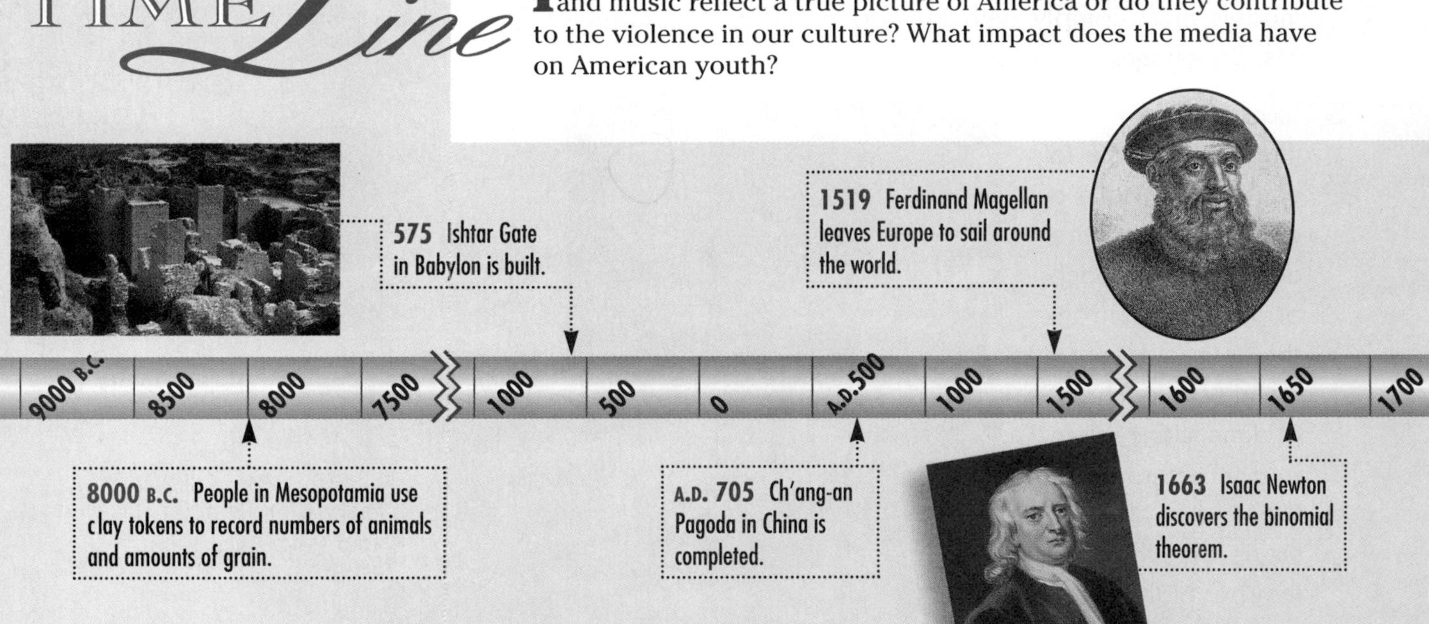

PEOPLE IN THE NEWS

The title of **Robert Rodriquez's** new book is *Rebel Without a Crew: Or How a 23-Year-Old Filmmaker with $7,000 Became a Hollywood Player*. It is the story of how the University of Texas film student from Austin, Texas, made a feature film on a very small budget. He used friends as actors, wrote the script, directed the 14-day shoot, and handled the camera work as a one-man crew. The film, *El Mariachi*, went on to win the Audience Award at the Sundance Film Festival, was released by Columbia, and is now on video.

Robert's second film, a full budget production, was *Desperado*, released in 1995. He told Columbia that he would sign a contract if he could stay in Texas, near his family and his inspiration. His advice to future film makers, "Grab your camera and just do it."

Chapter Project

- Pick five of your favorite movies. List five hints for each movie that will help your classmates to guess the names of the movies.
- Exchange your list of hints with a classmate. Try to guess your classmate's favorite movies. Explain how each hint helped you to eliminate some movies and to concentrate on others. Did you guess your classmate's favorite movies correctly?
- Explain how guessing can help when factoring polynomials.
- List five polynomials for one of your classmates to factor. Make sure that one of your polynomials cannot be factored.
- Exchange your polynomials with a classmate and factor the polynomials on the list you receive.

1995 *Waterworld*, which cost a record $175,000,000 to produce, is released.

1850 1900 1910 1920 1930 1940 1950 1960 1970 1980 1990 2000

1939 Marian Anderson gives a concert for 75,000 at the Lincoln Memorial.

1975 The VHS format (Video Home System) is launched by the Japanese company JVC.

10–1 Factors and Greatest Common Factors

What YOU'LL LEARN

- To find prime factorizations of integers, and
- to find greatest common factors (GCF) for sets of monomials.

Why IT'S IMPORTANT

You can use factors to solve problems involving packaging and gardening.

INTEGRATION

Geometry

Suppose you were asked to use grid paper to draw all of the possible rectangles with whole number dimensions that have areas of 12 square units each. The figure at the right shows one possible drawing.

Rectangles A and B are both 3 by 4 and, therefore, can be considered the same. Likewise, Rectangles C and D and Rectangles E and F are considered the same.

Recall that when two or more numbers are multiplied to form a product, each number is a *factor* of the product. In the example above, 12 is expressed as the product of different pairs of whole numbers.

$12 = 3 \times 4 \qquad 12 = 2 \times 6 \qquad 12 = 12 \times 1$

$12 = 4 \times 3 \qquad 12 = 6 \times 2 \qquad 12 = 1 \times 12$

The whole numbers 1, 2, 3, 4, 6, and 12 are factors of 12.

Example  **Find the factors of 72.**

To find the factors of 72, list all the pairs of numbers whose product is 72.

$1 \times 72 \qquad 2 \times 36 \qquad 3 \times 24 \qquad 4 \times 18 \qquad 6 \times 12 \qquad 8 \times 9$

Therefore, the factors of 72, in increasing order, are 1, 2, 3, 4, 6, 8, 9, 12, 18, 24, 36, and 72.

Some whole numbers have exactly two factors, the number itself and 1. These numbers are called **prime numbers**. Whole numbers that have more than two factors are called **composite numbers**.

Definitions of Prime and Composite Numbers	**A prime number is a whole number, greater than 1, whose only factors are 1 and itself. A composite number is a whole number, greater than 1, that is not prime.**

0 and 1 are neither prime nor composite.

FYI

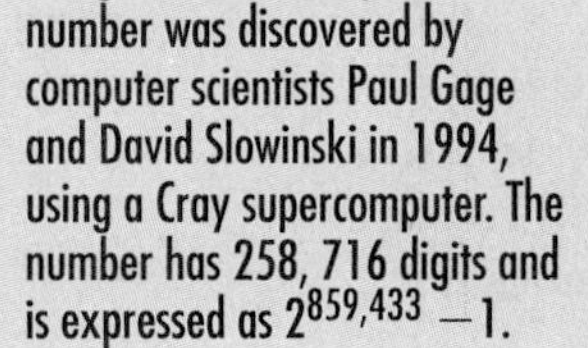

The greatest known prime number was discovered by computer scientists Paul Gage and David Slowinski in 1994, using a Cray supercomputer. The number has 258, 716 digits and is expressed as $2^{859,433} - 1$.

The number 6 is a factor of 12, but not a *prime factor* of 12, since 6 is not a prime number. When a whole number is expressed as a product of factors that are all prime numbers, the expression is called the **prime factorization** of the number. Thus, the prime factorization of 12 is $2 \cdot 2 \cdot 3$ or $2^2 \cdot 3$.

The prime factorization of every number is unique except for the order in which the factors are written. For example, $2 \cdot 3 \cdot 2$ is also a prime factorization of 12, but it is the same as $2 \cdot 2 \cdot 3$. This property of numbers is called the **unique factorization theorem**.

Example 2 Find the prime factorization of 140.

Method 1

$140 = 2 \cdot 70$ *The least prime factor of 140 is 2.*

$= 2 \cdot 2 \cdot 35$ *The least prime factor of 70 is 2.*

$= 2 \cdot 2 \cdot 5 \cdot 7$ *The least prime factor of 35 is 5.*

All the factors in the last row are prime. Thus, the prime factorization of 140 is $2 \cdot 2 \cdot 5 \cdot 7$ or $2^2 \cdot 5 \cdot 7$.

Method 2

Use a factor tree.

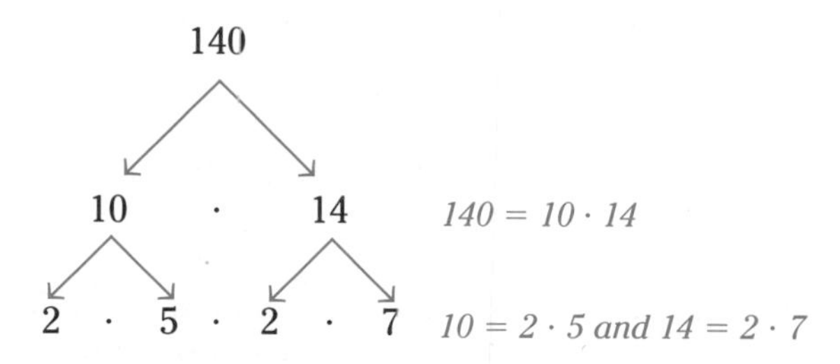

140 = 10 · 14

10 = 2 · 5 and 14 = 2 · 7

All of the factors in the last branch of the factor tree are prime. Thus, the prime factorization of 140 is $2 \cdot 2 \cdot 5 \cdot 7$ or $2^2 \cdot 5 \cdot 7$.

FYI

Computer scientists have been able to find the prime factors of a 155-digit number. This is particularly significant because some secret codes are based on the factors of large numbers.

A negative integer is factored completely when it is expressed as the product of -1 and prime numbers.

Example 3 Factor -150 completely.

$-150 = -1 \cdot 150$ *Express -150 as -1 times 150.*

$= -1 \cdot 2 \cdot 75$ *Find the prime factors of 150.*

$= -1 \cdot 2 \cdot 3 \cdot 25$

$= -1 \cdot 2 \cdot 3 \cdot 5 \cdot 5$ or $-1 \cdot 2 \cdot 3 \cdot 5^2$

A monomial is in **factored form** when it is expressed as the product of prime numbers and variables and no variable has an exponent greater than 1.

Example 4 Factor $45x^3y^2$.

$45x^3y^2 = 3 \cdot 15 \cdot x \cdot x \cdot x \cdot y \cdot y$

$= 3 \cdot 3 \cdot 5 \cdot x \cdot x \cdot x \cdot y \cdot y$

Two or more numbers may have some common factors. Consider the numbers 84 and 70, for example.

Factors of 84: 1, 2, 3, 4, 6, 7, 12, 14, 21, 28, 42, 84

Factors of 70: 1, 2, 5, 7, 10, 14, 35, 70

There are some factors that appear on both lists. The greatest of these numbers is 14, which is called the **greatest common factor (GCF)** of 84 and 70.

Definition of Greatest Common Factor	The greatest common factor of two or more integers is the greatest number that is a factor of all of the integers.

There is an easier way to find the GCF of numbers without having to find all of their factors. Look at the prime factorizations of the numbers, and multiply all the prime factors they have in common. *If there are no common prime factors, the GCF is 1.*

$$84 = 2 \cdot 2 \cdot 3 \cdot 7 \qquad 70 = 2 \cdot 5 \cdot 7$$

The integers 84 and 70 have 2 and 7 as common prime factors. The product of these common prime factors is 14, the GCF of 84 and 70.

Example 5 **Find the GCF of 54, 63, and 180.**

$54 = 2 \cdot 3 \cdot (3) \cdot (3)$ *Factor each number.*

$63 = (3) \cdot (3) \cdot 7$ *Circle the common factors.*

$180 = 2 \cdot 2 \cdot (3) \cdot (3) \cdot 5$

The GCF of 54, 63, and 180 is $3 \cdot 3$ or 9.

Example 6

APPLICATION
Packaging

A bakery packages its fat-free cookies in two sizes of boxes. One box contains 18 cookies, and the other contains 24 cookies. In order to keep the cookies fresh, the bakery plans to wrap a smaller number of cookies in cellophane before they are placed in the boxes. To save money, the bakery wants to use the same size cellophane packages for each box and to place the greatest possible number of cookies in each cellophane package.

a. How many cookies should the bakery place in each cellophane package?

b. How many cellophane packages will go in each size of box?

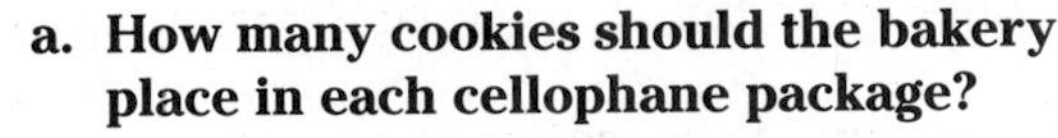

a. Find the GCF of 18 and 24.

$18 = (2) \cdot 3 \cdot (3)$

$24 = (2) \cdot 2 \cdot 2 \cdot (3)$

The bakery should put $2 \cdot 3$ or 6 cookies in each inner cellophane package.

b. The box of 18 cookies will contain $18 \div 6$ or 3 cellophane packages. The box of 24 cookies will contain $24 \div 6$ or 4 cellophane packages.

CAREER CHOICES

Pastry chefs are professionals who prepare bread and pastry goods for restaurants, institutions, and retail bakery shops. They do most of their work by hand, taking pride in their creations.

Apprenticeship programs include studying food preparation, decoration techniques, purchasing, and business mathematics.

For more information, contact:

The Educational Foundation of the National Restaurant Association
250 South Wacker Dr.
Suite 1400
Chicago, IL 60606

The GCF of two or more monomials is the product of their common factors, when each monomial is expressed in factored form.

Example  **Find the GCF of $12a^2b$ and $90a^2b^2c$.**

$12a^2b = (2) \cdot 2 \cdot (3) \cdot (a) \cdot (a) \cdot (b)$ *Factor each monomial.*

$90a^2b^2c = (2) \cdot 3 \cdot (3) \cdot 5 \cdot (a) \cdot (a) \cdot (b) \cdot b \cdot c$ *Circle the common factors.*

The GCF of $12a^2b$ and $90a^2b^2c$ is $2 \cdot 3 \cdot a \cdot a \cdot b$ or $6a^2b$.

CHECK FOR UNDERSTANDING

Communicating Mathematics

Study this lesson. Then complete the following.

1. **Draw** and label as many rectangles as possible with whole number dimensions that have an area of 48 square inches.
2. Is $2 \cdot 3^2 \cdot 4$ the prime factorization of 72? Why or why not?
3. If the GCF of two numbers is 1, must the numbers be prime? Explain.

4. How many prime numbers do you believe there are? Write a statement to support your opinion.

Guided Practice

Find the factors of each number.

5. 4
6. 56

State whether each number is *prime* or *composite*. If the number is composite, find its prime factorization.

7. 89
8. 39

Factor each expression completely. Do not use exponents.

9. -30
10. $22m^2n$

Find the GCF of the given monomials.

11. 4, 12
12. 10, 15
13. $24d^2$, $30c^2d$
14. 18, 35
15. $-20gh$, $36g^2h^2$
16. $30a^2$, $42a^3$, $54a^3b$

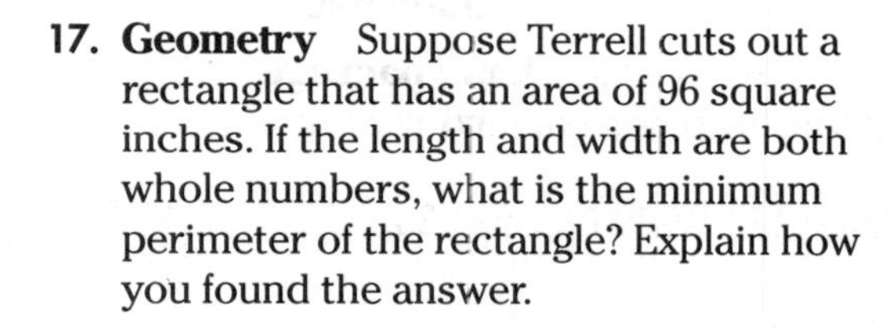

17. **Geometry** Suppose Terrell cuts out a rectangle that has an area of 96 square inches. If the length and width are both whole numbers, what is the minimum perimeter of the rectangle? Explain how you found the answer.

EXERCISES

Practice

Find the factors of each number.

18. 25
19. 67
20. 36
21. 80
22. 400
23. 950

State whether each number is *prime* or *composite*. If the number is composite, find its prime factorization.

24. 17
25. 63
26. 91
27. 97
28. 304
29. 1540

Factor each expression completely. Do not use exponents.

30. -70
31. -117
32. $66z^2$
33. $4b^3d^2$
34. $-102x^3y$
35. $-98a^2b$

Find the GCF of the given monomials.

36. 18, 36
37. 18, 45
38. 84, 96
39. 28, 75
40. −34, 51
41. 95, −304
42. $17a, 34a^2$
43. $21p^2q, 35pq^2$
44. $12an^2, 40a^4$
45. $-60r^2s^2t^2, 45r^3t^3$
46. 18, 30, 54
47. 24, 84, 168
48. $14a^2b^3, 20a^3b^2c, 35ab^3c^2$
49. $18x^2, 30x^3y^2, 54y^3$
50. $14a^2b^2, 18ab, 2a^3b^3$
51. $32m^2n^3, 8m^2n, 56m^3n^2$

Find each missing factor.

52. $42a^2b^5c = 7a^2b^3(\underline{\ ?\ })$
53. $-48x^4y^2z^3 = 4xyz(\underline{\ ?\ })$
54. $48a^5b^5 = 2ab^2(4ab)(\underline{\ ?\ })$
55. $36m^5n^7 = 2m^3n(6n^5)(\underline{\ ?\ })$

56. **Geometry** The area of a rectangle is 116 square inches. What are its possible whole number dimensions?

57. **Geometry** The area of a rectangle is 1363 square centimeters. If the measures of the length and width are both prime numbers, what are the dimensions of the rectangle?

58. **Number Theory** Check to see if your house number is a prime number and if the last four digits in your telephone number form a prime number. Explain how you decided.

59. **Number Theory** *Twin primes* are two consecutive odd numbers that are prime, such as 11 and 13. List the twin primes where both primes are less than 100.

Programming

60. Use the graphing calculator program below to find the GCF of two numbers.

```
PROGRAM:GCF
: Input "INTEGER",A       : Goto 4
: Input "INTEGER",B       : A-B→A
: A→E                     : Goto R
: B→F                     : Lbl 4
: Lbl R                   : B-A→B
: If A=B                  : Goto R
: Goto 5                  : Lbl 5
: If A<B                  : Disp "GCF IS", A
```

Use the program to find the GCF of each pair of numbers.

a. 896, 700
b. 1015, 3132
c. 567, 416
d. 486, 432
e. 891, 1701
f. 1105, 1445

Critical Thinking

61. **Geometry** Suppose the volume of a rectangular solid is $2b^3$ and the measure of each side is a monomial with integral coefficients.
    a. List the demensions of each such rectangular solid. (*Hint:* There are 6.)
    b. Draw and label each solid.
    c. Find the surface area of each solid if $b = 6$.
    d. What can you conclude about the surface areas of these solids, given that the volume remains constant?

Applications and Problem Solving

62. **Gardening** Marisela is planning to have 100 tomato plants in her garden. In what ways can she arrange them so that she has the same number of plants in each row, at least 5 rows of plants, and at least 5 plants in each row?

63. **Sports** A new athletic field is being sodded at Beck High School using 2-yard-by-2-yard squares of sod. If the length of the field is 70 yards longer than the width and its area is 6000 square yards, how many squares of sod will be needed?

Mixed Review

64. Find $(1.1x + y)^2$. (Lesson 9–8)

65. Simplify $\frac{12b^5}{4b^4}$. (Lesson 9–2)

66. Solve the system of equations by graphing. (Lesson 8–1)
$y = -x$
$y = 2x$

67. Graph the compound inequality $y > 2$ or $y < 1$. (Lesson 7–4)

68. Solve $16x < 96$. Check your solution. (Lesson 7–2)

69. Write the standard form of an equation of the line that passes through $(4, -2)$ and $(4, 8)$. (Lesson 6–2)

70. Graph $8x - y = 16$. (Lesson 5–4)

71. **Physics** Weights of 50 pounds and 75 pounds are placed on a lever. The two weights are 16 feet apart, and the lever is balanced. How far from the fulcrum is the 50-pound weight? (Lesson 4–8)

72. **Waves** The highest wave ever sighted and recorded was 112 feet high. This wave was brought on by a wind of 74 mph. Using ratios, determine how high a wave brought on by a 25-mph wind could reach. (Lesson 4–1)

73. Solve $9 = x + 13$. (Lesson 3–1)

74. Write a verbal expression for $z^7 + 2$. (Lesson 1–1)

Mathematics and SOCIETY

Number Sieve

The article below appeared in *Science News* on October 1, 1994.

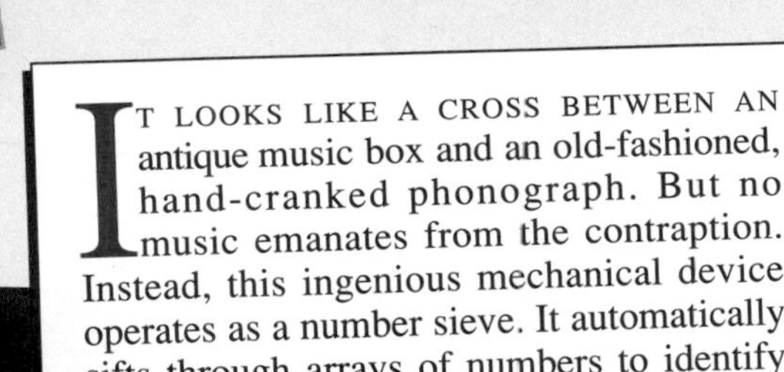

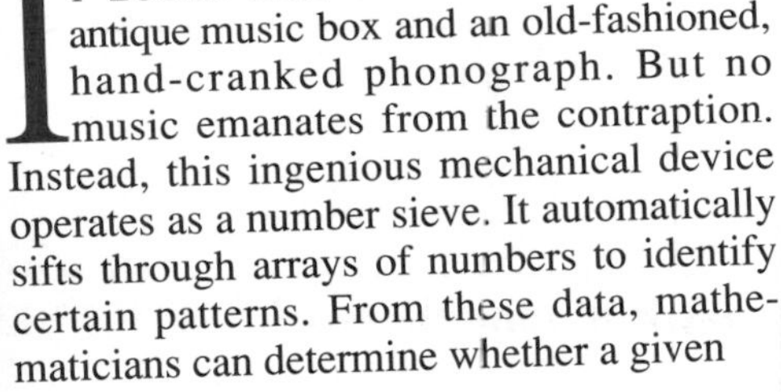

IT LOOKS LIKE A CROSS BETWEEN AN antique music box and an old-fashioned, hand-cranked phonograph. But no music emanates from the contraption. Instead, this ingenious mechanical device operates as a number sieve. It automatically sifts through arrays of numbers to identify certain patterns. From these data, mathematicians can determine whether a given number is a prime or the product of two or more primes multiplied together. Constructed 75 years ago, it also represents the first known, successful attempt to automate the factoring of whole numbers. Until three researchers tracked down the machine recently, few people knew of its existence. Now, this unique device can take its proper place in the history of computational number theory. ■

1. After the death of the machine's French inventor Eugène Olivier Carissan in 1925, the machine was given to an astronomer who put it away for safekeeping. Why do you think a machine so far ahead of its time did not find a greater use?

2. One of the main uses of prime numbers today is in cryptography, the coding and decoding of data and messages. Why are more sophisticated codes needed today than they were in the 1920s?

10–2A Factoring Using the Distributive Property

Materials: algebra tiles product mat

A Preview of Lesson 10–2

When two or more numbers are multiplied, these numbers are factors of the product. Sometimes you know the product of binomials and are asked to find the factors. This is called **factoring.** You can use algebra tiles to factor binomials.

Activity 1 Use algebra tiles to factor $2x + 8$.

Step 1 Model the polynomial $2x + 8$.

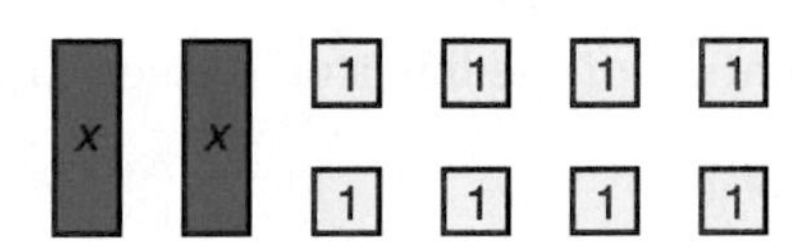

Step 2 Arrange the tiles into a rectangle. The total area of the tiles represents the product and its length and width represent the factors.

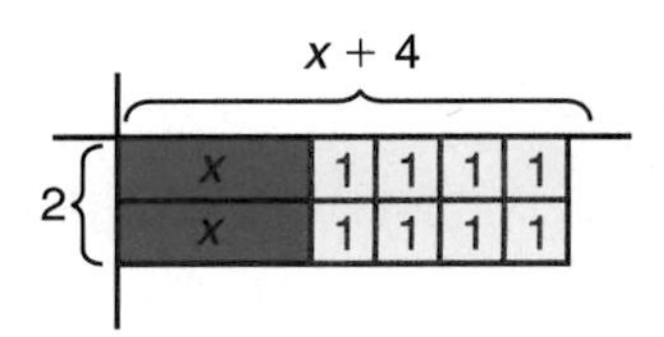

The rectangle has a width of 2 and a length of $x + 4$. Therefore, $2x + 8 = 2(x + 4)$.

Activity 2 Use algebra tiles to factor $x^2 - 3x$.

Step 1 Model the polynomial $x^2 - 3x$.

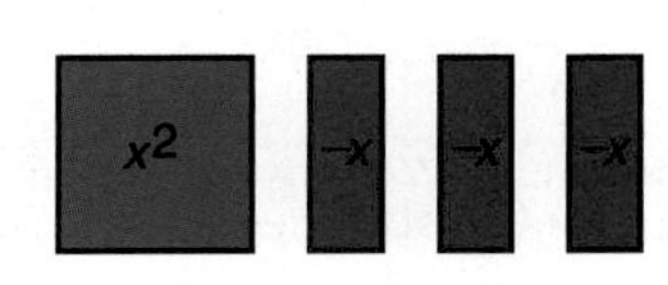

Step 2 Arrange the tiles into a rectangle.

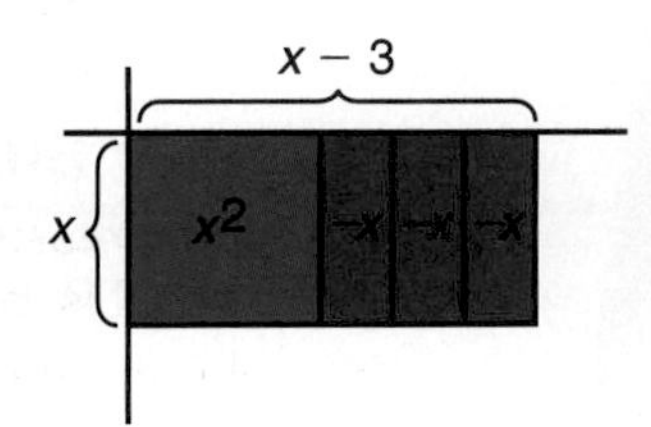

The rectangle has a width of x and a length of $x - 3$. Therefore, $x^2 - 3x = x(x - 3)$.

Model **Use algebra tiles to factor each binomial.**

1. $3x + 9$
2. $4x - 10$
3. $3x^2 + 4x$
4. $10 - 5x$

Draw **Tell whether each binomial can be factored. Justify your answer with a drawing.**

5. $2x + 3$
6. $3 - 9x$
7. $x^2 - 5x$
8. $3x^2 + 5$

Write

9. Write a paragraph that explains how you can determine whether a binomial can be factored. Include an example of one binomial that can be factored and one that cannot.

10–2 Factoring Using the Distributive Property

What YOU'LL LEARN

- To use the greatest common factor (GCF) and the distributive property to factor polynomials, and
- to use grouping techniques to factor polynomials with four or more terms.

Why IT'S IMPORTANT

You can use factoring to solve problems involving sports and construction.

Sports

Rugby is a contact sport in which each of two teams tries to get an oval ball behind its opponent's goal line or kick it over its opponent's goal. It is similar to American football. However, the action in the game is almost nonstop, and the players wear little protective gear.

There are two versions of rugby—Rugby Union and Rugby League. Rugby Union is popular in Australia, Canada, England, France, Ireland, Japan, New Zealand, Scotland, South Africa, and Wales. It is played on a rectangular field.

LOOK BACK

You can refer to Lesson 1-7 for information on the distributive property.

If the width of this field is represented by x, its length can be represented by $x + 75$ and the area of the field by $x(x + 75)$, or $x^2 + 75x$. If $x(x + 75) = x^2 + 75x$, then $x^2 + 75x = x(x + 75)$. *Why?*

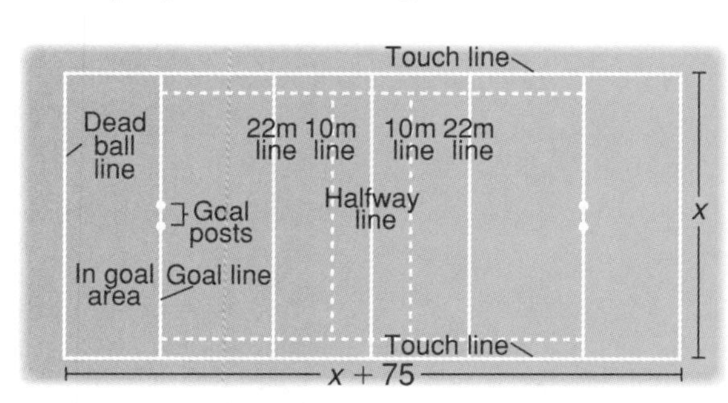

The expression $x(x + 75)$ is called the *factored form* of $x^2 + 75x$. A polynomial is in factored form, or **factored,** when it is expressed as the product of monomials and polynomials.

GLOBAL CONNECTIONS

Rugby is named for Rugby School in England where the game was first played in 1823. The game has spread throughout the British Empire and into Asia. The Rugby World Cup, the World Championship played in 1995, was won by the South African team.

In Chapter 9, you multiplied a polynomial by a monomial using the distributive property. You can also reverse this process and express a polynomial in factored form by using the distributive property.

Model	Multiplying Polynomials	Factoring Polynomials
sides 3 by $2a$, b; regions $6a$, $3b$	$3(2a + b) = 6a + 3b$	$6a + 3b = 3(a + 2b)$
sides $5x$ by $3x$, $-4y$; regions $15x^2$, $-20xy$	$5x(3x - 4y) = 15x^2 - 20xy$	$15x^2 - 20xy = 5x(3x - 4y)$
sides 3 by x^2, $5x$; regions $3x^2$, $15x$	$3(x^2 + 5x) = 3x^2 + 15x$	$3x^2 + 15x = 3(x^2 + 5x)$

Factoring a polynomial or finding the factored form of a polynomial means to find its *completely* factored form. The expression $3(x^2 + 5x)$ above is not considered completely factored since the polynomial $x^2 + 5x$ can be factored as $x(x + 5)$. The completely factored form of $3x^2 + 15x$ is $3x(x + 5)$.

Example 1 **Use the distributive property to factor each polynomial.**

a. $\mathbf{12mn^2 - 18m^2n^2}$

First find the GCF for $12mn^2$ and $18m^2n^2$.

$12mn^2 = 2 \cdot 2 \cdot 3 \cdot m \cdot n \cdot n$

$18m^2n^2 = 2 \cdot 3 \cdot 3 \cdot m \cdot m \cdot n \cdot n$ *The GCF is $2 \cdot 3 \cdot m \cdot n \cdot n$ or $6mn^2$.*

Notice that $12mn^2 = 6mn^2(2)$ and $18m^2n^2 = 6mn^2(3m)$. Then use the distributive property to express the polynomial as the product of the GCF and the remaining factor of each term.

$$12mn^2 - 18m^2n^2 = 6mn^2(2) - 6mn^2(3m)$$
$$= 6mn^2(2 - 3m) \quad \textit{Distributive property}$$

b. $\mathbf{20abc + 15a^2c - 5ac}$

$20abc = 2 \cdot 2 \cdot 5 \cdot a \cdot b \cdot c$

$15a^2c = 3 \cdot 5 \cdot a \cdot a \cdot c$

$5ac = 5 \cdot a \cdot c$ *The GCF is 5ac.*

$$20abc + 15a^2c - 5ac = 5ac(4b) + 5ac(3a) - 5ac(1)$$
$$= 5ac(4b + 3a - 1)$$

Factoring a polynomial can simplify computations.

Example 2

Construction

The Lopez family wants to build a swimming pool in the shape of the figure below. Although the family has not yet decided on the actual dimensions of the pool, they do know that they want to build a deck that is 4 feet wide around the pool.

a. Write an equation for the area of the deck.

b. If they decide to let *a* be 24 feet long, *b* be 6 feet long, and *c* be 10 feet, find the area of the deck.

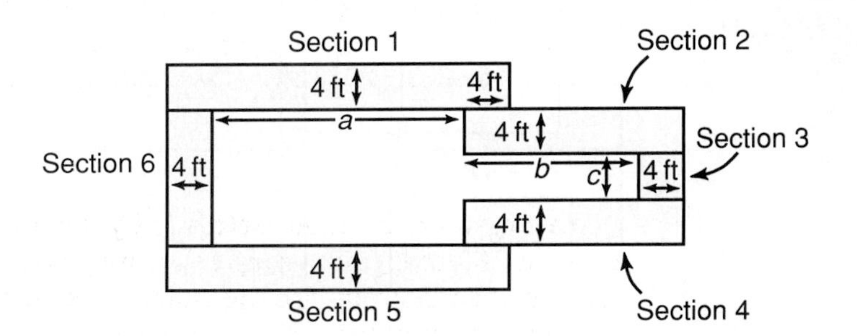

a. You can find the area of the deck by finding the sum of the areas of the 6 rectangular sections shown in the figure. The resulting expression can be simplified by first using the distributive property and then factoring.

Section 1 Section 2 Section 3 Section 4 Section 5 Section 6

$$A = 4(a + 4 + 4) + 4(b + 4) + 4c + 4(b + 4) + 4(a + 4 + 4) + 4(c + 4 + 4)$$
$$= 4a + 16 + 16 + 4b + 16 + 4c + 4b + 16 + 4a + 16 + 16 + 4c + 16 + 16$$
$$= 8a + 8b + 8c + 128 \quad \text{Combine like terms.}$$
$$= 8(a + b + c + 16) \quad \text{The GCF is 8.}$$

The area of the deck is $8(a + b + c + 16)$ square feet.
Would dividing the deck into different sections result in a different answer?

b. Replace a with 24, b with 6, and c with 10.

$$A = 8(24 + 6 + 10 + 16)$$
$$= 8(56) \text{ or } 448$$

The area of the deck would be 448 square feet.

Just as it is possible to use the distributive property to factor a polynomial into monomial and polynomial factors, it is possible to factor some polynomials containing four or more terms into the product of two polynomials. Consider $(3a + 2b)(4c + 7d) = 12ac + 21ad + 8bc + 14bd$. Here, the product of two binomials results in a polynomial with four terms. How can the process be reversed to factor the four-term polynomial into its two binomial factors?

Example 3 **Factor $12ac + 21ad + 8bc + 14bd$.**

$12ac + 21ad + 8bc + 14bd$

$= (12ac + 21ad) + (8bc + 14bd)$ *Apply the associative property, since a common factor of 3a appears in the first two terms and a common factor of 2b appears in the last two terms.*

$= 3a(4c + 7d) + 2b(4c + 7d)$ *Factor the first two terms and the last two terms.*

$= (3a + 2b)(4c + 7d)$ *4c + 7d is a common factor. Use the distributive property.*

LOOK BACK

You can refer to Lesson 9-6 for information on FOIL.

Check by using FOIL.

$$(3a + 2b)(4c + 7d) = \overset{F}{(3a)(4c)} + \overset{O}{(3a)(7d)} + \overset{I}{(2b)(4c)} + \overset{L}{(2b)(7d)}$$
$$= 12ac + 21ad + 8bc + 14bd \ ✔$$

This method is called **factoring by grouping**. It is necessary to group the terms and factor each group separately so that the remaining polynomial factors of each group are the same. This allows the distributive property to be applied a second time with a polynomial as the common factor.

Sometimes you can group the terms in more than one way when factoring a polynomial. For example, the polynomial in Example 3 could have been factored in the following way.

$$\begin{aligned} 12ac + 21ad + 8bc + 14bd &= (12ac + 8bc) + (21ad + 14bd) \\ &= 4c(3a + 2b) + 7d(3a + 2b) \\ &= (4c + 7d)(3a + 2b) \end{aligned}$$

The result is the same as in Example 3.

$-1(a - 3) = -a + 3$
$= 3 - a$

Recognizing binomials that are additive inverses is often helpful in factoring. For example, the binomials $3 - a$ and $a - 3$ are additive inverses since the sum of $3 - a$ and $a - 3$ is 0. Thus, $3 - a$ and $-a + 3$ are equivalent. What is the additive inverse of $5 - y$?

Example 4

Factor $15x - 3xy + 4y - 20$.

$$\begin{aligned} 15x - 3xy + 4y - 20 &= (15x - 3xy) + (4y - 20) \\ &= 3x(5 - y) + 4(y - 5) \\ &= 3x(-1)(y - 5) + 4(y - 5) \\ &= -3x(y - 5) + 4(y - 5) \\ &= (-3x + 4)(y - 5) \end{aligned}$$

$(5 - y)$ and $(y - 5)$ are additive inverses. $(5 - y) = (-1)(-5 + y)$ or $(-1)(y - 5)$

Check: $(-3x + 4)(y - 5) = (-3x)(y) + (-3x)(-5) + 4(y) + 4(-5)$
$= -3xy + 15x + 4y - 20$
$= 15x - 3xy + 4y - 20$ ✔

In summary, a polynomial can be factored by grouping if all of the following situations exist.

- There are four or more terms.
- Terms with common factors can be grouped together.
- The two common factors are identical or differ by a factor of -1.

CHECK FOR UNDERSTANDING

Communicating Mathematics

Study the lesson. Then complete the following.

1. a. **Express** $8d^2 - 14d$ as a product of factors in three different ways.
   b. Which of the three answers in part a is the completely factored form of $8d^2 - 14d$? Explain.

2. a. **Express** the area of the rectangle at the right by adding the areas of the smaller rectangles.
   b. **Express** the area as the product of the length and the width.
   c. What is the relationship between the expressions in parts a and b?

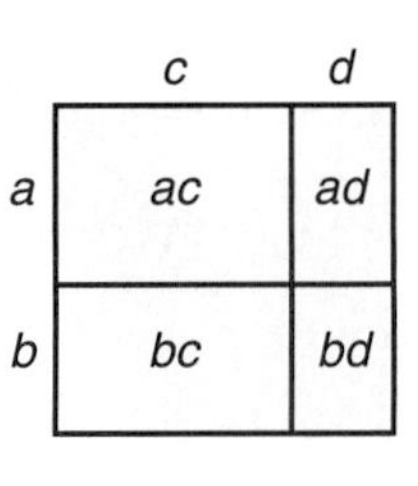

3. **List** the properties used to factor $4gh + 8h + 3g + 6$ by grouping.
4. **Group** the terms of $4gh + 8h + 3g + 6$ in pairs in two different ways so that the pairs of terms have a common monomial factor.
5. **Write** the additive inverse of $7p^2 - q$.

6. Use algebra tiles to factor $2x^2 - x$.

Guided Practice

Find the GCF of the terms in each expression.

7. $3y^2 + 12$
8. $5n - n^2$
9. $5a + 3b$
10. $6mn + 15m^2$
11. $12x^2y^2 - 8xy^2$
12. $4x^2y - 6xy^2$

Express each polynomial in factored form.

13. $a(x + y) + b(x + y)$
14. $3m(a - 2b) + 5n(a - 2b)$
15. $x(3a + 4b) - y(3a + 4b)$
16. $x^2(a^2 + b^2) + (a^2 + b^2)$

fabulous **FIRSTS**

Karch Kiraly (1961–)

Karch Kiraly is the first person to win an Olympic gold medal, win the World Championship of Beach Volleyball, and earn over $1,500,000 in pro volleyball. He also helped UCLA win 4 NCAA championships.

Complete. In Exercise 18, both blanks represent the same expression.

17. $20s + 12t = 4(5s + \underline{\ ?\ })$
18. $(6x^2 - 10xy) + (9x - 15y) = 2x(\underline{\ ?\ }) + 3(\underline{\ ?\ })$

Factor each polynomial.

19. $29xy - 3x$
20. $x^5y - x$
21. $3c^2d - 6c^2d^2$
22. $ay - ab + cb - cy$
23. $rx + 2ry + kx + 2ky$
24. $5a - 10a^2 + 2b - 4ab$
25. **Volleyball** Peta is scheduling the games for a volleyball league. To find the number of games she needs to schedule, she can use the equation $g = \frac{1}{2}n^2 - \frac{1}{2}n$, where g represents the number of games needed for each team to play each other team exactly once and n represents the number of teams.
    a. Write this equation in factored form.
    b. How many games are needed for 14 teams to play each other exactly once?
    c. How many games are needed for 7 teams to play each other exactly 3 times?

EXERCISES

Practice

Complete. In exercises with two blanks, both blanks represent the same expression.

26. $10g - 15h = 5(\underline{\ ?\ } - 3h)$
27. $8rst + 8rs^2 = \underline{\ ?\ }(t + s)$
28. $11p - 55p^2q = \underline{\ ?\ }(1 - 5pq)$
29. $(6xy - 15x) + (-8y + 20) = 3x(\underline{\ ?\ }) - 4(\underline{\ ?\ })$
30. $(a^2 + 3ab) + (2ac + 6bc) = a(\underline{\ ?\ }) + 2c(\underline{\ ?\ })$
31. $(20k^2 - 28kp) + (7p^2 - 5kp) = 4k(\underline{\ ?\ }) - p(\underline{\ ?\ })$

Factor each polynomial.

32. $9t^2 + 36t$
33. $14xz - 18xz^2$
34. $15xy^3 + y^4$
35. $17a - 41a^2b$
36. $2ax + 6xc + ba + 3bc$
37. $2my + 7x + 7m + 2xy$
38. $3m^2 - 5m^2p + 3p^2 - 5p^3$
39. $3x^3y - 9xy^2 + 36xy$
40. $5a^2 - 4ab + 12b^3 - 15ab^2$
41. $2x^3 - 5xy^2 - 2x^2y + 5y^3$
42. $12ax + 20bx + 32cx$
43. $4ax - 14bx + 35by - 10ay$
44. $3my - ab + am - 3by$
45. $28a^2b^2c^2 + 21a^2bc^2 - 14abc$
46. $6a^2 - 6ab + 3bc - 3ca$
47. $12mx - 8m + 6rx - 4r$
48. $2ax + bx - 6ay - 3by - bz - 2az$
49. $7ax + 7bx + 3at + 3bt - 4a - 4b$

Geometry

Write an expression in factored form for the area of each shaded region.

50. 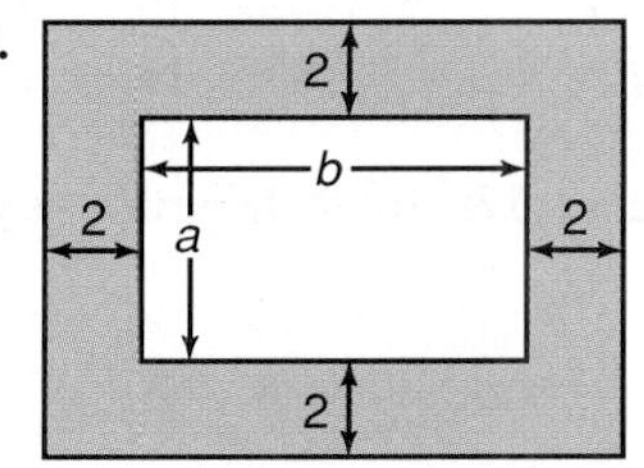

51. 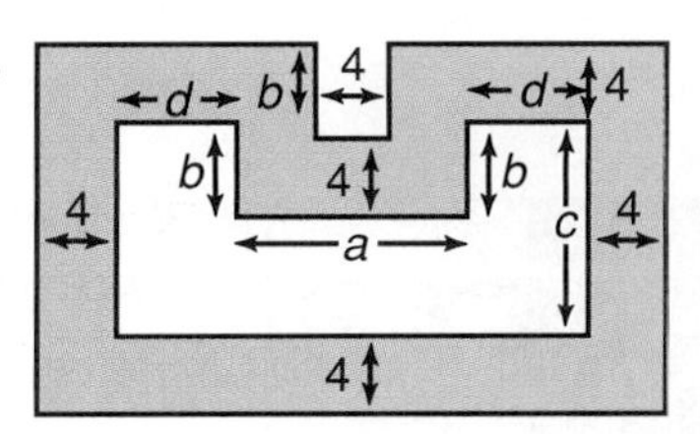

52. 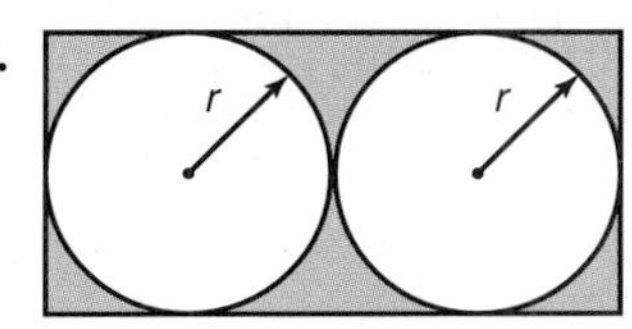

53. 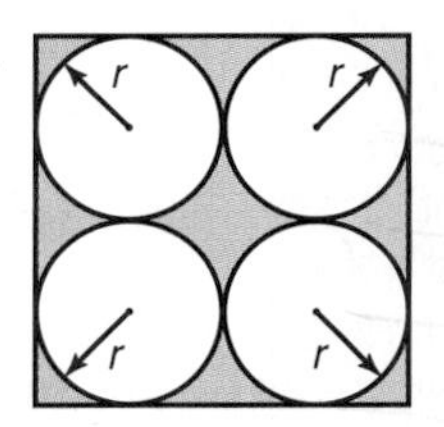

Find the dimensions of a rectangle having the given area if its dimensions can be represented by binomials with integral coefficients.

54. $(5xy + 15x - 6y - 18)$ cm^2
55. $(4z^2 - 24z - 18m + 3mz)$ cm^2

56. **Geometry** The perimeter of a square is $(12x + 20y)$ inches. Find the area of the square.

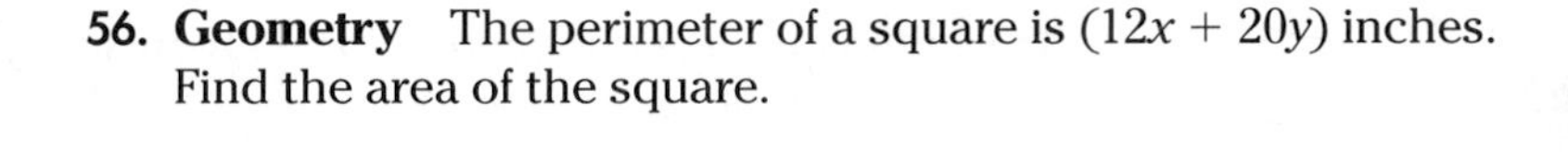

Critical Thinking

57. **Geometry** The perimeter of a rectangle is $(6a + 4b + 2ab + 12)$ centimeters. Find three possible expressions, in factored form, for the measure of its area.

Applications and Problem Solving

58. **Gardening** The length of Eduardo's garden is 5 feet more than twice its width w. This year, Eduardo decided to make the garden 4 feet longer and double its width. How much additional area did Eduardo add to his garden?

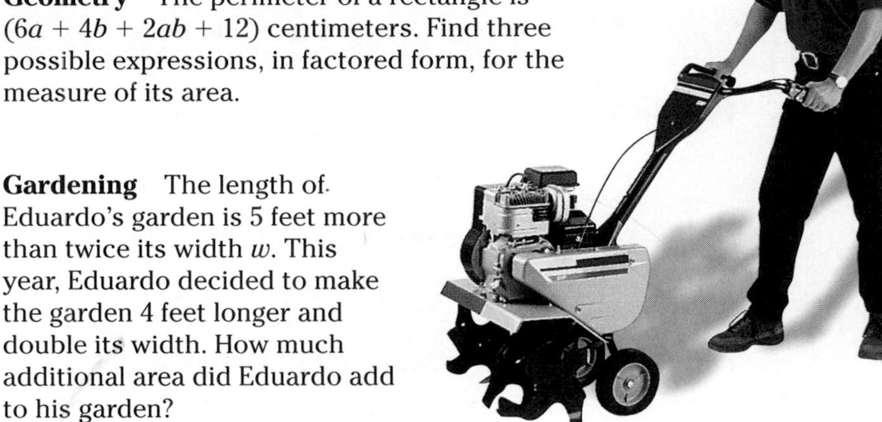

59. **Construction** A 4-foot wide stone path is to be built along each of the longer sides of a rectangular flower garden. The length of the longer side of the garden is 3 feet less than twice the length of the shorter side *s*. Write an expression, in factored form, to represent the measure of the total area of the garden and path.

60. **Rugby** Refer to the application at the beginning of the lesson.
 a. The width of a Rugby Union field is 69 meters. What is the length of the field?
 b. What is the area of a Rugby Union field?
 c. The length of a Rugby League field is 52 meters longer than its width. Write an expression for the area of the field.
 d. The width of a Rugby League field is 68 meters. Find the area of the field.
 e. Which type of rugby field has a greater area?

Mixed Review

61. **Music** Two musical notes played at the same time produce harmony. The closest harmony is produced by frequencies with the greatest GCF. A, C, and C sharp have frequencies of 220, 264, and 275, respectively. Which pair of these notes produces the closest harmony? (Lesson 10–1)

62. **Government** In 1990, the population of the United States was 248,200,000. The area of the United States is 3,540,000 square miles. (Lesson 9–3)
 a. If the population were equally spaced over the land, how many people would there be for each square mile?
 b. In 1990, the federal budget deficit was $220,000,000,000. How much would each American have had to pay in 1990 to erase the deficit?

63. **Geometry** The graphs of $3x + 2y = 1$, $y = 2$, and $3x - 4y = -29$ contain the sides of a triangle. Find the measure of the area of the triangle. (Hint: Use the formula $A = \frac{1}{2}bh$.) (Lesson 8–2)

64. **Statistics** The ten highest-paying occupations in America are shown in the table at the right. (Lesson 7–7)
 a. Make a box-and-whisker plot of the data.
 b. Name any outliers.

Occupation	Median Salary
Physician	$148,000
Dentist	93,000
Lobbyist	91,300
Management Consultant	61,900
Lawyer	60,500
Electrical Engineer	59,100
School Principal	57,300
Aeronautical Engineer	56,700
Airline Pilot	56,500
Civil Engineer	55,800

Source: Bureau of Labor Statistics

65. Solve $17.42 - 7.029z \geq 15.766 - 8.029z$. (Lesson 7–1)

66. **Geometry** Find the coordinates of the midpoint of the line segment whose endpoints are $A(5, -2)$ and $B(7, 3)$. (Lesson 6–7)

67. Solve $3a + 2b = 11$ if the domain is $\{-3, 0, 1, 2, 5\}$. (Lesson 5–3)

68. Fourteen is 50% less than what number? (Lesson 4–5)

69. Solve $4x + 3y = 7$ for y. (Lesson 3–6)

70. Solve $\frac{5}{2}x = -25$. (Lesson 3–2)

71. Find the value of $(2^5 - 5^2) + (4^2 - 2^4)$. (Lesson 1–6)

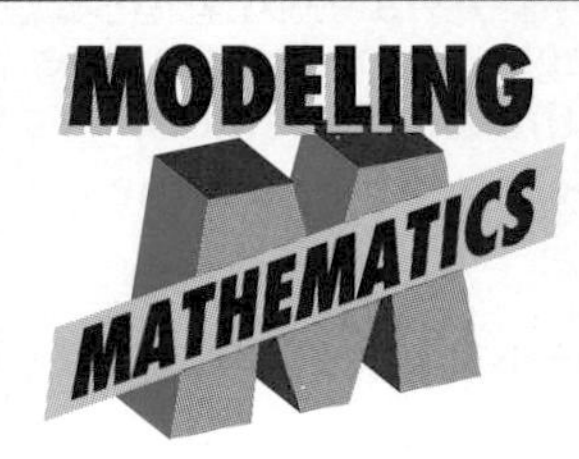

A Preview of Lesson 10–3

10–3A Factoring Trinomials

Materials: algebra tiles product mat

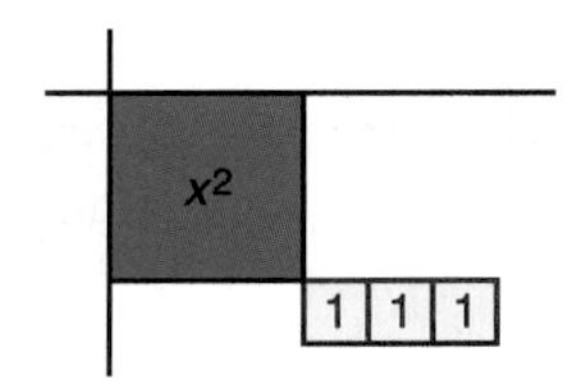

You can use algebra tiles to factor trinomials. If a rectangle cannot be formed to represent the trinomial, then the trinomial is not factorable.

Activity 1 Use algebra tiles to factor $x^2 + 4x + 3$.

Step 1 Model the polynomial $x^2 + 4x + 3$.

Step 2 Place the x^2-tile at the corner of the product mat. Arrange the 1-tiles into a 1-by-3 rectangular array as shown.

Step 3 Complete the rectangle with the x-tiles.

The rectangle has a width of $x + 1$ and a length of $x + 3$. Therefore, $x^2 + 4x + 3 = (x + 1)(x + 3)$.

You will need to use the guess-and-check strategy with many trinomials.

Activity 2 Use algebra tiles to factor $x^2 + 5x + 4$.

Step 1 Model the polynomial $x^2 + 5x + 4$.

Step 2 Place the x^2 tile at the corner of the product mat. Arrange the 1-tiles into a 2-by-2 rectangular array as shown. Try to complete the rectangle. Notice that there is an extra x-tile.

Step 3 Arrange the 1-tiles into a 1-by-4 rectangular array. This time you can complete the rectangle with the x-tiles.

The rectangle has a width of $x + 1$ and a length of $x + 4$. Therefore, $x^2 + 5x + 4 = (x + 1)(x + 4)$.

Activity 3 Use algebra tiles to factor $x^2 - 4x + 4$.

Step 1 Model the polynomial $x^2 - 4x + 4$.

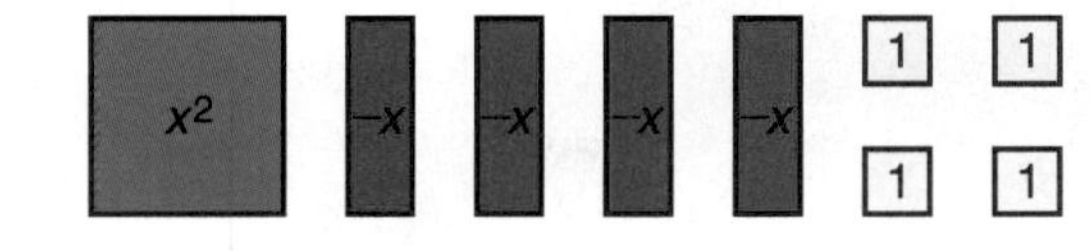

Step 2 Place the x^2-tile at the corner of the product mat. Arrange the 1-tiles into a 2-by-2 rectangular array as shown.

Step 3 Complete the rectangle with the x-tiles.

The rectangle has a width of $x - 2$ and a length of $x - 2$. Therefore, $x^2 - 4x + 4 = (x - 2)(x - 2)$.

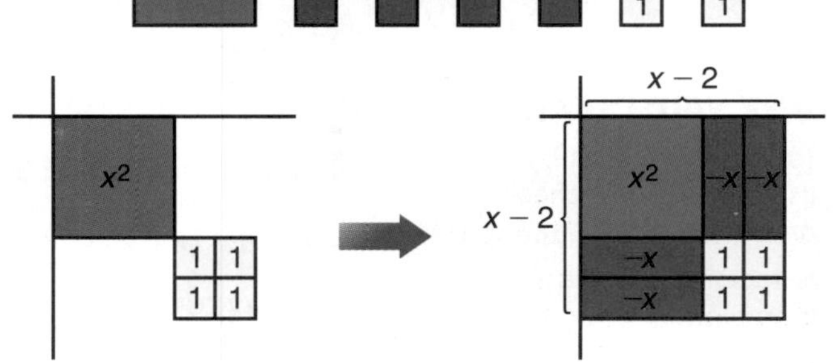

Activity 4 Use algebra tiles to factor $x^2 - x - 2$.

Step 1 Model the polynomial $x^2 - x - 2$.

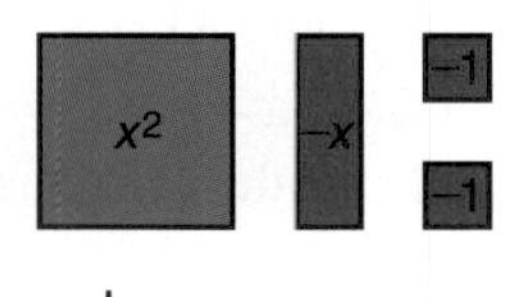

Step 2 Place the x^2-tile at the corner of the product mat. Arrange the 1-tiles into a 1-by-2 rectangular array as shown.

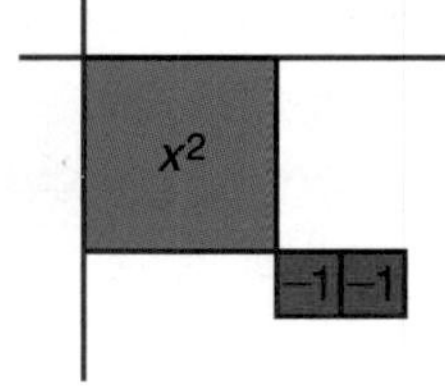

Step 3 Place the x-tile as shown. Recall that you can add zero-pairs without changing the value of the polynomial. In this case, add a zero pair of x-tiles.

The rectangle has a width of $x + 1$ and a length of $x - 2$. Therefore, $x^2 - x - 2 = (x + 1)(x - 2)$.

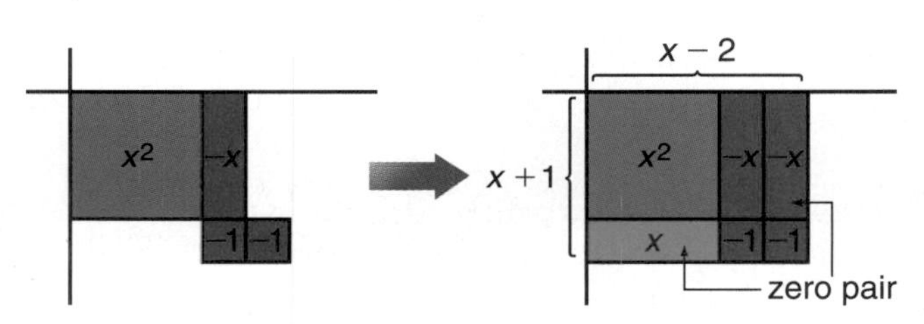

Model **Use algebra tiles to factor each trinomial.**

1. $x^2 + 6x + 5$
2. $x^2 + 5x + 6$
3. $x^2 + 7x + 12$
4. $x^2 - 6x + 9$
5. $x^2 - 3x + 2$
6. $x^2 - 6x + 8$
7. $x^2 + 4x - 5$
8. $x^2 - x - 6$

Draw **Tell whether each trinomial can be factored. Justify your answer with a drawing.**

9. $x^2 + 7x + 10$
10. $x^2 - 4x + 5$
11. $x^2 + 5x - 4$
12. $x^2 + 2x + 6$

Write

13. Write a paragraph that explains how you can determine whether a trinomial can be factored. Include an example of one trinomial that can be factored and one that cannot.

10-3

Factoring Trinomials

***What* YOU'LL LEARN**

- To solve problems by using guess and check, and
- to factor quadratic trinomials.

***Why* IT'S IMPORTANT**

You can use factoring to solve problems involving shipping and geometry.

INTEGRATION
Number Theory

The product of two consecutive odd integers is 3363. What are the numbers?

One way to find the two numbers is to use a problem-solving strategy called **guess and check**. To use this strategy, guess the answer to the problem, and then check whether the guess is correct. If the first guess is incorrect, guess and check again until you find the correct answer. Often, the results of one guess can help you make a better guess. Always keep an organized record of your guesses so you don't make the same guess twice.

Guess: Try 41 and 43.

Check: $41 \times 43 = 1763$
This product is considerably less than 3363.

Guess: Try two numbers greater than 41 and 43, such as 61 and 63.

Check: $61 \times 63 = 3843$
This product is greater than 3363.

Guess: Try 51 and 53.

Check: $51 \times 53 = 2703$
This product is less than 3363. Since the ones digit is not 0 or 5, do not use 55 as one of the numbers. *Why?*

Guess: Try 57 and 59.

Check: $57 \times 59 = 3363$

The two numbers are 57 and 59.

In Lesson 10–1, you learned that when two numbers are multiplied, each number is a factor of the product. Similarly, when two binomials are multiplied, each binomial is a factor of the product. You can use the guess-and-check strategy to find the factors of a trinomial. Consider the binomials $5x + 3$ and $2x + 7$. You can use the FOIL method to find their product.

$$\begin{aligned}(5x + 3)(2x + 7) &= \overset{F}{(5x)(2x)} + \overset{O}{(5x)(7)} + \overset{I}{(3)(2x)} + \overset{L}{(3)(7)} \\ &= 10x^2 + 35x + 6x + 21 \\ &= 10x^2 + (35 + 6)x + 21 \\ &= 10x^2 + 41x + 21\end{aligned}$$

Notice that $10 \cdot 21 = 210$ and $35 \cdot 6 = 210$.

The binomials $5x + 3$ and $2x + 7$ are factors of $10x^2 + 41x + 21$.

When using the FOIL method above, look at the product of the coefficients of the first and last terms, 10 and 21. Notice that this product, 210, is the same as the product of the coefficients of the two middle terms, 35 and 6. Their sum is the coefficient of the middle term of the final product.

You can use this pattern to factor quadratic trinomials, such as $3y^2 + 10y + 8$.

$3y^2 + 10y + 8$ — The product of 3 and 8 is 24.

$3y^2 + (\underline{?} + \underline{?})y + 8$

You need to find two integers whose *product is 24* and whose *sum is 10*. Use the guess-and-check strategy to find these numbers.

Factors of 24	Sum of Factors	
1, 24	$1 + 24 = 25$	*no*
2, 12	$2 + 12 = 14$	*no*
3, 8	$3 + 8 = 11$	*no*
4, 6	$4 + 6 = 10$	*yes*

$3y^2 + 10y + 8$

$= 3y^2 + (4 + 6)y + 8$ — *Select the factors 4 and 6.*

$= 3y^2 + 4y + 6y + 8$

$= (3y^2 + 4y) + (6y + 8)$ — *Group terms that have a common monomial factor.*

$= y(3y + 4) + 2(3y + 4)$ — *Factor.*

$= (y + 2)(3y + 4)$ — *Use the distributive property.*

Therefore, $3y^2 + 10y + 8 = (y + 2)(3y + 4)$. *Check by using FOIL.*

Example 1 **Factor $10x^2 - 27x + 18$.**

$10x^2 - 27x + 18$ — The product of 10 and 18 is 180.

$10x^2 + (\underline{?} + \underline{?})x + 18$ — Since the product is positive and the sum is negative, both factors of 180 must be negative. *Why?*

Factors of 180	Sum of Factors	
$-180, -1$	$-180 + (-1) = -181$	*no*
$-90, -2$	$-90 + (-2) = -92$	*no*
$-45, -4$	$-45 + (-4) = -49$	*no*
$-15, -12$	$-15 + (-12) = -27$	*yes* — *You can stop listing factors when you find a pair that works.*

$10x^2 - 27x + 18$

$= 10x^2 + [-15 + (-12)]x + 18$

$= 10x^2 - 15x - 12x + 18$

$= (10x^2 - 15x) + (-12x + 18)$

$= 5x(2x - 3) + (-6)(2x - 3)$ — *Factor the GCF from each group.*

$= (5x - 6)(2x - 3)$ — *Use the distributive property.*

Therefore, $10x^2 - 27x + 18 = (5x - 6)(2x - 3)$. *Check by using FOIL.*

Example 2 The area of a rectangle is $(a^2 - 3a - 18)$ square inches. This area is increased by adding 5 inches to both the length and the width. If the dimensions of the original rectangle are represented by binomials with integral coefficients, find the area of the new rectangle.

Geometry

To determine the area of the new rectangle, you must first find the dimensions of the original rectangle by factoring $a^2 - 3a - 18$. The coefficient of a^2 is 1. Thus, you must find two numbers whose product is $1 \cdot (-18)$ or -18 and whose sum is -3.

Original Rectangle
Area = $(a^2 - 3a - 18)$ in^2
? in.
? in.

Factors of −18	Sum of Factors	
$-18, 1$	$-18 + 1 = -17$	*no*
$-9, 2$	$-9 + 2 = -7$	*no*
$-6, 3$	$-6 + 3 = -3$	*yes*

The factors of −18 should be chosen so that exactly one factor in each pair is negative and that factor has the greater absolute value. Why?

$$\begin{aligned} a^2 - 3a - 18 &= a^2 + [(-6) + 3]a - 18 \\ &= a^2 - 6a + 3a - 18 \\ &= (a^2 - 6a) + (3a - 18) \\ &= a(a - 6) + 3(a - 6) \\ &= (a + 3)(a - 6) \end{aligned}$$

Check by using FOIL.

The dimensions of the original rectangle are $(a + 3)$ inches and $(a - 6)$ inches. Therefore, the dimensions of the new rectangle are $(a + 3) + 5$ or $(a + 8)$ inches and $(a - 6) + 5$ or $(a - 1)$ inches. Now, find an expression for the area of the new rectangle.

$$\begin{aligned} (a + 8)(a - 1) &= a^2 - a + 8a - 8 \\ &= a^2 + 7a - 8 \end{aligned}$$

The area of the new rectangle is $(a^2 + 7a - 8)$ square inches.

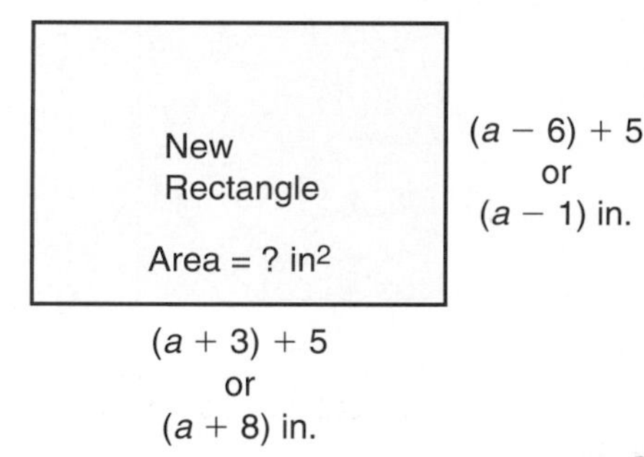

Let's study the factorization of $a^2 - 3a - 18$ from Example 2 more closely.

$$a^2 - 3a - 18 = (a + 3)(a - 6)$$

Notice that the sum of 3 and -6 is equal to -3, the coefficient of a in the trinomial. Also, the product of 3 and -6 is equal to -18, the constant term of the trinomial. This pattern holds for all trinomials whose quadratic term has a coefficient of 1.

Occasionally the terms of a trinomial will contain a common factor. In these cases, first use the distributive property to factor out the common factor, and then factor the trinomial.

Example 3 **Factor $14t - 36 + 2t^2$.**

First rewrite the trinomial so the terms are in descending order.

$$14t - 36 + 2t^2 = 2t^2 + 14t - 36$$
$$= 2(t^2 + 7t - 18)$$ *The GCF of the terms is 2. Use the distributive property.*

Now factor $t^2 + 7t - 18$. Since the coefficient of t^2 is 1, we need to find two factors of -18 whose sum is 7.

Factors of -18	Sum of Factors	
$18, -1$	$18 + (-1) = 17$	*no*
$9, -2$	$9 + (-2) = 7$	*yes*

The desired factors are 9 and -2 and $t^2 + 7t - 18 = (t + 9)(t - 2)$. Therefore, $2t^2 + 14t - 36 = 2(t + 9)(t - 2)$.

A polynomial that cannot be written as a product of two polynomials with integral coefficients is called a **prime polynomial**.

Example 4 **Factor $2a^2 - 11a + 7$.**

You must find two numbers whose product is $2 \cdot 7$ or 14 and whose sum is -11. Since the sum has to be negative, both factors of 14 have to be negative.

Factors of 14	Sum of Factors	
$-1, -14$	$-1 + (-14) = -15$	*no*
$-2, -7$	$-2 + (-7) = -9$	*no*

There are no factors of 14 whose sum is -11.

Therefore, $2a^2 - 11a + 7$ cannot be factored using integers. Thus, $2a^2 - 11a + 7$ is a prime polynomial.

You can use your knowledge of factoring to write polynomials that can be factored using integers.

Example 5 **Find all values of k so the trinomial $3x^2 + kx - 4$ can be factored using integers.**

For $3x^2 + kx - 4$ to be factorable, k must equal the sum of the factors of $3(-4)$ or -12.

Factors of -12	Sum of Factors (k)
$-12, 1$	$-12 + 1 = -11$
$12, -1$	$12 + (-1) = 11$
$-6, 2$	$-6 + 2 = -4$
$6, -2$	$6 + (-2) = 4$
$-4, 3$	$-4 + 3 = -1$
$4, -3$	$4 + (-3) = 1$

Therefore, the values of k are -11, 11, -4, 4, -1, and 1.

You can graph a polynomial and its factored form on the same axes to see if you have factored correctly. If the two graphs coincide, the factored form is probably correct.

EXPLORATION GRAPHING CALCULATORS

Suppose $x^2 - x + 6$ has been factored as $(x + 2)(x - 3)$.

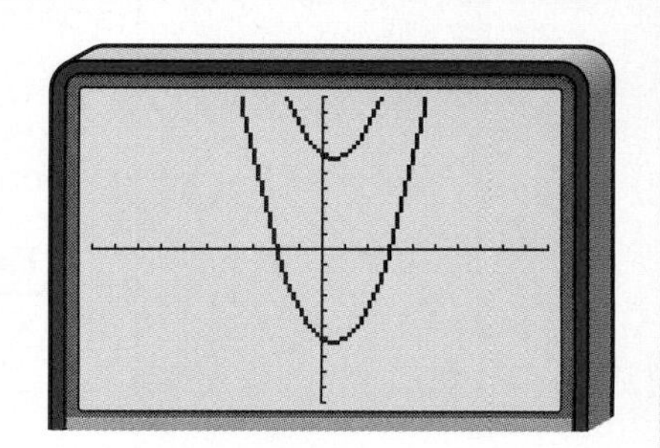

a. Press [Y=]. Enter $x^2 - x + 6$ for Y1 and $(x + 2)(x - 3)$ for Y2.

b. Press [ZOOM] 6. Notice that two different graphs appear. Therefore, $x^2 - x + 6 \neq (x + 2)(x - 3)$.

Your Turn

Determine whether each equation is a true statement. If it is not correct, state the correct factorization of the trinomial.

a. $x^2 - x - 2 = (x - 1)(x + 2)$ **b.** $x^2 - 2x - 3 = (x - 3)(x + 1)$

c. $x^2 - 5x + 4 = (x - 4)(x - 1)$ **d.** $2x^2 + 3x - 2 = (2x + 2)(x - 1)$

CHECK FOR UNDERSTANDING

Communicating Mathematics

Study this lesson. Then complete the following.

1. **Explain** why you should keep a record of your guesses when you are using the guess-and-check strategy.
2. **Write** an example of a prime polynomial.

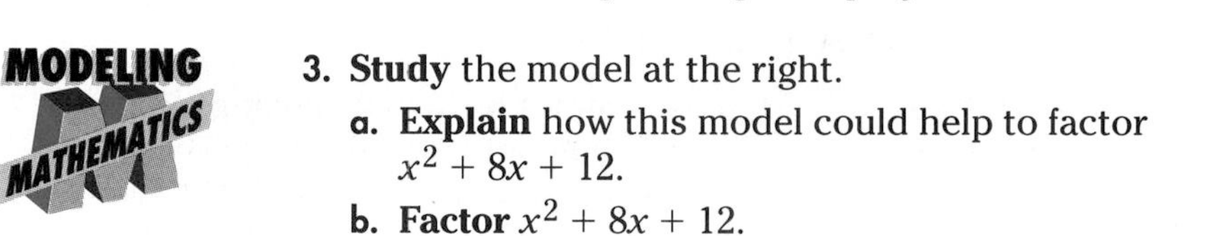

3. **Study** the model at the right.
   a. **Explain** how this model could help to factor $x^2 + 8x + 12$.
   b. **Factor** $x^2 + 8x + 12$.

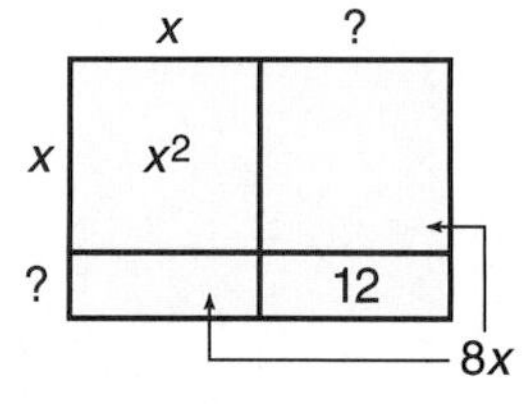

4. Use algebra tiles to factor $x^2 + 3x - 4$. Make a drawing of the algebra tiles.

Guided Practice

For each trinomial of the form $ax^2 + bx + c$, find two integers whose product is equal to ac and whose sum is equal to b.

5. $x^2 + 11x + 24$
6. $x^2 + 4x - 45$
7. $2x^2 + 13x + 20$
8. $3x^2 - 19x + 6$
9. $4x^2 - 8x + 3$
10. $5x^2 - 13x - 6$

Complete.

11. $r^2 - 5r - 14 = (r + 2)(r \underline{\ ?\ } 7)$
12. $2g^2 + 5g - 12 = (2g - 3)(g + \underline{\ ?\ })$

Factor each trinomial, if possible. If the trinomial cannot be factored using integers, write *prime*.

13. $t^2 + 7t + 12$
14. $c^2 - 13c + 36$
15. $2y^2 - 2y - 12$
16. $3d^2 - 12d + 9$
17. $2x^2 + 5x - 2$
18. $6p^2 + 15p - 9$

Find all values of k so each trinomial can be factored using integers.

19. $x^2 + kx + 14$
20. $2b^2 + kb - 3$

21. **Geometry** The area of a rectangle is $(3x^2 + 14x + 15)$ square meters. This area is reduced by decreasing both the length and width by 3 meters. If the dimensions of the original rectangle are represented by binomials with integral coefficients, find the area of the new rectangle.

EXERCISES

Practice

Complete.

22. $a^2 + a - 30 = (a - 5)(a \underline{\ ?\ } 6)$
23. $g^2 - 8g + 16 = (g - 4)(g \underline{\ ?\ } 4)$
24. $4y^2 - y - 3 = (\underline{\ ?\ } + 3)(y - 1)$
25. $6t^2 - 23t + 20 = (3t - 4)(2t - \underline{\ ?\ })$
26. $4x^2 + 4x - 3 = (2x - 1)(\underline{\ ?\ } + 3)$
27. $15g^2 + 34g + 15 = (5g + 3)(3g + \underline{\ ?\ })$

Factor each trinomial, if possible. If the trinomial cannot be factored using integers, write *prime*.

28. $b^2 + 7b + 12$
29. $m^2 - 14m + 40$
30. $z^2 - 5z - 24$
31. $t^2 - 2t + 35$
32. $s^2 + 3s - 180$
33. $2x^2 + x - 21$
34. $7a^2 + 22a + 3$
35. $2x^2 - 5x - 12$
36. $3c^2 - 3c - 5$
37. $4n^2 - 4n - 35$
38. $72 - 26y + 2y^2$
39. $10 + 19m + 6m^2$
40. $a^2 + 2ab - 3b^2$
41. $12r^2 - 11r + 3$
42. $15x^2 - 13xy + 2y^2$
43. $12x^3 + 2x^2 - 80x$
44. $5a^3b^2 + 11a^2b^2 - 36ab^2$
45. $20a^4b - 58a^3b^2 + 42a^2b^3$

Find all values of k so each trinomial can be factored using integers.

46. $r^2 + kr - 13$
47. $x^2 + kx + 10$
48. $2c^2 + kc + 12$
49. $3s^2 + ks - 14$
50. $x^2 + 8x + k, k > 0$
51. $n^2 - 5n + k, k > 0$

52. **Geometry** The area of a rectangle is $(6x^2 - 31x + 35)$ square inches. If the dimensions of the rectangle are all whole numbers, what is the minimum possible area of the rectangle?

53. **Geometry** The volume of a rectangular prism is $(15r^3 - 17r^2 - 42r)$ cubic centimeters. If the dimensions of the prism are represented by polynomials with integral coefficients, find the dimensions of the prism.

Graphing Calculator

Use a graphing calculator to determine whether each equation is a true statement. If it is not correct, state the correct factorization of the trinomial.

54. $x^2 - 2x - 15 = (x - 5)(x + 3)$
55. $2x^2 + x - 3 = (2x - 1)(x + 3)$
56. $3x^2 - 4x - 4 = (3x - 2)(x + 2)$
57. $x^2 - 6x + 9 = (x + 3)(x - 3)$

Critical Thinking

58. Complete each polynomial in three different ways so that the resulting polynomial can be factored. Then factor each polynomial.
    a. $x^2 + 8x + \underline{\ ?\ }$
    b. $x^2 + \underline{\ ?\ }x - 10$

Applications and Problem Solving

59. **Shipping** A shipping crate is to be built in the shape of a rectangular solid. The volume of the crate is $(45x^2 - 174x + 144)$ cubic feet where x is a positive integer. If the height of the crate is 3 feet, what is the minimum volume possible for this crate?

60. **Guess and Check** Place the digits 1, 2, 3, 4, 5, 6, 8, 9, 10, 12 on the dots at the right so that the sum of the integers on any line equals the sum on any other line.

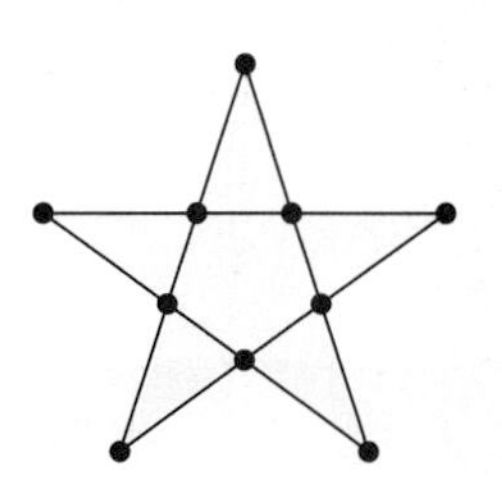

Mixed Review

61. **Finance** During the first hour of trading, John Sugarman sold x shares of stock that cost \$4 per share. During the next hour, he sold stock that cost \$8 per share. He sold 5 more shares during the first hour than the second hour. If he had sold only the stock that cost \$4 per share during the two hours, how many shares would he have needed to sell to have the same amount of total sales? (Lesson 10–2)

62. Find the degree of $7x^3 + 4xy + 3xz^3$. (Lesson 9–4)

63. Use elimination to solve the system of equations. (Lesson 8–3)
$2x = 4 - 3y$
$3y - x = -11$

64. Solve $|2y - 7| \geq -6$. (Lesson 7–6)

65. Write an equation of the line that is parallel to the graph of $2x + 3y = 1$ and passes through (4, 2). (Lesson 6–6)

66. Determine the slope of the line that passes through the points at $(-3, 6)$ and $(-5, 9)$. (Lesson 6–1)

67. State the domain and range of $\{(0, 2), (1, -2), (2, 4)\}$. (Lesson 5–2)

68. **Geography** There is three times as much water as land on Earth's surface. What percent of Earth is covered by water? (Lesson 4–4)

69. Find the supplement of 90°. (Lesson 3–4)

70. **Time** When it is noon in Richmond, Virginia, it is 2:00 A.M. the following morning in Kanagawa, Japan. Joel, who teaches English in Kanagawa, would like to call his mother in Richmond at 7:30 on the morning of her birthday, October 26. On what day and at what time would Joel have to call her? (Lesson 2–3)

71. Simplify $3(x + 2y) - 2y$. (Lesson 1–7)

SELF TEST

Find the GCF of the given monomials. (Lesson 10–1)

1. $50n^4, 40n^2p^2$
2. $15abc, 35a^2c, 105a$

Factor each polynomial, if possible. If the polynomial cannot be factored using integers, write *prime*. (Lessons 10–2 and 10–3)

3. $18xy^2 - 24x^2y$
4. $2ab + 2am - b - m$
5. $2q^2 - 9q - 18$
6. $t^2 + 5t - 20$
7. $3y^2 - 8y + 5$
8. $27m^2n^2 - 75mn$

9. **Guess and Check** Write an eight-digit number using the digits 1, 2, 3, and 4 each twice so that the 1s are separated by 1 digit, the 2s are separated by 2 digits, the 3s are separated by 3 digits, and the 4s are separated by 4 digits. (Lesson 10–3)

10. **Geometry** The area of a rectangle is $(x^2 - x - 6)$ square meters. The length and width are each increased by 9 meters. If the dimensions of the original rectangle are binomials with integral coefficients, find the area of the new rectangle. (Lesson 10–3)

10–4 Factoring Differences of Squares

What YOU'LL LEARN

- To identify and factor binomials that are the differences of squares.

Modeling

You have used algebra tiles to factor trinomials. You can also use algebra tiles to factor some binomials.

MODELING MATHEMATICS

Difference of Squares

Materials: algebra tiles product mat

Factor $x^2 - 9$.

Step 1

Model the polynomial $x^2 - 9$.

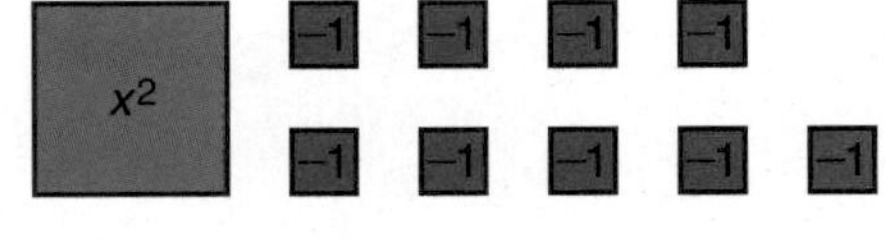

Step 2

Place the x^2-tile at the corner of the product mat. Arrange the 1-tiles into a 3-by-3 square.

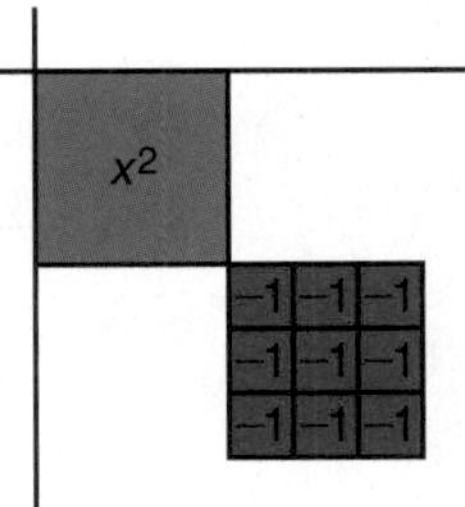

Step 3

Complete the rectangle using 3 zero pairs as shown.

The rectangle has a width of $x - 3$ and a length of $x + 3$. Therefore, $x^2 - 9 = (x - 3)(x + 3)$.

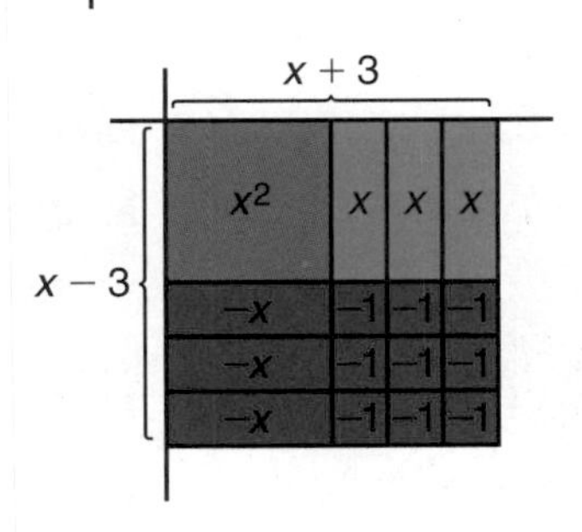

Your Turn

a. Use algebra tiles to factor each binomial.

$x^2 - 16$	$x^2 - 4$	$x^2 - 1$
$4x^2 - 9$	$9x^2 - 4$	$4x^2 - 1$

b. Binomials such as those in part a are called the *difference of squares*. Explain why you think this term applies to these binomials.

c. Study the factors of the binomials in part a. What do you notice about the signs of the factors? about the terms of the factors?

d. Use the pattern that you observe to factor $x^2 - 100$.

e. Use FOIL to check your answer in part d. Was your answer correct?

Why IT'S IMPORTANT

You can use factoring to solve problems involving geometry and number theory.

Recall that the product of the sum and difference of two binomials such as $n + 8$ and $n - 8$ is called the *difference of squares.*

$$(n + 8)(n - 8) = n^2 - 8n + 8n - 64 \quad \text{Use FOIL.}$$
$$= n^2 - 64 \quad \text{Note that this is the difference of two squares, } n^2 \text{ and } 64.$$

The Modeling Mathematics activity suggests the following rule for factoring the difference of squares.

Difference of Squares	$a^2 - b^2 = (a - b)(a + b) = (a + b)(a - b)$

You can use this rule to factor binomials that can be written in the form $a^2 - b^2$.

Example 1 **Factor each binomial.**

a. $\boldsymbol{m^2 - 81}$

$$m^2 - 81 = (m)^2 - (9)^2 \qquad \textit{m · m = m}^2 \textit{ and } 9 \cdot 9 = 81$$
$$= (m - 9)(m + 9) \qquad \textit{Use the difference of squares.}$$

b. $\boldsymbol{100s^2 - 25t^2}$

$$100s^2 - 25t^2 = 25(4s^2 - t^2) \qquad \textit{25 is the GCF.}$$
$$= 25[(2s)^2 - t^2] \qquad 2s \cdot 2s = 4s^2 \textit{ and } t \cdot t = t^2$$
$$= 25(2s - t)(2s + t) \qquad \textit{Use the difference of squares.}$$

c. $\boldsymbol{\frac{1}{9}x^2 - \frac{4}{25}y^2}$

$$\frac{1}{9}x^2 - \frac{4}{25}y^2 = \left(\frac{1}{3}x\right)^2 - \left(\frac{2}{5}y\right)^2 \qquad \textit{Why?}$$
$$= \left(\frac{1}{3}x - \frac{2}{5}y\right)\left(\frac{1}{3}x + \frac{2}{5}y\right) \qquad \textit{Check this result by using FOIL.}$$

Sometimes the terms of a binomial have common factors. If so, the GCF should always be factored out first. Occasionally, the difference of squares needs to be applied more than once or along with grouping in order to completely factor a polynomial.

Example  **Factor each polynomial.**

a. $\boldsymbol{20cd^2 - 125c^5}$

$$20cd^2 - 125c^5 = 5c(4d^2 - 25c^4) \qquad \textit{The GCF of } 20cd^2 \textit{ and } 125c^5 \textit{ is } 5c.$$
$$= 5c(2d - 5c^2)(2d + 5c^2) \qquad 2d \cdot 2d = 4d^2 \textit{ and } 5c^2 \cdot 5c^2 = 25c^4$$

b. $\boldsymbol{3k^4 - 48}$

$$3k^4 - 48 = 3(k^4 - 16) \qquad \textit{Why?}$$
$$= 3(k^2 - 4)(k^2 + 4) \qquad k^2 \cdot k^2 = k^4$$
$$= 3(k - 2)(k + 2)(k^2 + 4) \qquad k^2 + 4 \textit{ cannot be factored. Why not?}$$

c. $\boldsymbol{9x^5 + 11x^3y^2 - 100xy^4}$

To factor the trinomial, we need two numbers whose product is −900 and whose sum is 11.

$$9x^5 + 11x^3y^2 - 100xy^4 = x(9x^4 + 11x^2y^2 - 100y^4)$$
$$= x[(9x^4 - 25x^2y^2) + (36x^2y^2 - 100y^4)]$$
$$= x[x^2(9x^2 - 25y^2) + 4y^2(9x^2 - 25y^2)]$$
$$= x(x^2 + 4y^2)(9x^2 - 25y^2)$$
$$= x(x^2 + 4y^2)(3x - 5y)(3x + 5y)$$

The difference of squares can be used to multiply numbers mentally.

Example 3 **Show a method for finding the product of 37 and 43 mentally.**

Since $37 = 40 - 3$ and $43 = 40 + 3$, the product of $(37)(43)$ can be expressed as $(40 - 3)(40 + 3)$.

$$\begin{aligned}(43)(37) &= (40 + 3)(40 - 3)\\ &= 40^2 - 3^2\\ &= 1600 - 9 \text{ or } 1591\end{aligned}$$

Pythagoras

According to the Pythagorean theorem, the sum of the squares of the measures of the legs of a right triangle equals the square of the measure of the hypotenuse.

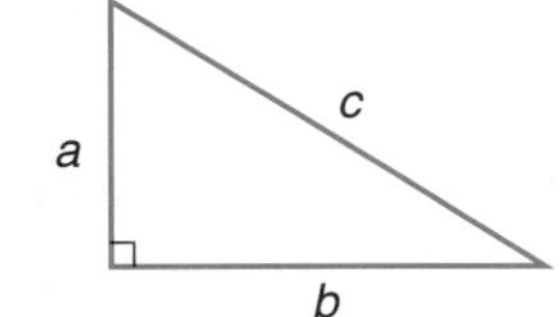

$$a^2 + b^2 = c^2$$

A **Pythagorean triple** is a group of three whole numbers that satisfy the equation $a^2 + b^2 = c^2$. For example, the numbers 3, 4, and 5 form a Pythagorean triple.

$$\begin{aligned}3^2 + 4^2 &\stackrel{?}{=} 5^2\\ 9 + 16 &\stackrel{?}{=} 25\\ 25 &= 25 \checkmark\end{aligned}$$

You can use the difference of squares to find Pythagorean triples.

Example  **Find a Pythagorean triple that includes 8 as one of its numbers.**

Number Theory

FYI

Finding Pythagorean triples has interested mathematicians since ancient times. Some Pythagorean triples have been found on Babylonian cuneiform tablets.

First find the square of 8. $8^2 = 64$

Factor 64 into two even factors or two odd factors. $64 = (2)(32)$

Find the mean of the two factors. $\frac{2 + 32}{2} = 17$

Complete the following statement.

$$\begin{aligned}(2)(32) &= (17 - \underline{\ ?\ })(17 + \underline{\ ?\ })\\ &= (17 - 15)(17 + 15)\\ &= 17^2 - 15^2\end{aligned}$$

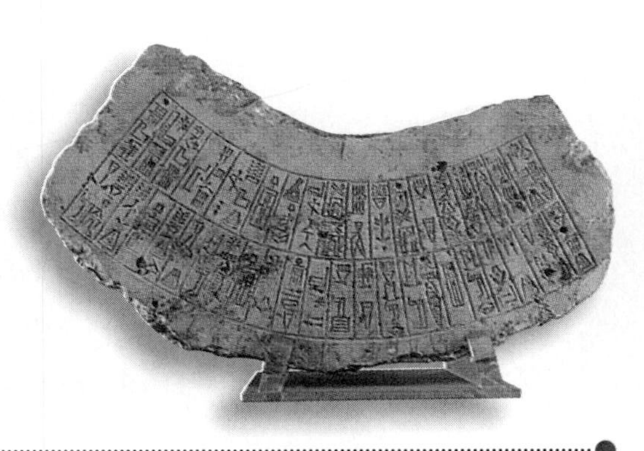

Therefore, $8^2 = 17^2 - 15^2$ or $8^2 + 15^2 = 17^2$. The numbers 8, 15, and 17 form a Pythagorean triple.

CHECK FOR UNDERSTANDING

Communicating Mathematics

Study this lesson. Then complete the following.

1. **Describe** a binomial that is the difference of two squares.
2. **Write** a polynomial that is the difference of two squares. Factor your polynomial.
3. **Explain** how to factor a difference of squares by using the method for factoring trinomials presented in Lesson 10-3.
4. **You Decide** Patsy says that $28f^2 - 7g^2$ can be factored using the difference of squares. Sally says it cannot. Who is correct? Explain.
5. **Show** how to use the difference of squares to find $\frac{15}{16} \cdot \frac{17}{16}$.

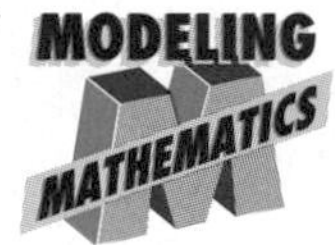

6. Use algebra tiles to factor $4 - x^2$.

Guided Practice

State whether each binomial can be factored as a difference of squares.

7. $p^2 - 49q^2$
8. $25a^2 - 81b^4$
9. $9x^2 + 16y^2$

Match each binomial with its factored form.

10. $4x^2 - 25$ a. $25(x - 1)(x + 1)$
11. $16x^2 - 4$ b. $(5x - 2)(5x + 2)$
12. $25x^2 - 4$ c. $(2x - 5)(2x + 5)$
13. $25x^2 - 25$ d. $4(2x - 1)(2x + 1)$

Factor each polynomial, if possible. If the polynomial cannot be factored, write *prime*.

14. $t^2 - 25$
15. $1 - 16g^2$
16. $2a^2 - 25$
17. $20m^2 - 45n^2$
18. $(a + b)^2 - c^2$
19. $x^4 - y^4$
20. Find the product of 17 and 23 mentally using difference of squares.
21. The difference of two numbers is 3. If the difference of their squares is 15, what is the sum of the numbers?

EXERCISES

Practice

Factor each polynomial, if possible. If the polynomial cannot be factored, write *prime*.

22. $w^2 - 81$
23. $4 - v^2$
24. $4q^2 - 9$
25. $100d^2 - 1$
26. $16a^2 - 25b^2$
27. $2z^2 - 98$
28. $9g^2 - 75$
29. $4t^2 - 27$
30. $8x^2 - 18$
31. $17 - 68k^2$
32. $25y^2 - 49z^4$
33. $36x^2 - 125y^2$
34. $-16 + 49h^2$
35. $16b^2c^4 + 25d^8$
36. $-9r^2 + 81$
37. $a^2x^2 - 0.64y^2$
38. $\frac{1}{16}x^2 - 25z^2$
39. $\frac{9}{2}a^2 - \frac{49}{2}b^2$
40. $(4p - 9q)^2 - 1$
41. $(a + b)^2 - (c + d)^2$
42. $25x^2 - (2y - 7z)^2$
43. $x^8 - 16y^4$
44. $a^6 - a^2b^4$
45. $a^4 + a^2b^2 - 20b^4$

Find each product mentally by using differences of squares.

46. 29×31
47. 24×26
48. 94×106

Find the dimensions of a rectangle with the same area as the shaded region in each drawing. Assume that the dimensions of the rectangle must be represented by binomials with integral coefficients.

49.

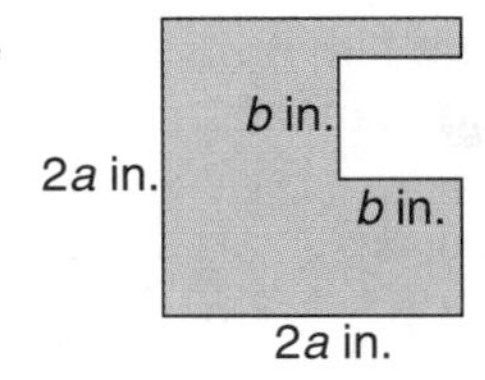

50.

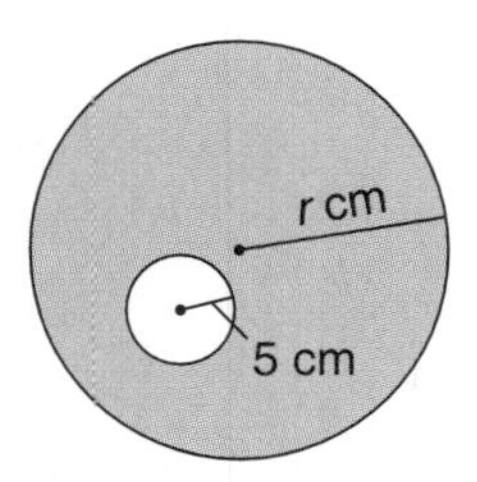

51.

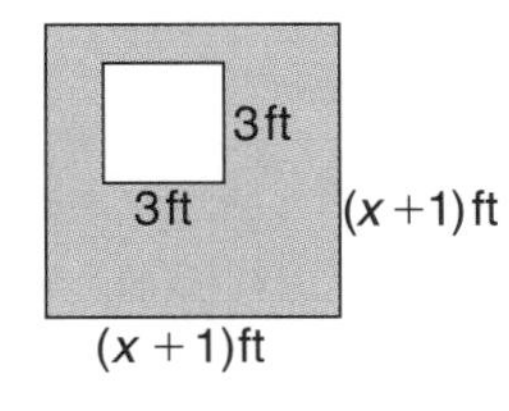

Find the dimensions of a rectangular solid having the given volume if each dimension can be written as a binomial with integral coefficients.

52. $(7mp^2 + 2np^2 - 7mr^2 - 2nr^2)$ cubic centimeters

53. $(5a^3 - 125ab^2 - 75b^3 + 3a^2b)$ cubic inches

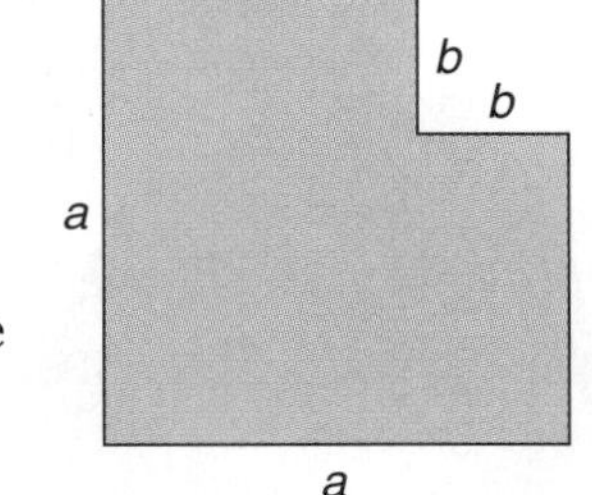

Critical Thinking

54. Show how to divide and rearrange the diagram at the right to show that $a^2 - b^2 = (a - b)(a + b)$. Make a diagram to show your reasoning.

Applications and Problem Solving

55. **Geometry** The side of a square is x centimeters long. The length of a rectangle is 5 centimeters longer than a side of the square, and the width of the rectangle is 5 centimeters shorter than the side of the square.
   a. Which has the greater area, the square or the rectangle?
   b. How much greater is that area?

56. **Number Theory** Find a Pythagorean triple that includes 7.

57. **Number Theory** Find a Pythagorean triple that includes 9.

58. **Geometry** Express the square of the length of the missing side of the triangle at the right as the product of two binomials.

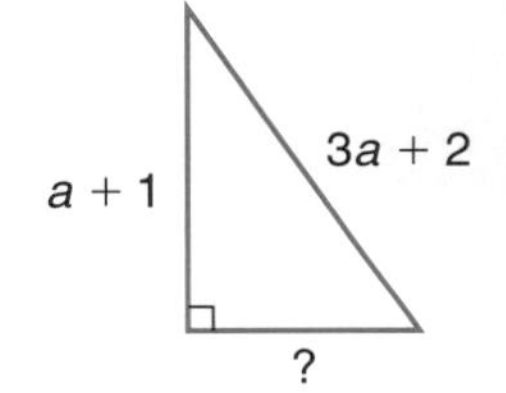

Mixed Review

59. **Guess and Check** Julie went to the corner store to buy four items. The clerk at the store had to use a calculator to add the four prices, since her cash register was broken. When the clerk figured Julie's bill, she mistakenly hit the multiplication key each time, instead of the plus key. Julie had already mentally computed her sum, so, realizing that the total was actually correct, she paid the clerk the amount of \$7.11. How much was each item that Julie bought? (Lesson 10–3)

60. Find $(n^2 + 5n + 3) + (2n^2 + 8n + 8)$. (Lesson 9–5)

61. Use elimination to solve the system of equations. (Lesson 8–4)
   $x + y = 20$
   $0.4x + 0.15y = 4$

62. **Probability** Use a tree diagram to find the probability of getting at least one tail when four fair coins are tossed. (Lesson 7–5)

63. Graph $6x - 3y = 6$. (Lesson 6–5)

64. Write an equation for the relation given in the chart below. (Lesson 5–6)

a	1	2	3	4	5	6
b	1	4	7	10	13	16

65. Graph $M(0, 3)$. (Lesson 5–1)

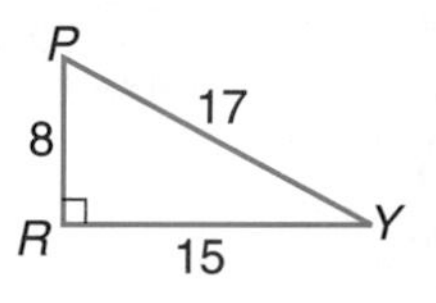

66. Trigonometry For the triangle at the right, find sin Y, cos Y, and tan Y to the nearest thousandth. (Lesson 4–3)

67. Statistics The populations in millions of the 50 states in 1993 are shown below. Find the mean, median, and mode population. (Lesson 3–7)

1.2	1.1	0.6	6.0	1.0	3.3	18.2	7.9	12.0
11.1	5.7	11.7	9.5	5.0	4.5	2.8	5.2	0.6
0.7	1.6	2.5	0.7	5.0	6.5	1.8	6.9	3.6
6.9	13.7	3.8	5.1	4.2	2.6	2.4	4.3	3.2
18.0	0.8	1.1	0.5	3.6	1.6	3.9	1.9	1.4
5.3	3.0	31.2	0.6	1.2				

68. In Mr. Tucker's algebra class, students can get extra points for finding the correct solution to the "Riddle of the Week." One week, Mr. Tucker posed this riddle. (Lesson 2–9)

Luis is 10 years older than his brother. Next year, he will be three times as old as his brother. How old is Luis now?

Josh's answer is: *Luis is 12 years old and his brother is 4.*

a. If the brother's age is represented by a, what is Luis' age?

b. What will be the brother's age next year?

c. Does Josh get the extra points for his answer? Why or why not?

Refer to the Investigation on pages 554–555.

Suppose your manager, Ms. Brown, also gave you certain specifications as to which bricks to use and how many of each kind to use.

1 Using one of each type of brick, is there a rectangular pattern you can make with just three bricks? Justify your answer.

2 Select two of one type of brick and one each of the other two types. Is there a way to arrange these bricks into a rectangular pattern? If so, is there more than one pattern? Sketch a drawing of the rectangular pattern(s) you found and label the size of each brick.

3 If there is not a way to arrange the bricks into a rectangular pattern, explain why not. Is there more than one possible choice of four bricks that will make a rectangular pattern?

4 Now, using at least one brick of each type, find all the different patterns (if any) there are to arrange the bricks into a rectangular pattern of five bricks.

5 Now try designs using at least one of each type for a pattern of six bricks, seven bricks, eight bricks, nine bricks, and ten bricks.

6 Draw a diagram of all of the possible patterns. Label the size of each brick. Describe how you developed a process for finding the patterns. Are there generalizations you can make about the number of bricks and rectangular arrangements? Justify any generalizations. Explain how you know you have all of the possible patterns.

Add the results of your work to your Investigation Folder.

10–5 Perfect Squares and Factoring

What YOU'LL LEARN

- To identify and factor perfect square trinomials.

Why IT'S IMPORTANT

You can use factoring to solve problems involving finance and construction.

INTEGRATION
Number Theory

Recall that the numbers 1, 4, 9, 16, and 25 are called *perfect square* numbers, since they can each be expressed as the square of an integer.

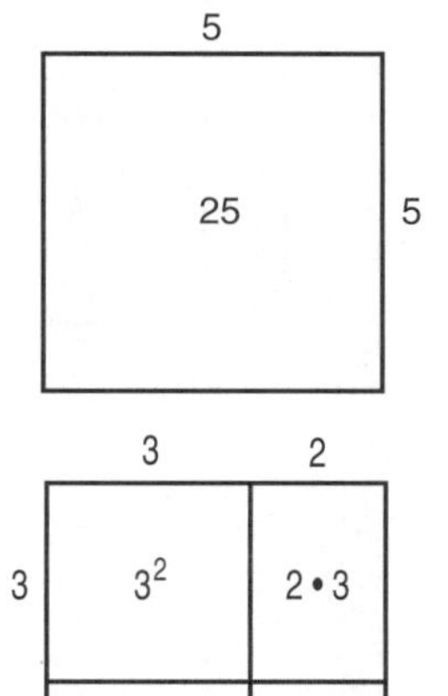

The equation $5^2 = 25$ can be modeled as the area of a square having a side of length 5 as shown at the right.

Suppose $(3 + 2)$ is substituted for 5. Then $(3 + 2)^2$ can be modeled by using a 5-by-5 square and divided it into four regions as shown at the right. The sum of the areas of the four regions equals the area of the square.

$$\begin{aligned}(3 + 2)^2 &= 3^2 + (2 \cdot 3) + (3 \cdot 2) + 2^2 \\ &= 3^2 + (3 \cdot 2) + (3 \cdot 2) + 2^2 \\ &= 3^2 + 2(3 \cdot 2) + 2^2\end{aligned}$$

The last line shows a very interesting relationship.

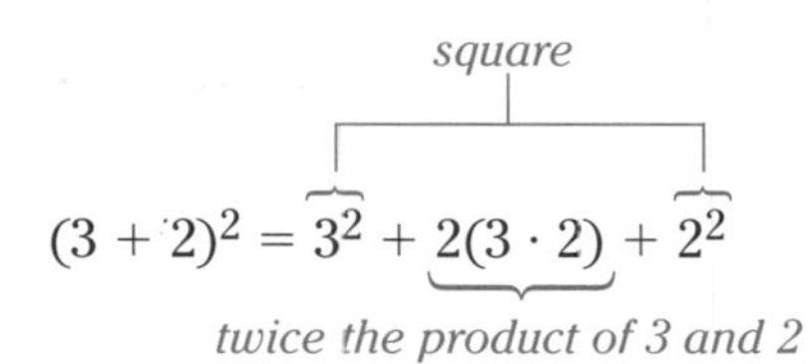

The square of this binomial is the sum of

- the square of the first term,
- twice the product of the first and second term, and
- the square of the second term.

This observation is generalized by the model at the right.

$(a + b)^2 = a^2 + 2ab + b^2$

LOOK BACK

You can refer to Lesson 9-8 for information on squares of sums and squares of differences.

To square $(a - b)$, write the binomial as $[a + (-b)]$ and then square this binomial.

$$\begin{aligned}(a - b)^2 &= [a + (-b)]^2 \\ &= a^2 + 2a(-b) + (-b)^2 \\ &= a^2 - 2ab + b^2\end{aligned}$$

Products of the form $(a + b)^2$ and $(a - b)^2$ are called perfect squares, and their expansions are **perfect square trinomials.**

Perfect Square Trinomials	$(a + b)^2 = a^2 + 2ab + b^2$ $(a - b)^2 = a^2 - 2ab + b^2$

These patterns can be used to factor trinomials.

Model	Squaring a Binomial	Factoring a Perfect Square Trinomial
v, 3 across; v, 3 down v^2	$3v$ $3v$	3^2	$(v + 3)^2 = v^2 + 2(v)(3) + 3^2$ $= v^2 + 6v + 9$	$v^2 + 6v + 9 = (v)^2 + 2(v)(3) + (3)^2$ $= (v + 3)^2$
$3p$, $-2q$ across; $3p$, $-2q$ down $(3p)^2$	$(-2q)(3p)$ $(3p)(-2q)$	$(-2q)^2$	$(3p - 2q)^2 = (3p)^2 + 2(3p)(-2q) + (-2q)^2$ $= 9p^2 - 12pq + 4q^2$	$9p^2 - 12pq + 4q^2 = (3p)^2 - 2(3p)(2q) + (2q)^2$ $= (3p - 2q)^2$

To determine whether a trinomial can be factored using these patterns, you must decide if it is a perfect square trinomial. In other words, you must determine if it can be written in the form $a^2 + 2ab + b^2$ or in the form $a^2 - 2ab + b^2$. For a trinomial to be in one of these forms, the following must be satisfied.

- The first term is a perfect square.
- The third term is a perfect square.
- The middle term is either 2 or -2 times the product of the square root of the first term and the square root of the last term.

Example  **Determine whether each trinomial is a perfect square trinomial. If so, factor it.**

a. $4y^2 + 36yz + 81z^2$

To determine whether $4y^2 - 36yz + 81z^2$ is a perfect square trinomial, answer each question.

- Is the first term a perfect square? $4y^2 \stackrel{?}{=} (2y)^2$ *yes*
- Is the last term a perfect square? $81z^2 \stackrel{?}{=} (9z)^2$ *yes*
- Is the middle term twice the product of $2y$ and $9z$? $36yz \stackrel{?}{=} 2(2y)(9z)$ *yes*

$4y^2 + 36yz + 81z^2$ is a perfect square trinomial.

$$4y^2 + 36yz + 81z^2 = (2y)^2 + 2(2y)(9z) + (9z)^2$$
$$= (2y + 9z)^2$$

b. $9n^2 + 49 - 21n$

First arrange the terms of $9n^2 + 49 - 21n$ so that the powers of n are in descending order.

$$9n^2 + 49 - 21n = 9n^2 - 21n + 49$$

- Is the first term a perfect square? $9n^2 \stackrel{?}{=} (3n)^2$ *yes*
- Is the last term a perfect square? $49 \stackrel{?}{=} (7)^2$ *yes*
- Is the middle term the product of -2, $3n$, and 7? $-21n \stackrel{?}{=} -2(3n)(7)$ *no*

$9n^2 - 21n + 49$ is not a perfect square trinomial.

Example 2 **Suppose the dimensions of a rectangle can be written as binomials with integral coefficients. Is the rectangle with the area of $(121x^2 - 198xy + 81y^2)$ square millimeters a square? If so, what is the measure of each side of the square?**

Geometry

Explore You know that the dimensions of the rectangle can be written as binomials with integral coefficients. The problem gives the area of a rectangle and asks whether it is a square. If it is a square, you need to find the dimension of each side.

Plan The rectangle is a square if $121x^2 - 198xy + 81y^2$ is a perfect square trinomial. You must answer three questions to determine if it is a perfect square trinomial. If it is a perfect square trinomial, you must factor it to find the measure of each side of the square.

Solve

- Is the first term a perfect square? $121x^2 \stackrel{?}{=} (11x)^2$ *yes*
- Is the last term a perfect square? $81y^2 \stackrel{?}{=} (9y)^2$ *yes*
- Is the middle term the product of -2, $11x$, and $9y$? $-198xy \stackrel{?}{=} -2(11x)(9y)$ *yes*

Since $121x^2 - 198xy + 81y^2$ is a perfect square trinomial, the rectangle is a square. To find the measure of each side, factor the trinomial.

$$\begin{aligned} 121x^2 - 198xy + 81y^2 &= (11x)^2 - 2(11x)(9y) + (9y)^2 \\ &= (11x - 9y)^2 \end{aligned}$$

The measure of each side is $(11x - 9y)$ millimeters.

Examine If each side of the square is $(11x - 9y)$ millimeters, then the area of the square is $(11x - 9y)^2$ square millimeters. Use FOIL to see if $(11x + 9y)^2$ equals $(121x^2 - 198xy + 81y^2)$.

$$\begin{aligned} (11x - 9y)^2 &= (11x - 9y)(11x - 9y) \\ &= (11x)(11x) + (11x)(-9y) + (-9y)(11x) + (-9y)(-9y) \\ &= 121x^2 - 99xy - 99xy + 81y^2 \\ &= 121x^2 - 198xy + 81y^2 \ \checkmark \end{aligned}$$

As you continue your study of mathematics, you will find that forming a perfect square trinomial can sometimes be a useful tool for solving problems.

Example  **Determine all values of k that make $25x^2 + kx + 49$ a perfect square trinomial.**

$$25x^2 + kx + 49 = (5x)^2 + kx + (7)^2$$

In order for this to be a perfect square trinomial, kx must equal either $2(5x)(7)$ or $-2(5x)(7)$. *Why?*

$kx = 2(5x)(7)$	or	$kx = -2(5x)(7)$
$kx = 70x$		$kx = -70x$
$k = 70$		$k = -70$

Check to see if $25x^2 + 70x + 49$ and $25x^2 - 70x + 49$ are perfect square trinomials.

In this chapter, you have learned various methods to factor different types of polynomials. The following chart summarizes these methods and can help you decide when to use a specific method.

Check for:	Number of Terms		
	Two	Three	Four or More
greatest common factor	✓	✓	✓
difference of squares	✓		
perfect square trinomials		✓	
trinomial that has two binomial factors		✓	
pairs of terms that have a common monomial factor			✓

Whenever there is a GCF other than 1, always factor it out first. Then, check the appropriate factoring methods in the order shown in the table. Use these methods to factor until all of the factors are prime.

Example 4 **Factor each polynomial.**

a. $\mathbf{4k^2 - 100}$

First check for a GCF. Then, since the polynomial has two terms, check for the difference of squares.

$4k^2 - 100 = 4(k^2 - 25)$ *The GCF is 4.*

$= 4(k - 5)(k + 5)$ *$k^2 - 25$ is the difference of squares since $k \cdot k = k^2$ and $5 \cdot 5 = 25$.*

Therefore, $4k^2 - 25$ is completely factored as $4(k - 5)(k + 5)$.

b. $\mathbf{9x^2 - 3x - 20}$

The polynomial has three terms. The GCF is 1. $9x^2 = (3x)^2$, but -20 is not a perfect square. The trinomial is not a perfect square trinomial.

Are there two numbers whose product is $9(-20)$ or -180 and whose sum is -3? Yes, the product of -15 and 12 is -180, and their sum is -3.

$$\begin{aligned} 9x^2 - 3x - 20 &= 9x^2 - 15x + 12x - 20 \\ &= (9x^2 - 15x) + (12x - 20) \\ &= 3x(3x - 5) + 4(3x - 5) \\ &= (3x + 4)(3x - 5) \end{aligned}$$

Therefore, $9x^2 - 3x - 20$ is completely factored as $(3x + 4)(3x - 5)$.

c. $\mathbf{4m^4n + 6m^3n - 16m^2n^2 - 24mn^2}$

Since the polynomial has four terms, first check for the GCF and then check for pairs of terms that have a common factor.

$$\begin{aligned} 4m^4n + 6m^3n - 16m^2n^2 - 24mn^2 &= 2mn(2m^3 + 3m^2 - 8mn - 12n) \\ &= 2mn[(2m^3 + 3m^2) + (-8mn - 12n)] \\ &= 2mn[m^2(2m + 3) + (-4n)(2m + 3)] \\ &= 2mn(m^2 - 4n)(2m + 3) \end{aligned}$$

Therefore, $4m^4n + 6m^3n - 16m^2n^2 - 24mn^2$ is completely factored as $2mn(m^2 - 4n)(2m + 3)$.

CHECK FOR UNDERSTANDING

Communicating Mathematics

Study the lesson. Then complete the following.

1. a. **Draw** a rectangle to show how to factor $4x^2 + 12x + 9$. Label the dimensions and the area of the rectangle.
   b. **Explain** why the name *perfect square trinomial* is appropriate for this trinomial.
2. a. **Write** a polynomial that is a perfect square trinomial.
   b. **Factor** your trinomial.
3. a. **Describe** the first step in factoring any polynomial.
   b. **Explain** why this step is important.
4. **You Decide** Robert says that $12a^4 - 8a^2 - 4$ is completely factored as $4(3a^2 + 1)(a^2 - 1)$. Samuel says that he can factor it further. Who is correct? Explain your answer.

5. **Assess Yourself** Describe the relationship between multiplying polynomials and factoring polynomials. Do you prefer to multiply polynomials or to factor polynomials? Explain.

Guided Practice

Complete.

6. $b^2 + 10b + 25 = (b + \underline{\ ?\ })^2$
7. $64a^2 - 16a + 1 = (\underline{\ ?\ } - 1)^2$
8. $81n^2 + 36n + 4 = (\underline{\ ?\ } + 2)^2$
9. $1 - 12c + 36c^2 = (1 - \underline{\ ?\ })^2$

Determine whether each trinomial is a perfect square trinomial. If so, factor it.

10. $t^2 + 18t + 81$
11. $4n^2 - 28n + 49$
12. $9y^2 + 30y - 25$
13. $16b^2 - 56bc + 49c^2$

Factor each polynomial, if possible. If the polynomial cannot be factored, write *prime*.

14. $15g^2 + 25$
15. $4a^2 - 36b^2$
16. $x^2 + 6x - 9$
17. $50g^2 + 40g + 8$
18. $9t^3 + 66t^2 - 48t$
19. $20a^2x - 4a^2y - 45xb^2 + 9yb^2$
20. a. Find the missing value that makes the following a perfect square trinomial.
    $$9x^2 + 24x + \underline{\ ?\ }$$
    b. Copy and complete the model for this trinomial.

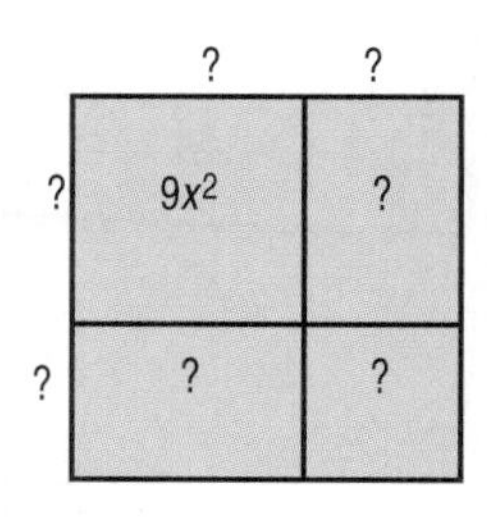

EXERCISES

Practice

Determine whether each trinomial is a perfect square trinomial. If so, factor it.

21. $r^2 - 8r + 16$
22. $d^2 + 50d + 225$
23. $49p^2 - 28p + 4$
24. $4y^2 + 12yz + 9z^2$
25. $49s^2 - 42st + 36t^2$
26. $25y^2 + 20yz - 4z^2$
27. $4m^2 + 4mn + n^2$
28. $81t^2 - 180t + 100$

Determine whether each trinomial is a perfect square trinomial. If so, factor it.

29. $2g^2 - 10g + 25$
30. $1 + 100h^2 + 20h$
31. $64b^2 - 72b + 81$
32. $9a^2 - 24a + 16$
33. $\frac{1}{4}a^2 + 3a + 9$
34. $\frac{4}{9}x^2 - \frac{16}{3}x + 16$

Factor each polynomial, if possible. If the polynomial cannot be factored, write *prime*.

35. $45a^2 - 32ab$
36. $c^2 - 5c + 6$
37. $v^2 - 30v + 225$
38. $m^2 - p^4$
39. $9a^2 + 12a - 4$
40. $3a^2b + 6ab + 9ab^2$
41. $3y^2 - 147$
42. $20n^2 + 34n + 6$
43. $18a^2 - 48a + 32$
44. $3m^3 + 48m^2n + 192mn^2$
45. $x^2y^2 - y^2 - z^2 + x^2z^2$
46. $5a^2 + 7a + 6b^2 - 4b$
47. $4a^3 + 3a^2b^2 + 8a + 6b^2$
48. $(x + y)^2 - (w - z)^2$
49. $0.7p^2 - 3.5pq + 4.2q^2$
50. $(x + 2y)^2 - 3(x + 2y) + 2$
51. $g^4 + 6g^3 + 9g^2 - 3g^2h - 18gh - 27h$
52. $12mp^2 - 15np^2 - 16m + 20np - 16mp + 20n$

Determine all values of k that make each of the following a perfect square trinomial.

53. $25t^2 - kt + 121$
54. $64x^2 - 16xy + k$
55. $ka^2 - 72ab + 144b^2$
56. $169n^2 + knp + 100p^2$

Geometry

57. The area of a circle is $(9y^2 + 78y + 169)\pi$ square centimeters. What is the diameter of the circle?

58. The volume of a rectangular prism is $(x^3y - 63y^2 + 7x^2 - 9xy^3)$ cubic inches. Find the dimensions of the prism, if its dimensions can be represented by binomials with integral coefficients.

LOOK BACK

You can refer to Lesson 2-8 to review square roots.

59. The length of a rectangle is 3 centimeters greater than the length of a side of a square. The width of the rectangle is one-half the length of the side of the square. If the area of the square is $(16x^2 - 56x + 49)$ square centimeters, what is the area of the rectangle?

60. The area of a square is $(81 - 90x + 25x^2)$ square meters. If x is a positive integer, what is the least possible perimeter measure for the square?

Critical Thinking

61. Consider the value of $\sqrt{a^2 - 2ab + b^2}$.
    a. Under what circumstances does the value equal $a - b$?
    b. Under what circumstances does the value equal $b - a$?
    c. Under what circumstances does the value equal $a - b$ and $b - a$?

Applications and Problem Solving

62. **Construction** The builders of an office complex are looking for a square lot. They found a vacant lot that was long enough. However, its length was 60 yards more than its width w, so it was not a square. It also did not have enough area; they needed 900 additional square yards. So they are still looking for a lot. Write an expression for the length of a side of the square lot they should be looking for.

63. **Investments** Tamara plans to invest some money in a certificate of deposit. After 2 years, the value of the certificate will be $p + 2pr + pr^2$, where p represents the amount of money invested and r represents the annual interest rate.
   a. If Tamara invests \$1000 at an annual interest rate of 8%, find the value of the certificate after 2 years.
   b. Factor the expression that represents the value of the certificate after 2 years.
   c. Suppose Tamara invests \$1000 at 7%. Use your expression in part b to find the value of the certificate after 2 years.
   d. Which form of the expression do you prefer to use to make your computations? Explain.

Mixed Review

64. Factor $45x^2 - 20y^2z^2$. (Lesson 10–4)
65. Simplify $2.5t(8t - 12) + 5.1(6t^2 + 10t - 20)$. (Lesson 9–6)
66. Simplify $(3a^2)(4a^3)$. (Lesson 9–1)
67. **Employment** Mike's parents allow him to work 30 hours a week. He would like to use this time to help out in his parents' hardware store, but it pays only \$5 per hour. He could mow lawns for \$7.50 per hour, but there is less than 20 hours of lawn work available. What is the maximum amount of time Mike can work in his parents' store and still make at least \$175 per week? (Lesson 8–5)
68. **Physical Science** A European-made hot tub is advertised to have a temperature of 35°C to 40°C, inclusive. What is the temperature range for the hot tub in degrees Fahrenheit? (*Hint:* Use $F = \frac{9}{5}C + 32$.) (Lesson 7–4)
69. **Statistics** The median incomes of American families since 1970 are shown in the table at the left. (Lesson 6–3)
   a. Make a scatter plot of the data.
   b. Can the data be approximated by a straight line? If so, graph the line and write an equation of this line.
   c. Estimate the median family income for this year.

Year	Median Income
1970	\$ 8734
1975	11,800
1980	17,710
1981	19,074
1982	20,191
1983	21,018
1984	22,415
1985	23,618
1986	24,897
1987	26,061
1988	27,225
1989	28,905
1990	29,943
1991	30,126
1992	30,786

Source: U.S. Census Bureau

70. Refer to Exercise 67 on page 586. Find the range and interquartile range of the state populations. (Lesson 5–7)
71. **Probability** If a card is selected at random from a deck of 52 cards, what are the odds of selecting a club? (Lesson 4–6)
72. **Geometry** ΔABC and ΔDEF are similar. If $a = 5$, $d = 11$, $f = 6$, and $e = 14$, find the missing measures. (Lesson 4–2)

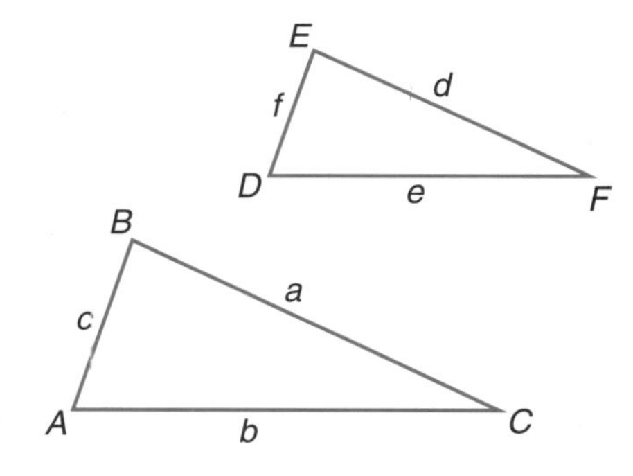

73. **National Landmarks** The Statue of Liberty and the pedestal on which it stands are 302 feet tall altogether. The pedestal is 2 feet shorter than the statue. How tall is the statue? (Lesson 3–3)
74. **Basketball** In 1962, Wilt Chamberlain set an NBA record by averaging 50.4 points per game for a few games. If he had been able to maintain this average over an 82-game season, how many total points would he have scored? Round to the nearest whole number. (Lesson 2–6)

10–6

Solving Equations by Factoring

What YOU'LL LEARN

- To use the zero product property to solve equations.

Why IT'S IMPORTANT

You can use equations to solve problems involving diving, history, and bridges.

APPLICATION

Diving

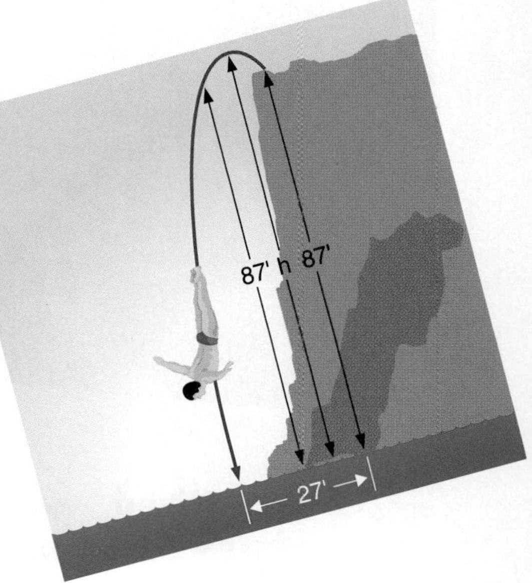

In Acapulco, Mexico, divers leap from La Quebrada, the "Break in the Rocks," diving headfirst into the Pacific Ocean 87 feet below. Because the base rocks extend out 21 feet from the starting point, the divers must jump outward 27 feet. The equation $h = 87 + 8t - 16t^2$ describes the height h (in feet) of a diver t seconds after leaping.

To find the time at which the diver reaches the highest point of the dive, you can find the time at which he will be 87 feet above the ocean on his way down. The diver springs upward at the beginning of the dive and will be at the highest point halfway between this time and the start of the dive.

$$87 = 87 + 8t - 16t^2$$

$$0 = 8t - 16t^2 \quad \textit{Subtract 87 from each side.}$$

$$0 = 8t(1 - 2t) \quad \textit{Factor } 8t - 16t^2.$$

To solve this equation, you need to find the values of t that make the product $8t(1 - 2t)$ equal to 0. Consider the following products.

$$8(0) = 0 \qquad 0(-17)(0) = 0 \qquad (29 - 11)(0) = 0 \qquad 0(2a - 3) = 0$$

Notice that in each case, *at least one* of the factors is zero. These examples illustrate the **zero product property.**

Zero Product Property	**For all numbers a and b, if $ab = 0$, then $a = 0$, $b = 0$, or both a and b equal 0.**

Thus, if an equation can be written in the form $ab = 0$, then the zero product property can be applied to solve that equation.

Use this information to solve the equation about the diver. If $0 = 8t(1 - 2t)$, then either $8t = 0$ or $(1 - 2t) = 0$.

$$8t = 0 \quad \text{or} \quad 1 - 2t = 0$$

$$t = 0 \qquad\qquad 1 = 2t$$

$$\frac{1}{2} = t$$

The diver will be 87 feet above the ocean at the start of the dive ($t = 0$) and $\frac{1}{2}$ second later. He will be at the highest point of the dive $\frac{1}{4}$ second after starting the dive. *You will solve more problems about this dive in Example 5.*

Example 1 **Solve each equation. Then check the solution.**

a. $(p - 8)(2p + 7) = 0$

If $(p - 8)(2p + 7) = 0$, then $p - 8 = 0$ or $2p + 7 = 0$. *Zero product property*

$p - 8 = 0$ or $2p + 7 = 0$

$p = 8$ $\quad\quad 2p = -7$

$\quad\quad p = -\frac{7}{2}$

Check: $(p - 8)(2p + 7) = 0$

$(8 - 8)[2(8) + 7] \stackrel{?}{=} 0$ or $\left(-\frac{7}{2} - 8\right)\left[2\left(-\frac{7}{2}\right) + 7\right] \stackrel{?}{=} 0$

$0(23) \stackrel{?}{=} 0$ $\quad\quad -\frac{23}{2}(0) \stackrel{?}{=} 0$

$0 = 0$ ✔ $\quad\quad 0 = 0$ ✔

The solution set is $\left\{8, -\frac{7}{2}\right\}$.

b. $t^2 = 9t$

Write the equation in the form $ab = 0$.

$t^2 = 9t$

$t^2 - 9t = 0$

$t(t - 9) = 0$ *Factor out the GCF, t.*

$t = 0$ or $t - 9 = 0$ *Zero product property*

$t = 9$

Check: $t^2 = 9t$

$(0)^2 \stackrel{?}{=} 9(0)$ or $(9)^2 \stackrel{?}{=} 9(9)$

$0 = 0$ ✔ $\quad\quad 81 = 81$ ✔

The solution set is $\{0, 9\}$.

To solve $t^2 = 9t$ in Example 1b, you may be tempted to divide each side of the equation by t. If you did, the solution would be 9. Since it is not possible for an equation to have two different solutions, which process is correct? Recall that division by 0 is undefined. But when you divide each side by t, t is unknown. Thus, you may actually be dividing by 0. In fact, 0 is one of the solutions of this equation. To avoid situations like this, keep in mind that you cannot divide each side of an equation by an expression containing a variable unless you know that the value of the expression is not 0.

Example 2 **Solve $m^2 + 144 = 24m$. Then check the solution.**

$m^2 + 144 = 24m$

$m^2 - 24m + 144 = 0$ *Rewrite the equation.*

$(m - 12)^2 = 0$ *Factor $m^2 - 24m + 144$ as a perfect square trinomial.*

$(m - 12)(m - 12) = 0$

$m - 12 = 0$ or $m - 12 = 0$

$m = 12$ $\quad\quad m = 12$

(continued on the next page)

Check: $m^2 + 144 = 24m$

$$(12)^2 + 144 \stackrel{?}{=} 24(12)$$
$$144 + 144 \stackrel{?}{=} 288$$
$$288 = 288 \checkmark$$

The solution set is $\{12\}$.

You can apply the zero product property to an equation that is written as the product of any number of factors equal to zero.

Example 3 **Solve $5b^3 + 34b^2 = 7b$.**

$5b^3 + 34b^2 = 7b$

$5b^3 + 34b^2 - 7b = 0$ *Arrange the terms so the powers of b are in descending order.*

$b(5b^2 + 34b - 7) = 0$ *Factor the GCF, b.*

$b(5b - 1)(b + 7) = 0$ *Factor $5b^2 + 34b - 7$.*

$b = 0$ or $5b - 1 = 0$ or $b + 7 = 0$

$5b = 1$ $\qquad b = -7$

$b = \frac{1}{5}$

The solution set is $\left\{0, \frac{1}{5}, -7\right\}$. *Check this result.*

If an object is launched from ground level, it reaches its maximum height in the air at the time halfway between the launch and impact times. Its height above the ground after t seconds is given by the formula $h = vt - 16t^2$. In this formula, h represents the height of the object in feet, and v represents the object's initial upward velocity in feet per second.

Example 4 **A flare is launched from a life raft with an initial upward velocity of 144 feet per second. How long will the flare stay aloft? What will be the maximum height attained by the flare?**

Rescue Missions

Explore You know the initial upward velocity of the flare is 144 feet per second. You need to determine the length of time the flare will be in the air and how high the flare will go.

Plan The flare will be in the air until the height is 0. Use the general formula $h = vt - 16t^2$ to determine how long the flare will be in the air. The flare will reach its maximum height halfway between the launch and impact times. Use the formula again to determine the height at the middle of its flight.

Solve $h = vt - 16t^2$

$0 = 144t - 16t^2$ *Replace v with 144.*

$0 = 16t(9 - t)$

$16t = 0$ or $9 - t = 0$

$t = 0$ $\qquad 9 = t$

Since 0 seconds is the launch time, the landing time is 9 seconds. The flare will be aloft for 9 seconds. It will reach its maximum height halfway through its flight time at $\frac{1}{2}(9)$ or 4.5 seconds.

$h = 144t - 16t^2$

$= 144(4.5) - 16(4.5)^2$ *Replace t with 4.5.*

$= 648 - 324$

$= 324$

The flare will reach its maximum height of 324 feet after 4.5 seconds.

Examine Check to see if the flare will actually be at a height of 0 feet after 9 seconds.

$0 \stackrel{?}{=} 144(9) - 16(9)^2$

$0 \stackrel{?}{=} 1296 - 1296$

$0 = 0$ ✔

The flare will be aloft for 9 seconds and the maximum height will be reached after 4.5 seconds.

Example 5 **Refer to the application at the beginning of the lesson.**

Diving

a. What is the diver's maximum height?

b. When will the diver enter the water?

a. The diver will reach the maximum height after $\frac{1}{4}$ second.

$h = 87 + 8t - 16t^2$

$= 87 + 8\left(\frac{1}{4}\right) - 16\left(\frac{1}{4}\right)^2$

$= 87 + 2 - 1$

$= 88$

The diver's maximum height is 88 feet. *Does this seem reasonable?*

b. When the diver reaches 88 feet, his upward motion has stopped and the diver begins to fall to the sea below. His velocity at this time is 0. Therefore, the vt term in the general formula $h = vt - 16t^2$ is also 0.

$88 = 16t^2$ *Why does 88 equal $16t^2$ instead of $-16t^2$?*

$5.5 = t^2$

$2.35 \approx t$

It takes the diver $\frac{1}{4}$ or 0.25 second to reach the maximum height and about another 2.35 seconds to reach the water. The diver will reach the water about 0.25 + 2.35 or 2.60 seconds after he starts his dive.

CHECK FOR UNDERSTANDING

Communicating Mathematics

Study the lesson. Then complete the following.

1. If the product of two or more factors is zero, what must be true of the factors?
2. **Describe** the type of equation that can be solved by using the zero-product property.
3. Can $(x + 3)(x - 5) = 0$ be solved by dividing each side of the equation by $x + 3$? Explain.
4. **You Decide** Diana says that if $(x + 2)(x - 3) = 8$, then $x + 2 = 8$ or $x - 3 = 8$. Caitlin disagrees. Who is correct?

Guided Practice

Solve each equation. Check your solutions.

5. $g(g + 5) = 0$
6. $(n - 4)(n + 2) = 0$
7. $5m = 3m^2$
8. $x^2 = 5x + 14$
9. $7r^2 = 70r - 175$
10. $a^3 - 29a^2 = -28a$
11. **Geometry** The dimensions of a rectangle are $(2x + 9)$ inches and $(2x - 1)$ inches. A square with side x inches is cut out of one of the corners. If the remaining area is 195 square inches, what is x?

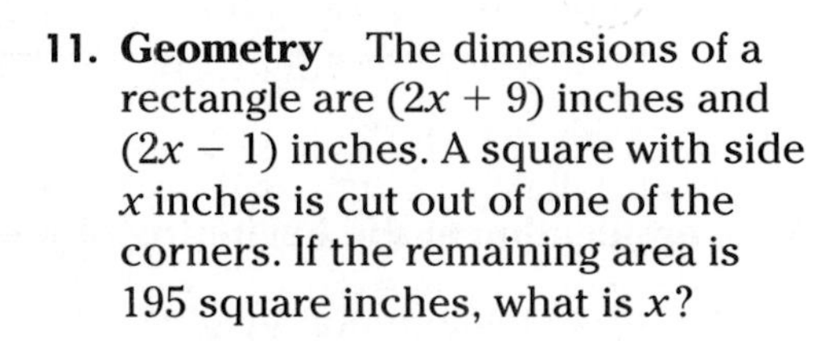

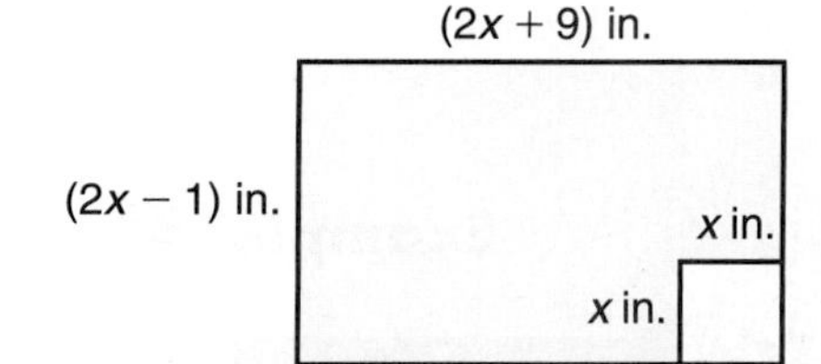

12. **Baseball** Nolan Ryan, the greatest strike-out pitcher in the history of baseball, had a fastball clocked at 103 miles per hour, which is 151 feet per second.
    a. If he threw the ball directly upward with the same velocity, how many seconds would it take for the ball to return to his glove? (*Hint:* Use the general formula $h = vt - 16t^2$.)
    b. How high above his glove would the ball travel?

Top Five List

First professional pitchers to throw a perfect game

1. Lee Pichmond, June 12, 1880
2. Monte Ward, June 17, 1880
3. Cy Young, May 5, 1904
4. Adrian Joss, Oct. 2, 1908
5. Charlie Robertson, Apr. 30, 1922

EXERCISES

Practice

Solve each equation. Check your solutions.

13. $x(x - 24) = 0$
14. $(q + 4)(3q - 15) = 0$
15. $(2x - 3)(3x - 8) = 0$
16. $(4a + 5)(3a - 7) = 0$
17. $a^2 + 13a + 36 = 0$
18. $x^2 - x - 56 = 0$
19. $y^2 - 64 = 0$
20. $5s - 2s^2 = 0$
21. $3z^2 = 12z$
22. $m^2 - 24m = -144$
23. $6q^2 + 5 = -17q$
24. $5b^3 + 34b^2 = 7b$
25. $\frac{x^2}{12} - \frac{2x}{3} - 4 = 0$
26. $t^2 - \frac{t}{6} = \frac{35}{6}$
27. $n^3 - 81n = 0$
28. $(x + 8)(x + 1) = -12$
29. $(r - 1)(r - 1) = 36$
30. $(3y + 2)(y + 3) = y + 14$

31. **Number Theory** Find two consecutive even integers whose product is 168.

32. **Number Theory** Find two consecutive odd integers whose product is 1023.

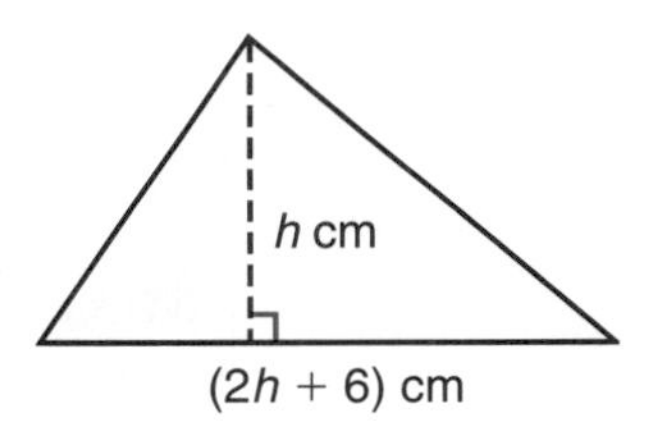

33. **Geometry** The triangle at the right has an area of 40 square centimeters. Find the height h of the triangle.

Critical Thinking

34. Write an equation with integral coefficients that has $\{-3, 0, 7\}$ as its solution set.

35. Consider the equations $a^2 + 5a = 6$ and $|2x + 5| = 7$.
   a. Solve each equation.
   b. What is the relationship between the two equations? What does that mean?

Applications and Problem Solving

36. **Gardening** LaKeesha has enough bricks to make a 30-foot-long border around the rectangular vegetable garden she is planning. The booklet she got from the plant nursery when she bought the seeds says the plants will need space to grow, and it advises that the seeds should be planted in an area of 54 square feet. What should the dimensions of her garden be?

37. **History** During the late 16th century, Galileo was a professor at the University of Pisa, Italy. During this time, he demonstrated that objects of different weights fell at the same velocity by dropping two objects of different weights from the top of the Leaning Tower of Pisa.
   a. If he dropped the objects from a height of 180 feet, how long did it take them to hit the ground?
   b. Research Galileo's life. Why was he criticized for experimenting with these weights? What other belief of his led to his trial by the Inquisition of 1633?

38. **Fountains** The tallest fountain in the world is at Fountain Hills, Arizona. When all three pumps are working, the nozzle speed of the water is 146.7 miles per hour (215.16 feet per second). It is claimed the water reaches a height of 625 feet. Do you agree or disagree with this claim? Explain.

FYI

Superman first appeared in a comic book in 1938 and in a newspaper strip in 1939. Jerry Seigel wrote these comics and Joe Shuster provided the artwork.

39. **Superheroes** A meteorite is headed towards Metropolis when Superman intercepts it. He takes it to the top of the *Daily Planet* (180 feet high) and tosses the meteorite into space with an upward velocity of 2400 feet per second. What height will the meteorite attain before returning to Earth?

40. **Bridges** The chart at the right compares the highest bridge in the world at Royal Gorge, Colorado with the highest railroad bridge in the world at Kolasin, Yugoslavia. If Elva accidentally drops her keys from the Royal Gorge Bridge, will the keys hit the Arkansas River below within 8 seconds?

High Bridges

Bridge	Height (ft)
Royal Gorge, Colorado	880
Kolasin, Yugoslavia	650

41. **Oil Leases** A drilling company has the oil rights for a rectangular piece of land that is 6 kilometers by 5 kilometers. According to a lease agreement, the company can only dig their wells in the inner two-thirds of the land. A uniform strip around the edge of the land must remain untouched. What is the width of the strip of land that must remain untouched?

Mixed Review

42. Factor $100x^2 + 20x + 1$. (Lesson 10–5)

43. Find $(5q + 2r)(8q - 3r)$. (Lesson 9–7)

44. Solve $9x + 4 < 7 - 13x$. (Lesson 7–3)

45. Determine the x- and y-intercepts of the graph of $2x - 7y = 28$. (Lesson 6–4)

46. **Patterns** Copy and complete the table below. (Lesson 5–5)

s	4	2	0	−2	−4
$r(s)$	19	11	3		

47. **World Records** The Huey P. Long Bridge in Metairie, Louisiana, is the longest railroad bridge in the world. If you were traveling on a train going 60 miles per hour across the 22,996-foot bridge, how long would it take you to cross it? (Lesson 4–7)

48. **Consumerism** Julio is offered two payment plans when he buys a sofa. Under one plan, he pays \$400 down and x dollars per month for 9 months. Under the other plan, he pays no money down and $x + 25$ dollars per month for 12 months. How much does the sofa cost? (Lesson 3–5)

WORKING ON THE In·ves·ti·ga·tion

Refer to the Investigation on pages 554–555.

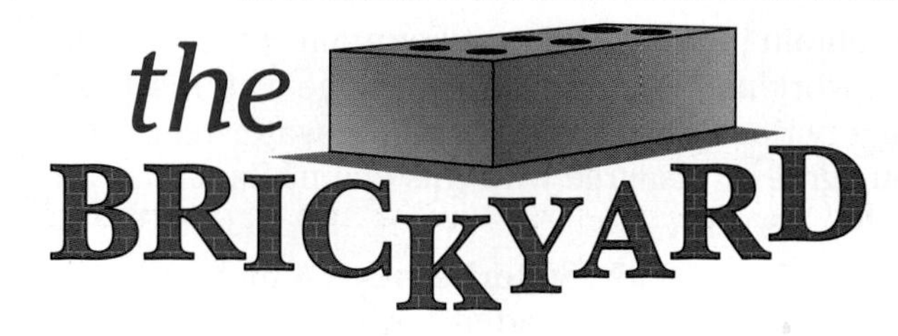

The perimeter P of a square is given by the formula $P = 4s$, where s is the length of the side of the square. The formula for the area A of a square is $A = s^2$. The perimeter of a rectangle is found using $P = 2\ell + 2w$, where ℓ is the length of the rectangle and w is the width. The formula for the area of the rectangle is $A = \ell w$.

1. Suppose the large square brick has a side x units long and the small square brick has a side y units long. What are the dimensions of the rectangular brick? What is the area of each of the bricks? Draw a diagram to illustrate the dimensions of each brick.
2. Refer to the drawings of all the rectangular brick patterns you found using 4, 5, 6, 7, 8, 9, and 10 bricks. Calculate the perimeter and area in terms of x and y of all of the patterns you discovered.
3. Create a table to record the linear dimensions, the perimeter, and the area of the brick patterns you found. Use the table to record the dimensions, perimeter, and area in terms of x and y. The table should be set up like the one below.

Add the results of your work to your Investigation Folder.

# of bricks in pattern	length of pattern	width of pattern	perimeter of pattern	area of pattern
4				
5				
6				

CHAPTER 10 HIGHLIGHTS

VOCABULARY

After completing this chapter, you should be able to define each term, property, or phrase and give an example or two of each.

Algebra

composite numbers (p. 558)

difference of squares (p. 581)

factored form (p. 559)

factoring or factored (pp. 564, 565)

factoring by grouping (p. 567)

greatest common factor (GCF) (p. 559)

perfect square trinomials (p. 587)

prime factorization (p. 558)

prime numbers (p. 558)

prime polynomial (p. 577)

Pythagorean triple (p. 583)

unique factorization theorem (p. 558)

zero product property (p. 594)

Problem Solving

guess and check (p. 574)

UNDERSTANDING AND USING THE VOCABULARY

State whether each sentence is *true* or *false*. If false, replace the underlined word or number to make a true sentence.

1. The number 27 is an example of a <u>prime</u> number.
2. <u>$2x$</u> is the greatest common factor (GCF) of $12x^2$ and $14xy$.
3. <u>66</u> is an example of a perfect square.
4. 61 is a <u>factor</u> of 183.
5. The prime factorization for 48 is <u>$3 \cdot 4^2$</u>.
6. $x^2 - 25$ is an example of a <u>perfect square trinomial</u>.
7. The number 35 is an example of a <u>composite</u> number.
8. <u>$x^2 - 3x - 70$</u> is an example of a prime polynomial.
9. The <u>unique factorization theorem</u> allows you to solve equations.
10. <u>$(b - 7)(b + 7)$</u> is the factorization of a difference of squares.

CHAPTER 10

STUDY GUIDE AND ASSESSMENT

SKILLS AND CONCEPTS

OBJECTIVES AND EXAMPLES

Upon completing this chapter, you should be able to:

- find the prime factorization of integers (Lesson 10–1)

Find the prime factorization of 180.

$$180 = 2 \cdot 90$$
$$= 2 \cdot 2 \cdot 45$$
$$= 2 \cdot 2 \cdot 3 \cdot 15$$
$$= 2 \cdot 2 \cdot 3 \cdot 3 \cdot 5$$

The prime factorization of 180 is $2 \cdot 2 \cdot 3 \cdot 3 \cdot 5$ or $2^2 \cdot 3^2 \cdot 5$.

REVIEW EXERCISES

Use these exercises to review and prepare for the chapter test.

State whether each number is *prime* or *composite*. If the number is composite, find its prime factorization.

11. 28

12. 33

13. 150

14. 301

15. 83

16. 378

- find the greatest common factors (GCF) for sets of monomials (Lesson 10–1)

Find the GCF of $15x^2y$ and $45xy^2$.

$15x^2y = 3 \cdot 5 \cdot x \cdot x \cdot y$

$45xy^2 = 3 \cdot 3 \cdot 5 \cdot x \cdot y \cdot y$

The GCF is $3 \cdot 5 \cdot x \cdot y$ or $15xy$.

Find the GCF of the given monomials.

17. 35, 30

18. 12, 18, 40

19. $12ab$, $-4a^2b^2$

20. $16mrt$, $30m^2r$

21. $20n^2$, $25np^5$

22. $60x^2y^2$, $35xz^3$

23. $56x^3y$, $49ax^2$

24. $6a^2$, $18b^2$, $9b^3$

- use the greatest common factor (GCF) and the distributive property to factor polynomials (Lesson 10–2)

Factor $12a^2 - 8ab$.

$$12a^2 - 8ab = 4a(3a) - 4a(2b)$$
$$= 4a(3a - 2b)$$

Factor each polynomial.

25. $13x + 26y$

26. $6x^2y + 12xy + 6$

27. $24a^2b^2 - 18ab$

28. $26ab + 18ac + 32a^2$

29. $36p^2q^2 - 12pq$

30. $a + a^2b + a^3b^3$

- use grouping techniques to factor polynomials with four or more terms (Lesson 10–2)

Factor $2x^2 - 3xz - 2xy + 3yz$.

$$2x^2 - 3xz - 2xy + 3yz = (2x^2 - 3xz) + (-2xy + 3yz)$$
$$= x(2x - 3z) - y(2x - 3z)$$
$$= (x - y)(2x - 3z)$$

Factor each polynomial.

31. $a^2 - 4ac + ab - 4bc$

32. $4rs + 12ps + 2mr + 6mp$

33. $16k^3 - 4k^2p^2 - 28kp + 7p^3$

34. $dm + mr + 7r + 7d$

35. $24am - 9an + 40bm - 15bn$

36. $a^3 - a^2b + ab^2 - b^3$

OBJECTIVES AND EXAMPLES

- factor quadratic trinomials (Lesson 10–3)

$a^2 - 3a - 4 = (a + 1)(a - 4)$

$$\begin{aligned} 4x^2 - 4xy - 15y^2 &= 4x^2 + (-10 + 6)xy - 15y^2 \\ &= 4x^2 - 10xy + 6xy - 15y^2 \\ &= (4x^2 - 10xy) + (6xy - 15y^2) \\ &= 2x(2x - 5y) + 3y(2x - 5y) \\ &= (2x - 5y)(2x + 3y) \end{aligned}$$

REVIEW EXERCISES

Factor each trinomial, if possible. If the trinomial cannot be factored using integers, write *prime*.

37. $y^2 + 7y + 12$
38. $x^2 - 9x - 36$
39. $6z^2 + 7z + 3$
40. $b^2 + 5b - 6$
41. $2r^2 - 3r - 20$
42. $3a^2 - 13a + 14$

- identify and factor binomials that are the differences of squares (Lesson 10–4)

$$\begin{aligned} a^2 - 9 &= (a)^2 - (3)^2 \\ &= (a - 3)(a + 3) \end{aligned}$$

$$\begin{aligned} 3x^3 - 75x &= 3x(x^2 - 25) \\ &= 3x(x - 5)(x + 5) \end{aligned}$$

Factor each polynomial, if possible. If the polynomial cannot be factored using integers, write *prime*.

43. $b^2 - 16$
44. $25 - 9y^2$
45. $16a^2 - 81b^4$
46. $2y^3 - 128y$
47. $9b^2 - 20$
48. $\frac{1}{4}n^2 - \frac{9}{16}r^2$

- identify and factor perfect square trinomials (Lesson 10–5)

$$\begin{aligned} &16z^2 - 8z + 1 \\ &\quad = (4z)^2 - 2(4z)(1) + (1)^2 \\ &\quad = (4z - 1)^2 \end{aligned}$$

$$\begin{aligned} 9x^2 + 24xy + 16y^2 &= (3x)^2 - 2(3x)(4y) + (4y)^2 \\ &= (3x + 4y)^2 \end{aligned}$$

Factor each polynomial, if possible. If the polynomial cannot be factored using integers, write *prime*.

49. $a^2 + 18a + 81$
50. $9k^2 - 12k + 4$
51. $4 - 28r + 49r^2$
52. $32n^2 - 80n + 50$
53. $6b^3 - 24b^2g + 24bg^2$
54. $49m^2 - 126m + 81$
55. $25x^2 - 120x + 144$

OBJECTIVES AND EXAMPLES

- use the zero product property to solve equations (Lesson 10–6)

Solve $b^2 - b - 12 = 0$.

$b^2 - b - 12 = 0$

$(b - 4)(b + 3) = 0$

If $(b - 4)(b + 3) = 0$, then $(b - 4) = 0$ or $(b + 3) = 0$.

$b - 4 = 0$ or $b + 3 = 0$

$b = 4$ $\qquad$ $b = -3$

The solution set is $\{4, -3\}$.

REVIEW EXERCISES

Solve each equation. Check your solution.

56. $y(y + 11) = 0$

57. $(3x - 2)(4x + 7) = 0$

58. $2a^2 - 9a = 0$

59. $n^2 = -17n$

60. $\frac{3}{4}y = \frac{1}{2}y^2$

61. $y^2 + 13y + 40 = 0$

62. $2m^2 + 13m = 24$

63. $25r^2 + 4 = -20r$

APPLICATIONS AND PROBLEM SOLVING

64. Geometry The measure of the area of a rectangle is $4m^2 - 3mp + 3p - 4m$. If the dimensions of the rectangle are represented by polynomials with integral coefficients, find the dimensions of the rectangle. (Lesson 10–2)

65. Guess and Check Numero Uno says, "I am thinking of a three-digit number. If you multiply the digits together and then multiply the result by 4, the answer is the number I'm thinking of. What is my number?" (Lesson 10–3)

66. Photography To get a square photograph to fit into a rectangular frame, Li-Chih had to trim a 1-inch strip from one pair of opposite sides of the photo and a 2-inch strip from the other two sides. In all, he trimmed off 64 square inches. What were the original dimensions of the photograph? (Lesson 10–4)

67. Guess and Check Fill in each box below with a digit from 1 to 6 to make this multiplication work. Use each digit exactly once. (Lesson 10–3)

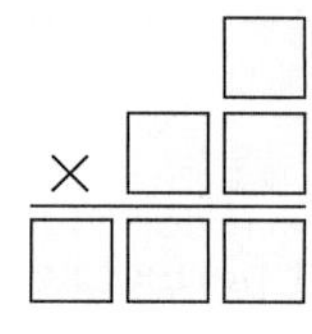

68. Geometry The measure of the area of a rectangle is $16x^2 - 9$. Find the measure of its perimeter. (Lesson 10–4)

69. Number Theory The product of two consecutive odd integers is 99. Find the integers. (Lesson 10–6)

A practice test for Chapter 10 is provided on page 796.

ALTERNATIVE ASSESSMENT

COOPERATIVE LEARNING PROJECT

Art and Framing In this project, you will determine what size mat and frame will be the most visually stimulating for an abstract print that you just bought on your trip to The Chicago Institute of Art. The print is 9 inches by 12 inches and can be cropped an inch on either side and not spoil its effect.

You called a friend to get some advice on how to mat and frame your print. This friend majored in art and works in the art industry, but he also enjoys puzzling you. He gave you two problems and told you to solve them and decide which you think would be the most appropriate dimensions for your mat and frame. Here are the two problems.

- The inside rectangle has a width four inches less than its length. The length of the larger rectangle is twice its width. Find the dimensions of each rectangle if the matted area is twice as much as the inside rectangle area and the perimeter of the inside rectangle is 32 inches less than the perimeter of the larger rectangle.
- The inside rectangle has a width four inches less than its length. The length of the larger rectangle is twice its width. Find the dimensions of each rectangle if the matted area is seven times as much in square inches as the perimeter of the smaller rectangle is in inches, and the width of the larger rectangle is two inches more than the length of the smaller rectangle.

Which problem will give you the dimensions that will best suit your print? What are the dimensions? Will you have to crop your print? If so, how much?

Follow these steps to determine the appropriate dimensions for the mat and frame.

- Illustrate each of the problems.
- Label each of the drawings with as much information as possible.
- Develop an equation for each problem that describes the situation.
- Investigate your answers and determine the appropriate dimensions for the mat and frame.
- If possible, find a 9-inch by 12-inch photograph or advertisement from a magazine. Have two color photocopies made of it and frame each with a cardboard frame that meets the specifications.
- Write a paragraph describing the two problems and how you determined the solution.

THINKING CRITICALLY

- A *nasty* number is a positive integer with at least four different factors such that the difference between the numbers in one pair of factors equals the sum of the numbers in another pair. The first nasty number is 6 since $6 = 6 \cdot 1 = 2 \cdot 3$ and $6 - 1 = 2 + 3$. Find the next five nasty numbers. (*Hint*: They are all multiples of 6.)
- Write an equation with integral coefficients that has $\left\{\frac{2}{3}, -1\right\}$ as the solution.

PORTFOLIO

When using the guess-and-check strategy for solving a math problem, if a first guess does not work, you must make a second guess. How do you make a second guess when the first one is incorrect? In making this second guess, you must know how and why the first guess was incorrect. Find a problem from your work in this chapter in which you used the guess-and-check strategy. Write a step-by-step description of why you chose each successive guess after the previous guess for that problem. Place this in your portfolio.

SELF EVALUATION

Having a strategy before solving a problem is helpful. It helps you to focus on the problem and the solution and to be organized in the solution of your problem. Using strategies or check lists are fundamental skills used in critical thinking.

Assess yourself. Do you determine a strategy for solving a problem before you delve into it? Do you stay with the initial plan or do you reevaluate after a time period and try a new strategy? Do you keep a record of your strategy and attempts so that you can refer back to it to determine if you are still on track? Give an example of a math problem in which a strategy is beneficial in order to keep track of what has been attempted. Also, give an example of a daily life problem in which you used a strategy to help organize your plan for a solution.

CUMULATIVE REVIEW

CHAPTERS 1–10

SECTION ONE: MULTIPLE CHOICE

There are nine multiple-choice questions in this section. After working each problem, write the letter of the correct answer on your paper.

1. The square of a number subtracted from 8 times the number is equal to twice the number. Find the number.

   A. 0 or 6
   B. 3
   C. 2
   D. 0 or 10

2. Choose the open sentence that represents the range of acceptable diameters for a lawn mower bolt that will work properly only if its diameter differs from 2 cm by no more than 0.04 cm.

   A. $|d - 0.04| \geq 2$
   B. $|d| < 1.96$
   C. $|d - 2| \leq 0.04$
   D. $|d| > 2.04$

3. Choose an expression for the area of the shaded region shown below.

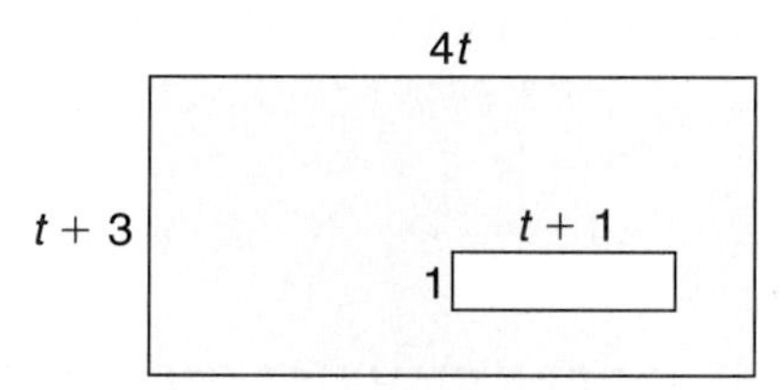

   A. $(4t)(t + 3)$
   B. $(4t - t)(t + 3) - (t + 1)$
   C. $(t + 1)(t + 3) - (4t)(t)$
   D. $(4t)(t + 3) - (t + 1)$

4. Choose the equivalent equation for $A = \frac{1}{2}h(a + b)$.

   A. $h = \frac{2A}{a + b}$
   B. $h = 2A - (a + b)$
   C. $h = \frac{\frac{1}{2}(a + b)}{A}$
   D. $h = \frac{A - b}{2a}$

5. Bob and Vicki took a trip to Zuma Beach. On the way there, their average speed was 42 miles per hour. On the way home, their average speed was 56 miles per hour. If their total travel time was 7 hours, find the distance to the beach.

   A. 126 miles
   B. 168 miles
   C. 98 miles
   D. 294 miles

6. Choose which statement is true.

   A. The range is the difference between the greatest value and the lower quartile in a set of data.
   B. An outlier will only affect the mean of a set of data.
   C. Quartiles are values that divide a set of data into equal halves.
   D. The interquartile range is the sum of the upper and lower quartiles in a set of data.

7. Choose the prime polynomial.

   A. $y^2 + 12y + 27$
   B. $6x^2 - 11x + 4$
   C. $h^2 + 5h - 8$
   D. $9k^2 + 30km + 25m^2$

8. Choose the equation of the line that passes through the points at (9, 5) and (−3, −4).

   A. $y = -\frac{1}{12}x + \frac{23}{4}$
   B. $y = \frac{4}{3}x$
   C. $y = -\frac{7}{2}x - \frac{3}{4}$
   D. $y = \frac{3}{4}x - \frac{7}{4}$

9. Carrie's bowling scores for four games are $b + 2$, $b + 3$, $b - 2$, and $b - 1$. What must her score be on her fifth game to average $b + 2$?

   A. $b + 8$
   B. b
   C. $b - 2$
   D. $b + 5$

SECTION TWO: SHORT ANSWER

This section contains ten questions for which you will provide short answers. Write your answer on your paper.

10. The measure of the perimeter of a square is $20m + 32p$. Find the measure of its area.

11. At a bake sale, cakes cost twice as much as pies. Pies were $4 more than triple the price of cookies. Darin bought a cake, three pies, and four cookies for $24.75. What is the price of each item?

12. Factor $3g^2 - 10gh - 8h^2$.

13. On the first day of school, 264 school notebooks were sold. Some sold for 95¢ each, and the rest sold for $1.25 each. How many of each were sold if the total sales were $297?

14. A rectangular photograph is 8 centimeters wide and 12 centimeters long. The photograph is enlarged by increasing the length and width by an equal amount. If the area of the new photograph is 69 square centimeters greater than the area of the original photograph, what are the dimensions of the new photograph?

15. Linda plans to spend at most $50 on shorts and blouses. She bought 2 pairs of shorts for $14.20 each. How much can she spend on blouses?

16. The difference of two numbers is 3. If the difference of their squares is 15, what is the sum of the numbers?

17. **Geometry** The length of a rectangle is eight times its width. If the length was decreased by 10 meters and the width was decreased by 2 meters, the area would be decreased by 162 square meters. Find the original dimensions.

18. Ben's car gets between 18 and 21 miles per gallon of gasoline. If his car's tank holds 15 gallons, what is the range of distance that Ben can drive his car on one tank of gasoline?

19. Find the measure of the third side of a triangle if the perimeter of the triangle is $8x^2 + x + 15$ and the measures of two of its sides can be represented by the expressions $2x^2 - 5x + 7$ and $x^2 - x + 11$.

SECTION THREE: OPEN-ENDED

This section contains two open-ended problems. Demonstrate your knowledge by giving a clear, concise solution to each problem. Your score on these problems will depend on how well you do the following.

- Explain your reasoning.
- Show your understanding of the mathematics in an organized manner.
- Use charts, graphs, and diagrams in your explanation.
- Show the solution in more than one way or relate it to other situations.
- Investigate beyond the requirements of the problem.

20. A tinsmith is going to make a box out of tin by cutting a square from each corner and folding up the sides. The box needs to be 3 inches high (so he will be cutting out 3-inch squares), and it needs to be twice as long as it is wide, so that it can hold two smaller square cardboard boxes. Finally, its volume needs to be 1350 cubic inches. What should the dimensions be?

21. The length of a rectangle is five times its width. If the length was increased by 7 meters and the width was decreased by 4 meters, the area would be decreased by 132 square meters. Find the original dimensions.

CHAPTER

11 Exploring Quadratic and Exponential Functions

Objectives

In this chapter, you will:

- find the equation of the axis of symmetry and the coordinates of the vertex of a parabola,
- graph quadratic and exponential functions,
- use estimation to find roots of quadratic equations by graphing,
- find roots of quadratic equations by using the quadratic formula,
- solve problems by looking for and using a pattern, and
- solve problems involving growth and decay.

Reaching Out to Help Others

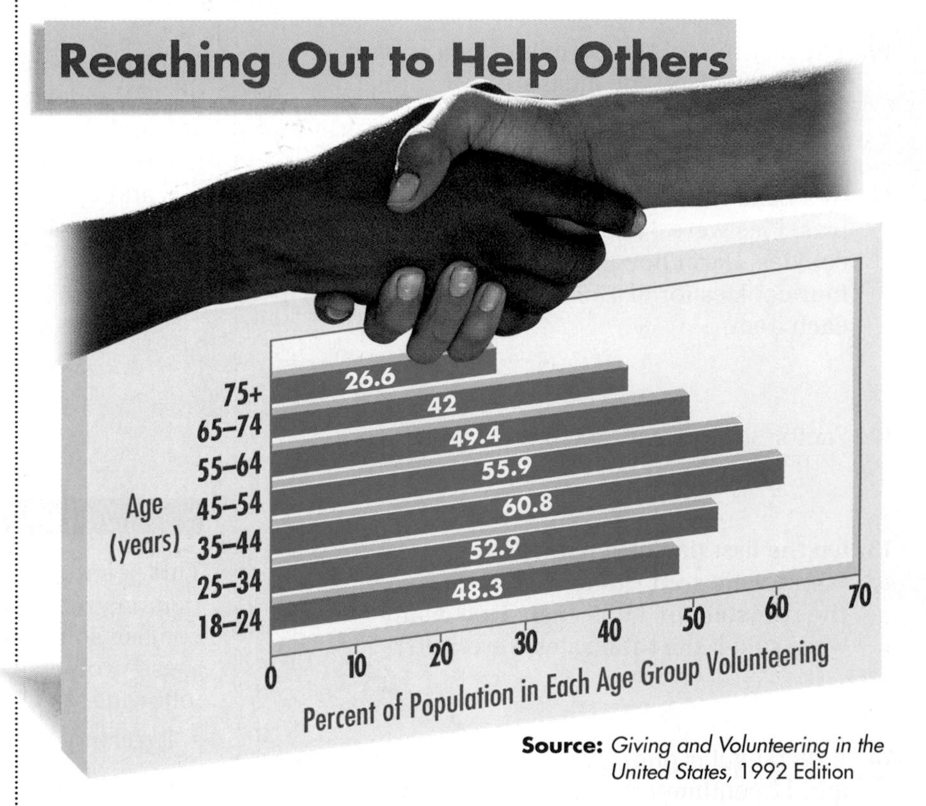

Source: *Giving and Volunteering in the United States,* 1992 Edition

Dr. Robert Coles, a psychiatrist and Harvard professor, calls it "the call of service," men and women who volunteer their time to help others. Many teens would like to get involved in volunteer service, but may not know where to start. Some opportunities may be available in your own community—helping senior citizens, working as a Big Brother or Big Sister, or volunteering at a food pantry. Volunteering will help others and may help you feel great, too.

TIME *Line*

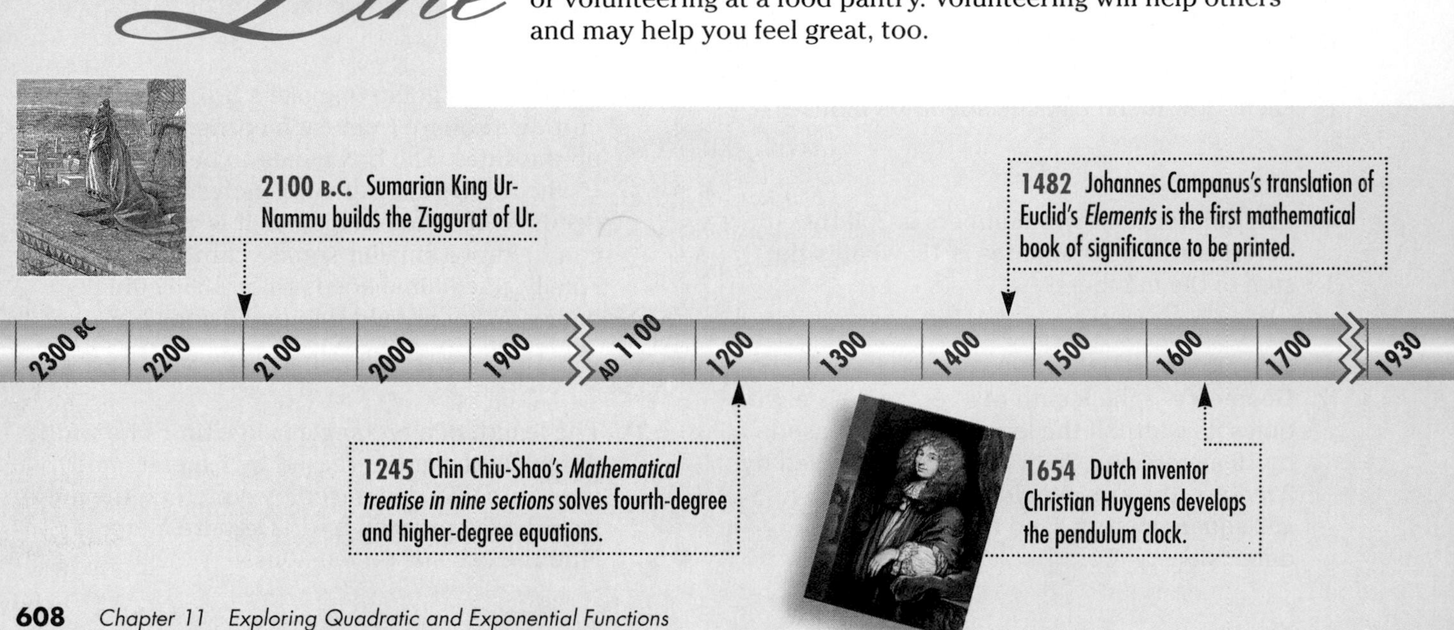

PEOPLE IN THE NEWS

A perfect example of a teenage volunteer is **Liz Alvarez** of St. Petersburg, Florida. She has been a community volunteer at St. Anthony's Hospital, giving more than 600 hours of her time to help the staff. She answers phones in the pastoral care department, does filing as an administrative assistant, and carries messages for staff members. She helps patients by delivering items to their rooms, works in the gift shop, and assists other volunteers.

She enjoys her time spent volunteering, is happy to help others, and believes that the experience is very rewarding. She encourages other teens to investigate the opportunities in their communities and join a volunteer team.

Chapter Project

Organizations in every community study the characteristics of those who volunteer their time. In this manner, they know what population to target when seeking volunteers for a particular project or event. Survey companies, such as the Gallup Organization, Inc., often provide statistics for these organizations.

One of the characteristics the Gallup Organization studied was the income of volunteers. They found the following information on income brackets and the percent of the people in each bracket who volunteer.

Income ($)	Percent of Population Volunteering
Under 10,000	31.6
10,000–19,999	37.9
20,000–29,999	51.3
30,000–39,999	56.4
40,000–49,999	67.4
50,000–59,999	67.7
60,000–74,999	55.0
75,000–99,999	62.8
100,000 +	73.7

Analyze these findings.

- Make a graph of the data.
- What type of behavior do these data present?
- Why do you think this behavior exists?
- Research in an almanac or statistical abstract to find the average income for the age groups listed in the graph on the previous page.
- Use your research, the graph on the previous page, and the table to make a conjecture about the characteristics of the average volunteer.

1940 U.S. photographer Helen Levitt chronicles life in the streets of New York City with her print *Children.*

1981 The Peace Corps becomes an independent agency, sending volunteers to improve living conditions in developing countries.

1940 1945 1950 1955 1960 1965 1970 1975 1980 1985 1990 1995 2000

1961 American soprano Leontyne Price makes her Metropolitan Opera debut.

1991 Advances in music technology enable Natalie Cole to produce a duet album with her late father Nat "King" Cole.

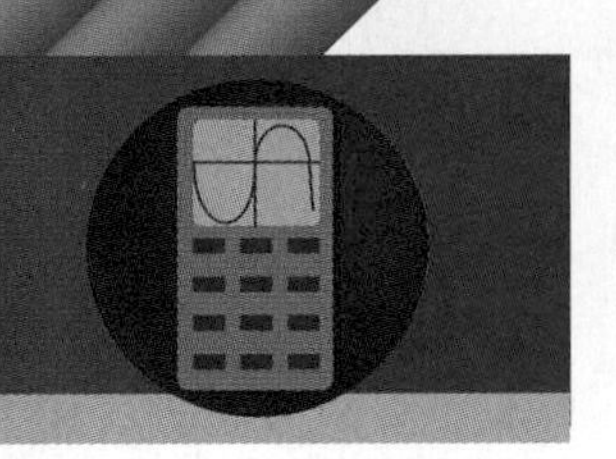

11–1A Graphing Technology Quadratic Functions

A Preview of Lesson 11–1

Equations in the form $y = ax^2 + bx + c$ are called **quadratic functions,** and their graphs are called **parabolas.** A parabola is a U-shaped curve that can open upward or downward. The maximum or minimum point of a parabola is called its **vertex.** You can use a graphing calculator to graph a quadratic function and find the coordinates of its vertex.

Sometimes the coordinates of the vertex may not be integers. Use the ZOOM feature several times until you get coordinates that are fairly consistent (within three decimal places).

Example

Graph $y = \frac{1}{4}x^2 - 4x - 2$ and locate its vertex. Use the integer window.

A quadratic function is entered into the Y= list in the same way that you entered linear functions.

Enter: Y= .25 X,T,θ x^2 – 4 X,T,θ – 2 *Enters the function.*

ZOOM 6 ZOOM 8 ENTER *Selects the integer window.*

The parabola opens upward. The vertex is a minimum point.

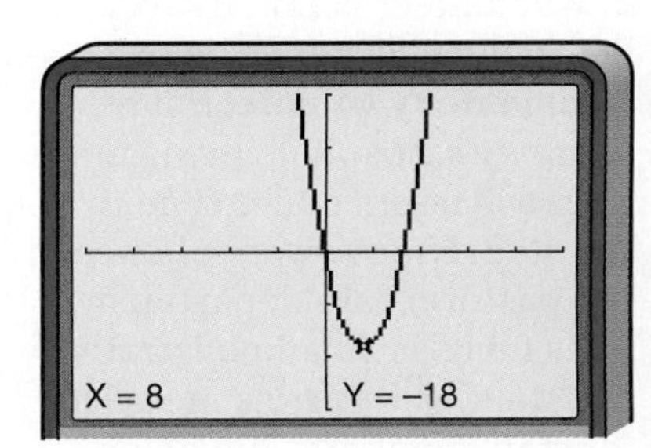

Method 1: Use TRACE.

Press TRACE and use the left and right arrow keys to move the cursor to the vertex. Watch the coordinates at the bottom of the screen as you move the cursor. The vertex will be the point of the least y value. *Why?*

Method 2: Use CALC.

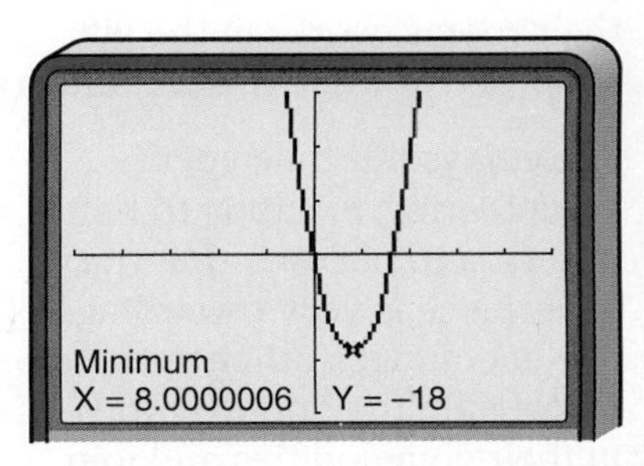

Since the parabola opens upward, the vertex is a minimum point. Press 2nd CALC 3.

- A "Lower Bound?" prompt appears. Move the cursor to a location left of where you think the vertex is and press ENTER.
- An "Upper Bound?" prompt appears. Move the cursor to a location right of where you think the vertex is and press ENTER.
- When the "Guess?" prompt appears, move the cursor to your choice for the vertex and press ENTER. The screen will give you an approximation of the vertex coordinates.

You can also press ENTER at the Guess? prompt without entering a guess.

The coordinates of the vertex are (8, –18).

EXERCISES

Graph each function. Make a sketch of the graph and note the ordered pair representing the vertex on the graph.

1. $y = x^2 + 16x + 59$
2. $y = 12x^2 + 18x + 10$
3. $y = x^2 - 10x + 25$
4. $y = -2x^2 - 8x - 1$
5. $y = 2(x - 10)^2 + 14$
6. $y = -0.5x^2 - 2x + 3$

11–1 Graphing Quadratic Functions

What YOU'LL LEARN

- To find the equation of the axis of symmetry and the coordinates of the vertex of a parabola, and
- to graph quadratic functions.

Why IT'S IMPORTANT

You can graph quadratic functions to solve problems involving fireworks and football.

Landmarks

The Gateway Arch of the Jefferson National Expansion Memorial in St. Louis, Missouri, is shaped like an upside-down U. This shape is actually a *catenary,* which resembles a geometric shape called a **parabola.** The shape of the arch can be approximated by the graph of the function $f(x) = -0.00635x^2 + 4.0005x - 0.07875$, where $f(x)$ is the height of the arch in feet and x is the horizontal distance from one base.

This type of function is an example of a **quadratic function.** A quadratic function can be written in the form $f(x) = ax^2 + bx + c$, where $a \neq 0$. Notice that this function has a degree of 2 and the exponents are positive.

Definition of a Quadratic Function	**A quadratic function is a function that can be described by an equation of the form $y = ax^2 + bx + c$, where $a \neq 0$.**

Since distance and height are involved, the domain and range must both be positive.

To graph a quadratic equation, you can use a table of values. The table at the right shows the distance (in 35-foot increments) from one base of the arch and the height of the arch at each increment (rounded to the nearest foot). Graph the ordered pairs and connect them with a smooth curve. The graph of $f(x) = -0.00635x^2 + 4.0005x - 0.07875$ is shown below.

x	f(x)
0	0
35	132
70	249
105	350
140	436
175	506
210	560
245	599
280	622
315	630
350	622
385	599
420	560
455	506
490	436
525	350
560	249
595	132
630	0

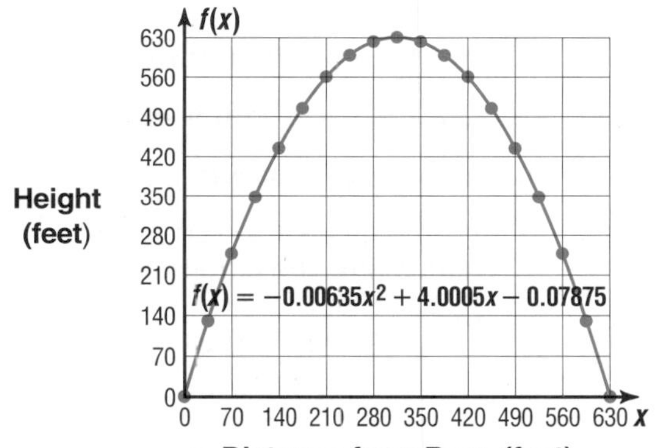

GLOBAL CONNECTIONS

The arch as an architectural form dates back to Egypt and Greece, but was first used extensively by the Romans in building bridges and aqueducts, like the Pont du Gard at Nimes, France. The architects in medieval times used majestic pointed arches in the construction of gothic cathedrals.

Notice that the value of a in this function is negative and the curve opens downward. The greatest value of $f(x)$ seems to be 630 feet, which occurs when the distance from the base is 315 feet. The point at (315, 630) would be the **vertex** of the parabola. For a parabola that opens downward, the vertex is a **maximum** point of the function. If the parabola opens upward, the vertex is a **minimum** point of the function.

Parabolas possess a geometric property called **symmetry.** Symmetrical figures are those in which the figure can be folded and each half matches the other exactly. The following activity explores the symmetry of a parabola.

MODELING MATHEMATICS

Symmetry of Parabolas

Materials: 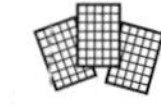 grid paper

Your Turn

a. Graph $y = x^2 - 6x + 5$ on grid paper.

b. Hold your paper up to the light and fold the parabola in half so the two sides match exactly.

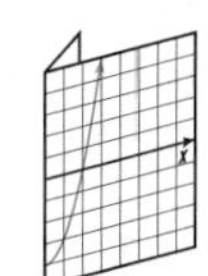

c. Unfold the paper. Which point on the parabola lies on the fold line?

d. Write an equation to describe the fold line

e. Write a few sentences to describe the symmetry of a parabola based on your findings in this activity.

The fold line in the activity above is called the **axis of symmetry** for the parabola. Each point on the parabola that is on one side of the axis of symmetry has a corresponding point on the parabola on the other side of the axis. The vertex is the only point on the parabola that is on the axis of symmetry. The equation for the axis of symmetry can be determined from the equation of the parabola.

Equation of the Axis of Symmetry of a Parabola	**The equation of the axis of symmetry for the graph of $y = ax^2 + bx + c$, where $a \neq 0$, is $x = -\frac{b}{2a}$.**

You can determine a lot of information about a parabola from its equation.

Example 

Given the equation $y = x^2 - 4x + 5$,

a. find the equation of the axis of symmetry,

b. find the coordinates of the vertex of the parabola, and

c. graph the equation.

a. In the equation $y = x^2 - 4x + 5$, $a = 1$ and $b = -4$. Substitute these values into the equation of the axis of symmetry.

$$x = -\frac{b}{2a}$$

$$= -\frac{-4}{2(1)} \text{ or } 2$$

b. Since the equation for the axis of symmetry is $x = 2$ and the vertex lies on the axis, the x-coordinate for the vertex is 2.

$$y = x^2 - 4x + 5$$

$$= 2^2 - 4(2) + 5 \quad \textit{Replace } x \textit{ with 2.}$$

$$= 4 - 8 + 5 \text{ or } 1$$

The coordinates of the vertex are (2, 1).

You can use the TABLE feature on a graphing calculator to help you find ordered pairs that satisfy an equation.

c. You can use the symmetry of the parabola to help you draw its graph. Draw a coordinate plane. Graph the vertex and axis of symmetry. Choose a value less than 2, say 0, and find the y-coordinate that satisfies the equation.

$y = x^2 - 4x + 5$

$\quad = 0^2 - 4(0) + 5$ or 5

Graph (0, 5). Since the graph is symmetrical, you can find another point on the other side of the axis of symmetry. The point at (0, 5) is 2 units left of the axis. Go 2 units right of the axis and plot a point at (4, 5). Repeat this for several other points. Then sketch the parabola.

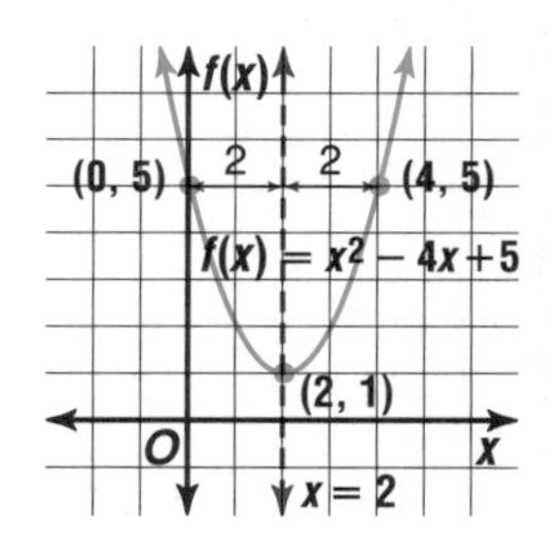

Check: Does (4, 5) satisfy the equation?

$y = x^2 - 4x + 5$

$5 \stackrel{?}{=} 4^2 - 4(4) + 5$

$5 \stackrel{?}{=} 16 - 16 + 5$

$5 = 5$ ✓

The point (4, 5) satisfies the equation $y = x^2 - 4x + 5$ and is part of the graph.

The graph of an equation describing the height of an object propelled into the air can be an example of a quadratic function. Because gravity overtakes the initial force of the object, the object falls back to Earth.

Example 2

Pyrotechnics

F Y I

The largest firework ever exploded was a 1543-pound shell called the Universe I Part II. It was used as part of the Lake Toya Festival in Hokkaido, Japan, on July 15, 1988, and had a burst 3937 feet in diameter.

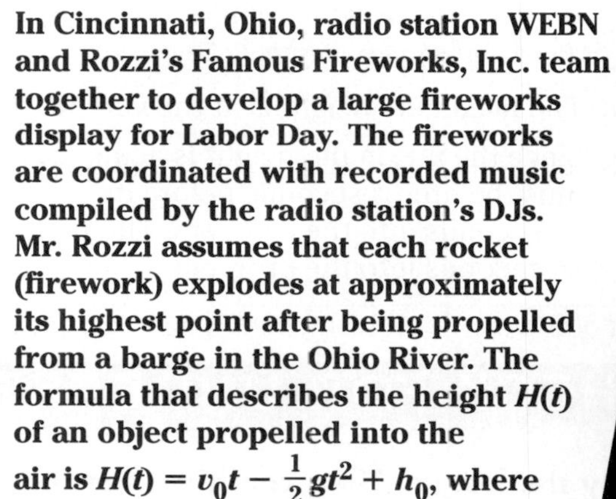

In Cincinnati, Ohio, radio station WEBN and Rozzi's Famous Fireworks, Inc. team together to develop a large fireworks display for Labor Day. The fireworks are coordinated with recorded music compiled by the radio station's DJs. Mr. Rozzi assumes that each rocket (firework) explodes at approximately its highest point after being propelled from a barge in the Ohio River. The formula that describes the height $H(t)$ of an object propelled into the air is $H(t) = v_0t - \frac{1}{2}gt^2 + h_0$, where v_0 represents the initial velocity in m/s, t represents time in seconds, g represents the acceleration of gravity (about 9.8 m/s^2), and h_0 is the initial height of the object at the time it is launched. A certain rocket has an initial velocity of 39.2 m/s and is launched 1.6 meters above the surface of the water.

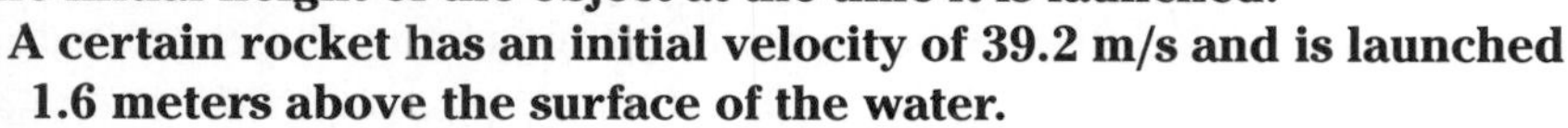

a. Graph the equation representing the height of the rocket.

b. What is the maximum height that the rocket achieves?

c. One of these rockets is scheduled to explode 2 minutes and 28 seconds into the program. When should the rocket be fired from the barge?

(continued on the next page)

a. Use the given velocity and the force of gravity to determine the formula.

$H(t) = v_0t - \frac{1}{2}gt^2 + h_0$

$= 39.2t - \frac{1}{2}(9.8)t^2 + 1.6$ *Replace v_0, g, and h_0 with the given values.*

$= 39.2t - 4.9t^2 + 1.6$ *Simplify.*

The values of a and b are -4.9 and 39.2, respectively.

$H(t)$ corresponds to $f(x)$ and t corresponds to x.

Find the equation of the axis of symmetry to help you locate the vertex.

$t = -\frac{b}{2a}$

$= -\frac{39.2}{2(-4.9)}$ or 4 seconds

Use a calculator to find $H(t)$ when $t = 4$.

$H(4) = 39.2(4) - 4.9(4)^2 + 1.6$ or 80 meters

The vertex of the graph is at (4, 80).

We can use a table of values to find other points that satisfy the equation.

The graph shows the height of the rocket at any given time. It does not show the path that the rocket took.

t	$H(t)$
0	1.6
1	35.9
2	60.4
3	75.1
4	80.0
5	75.1
6	60.4
7	35.9
8	1.6
9	−42.5

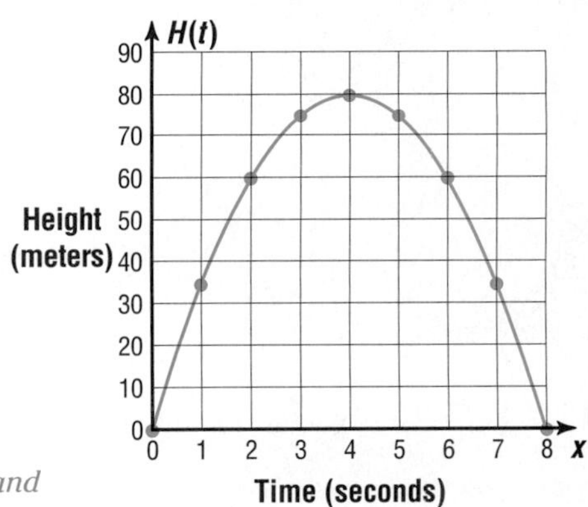

Since time is the independent variable and height is the dependent variable, the domain and range must both be positive.

b. The maximum height is at the vertex. This height is 80 meters.

c. Since the maximum height is achieved 4 seconds after launch, the rocket must be fired 4 seconds before its scheduled explosion at 2 minutes 28 seconds into the program. The rocket should be fired at 2 minutes 24 seconds into the program.

CHECK FOR UNDERSTANDING

Communicating Mathematics

Study the lesson. Then complete the following.

1. **Determine** the distance between the bases of the Gateway Arch.
2. **Explain** how you can use symmetry to help you graph a parabola.
3. **Write** a sentence to explain how you can write the equation of the axis of symmetry if you know the coordinates of the vertex of a parabola.
4. **You Decide** Lisa says that $y = -x^2$ is the same as $y = (-x)^2$. Angie says that they are different. Who is correct? Explain and include graphs.
5. How can you determine if the vertex is a maximum or minimum without graphing the equation first?

MODELING MATHEMATICS

6. Graph $y = -x^2 + 4x - 4$. How does this graph differ from the one in the activity on page 612? Use paper folding to determine the axis of symmetry.

Guided Practice

Write the equation of the axis of symmetry and find the coordinates of the vertex of the graph of each equation. State if the vertex is a maximum or minimum. Then graph the equation.

7. $y = x^2 + 2$
8. $y = -2x^2$
9. $y = x^2 + 4x - 9$
10. $y = x^2 - 14x + 13$
11. $y = -x^2 + 5x + 6$
12. $y = -2x^2 + 4x + 6.5$

13. Which equation describes the graph at the right?
    a. $f(x) = x^2 - 6x + 9$
    b. $f(x) = -x^2 + 6x + 9$
    c. $f(x) = x^2 + 6x + 9$

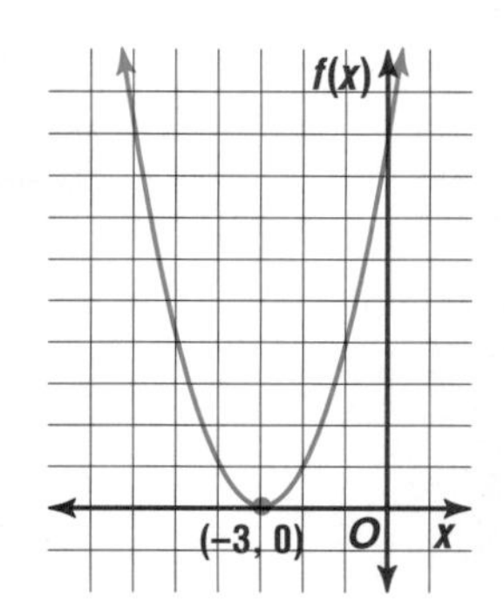

14. **Tennis** A tennis ball is propelled upward from the face of a racket at 40 ft/s. The racket face is 3 feet from the ground when it makes contact with the ball.
    a. If the force of gravity is 32 ft/s^2, at what time after hitting the ball is it at its highest point?
    b. How high does it go?

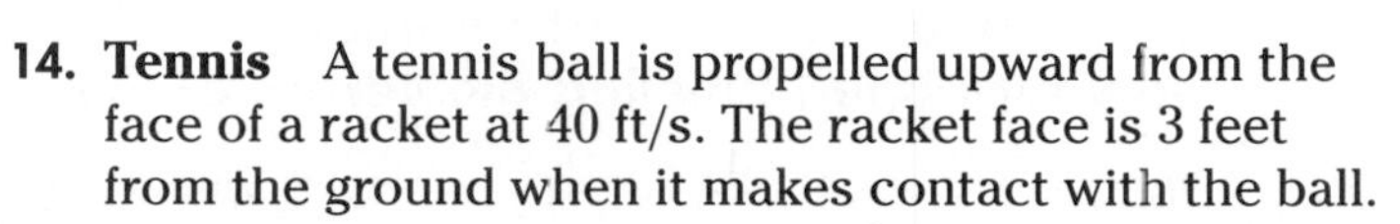

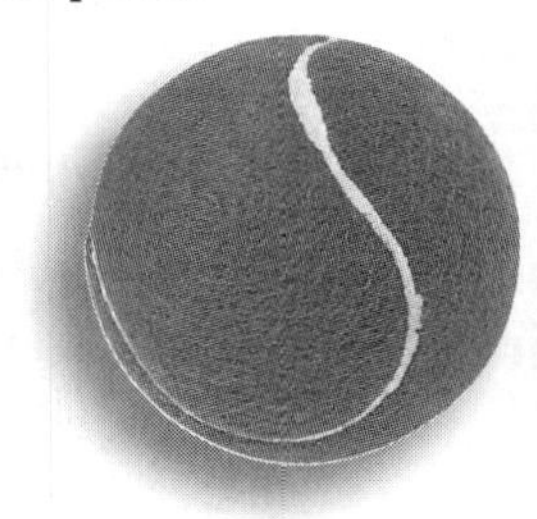

EXERCISES

Practice

Write the equation of the axis of symmetry and find the coordinates of the vertex of the graph of each equation. State if the vertex is a maximum or minimum. Then graph the equation.

15. $y = 4x^2$
16. $y = -x^2 + 4x - 1$
17. $y = x^2 + 2x + 18$
18. $y = x^2 - 3x - 10$
19. $y = x^2 - 5$
20. $y = 4x^2 + 16$
21. $y = 2x^2 + 12x - 11$
22. $y = 3x^2 + 24x + 80$
23. $y = x^2 - 25$
24. $y = 15 - 6x - x^2$
25. $y = -3x^2 - 6x + 4$
26. $y = 5 + 16x - 2x^2$
27. $y = 3(x + 1)^2 - 20$
28. $y = -(x - 2)^2 + 1$
29. $y = \frac{2}{3}(x + 1)^2 - 1$

Match each equation with its graph.

30. $f(x) = \frac{1}{2}x^2 + 1$
31. $f(x) = -\frac{1}{2}x^2 + 1$
32. $f(x) = \frac{1}{2}x^2 - 1$

a.
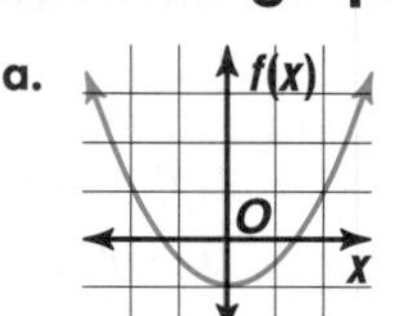

b.

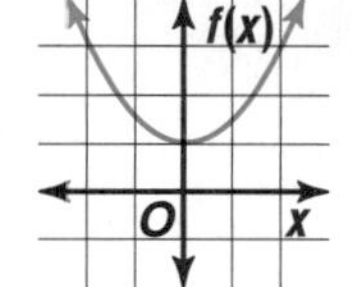

c.
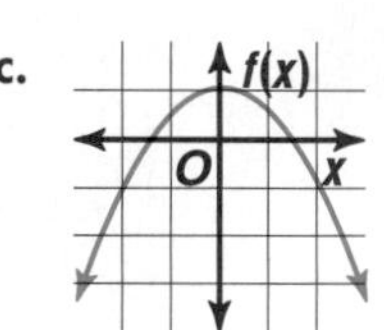

Graph each equation.

33. $y + 2 = x^2 - 10x + 25$
34. $y + 1 = 3x^2 + 12x + 12$
35. $y + 3 = -2(x - 4)^2$
36. $y - 5 = \frac{1}{3}(x + 2)^2$
37. What is the equation of the axis of symmetry of a parabola if its x-intercepts are -6 and 4?

38. The vertex of a parabola is at $(-4, -3)$. If one x-intercept is -11, what is the other x-intercept?

39. Two points on a parabola are at $(-8, 7)$ and $(12, 7)$. What is the equation of the axis of symmetry?

Graphing Calculator

You can draw the axis of symmetry on a graphed parabola by using the VERTICAL command on the DRAW menu. Press 2nd DRAW 4 and use the arrow keys to move the line into place. To graph the axis of symmetry before the graph is drawn, enter the equation in the Y= list and press 2nd DRAW 4 (the x value through which the line will pass) ENTER. Graph each equation and its axis of symmetry. Make a sketch of each graph on your paper, labeling the vertex of the graph.

40. $y = 8 - 4x - x^2$

41. $y = 20x^2 + 44x + 150$

42. $y = 0.023x^2 + 12.33x - 66.98$

43. $y = -78.23x^2 - 23.76x + 88.34$

Critical Thinking

44. Graph $y = x^2 + 2$ and $x + y = 8$ on the same coordinate plane. What are the coordinates of the points they have in common? Explain how you determined these points.

Applications and Problem Solving

45. **College** The cost of a college education is increasing every year. Many parents are starting college funds for their children before they are even born. The average tuition and fees for public college during the years 1970–1993 can be estimated using the function $U(t) = 2.97t^2 - 6.78t + 329.96$, where $U(t)$ represents tuition and fees (in dollars) for one year and t represents the number of years after 1970.

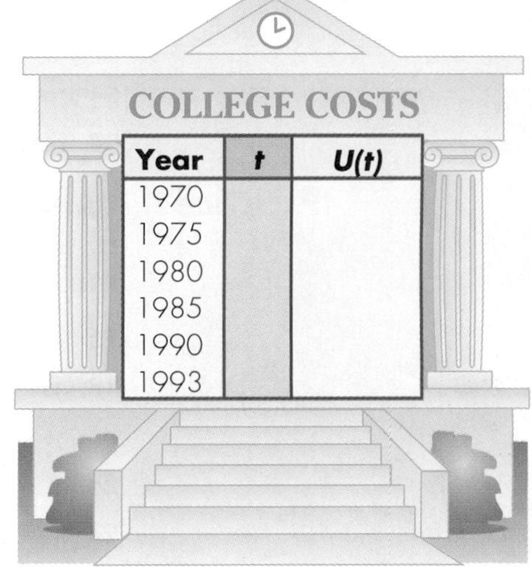

COLLEGE COSTS

Year	t	$U(t)$
1970		
1975		
1980		
1985		
1990		
1993		

   a. Copy and complete the table at the right.
   b. Determine the domain and range values for which this function makes sense.
   c. Graph this function.
   d. Assume that this function is a model for all years after 1993. How much will the tuition and fees for your first year of college be if you attend a public college?

46. **Football** When a football player punts a football, he hopes for a long "hang time," the total amount of time the ball stays in the air. Any hang time of more than about 4.5 seconds is usually good. Manuel is the punter for his high school team. He can kick the ball with an upward velocity of 80 ft/s, and his foot meets the ball 2 feet off the ground.
   a. Write a quadratic equation to describe the height of the football at any given time t. Use 32 ft/s^2 as the acceleration of gravity.
   b. How high is the ball after 1 second? 2 seconds? 3 seconds?
   c. What is Manuel's hang time?

Mixed Review

47. **Agriculture** A field is 1.2 kilometers long and 0.9 kilometers wide. A farmer begins plowing the field by starting at the outer edge and going all the way around the field. When he stops for lunch, a strip of uniform width has been plowed on all sides of the field and half the field is plowed. What is the width of the strip? (Lesson 10–7)

48. Find $(x - 4)(x - 8)$. (Lesson 9–7)

49. **Astronomy** Mars, located 227,920,000 kilometers from the sun, has a diameter of 6.79×10^3 kilometers. (Lesson 9–3)
   a. Write the diameter of Mars in decimal notation.
   b. Write the distance Mars is from the sun in scientific notation.

50. Use elimination to solve the system of equations. (Lesson 8–4)

$3x + 4y = -25$

$2x - 3y = 6$

51. **Meteorology** The table below lists the record 24-hour precipitation for each state as of 1990. (Lesson 7–7)

State	Inches	State	Inches	State	Inches	State	Inches	State	Inches
AL	20.33	HI	38.00	MA	18.15	NM	11.28	SD	8.00
AK	15.20	ID	7.17	MI	9.78	NY	11.17	TN	11.00
AZ	11.40	IL	16.54	MN	10.84	NC	22.22	TX	43.00
AR	14.06	IN	10.50	MS	15.68	ND	8.10	UT	6.00
CA	26.12	IA	16.70	MO	18.18	OH	10.51	VT	8.77
CO	11.08	KS	12.59	MT	11.50	OK	15.50	VA	27.00
CT	12.77	KY	10.40	NE	13.15	OR	10.17	WA	12.00
DE	8.50	LA	22.00	NV	7.40	PA	34.50	WV	19.00
FL	38.70	ME	8.05	NH	10.38	RI	12.13	WI	11.72
GA	18.00	MD	14.75	NJ	14.81	SC	13.25	WY	6.06

Source: National Climatic Data Center

   a. Make a box-and-whisker plot of the data.
   b. Are there any outliers? If so, list them.

52. Graph $y = 3x + 4$. (Lesson 5–5)

53. Solve $3x = -15$. (Lesson 3–2)

54. **Patterns** Complete the pattern: 3, 6, 12, 24, __?__, __?__. (Lesson 1–2)

WORKING ON THE

In·ves·ti·ga·tion

Refer to the Investigation on pages 554–555.

Examine the table you created in Lesson 10–7 that included the length, width, perimeter, and area of the rectangular brick patterns. Measure each of your models in millimeters to determine the values of x and y and record the measures.

1 Graph the data, plotting the length of the pattern on the horizontal axis and the area of the pattern on the vertical axis. What kind of a graph is it? What relationship does it show?

2 Make another graph, plotting the perimeter of the pattern on the horizontal axis and the area of the pattern on the vertical axis. What kind of a graph is it? What kind of relationship does it show?

3 What is the relationship between the measures of the area, length, and width? Why are there some rectangular patterns that have the same area but different perimeters?

Add the results of your work to your Investigation Folder.

11–1B Graphing Technology Parent and Family Graphs

An Extension of Lesson 11–1

A family of graphs is a group of graphs that have at least one characteristic in common. In Lesson 6–5A, you learned about families of linear graphs that shared the same slope or y-intercept. Families of parabolas often fall into two categories—those that have the same vertex and those that have the same shape. Graphing calculators make it easy to study the characteristics of families of parabolas.

The parent function in each of these families is $y = x^2$.

Example 1 **Graph each group of equations on the same screen. Compare and contrast the graphs.**

a. $y = x^2, y = 2x^2, y = 3x^2$

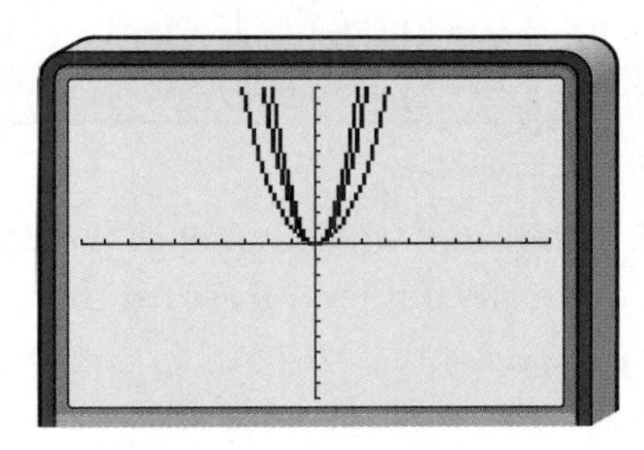

Each graph opens upward and has its vertex at the origin. The graphs of $y = 2x^2$ and $y = 3x^2$ are narrower than the graph of $y = x^2$.

b. $y = x^2, y = 0.5x^2, y = 0.3x^2$

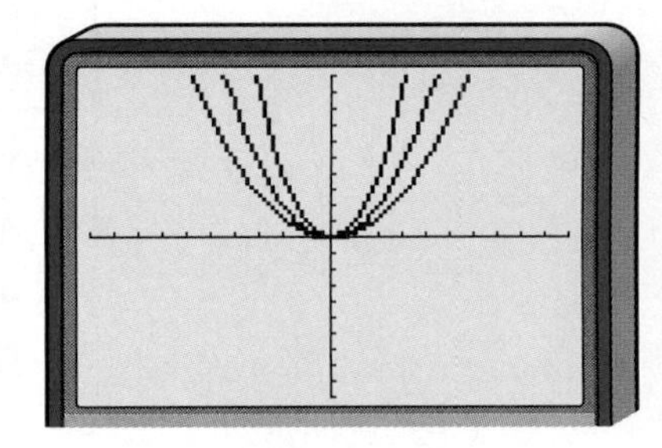

Each graph opens upward and has its vertex at the origin. The graphs of $y = 0.5x^2$ and $y = 0.3x^2$ are wider than the graph of $y = x^2$.

How does the value of a in $y = ax^2$ affect the shape of the graph?

c. $y = x^2, y = x^2 + 2,$
$y = x^2 - 3, y = x^2 - 5$

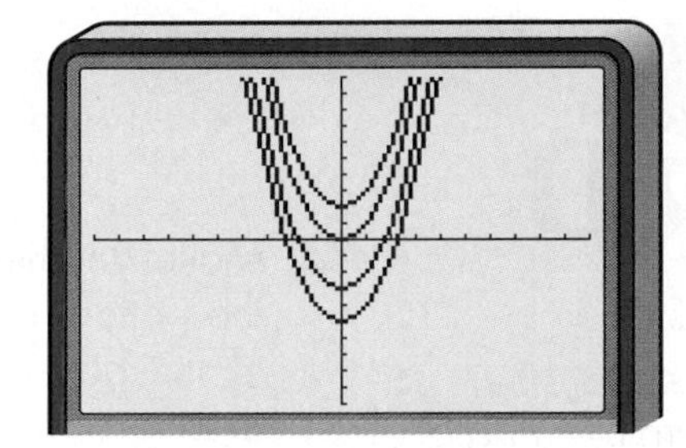

Each graph opens upward and has the same shape as $y = x^2$. However, each parabola has a different vertex, located along the y-axis. *How does the value of the constant affect the placement of the graph?*

d. $y = x^2, y = (x - 2)^2,$
$y = (x + 3)^2, y = (x + 1)^2$

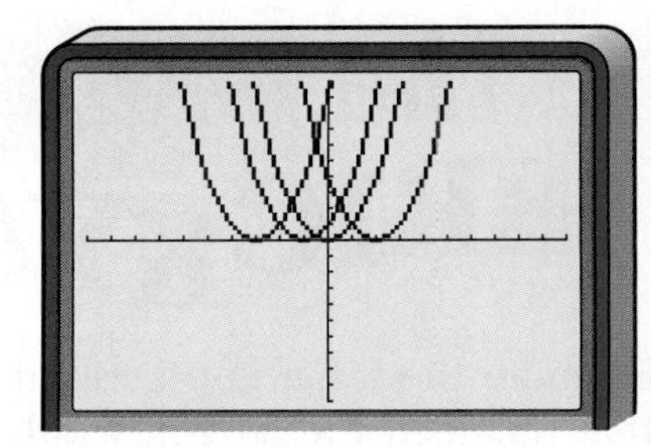

Each graph opens upward and has the same shape as $y = x^2$. However, each parabola has a different vertex, located along the x-axis. *How is the location of the vertex related to the equation of the graph?*

When analyzing or comparing the shapes of various graphs on different screens, it is important to compare the graphs using the same parameters. That is, the window used to compare the graphs should be the same, with the same scale factor. Suppose we graph the same equation using a different window for each. How will this affect the appearance of the graph?

Example 2 **Graph $y = x^2 - 5$ in each viewing window. What conclusions can you draw about the appearance of a graph in the window used?** *The scale is 1 unless otherwise noted.*

a. standard viewing window

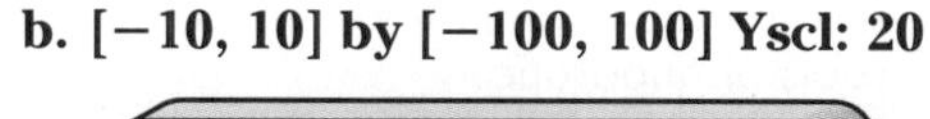

b. [−10, 10] by [−100, 100] Yscl: 20

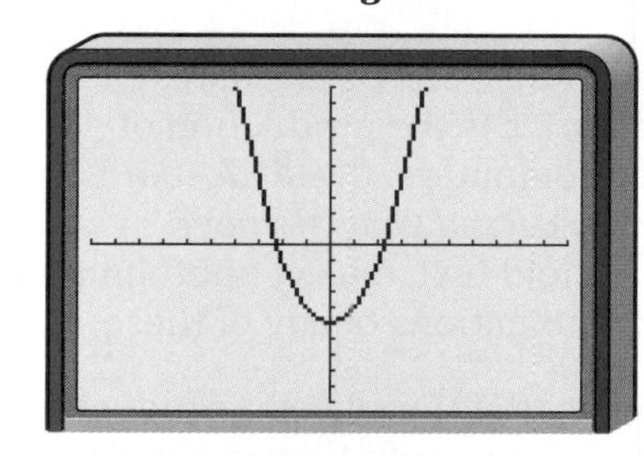

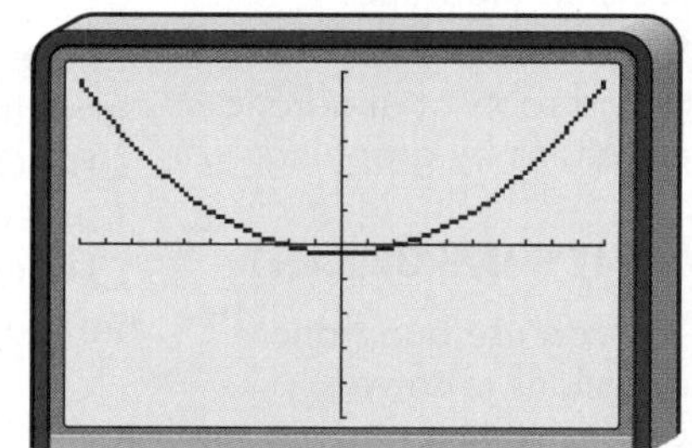

c. [−50, 50] Xscl: 5 by [−10, 10]

d. [−0.5, 0.5] Xscl: 0.1 by [−10, 10]

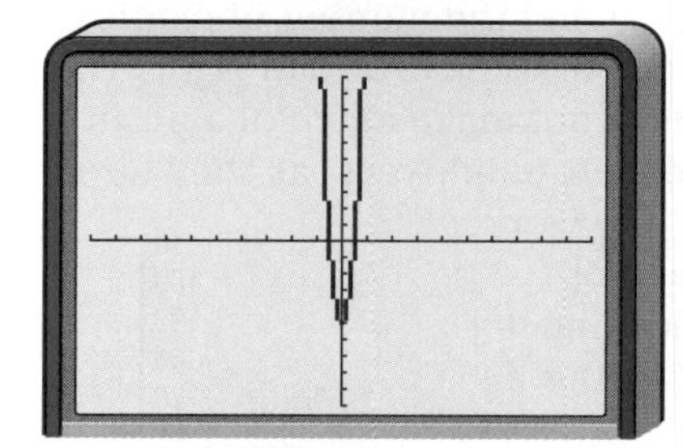

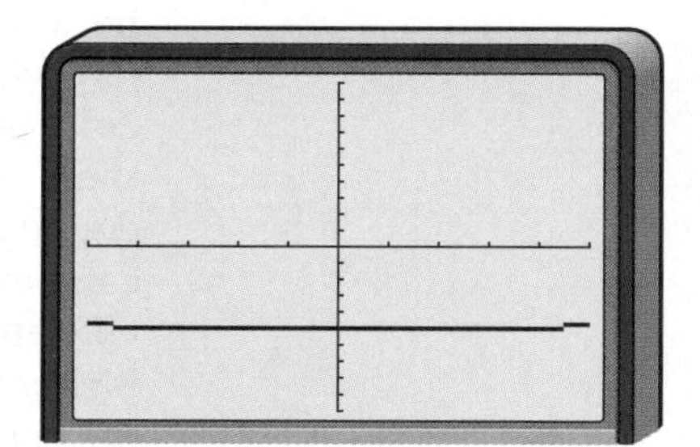

The window greatly affects the appearance of the parabola. Without knowing the window, graph b might be of the family $y = ax^2$, where $0 < a < 1$. Graph c makes the graph look like a member of $y = ax^2$, where $a > 1$. Graph d looks more like a line. However, all are graphs of the same equation.

EXERCISES

Graph each group of equations on the same screen. Make a sketch of the screen on grid paper, and compare and contrast the graphs.

1. $y = -x^2$
   $y = -2x^2$
   $y = -5x^2$
2. $y = -x^2$
   $y = -0.3x^2$
   $y = -0.7x^2$
3. $y = -x^2$
   $y = -(x + 4)^2$
   $y = -(x - 8)^2$
4. $y = -x^2$
   $y = -x^2 + 6$
   $y = -x^2 - 4$

Use the families of graphs that have appeared in this lesson to predict the appearance of the graph of each equation. Then sketch the graph.

5. $y = 4x^2$
6. $y = x^2 - 6$
7. $y = -0.1x^2$
8. $y = (x + 1)^2$
9. Describe how each change in the equation of $y = x^2$ would affect the graph of $y = x^2$. Be sure to consider all values of a and b.
   a. $y = ax^2$
   b. $y = x^2 + a$
   c. $y = (x + a)^2$
   d. $y = (x + a)^2 + b$

11–2 Solving Quadratic Equations by Graphing

What YOU'LL LEARN

- To use estimation to find roots of quadratic equations, and
- to find roots of quadratic equations by graphing.

Why IT'S IMPORTANT

You can use quadratic equations to solve problems involving architecture and number theory.

Technology

In the United States, one of the fastest growing industries is the production of CD-ROMs for computers. CD-ROM stands for *Compact Disc-Read Only Memory*. CD-ROMs can hold text, music, photographic images, or combinations of any of these.

Any company that produces a product to sell finds that the profit they can make depends on the number of employees they have (among other things). The relationship between profit and the number of employees looks like the parabola drawn at right. The company does not make much of a profit if there are too few employees; as they hire more employees, the work can be done more efficiently, leading to higher profits. But if the company hires too many employees, it may not have room for them or enough work for them to do, yet they still have to be paid, causing lower profits.

Since number of employees is the independent variable and profit is the dependent variable, the domain and range must both be positive.

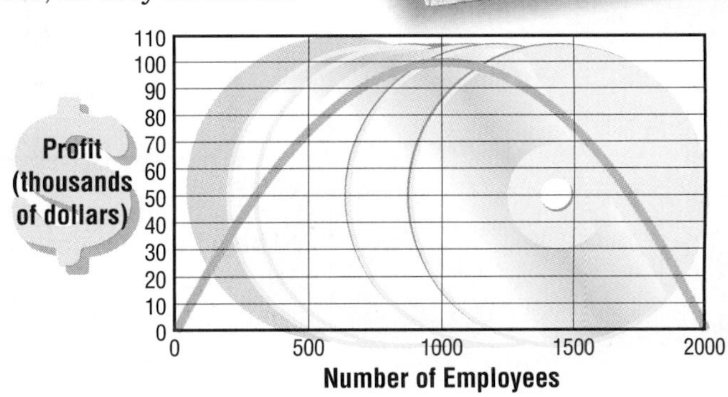

Suppose a company that produces CD-ROMs can express its profit as $P(x) = -0.1x^2 + 200x$, where x is the number of employees in the company. As the personnel manager of the company, it is your job to determine the least number of employees the company should have in order to reach its goal of having $75,000 in profit. *You will solve this problem in Exercise 4.*

A **quadratic equation** is an equation in which the value of the related quadratic function is 0. That is, for the quadratic equation $0 = x^2 + 6x - 7$, the related quadratic function is $f(x) = x^2 + 6x - 7$, where $f(x) = 0$. You have used factoring to solve equations like $x^2 + 6x - 7 = 0$. You can also use graphing to estimate the solutions to this equation.

The solutions of a quadratic equation are called the **roots** of the equation. The roots of a quadratic equation can be found by finding the x-intercepts or **zeros** of the related quadratic function.

Example 1

Solve $x^2 - 2x - 3 = 0$ by graphing. Check by factoring.

Graph the related function $f(x) = x^2 - 2x - 3$. The equation of the axis of symmetry is $x = 1$, and the coordinates of the vertex are $(1, -4)$. Make a table of values to find other points to sketch the graph of $f(x) = x^2 - 2x - 3$.

x	f(x)
−1	0
0	−3
1	−4
2	−3
3	0

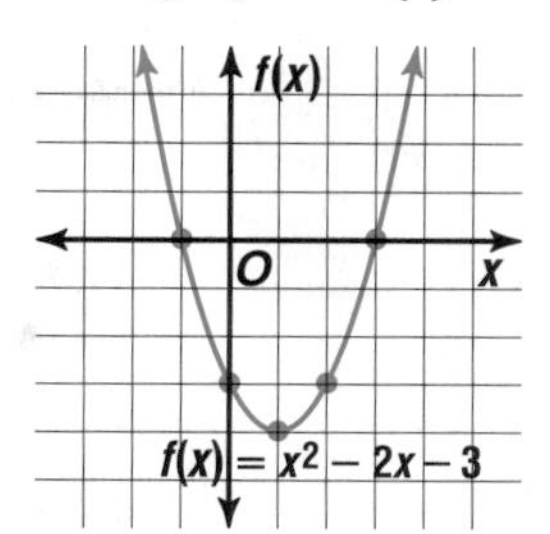

To solve the equation $x^2 - 2x - 3 = 0$, we need to know where the value of $f(x)$ is 0. On the graph, this occurs at the x-intercepts. The x-intercepts of the parabola appear to be -1 and 3.

LOOK BACK

You can refer to Lesson 10-6 for more information on the zero product property and solving equations by factoring.

Check: Solve by factoring.

$$x^2 - 2x - 3 = 0$$

$$(x - 3)(x + 1) = 0$$

Set each factor equal to 0.

$x - 3 = 0$ *Zero product property*

$x = 3$ *Add 3 to each side.*

$x + 1 = 0$ *Zero product property*

$x = -1$ *Add −1 to each side.*

The solutions of the equation are 3 and -1.

In Example 1, the zeros of the function were integers. Usually the zeros of a quadratic function are not integers. In these cases, use estimation to approximate the roots of the equation.

Example 2

Solve $x^2 + 9x + 5 = 0$ by graphing. If integral roots cannot be found, estimate the roots by stating the consecutive integers between which the roots lie.

Use a table of values to graph the related function $f(x) = x^2 + 9x + 5$.

Notice in the table of values that the value of the function changes from negative to positive between the x values of −9 and −8 and −1 and 0.

x	f(x)
−9	5
−8	−3
−7	−9
−6	−13
−5	−15
−4	−15
−3	−13
−2	−9
−1	−3
0	5

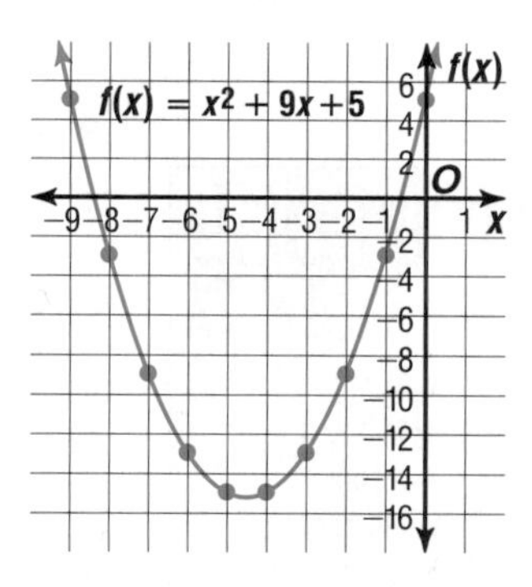

The x-intercepts of the graph are between -9 and -8 and between -1 and 0. So, one root of the equation is between -9 and -8, and the other root is between -1 and 0.

You can use a graphing calculator to find a more accurate estimate for the root of a quadratic equation than you can by using paper and pencil. One way to estimate the root is to ZOOM IN on the zero. Another method is to use the ROOT feature on the CALC menu.

EXPLORATION GRAPHING CALCULATORS

Use a graphing calculator to solve $3x^2 - 6x - 2 = 0$ to the nearest hundredth.

Graph $y = 3x^2 - 6x - 2$ in the standard viewing window. To use the ROOT feature, you must use the cursor to define the interval in which the calculator will look. You define the interval in the same way the MAXIMUM and MINIMUM intervals are defined.

Enter: 2nd CALC 2

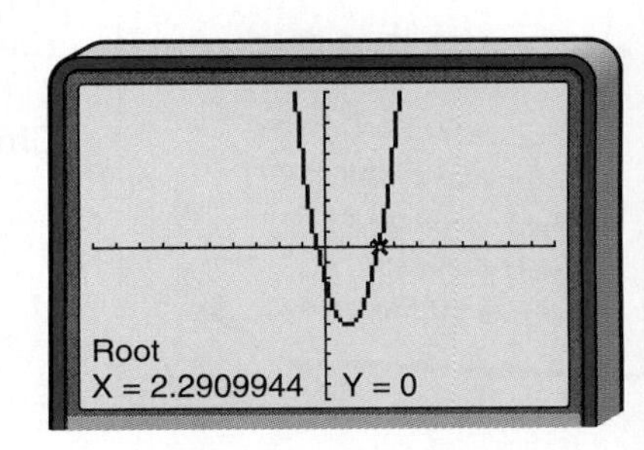

Now use the arrow keys to move the cursor to the left of one of the x-intercepts and press ENTER to define the lower bound. Then use the arrow keys to move the cursor to the right of that x-intercept and press ENTER to define the upper bound.

The y-coordinate value may appear as a decimal in scientific notation, such as −1E−12 rather than 0.

Press ENTER and the approximate coordinates of the root will appear. Repeat this process to find the coordinates of the other root.

Your Turn

Use a graphing calculator to estimate the roots of each equation.

a. $x^2 + 2x - 9 = 0$ **b.** $7.5x^2 - 9.5 = 0$ **c.** $6x^2 + 5x + 5 = 0$

d. Use a graphing calculator to solve $x^2 + 9x + 5 = 0$. Compare your results to those in Example 2. Which method of solution do you prefer?

Quadratic equations always have two roots. However, these roots may not be two distinct numbers.

Example  **Solve $x^2 - 12x + 36 = 0$ by graphing.**

Graph the related function $f(x) = x^2 - 12x + 36$.

x	$f(x)$
4	4
5	1
6	0
7	1
8	4

The equation of the axis of symmetry is $x = 6$. The vertex of the parabola is at (6, 0).

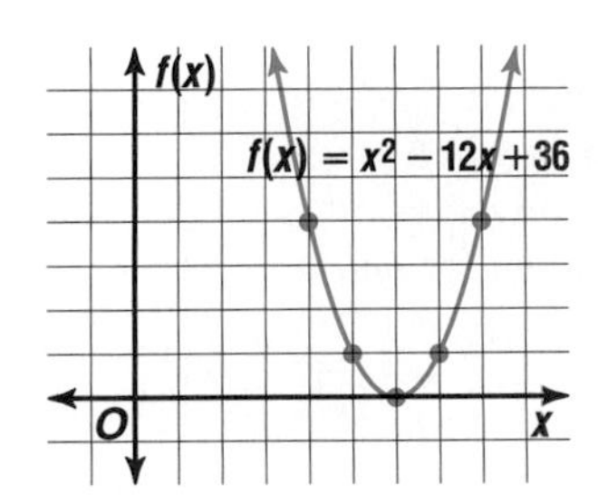

TECHNOLOGY Tip

If you graph a function with a graphing calculator and it appears that the vertex of a parabola is its x-intercept, use the ZOOM IN feature several times. You may find that the parabola actually does cross the x-axis twice.

Notice that the vertex of the parabola is the x-intercept. Thus, one solution is 6. What is the other solution?

Try solving by factoring.

$x^2 - 12x + 36 = 0$

$(x - 6)(x - 6) = 0$

Set each factor equal to 0.

$x - 6 = 0$ $\quad x - 6 = 0$ *Zero product property*

$x = 6$ $\quad x = 6$ *Add 6 to each side.*

There are two identical roots to this equation. So, there is only one distinct root. The solution for $x^2 - 12x + 36 = 0$ is 6.

Thus far, we have seen that quadratic equations can have two distinct real roots or one distinct real root. Is it possible that there may be no real roots?

Example 4 **Solve $x^2 + 2x + 5 = 0$ by graphing.**

Graph the function $f(x) = x^2 + 2x + 5$.

x	$f(x)$
−3	8
−2	5
−1	4
0	5
1	8

The equation of the axis of symmetry is $x = -1$. The vertex of the parabola is at $(-1, 4)$.

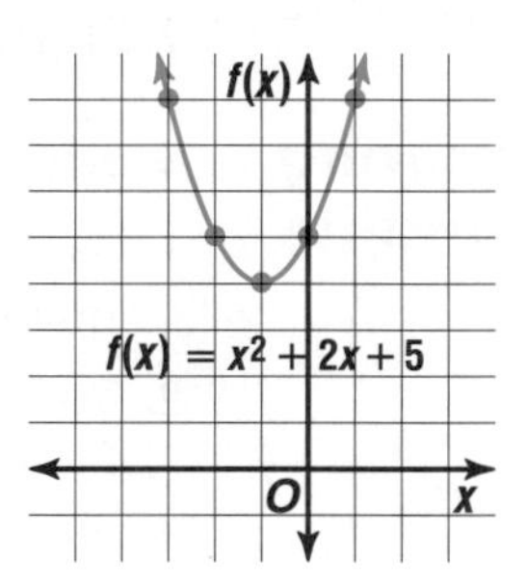

This graph has no x-intercept. Thus, there are no real number solutions for this equation.

The symbol ∅, indicating an empty set, is often used to represent no real solution.

Quadratic equations can be used to solve number problems.

Example 5 **Two numbers have a sum of 4. What are the numbers if their product is −12?**

INTEGRATION

Number Theory

Explore Let n represent one of the numbers. Then the other number is $4 - n$.

Plan A function that describes the product of these two numbers is $f(n) = n(4 - n)$ or $f(n) = -n^2 + 4n$. Find the value of n if $f(n)$ equals −12.

Solve Solve $f(n) = -n^2 + 4n$ if $f(n) = -12$.

$f(n) = -n^2 + 4n$

$-12 = -n^2 + 4n$ $\quad$ *$f(n) = -12$*

$0 = -n^2 + 4n + 12$ $\quad$ *Rewrite the equation so one side is 0.*

Graph the related function $f(n) = -n^2 + 4n + 12$.

x	y
−3	−9
−2	0
−1	7
0	12
1	15
2	16
3	15
4	12
5	7
6	0
7	−9

The equation of the axis of symmetry is $n = 2$. The vertex of the parabola is at $(2, 16)$.

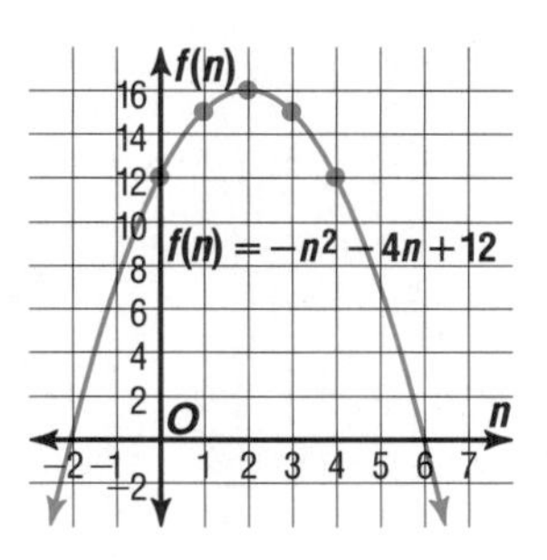

(continued on the next page)

The n-intercepts of the graph are -2 and 6. Use these values of n to find the value of the other number $4 - n$.

If $n = -2$, then $4 - n = 4 - (-2)$ or 6.

If $n = 6$, then $4 - n = 4 - 6$ or -2.

So the numbers are 6 and -2.

Examine Test to see if the numbers satisfy the problem.

The sum of the numbers is 4. The product of the numbers is -12.

$-2 + 6 \stackrel{?}{=} 4$ $-2(6) \stackrel{?}{=} -12$

$4 = 4$ ✓ $-12 = -12$ ✓

The two numbers are -2 and 6.

CHECK FOR UNDERSTANDING

Communicating Mathematics

Study the lesson. Then complete the following.

1. **Explain** why the x-intercepts of a quadratic function can be used to solve a quadratic equation.
2. **You Decide** Joshua says he likes to solve a quadratic equation by factoring rather than graphing. Hanna says that she likes graphing because she can always get an answer. Who is correct? Give examples to support your answer.
3. What is the related function you would use to solve $x^2 + 9x + 2 = 3x - 4$ by graphing?
4. Refer to the application at the beginning of the lesson. What is the least number of employees that would help the company make its goal of $75,000 profit?

5. **Draw** an example of each type of situation that may occur when using graphing to find the solutions of a quadratic equation. Identify the number of real roots of the quadratic function in each situation.

Guided Practice

Determine the number of real roots for each quadratic equation whose related function is graphed below.

6. 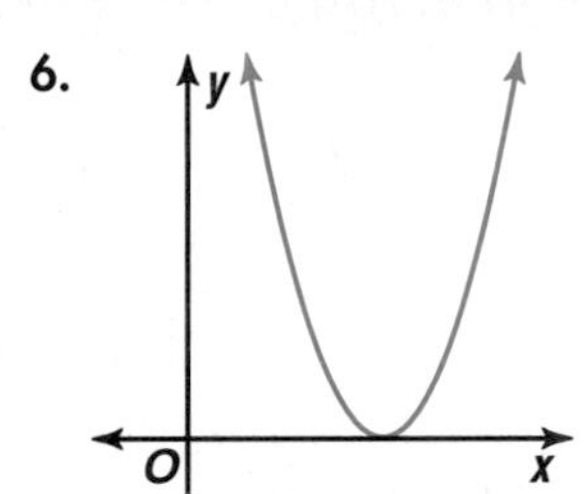

7. 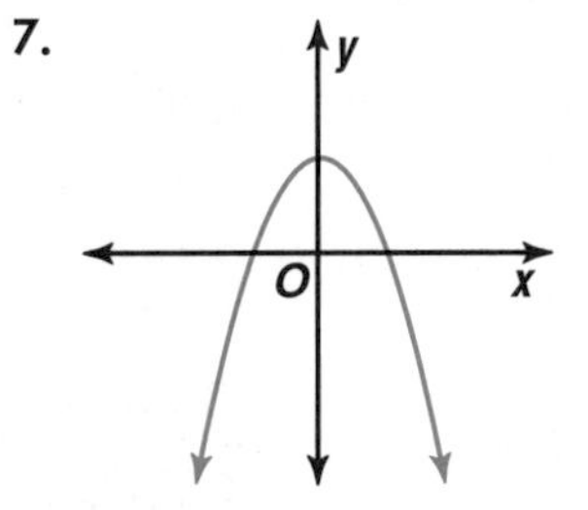

8. 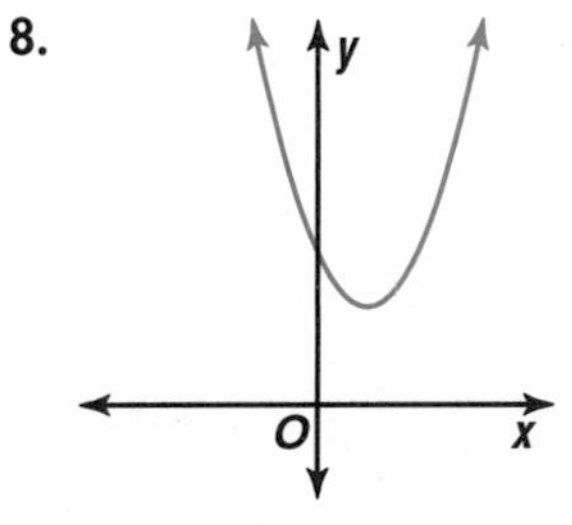

9. State the real roots of the quadratic equation whose related function is graphed at the right.

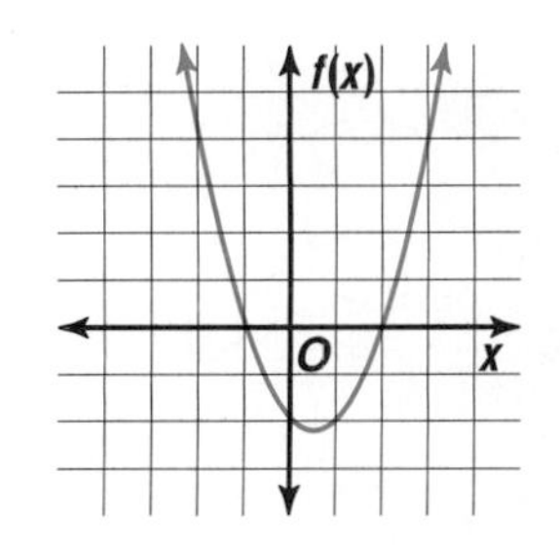

Solve each equation by graphing. If integral roots cannot be found, state the consecutive integers between which the roots lie.

10. $x^2 - 7x + 6 = 0$
11. $c^2 - 5c - 24 = 0$
12. $5n^2 + 2n + 6 = 0$
13. $w^2 - 3w = 5$
14. $b^2 - b + 4 = 0$
15. $a^2 - 10a = -25$

16. **Number Theory** Use a quadratic equation to find two real numbers whose sum is 5 and whose product is -24.

EXERCISES

Practice

State the real roots of each quadratic equation whose related function is graphed below.

17. 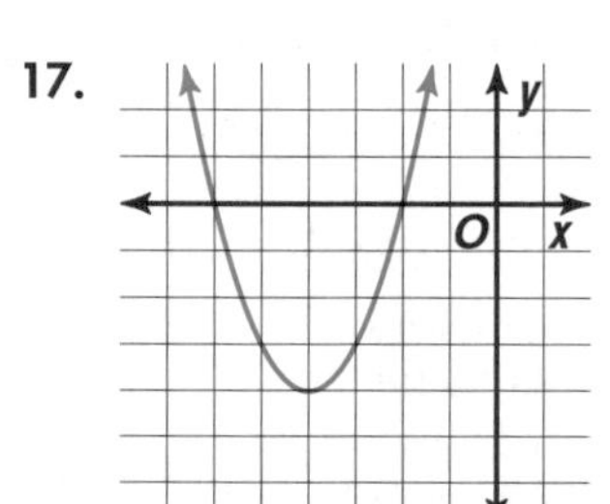

18. 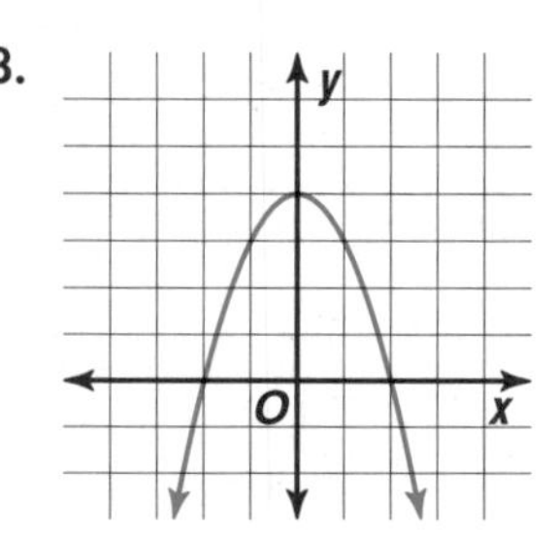

19. 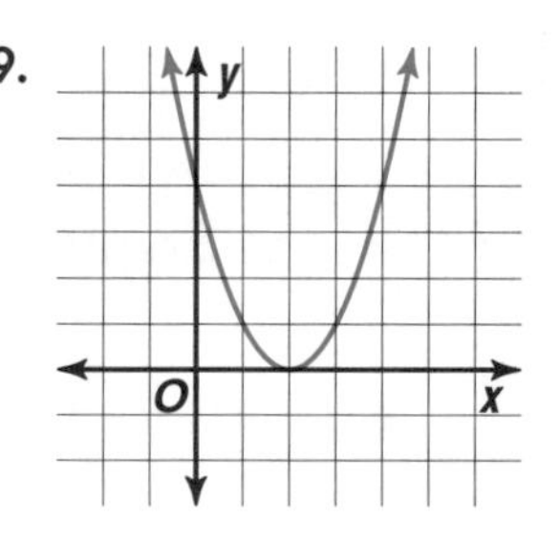

Solve each equation by graphing. If integral roots cannot be found, state the consecutive integers between which the roots lie.

20. $x^2 + 7x + 12 = 0$
21. $x^2 - 16 = 0$
22. $a^2 + 6a + 7 = 0$
23. $x^2 + 6x + 9 = 0$
24. $r^2 + 4r - 12 = 0$
25. $c^2 + 3 = 0$
26. $2c^2 + 20c + 32 = 0$
27. $3x^2 + 9x - 12 = 0$
28. $2x^2 - 18 = 0$
29. $p^2 + 16 = 8p$
30. $w^2 - 10w = -21$
31. $a^2 - 8a = 4$
32. $m^2 - 2m = -2$
33. $12n^2 - 26n = 30$
34. $4x^2 - 35 = -4x$

The roots of a quadratic equation are given. Graph the related quadratic function if it has the indicated maximum or minimum point.

35. roots: $0, -6$
    maximum point: $(-3, 4)$
36. roots: $-2, -6$
    minimum point : $(-4, -2)$
37. roots: no real roots
    minimum point: $(2, 5)$
38. roots: $-4 < x < -3,\ 1 < x < 2$
    maximum point: $(-1, 6)$

Number Theory

Use a quadratic equation to determine the two numbers that satisfy each situation.

39. Their difference is 4 and their product is 32.
40. Their sum is 9 and their product is 20.
41. Their sum is 4 and their product is 5.
42. They differ by 2. The sum of their squares is 130.

Estimate the *y*-intercepts of each quadratic equation by graphing.

43. $x = -0.75y^2 - 6y - 9$
44. $x = y^2 - 4y + 1$
45. $x = 3y^2 + 2y + 4$

Graphing Calculator

Use a graphing calculator to solve each equation to the nearest hundredth.

46. $4x^2 - 11 = 0$ **47.** $-2x^2 - x - 3 = 0$ **48.** $x^2 + 22x + 121 = 0$

49. $6x^2 - 12x + 3 = 0$ **50.** $5x^2 + 4x - 7 = 0$ **51.** $-4x^2 + 7x + 8 = 0$

Use a graphing calculator to find values for *k* for each equation so that it will have (a) one distinct root, (b) two real roots, and (c) no real roots.

52. $x^2 + 3x + k = 0$ **53.** $kx^2 + 4x - 2 = 0$

Critical Thinking

54. Suppose the value of a quadratic function is negative when $x = 10$ and positive when $x = 11$. Explain why it is reasonable to assume that the related equation has a solution between 10 and 11.

Applications and Problem Solving

55. Architecture A painter is hired to paint an art gallery whose walls are sculptured with arches that can be represented by the quadratic function $f(x) = -x^2 - 4x + 12$. The wall space under each arch is to be painted a different color from the arch itself. The painter can use the formula $A = \frac{2}{3}bh$ to estimate the area under a parabola, where b is the length of a horizontal segment connecting two points on the parabola and h is the height from that segment to the vertex. Suppose the horizontal segment is represented by the floor, which is the x-axis, and each unit represents 1 foot.

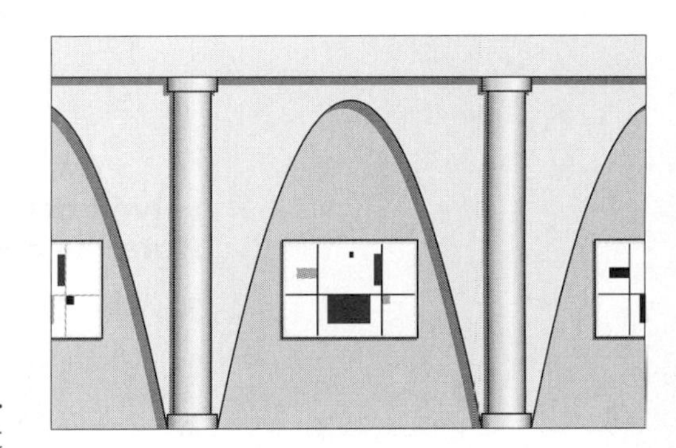

a. Graph the quadratic function and determine its x-intercepts.

b. What is the length of the segment along the floor?

c. What is the height of the arch?

d. How much wall space is there under each arch?

e. How much would the paint cost to paint the walls under 12 arches if the paint is \$27/gallon, she applies two coats, and the manufacturer states that each gallon will cover 200 ft^2? *Remember you cannot buy part of a gallon.*

Top Five List

Most-Visited U.S. National Parks in 1993

Park	Visitors
1. Great Smoky Mountains	9,283,848
2. Grand Canyon	4,575,602
3. Yosemite	3,839,645
4. Yellowstone	2,912,193
5. Rocky Mountains	2,780,342

56. Make a Drawing Banneker Park has set aside a section of the park as a nature preserve with an observation deck and telescopes so that people can observe the wildlife and plants up close. To accommodate the increased number of visitors they expect, the park commission has received funding to double the parking area. The current lot is 64 yards by 96 yards, and they are going to add strips of equal width to the end and side of the lot to create a larger rectangle that is twice the size of the original.

a. Make a drawing of the lot and the proposed additions.

b. Write a quadratic equation to find x, the width of the strips, so that the area of the parking lot is doubled.

c. How wide are the strips to be added?

d. What are the dimensions of the new parking lot?

Mixed Review

57. Find the equation of the axis of symmetry and the coordinates of the vertex of the graph of $y = -3x^2 + 4$. (Lesson 11–1)

58. Solve $81x^3 + 36x^2 = -4x$. (Lesson 10–6)

59. **Geometry** The area of a rectangle is $(8x^2 - 10x + 3)$ square meters. What are the dimensions of the rectangle? (Lesson 10–3)

60. Find $(x - 2y)^3$. (Lesson 9–8)

61. Find the degree of $6x^2y + 5x^3y^2z - x + x^2y^2$. (Lesson 9–4)

62. Use graphing to solve the system of equations. (Lesson 8–1)
$x + y = 3$
$x + y = 4$

63. Solve $|3x + 4| < 8$. (Lesson 7–6)

64. Graph $y = -x + 6$. (Lesson 6–5)

65. **Commerce** Find the sale price of an item originally marked as $33 with 25% off. (Lesson 4–4)

66. Find $-4 + 6 + (-10) + 8$. (Lesson 2–3)

WORKING ON THE In·ves·ti·ga·tion

Refer to the Investigation on pages 554–555.

the BRICKYARD

In order to best use the excess inventory of bricks, you are going to investigate the possible combinations that will use up the inventory at a steady rate.

1 Suppose you have four large square bricks and three small square bricks. How many rectangular bricks would you need to create a rectangular pattern? Is there more than one pattern that can be made if you change the number of rectangular bricks?
- What are the dimensions, perimeters, and areas of the rectangular patterns formed?
- Use the variables x and y as the length of the large square and the length of the small square, respectively. List the dimensions, perimeter, and area of each pattern in terms of x and y.
- Explain your findings. How do the perimeters and areas of these patterns relate in this matter?

2 If you had three large square bricks, five rectangular bricks, and two small square bricks, what size rectangular patterns can you make? Express the dimensions, perimeter, and area of each pattern in terms of x and y.

3 If you had two large bricks, six rectangular bricks, and four small bricks, what size rectangular patterns can you make? Express the dimensions, perimeter, and area of each pattern in terms of x and y.

4 Explain the method you used to solve these problems and any generalizations you found.

Add the results of your work to your Investigation Folder.

11–3 Solving Quadratic Equations by Using the Quadratic Formula

What YOU'LL LEARN
- To solve quadratic equations by using the quadratic formula.

Why IT'S IMPORTANT

You can use quadratic equations to solve problems involving hydraulics and civics.

Civics

The number of citizens (in millions) voting in each presidential election since 1824 can be approximated by the quadratic function $V(t) = 0.0046t^2 - 0.185t + 3.30$, where t represents the number of years since 1824. For her history project, Marcela Ruiz needed to determine in what year the number of voters in a presidential election was approximately 55 million. She used the function above, replacing $V(t)$ with 55.

$$V(t) = 0.0046t^2 - 0.185t + 3.30$$
$$55 = 0.0046t^2 - 0.185t + 3.30$$
$$0 = 0.0046t^2 - 0.185t - 51.7$$

Marcela knew that she could use her graphing calculator to estimate the solutions of this equation, but she wondered how she could get a good estimate of t if she didn't have a graphing calculator. *You will estimate these solutions in Example 4.*

You can use the **quadratic formula** to solve any quadratic equation.

The Quadratic Formula	The solutions of a quadratic equation in the form $ax^2 + bx + c = 0$, where $a \neq 0$, are given by the formula $x = \frac{-b \pm \sqrt{b^2 - 4ac}}{2a}$.

The quadratic formula can be used to solve any quadratic equation involving any variable.

Example  **Use the quadratic formula to solve each equation.**

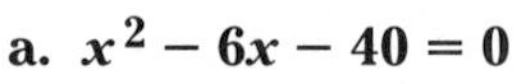

a. $x^2 - 6x - 40 = 0$

In the equation, $a = 1$, $b = -6$, and $c = -40$. Substitute these values into the quadratic formula.

The symbol ± means to evaluate the expression first using + and then evaluate it again using −. This provides for the two solutions of the equation.

$$x = \frac{-b \pm \sqrt{b^2 - 4ac}}{2a}$$
$$= \frac{-(-6) \pm \sqrt{(-6)^2 - 4(1)(-40)}}{2(1)} \quad a = 1, b = -6, \text{ and } c = -40$$
$$= \frac{6 \pm \sqrt{36 + 160}}{2}$$
$$= \frac{6 \pm \sqrt{196}}{2}$$
$$= \frac{6 \pm 14}{2}$$

You can also check the solution to any equation by substituting each value into the original equation.

$x = \frac{6 + 14}{2}$ or $x = \frac{6 - 14}{2}$

$= \frac{20}{2}$ or 10 $\qquad = -\frac{8}{2}$ or -4

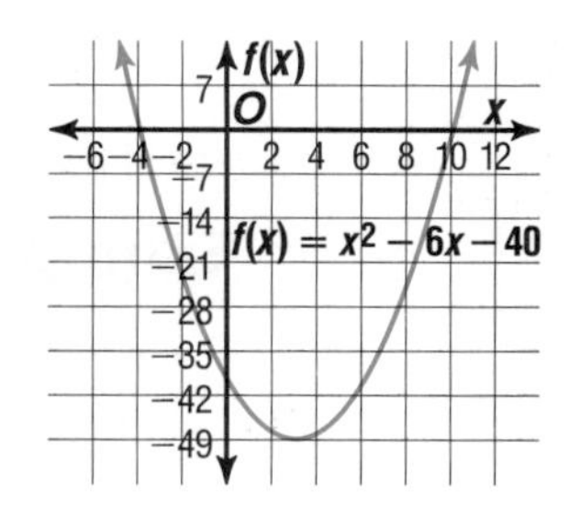

Check: Solve by graphing the related function $f(x) = x^2 - 6x - 40$. The x-intercepts appear to be 10 and -4. This agrees with the algebraic solution.

The solutions are 10 and -4.

b. $\mathbf{y^2 - 6y + 9 = 0}$

$$y = \frac{-b \pm \sqrt{b^2 - 4ac}}{2a}$$

$$= \frac{-(-6) \pm \sqrt{(-6)^2 - 4(1)(9)}}{2(1)} \quad \textit{a = 1, b = −6, and c = 9}$$

$$= \frac{6 \pm \sqrt{36 - 36}}{2}$$

$$= \frac{6 \pm \sqrt{0}}{2}$$

$y = \frac{6 + 0}{2}$ or $y = \frac{6 - 0}{2}$

$= 3 \qquad = 3$

Check: Solve by factoring.

$y^2 - 6y + 9 = 0$

$(y - 3)(y - 3) = 0$

$y - 3 = 0 \qquad y - 3 = 0$ *Zero product property*

$y = 3 \qquad y = 3$ *Add 3 to each side.*

There is one distinct solution, 3.

LOOK BACK

You can refer to Lesson 2-8 for more information on irrational numbers.

Sometimes when you use the quadratic formula, you find the solutions are irrational numbers. It is helpful to use a calculator to estimate the values of the solutions in this case.

Example 2 **Use the quadratic formula to solve $2n^2 - 7n - 3 = 0$.**

$$n = \frac{-b \pm \sqrt{b^2 - 4ac}}{2a}$$

$$= \frac{-(-7) \pm \sqrt{(-7)^2 - 4(2)(-3)}}{2(2)} \quad \textit{a = 2, b = −7, and c = −3}$$

$$= \frac{7 \pm \sqrt{49 + 24}}{4}$$

$$= \frac{7 \pm \sqrt{73}}{4}$$

$\sqrt{73}$ is an irrational number. We can approximate the solutions by using a calculator to find a decimal value for $\sqrt{73}$.

$n = \frac{7 + \sqrt{73}}{4} \approx 3.886 \qquad n = \frac{7 - \sqrt{73}}{4} \approx -0.386$

The two solutions are approximately -0.386 and 3.886.

When we solved quadratic equations by graphing, we found that some quadratic equations have no real solutions. How does the quadratic formula work in this situation?

Example 3 **Use the quadratic formula to solve $z^2 - 5z + 12 = 0$.**

$$z = \frac{-b \pm \sqrt{b^2 - 4ac}}{2a}$$

$$= \frac{-(-5) \pm \sqrt{(-5)^2 - 4(1)(12)}}{2(1)} \quad a = 1,\ b = -5, \text{ and } c = 12$$

$$= \frac{5 \pm \sqrt{25 - 48}}{2}$$

$$= \frac{5 \pm \sqrt{-23}}{2}$$

Since there is no real number that is the square root of a negative number, this equation has no real solutions.

It is often helpful to use a calculator when using the quadratic formula to solve real-world problems. If there are no real solutions for the equation, the calculator will give you an error message.

Example 4 **Refer to the connection at the beginning of the lesson. Determine in what year the number of people voting in a presidential election was approximately 55 million.**

Civics

Use a scientific calculator and the quadratic formula to find values for t. The values for a, b, and c are 0.0046, −0.185, and −51.7, respectively. Find the value of $\sqrt{b^2 - 4ac}$ and store it in the calculator's memory.

Enter: [( ] .185 [+/−] [x^2] [−] 4 [×] .0046 [×] 51.7 [+/−] [ )] [$\sqrt{x}$] [STO] *0.992726044*

Now evaluate the quadratic formula.

Enter: [( ] [( ] .185 [+/−] [ )] [+/−] [+] [RCL] [ )] [÷] [( ] 2 [×] .0046 [ )] [=] *128.0137005*

Enter: [( ] [( ] .185 [+/−] [ )] [+/−] [−] [RCL] [ )] [÷] [( ] 2 [×] .0046 [ )] [=] *−87.79630922*

The negative root has no meaning in this problem.

Since t represents the number of years since 1824, add 128 to 1824: 128 + 1824 = 1952.

So, 1952 was the year in which approximately 55 million people voted.

CHECK FOR UNDERSTANDING

Communicating Mathematics

Study the lesson. Then complete the following.

1. **Explain** how you get two solutions when using the quadratic formula.
2. **Explain** what happens in the quadratic formula when there are no real solutions for the equation.
3. Refer to the connection at the beginning of the lesson.
   a. Predict how many will vote in the year 2000.
   b. Describe the domain and range of this relation.

4. **Assess Yourself** You have learned to use graphing, factoring, and the quadratic formula to solve quadratic equations. Which method do you prefer and why?

Guided Practice

State the values of *a*, *b*, and *c* for each quadratic equation. Then solve the equation by using the quadratic formula. Approximate irrational roots to the nearest hundredth.

5. $x^2 + 3x - 18 = 0$
6. $14 = 12 - 5x - x^2$
7. $4x^2 - 2x + 15 = 0$
8. $x^2 = 25$

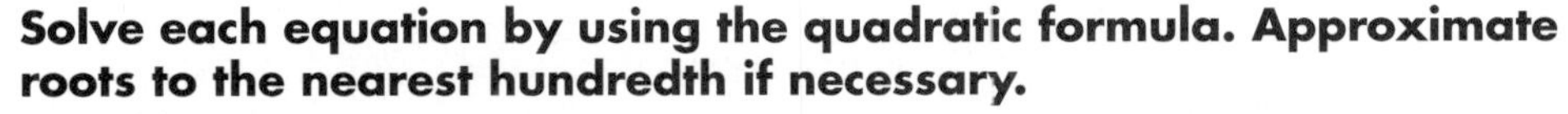

Solve each equation by using the quadratic formula. Approximate roots to the nearest hundredth if necessary.

9. $4x^2 + 2x - 17 = 0$
10. $3b^2 + 5b + 11 = 0$
11. $x^2 + 7x + 6 = 0$
12. $z^2 - 13z = 32$

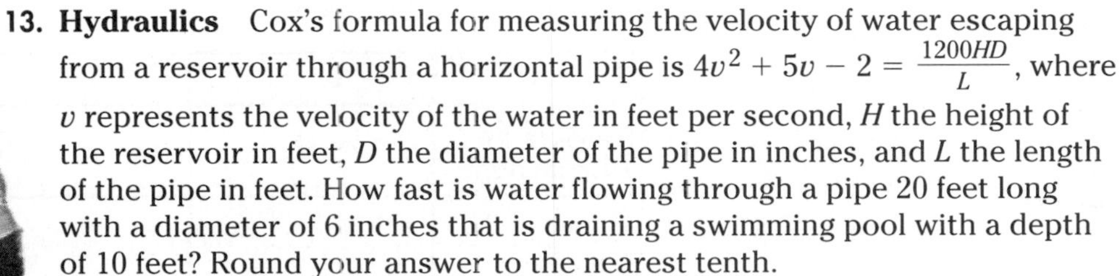

13. **Hydraulics** Cox's formula for measuring the velocity of water escaping from a reservoir through a horizontal pipe is $4v^2 + 5v - 2 = \frac{1200HD}{L}$, where v represents the velocity of the water in feet per second, H the height of the reservoir in feet, D the diameter of the pipe in inches, and L the length of the pipe in feet. How fast is water flowing through a pipe 20 feet long with a diameter of 6 inches that is draining a swimming pool with a depth of 10 feet? Round your answer to the nearest tenth.

EXERCISES

Practice

Solve each equation by using the quadratic formula. Approximate irrational roots to the nearest hundredth.

14. $x^2 - 2x - 24 = 0$
15. $a^2 + 10a + 12 = 0$
16. $c^2 + 12c + 20 = 0$
17. $5y^2 - y - 4 = 0$
18. $r^2 + 25 = 0$
19. $3b^2 - 7b - 20 = 0$
20. $y^2 + 12y + 36 = 0$
21. $2r^2 + r - 14 = 0$
22. $2x^2 + 4x = 30$
23. $2x^2 - 28x + 98 = 0$
24. $24x^2 - 14x = 6$
25. $6x^2 + 15 = -19x$
26. $12x^2 = 48$
27. $x^2 + 6x = 36 + 6x$
28. $1.34a^2 - 1.1a = -1.02$
29. $3m^2 - 2m = 1$
30. $24a^2 - 2a = 15$
31. $2w^2 = -(7w + 3)$
32. $a^2 - \frac{3}{5}a + \frac{2}{25} = 0$
33. $-2x^2 + 0.7x = -0.3$
34. $2y^2 - \frac{5}{4}y = \frac{1}{2}$

Without graphing, determine the *x*-intercepts of the graph of each function to the nearest tenth.

35. $f(x) = 2x^2 - 5x + 2$
36. $f(x) = 4x^2 - 9x + 4$
37. $f(x) = 13x^2 - 16x - 4$

Use the quadratic formula to determine values for *a*, *b*, and *c* if the given numbers are solutions of a quadratic equation. Then write the equation.

38. $-1 \pm \sqrt{3}$ **39.** $\frac{-5 \pm \sqrt{2}}{2}$ **40.** $\frac{4 \pm \sqrt{29}}{2}$

Programming

41. The graphing calculator program at the right determines what type of solutions a quadratic equation will have and then prints decimal approximations of the solutions if they exist.

```
PROGRAM:SOLUTIONS
: Prompt A, B, C
: B²−4AC→D
: If D<0
: Then
: Disp "NO REAL SOLUTIONS"
: Stop
: End
: If D=0
: Then
: Disp "1 DISTINCT SOLUTION:"
  −B/2A
: Else
: Disp "2 REAL SOLUTIONS",
  (−B+√ D)/2A, (−B−√ D)/2A
: End
```

Use the program to find the solutions of each equation.

a. $x^2 - 11x + 10 = 0$
b. $3x^2 - 2x + 1 = 0$
c. $4x^2 + 4x + 1 = 0$
d. $7x^2 + 2x - 5 = 0$

Critical Thinking

42. The expression $b^2 - 4ac$ is called the **discriminant** of a quadratic equation. The discriminant can help you determine what type of solutions to expect when you solve a quadratic equation. Copy and complete the table below.

Equation	$x^2 - 4x + 1 = 0$	$x^2 + 6x + 11 = 0$	$x^2 - 4x + 4 = 0$
Value of the Discriminant			
Graph of the Equation			
Number of *x*-intercepts			
Number of Real Solutions			

43. Use the results of Examples 1–3 and the table above to describe the discriminant of a quadratic equation for each type of solution.

a. two irrational solutions
b. two noninteger rational solutions
c. two integral solutions
d. 1 distinct integral solution

Applications and Problem Solving

44. Government Between 1980 and 1993, the income (billions of dollars) received by the federal government can be modeled by the quadratic function $I(t) = 0.26t^2 + 49.94t + 511.4$, where t represents the number of years since 1980.

a. Determine the domain and range values for which this function makes sense.
b. Determine the income in 1993.
c. Assume that the pattern continues to hold. What is the projected federal income for the year 2000?
d. Determine in which year the federal income was \$1000 billion or \$1 trillion.

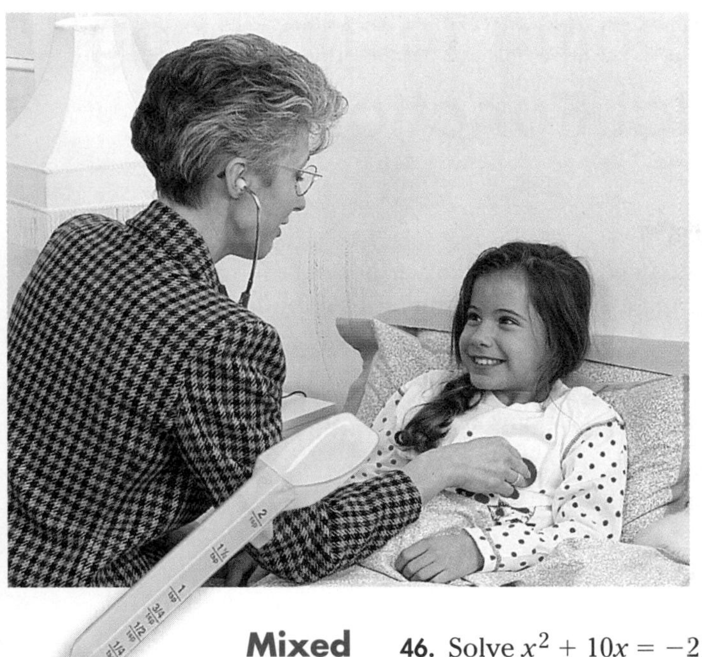

45. **Medicine** Two rules that govern the amount of medicine you should give a child if you know the adult dosage are Young's rule, $c = \frac{ad}{a + 12}$, and Cowling's rule, $c = \frac{(a + 1)d}{24}$. In both formulas, a represents the age of the child (in years), d represents the amount of the adult dosage, and c represents the amount of the child's dosage.
    a. The adult dosage for a drug is 30 mg/day. Calculate the dosage for a 6-year-old using each rule.
    b. Write a quadratic equation that represents the age(s) at which the two rules give the same dosage. Then solve the equation.

Mixed Review

46. Solve $x^2 + 10x = -21$ by graphing. (Lesson 11–2)
47. Solve $4s^2 = -36s$. (Lesson 10–6)
48. Find $3a^2(a - 4) + 6a(3a^2 + a - 7) - 4(a - 7)$. (Lesson 9–6)
49. **Number Theory** The sum of two numbers is 42. Their difference is 6. Find the numbers. (Lesson 8–3)
50. Graph the solution set for $b > 5$ or $b \leq 0$. (Lesson 7–4)
51. Determine if the following relation is a function. (Lesson 5–5)
    $\{(3, 2), (-3, -2), (-4, -2), (4, -2)\}$
52. Solve $3x + 8 = 2x - 4$. (Lesson 3–5)
53. **Weather** The temperature at 8:00 A.M. was 36°F. A cold front went through that evening, and by 3:00 A.M. the next day, the temperature had fallen 40°. What was the temperature at 3:00 A.M.? (Lesson 2–1)

SELF TEST

Write the equation of the axis of symmetry and find the coordinates of the vertex of the graph of each equation. Then graph the equation. (Lesson 11–1)

1. $y = x^2 - x - 6$
2. $y = 2x^2 + 3$
3. $y = -x^2 + 7$

4. **Physics** The height h in feet of an experimental rocket t seconds after blast-off is given by the formula $h = -16t^2 + 2320t + 125$. (Lesson 11–1)
    a. Approximately how long after blast-off does the rocket reach a height of 84,225 feet?
    b. How much longer from this height does it take the rocket to reach its maximum height?

Solve each equation by graphing. If integral roots cannot be found, state the consecutive integers between which the roots lie. (Lesson 11–2)

5. $x^2 = 81$
6. $4x^2 = 35 - 4x$
7. $6x^2 + 36 = 0$

Solve each equation by using the quadratic formula. Approximate irrational roots to the nearest hundredth. (Lesson 11–3)

8. $a^2 + 7a = -6$
9. $y^2 + 6y + 10 = 0$
10. $z^2 - 13z - 32 = 0$

11–4A Graphing Technology Exponential Functions

A Preview of Lesson 11–4

Graphing calculators can be used to graph many types of functions easily so patterns in the functions can be studied. This includes **exponential functions** of the form $y = a^x$, where $a > 0$ and $a \neq 1$.

Example **Graph each equation in the standard viewing window. Describe the graph.**

a. $y = 3^x$

Enter the equation in the Y= list.

Enter: Y= 3 ^ X,T,θ ZOOM 6

Notice that the graph increases rapidly as x becomes greater. The graph passes through the point at (0, 1). The domain of the function is all real numbers, and the range is all positive real numbers.

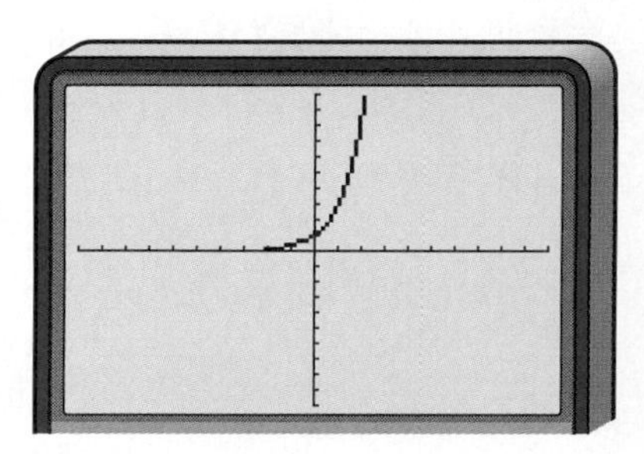

b. $y = \left(\frac{1}{3}\right)^x$

Enter: Y= (1 ÷ 3) ^ X,T,θ GRAPH

The graph decreases as x increases. The graph passes through the point at (0, 1). The domain is all real numbers, and the range is all positive real numbers.

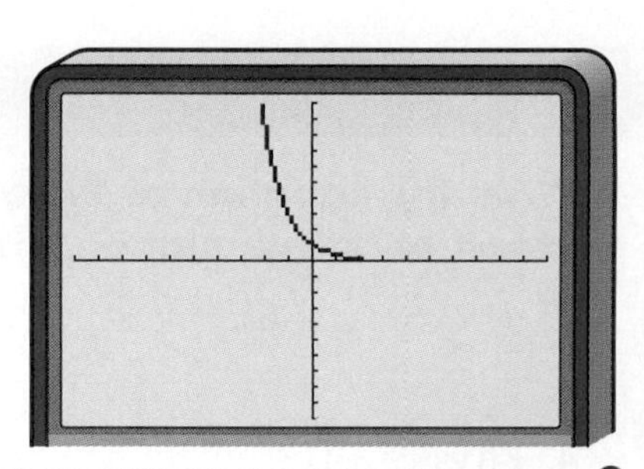

EXERCISES

Use a graphing calculator to graph each exponential equation. Sketch the graphs on a separate piece of paper.

1. $y = 2^x$ **2.** $y = 5^x$ **3.** $y = 0.1^x$

4. $y = \left(\frac{2}{3}\right)^x$ **5.** $y = 0.25^x$ **6.** $y = 1.6^x$

7. $y = 0.2^x$ **8.** $y = 0.5^{-x}$ **9.** $y = 10^x$

10. Solve $1.2^x = 10$ graphically. Explain how you solved the equation and write the solution accurately to the nearest hundredth.

11-4 Exponential Functions

What YOU'LL LEARN

- To graph exponential functions,
- to determine if a set of data displays exponential behavior, and
- to solve exponential equations.

Why IT'S IMPORTANT

You can use exponential equations to solve problems involving biology and archaeology.

Folklore

A wise man asked his ruler to provide rice for feeding his people. Rather than receiving a constant daily supply of rice, the wise man asked the ruler to give him 2 grains of rice for the first square on a chessboard, 4 grains of rice for the second, 8 grains of rice for the third, 16 grains of rice for the fourth, and so on, doubling the amount of rice with each square of the board. How many grains of rice will he receive for the last (64th) square on the chessboard?

You could make a table and look for a pattern to determine how many grains of rice he received for each square of the chessboard.

Square	Grains	Pattern	Square	Grains	Pattern
1	2	2^1	19	524,288	2^{19}
2	4	2^2	20	1,048,576	2^{20}
3	8	2^3	21	2,097,152	2^{21}
4	16	2^4	22	4,194,304	2^{22}
5	32	2^5	23	8,388,608	2^{23}
6	64	2^6	24	16,777,216	2^{24}
7	128	2^7	25	33,554,432	2^{25}
8	256	2^8	26	67,108,864	2^{26}
9	512	2^9	27	134,217,728	2^{27}
10	1024	2^{10}	28	268,435,456	2^{28}
11	2048	2^{11}	29	536,870,912	2^{29}
12	4096	2^{12}	30	1,073,741,824	2^{30}
13	8192	2^{13}	31	2,147,483,648	2^{31}
14	16,384	2^{14}	32	4,294,967,296	2^{32}
15	32,768	2^{15}	33	8,589,934,592	2^{33}
16	65,536	2^{16}	34	17,179,869,184	2^{34}
17	131,072	2^{17}	35	34,359,738,368	2^{35}
18	262,144	2^{18}	36	68,719,476,736	2^{36}

FYI

There are approximately 24,000 grains of rice in 1 pound of rice.

Notice that when only 36 of the 64 squares are calculated, there are over 68 billion grains of rice. How many grains would there be for square 64? *You will answer this question in Exercise 1.*

Study the pattern column. Notice that the exponent number matches the number of the square on the chessboard. So we can write an equation to describe y, the number of grains of rice for any given square x as $y = 2^x$. This type of function, in which the variable is the exponent, is called an **exponential function.**

Definition of Exponential Function	**An exponential function is a function that can be described by an equation of the form $y = a^x$, where $a > 0$ and $a \neq 1$.**

You can use paper-folding to illustrate an exponential function.

Modeling an Exponential Function

Materials: ☐ large piece of paper

Your Turn

a. Fold a large rectangular piece of paper in half. Unfold it and record how many sections are formed by the creases. Refold the paper.

b. Fold the paper in half again. Record how many sections are formed by the creases. Refold the paper.

c. Continue folding in half and recording the number of sections until you can no longer fold the paper.

d. How many folds could you make?

e. How many sections were formed?

f. What exponential function is modeled by the folds and sections created?

As with other functions, you can use ordered pairs to graph an exponential function. Use a table of values and a calculator to find ordered pairs that satisfy $y = 2^x$. While the negative values of x have no meaning in the rice problem, they should be included in the graph of the function. Connect the points to form a smooth curve.

x	y
−5	0.03125
−4	0.0625
−3	0.125
−2	0.25
−1	0.5
0	1
1	2
2	4
3	8
4	16

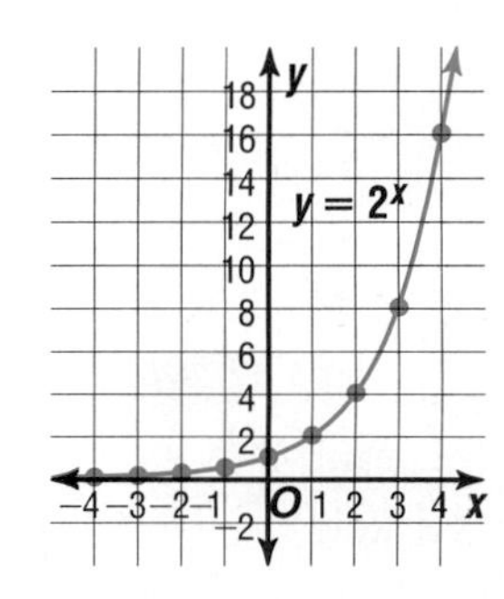

Notice that the graph has a y-intercept of 1. Does it have an x-intercept? *You will answer this question in Exercise 2.*

The graph shown above represents all real values of x and their corresponding values of y for $y = 2^x$.

You can use a scientific calculator to help you find ordered pairs to graph other exponential functions. For example, suppose $y = 3^x$ and $x = -2$.

Enter: 3 [y^x] 2 [+/−] [=] 0.111111111

Example 1 **Graph each function. State the y-intercept of each graph.**

a. $y = 3^x$

x	y
−3	0.037
−2	0.111
−1	0.333
0	1
1	3
2	9
3	27

The y-intercept is 1.

b. $y = \left(\frac{1}{3}\right)^x$

x	y
−3	27
−2	9
−1	3
0	1
1	0.333
2	0.111
3	0.037
4	0.012

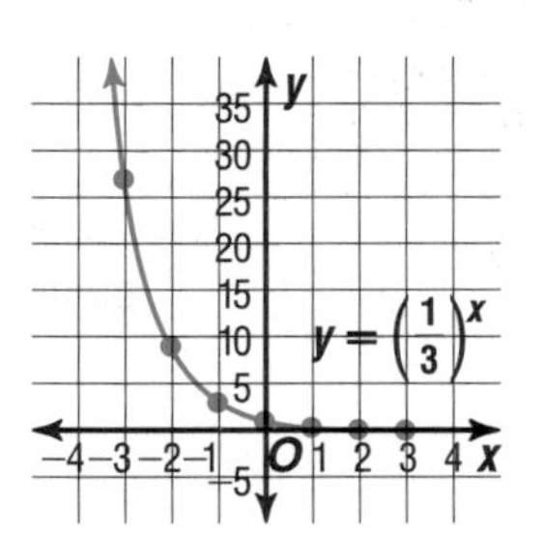

The y-intercept is 1.

Notice that the graph of $y = \left(\frac{1}{a}\right)^x$ decreases rapidly as x increases.

c. $y = 3^x - 7$

x	y
−3	−6.96
−2	−6.89
−1	−6.67
0	−6
1	−4
2	2
3	20
4	74

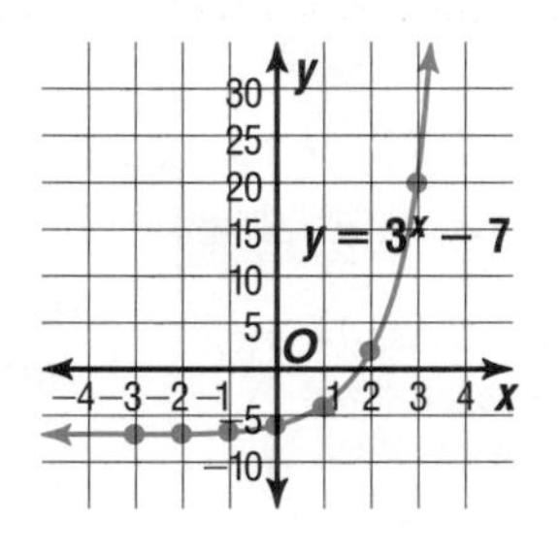

The y-intercept is −6.

LOOK BACK

You can refer to Lesson 10-1 for more information on factors.

How do you know if a set of data is exponential? One method is to observe the shape of the graph. But the graph of an exponential function may resemble part of the graph of a quadratic function. Another way is to **look for a pattern** in the data.

Example  **Determine whether each set of data displays exponential behavior.**

PROBLEM SOLVING

Look for a Pattern

You could also use a graphing calculator to make a scatter plot of the data to observe the patterns.

a.

x	0	5	10	15	20	25
y	800	400	200	100	50	25

Method 1: Look for a Pattern

The domain values are at regular intervals of 5. Let's see if there is a common factor among the range values.

800 → 400 → 200 → 100 → 50 → 25, each step $\times\frac{1}{2}$

Since the domain values are at regular intervals and the range values have a common factor, the data are probably exponential. The equation for the data probably involves $\left(\frac{1}{2}\right)^x$.

Method 2: Graph the data.

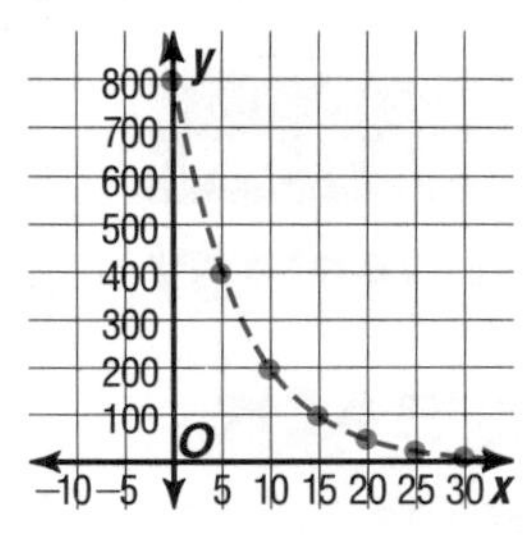

The graph shows a rapidly decreasing value of y as x increases. This is a characteristic of exponential behavior.

b.

x	0	5	10	15	20	25
y	3	6	9	12	15	18

Method 1: Look for a Pattern

The domain values are at regular intervals of 5. The range values have a common difference of 3.

3 6 9 12 15 18

+3 +3 +3 +3 +3

These data do not display exponential behavior, but rather linear behavior.

Method 2: Graph the data.

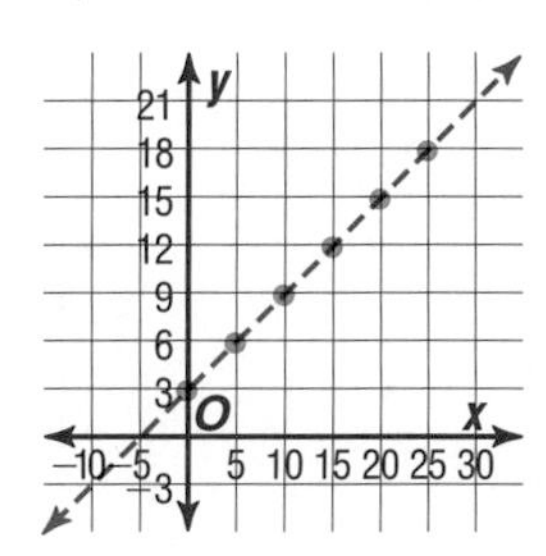

This is the graph of a line, not an exponential function.

F Y I

Radiocarbon, or carbon-14, decays into nitrogen. It was first used in the late 1940s by Williard F. Libby to determine the age of an artifact. It has been used to date objects as old as 50,000 years.

Exponential functions are often used to describe real-life situations. In the late 1890s, Henri Becquerel discovered that fossils contain naturally-occurring radioactive atoms of carbon-14. When an organism is alive, it absorbs carbon-14 from the sun. When the organism dies, no more carbon-14 is absorbed, and the initial amount of carbon-14 begins to decay gradually into other elements.

The **half-life** of a radioactive element is defined as the time that it takes for one-half a quantity of the element to decay. Radioactive carbon-14 has a half-life of 5730 years. This means that after 5730 years, half the original amount of carbon-14 has decayed. In another 5730 years, half of the remaining half will decay, and so on. This pattern of decay can be described by $a = 0.5^t$, where a is the decay factor and t is the number of half-lives.

Study the chart to see how 64 grams of carbon-14 decays over several half-lives. Notice its relationship to the exponential function $a = 0.5^t$.

Half-life	Grams of Carbon-14 left
0	64 $= 64 \times 0.5^0$
1	32 $= 64 \times 0.5^1$
2	16 $= 64 \times 0.5^2$
3	8 $= 64 \times 0.5^3$
4	4 $= 64 \times 0.5^4$
5	2 $= 64 \times 0.5^5$
6	1 $= 64 \times 0.5^6$
7	0.5 $= 64 \times 0.5^7$
8	0.25 $= 64 \times 0.5^8$
9	0.125 $= 64 \times 0.5^9$

Example 3

APPLICATION

Archaeology

If the original concentration of carbon-14 in a living organism was 256 grams, determine the concentration of carbon-14 remaining in a fossil for each situation.

a. The organism lived 1000 years ago.

b. The organism lived 10,000 years ago.

LOOK BACK

You can refer to Lesson 9-3 for more information on numbers written in scientific notation.

a. First determine how many carbon-14 half-lives there are in 1 thousand years.

$\frac{1000}{5730} \approx 0.1745$ half-lives

Then determine the value of the disintegration factor when $t = 0.1745$.

$a = 0.5^t$

$= 0.5^{0.1745}$

≈ 0.886 *Use a calculator.*

Multiply the original amount of carbon-14 by this factor.
256 grams $\times$ 0.886 = 226.8 grams

b. Find how many half-lives there are in 10,000 years.

$\frac{10{,}000}{5730} \approx 1.75$ half-lives

Determine the value of the disintegration factor when $t = 1.75$.

$a = 0.5^{1.75}$

≈ 0.297

Multiply the original amount of carbon-14 by this factor.
256 grams $\times$ 0.297 $\approx$ 76.0 grams

You can use algebra to solve equations involving exponential expressions by using the following rule.

Property of Equality for Exponential Functions

Suppose a is a positive number other than 1. Then $a^{x_1} = a^{x_2}$ if and only if $x_1 = x_2$.

The skills you learned when solving quadratic equations can be helpful when solving some exponential equations.

Example 4 **Solve $64^3 = 4^{x^2}$.**

LOOK BACK

You can refer to Lesson 9-1 for more information about properties of exponents.

Explore The two quantities do not have the same base, or value for a. However, 64 is a power of 4.

Plan In order to use the property of equality, we must rewrite the terms so that they have the same base. Then we can use the property of equality for exponential functions.

Solve

$64^3 = 4^{x^2}$

$(4^3)^3 = 4^{x^2}$ *$64 = 4 \cdot 4 \cdot 4$ or 4^3*

$4^9 = 4^{x^2}$ *Product of powers*

$9 = x^2$ *Property of equality for exponential functions*

$x^2 - 9 = 0$ *Rewrite the equation in standard form.*

$(x + 3)(x - 3) = 0$ *Factor.*

$x + 3 = 0$ $\quad$ $x - 3 = 0$ *Zero product property*

$x = -3$ $\quad$ $x = 3$

Examine Check each solution by substituting it into the original equation.

$64^3 = 4^{x^2}$ $\quad$ $64^3 = 4^{x^2}$

$64^3 \stackrel{?}{=} 4^{3^2}$ *$x = 3$* $\quad$ $64^3 \stackrel{?}{=} 4^{(-3)^2}$ *$x = -3$*

$262{,}144 \stackrel{?}{=} 4^9$ $\quad$ $262{,}144 \stackrel{?}{=} 4^9$

$262{,}144 = 262{,}144$ ✔ $\quad$ $262{,}144 = 262{,}144$ ✔

The solutions are 3 and -3.

CHECK FOR UNDERSTANDING

Communicating Mathematics

Study the lesson. Then complete the following.

1. **Refer** to the application at the beginning of the lesson.
   a. Use a calculator to determine how many grains there would be for the 64th square of the chessboard.
   b. How many tons of rice is this? (*Hint:* Recall that 1 T = 2000 lb.)
2. a. **Determine** whether the graph of $y = 2^x$ has an x-intercept.
   b. **Describe** your method.
   c. Is this true of all exponential functions?
3. a. **Determine** whether the graph of $y = 2^x$ has a vertex.
   b. **Describe** your method for answering part a.
   c. Is this true for all exponential functions?
4. **Explain** why $a \neq 1$ in the definitions and properties involving exponential functions.
5. **Write** a paragraph explaining why you think a graphing calculator might be a good tool to have when studying exponential functions.

6. Refer to the Modeling Mathematics activity on page 636.
   a. The area of the large rectangle is 1. Find the area of each section after each set of folds. Record your findings in a table.
   b. Compare the number of folds to the area of each section. What pattern do you see?
   c. Write an exponential function relating the number of folds to the area of each section.

Guided Practice

Use a calculator to determine the approximate value of each expression to the nearest hundredth.

7. $3^{1.5}$
8. $3^{-0.9}$
9. $3^{2.3}$

Graph each function. State the y-intercept.

10. $y = 0.5^x$
11. $y = 2^x + 6$
12. Determine if the data in the table below displays exponential behavior. Describe the behavior.

x	0	1	2	3	4	5
y	1	6	36	216	1296	7776

Solve each equation.

13. $5^{3y+4} = 5^y$
14. $2^5 = 2^{2x-1}$
15. $3^x = 9^{x+1}$
16. **Biology** Suppose $B = 100 \cdot 2^t$ represents the number of bacteria B in a petri dish after t hours if you began with 100 bacteria. How long would it take to obtain 1000 bacteria?

EXERCISES

Practice

Use a calculator to determine the approximate value of each expression to the nearest hundredth.

17. $4^{1.7}$
18. $10^{-0.5}$
19. $\left(\frac{2}{3}\right)^{-1.2}$
20. $\left(\frac{1}{3}\right)^{4.1}$
21. $50(3^{-0.6})$
22. $10(3^{-1.8})$
23. $0.4(3^{0.7})$
24. $20(0.25^{-2.7})$

Graph each function. State the y-intercept.

25. $y = 2^x + 4$
26. $y = 2^{x+4}$
27. $y = 3\left(\frac{1}{3}\right)^x$
28. $y = 2 \cdot 3^x$
29. $y = 4^x$
30. $y = \left(\frac{1}{4}\right)^x$

Determine if the data in each table display exponential behavior. Explain why or why not.

31.

x	y
−2	−5
−1	−2
0	1
1	4

32.

x	y
0	1
1	0.5
2	0.25
3	0.125

33.

x	y
−1	−0.5
0	1.0
1	−2.0
2	4.0

Solve each equation.

34. $5^{3x} = 5^{-3}$
35. $2^{x+3} = 2^{-4}$
36. $5^x = 5^{3x+1}$
37. $10^x = 0.001$
38. $2^{2x} = \frac{1}{8}$
39. $\left(\frac{1}{6}\right)^q = 6^{q-6}$
40. $16^{x-1} = 64^x$
41. $81^x = 9^{x^2-3}$
42. $4^{x^2-2x} = 8^{x^2+1}$

Graphing Calculator

As with linear graphs and quadratic graphs, exponential graphs can form families of graphs. Graph each set of equations on the same screen. Sketch the graphs and discuss any similarities or differences.

43. $y = 3^x$
$y = 3^{x+4}$
$y = 3^{x-2}$

44. $y = \left(\frac{1}{3}\right)^x$
$y = \left(\frac{1}{3}\right)^x + 5$
$y = \left(\frac{1}{3}\right)^x - 3$

45. $y = 2^x$
$y = 2^{x-7}$
$y = 2^{x-2}$

46. $y = 6^x$
$y = 6^{3x}$
$y = 6^{8x}$

47. a. Use a graphing calculator to solve $2.5^x = 10$. Write the solution to the nearest hundredth.
b. Explain why you cannot solve this equation algebraically like you did the equation in Exercise 41.

Critical Thinking

48. Refer to the equations in Example 1. Use a calculator to find additional values to complete the following.
a. For $y = 3^x$, as x decreases, the value of y approaches what number?
b. For $y = \left(\frac{1}{3}\right)^x$, as x increases, the value of y approaches what number?
c. For $y = 3^x - 7$, as x decreases, the value of y approaches what number?
d. For any equation $y = a^x + c$, where $a > 1$, what value does y approach as x decreases?
e. For any equation $y = a^x + c$, where $0 < a < 1$, what value does y approach as x increases?

Applications and Problem Solving

49. **Biology** Mitosis is a process of cell reproduction in which one cell divides into two identical cells. *E. coli* is a fast-growing bacteria that is often responsible for food poisoning in uncooked meat. It can reproduce itself in 15 minutes. If you begin with 100 *e. coli* bacteria, how many bacteria will there be in 1 hour?

50. **Currency** In the United States between 1910 and 1994, the amount of currency in circulation $M(t)$ (billions of dollars) can be approximated by the function $M(t) = 2.08(1.06)^t$, where t represents the number of years since 1910.
    a. Determine $M(t)$ for the years 1920, 1950, 1980, and 2000.
    b. Determine the amount of currency in circulation in the years 1920, 1950, 1980, and 2000.

Mixed Review

51. Solve $2x^2 + 3 = -7x$ by using the quadratic formula. Check your solution by factoring. (Lesson 11–3)

52. **Geometry** A rectangle has an area of $(16p^2 - 40pr + 25r^2)$ square kilometers. (Lesson 10–5)
    a. Find the dimensions of the rectangle.
    b. Sketch the rectangle, labeling its dimensions.

53. Factor $\frac{4}{5}a^2b - \frac{3}{5}ab^2 - \frac{1}{5}ab$. (Lesson 10–2)

54. **City Planning** A section of Lithopolis is shaped like a trapezoid with an area of 81 square miles. The distance between Union Street and Lee Street is 9 miles. The length of Union Street is 14 miles less than 3 times the length of Lee Street. Find the length of Lee Street. Use $A = \frac{h(a + b)}{2}$. (Lesson 9–8)

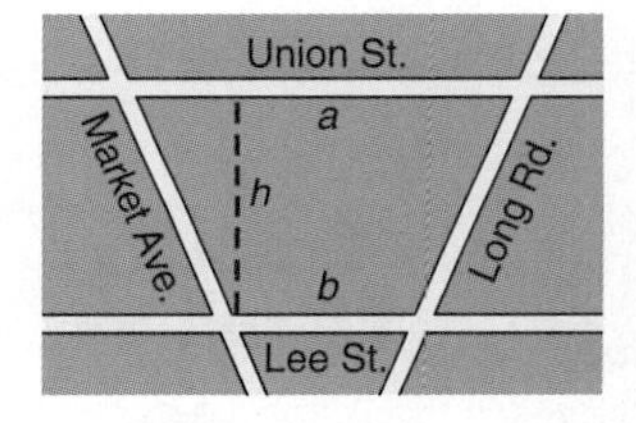

55. **Basketball** On December 13, 1983, the Denver Nuggets and the Detroit Pistons broke the record for the highest score in a professional basketball game. The two teams scored a total of 370 points. If the Nuggets scored 2 points less than the Pistons, what was the Nuggets' final score? (Lesson 9–5)

56. Solve the system of inequalities by graphing. (Lesson 8–5)
    $y \leq 2x + 2$
    $y \geq -x - 1$

57. **Astronomy** The table at the right shows the relationship between the distance from the sun, measured in millions of miles, and the time to complete an orbit, measured in Earth years. (Lesson 6–3)
    a. Make a scatter plot of these data.
    b. Draw a best-fit line and write an equation for the line.
    c. Suppose a tenth planet was discovered at a distance of 4.1 billion miles from the sun. Use the scatter plot to estimate how long it would take it to orbit the sun.

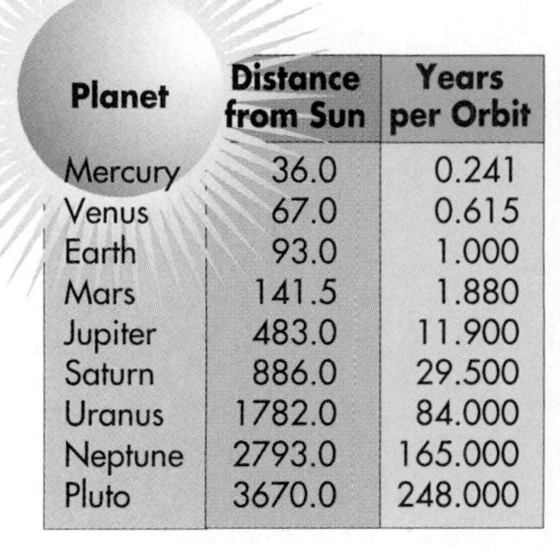

Planet	Distance from Sun	Years per Orbit
Mercury	36.0	0.241
Venus	67.0	0.615
Earth	93.0	1.000
Mars	141.5	1.880
Jupiter	483.0	11.900
Saturn	886.0	29.500
Uranus	1782.0	84.000
Neptune	2793.0	165.000
Pluto	3670.0	248.000

fabulous FIRSTS

Clair C. Patterson (1922–1995)

Geochemist Clair Patterson, a professor at the California Institute of Technology, was the first scientist to establish Earth's age based on analysis of radioactive isotopes and their daughter products. In 1953, he established Earth's age to be 4.6 billion years.

58. **Geometry** Triangle ABC is similar to triangle ADE in the figure at the right. Find the value of s. (Lesson 4–2)

59. Find $(-2)(3)(-10)$. (Lesson 2–6)

11-5 Growth and Decay

What YOU'LL LEARN

- To solve problems involving growth and decay.

Why IT'S IMPORTANT

You can solve growth and decay problems to learn more about demographics and energy.

Demographics

How quickly has the population of your state grown in the 20th century? Both California and Nebraska have grown in population during this century by a constant percent. The graph at the right shows the population of each state where t represents the number of years since 1900. Since 1900, the population in Nebraska has grown at a rate of 0.4% per year, while the population of California has grown at a rate of 3% per year.

The exponential functions that model the growth of each state are given below.

California: $y = 1.77(1.03)^t$ *The population in 1900 was 1.77 million.*

Nebraska: $y = 1.14(1.004)^t$ *The population in 1900 was 1.14 million.*

Which state exhibits the more rapid growth? What will the population of each state be in the year 2000? *You will answer these questions in Exercise 1.*

The equations for the two states' populations are variations of the equation $y = C(1 + r)^t$. This is the **general equation for exponential growth** in which the initial amount C increases by the same percent r over a given period of time t. This equation can be applied to many kinds of growth applications.

One of these applications is monetary growth. When solving problems involving compound interest, the growth equation becomes $A = P\left(1 + \frac{r}{n}\right)^{nt}$, where A is the amount of the investment over a period of time, P is the principal (initial amount of the investment), r is the annual rate of interest expressed as a decimal, n is the number of times that the interest is compounded each year, and t is the number of years that the money is invested.

CAREER CHOICES

Urban, or **city, planners** use statistics on demographics, traffic patterns, and economics to help officials make decisions on social, economic, and environmental issues in their communities. A bachelors degree in urban or regional planning is required, and a masters degree in civil engineering or planning is helpful for advancement.

For further information, contact:

American Planning Association
1776 Massachusetts Ave., NW
Washington, DC 20036

Example 1

Finance

In the spring of 1994, Mr. and Mrs. Mitzu had \$10,000 they wished to place in a bank certificate of deposit toward their retirement in the year 2004. The interest rate at that time was 2.5% compounded monthly. However, there were seven increases in the prime rate in a year so that in the spring of 1995, the interest rate had risen to 5.5%.

a. Determine the amount of their investment after 10 years if they invested the principal and let it remain at the 2.5% rate.

b. Determine the amount of the investment after 9 years if they waited and invested the principal at the 5.5% rate.

c. What are the best options for their investment?

(continued on the next page)

a. The interest rate r as a decimal is 0.025 and $n = 12$. *Why?*

$$A = P\left(1 + \frac{r}{n}\right)^{nt}$$

$$= 10{,}000\left(1 + \frac{0.025}{12}\right)^{12 \cdot 10} \quad P = 10{,}000,\ r = 0.025,\ n = 12, \text{ and } t = 10$$

Use a calculator.

Enter: 10000 [×] [(] 1 [+] [(] .025  12 

[y^x] 120 [=] 12836.91542

The amount of the account after 10 years at 2.5% is about \$12,836.92.

b. The interest rate as a decimal is 0.055, and t is 9 years.

$$A = P\left(1 + \frac{r}{n}\right)^{nt}$$

$$= 10{,}000\left(1 + \frac{0.055}{12}\right)^{12 \cdot 9} \quad P = 10{,}000,\ r = 0.055,\ n = 12, \text{ and } t = 9$$

$$\approx 16{,}386.44$$

c. If they waited until the rate went up, they would have had more money than leaving the money at the initial rate. However, if it is possible to reinvest the money each year, they could have invested the money for one year at 2.5% and then reinvested it the next year at 5.5%. That would yield even more money for their retirement in 2004.

A variation of the growth equation can be used as the **general equation for exponential decay.** In the formula $A = C(1 - r)^t$, A represents the final amount, C is the initial amount, r is the rate of decay, and t denotes time.

Example 2

APPLICATION

Demographics

The cities listed at the right have experienced declining populations since 1970.

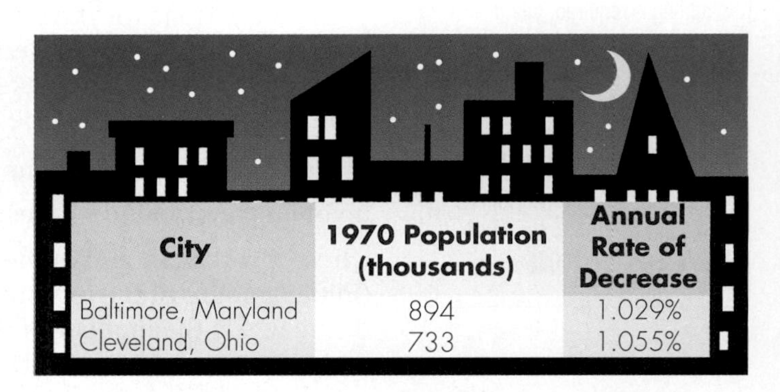

City	1970 Population (thousands)	Annual Rate of Decrease
Baltimore, Maryland	894	1.029%
Cleveland, Ohio	733	1.055%

a. If t represents the number of years since 1970 and C represents the 1970 population, write an exponential decay equation for each city.

b. Assume that each city maintains the same decrease rate into the next century. Calculate the population of each city in the year 2070.

c. How do the projected populations of the cities in 2070 compare to the actual populations in 1970?

a. Baltimore

$C = 894$, $r = 0.01029$

$A = C(1 - r)^t$

$A = 894(1 - 0.01029)^t$

$A = 894(0.98971)^t$

Cleveland

$C = 733$, $r = 0.01055$

$A = C(1 - r)^t$

$A = 733(1 - 0.01055)^t$

$A = 733(0.98945)^t$

b. The year 2070 is 100 years later. Evaluate each decay equation for $t = 100$.

Baltimore	**Cleveland**
$A = 894(0.98971)^t$	$A = 733(0.98945)^t$
$A = 894(0.98971)^{100}$	$A = 733(0.98945)^{100}$
$A \approx 317.78$	$A \approx 253.80$

c. In 1970, the difference in the populations of Baltimore and Cleveland was 161,000. In 2070, the difference is only about 64,000. If the populations would continue to decrease at the same rates, they would grow even closer together.

Sometimes items decrease in value. For example, as equipment gets older, it *depreciates*. You can use the decay formula to determine the value of an item at a given time.

Example 3

Consumerism

Hogan Blackburn is considering the purchase of a new car. He is faced with the decision to lease or buy. If he leases the car, he pays \$369 a month for 2 years and then has the option to buy the car for \$13,642. The price of the car now is \$16,893.

a. If the car depreciates at 17% per year, how will the depreciated price compare with the buyout price of the lease?

b. At the end of the lease, the dealer offers a loan for 3 years at a rate of \$435.79 per month. If he buys the car now, he will pay \$464.56 monthly for 4 years. Which is the best way to go if he plans to keep the car for at least 5 years?

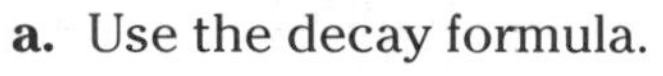

a. Use the decay formula.

$A = C(1 - r)^t$

$= 16{,}893(1 - 0.17)^2$ *$r = 0.17$, $C = 16{,}893$, and $t = 2$*

$= 11{,}637.59$

The depreciated value is \$2000 less than the buyout price.

b. Calculate the cost of each possibility over a 5-year period.

Lease and Buy: \$369(24 months) + \$435.79(36 months) = \$24,544.44

Buy: \$464.56(48 months) = \$22,298.88

The best way to go may depend on Hogan's financial status.

- Overall, the buy option costs less and lasts only 4 years while the lease and buy option costs more and lasts for 5 years.
- However, if Hogan really wants a new car and cannot afford the larger monthly payment now, the lease option may be best. However, he will eventually have to make the larger payment if he keeps the car.

You can use a spreadsheet to quickly evaluate different values for any exponential growth or decay formula.

EXPLORATION — SPREADSHEETS

Recall that in a spreadsheet you can refer to each cell by its name (A1 means column A, row 1). Suppose we set up a spreadsheet to evaluate the formula for monetary exponential growth.

- You can enter labels into the first row to identify the data in each column. Let column A contain the values of P, column B the values of r, column C the values of n, column D the value of t, and column E the values of A.

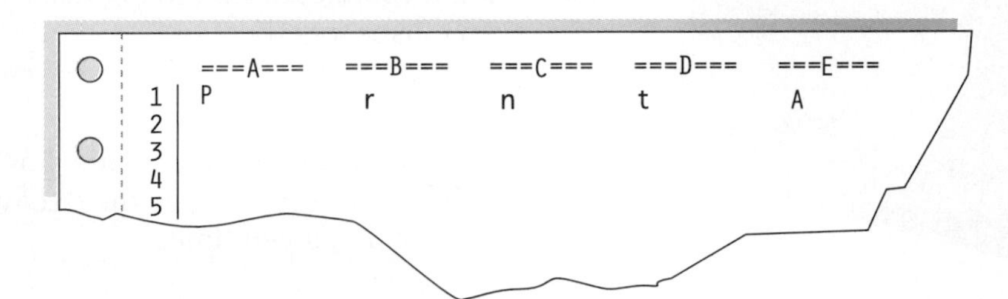

- You can enter a formula into cell E2 to evaluate the formula given the values entered into cells A2 through D2. The formula is A2 *(1 + (B2/C2))^(C2*D2).
- Copy this formula into the other cells in column E.

Your Turn

a. Use the spreadsheet to evaluate the data in Example 1.

b. How would you change the spreadsheet to evaluate the general formula for decay?

LOOK BACK

Refer to the Exploration in Lesson 6-1 for more information about spreadsheets.

CHECK FOR UNDERSTANDING

Communicating Mathematics

Study the lesson. Then complete the following.

1. Refer to the application at the beginning of the lesson.
   a. Which state exhibits more rapid growth?
   b. What will the population of each state be in the year 2000?
2. **Explain** how you could tell the difference between an exponential graph that shows 0.7% growth versus one that shows 7% growth.

Guided Practice

3. **State** the value of n in the monetary growth formula for each period of compounded interest.
   a. annually **b.** semi-annually **c.** quarterly **d.** daily

Determine whether each exponential equation represents growth or decay.

4. $y = 10(1.03)^x$
5. $y = 10(0.50)^x$
6. $y = 10(0.75)^x$

7. **History** In 1626, Peter Minuit, governor of the colony of New Netherland, bought the island of Manhattan from the Indians for beads, cloth, and trinkets worth 60 Dutch guilders ($24). If that $24 had been invested at 6% per year compounded annually, how much money would there be in the year 2000?

8. **Farming** Many self-employed individuals such as farmers can depreciate the value of the machinery they buy as a part of their income tax returns. Suppose a tractor valued at $50,000 depreciates 10% per year. Make a table to determine in how many years the value of the tractor would be less than $25,000.

EXERCISES

Applications and Problem Solving

Finance Determine the final amount from the investment for each situation.

	Initial Amount	Annual Interest Rate	Time	Type of Compounding
9.	$400	7.25%	7 years	quarterly
10.	$500	5.75%	25 years	monthly
11.	$10,000	6.125%	18 months	daily
12.	$250	10.3%	40 years	monthly

13. **Demographics** The population in 1994 and the growth rate for four countries is given below.

Country	Continent	Growth Rate	1994 Population
Ethiopia	Africa	2.5%	58.7 million
India	Asia	1.9%	919.9 million
Colombia	S. America	2.1%	35.6 million
Singapore	Asia	1.3%	2.9 million

   a. Write an exponential equation for each country's growth.
   b. Compute the estimated population for each country in the year 2000.
   c. Exclude the data for India. Make a double-bar graph to show how the population in the other three countries in 1994 compares with the estimate for the year 2000.

14. **Energy** The Environmental Protection Agency (EPA) has called for businesses to find cleaner sources of energy as we approach the 21st century. Coal is not considered to be a clean source of energy. In 1950, the use of coal by residential and commercial users was 114.6 million tons. Since then, the use of coal has decreased by 6.6% per year.
   a. Write an equation to represent the use of coal since 1950.
   b. Suppose the use of coal continues to decrease at the same rate. Use a calculator to estimate in what year the use of coal will end.

15. **Insurance** The total amount of life insurance sold in 1950 was \$231.5 billion. Since then, the annual increase has been estimated at 9.63%.
    a. Write an equation to represent the amount of insurance sold annually since 1950.
    b. Find the estimated amount of life insurance that will be sold in the year 2010.

16. **Wildlife** In 1980, there were 1.2 million elephants living in Africa. Because the natural grazing lands for the elephant are disappearing due to increased population and cultivation of the land, the number of elephants in Africa has decreased by about 6.8% per year.
    a. Write an equation to represent the population of elephants in Africa.
    b. In what year did the population of African elephants drop to less than half of the number in 1980?
    c. What factors might affect the rate of decline in elephant population?

17. **Savings** Sheena is investing her \$5000 inheritance in a saving certificate that matures in 4 years. The interest rate is 8.25% compounded quarterly.
    a. Determine the balance in the account after 4 years.
    b. Her friend, LaDonna, invests the same amount of money at the same interest rate but her bank compounds daily. Determine how much she will have after 4 years.
    c. What is the difference in the amount Sheena and LaDonna have after 4 years?
    d. Which type of compounding appears to be more profitable?

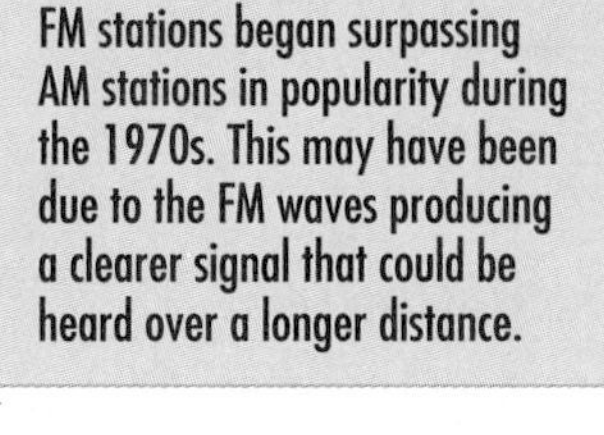

FYI

FM stations began surpassing AM stations in popularity during the 1970s. This may have been due to the FM waves producing a clearer signal that could be heard over a longer distance.

18. **Radio** FM broadcast frequencies range from 88 to 108 MHz in tenths. Before digital displays existed, you turned a knob and a bar slide from across the display or *dial* to find the station you wanted to hear. The equation that relates the MHz reading to position of the slide is $f(d) = 88(1.0137)^d$, where d is the distance from the left of the dial. Suppose someone didn't know the number of their favorite station, but knew it was about halfway across the dial. If the dial is 15 cm long, what station is their favorite ?

19. **Population** Research to find the population in your community for the past 50 years.
    a. Make a graph to show the population change.
    b. Does the population of your community show growth or decay?
    c. Estimate what the population of your community will be when you are 50 years old.

Graphing Calculator

20. **Radioactivity** A formula for examining the decay of radioactive materials is $y = Ne^{kt}$, where N is the beginning amount in grams, $e \approx 2.72$, k is a negative constant for the substance, and t is the number of years. Use a graphing calculator to estimate each of the following.
    a. How long will it take 250 grams of a radioactive substance to reduce to 50 grams if $k = -0.08042$?
    b. In 10 years, 200 grams of a radioactive substance is reduced to 100 grams. Find an estimate for the constant k for this substance.

Critical Thinking

21. The general exponential formula for growth or decay is $y = Ca^x$. Determine what values of a describe growth or decay. What does x usually represent?

22. Consider the equation $y = C(1 + r)^x$. How do you determine the y-intercept of the graph of this equation without graphing or evaluating the function for values of x?

Mixed Review

23. Solve $3^y = 3^{3y+1}$. (Lesson 11–4)

24. Find $(n^2 + 5n + 3) - (2n^2 + 8n + 8)$. (Lesson 9–5)

25. Simplify $\frac{-6r^3s^5}{18r^{-7}s^5t^{-2}}$. (Lesson 9–2)

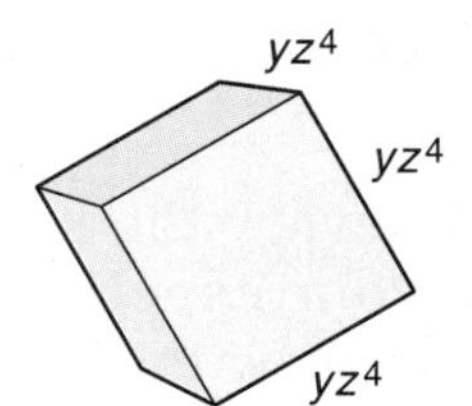

26. **Geometry** Find the volume of the cube shown at the right. (Lesson 9–1)

27. **Geometry** Write the equation of a line that passes through $(-2, 7)$ and is perpendicular to the line whose equation is $2x - 5y = 3$. Use slope-intercept form. (Lesson 6–6)

Minimizing Computers

The following excerpt appeared in an article in *The New York Times* on November 22, 1994.

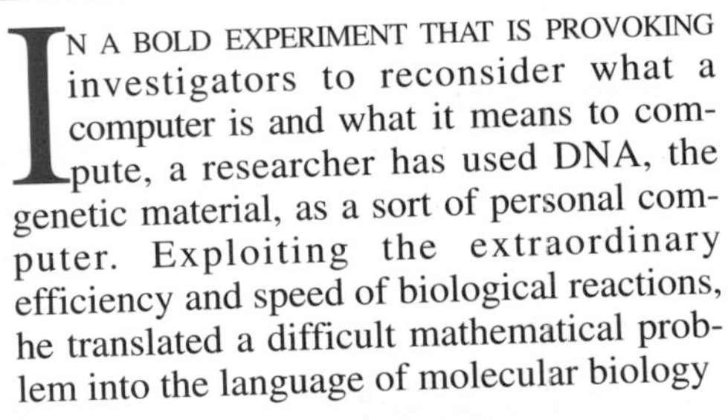

In a bold experiment that is provoking investigators to reconsider what a computer is and what it means to compute, a researcher has used DNA, the genetic material, as a sort of personal computer. Exploiting the extraordinary efficiency and speed of biological reactions, he translated a difficult mathematical problem into the language of molecular biology and solved it by carrying out a reaction in one-fiftieth of a teaspoon of solution in a test tube....Molecular computers can perform more than a trillion operations per second, which makes them a thousand times as fast as the fastest supercomputer. They are a billion times as energy efficient as conventional computers. And they can store information in a trillionth of the space required by ordinary computers. ■

1. Does the idea of a molecular computer in a test tube surprise you? Why or why not?
2. If biological systems can have computational abilities, what effects might this have on computer scientists, programmers, and mathematicians?
3. One class of problems that molecular computers could help solve involves finding one desired solution or path out of a huge number of possibilities. How could this be used if the problem was a defective gene causing the rate of a deadly disease to grow exponentially?

CLOSING THE

In·ves·ti·ga·tion

the BRICKYARD

Refer to the Investigation on pages 554–555.

Review the knowledge you have gained from your experiments working with the bricks. Review the instructions given to you by your manager as you begin to close this Investigation.

> Please create several patio designs that will utilize these bricks. Submit at least three different plans, explaining the materials required for each patio. I am anxious to see the different ways in which these bricks can be arranged to form rectangular patios. Is there a general formula or pattern we can use to design these in the future? I look forward to your report on helping us solve our inventory problems.

Analyze

You have conducted experiments and organized your data in various ways. It is now time to analyze your findings and state your conclusions.

1 Look over your data and complete a chart like the one on page 555 for each design. Use actual measurements.

2 What information does this chart reflect? Does it give information that can be used to generalize a method for forming future brick patio patterns? Explain.

PORTFOLIO ASSESSMENT

You may want to keep your work on this Investigation in your portfolio.

Write

The report to your manager should explain your process for investigating these rectangular brick patterns and what you found from your investigations.

3 What size rectangular brick patterns are possible? Draw sketches of the possible patterns. Describe the numbers of bricks used and include the dimensions, perimeter, and area of each pattern in terms of x and y.

4 Write procedures or generalizations that may be followed to find rectangular patterns in the following situations.

- You have a certain number of large squares and small squares. How many rectangular tiles are necessary to create a rectangular pattern?
- How can you find the dimensions of a pattern given the type and number of bricks available?
- How can you find the different possible patterns for any set number of bricks?
- How do you know that no possible rectangular pattern can be made given a set of the three types of bricks?

5 Summarize your findings and give recommendations for possible brick patterns. Explain methods for exploring more patterns in the future.

CHAPTER 11 HIGHLIGHTS

VOCABULARY

After completing this chapter, you should be able to define each term, property, or phrase and give an example or two of each.

Algebra

axis of symmetry (p. 612)

discriminant (p. 632)

exponential function (pp. 634, 635)

general equation for exponential decay (p. 644)

general equation for exponential growth (p. 643)

half-life (p. 638)

maximum (p. 611)

minimum (p. 611)

parabola (pp. 610, 611)

quadratic equation (p. 620)

quadratic formula (p. 628)

quadratic function (pp. 610, 611)

roots (p. 620)

symmetry (p. 612)

vertex (pp. 610, 611)

zeros (p. 620)

Problem Solving

look for a pattern (p. 637)

UNDERSTANDING AND USING THE VOCABULARY

Choose the letter of the term that best matches each equation or phrase.

1. $y = C(1 + r)^t$
2. $f(x) = ax^2 + bx + c$
3. a geometric property of parabolas
4. $x = \frac{-b}{2a}$
5. $y = a^x$
6. maximum or minimum point of a parabola
7. $A = C(1 - r)^t$
8. solutions of a quadratic equation
9. $x = \frac{-b \pm \sqrt{b^2 - 4ac}}{2a}$
10. the graph of a quadratic function

a. equation of axis of symmetry
b. exponential decay formula
c. exponential function
d. exponential growth formula
e. parabola
f. quadratic formula
g. quadratic function
h. roots
i. symmetry
j. vertex

SKILLS AND CONCEPTS

OBJECTIVES AND EXAMPLES

Upon completing this chapter, you should be able to:

- find the equation of the axis of symmetry and the coordinates of the vertex of a parabola (Lesson 11–1)

In the equation $y = x^2 - 8x + 12$, $a = 1$ and $b = -8$.

The equation of the axis of symmetry is

$x = \frac{-b}{2a} = \frac{-(-8)}{2(1)}$ or 4.

Use the value $x = 4$ to find the coordinates of the vertex.

$y = x^2 - 8x + 12$

$= (4)^2 - 8(4) + 12$

$= 16 - 32 + 12$ or -4

The coordinates of the vertex are $(4, -4)$.

REVIEW EXERCISES

Use these exercises to review and prepare for the chapter test.

Write the equation of the axis of symmetry and find the coordinates of the vertex of the graph of each equation.

11. $y = -3x^2 + 4$
12. $y = x^2 - 3x - 4$
13. $y = 3x^2 + 6x - 17$
14. $y = 3(x + 1)^2 - 20$
15. $y = x^2 + 2x$

- graph quadratic functions (Lesson 11–1)

Graph $y = x^2 - 8x + 12$. Use the information above.

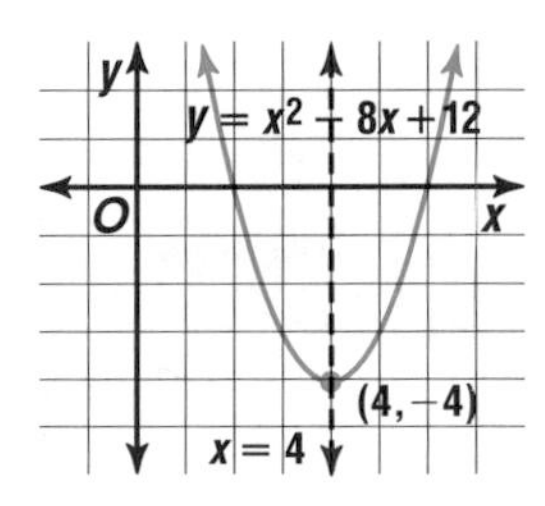

Using the results from Exercises 11–15, graph each equation.

16. $y = -3x^2 + 4$
17. $y = x^2 - 3x - 4$
18. $y = 3x^2 + 6x - 17$
19. $y = 3(x + 1)^2 - 20$
20. $y = x^2 + 2x$

- find roots of quadratic equations by graphing (Lesson 11–2)

Based on the graph of $y = x^2 - 8x + 12$ shown above, the roots of the equation $x^2 - 8x + 12 = 0$ are 2 and 6. This is because $y = 0$ at the x-intercepts, which appear to be at 2 and 6.

Substitute these values into the original equation.

$x^2 - 8x + 12 = 0$	$x^2 - 8x + 12 = 0$
$(2)^2 - 8(2) + 12 = 0$	$(6)^2 - 8(6) + 12 = 0$
$4 - 16 + 12 = 0$	$36 - 48 + 12 = 0$
$0 = 0$ ✓	$0 = 0$ ✓

The solutions of the equation are 2 and 6.

Solve each equation by graphing. If integral roots cannot be found, state the consecutive integers between which the roots lie.

21. $x^2 - x - 12 = 0$
22. $x^2 + 6x + 9 = 0$
23. $x^2 + 4x - 3 = 0$
24. $2x^2 - 5x + 4 = 0$
25. $x^2 - 10x = -21$
26. $6x^2 - 13x = 15$

OBJECTIVES AND EXAMPLES

- solve quadratic equations by using the quadratic formula (Lesson 11–3)

Solve $2x^2 + 7x - 15 = 0$.

In the equation, $a = 2$, $b = 7$, and $c = -15$. Substitute these values into the quadratic formula.

$$x = \frac{-(7) \pm \sqrt{(7)^2 - 4(2)(-15)}}{2(2)}$$

$$= \frac{-7 \pm \sqrt{169}}{4}$$

$$x = \frac{-7 + 13}{4} \quad \text{or} \quad x = \frac{-7 - 13}{4}$$

$$= \frac{3}{2} \qquad\qquad = -5$$

REVIEW EXERCISES

Solve each equation by using the quadratic formula. Approximate irrational roots to the nearest hundredth.

27. $x^2 - 8x = 20$
28. $r^2 + 10r + 9 = 0$
29. $4p^2 + 4p = 15$
30. $2y^2 + 3 = -8y$
31. $9k^2 - 13k + 4 = 0$
32. $9a^2 + 25 = 30a$
33. $-a^2 + 5a - 6 = 0$
34. $-2d^2 + 8d + 3 = 3$
35. $21a^2 + 5a - 7 = 0$
36. $2m^2 = \frac{17}{6}m - 1$

- graph exponential functions (Lesson 11–4)

Graph $y = 2^x - 3$.

x	y
−3	−2.875
−2	−2.75
−1	−2.5
0	−2
1	−1
2	1
3	5

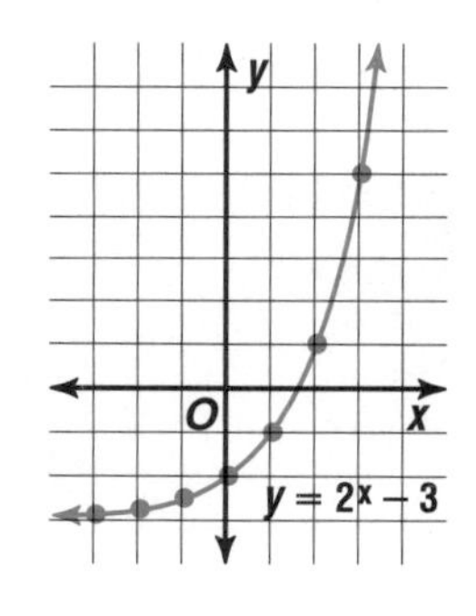

The y-intercept is -2.

Graph each function. State the y-intercept.

37. $y = 3^x + 6$
38. $y = 3^{x+2}$
39. $y = 2^x$
40. $y = 2\left(\frac{1}{2}\right)^x$

- solve exponential equations (Lesson 11–4)

Solve $25^{b+4} = \left(\frac{1}{5}\right)^{2b}$.

$$25^{b+4} = \left(\frac{1}{5}\right)^{2b}$$

$$(5^2)^{b+4} = (5^{-1})^{2b}$$

$$5^{2b+8} = 5^{-2b}$$

$$2b + 8 = -2b$$

$$8 = -4b$$

$$-2 = b$$

The solution is -2.

Solve each equation.

41. $3^{4x} = 3^{-12}$
42. $7^x = 7^{4x+9}$
43. $\left(\frac{1}{3}\right)^t = 27^{t+8}$
44. $0.01 = \left(\frac{1}{10}\right)^{4r}$
45. $64^{y-3} = \left(\frac{1}{16}\right)^{y^2}$

OBJECTIVES AND EXAMPLES

- solve problems involving growth and decay (Lesson 11–5)

Find the final amount from an investment of \$1500 invested at an interest rate of 7.5% compounded quarterly for 10 years.

$$A = P\left(1 + \frac{r}{n}\right)^{nt}$$

$$= 1500\left(1 + \frac{0.075}{4}\right)^{4 \cdot 10}$$

$$= 3153.523916$$

The amount of the account is about \$3153.52.

REVIEW EXERCISES

Determine the final amount from the investment for each situation.

	Initial Amount	Annual Interest Rate	Time	Type of Compounding
46.	\$2000	8%	8 years	quarterly
47.	\$5500	5.25%	15 years	monthly
48.	\$15,000	7.5%	25 years	monthly
49.	\$500	9.75%	40 years	daily

APPLICATIONS AND PROBLEM SOLVING

50. Archery The height h, in feet, that a certain arrow will reach t seconds after being shot directly upward is given by the formula $h = 112t - 16t^2$. What is the maximum height for this arrow? (Lesson 11–1)

51. Physics A projectile is shot vertically up in the air. Its distance s, in feet, after t seconds is given by the equation $s = 96t - 16t^2$. Find the values of t when s is 96 feet. (Lesson 11–3)

52. Finance Kevin deposited \$1400 for 8 years at $6\frac{1}{2}$% interest compounded quarterly. How much will he have at the end of 8 years? (Lesson 11–5)

53. Diving Wyatt is diving from a 10-meter platform. His height h in meters above the water when he is x meters away from the platform is given by the formula $h = -x^2 + 2x + 10$. Approximately how far away from the platform is he when he enters the water? (Lesson 11–2)

54. Number Theory Find a number whose square is 168 greater than 2 times the number. (Lesson 11–3)

55. Decision Making Juanita wants to buy a new computer but she only has \$500. She decides to wait a year and invest her money. Should she put it in a 1-year CD with a rate of 8% compounded monthly or in a savings account with a rate of 6% compounded daily? Explain your answer. (Lesson 11–5)

A practice test for Chapter 11 is provided on page 797.

ALTERNATIVE ASSESSMENT

COOPERATIVE LEARNING PROJECT

Sightseeing Tours In this project, you will model a business's profit. The Wash Student Tour Company offers one week tours of Washington, D.C. in small groups. While some of Wash's cost per person go down as the number of people on the tour increases, other costs go up because they must reserve rooms in another motel and rent extra vans. Wash has a function that enables them to predict their profit per student. If x is the number of students on the tour, and $f(x)$ is the profit (in dollars) per student, then $f(x) = -0.6x^2 + 18x - 45$.

Write a summary, using this model that describes in detail the profit structure for the Tour Company. Find the number of students that will give Wash the largest profit per student. What is the maximum profit? What does this represent? The company will offer tours as long as they do not lose money. What is the least or greatest number of students they should accept?

Follow these steps to accomplish your task.

- Substitute various numbers of students into the function to determine the profit for each.
- Substitute various amounts of profits into the function to determine the number of students that must be on the tour.
- Graph the function.
- Using the model, discuss terms such as profit, loss, break even, and maximum or minimum.
- Write a summary.

THINKING CRITICALLY

- If the value of a quadratic function is negative when $x = 1$ and positive when $x = 2$, explain what this means in terms of the roots and why.
- In the quadratic equation $ax^2 - bx + c = 0$, if $ac < 0$, what must be true about the nature of the roots of the equation?

PORTFOLIO

Use the quadratic formula and factoring to solve several quadratic equations. As you solve each quadratic equation in both ways, compare the similarities and the differences, if there are any. Write a description of how the quadratic formula can be used to determine whether or not a quadratic polynomial is factorable. Place this in your portfolio.

SELF EVALUATION

While the solution to a problem may be unforeseen, a good problem solver can use the details of the problem to plan a method of solution. Those details can often be part of the solution or preliminary steps needed to arrive at the solution.

Assess yourself. Do you pay attention to detail? Do you organize and evaluate as you go through a problem? Once you have a solution, do you analyze the solution by looking at all options or do you only look for the most obvious? Give an example of a math-related problem and a daily life problem in which detail was essential and explain how it helped.

LONG-TERM PROJECT

In·ves·ti·ga·tion

A Growing Concern

MATERIALS NEEDED

- calculator
- cardboard
- construction paper
- flashlight
- glue
- markers
- modeling clay
- paint
- ruler
- scissors
- duct tape

You have a small landscape company that specializes in residential landscape design. One of your clients is the Sanchez family. Mr. and Dr. Sanchez have three children, ages 5, 11, and 15. They have just moved into a home on a lot that is one third of an acre (1 acre = 43,560 ft^2). Their one-story home and garage occupy 3425 square feet.

Dr. Sanchez loves to garden, and Mr. Sanchez likes to swim laps to keep in shape. So, the Sanchez family is interested in a backyard pool, hot tub, deck and/or patio, and a fairly good-sized lawn. They also want to leave room to later construct a play area for their youngest child and a garden for Dr. Sanchez. The front yard was fully landscaped by the construction company that built the house.

The family prefers a low-maintenance landscape, which consists mainly of lawn care. Because of the dry climate, daily watering must be done with a sprinkler system. They are also interested in concrete walkways, to match the driveway and the path to the front door.

For this job, the Sanchez family will request bids from several suppliers. They have asked your company to construct a model of your design, along with a bid, for them to view.

The Sanchezes are interested in a reasonably low price, but will choose a higher bid if they prefer the design and features. In either case, they plan to spend no more than $65,000.

When calculating bids, you must consider the cost of supplies, materials, labor, and a profit margin. Use the company's labor and material tables to estimate costs. Labor costs are determined either by the entire job or by the hour. This cost is dependent on the individual job. The profit margin is 20% of the total cost of materials, supplies, and labor.

Your design team consists of three people. Make an Investigation Folder in which you can store all of your work on this Investigation for future use.

THE PROPERTY

Use the dimensions given in the diagram below and your knowledge of geometry to design and create a *rough draft* of a landscaping plan for the Sanchez family's backyard. Be sure to include a pool, hot tub, lawn area, flower beds, trees, deck and/or patio, and concrete walkways in your design. Think about the dimensions of all of the features in your design. Be sure to leave ample space for a play area and garden to be developed later.

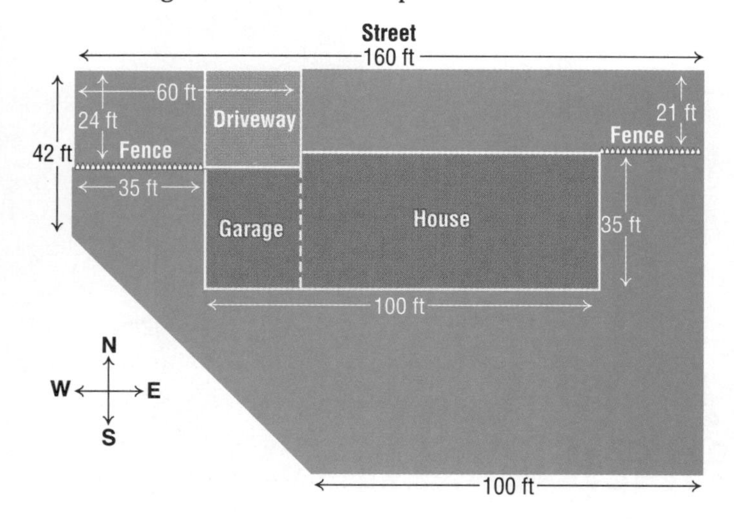

LABOR AND MATERIALS

POOL

Labor

digging: $1/cu ft

pool construction: $10/ft^2 of surface area

plumbing, filter, heating insulation: 8 h @ $35/h

Materials

pool construction: $18/ft^2 of surface area

plumbing and filter: $1250

heater: $1600

HOT TUB

Labor

digging: $1/cu ft

hot tub construction: $10/ft^2 of tub surface area

plumbing, filter, heating insulation: 8 h @ $35/h (3 hours if installed with a pool)

Materials

tub construction: $18/ft^2 of tub surface area

plumbing and filter: $1250

plumbing without filter: $250

heater: $1600 (pool and hot tub can share heater and filter)

DECK

Labor

8 ft^2 of deck per hour @ $25/h

Materials

redwood deck materials (4 ft × 1 ft): $12

PLANTS and TREES

Labor

plants/trees: 4 per hour @ $20/h

grass seed: 125 ft^2 per hour @ $20/h

sprinkler installation: 20 ft per hour @ $20/h

Materials

1 plant: $6.25

1 tree: $22.50

soil preparation: $1.75 per ft^2 of seeded area

sprinkler: 14 ft of sprinkler for every 10 ft^2 of lawn @ $1.50/ft

PATIO and WALKWAYS

Labor

12 ft^2 of patio or walkway per hour @ $22/h

Materials

concrete: $16/10 ft^2

You will continue working on this Investigation throughout Chapters 12 and 13.

Be sure to keep your designs, models, charts, and other materials in your Investigation Folder.

A Growing Concern Investigation

Working on the Investigation Lesson 12–1, p. 665

Working on the Investigation Lesson 12–7, p. 695

Working on the Investigation Lesson 13–1, p. 718

Working on the Investigation Lesson 13–5, p. 741

Closing the Investigation End of Chapter 13, p. 748

CHAPTER 12

Exploring Rational Expressions and Equations

Objectives

In this chapter, you will:

- simplify rational expressions,
- add, subtact, multiply, and divide rational expressions,
- divide polynomials, and
- make organized lists to solve problems.

Source: U.S. National Endowment for the Arts, *Annual Report*

The term "starving artist" is not without merit. Many artists often have other jobs to survive while trying to pursue their careers as artists. The National Endowment for the Arts is authorized to assist individuals and nonprofit organizations financially in a wide range of artistic endeavors. Some of the artistic forms that are funded are music, museums, theater, dance, media arts, and visual arts.

TIME Line

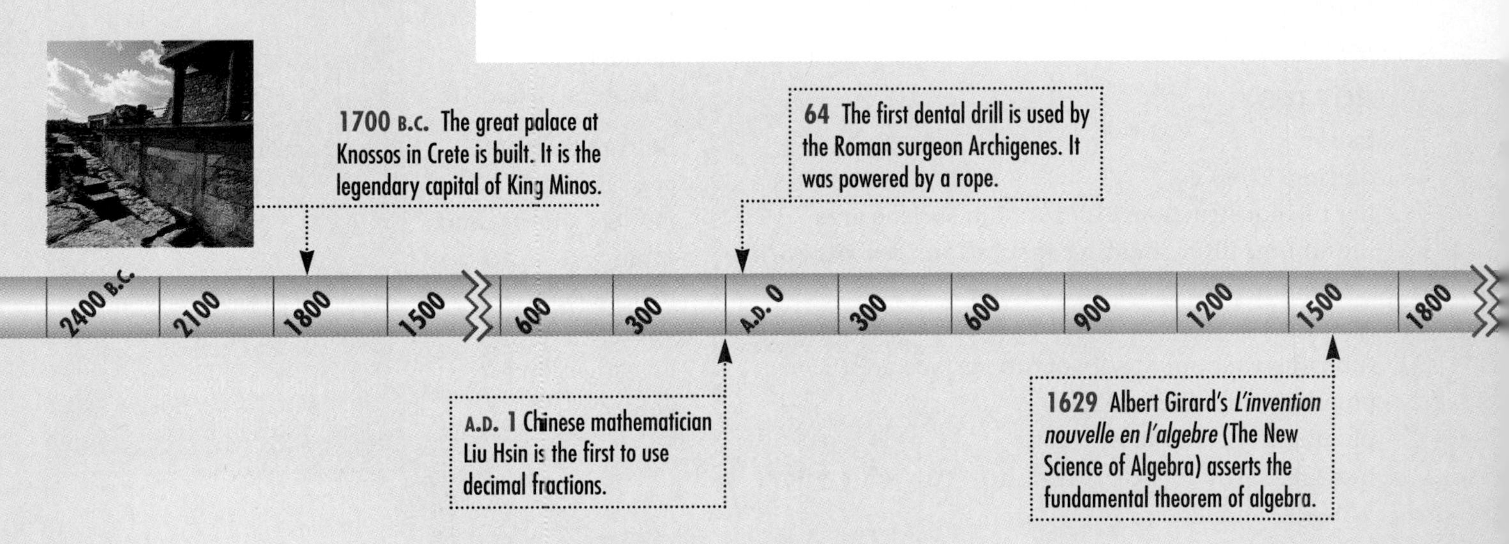

PEOPLE IN THE NEWS

The expression of energy, vibrancy, and harmony are just some of the thoughts that flow through the creativity of **Joe Maktima** of Flagstaff, Arizona. The themes of his work in acrylic paints and mediums are deeply rooted in his pueblo culture. Joe began his artistic endeavors in high school as a hobby, but it was winning the Best of Show Award in a national competition of Native American high school art students that gave him the confidence to become a professional artist.

Chapter Project

Native American art also includes blanket weaving. Many of the designs include pictorial representations of everyday objects or characters from folklore about their ancestors.

- Design a 64″ × 80″ blanket. Choose either objects that represent you or some part of your personal history for the blanket's design.
- Make a drawing to represent your design. Include a scale expressed as a rational number to show the actual size of your blanket.
- Use yarn or construction paper to weave a portion of your design.
- Calculate how much yarn in each color you would need to create a real blanket.
- Estimate the cost of your blanket in materials and labor.

Suppose a blanket is priced at $375. How does this price compare with your estimate?

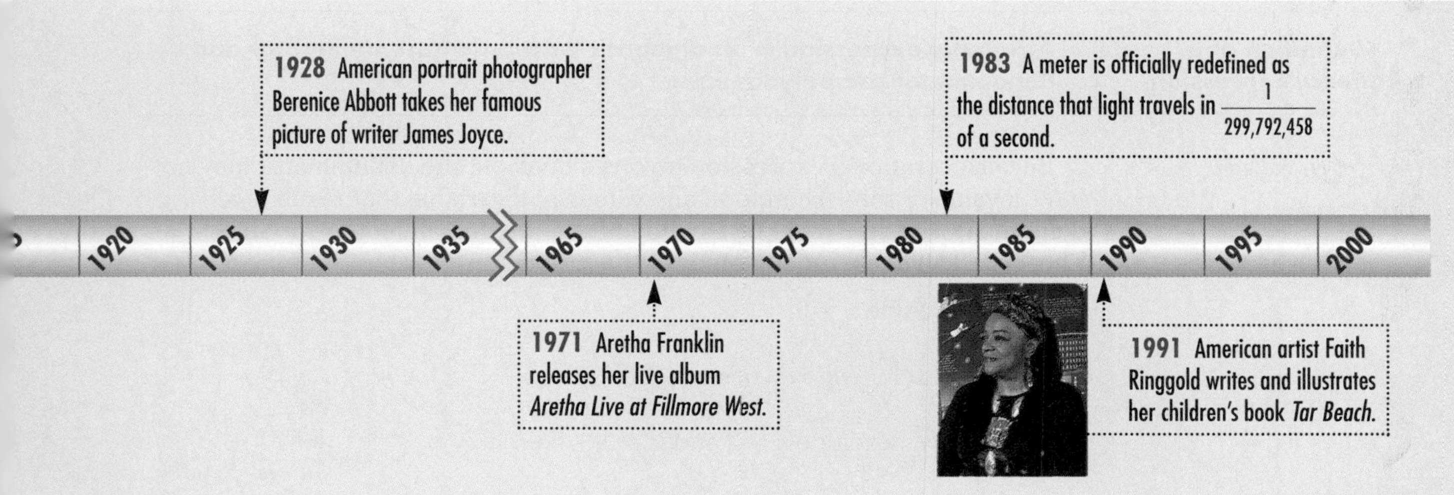

1928 American portrait photographer Berenice Abbott takes her famous picture of writer James Joyce.

1983 A meter is officially redefined as the distance that light travels in $\frac{1}{299,792,458}$ of a second.

1971 Aretha Franklin releases her live album *Aretha Live at Fillmore West.*

1991 American artist Faith Ringgold writes and illustrates her children's book *Tar Beach.*

12–1 Simplifying Rational Expressions

What YOU'LL LEARN

- To simplify rational expressions, and
- to identify values excluded from the domain of a rational expression.

Why IT'S IMPORTANT

You can use rational expressions to solve problems involving physics and carpentry.

Physical Science

Many bicyclists carry small tool kits in case they need to make adjustments to their bicycles while on the road. A wrench, for example, might be needed to tighten the bolts that keep the seat aligned. The force applied to one end of the wrench is multiplied so that the bolt at the other end of the wrench is made very secure.

A wrench is an example of a simple machine called a *lever*. In Lesson 7–2, you learned that to calculate the *mechanical advantage* (MA) of a lever, you find the ratio of the length of the effort arm to the length of the resistance arm. MA is usually expressed as a decimal value.

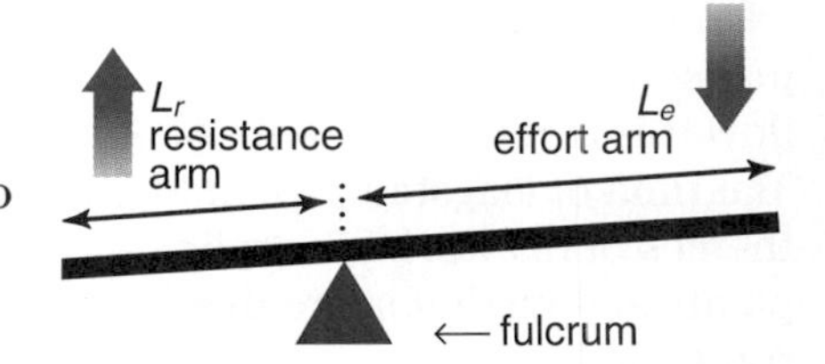

$$MA = \frac{\text{length of effort arm}}{\text{length of resistance arm}} = \frac{L_e}{L_r}$$

Suppose the total length of a lever is 6 feet. If the length of the resistance arm of the lever is x feet, then the length of the effort arm is $(6 - x)$ feet. The function $f(x) = \frac{6 - x}{x}$ represents the mechanical advantage of the lever. The table and graph at the right represent this function. They illustrate how varying the length of the resistance arm affects the mechanical advantage. Notice that, as x increases, the mechanical advantage $f(x)$ decreases. The expression that defines this function, $\frac{6-x}{x}$, is an example of a **rational expression.**

x	f(x)
0	undefined
1	5.0
2	2.0
3	1.0
4	0.5
5	0.2
6	0.0

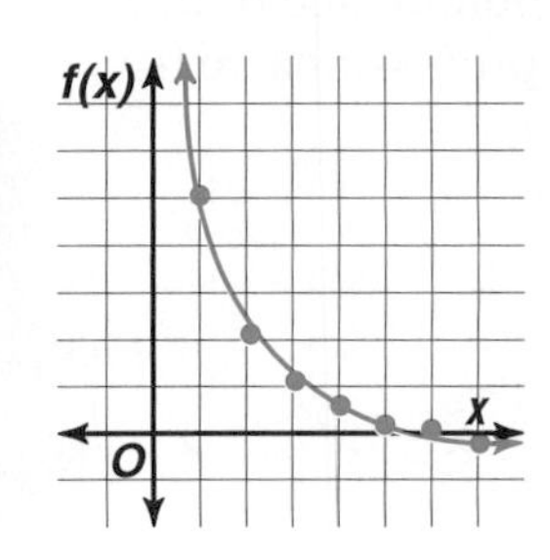

Definition of a Rational Expression	A rational expression is an algebraic fraction whose numerator and denominator are polynomials.

$f(x) = \frac{6 - x}{x}$ is called a rational function.

Because a rational expression involves division, the denominator may not have a value of zero. Therefore, any values of a variable that result in a denominator of zero must be excluded from the domain of the variable. These are called **excluded values** of the rational expression.

For $\frac{6 - x}{x}$, exclude $x = 0$.

For $\frac{5m + 3}{m + 6}$, exclude $m = -6$, since $-6 + 6 = 0$.

For $\frac{x^2 - 5}{x^2 - 5x + 6}$, exclude $x = 2$ and $x = 3$. *Why?*

Example 1 **For each rational expression, state the values of the variable that must be excluded.**

a. $\frac{7b}{b+5}$

Exclude the values for which $b + 5 = 0$.

$b + 5 = 0$

$b = -5$

Therefore, b cannot equal -5.

b. $\frac{r^2 + 32}{r^2 + 9r + 8}$

Exclude the values for which $r^2 + 9r + 8 = 0$.

$r^2 + 9r + 8 = 0$

$(r + 1)(r + 8) = 0$

$r = -1$ or $r = -8$ *Zero product property*

Therefore, r cannot equal -1 or -8.

To simplify a rational expression, you must eliminate any common factors of the numerator and denominator. To do this, use their greatest common factor (GCF). Remember that $\frac{ab}{ac} = \frac{a}{a} \cdot \frac{b}{c}$ and $\frac{a}{a} = 1$. So, $\frac{ab}{ac} = 1 \cdot \frac{b}{c}$ or $\frac{b}{c}$.

Example 2 **Simplify $\frac{9x^2yz}{24xyz^2}$. State the excluded values of x, y, and z.**

$$\frac{9x^2yz}{24xyz^2} = \frac{(3xyz)(3x)}{(3xyz)(8z)} \quad \textit{Factor; the GCF is 3xyz.}$$

$$= \frac{\overset{1}{\cancel{(3xyz)}}(3x)}{\underset{1}{\cancel{(3xyz)}}(8z)} \quad \textit{Divide by the GCF.}$$

$$= \frac{3x}{8z}$$

Exclude the values for which $24xyz^2 = 0$: $x = 0$, $y = 0$, or $z = 0$. Therefore, neither x, y, nor z can equal 0.

You can use the same procedure to simplify a rational expression in which the numerator and denominator are polynomials.

Example 3 **Simplify $\frac{a+3}{a^2 + 4a + 3}$. State the excluded values of a.**

$$\frac{a+3}{a^2 + 4a + 3} = \frac{a+3}{(a+1)(a+3)} \quad \textit{Factor the denominator.}$$

$$= \frac{\overset{1}{\cancel{a+3}}}{(a+1)\underset{1}{\cancel{(a+3)}}} \quad \textit{The GCF is } a + 3.$$

$$= \frac{1}{a+1}$$

Exclude the values for which $a^2 + 4a + 3 = 0$.

$a^2 + 4a + 3 = 0$

$(a + 1)(a + 3) = 0$

$a = -1$ or $a = -3$

Therefore, a cannot equal -1 or -3.

EXPLORATION CALCULATORS

You can use a scientific or graphing calculator to evaluate rational expressions for given values of the variables. When you do this, you must be careful to use parentheses to group both the numerator and the denominator of the expression. For example, to evaluate $\frac{x^2 - 6x + 8}{x - 2}$ when $x = -3$, use a key sequence like this on a scientific calculator.

Enter: (3 +/− x^2 − 6 × 3 +/− + 8) ÷ (3

 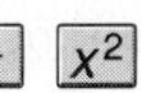 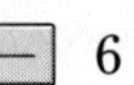   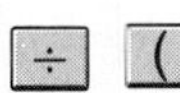

+/− − 2) = −7

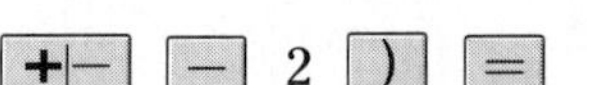

The key sequence for a graphing calculator is very similar.

Enter: (((−) 3) x^2 − 6 × (−) 3 + 8) ÷

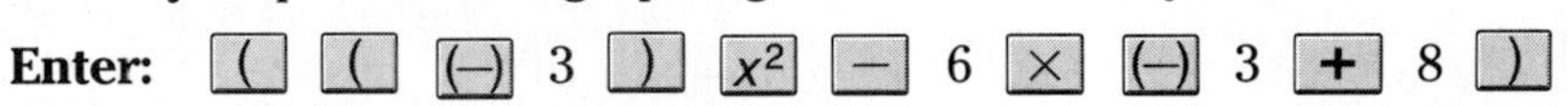  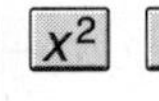  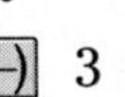 

((−) 3 − 2) ENTER −7

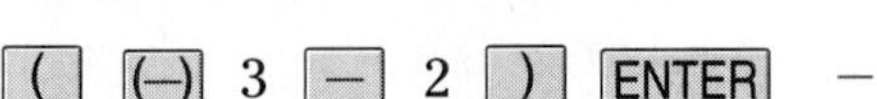

Your Turn

a. Copy and complete the table below.

x	−3	−2	−1	0	1	2	3
$\frac{x^2 - 6x + 8}{x - 2}$	−7						
$x - 4$							

b. For which value(s) of x is the value of $\frac{x^2 - 6x + 8}{x - 2}$ equal to the value of $x - 4$? Why does this result make sense?

c. For which value(s) of x is the value of $\frac{x^2 - 6x + 8}{x - 2}$ *not* equal to the value of $x - 4$? Why does this result make sense?

Example 4

CONNECTION
Physical Science

To pry the lid off a paint can, a screwdriver that is 20.5 cm long is used as a lever. It is placed so that 0.5 cm of its length extends inward from the rim of the can. Then a force of 5 pounds is applied at the end of the screwdriver. What is the force placed on the lid?

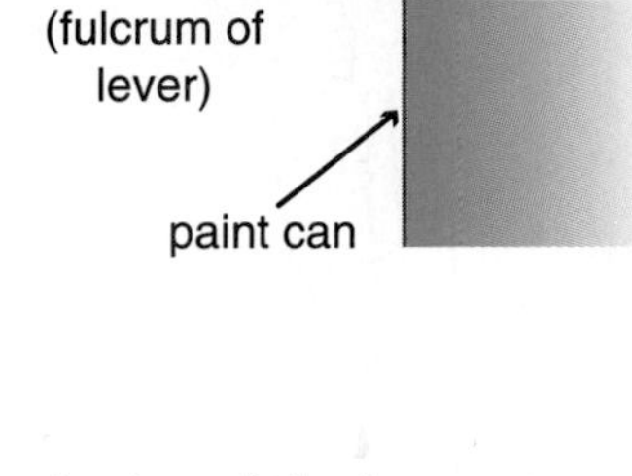

Let s represent the total length of the screwdriver.

Let r represent the length that extends inward from the rim. This is the length of the resistance arm of the lever.

Then $s - r$ represents the length that extends outward from the rim. This is the length of the effort arm of the lever.

Use the formula given in the connection at the beginning of the lesson to write an expression for mechanical advantage.

$$MA = \frac{\text{length of effort arm}}{\text{length of resistance arm}} \text{ or } \frac{s - r}{r}$$

Now evaluate the expression for the given values.

$$\frac{s-r}{r} = \frac{20.5-0.5}{0.5} \quad s = 20.5 \text{ and } r = 0.5$$

$$= \frac{20}{0.5} \text{ or } 40 \quad \text{Simplify.}$$

The mechanical advantage is 40. The force placed on the lid is the product of this number and the force applied at the end of the screwdriver.

$40 \cdot 5$ pounds $= 200$ pounds

The force placed on the lid is 200 pounds.

CHECK FOR UNDERSTANDING

Communicating Mathematics

Study the lesson. Then complete the following.

1. **Explain** why $x = 2$ is excluded from the domain of $f(x) = \frac{x^2 + 7x + 12}{x - 2}$.
2. **Estimate** the mechanical advantage of the lever shown on page 660 using the graph if x is 1.5.
3. **Explain** how you would determine the values to be excluded from the domain of $f(x) = \frac{x + 5}{x^2 + 6x + 5}$.
4. **Write** a rational expression involving one variable for which the excluded values are -2 and 7.
5. **Write** the meaning of the term *mechanical advantage* in your own words.

Guided Practice

For each expression, find the GCF of the numerator and the denominator. Then simplify. State the excluded values of the variables.

6. $\frac{13a}{14ay}$
7. $\frac{-7a^2b^3}{21a^5b}$
8. $\frac{a(m+3)}{a(m-2)}$
9. $\frac{3b}{b(b+5)}$
10. $\frac{(r+s)(r-s)}{(r-s)(r-s)}$
11. $\frac{m-3}{m^2-9}$
12. Evaluate $\frac{x^2 + 7x + 12}{x + 3}$ if $x = 2$.
13. **Landscaping** To clear land for a garden, Chang needs to move some large rocks. He plans to use a 6-foot-long pinch bar as a lever. He positions it next to each rock as shown at the right.
    a. Calculate the mechanical advantage.
    b. If Chang can apply a force of 150 pounds to the effort arm, what is the greatest weight he can lift?

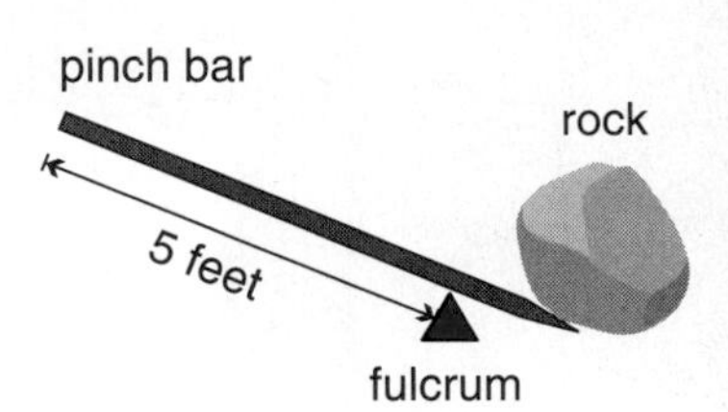

EXERCISES

Practice

Simplify each rational expression. State the excluded values of the variables.

14. $\frac{15a}{39a^2}$

15. $\frac{35y^2z}{14yz^2}$

16. $\frac{28a^2}{49ab}$

17. $\frac{56x^2y}{70x^3y}$

18. $\frac{4a}{3a + a^2}$

19. $\frac{y + 3y^2}{3y + 1}$

20. $\frac{x^2 - 9}{2x + 6}$

21. $\frac{y^2 - 49x^2}{y - 7x}$

22. $\frac{x + 5}{x^2 + x - 20}$

23. $\frac{a - 3}{a^2 - 7a + 12}$

24. $\frac{3x - 15}{x^2 - 7x + 10}$

25. $\frac{x + 4}{x^2 + 8x + 16}$

26. $\frac{x^2 - 2x - 15}{x^2 - x - 12}$

27. $\frac{a^2 + 4a - 12}{a^2 + 2a - 8}$

28. $\frac{x^2 - 36}{x^2 + x - 30}$

29. $\frac{b^2 - 3b - 4}{b^2 - 13b + 36}$

30. $\frac{14x^2 + 35x + 21}{12x^2 + 30x + 18}$

31. $\frac{4x^2 + 8x + 4}{5x^2 + 10x + 5}$

Calculator

Use a calculator to evaluate each expression for the given values.

32. $\frac{x^2 - x}{3x}, x = -3$

33. $\frac{x^4 - 16}{x^4 - 8x^2 + 16}, x = -1$

34. $\frac{x + y}{x^2 + 2xy + y^2}, x = 3, y = 2$

35. $\frac{x^3y^3 + 5x^3y^2 + 6x^3y}{xy^5 + 5xy^4 + 6xy^3}, x = 1, y = -2$

Critical Thinking

36. Explain why $\frac{m^2 - 16}{m + 4}$ is not the same as $m - 4$.

Applications and Problem Solving

37. **Carpentry** A house mover can lift a house from its foundation using a *jackscrew* like the one shown at the right. As shown, the effort force is applied in a circular motion. The vertical distance that the jackscrew moves in one turn is called its *pitch*. You can use the following formula to calculate the mechanical advantage of the jackscrew.

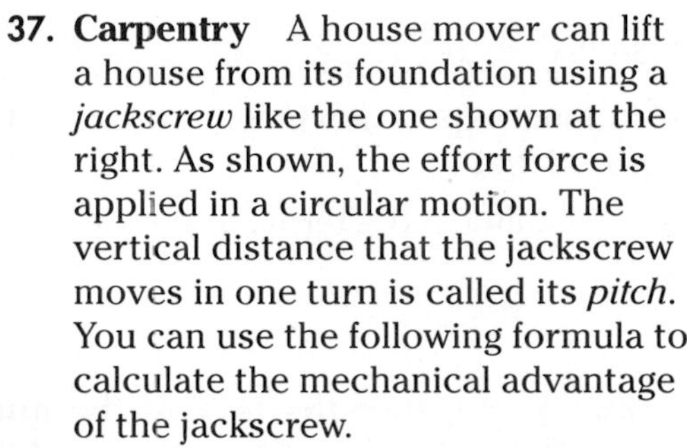

$$MA = \frac{\text{circumference of the circle}}{\text{pitch of the screw}}$$

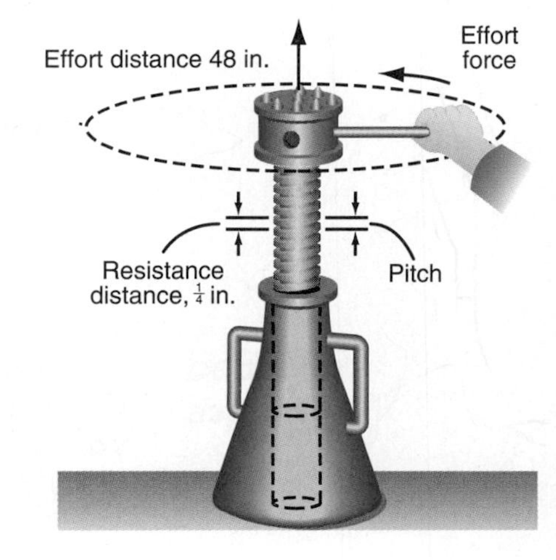

a. Calculate the mechanical advantage for the jackscrew shown above.

b. Suppose the effort force is 20 pounds. What is the greatest weight the jackscrew can lift?

38. **Aeronautics** Aircraft engineers can use the following formula to calculate the atmospheric pressure P in pounds per square inch when an aircraft is flying at an altitude a in feet.

$$P = \frac{-9.05\left[\left(\frac{a}{1000}\right)^2 - \frac{65a}{1000}\right]}{\left(\frac{a}{1000}\right)^2 + 40\left(\frac{a}{1000}\right)}$$

a. Calculate the atmospheric pressure outside a plane flying at an altitude of 20,000 feet.

b. Calculate the atmospheric pressure outside a plane flying at an altitude of 40,000 feet.

c. Is your answer to part b twice that of part a? By how much do they differ?

FYI

Because altitude affects things like boiling temperatures, recipe cooking temperatures and times must be altered so that a cake, for example, will turn out correctly. Look on a box of cake mix to find the high altitude directions.

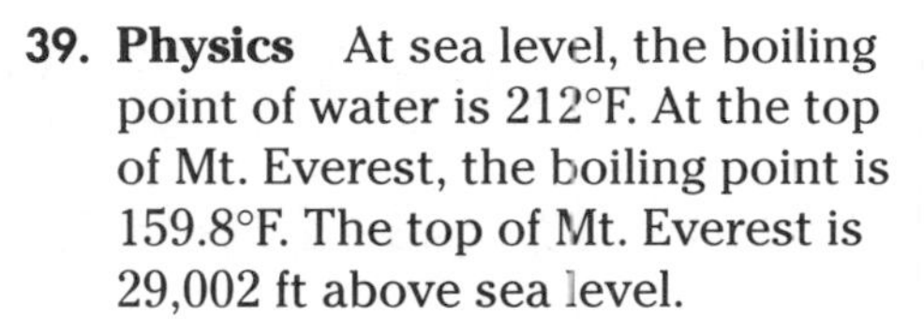

39. Physics At sea level, the boiling point of water is 212°F. At the top of Mt. Everest, the boiling point is 159.8°F. The top of Mt. Everest is 29,002 ft above sea level.

a. How many degrees does the boiling point drop for every mile up from sea level? Write the solution as a ratio involving units.

b. Simplify the expression you wrote in part a.

c. Draw a graph that represents the boiling point in degrees Fahrenheit as a function of altitude in miles.

Mixed Review

40. Finance In the compound interest formula $A = P\left(1 + \frac{r}{n}\right)^{nt}$, A represents the value of the investment in the future, P is the amount of the original investment, r is the annual interest rate, t is the number of years of the investment, and n is the number of times the interest is compounded each year. Find the total amount after \$2500 is invested for 18 years at a rate of 6%, compounded quarterly. (Lesson 11–5)

41. Number Theory The square of a number decreased by 121 is 0. Find the number. (Lesson 10–7)

42. Find $3x(-5x^2 - 2x + 7)$. (Lesson 9–6)

43. Graph the system of inequalities. (Lesson 8–5)

$y \geq x - 2$

$y \leq 2x - 1$

44. Solve $4 + |x| = 12$. (Lesson 7–6)

45. Graph $y = 3x + 2$. (Lesson 5–4)

46. Solve $\frac{x}{4} + 7 = 6$. (Lesson 3–3)

WORKING ON THE **In·ves·ti·ga·tion**

Refer to the Investigation on pages 656–657.

A Growing Concern

1 Use graph paper to make a scale drawing of the house, garage, driveway, boundary lines, and fences. This will serve as your template to which you will add the other details.

2 Develop a detailed plan for the pool and hot tub locations. Indicate the dimensions of both the pool and the hot tub. Explain why you chose the size, location, and orientation for the pool.

3 Figure the costs for materials, labor, and profit margin for construction of the pool and hot tub. Justify your costs and make a case for building the pool and hot tub as you have designed them.

4 Add the pool and hot tub to the scale drawing, indicating their measurements.

Add the results of your work to your Investigation Folder.

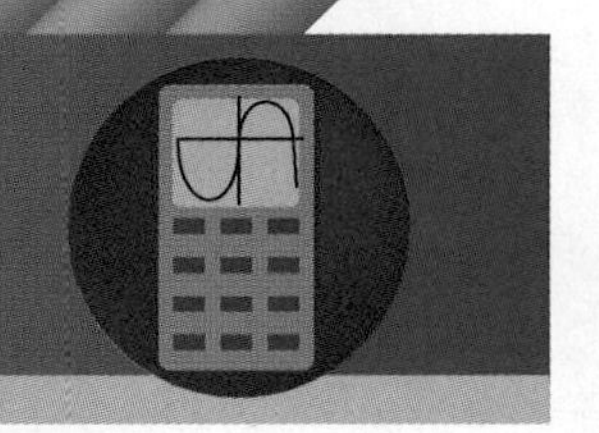

12–1B Graphing Technology Rational Expressions

An Extension of Lesson 12–1

When simplifying rational expressions, you can use a graphing calculator to support your answer. You can also use a calculator to find the excluded values.

Example **Simplify** $\frac{3x^2 - 8x + 5}{x^2 - 1}$.

$$\frac{3x^2 - 8x + 5}{x^2 - 1} = \frac{(3x - 5)(x - 1)}{(x + 1)(x - 1)} \quad \textit{Factor the numerator and denominator.}$$

$$= \frac{3x - 5}{x + 1} \quad \textit{Divide the common factor } (x - 1).$$

When $x = -1$ or $x = 1$, $x^2 - 1 = 0$. Therefore, x cannot equal -1 or 1.

Now use a graphing calculator. Graph $y = \frac{3x^2 - 8x + 5}{x^2 - 1}$ using the window $[-4.7, 4.7]$ by $[-10, 10]$ with scale factors of 1.

Enter:

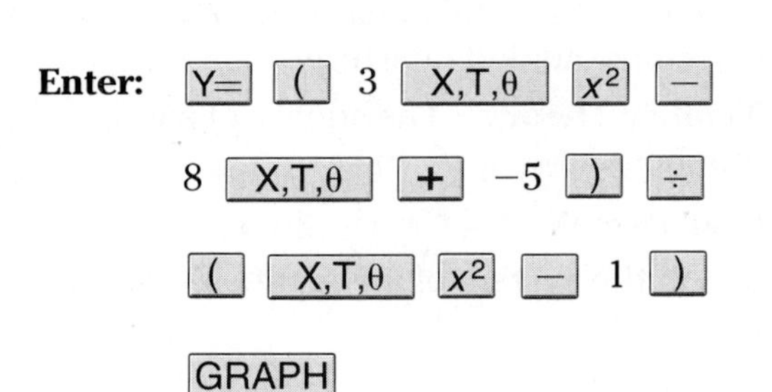

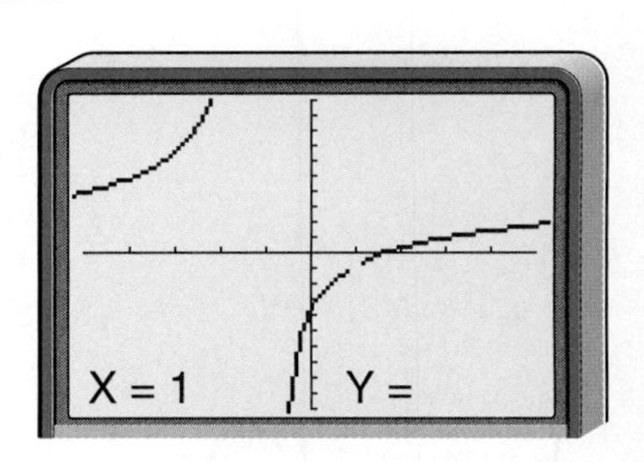

Use the TRACE key to observe the values of x and y. Notice that at $x = -1$ and $x = 1$, no value for y appears in the window. There are also "holes" in the graph at those points. These are the excluded values.

Now enter $y = \frac{3x - 5}{x + 1}$ as Y2 and observe the graphs.

Enter: Y= (3 X,T,θ — 5)
÷ (X,T,θ + 1)
GRAPH

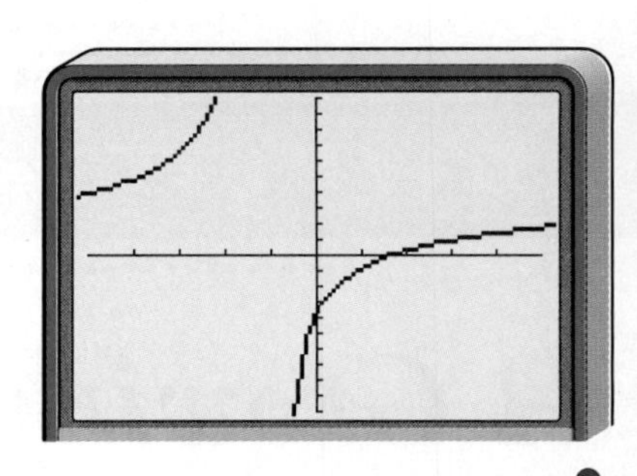

Notice that the two graphs appear to be identical. This means that the simplified expression is probably correct.

EXERCISES

Simplify each expression. Then verify your answer graphically. Name the excluded values.

1. $\frac{x^2 - 25}{x^2 + 10x + 25}$ **2.** $\frac{3x + 6}{x^2 + 7x + 10}$ **3.** $\frac{2x - 9}{4x^2 - 18x}$

4. $\frac{2x^2 - 4x}{x^2 + x - 6}$ **5.** $\frac{x^2 - 9x + 8}{x^2 - 16x + 64}$ **6.** $\frac{3x^2}{12x^2 + 192x}$

7. $\frac{-x^2 + 6x - 9}{x^2 - 6x + 9}$ **8.** $\frac{5x^2 + 10x + 5}{3x^2 + 6x + 3}$ **9.** $\frac{25 - x^2}{x^2 + x - 30}$

12-2 Multiplying Rational Expressions

What YOU'LL LEARN
- To multiply rational expressions.

Why IT'S IMPORTANT

You can use rational expressions to solve problems involving traffic and money.

Traffic

On the Sunday after Thanksgiving in 1994, westbound traffic was backed up for 13 miles at one exit of the Massachusetts Turnpike. Assume that each vehicle occupied an average of 30 feet of space in a lane and that the turnpike has three lanes. You can use the expression below to estimate the number of vehicles involved in the backup.

$$3 \text{ lanes} \cdot \left(\frac{13 \text{ miles}}{\text{lane}}\right)\left(\frac{5280 \text{ feet}}{\text{mile}}\right)\left(\frac{1 \text{ vehicle}}{30 \text{ feet}}\right)$$ *This expression will be simplified in Example 3.*

LOOK BACK

You can refer to Lesson 2-6 for information about multiplying rational numbers.

This multiplication is similar to the multiplication of rational expressions. Recall that to multiply rational numbers that are expressed as fractions, you multiply numerators and multiply denominators. You can use this same method to multiply rational expressions. From this point on, you may assume that no denominator of a rational expression has a value of 0.

rational numbers

$$\frac{4}{5} \cdot \frac{3}{7} = \frac{4 \cdot 3}{5 \cdot 7} = \frac{12}{35}$$

rational expressions

$$\frac{3}{m} \cdot \frac{k}{4} = \frac{3 \cdot k}{m \cdot 4} = \frac{3k}{4m}$$

Example 1 Find $\frac{3x^2y}{2rs} \cdot \frac{24r^2s}{15xy^2}$.

Method 1: Divide by the common factors after multiplying.

$$\frac{3x^2y}{2rs} \cdot \frac{24r^2s}{15xy^2} = \frac{72r^2sx^2y}{30rsxy^2}$$

$$= \frac{\overset{1}{\cancel{6rsxy}}(12rx)}{\underset{1}{\cancel{6rsxy}}(5y)} \quad \textit{The GCF is 6rsxy.}$$

$$= \frac{12rx}{5y}$$

Method 2: Divide by the common factors before multiplying.

$$\frac{3x^2y}{2rs} \cdot \frac{24r^2s}{15xy^2} = \frac{\overset{1\,x\,1}{\cancel{3}\cancel{x^2}\cancel{y}}}{\underset{1\,1\,1}{\cancel{2}\cancel{r}\cancel{s}}} \cdot \frac{\overset{12\,r\,1}{\cancel{24}\cancel{r^2}\cancel{s}}}{\underset{5\,1\,y}{\cancel{15}\cancel{x}\cancel{y^2}}}$$

$$= \frac{12rx}{5y}$$

Sometimes you must factor a quadratic expression before you can simplify a product of rational expressions.

Example 2 **Find each product.**

a. $\frac{m+3}{5m} \cdot \frac{20m^2}{m^2+8m+15}$

$$\frac{m+3}{5m} \cdot \frac{20m^2}{m^2+8m+15} = \frac{m+3}{5m} \cdot \frac{20m^2}{(m+3)(m+5)}$$ *Factor the denominator.*

$$= \frac{\overset{4}{\cancel{20}}\overset{m}{\cancel{m^2}}\overset{1}{\cancel{(m+3)}}}{\underset{1}{\cancel{5}}\underset{1}{\cancel{m}}\underset{1}{\cancel{(m+3)}}(m+5)}$$ *The GCF is 5m(m + 3).*

$$= \frac{4m}{m+5}$$

b. $\frac{3(x+5)}{4(2x^2+11x+12)} \cdot (2x+3)$

$$\frac{3(x+5)}{4(2x^2+11x+12)} \cdot (2x+3) = \frac{3(x+5)(2x+3)}{4(2x^2+11x+12)}$$

$$= \frac{3(x+5)(2x+3)}{4(2x+3)(x+4)}$$ *Factor the denominator.*

$$= \frac{3(x+5)\overset{1}{\cancel{(2x+3)}}}{4\underset{1}{\cancel{(2x+3)}}(x+4)}$$ *The GCF is (2x + 3).*

$$= \frac{3(x+5)}{4(x+4)}$$

$$= \frac{3x+15}{4x+16}$$ *Simplify the numerator and the denominator.*

When you multiply fractions that involve units of measure, you can divide by the units in the same way that you divide by variables. Recall that this process is called *dimensional analysis.*

Example 3 **APPLICATION** **Traffic**

Refer to the application at the beginning of the lesson. About how many vehicles were involved in the backup?

$$3 \text{ lanes} \cdot \left(\frac{13 \text{ miles}}{\text{lane}}\right)\left(\frac{5280 \text{ feet}}{\text{mile}}\right)\left(\frac{1 \text{ vehicle}}{30 \text{ feet}}\right)$$

$$= \frac{\overset{1}{\cancel{3}}\ \cancel{\text{lanes}}}{1} \cdot \left(\frac{13\ \cancel{\text{miles}}}{\cancel{\text{lane}}}\right)\left(\frac{\overset{528}{\cancel{5280}}\ \cancel{\text{feet}}}{\cancel{\text{mile}}}\right)\left(\frac{1 \text{ vehicle}}{\underset{1}{\cancel{30}}\ \cancel{\text{feet}}}\right)$$ *Divide by common units.*

$= 6864$ vehicles

The backup involved about 6900 vehicles.

CHECK FOR UNDERSTANDING

Communicating Mathematics

Study the lesson. Then complete the following.

1. **Write** two rational expressions whose product is $\frac{xy}{x-4}$.
2. **You Decide** The work of two students who simplified $\frac{4x+8}{4x-8} \cdot \frac{x^2+2x-8}{2x+8}$ is shown below. Which one is correct, and why?

   Matt: $\frac{\overset{1}{\cancel{4x}}+\overset{1}{\cancel{8}}}{\underset{1}{\cancel{4x}}-\underset{1}{\cancel{8}}} \cdot \frac{\overset{1}{\cancel{x^2}}+2\overset{1}{\cancel{x}}-\cancel{8}}{\underset{1}{\cancel{2x}}+\underset{1}{\cancel{8}}} = x^2$ Angie: $\frac{\overset{1}{\cancel{4}}(x+2)}{\underset{1}{\cancel{4}}\underset{1}{\cancel{(x-2)}}} \cdot \frac{\overset{1}{\cancel{(x-2)}}\overset{1}{\cancel{(x+4)}}}{2\underset{1}{\cancel{(x+4)}}} = \frac{x+2}{2}$

3. Write a paragraph explaining which is better—to simplify an expression first and then multiply, or to multiply and then simplify. Include examples to support your answers.

Guided Practice

Find each product. Assume that no denominator has a value of 0.

4. $\frac{12m^2y}{5r^2} \cdot \frac{r^2}{my}$

5. $\frac{16xy^2}{3m^2p} \cdot \frac{27m^3p}{32x^2y^3}$

6. $\frac{x-5}{r} \cdot \frac{r^2}{(x-5)(y+3)}$

7. $\frac{x+4}{4y} \cdot \frac{16y}{x^2+7x+12}$

8. $\frac{a^2+7a+10}{a+1} \cdot \frac{3a+3}{a+2}$

9. $\frac{4(x-7)}{3(x^2-10x+21)} \cdot (x-3)$

10. $3(x+6) \cdot \frac{x+3}{9(x^2+7x+6)}$

11. a. Multiply $\frac{2.54 \text{ centimeters}}{1 \text{ inch}} \cdot \frac{12 \text{ inches}}{1 \text{ foot}} \cdot \frac{3 \text{ feet}}{1 \text{ yard}}$. Then simplify.
    b. What does the simplified expression represent?

12. **Bicycling** Marisa says that bicycling at the rate of 20 miles per hour is equivalent to a rate of 1760 feet per minute. Is she correct? Write a product involving units of measure to justify your answer.

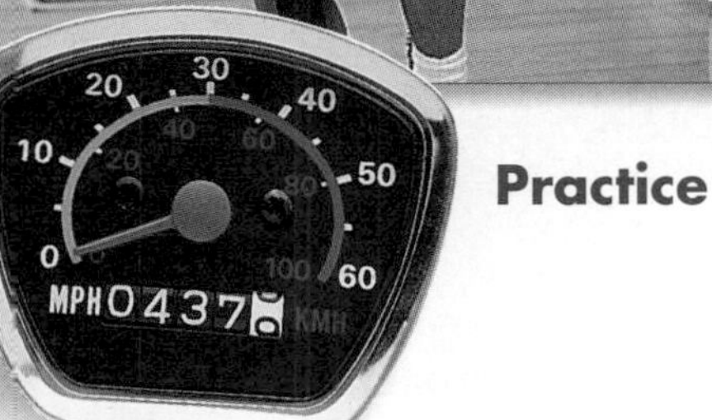

EXERCISES

Practice

Find each product. Assume that no denominator has a value of 0.

13. $\frac{7a^2}{5} \cdot \frac{15}{14a}$

14. $\frac{3m^2}{2m} \cdot \frac{18m^2}{9m}$

15. $\frac{10r^3}{6x^3} \cdot \frac{42x^2}{35r^3}$

16. $\frac{7ab^3}{11r^2} \cdot \frac{44r^3}{21a^2b}$

17. $\frac{64y^2}{5y} \cdot \frac{5y}{8y}$

18. $\frac{2a^2}{b} \cdot \frac{5bc}{6a}$

19. $\frac{m+4}{3m} \cdot \frac{4m^2}{m^2+9m+20}$

20. $\frac{m^2+8m+15}{a+b} \cdot \frac{7a+14b}{m+3}$

21. $\frac{5a+10}{10m^2} \cdot \frac{4m^3}{a^2+11a+18}$

22. $\frac{6r+3}{r+6} \cdot \frac{r^2+9r+18}{2r+1}$

23. $2(x+1) \cdot \frac{x+4}{x^2+5x+4}$

24. $4(a+7) \cdot \frac{12}{3(a^2+8a+7)}$

25. $\frac{x^2-y^2}{12} \cdot \frac{36}{x+y}$

26. $\frac{3a+9}{a} \cdot \frac{a^2}{a^2-9}$

27. $\frac{9}{3+2x} \cdot (12+8x)$

28. $(3x+3) \cdot \frac{x+4}{x^2+5x+4}$

29. $\frac{4x}{9x^2-25} \cdot (3x+5)$

30. $(b^2+12b+11) \cdot \frac{b+9}{b^2+20b+99}$

31. $\frac{4x+8}{x^2-25} \cdot \frac{x-5}{5x+10}$

32. $\frac{a^2-a-6}{a^2-9} \cdot \frac{a^2+7a+12}{a^2+4a+4}$

Multiply. Explain what each expression represents.

33. $\frac{32 \text{ feet}}{1 \text{ second}} \cdot \frac{60 \text{ seconds}}{1 \text{ minute}} \cdot \frac{60 \text{ minutes}}{1 \text{ hour}} \cdot \frac{1 \text{ mile}}{5280 \text{ feet}}$

34. $10 \text{ feet} \cdot 18 \text{ feet} \cdot 3 \text{ feet} \cdot \frac{1 \text{ yard}^3}{27 \text{ feet}^3}$

Simplify each expression.

35. **Heat** 20 grams at 540 Calories per gram

36. **Metal Refining** 4.025 grams per amp-hour at 2 amps for 5 hours

Critical Thinking

37. Find two different pairs of rational expressions whose product is $\frac{6x^2-6x-36}{x^2+3x-28}$.

Applications and Problem Solving

38. Traffic Refer to Example 3. Suppose there are eight toll collectors at the exit and it takes each an average of 24 seconds to collect the toll from one vehicle.

a. Write a product involving units to estimate the time it would take in hours to collect tolls from all the vehicles in the backup.

b. Simplify the product.

39. Money The chart at the right shows a recent foreign exchange rate table. It indicates how much of another country's currency you would receive in exchange for one American dollar at the time these rates were in effect.

Dollar Exchange

Country (Currency)	Equivalent to American Dollars
Canada (dollar)	0.7444
Hong Kong (dollar)	0.1292
Israel (shekel)	0.3281
France (franc)	0.1981
Mexico (peso)	0.1597

a. Write a rational expression involving units of measure that indicates how many American dollars you would receive for one French franc. Write another expression that indicates how many French francs you would receive for one American dollar.

b. Write a product involving units of measure that indicates how many French francs you would receive for 12,500 Mexican pesos. Then find the product.

c. Explain how to determine an exchange rate between Canada's dollar and Hong Kong's dollar.

Mixed Review

40. Cooking The formula $t = \frac{40(25 + 1.85a)}{50 - 1.85a}$ relates the time t in minutes that it takes to bake an average-size potato in an oven that is at an altitude of a thousands of feet. (Lesson 12–1)

a. What is the value of a for an altitude of 4500 feet?

b. Calculate the time is takes to bake a potato at an altitude of 3500 feet.

c. Calculate the time it takes to bake a potato at an altitude of 7000 feet.

d. The altitude in part c is twice that of part b. How do your baking times compare for those two altitudes?

41. Find the value of y for $y = 3^{2x}$ if $x = 3$. (Lesson 11–4)

42. Solve $5x^2 + 30x + 45 = 0$ by factoring. (Lesson 10–6)

43. Find the degree of $6x^2yz + 5xyz - x^3$. (Lesson 9–4)

44. Solve the system of equations. (Lesson 8–3)

$5 = 2x - 3y$

$-1 = -4x + 3y$

45. Find the slope of the line that passes through $(-3, 5)$ and $(8, 5)$. (Lesson 5–1)

46. What is 45% of \$1567 to the nearest dollar? (Lesson 4–4)

12–3

Dividing Rational Expressions

What YOU'LL LEARN

- To divide rational expressions.

Why IT'S IMPORTANT

You can use rational expressions to solve problems involving construction and railroads.

Parades

A crowd watching the Tournament of Roses Parade in Pasadena, California, fills the sidewalks along the route for about 5.5 miles on each side. Suppose these sidewalks are 10 feet wide and each person occupies an average of 4 square feet of space. To estimate the number of people along this part of the route, you can use dimensional analysis and the following expression.

$$\left(5.5 \text{ miles} \cdot \frac{5280 \text{ ft}}{1 \text{ mile}} \cdot 10 \text{ feet}\right) \div \frac{4 \text{ ft}^2}{1 \text{ person}}$$ *This expression will be simplified in Example 4.*

LOOK BACK

You can refer to Lesson 2-7 for information about dividing rational numbers.

This division is similar to the division of rational expressions. Recall that to divide rational numbers that are expressed as fractions you multiply by the reciprocal of the divisor. You can use this same method to divide algebraic rational expressions.

rational numbers

$$\frac{3}{2} \div \frac{2}{5} = \frac{3}{2} \cdot \frac{5}{2} \quad \text{The reciprocal of } \frac{2}{5} \text{ is } \frac{5}{2}.$$
$$= \frac{15}{4}$$

rational expressions

$$\frac{a}{b} \div \frac{c}{d} = \frac{a}{b} \cdot \frac{d}{c} \quad \text{The reciprocal of } \frac{c}{d} \text{ is } \frac{d}{c}.$$
$$= \frac{ad}{bc}$$

Example 1 Find $\frac{2a}{a+3} \div \frac{a+7}{a+3}$.

$$\frac{2a}{a+3} \div \frac{a+7}{a+3} = \frac{2a}{a+3} \cdot \frac{a+3}{a+7} \quad \textit{The reciprocal of } \frac{a+7}{a+3} \textit{ is } \frac{a+3}{a+7}.$$
$$= \frac{2a}{\cancel{a+3}_{1}} \cdot \frac{\overset{1}{\cancel{a+3}}}{a+7} \quad \textit{Divide by the common factor } a + 3.$$
$$= \frac{2a}{a+7}$$

Sometimes a quotient of rational expressions involves a divisor that is a binomial.

Example 2 Find $\frac{2m+6}{m+5} \div (m+3)$.

$$\frac{2m+6}{m+5} \div (m+3) = \frac{2m+6}{m+5} \cdot \frac{1}{m+3} \quad \textit{The reciprocal of } m+3 \textit{ is } \frac{1}{m+3}.$$
$$= \frac{2\overset{1}{\cancel{(m+3)}}}{(m+5)\underset{1}{\cancel{(m+3)}}} \quad \textit{Divide by the common factor } m + 3.$$
$$= \frac{2}{m+5}$$

Sometimes you must factor before you can simplify a quotient of rational expressions.

Example 3 Find $\dfrac{x}{x+2} \div \dfrac{x^2}{x^2+5x+6}$.

$$\frac{x}{x+2} \div \frac{x^2}{x^2+5x+6} = \frac{x}{x+2} \cdot \frac{x^2+5x+6}{x^2}$$

$$= \frac{\overset{1}{\cancel{x}}}{\underset{1}{\cancel{x+2}}} \cdot \frac{\overset{1}{\cancel{(x+2)}}(x+3)}{\underset{x}{\cancel{x^2}}} \quad \textit{Divide by common factors.}$$

$$= \frac{x+3}{x}$$

Example 4

Parades

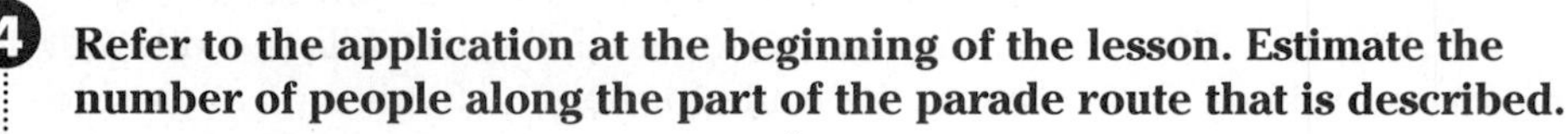

Refer to the application at the beginning of the lesson. Estimate the number of people along the part of the parade route that is described.

$$\left(5.5 \text{ miles} \cdot \frac{5280 \text{ ft}}{1 \text{ mile}} \cdot 10 \text{ feet}\right) \div \frac{4 \text{ ft}^2}{1 \text{ person}} \quad \textit{The reciprocal of } \frac{4 \text{ ft}^2}{1 \text{ person}} \textit{ is } \frac{1 \text{ person}}{4 \text{ ft}^2}.$$

$$= \left(5.5 \text{ miles} \cdot \frac{5280 \text{ ft}}{1 \text{ mile}} \cdot 10 \text{ feet}\right) \cdot \frac{1 \text{ person}}{4 \text{ ft}^2}$$

$$= \frac{5.5 \cancel{\text{ miles}}}{1} \cdot \frac{\overset{1320}{\cancel{5280 \text{ ft}}}}{1 \cancel{\text{ mile}}} \cdot \frac{10 \cancel{\text{ ft}}}{1} \cdot \frac{1 \text{ person}}{\underset{1}{\cancel{4 \text{ ft}^2}}} \quad \textit{Divide by common units.}$$

$$= 72{,}600 \text{ people}$$

This means that there were about 72,600 people along *each side*. Since there were two sides, the total number of people along the route was about 2(72,600), or 145,200.

CHECK FOR UNDERSTANDING

Communicating Mathematics

Study the lesson. Then complete the following.

1. **Analyze** the solution shown below.

$$\frac{a^2-9}{3a} \div \frac{a+3}{a-3} = \frac{3a}{a^2-9} \cdot \frac{a+3}{a-3}$$

$$= \frac{3a}{\underset{1}{\cancel{(a+3)}}(a-3)} \cdot \frac{\overset{1}{\cancel{(a+3)}}}{a-3}$$

$$= \frac{3a}{(a-3)^2}$$

What error was made?

2. **Write** two rational expressions whose quotient is $\frac{10r}{d^2}$.

3. An expression for calculating the mass of 1 cubic meter of a substance is given below. Write a procedure for simplifying the expression.

$$\frac{5.96 \text{ grams}}{\text{centimeter}^3} \cdot \frac{1 \text{ kilogram}}{1000 \text{ grams}} \cdot \frac{100^3 \text{ centimeters}^3}{1 \text{ meter}^3} \cdot 1 \text{ meter}^3$$

Guided Practice

Find the reciprocal of each expression.

4. $\frac{m^2}{3}$
5. $\frac{x}{5}$
6. $\frac{-9}{4y}$
7. $\frac{x^2-9}{y+3}$
8. $m - 3$
9. $x^2 + 2x + 5$

Find each quotient. Assume that no denominator has a value of 0.

10. $\frac{x}{x+7} \div \frac{x-5}{x+7}$

11. $\frac{m^2+3m+2}{4} \div \frac{m+1}{m+2}$

12. $\frac{5a+10}{a+5} \div (a+2)$

13. $\frac{x^2+7x+12}{x+6} \div (x+3)$

Simplify each dimensional expression. State what you think the expression represents.

14. $(8 \text{ feet} \cdot 3 \text{ feet} \cdot 12 \text{ feet}) \div \frac{27 \text{ feet}^3}{1 \text{ yard}^3}$

15. $(12 \text{ inches} \cdot 18 \text{ inches} \cdot 4 \text{ inches}) \div \frac{1728 \text{ inches}^3}{1 \text{ feet}^3}$

16. **Railroading** Jaheem is a railroad buff who likes to watch trains. At the railroad crossing near his house in Menominee Falls, Wisconsin, a Chicago & North Western freight train passes by at 40 miles per hour. Suppose each railroad car is 48 feet long.
    a. Write an expression involving units that represents the number of cars that pass by him in one minute.
    b. Find how many railroad cars would pass by him per minute.

EXERCISES

Practice

Find each quotient. Assume that no denominator has a value of 0.

17. $\frac{a}{a+3} \div \frac{a+11}{a+3}$

18. $\frac{m+7}{m} \div \frac{m+7}{m+3}$

19. $\frac{a^2b^3c}{m^2y^2} \div \frac{a^2bc^3}{m^3y^2}$

20. $\frac{5x^2}{7} \div \frac{10x^3}{21}$

21. $\frac{3m+15}{m+4} \div \frac{3m}{m+4}$

22. $\frac{3x}{x+2} \div (x-1)$

23. $\frac{4z+8}{z+3} \div (z+2)$

24. $\frac{x+3}{x+1} \div (x^2+5x+6)$

25. $\frac{2x+4}{x^2+11x+18} \div \frac{x+1}{x^2+14x+45}$

26. $\frac{k+3}{m^2+4m+4} \div \frac{2k+6}{m+2}$

27. What is the quotient when $\frac{2x+6}{x+5}$ is divided by $\frac{2}{x+5}$?

28. Find the quotient when $\frac{m-8}{m+7}$ is divided by $m^2 - 7m - 8$.

Find each quotient. Assume that no denominator has a value of 0.

29. $\frac{x^2+5x+6}{x^2-x-12} \div \frac{x+2}{x^2+x-20}$

30. $\frac{m^2+m-6}{m^2+8m+15} \div \frac{m^2-m-2}{m^2+9m+20}$

31. $\frac{2x^2+7x-15}{x+2} \div \frac{2x-3}{x^2+5x+6}$

32. $\frac{t^2-2t-8}{w-3} \div \frac{t-4}{w^2-7w+12}$

Simplify each expression. State what you think the expression represents.

33. $\left(\frac{60 \text{ miles}}{1 \text{ hour}} \cdot \frac{5280 \text{ feet}}{1 \text{ mile}} \div \frac{60 \text{ minutes}}{1 \text{ hour}}\right) \div \frac{60 \text{ seconds}}{1 \text{ minute}}$

34. $\frac{23.75 \text{ inches}}{1 \text{ revolution}} \cdot \frac{33\frac{1}{3} \text{ revolutions}}{1 \text{ minute}} \cdot 16.5 \text{ minutes}$

35. $(5 \text{ feet} \cdot 16.5 \text{ feet} \cdot 9 \text{ feet}) \div \frac{27 \text{ feet}^3}{1 \text{ yard}^3}$

36. $\left[\left(\frac{60 \text{ kilometers}}{1 \text{ hour}} \cdot \frac{1000 \text{ meters}}{1 \text{ kilometer}}\right) \div \frac{60 \text{ minutes}}{1 \text{ hour}}\right] \div \frac{60 \text{ seconds}}{1 \text{ minute}}$

Critical Thinking

37. **Geometry** The area of a rectangle is $\frac{x^2 - y^2}{2}$, and its length is $2x + 2y$. Find the width.

Applications and Problem Solving

38. **Construction** A construction supervisor needs to determine how many truckloads of earth must be removed from a site before a foundation can be poured. The bed of the truck has the shape shown at the right.

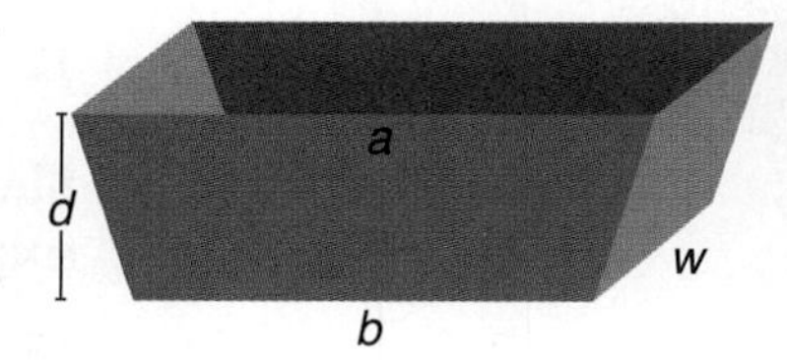

   a. Use the formula $V = \frac{d(a + b)}{2} \cdot w$ to write an expression involving units that represents the volume of the truck bed in cubic yards if $a = 18$ ft, $b = 15$ ft, $w = 9$ ft, and $d = 5$ ft.

   b. There are 20,000 cubic yards of earth that must be removed from the excavation site. Write an expression involving units that represents the number of truckloads that will be required to remove all the earth.

FYI

Phonograph records are played on a turntable that rotates at a constant speed. Each side of the record has a single groove, or *track*. As the record turns, the stylus of a *tone arm* vibrates in response to variations in the track. The vibration is converted to an electrical signal, which in turn is converted into sound.

39. **Railroads** The table at the right shows data about railroads in the United States.

Year	Miles of Track	Number of Freight Cars
1980	290,000	1,168,000
1985	257,000	867,000
1990	239,000	659,000
1992	227,000	605,000

   a. If one freight car is 48 feet long on average, how many miles long would a train without a locomotive be if all the freight cars for 1992 were connected?

   b. How many trains whose length is the answer to part a would it take to occupy all of the track in 1992?

   c. Repeat parts a and b for the year 1980.

40. **Music** Many old phonograph records turn at a rate of $33\frac{1}{3}$ revolutions per minute. Suppose a record of this type plays for 16.5 minutes and the average of the radii of the grooves on the record is $3\frac{3}{4}$ inches.

   a. Write an expression involving units that represents how many inches the needle travels while playing the record.

   b. Find the distance the needle travels.

Mixed Review

41. Find $\frac{2m + 3}{4} \cdot \frac{32}{(2m + 3)(m - 5)}$. (Lesson 12–2)

42. Solve $x^2 + 6x + 8 = 0$ by graphing. (Lesson 11–2)

43. Factor $16a^2 - 24ab^2 + 9b^4$. (Lesson 10–5)

44. **Health** If your heart beats once every second and you live to be 78 years old, your heart will have beat about 2,460,000,000 times. Write this number in scientific notation. (Lesson 9–3)

45. Solve $-3x + 6 > 12$. (Lesson 7–3)

46. Find $-3 + 4 - 10$. (Lesson 2–3)

47. Use the numbers 7 and 2 to write a mathematical sentence illustrating the commutative property of addition. (Lesson 1–8)

12–4 Dividing Polynomials

What YOU'LL LEARN

- To divide polynomials by monomials, and
- to divide polynomials by binomials.

Why IT'S IMPORTANT

You can use polynomials to solve problems involving science and interior design.

Interior Design

Tomi wants to put a decorative border waist-high around his dining room. The perimeter of the room is 52 feet, and the widths of the two windows and two doorways total $12\frac{3}{4}$ feet. To find the number of yards of border needed, he can use the expression $\frac{52 \text{ ft} - 12\frac{3}{4} \text{ ft}}{3 \text{ ft/yd}}$. The expression $\frac{52 \text{ ft}}{3 \text{ ft/yd}} - \frac{12\frac{3}{4} \text{ ft}}{3 \text{ ft/yd}}$ could also be used. In each case, each term of the numerator was divided by the denominator. *This expression will be simplified in Example 4.*

12 ft
14 ft
34.5 in.
34.5 in.
42 in.
42 in.

To divide a polynomial by a monomial, you divide each term of the polynomial by the monomial.

Example 1 **Find each quotient.**

a. $(3r^2 - 5) \div 12r$

$$(3r^2 - 5) \div 12r = \frac{3r^2 - 5}{12r} \quad \textit{Write as a rational expression.}$$

$$= \frac{3r^2}{12r} - \frac{5}{12r} \quad \textit{Divide each term by } 12r.$$

$$= \frac{\overset{1}{\cancel{3}}\overset{r}{\cancel{r^2}}}{\underset{4}{\cancel{12}}\underset{1}{\cancel{r}}} - \frac{5}{12r} \quad \textit{Divide by the common factors.}$$

$$= \frac{r}{4} - \frac{5}{12r}$$

b. $(9n^2 - 15n + 24) \div 3n$

$$(9n^2 - 15n + 24) \div 3n = \frac{9n^2 - 15n + 24}{3n}$$

$$= \frac{9n^2}{3n} - \frac{15n}{3n} + \frac{24}{3n}$$

$$= \frac{\overset{3}{\cancel{9}}\overset{n}{\cancel{n^2}}}{\underset{1}{\cancel{3n}}} - \frac{\overset{5}{\cancel{15n}}}{\underset{1}{\cancel{3n}}} + \frac{\overset{8}{\cancel{24}}}{\underset{1}{\cancel{3}}n} \quad \textit{Divide by the common factors.}$$

$$= 3n - 5 + \frac{8}{n}$$

Recall from Lesson 12–3 that, when you can factor, some divisions can be performed easily, as shown below.

$$(a^2 + 7a + 12) \div (a + 3) = \frac{a^2 + 7a + 12}{(a + 3)}$$

$$= \frac{\overset{1}{\cancel{(a+3)}}(a + 4)}{\underset{1}{\cancel{(a+3)}}} \quad \textit{Factor the dividend.}$$

$$= a + 4$$

You can use algebra tiles to model some quotients of polynomials.

MODELING MATHEMATICS

Dividing Polynomials

Materials: 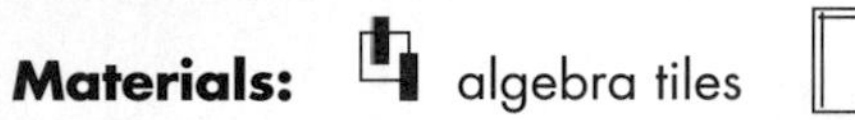 algebra tiles, product mat

Use algebra tiles to find $(x^2 + 2x - 8) \div (x + 4)$.

a. Model the polynomial $x^2 + 2x - 8$.

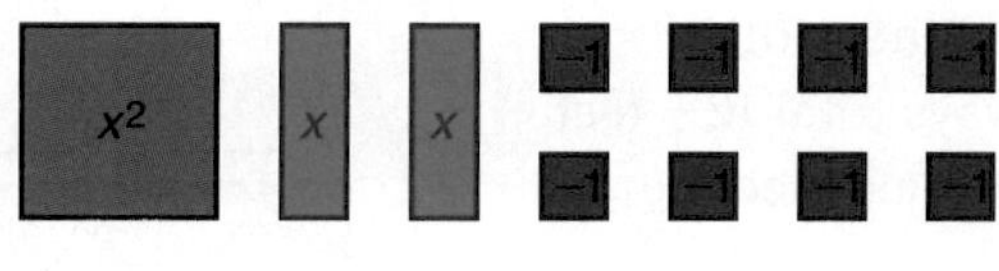

$x^2 + 2x + (-8)$

b. Place the x^2-tile at the corner of the mat. Arrange four of the 1-tiles as shown at the right, to make a length of $x + 4$.

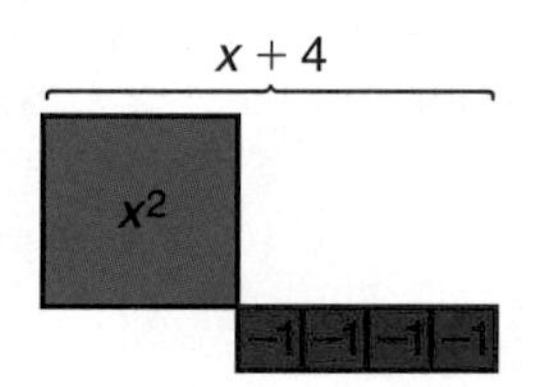

c. Use the remaining tiles to make a rectangular array. Recall that you can add zero-pairs without changing the value of the polynomial.

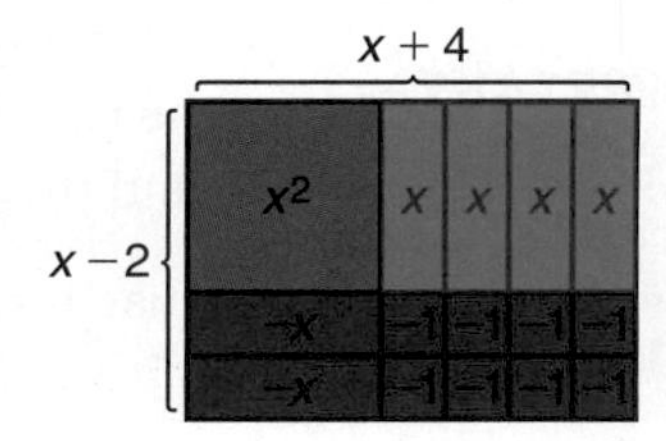

The width of the array, $x - 2$, is the quotient.

Your Turn

Use algebra tiles to find each quotient.

a. $(x^2 + 2x - 8) \div (x + 4)$

b. $(x^2 + 3x - 10) \div (x + 5)$

c. $(x^2 - 6x + 9) \div (x - 3)$

d. $(x^2 - 9) \div (x + 3)$

e. What happens when you try to model $(x^2 + 5x + 9) \div (x + 3)$? What do you think your result means?

When you cannot factor, you can use a long division process similar to the one you used in arithmetic. The division $(x^2 + 7x + 12) \div (x + 2)$ is shown below.

Step 1: To find the first term of the quotient, divide the first term of the dividend, x^2, by the first term of the divisor, x.

$$\begin{array}{r l}
x & \\
x + 2\overline{)x^2 + 7x + 12} & \quad x^2 \div x = x \\
\underline{(-)\ x^2 + 2x} & \quad \textit{Multiply } x \textit{ and } x + 2. \\
5x & \quad \textit{Subtract.}
\end{array}$$

Step 2: To find the next term of the quotient, divide the first term of the partial dividend, $5x$, by the first term of the divisor, x.

$$\begin{array}{r l}
x + \ 5 & \quad 5x \div x = 5. \\
x + 2\overline{)x^2 + 7x + 12} & \\
\underline{(-)\ x^2 + 2x} & \\
5x + 12 & \quad \textit{Bring down the 12.} \\
\underline{(-)\ 5x + 10} & \quad \textit{Multiply 5 and } x + 2. \\
2 & \quad \textit{Subtract.}
\end{array}$$

Therefore, the quotient when $x^2 + 7x + 12$ is divided by $x + 2$ is $x + 5$ with a remainder of 2. Since there is a nonzero remainder, the divisor is not a factor of the dividend. The remainder can be expressed as shown below.

$$\underbrace{(x^2 + 7x + 12)}_{\text{dividend}} \div \underbrace{(x + 2)}_{\text{divisor}} = \underbrace{x + 5 + \frac{2}{x + 2}}_{\text{quotient}}$$

Example 2 **Find $(3k^2 - 7k - 6) \div (3k + 2)$.**

Method 1: Long Division

$$\begin{array}{r} k - 3 \\ 3k + 2\overline{)3k^2 - 7k - 6} \\ (-)\ \underline{3k^2 + 2k} \\ -9k - 6 \\ (-)\ \underline{-9k - 6} \\ 0 \end{array}$$

Method 2: Factoring

$$\frac{3k^2 - 7k - 6}{3k + 2} = \frac{\overset{1}{\cancel{(3k + 2)}}(k - 3)}{\underset{1}{\cancel{3k + 2}}}$$

$$= k - 3$$

The quotient is $k - 3$ with a remainder of 0.

When the dividend is an expression like $s^3 + 9$, there is no s^2 term or s term. In such situations, you must rename the dividend using 0 as the coefficient of the missing terms.

Example 3 **Find $(s^3 + 9) \div (s - 3)$.**

$$\begin{array}{r} s^2 + 3s + 9 \\ s - 3\overline{)s^3 + 0s^2 + 0s + 9} \\ (-)\ \underline{s^3 - 3s^2} \\ 3s^2 + 0s \\ (-)\ \underline{3s^2 - 9s} \\ 9s + 9 \\ (-)\ \underline{9s - 27} \\ 36 \end{array} \quad s^3 + 9 = s^3 + 0s^2 + 0s + 9$$

Therefore, $(s^3 + 9) \div (s - 3) = s^2 + 3s + 9 + \frac{36}{s - 3}$.

Interior Design

Example 4 **Refer to the application at the beginning of the lesson. If the border comes in 5-yard rolls, how many rolls of border should Tomi buy?**

First find the total number of yards needed.

$$\frac{52 \text{ ft} - 12\frac{3}{4} \text{ ft}}{3 \text{ ft/yd}} = \frac{52 \text{ ft}}{3 \text{ ft/yd}} - \frac{12\frac{3}{4} \text{ ft}}{3 \text{ ft/yd}}$$

$$= \frac{52 \cancel{\text{ft}}}{1} \cdot \frac{1 \text{ yd}}{3 \cancel{\text{ft}}} - \frac{\overset{17}{\cancel{51}}}{4} \cancel{\text{ft}} \cdot \frac{1 \text{ yd}}{\underset{1}{\cancel{3}} \cancel{\text{ft}}}$$

$$= 17\frac{1}{3} \text{ yd} - 4\frac{1}{4} \text{ yd or } 13\frac{1}{12} \text{ yd}$$

Two rolls are 10 yards, which is not enough. So, Tomi should buy 3 rolls of border.

You can use a graphing calculator to compare a rational expression with the quotient that results when you divide its numerator by its denominator.

EXPLORATION — GRAPHING CALCULATORS

Consider the following rational expression and its quotient.

rational expression *quotient*

$$\frac{2x}{x+5} = 2 - \frac{10}{x+5}$$

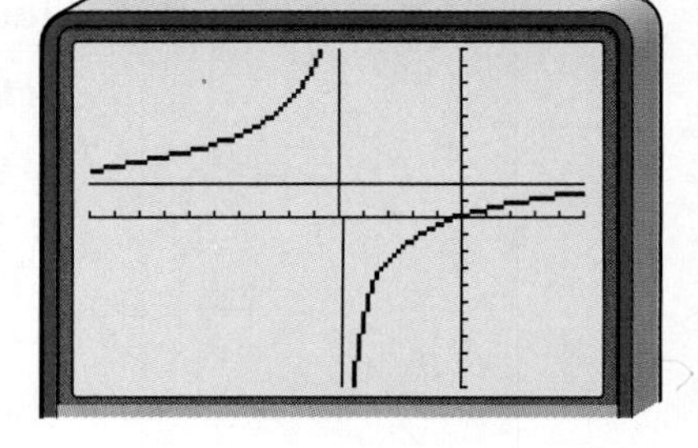

Use a graphing calculator to graph $y = \frac{2x}{x+5}$ and $y = 2$ on the same screen.

Use the viewing window $[-15, 5]$ by $[-10, 10]$.

Enter: Y= 2 X,T,θ ÷ (X,T,θ + 5) ENTER 2 GRAPH

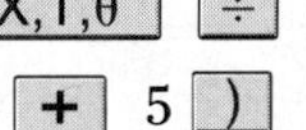

Your Turn

a. What can you say about the two graphs from examining the calculator screen?

b. Change Xmin and Xmax to -47 and 47. What can you say about the value of $\frac{2x}{x+5}$ and 2 as x gets larger and larger?

c. Use long division to find the quotient for $\frac{3x}{x-5}$.

d. What conclusion can you draw about the graph of $y = \frac{3x}{x-5}$ and the graph of $y = 3$?

CHECK FOR UNDERSTANDING

Communicating Mathematics

Study the lesson. Then complete the following.

1. Copy the division below and label the dividend, divisor, quotient, and remainder.

$$\frac{2x^2 - 11x - 20}{2x + 3} = x - 7 + \frac{1}{2x + 3}$$

2. **Explain** the meaning of a remainder of 0 in a long division of a polynomial by a binomial.
3. Refer to the application at the beginning of the lesson. Suppose Tomi decided to place the border at the bottom of the room above the baseboard, but below the windows. How many rolls of border should he buy?

4. **a.** Use algebra tiles to model $2x^2 - 9x + 9$.
   b. Which of the following divisors of $2x^2 - 9x + 9$ results in a remainder of 0?
   $x + 3$ $x - 3$ $2x - 3$ $2x + 3$

Guided Practice

Find each quotient.

5. $(9b^2 - 15) \div 3$
6. $(a^2 + 5a + 13) \div 5a$
7. $(t^2 + 6t - 7) \div (t + 7)$
8. $(s^2 + 11s + 18) \div (s + 2)$
9. $\frac{2m^2 + 7m + 3}{m + 2}$
10. $\frac{3r^2 + 11r + 7}{r + 5}$

11. **Geometry** Find the length of a rectangle if its area is $(10x^2 + 29x + 21)$ square meters and its width is $(5x + 7)$ meters.

EXERCISES

Practice

Find each quotient.

12. $(x^3 + 2x^2 - 5) \div 2x$
13. $(b^2 + 9b - 7) \div 3b$
14. $(3a^2 + 6a + 2) \div 3a$
15. $(m^2 + 7m - 28) \div 7m$
16. $(9xy^2 - 15xy + 3) \div 3xy$
17. $(a^3 + 8a - 21) \div (a - 2)$
18. $(2b^2 + 3b - 5) \div (2b - 1)$
19. $(m^2 + 4m - 23) \div (m + 7)$
20. $(2x^2 - 7x - 16) \div (2x + 3)$
21. $(2x^2 - 8x - 41) \div (x - 7)$
22. $\frac{14a^2b^2 + 35ab^2 + 2a^2}{7a^2b^2}$
23. $\frac{12m^3k + 16mk^3 - 8mk}{4mk}$
24. $\frac{3r^2 + 20r + 11}{r + 6}$
25. $\frac{a^2 + 10a + 20}{a + 3}$
26. $\frac{4m^2 + 8m - 19}{2m + 7}$
27. $\frac{6x^2 + 5x + 15}{2x + 3}$
28. $\frac{y^2 - 19y + 9}{y - 4}$
29. $\frac{4t^2 + 17t - 1}{4t + 1}$
30. Find the quotient when $x^2 + 9x + 15$ is divided by $x + 3$.
31. What is the quotient when $56x^3 + 32x^2 - 63x$ is divided by $7x$?

Programming

32. The graphing calculator program at the right will help you find the quotient and remainder when you divide a polynomial of the form $ax^2 + bx + c$ by a binomial of the form $x - r$. You enter the values of a, b, c, and r when prompted.

```
PROGRAM: POLYDIV
: Disp "ENTER A, B, C"
: Input A: Input B:
  Input C
: Disp "ENTER R"
: Input R
: Disp "COEFFICIENTS"
: Disp "OF QUOTIENT:"
: Disp A
: Disp B+A*R
: Disp "REMAINDER:"
: Disp C+B*R+A*R^2
```

Run the program to find each quotient and remainder.

a. $(7x^2 + 5x - 3) \div (x + 2)$
b. $(x^2 - 14x - 25) \div (3x + 4)$

33. Modify the graphing calculator program given above so that you can use it to find each quotient.
    a. $(x^3 + 2x^2 - 4x - 8) \div (x - 2)$
    b. $(20t^3 - 27t^2 + t - 6) \div (4t - 3)$
    c. $(2a^3 + 9a^2 + 5a - 12) \div (a + 3)$

Critical Thinking

34. Find the value of k if $x + 7$ is a factor of $x^2 - 2x - k$.
35. Find the value of k if $2m - 3$ is a factor of $2m^2 + 7m + k$.
36. Find the value of k if the remainder is 15 when $x^3 - 7x^2 + 4x + k$ is divided by $x - 2$.

Applications and Problem Solving

silver

raw copper

37. Environment Due to concerns about air pollution, it is important to invest money on the reduction of pollutants. The equation $C = \frac{120{,}000p}{1 - p}$ models the expenditure C in dollars needed to reduce pollutants by p percent where p is the percent in decimal form. If a utility company wants to remove 80% of the pollutants their equipment emits, what expenditure must they make?

38. Science The *density* of a material is its mass per unit volume. A 2.48 g block of copper occupies 0.28 cm^3. For example, the density of copper is found from the quotient $\frac{2.48 \text{ g}}{0.28 \text{ cm}^3} \approx 8.9 \text{ g/cm}^3$.

a. Make a table of densities for the materials above.

b. Make a line plot of the densities computed in part a. Use densities rounded to the nearest whole number.

c. Interpret the line plot made in part b.

Material	Mass(g)	Volume (cm^3)
aluminum	4.15	1.54
gold	2.32	0.12
silver	6.30	0.60
steel	7.80	1.00
iron	15.20	1.95
copper	2.48	0.28
blood	4.35	4.10
lead	11.30	1.00
brass	17.90	2.08
concrete	40.00	20.00

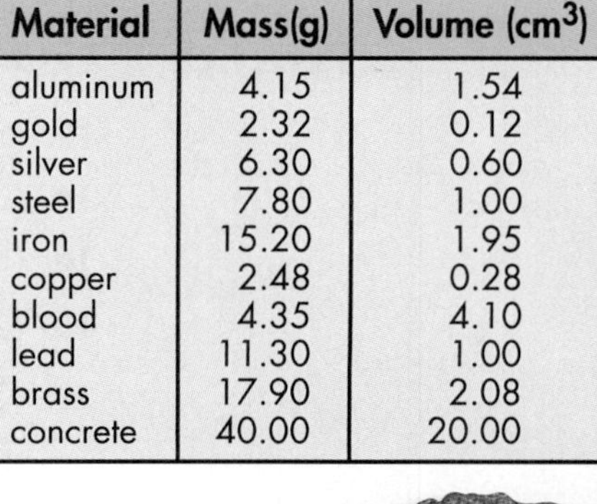

gold

Mixed Review

39. Find $\frac{x^2 - 16}{16 - x^2} \div \frac{7}{x}$. (Lesson 12–3)

40. Graph $y = -x^2 + 2x + 3$. (Lesson 11–1)

41. Factor $3x^2 - 6x - 105$. (Lesson 10–3)

42. Geometry Find the area of a rectangle if the length is $(2x + y)$ units and the width is $(x + y)$ units. (Lesson 9–7)

43. Solve the system of equations by graphing. (Lesson 8–1)

$y = 2x + 1$

$y = -2x + 5$

44. Write an equation for the relation $\{(2, 4), (3, 6), (-2, -4)\}$. (Lesson 5–6)

45. Graph $\{-2, -1, 4, 5\}$ on a number line. (Lesson 2–1)

SELF TEST

Simplify each rational expression. (Lesson 12–1)

1. $\frac{25x^3y^4}{36x^2y^5}$

2. $\frac{4x^2 - 9}{2x^2 + 13x - 15}$

Find each product or quotient. (Lessons 12–2 and 12–3)

3. $\frac{x^2 - 16}{x^2 + 5x + 6} \cdot \frac{4x^2 + 2x - 3}{x^2 - 5x + 4}$

4. $\frac{2x^2 - 5x + 2}{x^2 - 5x + 6} \div \frac{2x^2 + 9x - 5}{x^2 - 4x + 3}$

5. $(3x - 2) \cdot \frac{x - 5}{3x^2 + 10x - 8}$

6. $\frac{7x^2 + 36x + 5}{x - 5} \div (7x + 1)$

Find each quotient. (Lesson 12–4)

7. $(4x^2 - 18x + 20) \div (2x - 4)$

8. $\frac{3x^2 - 6x - 4}{x - 2}$

9. **Geometry** A rectangular field has an area of $12x^2 + 20x - 8$ square units and a width of $x + 2$ units. Find its length in terms of x. (Lesson 12–4)

10. **Travel** On the first day of a trip, Manuel drove 440 miles in 8 hours. The second day, he drove at the same speed but for only 6 hours. How far did he drive on the second day? (Lesson 12–2)

12-5 Rational Expressions with Like Denominators

What YOU'LL LEARN

- To add and subtract rational expressions with like denominators.

Why IT'S IMPORTANT

You can use rational expressions to solve problems involving history and geography.

History

At 11:40 P.M. on the night of April 14, 1912, the lookout on the bridge of the *Titanic* spotted an iceberg directly ahead. Despite their heroic efforts, the crew could not steer clear of the iceberg, and a 300-foot gash was ripped in the side of the ship. The *Titanic* sank less than three hours later, and 1500 lives were lost.

The New York Times.

TITANIC SINKS FOUR HOURS AFTER HITTING ICEBERG; 866 RESCUED BY CARPATHIA, PROBABLY 1250 PERISH; ISMAY SAFE, MRS. ASTOR MAYBE, NOTED NAMES MISSING

The Lost Titanic Being Towed Out of Belfast Harbor.

PARTIAL LIST OF THE SAVED.

CAPT. E. J. SMITH

The iceberg that the *Titanic* hit has been described as enormous. However, many people are surprised to learn that only about $\frac{1}{8}$ of any iceberg is visible above the surface of the water. This means the part of an iceberg that is submerged is about $1 - \frac{1}{8}$.

$$1 - \frac{1}{8} = \frac{8}{8} - \frac{1}{8} = \frac{8-1}{8} \text{ or } \frac{7}{8}$$

That is, about $\frac{7}{8}$, or 87.5% of the iceberg is submerged.

This example illustrates that, to add or subtract fractions with like denominators, you add or subtract the numerators and then write the sum or difference over the common denominator. You can use this same method to add or subtract rational expressions with like denominators.

rational numbers	*rational expressions*
$\frac{1}{9} + \frac{4}{9} = \frac{1+4}{9} = \frac{5}{9}$	$\frac{4}{y} + \frac{7}{y} = \frac{4+7}{y} = \frac{11}{y}$
$\frac{6}{7} - \frac{2}{7} = \frac{6-2}{7} = \frac{4}{7}$	$\frac{9}{5z} - \frac{6}{5z} = \frac{9-6}{5z} = \frac{3}{5z}$

Example 1 Find $\frac{a}{15m} + \frac{2a}{15m}$.

$\frac{a}{15m} + \frac{2a}{15m} = \frac{a+2a}{15m}$ *The common denominator is 15m.*

$= \frac{3a}{15m}$ *Add numerators.*

$= \frac{\overset{1}{\cancel{3}}a}{\underset{5}{\cancel{15}}m}$ *Divide by the common factors.*

$= \frac{a}{5m}$

LOOK BACK

You can refer to Lesson 2-5 for information about adding and subtracting rational numbers.

Sometimes the denominators of rational expressions are binomials. As long as each rational expression in a sum or difference has exactly the same binomial as its denominator, the process of adding or subtracting is the same.

Example  **Find $\frac{4}{x+3} - \frac{1}{x+3}$.**

$\frac{4}{x+3} - \frac{1}{x+3} = \frac{4-1}{x+3}$ *The common denominator is x + 3.*

$= \frac{3}{x+3}$ *Subtract numerators.*

Remember that, to subtract a polynomial, you add its additive inverse.

Example 3 **Find $\frac{2m+3}{m-4} - \frac{m-2}{m-4}$.**

$$\frac{2m+3}{m-4} - \frac{m-2}{m-4} = \frac{(2m+3)-(m-2)}{m-4}$$

$$= \frac{2m+3+[-(m-2)]}{m-4}$$ *The additive inverse of $m-2$ is $-(m-2)$.*

$$= \frac{2m+3-m+2}{m-4}$$

$$= \frac{m+5}{m-4}$$

Sometimes you must factor in order to simplify a sum or difference of rational expressions. Also a factor may need to be rewritten as its additive inverse to recognize the common denominator.

Example 4 **Find $\frac{7k+2}{4k-3} + \frac{8-k}{3-4k}$.**

The denominator $3-4k$ is the same as $-(-3+4k)$ or $-(4k-3)$. Rewrite the second expression so that the denominator is the same as the first expression.

$$\frac{7k+2}{4k-3} + \frac{8-k}{3-4k} = \frac{7k+2}{4k-3} - \frac{8-k}{4k-3}$$

$$= \frac{7k+2-(8-k)}{4k-3}$$

$$= \frac{8k-6}{4k-3}$$

$$= \frac{2(4k-3)}{(4k-3)}$$ *Factor the numerator.*

$$= \frac{2\overset{1}{\cancel{(4k-3)}}}{\underset{1}{\cancel{(4k-3)}}}$$ *Divide by the common factors.*

$$= 2$$

Example 5 **Find an expression for the perimeter of rectangle *ABCD*.**

$$P = 2\ell + 2w$$

$$= 2\left(\frac{9r}{2r+6s}\right) + 2\left(\frac{5r}{2r+6s}\right)$$

$$= \frac{2(9r)+2(5r)}{2r+6s}$$

$$= \frac{18r+10r}{2r+6s}$$

$$= \frac{28r}{2r+6s}$$

$$= \frac{28r}{2(r+3s)}$$ *Factor the denominator.*

$$= \frac{\overset{14}{\cancel{28}}r}{\underset{1}{\cancel{2}}(r+3s)}$$ *Divide by the common factors.*

$$= \frac{14r}{r+3s}$$

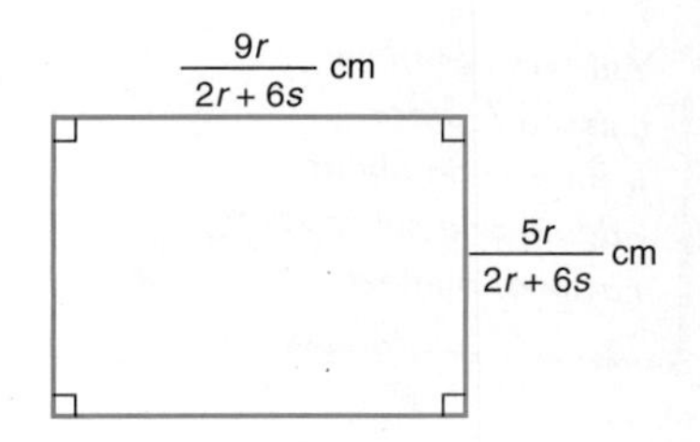

The perimeter of rectangle *ABCD* is $\left(\frac{14r}{r+3s}\right)$ cm.

CHECK FOR UNDERSTANDING

Communicating Mathematics

Study the lesson. Then complete the following.

1. **Compare and contrast** two rational expressions whose sum is 0 with two rational expressions whose difference is 0.
2. **Summarize** the procedure for adding or subtracting two rational expressions whose denominators are the same.
3. **You Decide** Abigail wrote $\frac{7x - 3}{2x + 11} + \frac{3x - 9}{2x + 11} = \frac{10x - 12}{4x + 22}$. What mistake did she make?

Guided Practice

Find each sum or difference. Express in simplest form.

4. $\frac{5x}{7} + \frac{2x}{7}$
5. $\frac{3}{x} + \frac{7}{x}$
6. $\frac{7}{3m} - \frac{4}{3m}$
7. $\frac{3}{a + 2} + \frac{7}{a + 2}$
8. $\frac{2m}{m + 3} - \frac{-6}{m + 3}$
9. $\frac{3x}{x + 4} - \frac{-12}{x + 4}$
10. If the sum of $\frac{2x - 1}{3x + 2}$ and another rational expression with denominator $3x + 2$ is $\frac{5x - 1}{3x + 2}$, what is the numerator of the second rational expression?

EXERCISES

Practice

Find each sum or difference. Express in simplest form.

11. $\frac{m}{3} + \frac{2m}{3}$
12. $\frac{3y}{11} - \frac{8y}{11}$
13. $\frac{5a}{12} - \frac{7a}{12}$
14. $\frac{4}{3z} + \frac{-7}{3z}$
15. $\frac{x}{2} - \frac{x - 4}{2}$
16. $\frac{a + 3}{6} - \frac{a - 3}{6}$
17. $\frac{2}{x + 7} + \frac{5}{x + 7}$
18. $\frac{2x}{x + 1} + \frac{2}{x + 1}$
19. $\frac{y}{y - 2} + \frac{2}{2 - y}$
20. $\frac{4m}{2m + 3} + \frac{5}{2m + 3}$
21. $\frac{-5}{3x - 5} + \frac{3x}{3x - 5}$
22. $\frac{3r}{r + 5} + \frac{15}{r + 5}$
23. $\frac{2x}{x + 2} + \frac{2x}{x + 2}$
24. $\frac{2m}{m - 9} + \frac{18}{9 - m}$
25. $\frac{2y}{y + 3} + \frac{-6}{y + 3}$
26. $\frac{3m}{m - 2} - \frac{5}{m - 2}$
27. $\frac{4x}{2x + 3} - \frac{-6}{2x + 3}$
28. $\frac{4t - 1}{1 - 4t} + \frac{2t + 3}{1 - 4t}$
29. If $\frac{3x - 100}{2x + 5}$ is added to the sum of $\frac{11x - 5}{2x + 5}$ and $\frac{11x + 12}{2x + 5}$, what is the result?
30. If $\frac{-3b + 4}{2b + 12}$ is subtracted from the sum of $\frac{b - 15}{2b + 12}$ and $\frac{-3b + 12}{2b + 12}$, what is the result?

Geometry

Find an expression for the perimeter of each rectangle.

31.

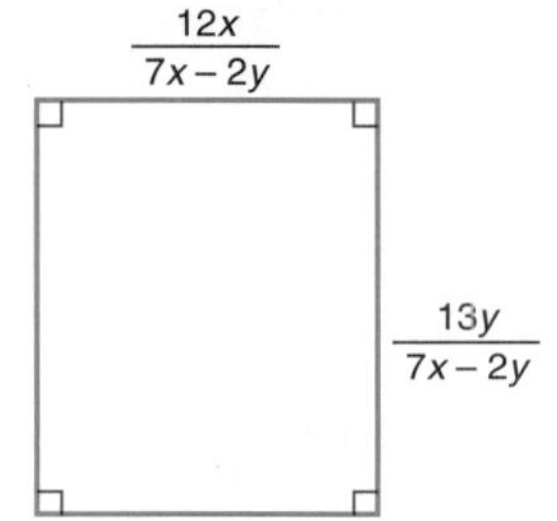

32.

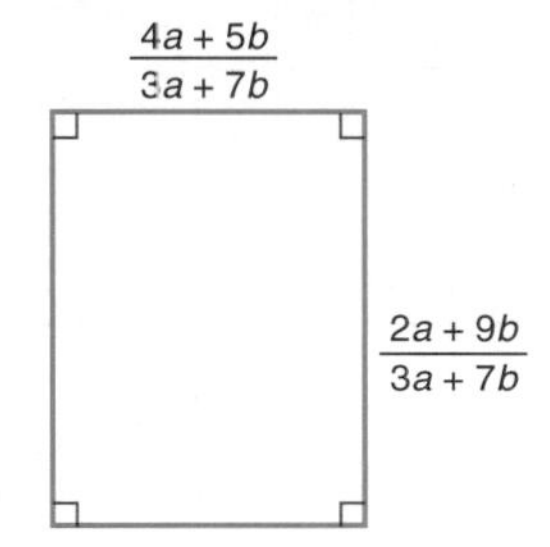

Critical Thinking

33. Which of the following rational numbers is not equivalent to the others?

a. $\frac{-3}{x-2}$ **b.** $-\frac{3}{2-x}$ **c.** $-\frac{3}{x-2}$ **d.** $\frac{3}{2-x}$

Applications and Problem Solving

34. Icebergs Water weighs 62.4 pounds per cubic foot. One cubic foot of water contains 7.48 gallons. Each cubic foot of ice yields 0.89 times as many gallons of water as a cubic foot of water.

a. How many gallons of water does one cubic foot of ice contain?

b. Some people have suggested moving icebergs from the North Atlantic to parts of the world that have little fresh water. If an iceberg is 1 mile wide, 2 miles long, and 800 feet thick, about how much fresh water would such an iceberg yield?

35. Utilities A person in the United States on the average uses 168 gallons of water each day. Suppose that 25% of the iceberg in Exercise 34 is lost as it is being moved to another location. How many days would the iceberg sustain the Atlanta, Georgia, metropolitan area, which has a population of about 3 million people?

Mixed Review

36. Engineering The Canadian Pacific Railroad constructed a track in the form of a circle so the train will spiral over itself as it passes through a mountain tunnel. This reduces the degree of incline for the train to climb. If we assume that the spiral is a circle one mile long, what would be the diameter of the circle in feet? (Lesson 12–4)

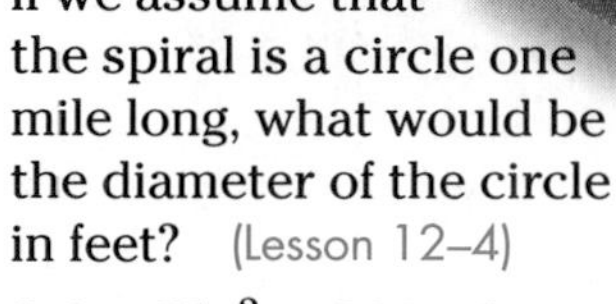

37. Solve $25x^2 = 36$ by factoring. (Lesson 10–6)

38. Geometry Find the perimeter of quadrilateral *MATH* shown at the right. (Lesson 9–5)

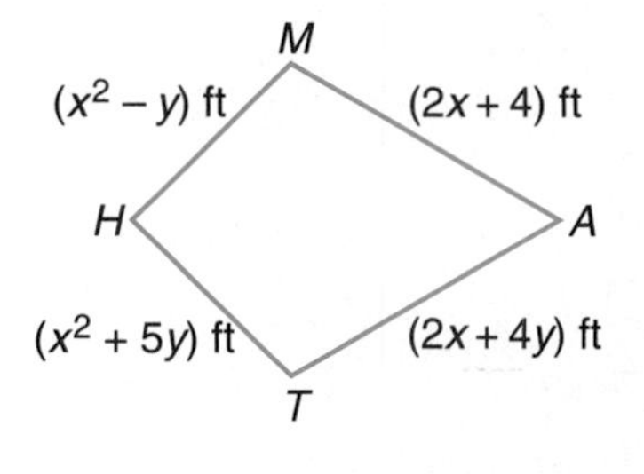

39. Use substitution to solve the system of equations. (Lesson 8–2)

$x + 2y = 5$

$x - 2y = -11$

40. Astronomy A *main sequence star* is a star that is in the neighborhood of the sun. The table below gives the names of the seven main sequence stars and the sun, their surface temperatures in thousands of °C, and their radii in multiples of the sun's radius. Draw a scatter plot with the temperatures on the horizontal axis and the radii on the vertical axis. (Lesson 6–3)

Stars	MU-1 Scorpii	Sirius A	Altair	Polycon A	Sun	61 Cygni A	Kueger 60	Barnard's Star
Surface Temp.	20	10.2	7.3	6.8	5.9	2.6	2.8	2.7
Radius	5.2	1.9	1.6	2.6	1.0	0.7	0.35	0.15

41. Probability A student rolls a die three times. What is the probability that each roll is a 1? (Lesson 4–6)

12–6 Rational Expressions with Unlike Denominators

What YOU'LL LEARN

- To add and subtract rational expressions with unlike denominators, and
- to make an organized list of possibilities to solve problems.

Why IT'S IMPORTANT

You can use rational expressions to solve problems involving astronomy and automobile maintenance.

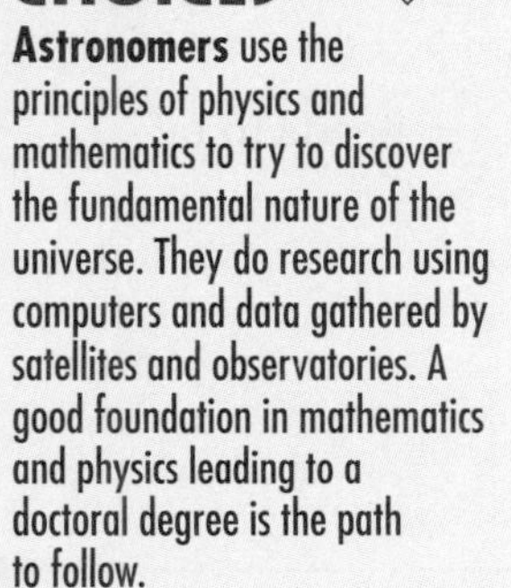

Astronomers use the principles of physics and mathematics to try to discover the fundamental nature of the universe. They do research using computers and data gathered by satellites and observatories. A good foundation in mathematics and physics leading to a doctoral degree is the path to follow.

For more information, contact:

American Astronomical Society
Education Office
University of Texas
Dept. of Astronomy
Austin, TX 78712-1083

Astronomy

Mars, Jupiter, and Saturn revolve around the sun approximately every 2 years, 12 years, and 30 years, respectively. The planets are said to be in *conjunction*, or alignment, when they appear close to one another in Earth's night sky. The last time this happened was in 1982. When will it happen again?

Mars will go through six revolutions for each one of Jupiter's. Mars will go through fifteen revolutions for each one of Saturn's. The times the planets align is related to the common multiples of 2, 12, and 30. In fact, the least number of years that will pass until the next alignment is the **least common multiple (LCM)** of these numbers. The least common multiple is the least number that is a common multiple of two or more numbers.

There are two methods you can use to find a least common multiple.

Method 1: Make an organized list.

You can often solve problems by listing the possibilities in an organized way. Use a systematic approach so you don't omit any important items.

Multiples of 2	Multiples of 12	Multiples of 30
$2 \cdot 0 = 0$	$12 \cdot 0 = 0$	$30 \cdot 0 = 0$
$2 \cdot 1 = 2$	$12 \cdot 1 = 12$	$30 \cdot 1 = 30$
$2 \cdot 2 = 4$	$12 \cdot 2 = 24$	$30 \cdot 2 = \mathbf{60}$
⋮	⋮	
$2 \cdot 30 = \mathbf{60}$	$12 \cdot 5 = \mathbf{60}$	

Compare the multiples in the table. Other than 0, the least number that is common to all three lists of multiples is 60. So, the LCM of 2, 12, and 30 is 60.

Method 2: Use prime factorization.

Find the prime factorization of each number.

$$2 = 2 \qquad 12 = 2 \cdot 2 \cdot 3 \qquad 30 = 2 \cdot 3 \cdot 5$$

Use each prime factor the greatest number of times it appears in any of the factorizations.

2 appears twice as a factor of 12. All other factors appear once.
Thus, the LCM of 2, 12, and 30 is $2 \cdot 2 \cdot 3 \cdot 5$ or 60.

Using either method, the LCM is 60. This means that the planets will be in alignment approximately 60 years after 1982, or about the year 2042.

You can use the same methods to find the LCM of two or more polynomials.

Example 1 **Find the LCM of $15a^2b^2$ and $24a^2b$.**

PROBLEM SOLVING
List the Possibilities

List the multiples of each coefficient and each variable expression.

15: 15, 30, 45, 60, 75, 90, 105, **120**, 135
24: 24, 48, 72, 96, **120**, 144
a^2: $\mathbf{a^2}$
b^2: $\mathbf{b^2}$, b^4, b^6
b: b, $\mathbf{b^2}$, b^3
LCM $= 120a^2b^2$

Example 2 **Find the LCM of $x^2 - x - 6$ and $x^2 + 2x - 15$.**

$x^2 - x - 6 = (x - 3)(x + 2)$ *Factor each expression.*

$x^2 + 2x - 15 = (x + 5)(x - 3)$

$\text{LCM} = (x - 3)(x + 2)(x + 5)$

To add or subtract fractions with unlike denominators, first rename the fractions so the denominators are alike. Any common denominator could be used. However, the computation usually is easier if you use the **least common denominator (LCD)**. Recall that the least common denominator is the LCM of the denominators.

You usually will use the following steps to add or subtract rational expressions with unlike denominators.

1. Find the LCD.
2. Change each rational expression into an equivalent expression with the LCD as the denominator.
3. Add or subtract as with rational expressions with like denominators.
4. Simplify if necessary.

Example 3 **Find $\frac{7}{3m} + \frac{5}{6m^2}$.**

$3m = 3 \cdot m$ *Factor each denominator.*

$6m^2 = 2 \cdot 3 \cdot m \cdot m$

$\text{LCM} = 2 \cdot 3 \cdot m \cdot m$, or $6m^2$

Since the denominator of $\frac{5}{6m^2}$ is already $6m^2$, only $\frac{7}{3m}$ needs to be renamed.

$$\frac{7}{3m} + \frac{5}{6m^2} = \frac{7(2m)}{3m(2m)} + \frac{5}{6m^2}$$ *Multiply $\frac{7}{3m}$ by $\frac{2m}{2m}$. Why?*

$$= \frac{14m}{6m^2} + \frac{5}{6m^2}$$

$$= \frac{14m + 5}{6m^2}$$ *Add the numerators.*

To combine rational expressions whose denominators are polynomials, follow the same procedure you use to combine expressions whose denominators are monomials.

Example 4 **Find each sum or difference.**

a. $\frac{s}{s + 3} + \frac{3}{s - 4}$

Since the denominators are $(s + 3)$ and $(s - 4)$, there are no common factors. The LCD is $(s + 3)(s - 4)$.

$$\frac{s}{s+3} + \frac{3}{s-4} = \frac{s}{s+3} \cdot \frac{s-4}{s-4} + \frac{3}{s-4} \cdot \frac{s+3}{s+3}$$

$$= \frac{s(s-4)}{(s+3)(s-4)} + \frac{3(s+3)}{(s+3)(s-4)}$$

$$= \frac{s^2 - 4s}{(s+3)(s-4)} + \frac{3s + 9}{(s+3)(s-4)}$$

$$= \frac{s^2 - 4s + 3s + 9}{(s+3)(s-4)}$$

$$= \frac{s^2 - s + 9}{(s+3)(s-4)}$$

b. $\dfrac{n-4}{(2-n)^2} - \dfrac{n-5}{n^2+n-6}$

$$\frac{n-4}{(2-n)^2} - \frac{n-5}{n^2+n-6}$$

$$= \frac{n-4}{(2-n)^2} - \frac{n-5}{(n-2)(n+3)} \quad \textit{Factor the denominators.}$$

$$= \frac{n-4}{(n-2)(n-2)} - \frac{n-5}{(n-2)(n+3)} \quad (2-n)^2 = (n-2)^2$$

$$= \frac{n-4}{(n-2)(n-2)} \cdot \frac{(n+3)}{(n+3)} - \frac{n-5}{(n-2)(n+3)} \cdot \frac{(n-2)}{(n-2)} \quad \textit{Rename each fraction using the LCD.}$$

$$= \frac{\left(n^2+3n-4n-12\right) - \left(n^2-2n-5n+10\right)}{(n-2)(n-2)(n+3)} \quad \textit{Subtract numerators}$$

$$= \frac{n^2-n-12-n^2+7n-10}{(n-2)^2(n+3)}$$

$$= \frac{6n-22}{(n-2)^2(n+3)}$$

CHECK FOR UNDERSTANDING

Communicating Mathematics

Study the lesson. Then complete the following.

1. **You Decide** Kaylee simplified $\frac{16}{64}$ as $\frac{1\not{6}}{\not{6}4} = \frac{1}{4}$. Is her answer correct? Was her method of simplifying correct? Explain.
2. **Describe** a situation in which the LCD of two or more rational expressions is equal to the denominator of one of the rational expressions.

Guided Practice

Find the LCD for each pair of rational expressions.

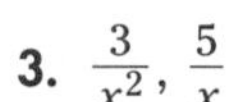

3. $\frac{3}{x^2}, \frac{5}{x}$
4. $\frac{4}{a^2b}, \frac{3}{ab^2}$
5. $\frac{4}{15m^2}, \frac{7}{18mb^2}$
6. $\frac{6}{a+6}, \frac{7}{a+7}$
7. $\frac{5}{x-3}, \frac{4}{x+3}$
8. $\frac{9}{2x-8}, \frac{10}{x-4}$

Find each sum or difference.

9. $\frac{7}{15m^2} + \frac{3}{5m}$
10. $\frac{2}{x+3} + \frac{3}{x-2}$
11. $3x + 6 - \frac{9}{x+2}$
12. $\frac{m+2}{m^2+4m+3} - \frac{6}{m+3}$
13. $\frac{11}{3y^2} - \frac{7}{6y}$
14. $\frac{4}{2g-7} + \frac{5}{3g+1}$
15. **Parades** At the Veteran's Day parade, the local members of the Veterans of Foreign Wars (VFW) found that they could arrange themselves in rows of 6, 7, or 8, with no one left over. What was the least number of VFW members in the parade?

EXERCISES

Practice

Find each sum or difference.

16. $\frac{m}{4} + \frac{3m}{5}$
17. $\frac{x}{7} - \frac{2x}{9}$
18. $\frac{m+1}{m} + \frac{m-3}{3m}$
19. $\frac{7}{x} + \frac{3}{xyz}$
20. $\frac{7}{6a^2} - \frac{5}{3a}$
21. $\frac{2}{st^2} - \frac{3}{s^2t}$

Find each sum or difference.

22. $\frac{3}{7m} + \frac{4}{5m^2}$

23. $\frac{3}{z+5} + \frac{4}{z-4}$

24. $\frac{d}{d+4} + \frac{3}{d+3}$

25. $\frac{k}{k+5} - \frac{2}{k+3}$

26. $\frac{3}{y-3} - \frac{y}{y+4}$

27. $\frac{10}{3r-2} - \frac{9}{r-5}$

28. $\frac{4}{3a-6} + \frac{a}{2+a}$

29. $\frac{5}{2m-3} - \frac{m}{6-4m}$

30. $\frac{b}{3b+2} + \frac{2}{9b+6}$

31. $\frac{w}{5w+2} - \frac{4}{15w+6}$

32. $\frac{h-2}{h^2+4h+4} + \frac{h-2}{h+2}$

33. $\frac{n+2}{n^2+4n+3} - \frac{6}{n+3}$

34. $\frac{a}{5-a} - \frac{3}{a^2-25}$

35. Find the difference when $\frac{2}{t+3}$ is subtracted from $\frac{3}{10t-9}$.

36. Find the sum of $\frac{2y}{y^2+7y+12}$ and $\frac{y+2}{y+4}$.

37. Find the sum of $\frac{2}{v+4}$, $\frac{v}{v-1}$, and $\frac{5v}{v^2+3v-4}$.

38. Find the difference when $\frac{2}{a+1}$ is subtracted from $\frac{6}{a-2}$. Then find the difference when $\frac{6}{a-2}$ is subtracted from $\frac{2}{a+1}$. How are the differences related?

Critical Thinking

39. Copy and complete.

a. $15 \cdot 24 = \underline{\ ?\ }$
GCF of 15 and 24 = $\underline{\ ?\ }$
LCM of 15 and 24 = $\underline{\ ?\ }$
GCF · LCM = $\underline{\ ?\ }$

b. $18 \cdot 30 = \underline{\ ?\ }$
GCF of 18 and 30 = $\underline{\ ?\ }$
LCM of 18 and 30 = $\underline{\ ?\ }$
GCF · LCM = $\underline{\ ?\ }$

c. Write a rule that describes the relationship between the GCF, the LCM, and the product of the two numbers.

d. Use your rule to describe how to find the LCM of two numbers if you know their GCF.

Applications and Problem Solving

40. **List the Possibilities** Doug Paulsen, the choreographer of a Broadway musical, has asked the producer of the show to hire enough dancers so they can be arranged in groups of exactly three, six, or seven, with no dancer left out. What is the least number of dancers required?

41. **Automobiles** Car owners need to follow a regular maintenance schedule to keep their cars running smoothly. The table below shows several of the checkups that should be performed on a regular basis, according to the Spring, 1995, issue of *Know-How* magazine.

If all these inspections and services are performed on April 20, 1996, and the owner follows the recommendations shown in the table, what is the next date on which they all should be performed again?

Inspection or Service	Frequency
engine oil and oil filter change	every 3000 miles (about 3 months)
transmission fluid level check	every oil change
brake system inspection	every oil change
chassis lubrication	every other oil change
power steering pump fluid level check	twice a year
tire and wheel rotation and inspection	every 15,000 miles

42. **Travel** Jaheed Toliver spent, in this order, a third of his life in the United States, a sixth of his life in Kenya, twelve years in Saudi Arabia, half the remainder in Australia, and as long in Canada as he spent in Hong Kong. How many years did Jaheed live if he spent his 45th birthday in Saudi Arabia and he spent a whole number of years in each country?

43. Find $\frac{8z + 3}{3z + 4} - \frac{2z - 5}{3z + 4}$. (Lesson 12–5)

Mixed Review

44. Simplify $\frac{x^2 + 7x + 6}{3x^2 + x - 2}$. State the excluded values of x. (Lesson 12–1)
45. Solve $2x^2 - 3x - 4 = 0$ by using the quadratic formula. (Lesson 11–3)
46. Factor $3x^2y + 6xy + 9y^2$. (Lesson 10–1)
47. Find $(3a^2b)(-5a^4b^2)$. (Lesson 9–1)
48. Graph the solution set of $3x > -15$ and $2x \leq 6$. (Lesson 7–4)
49. Solve $3n - 12 = 5n - 20$. (Lesson 3–5)
50. **Demographics** The populations of the capitals of some southern states are listed in the table below. Make a stem-and-leaf plot of the populations. (Lesson 1–4)

Capital	Population (thousands)	Capital	Population (thousands)
Atlanta, GA	394	Montgomery, AL	188
Austin, TX	466	Nashville, TN	488
Baton Rouge, LA	220	Oklahoma City, OK	445
Columbia, SC	98	Raleigh, NC	208
Frankfort, KY	26	Richmond, VA	203
Jackson, MS	197	Tallahassee, FL	125
Little Rock, AR	176		

Reading the Labels

The excerpt below appeared in an article in *Aging Magazine*, issue number 366, 1994.

BY THIS SUMMER, ALMOST ALL PACKAGED foods found in your local supermarket included new, improved labels that make it easier to understand the nutritional information contained on them and to compare the nutritional value of different products. This revolution in food labeling is the result of years of work by consumer advocates who pressed for more complete, accurate, and pertinent information on food labels....The percent of daily fat consumption should be no more than 30 percent of calories....The labels also give percentages of daily allowances for sodium, sugar, fiber, and protein. ■

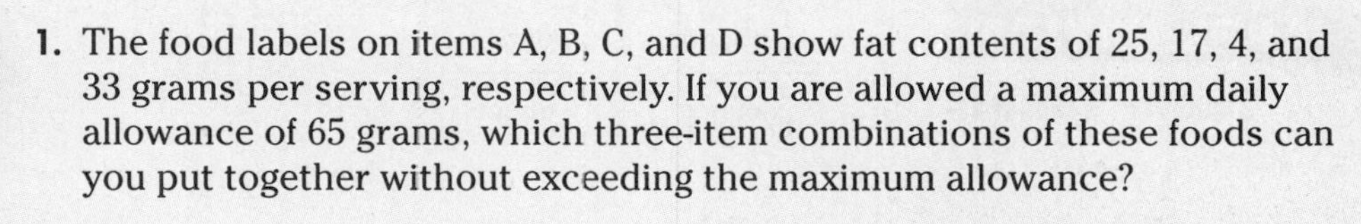

1. The food labels on items A, B, C, and D show fat contents of 25, 17, 4, and 33 grams per serving, respectively. If you are allowed a maximum daily allowance of 65 grams, which three-item combinations of these foods can you put together without exceeding the maximum allowance?

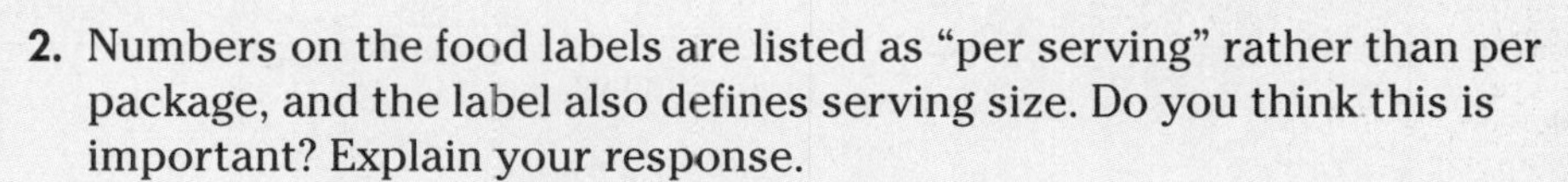

2. Numbers on the food labels are listed as "per serving" rather than per package, and the label also defines serving size. Do you think this is important? Explain your response.
3. When you shop for food, do you use the nutritional data on the labels in deciding what to buy? Why or why not?

12–7 Mixed Expressions and Complex Fractions

***What* YOU'LL LEARN**
- To simplify mixed expressions and complex fractions.

***Why* IT'S IMPORTANT**

You can use rational expressions to solve problems involving acoustics and statistics.

APPLICATION
Table Tennis

A Ping-Pong™ ball weighs about $\frac{1}{10}$ of an ounce. How many Ping-Pong balls weigh $1\frac{1}{2}$ pounds altogether? *This problem will be solved in Example 2.*

A number like $1\frac{1}{2}$ is a mixed number. Expressions like $a + \frac{b}{c}$ and $4 + \frac{x+y}{x-5}$ are **mixed expressions.** Changing mixed expressions to rational expressions is similar to changing mixed numbers to simple fractions (improper fractions).

mixed number to improper fraction

$$5\frac{4}{7} = 5 + \frac{4}{7} = \frac{5(7) + 4}{7} = \frac{35 + 4}{7} = \frac{39}{7}$$

mixed expression to rational expression

$$4 + \frac{x+y}{x-5} = \frac{4(x-5)}{x-5} + \frac{x+y}{x-5} = \frac{4(x-5) + (x+y)}{(x-5)} = \frac{4x - 20 + x + y}{x-5} = \frac{5x + y - 20}{x-5}$$

Example 1 **Simplify $7 + \frac{y-3}{y+4}$.**

$7 + \frac{y-3}{y+4} = \frac{7(y+4)}{y+4} + \frac{y-3}{y+4}$ *The LCD is y + 4.*

$= \frac{7(y+4) + y - 3}{y+4}$ *Add the numerators.*

$= \frac{7y + 28 + y - 3}{y+4}$

$= \frac{8y + 25}{y+4}$ *Simplify.*

Now let's solve the Ping-Pong ball problem.

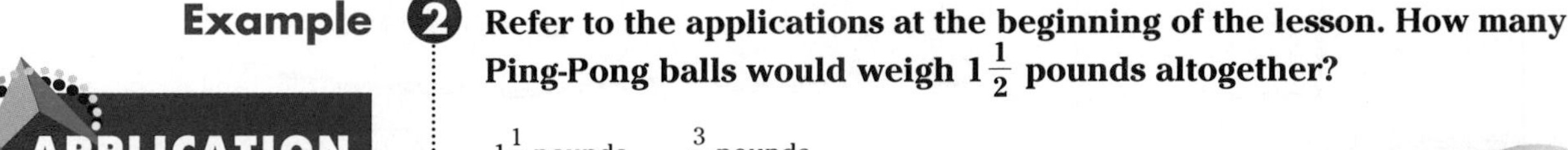

Example 2 **Refer to the applications at the beginning of the lesson. How many Ping-Pong balls would weigh $1\frac{1}{2}$ pounds altogether?**

APPLICATION
Table Tennis

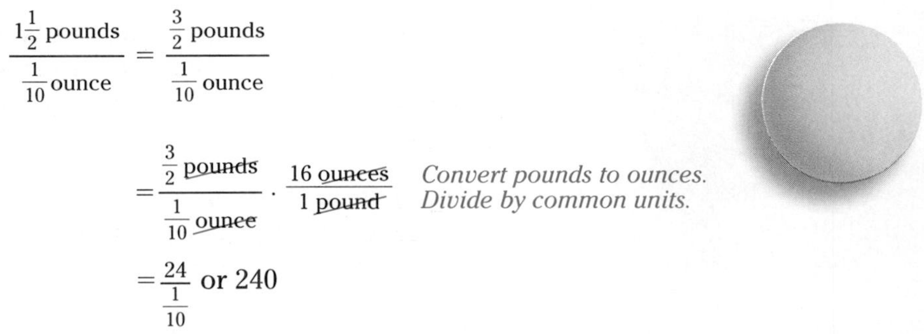

$$\frac{1\frac{1}{2}\text{ pounds}}{\frac{1}{10}\text{ ounce}} = \frac{\frac{3}{2}\text{ pounds}}{\frac{1}{10}\text{ ounce}}$$

$= \frac{\frac{3}{2}\text{ pounds}}{\frac{1}{10}\text{ ounce}} \cdot \frac{16\text{ ounces}}{1\text{ pound}}$ *Convert pounds to ounces. Divide by common units.*

$= \frac{24}{\frac{1}{10}}$ or 240

It would take 240 Ping-Pong balls to weigh $1\frac{1}{2}$ pounds.

Recall that if a fraction has one or more fractions in the numerator or denominator, it is called a *complex fraction*. Some complex fractions are shown below.

$$\frac{5\frac{1}{2}}{3\frac{3}{4}} \qquad \frac{9}{\frac{x}{y}} \qquad \frac{\frac{x+y}{y}}{\frac{x-y}{x}} \qquad \frac{\frac{1}{a}-\frac{1}{b}}{\frac{1}{a}+\frac{1}{b}}$$

You simplify an algebraic complex fraction in the same way you simplify a numerical complex fraction.

numerical

$$\frac{\frac{11}{2}}{\frac{15}{4}} = \frac{11}{2} \div \frac{15}{4}$$

$$= \frac{11}{2} \cdot \frac{4}{15} \quad \textit{The reciprocal of } \frac{15}{4} \textit{ is } \frac{4}{15}.$$

$$= \frac{22}{15}$$

algebraic

$$\frac{\frac{a}{b}}{\frac{c}{d}} = \frac{a}{b} \div \frac{c}{d}$$

$$= \frac{a}{b} \cdot \frac{d}{c} \quad \textit{The reciprocal of } \frac{c}{d} \textit{ is } \frac{d}{c}.$$

$$= \frac{ad}{bc}$$

Simplifying a Complex Fraction	Any complex fraction $\frac{\frac{a}{b}}{\frac{c}{d}}$, where $b \neq 0$, $c \neq 0$, and $d \neq 0$, can be expressed as $\frac{ad}{bc}$.

Example 3 Simplify each rational expression.

a. $\dfrac{1+\frac{4}{a}}{\frac{a}{6}+\frac{2}{3}}$

Simplify the numerator and denominator separately. Then divide.

$$\frac{1+\frac{4}{a}}{\frac{a}{6}+\frac{2}{3}} = \frac{\frac{1}{1}\cdot\frac{a}{a}+\frac{4}{a}}{\frac{a}{6}+\frac{2}{3}\cdot\frac{2}{2}} \quad \textit{The LCD of the numerator is a. The LCD of the denominator is 6.}$$

$$= \frac{\frac{a+4}{a}}{\frac{a+4}{6}} \quad \textit{Add to simplify both numerator and denominator.}$$

$$= \frac{a+4}{a} \div \frac{a+4}{6} \quad \textit{Rewrite as a division sentence.}$$

$$= \frac{\overset{1}{\cancel{a+4}}}{a} \cdot \frac{6}{\underset{1}{\cancel{a+4}}} \quad \textit{The reciprocal of } \frac{a+4}{6} \textit{ is } \frac{6}{a+4}.$$

$$= \frac{6}{a} \quad \textit{Divide by common factors.}$$

b. $\dfrac{m-\frac{m+5}{m-3}}{m+1}$

$$\frac{m-\frac{m+5}{m-3}}{m+1} = \frac{\frac{m(m-3)}{(m-3)}-\frac{m+5}{m-3}}{m+1} \quad \textit{The LCD of the numerator is } m-3. \textit{ The LCD of the denominator is } m+1.$$

$$= \frac{\frac{m^2-3m-m-5}{m-3}}{m+1} \quad \textit{Subtract to simplify the numerator.}$$

(continued on the next page)

$$= \frac{\frac{m^2 - 4m - 5}{m - 3}}{m + 1}$$ *Simplify.*

$$= \frac{\frac{(m + 1)(m - 5)}{m - 3}}{m + 1}$$ *Factor the numerator.*

$$= \frac{(m + 1)(m - 5)}{m - 3} \div (m + 1)$$ *Rewrite as a division sentence.*

$$= \frac{(m + 1)(m - 5)}{m - 3} \cdot \frac{1}{(m + 1)}$$ *The reciprocal of $m + 1$ is $\frac{1}{m + 1}$.*

$$= \frac{\overset{1}{\cancel{(m + 1)}}(m - 5)}{m - 3} \cdot \frac{1}{\underset{1}{\cancel{(m + 1)}}}$$ *Divide by the common factors.*

$$= \frac{m - 5}{m - 3}$$

You can use a graphing calculator to check the work you did in Example 3.

EXPLORATION — GRAPHING CALCULATORS

When you graph a function that involves rational expressions, remember to enclose every numerator and denominator in parentheses.

Your Turn

a. Graph $y = \frac{x - \frac{x + 5}{x - 3}}{x + 1}$. Use the window [−1.4, 8] by [−5, 5].

b. What are the excluded values of x in the expression in part a?

c. Graph $y = \frac{x - 5}{x - 3}$ on the same display used in part a.

d. What are the excluded values of x in the expression in part c?

e. Except for excluded values, do the graphs appear the same?

Example  **Simplify** $\frac{a - 2 + \frac{3}{a + 2}}{a + 1 - \frac{10}{a + 4}}$.

$$\frac{a - 2 + \frac{3}{a + 2}}{a + 1 - \frac{10}{a + 4}} = \frac{\frac{(a - 2)(a + 2)}{a + 2} + \frac{3}{a + 2}}{\frac{(a + 1)(a + 4)}{(a + 4)} - \frac{10}{a + 4}}$$ *The LCD of the numerator is $a + 2$.* *The LCD of the denominator is $a + 4$.*

$$= \frac{\frac{a^2 - 4 + 3}{a + 2}}{\frac{a^2 + 5a + 4 - 10}{(a + 4)}}$$ *Add to simplify the numerator.* *Subtract to simplify the denominator.*

$$= \frac{\frac{a^2 - 1}{(a + 2)}}{\frac{a^2 + 5a - 6}{(a + 4)}}$$ *Simplify.*

$$= \frac{\frac{(a + 1)(a - 1)}{(a + 2)}}{\frac{(a + 6)(a - 1)}{(a + 4)}}$$ *Factor to simplify the numerator and denominator.*

$$= \frac{(a + 1)\overset{1}{\cancel{(a - 1)}}}{(a + 2)} \cdot \frac{(a + 4)}{(a + 6)\underset{1}{\cancel{(a - 1)}}}$$ *Multiply by the reciprocal.*

$$= \frac{a^2 + 5a + 4}{a^2 + 8a + 12}$$ *Multiply.*

CHECK FOR UNDERSTANDING

Communicating Mathematics

Study the lesson. Then complete the following.

1. **Determine** the simplified form of $\dfrac{\frac{3}{(x+1)(x+2)(x+3)}}{\frac{4}{(x+3)(x+2)(x+1)}}$ mentally.
2. a. What is the LCD for the expression $3 + \frac{x}{2} + \frac{4}{x} + \frac{x^2 + 3x + 15}{x - 2}$?

   b. Write the expression in simplified form.

Guided Practice

Write each mixed expression as a rational expression.

3. $8 + \frac{3}{x}$
4. $5 + \frac{8}{3m}$
5. $3m + \frac{m+1}{2m}$

Simplify.

6. $\dfrac{4\frac{1}{3}}{5\frac{4}{7}}$
7. $\dfrac{6\frac{2}{5}}{3\frac{5}{9}}$
8. $\dfrac{\frac{3}{x}}{\frac{x}{3}}$
9. $\dfrac{\frac{5}{y}}{\frac{10}{y^2}}$
10. $\dfrac{\frac{x+4}{x-2}}{\frac{x+5}{x-2}}$
11. $\dfrac{\frac{a-b}{a+b}}{\frac{3}{a+b}}$
12. Nakita performed an operation on $\frac{x+2}{3x-1}$ and $\frac{2x^2-8}{3x-1}$ and got $\frac{1}{2(x-2)}$. What operation was it? Explain.

EXERCISES

Practice

Write each mixed expression as a rational expression.

13. $3 + \frac{6}{x+3}$
14. $11 + \frac{a-b}{a+b}$
15. $3 - \frac{4}{2x+1}$
16. $3 + \frac{x-4}{x+y}$
17. $5 + \frac{r-3}{r^2-9}$
18. $3 + \frac{x^2+y^2}{x^2-y^2}$

Simplify.

19. $\dfrac{7\frac{2}{3}}{5\frac{3}{4}}$
20. $\dfrac{6\frac{1}{7}}{8\frac{3}{5}}$
21. $\dfrac{\frac{a^3}{b}}{\frac{a^2}{b^2}}$
22. $\dfrac{\frac{x^2y^2}{a}}{\frac{x^2y}{a^3}}$
23. $\dfrac{2+\frac{5}{x}}{\frac{x}{3}+\frac{5}{6}}$
24. $\dfrac{4+\frac{3}{y}}{\frac{3}{8}+\frac{y}{2}}$
25. $\dfrac{a-\frac{15}{a-2}}{a+3}$
26. $\dfrac{x+\frac{35}{x+12}}{x+7}$
27. $\dfrac{\frac{x^2-4}{x^2+5x+6}}{x-2}$
28. $\dfrac{m+\frac{3m+7}{m+5}}{m+1}$
29. $\dfrac{m+5+\frac{2}{m+2}}{m+1+\frac{6}{m+6}}$
30. $\dfrac{a+1+\frac{3}{a+5}}{a+1+\frac{3}{a-1}}$
31. $\dfrac{y+6+\frac{3}{y+2}}{y+11+\frac{48}{y-3}}$
32. $\dfrac{x+3+\frac{4}{x-2}}{x-1-\frac{2}{x+3}}$
33. $\dfrac{t+1+\frac{1}{t+1}}{1-t-\frac{1}{t+1}}$

34. What is the quotient when $b + \frac{1}{b}$ is divided by $a + \frac{1}{a}$?

35. What is the product when $\frac{2b^2}{5c}$ is multiplied by the quotient of $\frac{4b^3}{2c}$ and $\frac{7b^3}{8c^2}$?

36. Write $1 + \cfrac{1}{1 + \cfrac{1}{1 + \cfrac{1}{1 + \cfrac{1}{x}}}}$ in simplest form.

Graphing Calculator

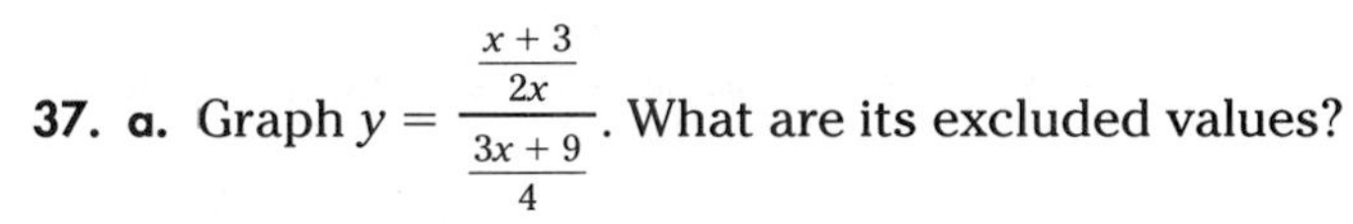

37. a. Graph $y = \dfrac{\frac{x+3}{2x}}{\frac{3x+9}{4}}$. What are its excluded values?

b. Graph $\frac{x+3}{2x} \cdot \frac{4}{3x+9}$ on the same display used in part a. What are its excluded values?

c. Except for the excluded values, are the graphs the same?

Critical Thinking

38. Simplify $\dfrac{3}{1 - \frac{3}{3+y}} - \dfrac{3}{\frac{3}{3-y} - 1}$.

Applications and Problem Solving

39. **Acoustics** If a train is moving toward you at v miles per hour and blowing its whistle at a frequency of f, then you hear it as though it were blowing its whistle with a frequency h, which is defined as $h = \dfrac{f}{1 - \frac{v}{s}}$, where s is the speed of sound.

a. Simplify the right side of this formula.

b. Suppose a train whistle blows at 370 cycles per second (at the same frequency as the first F sharp (F#) above middle C on the piano). The train is moving toward you at a speed of 80 miles per hour. The speed of sound is 760 miles per hour. Find the frequency of the sound as you hear it.

c. The note F# and the notes to its right on the piano are listed with their frequencies in the table.

Note	F#	G	G#	A	A#	B	C	C#
Frequency	370.0	392.0	415.3	440.0	466.1	493.8	523.2	554.3

By approximately how many notes did the sound rise in part b?

d. Find the frequency of the same whistle as you would hear it from an approaching TGV, the French train that is the fastest in the world ($v = 236$ miles per hour). By approximately how many notes would it rise?

FYI

Have you noticed that the pitch (frequency) of a train whistle or a horn from a speeding car gets higher as it approaches you and lower as it moves away from you? This effect is known as the *Doppler effect,* named for the Austrian scientist Christian Doppler, who discovered it in 1842.

40. **Statistics** In 1993, New Jersey was the most densely populated state, and Alaska was the least densely populated. The population of New Jersey was 7,879,000, and the population of Alaska was 599,000. The land area of New Jersey is about 7419 square miles, and the land area of Alaska is about 570,374 square miles. How many more people were there per square mile in New Jersey than in Alaska in 1993?

Mixed Review

41. Find $\frac{4x}{2x+6} + \frac{3}{x+3}$. (Lesson 12–6)

42. Find $\frac{b-5}{b^2-7b+10} \cdot \frac{b-2}{3}$. (Lesson 12–2)

43. Factor $4x^2 - 1$. (Lesson 10–4)

44. Find $x^4y^5 \div xy^3$. (Lesson 9–2)

45. Solve the system of equations. (Lesson 8–4)

$3a + 4b = -25$
$2a - 3b = 6$

46. **Geometry** A quadrilaterial has vertices $A(-1, -4)$, $B(2, -1)$, $C(5, -4)$, and $D(2, -7)$. (Lesson 6–6)
    a. Graph the quadrilateral and determine the relationship, if any, among its sides.
    b. What type of quadrilateral is $ABCD$?

47. Solve $P = 2\ell + 2w$ for w. (Lesson 3–6)

48. Find $\sqrt{225}$. (Lesson 2–8)

WORKING ON THE

In·ves·ti·ga·tion

Refer to the Investigation on pages 656–657.

A Growing Concern

1. Determine where the future play area and future garden should be placed. Mark off this space on your scale drawing. Justify your reasons for placing them where you did.
2. Develop a detailed plan for the deck and/or patio and walkways for the Sanchez family. Indicate the dimensions. Explain why you chose the placement and the type of materials for each place.
3. Figure the cost of materials, labor, and profit margin of the construction of the deck and/or patio. Justify your costs and make a case for building a deck and/or patio and the walkways as you designed them.
4. Be sure to add the deck and/or patio and the walkways to your drawing.

Add the results of your work to your Investigation Folder.

12-8 Solving Rational Equations

What YOU'LL LEARN

- To solve rational equations.

Why IT'S IMPORTANT

You can use rational equations to solve problems involving work, psychology, and electronics.

Work

Tiko and Julio have decided to start a lawn care service in their neighborhood. In order to schedule their clients in an organized manner, they compared notes on how long it took each of them to mow the same yard. Tiko could mow and trim Mrs. Harris's lawn in 3 hours, while it took Julio 2 hours. How long would it take if they both worked together?

You can answer this question by solving a **rational equation**. A rational equation is an equation that contains rational expressions.

Explore Since it takes Tiko 3 hours to do the yard, he can finish $\frac{1}{3}$ of the yard in 1 hour. At his rate, Julio can finish $\frac{1}{2}$ of the yard in an hour. Use the following formula.

$$\underbrace{r}_{\text{rate of work}} \cdot \underbrace{t}_{\text{time}} = \underbrace{w}_{\text{work done}}$$

Plan In t hours, Tiko can do $t \cdot \frac{1}{3}$ or $\frac{t}{3}$ of the job and Julio can do $t \cdot \frac{1}{2}$ or $\frac{t}{2}$ of the job. Thus, $\frac{t}{3} + \frac{t}{2} = 1$, where 1 represents the finished job.

Solve

$$\frac{t}{3} + \frac{t}{2} = 1$$

$$6\left(\frac{t}{3} + \frac{t}{2}\right) = 6(1) \quad \textit{Multiply each side by the LCD, 3(2) or 6.}$$

$$2t + 3t = 6 \quad \textit{Use the distributive property.}$$

$$5t = 6$$

$$t = \frac{6}{5} \text{ or } 1\frac{1}{5} \quad \textit{Check this solution.}$$

Examine If Tiko and Julio work for $1\frac{1}{5}$ hours, will the whole yard get done?

$$\text{Tiko's } rt + \text{Julio's } rt \stackrel{?}{=} 1 \rightarrow \frac{1}{3} \cdot \frac{6}{5} + \frac{1}{2} \cdot \frac{6}{5} \stackrel{?}{=} 1$$

$$\frac{2}{5} + \frac{3}{5} \stackrel{?}{=} 1$$

$$1 = 1$$

Working together, Tiko and Julio can finish the yard in $1\frac{1}{5}$ hours.

Example 1 **Solve $\frac{10}{3y} - \frac{5}{2y} = \frac{1}{4}$.**

$$\frac{10}{3y} - \frac{5}{2y} = \frac{1}{4}$$

$$12y\left(\frac{10}{3y} - \frac{5}{2y}\right) = 12y\left(\frac{1}{4}\right) \quad \textit{Multiply each side by the LCD, 12y.}$$

$$12y \cdot \frac{10}{3y} - 12y \cdot \frac{5}{2y} = 12y \cdot \frac{1}{4} \quad \textit{Use the distributive property.}$$

$$40 - 30 = 3y$$

$$10 = 3y$$

$$y = \frac{10}{3} \text{ or } 3\frac{1}{3}$$

Check: $\dfrac{10}{3\left(3\frac{1}{3}\right)} - \dfrac{5}{2\left(3\frac{1}{3}\right)} = \dfrac{1}{4}$

$$\frac{10}{10} - \frac{15}{20} \stackrel{?}{=} \frac{1}{4}$$

$$\frac{20}{20} - \frac{15}{20} \stackrel{?}{=} \frac{1}{4}$$

$$\frac{1}{4} = \frac{1}{4} \quad ✔$$

The solution is $3\frac{1}{3}$.

Multiplying each side of an equation by the LCD of two rational expressions can yield results that are not solutions to the original equation. Such solutions are called **extraneous solutions** or "false" solutions.

Example 2 **Solve $\dfrac{x}{x-1} + \dfrac{2x-3}{x-1} = 2$.**

$$\frac{x}{x-1} + \frac{2x-3}{x-1} = 2$$

$$(x-1)\left(\frac{x}{x-1} + \frac{2x-3}{x-1}\right) = (x-1)2 \quad \textit{Multiply each side by the LCD, } x-1.$$

$$(x-1)\left(\frac{x}{x-1}\right) + (x-1)\left(\frac{2x-3}{x-1}\right) = (x-1)2$$

$$x + 2x - 3 = 2x - 2$$

$$3x - 3 = 2x - 2$$

$$x = 1$$

The number 1 is not a solution, since 1 is an excluded value for x. Thus, the equation has no solution.

GLBAL CONNECTIONS

Algonquin Indians named the Mississippi River the "Father of Waters." Their phrase *Misi Sipi* meant big water. The Mississippi River drains one eighth of the North American continent, discharging 350 trillion gallons of water a day.

You can use the distance formula to solve real-world problems.

Example 3

Commerce

A grain barge operates between Minneapolis, Minnesota, and New Orleans, Louisiana, along the Mississippi River. The maximum speed of the barge in still water is 8 miles per hour. At this rate, a 30-mile trip downstream (with the current) takes as much time as an 18-mile trip upstream (against the current). What is the speed of the current?

Explore Let c = the speed of the current. The speed of the barge traveling downstream is 8 miles per hour plus the speed of the current, that is, $(8 + c)$ miles per hour. The speed of the barge traveling upstream is 8 miles per hour minus the speed of the current, that is, $(8 - c)$ miles per hour.

Plan To represent time t, solve $d = rt$ for t. Thus, $t = \frac{d}{r}$.

	d	r	$t = \frac{d}{r}$
downstream	30	$8 + c$	$\frac{30}{8+c}$
upstream	18	$8 - c$	$\frac{18}{8-c}$

(continued on the next page)

Longest Rivers in North America

River	Miles
1. Mackenzie-Peace	2635
2. Missouri-Red Rock	2540
3. Mississippi	2348
4. Missouri	2315
5. Yukon	1979

Solve

$$\frac{30}{8+c} = \frac{18}{8-c}$$

$$30(8-c) = 18(8+c) \quad \textit{Find the cross products.}$$

$$240 - 30c = 144 + 18c$$

$$96 = 48c$$

$$c = 2$$

Examine Check the value to see if it makes sense.
The barge goes downstream at $(8 + 2)$ or 10 mph.
A 30-mile trip would take $30 \div 10$ or 3 hours.
The barge goes upstream at $(8 - 2)$ or 6 mph.
An 18-mile trip takes $18 \div 6$ or 3 hours.
Both trips take the same amount of time, so the speed of the current must be 2 miles per hour.

Electricity can be described as the flow of electrons through a conductor, such as a copper wire. Electricity flows more freely through some conductors than others. The force opposing the flow is called *resistance*. The unit of resistance commonly used is the *ohm*.

Resistances can occur one after another, that is, *in series*. Resistances can also occur in branches with the conductor going in the same direction, or *in parallel*.

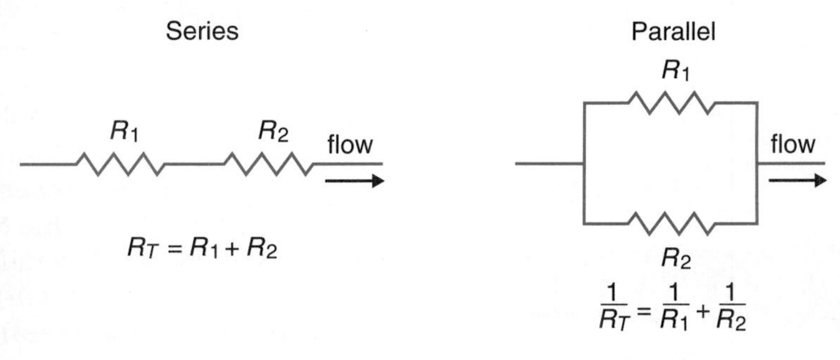

Example 4

Electronics

Assume that $R_1 = 4$ ohms and $R_2 = 3$ ohms. Compute the total resistance of the conductor when the resistances are in series and in parallel.

series

$$R_T = R_1 + R_2$$

$$= 4 + 3 \text{ or } 7$$

parallel

$$\frac{1}{R_T} = \frac{1}{R_1} + \frac{1}{R_2}$$

$$\frac{1}{R_T} = \frac{1}{4} + \frac{1}{3}$$

$$\frac{1}{R_T} = \frac{7}{12}$$

$$12 = 7R_T \quad \textit{Cross products}$$

$$R_T = \frac{12}{7} \text{ or } 1\frac{5}{7}$$

A circuit, or path for the flow of electrons, often has some resistances connected in series and others in parallel.

Example 5

Electronics

A parallel circuit has one branch in series as shown at the right. Given that the total resistance is 2.25 ohms, $R_1 = 3$ ohms, and $R_2 = 4$ ohms, find R_3.

$$\frac{1}{R_T} = \frac{1}{R_1} + \frac{1}{R_2 + R_3}$$ *The total resistance of the branch in series is $R_2 + R_3$.*

$$\frac{1}{2.25} = \frac{1}{3} + \frac{1}{4 + R_3}$$ *$R_T = 2.25$ and $R_1 = 3$*

$$\frac{1}{2.25} - \frac{1}{3} = \frac{1}{4 + R_3}$$

$$\frac{4}{9} - \frac{3}{9} = \frac{1}{4 + R_3}$$

$$\frac{1}{9} = \frac{1}{4 + R_3}$$

$$4 + R_3 = 9$$ *Find the cross products.*

$$R_3 = 5$$

Thus, R_3 is 5 ohms.

CHECK FOR UNDERSTANDING

Communicating Mathematics

Study the lesson. Then complete the following.

1. Refer to the application at the beginning of the lesson. How would the solution be different if Julio takes 6 hours to complete the lawn?
2. **You Decide** Antoinette solved $\frac{2m}{1 - m} + \frac{m + 3}{m^2 - 1} = 1$ and claimed that 1 and $-\frac{4}{3}$ were the solutions. Joel says that $-\frac{4}{3}$ is the only solution. Who is correct, and why?
3. **Define** a rational equation and distinguish it from a linear equation.

4. **Assess Yourself**
   a. Describe an activity that you and a friend do together that each can complete separately. Estimate the time it would take for each of you to complete it working alone.
   b. Use the math you have learned to find out how long it would take to complete the activity if the two of you worked together.

Guided Practice

Solve each equation.

5. $\frac{1}{4} + \frac{4}{x} = \frac{1}{x}$
6. $\frac{1}{5} + \frac{3}{2y} = \frac{3}{3y}$
7. $\frac{4}{x + 5} = \frac{4}{3(x + 2)}$
8. $\frac{x}{2} = \frac{3}{x + 1}$
9. $\frac{a - 1}{a + 1} - \frac{2a}{a - 1} = -1$
10. $\frac{w - 2}{w} - \frac{w - 3}{w - 6} = \frac{1}{w}$
11. **Work** Olivia can wash and wax her car, vacuum the interior, and wash the insides of the windows in 5 hours. What part of the job can she do in
    a. 1 hour? b. 3 hours? c. x hours?

12. **Recreation** Sally and her brother rented a boat to go fishing in Jones Creek. The maximum speed of the boat in still water was 3 miles per hour. At this rate, a 9-mile trip downstream (with the current) took the same amount of time as a 3-mile trip upstream. Let c = the speed of the current. Copy and complete the table below.

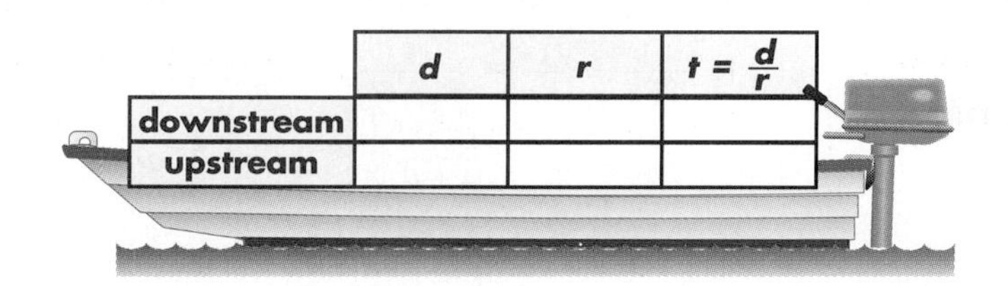

	d	r	$t = \frac{d}{r}$
downstream			
upstream			

   a. Write an equation that represents the conditions in the problem.
   b. Find the speed of the current.

Electronics: Exercises 13–15 refer to the diagram below.

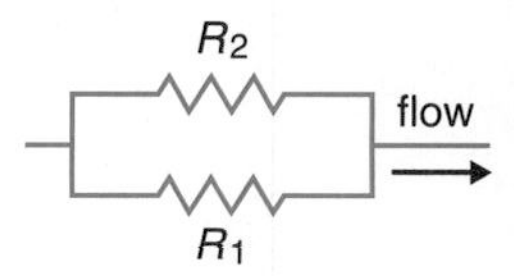

13. Find the total resistance, R_T, given that $R_1 = 8$ ohms and $R_2 = 6$ ohms.
14. Find R_1, given that R_T is $2.\overline{2}$ ohms and $R_2 = 5$ ohms.
15. Find R_1 and R_2, given that the total resistance is $2.\overline{6}$ ohms and R_1 is twice as great as R_2.

EXERCISES

Practice

Solve each equation.

16. $\frac{1}{4} + \frac{3}{x} = \frac{1}{x}$
17. $\frac{1}{5} - \frac{4}{3m} = \frac{2}{m}$
18. $x + 3 = -\frac{2}{x}$
19. $\frac{m+1}{m} + \frac{m+4}{m} = 6$
20. $\frac{x}{x+1} + \frac{5}{x-1} = 1$
21. $\frac{m-1}{m+1} - \frac{2m}{m-1} = -1$
22. $\frac{-4}{a+1} + \frac{3}{a} = 1$
23. $\frac{3x}{10} - \frac{1}{5x} = \frac{1}{2}$
24. $\frac{b}{4} + \frac{1}{b} = \frac{-5}{3}$
25. $\frac{-4}{n} = 11 - 3n$
26. $\frac{x-3}{x} = \frac{x-3}{x-6}$
27. $\frac{7}{a-1} = \frac{5}{a+3}$
28. $\frac{3}{r+4} - \frac{1}{r} = \frac{1}{r}$
29. $\frac{3}{x} + \frac{4x}{x-3} = 4$
30. $\frac{1}{4m} + \frac{2m}{m-3} = 2$
31. $\frac{a-2}{a} - \frac{a-3}{a-6} = \frac{1}{a}$
32. $\frac{x+3}{x+5} + \frac{2}{x-9} = \frac{5}{2x+10}$
33. $\frac{-1}{w+2} = \frac{w^2-7w-8}{3w^2+2w-8}$

Electronics: Refer to the diagram at the right.

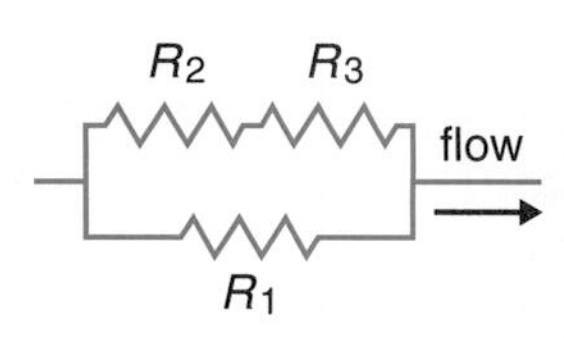

34. Find R_T, given that $R_1 = 5$ ohms, $R_2 = 4$ ohms, and $R_3 = 3$ ohms.
35. Find R_1, given that $R_T = 2\frac{10}{13}$ ohms, $R_2 = 3$ ohms, and $R_3 = 6$ ohms.
36. Find R_2, given that $R_T = 3.5$ ohms, $R_1 = 5$ ohms, and $R_3 = 4$ ohms.

Electronics: Solve each formula for the variable indicated.

37. $\frac{1}{R_T} = \frac{1}{R_1} + \frac{1}{R_2}$, for R_1

38. $I = \frac{E}{r + R}$, for R

39. $I = \frac{nE}{nr + R}$, for n

40. $I = \frac{E}{\frac{r}{n} + R}$, for r

Critical Thinking

41. What number would you add to both the numerator and denominator of $\frac{4}{11}$ to make a fraction equivalent to $\frac{2}{3}$?

42. Refer to the diagram at the right.
   a. Write an equation for the total resistance for the diagram.
   b. Find the total resistance, given that R_1 = 5 ohms, R_2 = 4 ohms, and R_3 = 6 ohms.

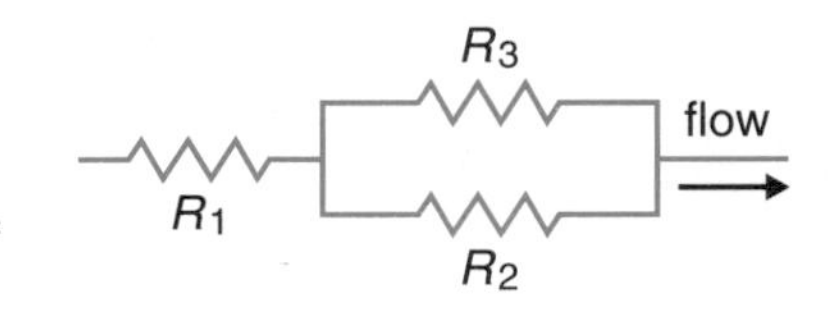

Applications and Problem Solving

43. **Electricity** Eight lights on a decorated tree are connected in series. Each has a resistance of 12 ohms. What is the total resistance?

44. **Electricity** Three appliances are connected in parallel: a lamp with a resistance of 60 ohms, an iron with a resistance of 20 ohms, and a heating coil with a resistance of 80 ohms. Find the total resistance.

45. **Air Travel** The flying distance from Honolulu, Hawaii, to San Francisco, California, is approximately 2400 miles. The air speed of a 747-100 airplane is 520 miles per hour. A strong tailwind blows at 120 miles per hour. At 1000 miles into the trip (1400 miles from San Francisco), the aircraft loses power in one engine. As a navigator for the aircraft, you must answer the following questions.
   a. How long would it take to return to Honolulu?
   b. How long would it take to continue on to San Francisco?
   c. What is the point-of-no-return? That is, at what point into the flight (in miles) would it be quicker to continue the flight to San Francisco than return to Honolulu?

46. **Bicycling** The Lake Pontchartrain Causeway in Louisiana is 24 miles long.
   a. Suppose Todd starts cycling at one end at 20 miles per hour and Kristie starts at the other end at 16 miles per hour. How long would it take them to meet? At what distance from either end would they meet?
   b. Todd's rate with no wind is 20 miles per hour, while Kristie's is 16 miles per hour. If Todd cycles against the wind, while Kristie cycles with the wind, and they meet at the midpoint of the bridge, what is the wind speed?

   c. The causeway consists of two parallel bridges. Assume Todd starts to bicycle from one end in one direction at 20 miles per hour. Kristie starts from the other end on the parallel bridge, bicycling at 16 miles per hour in the opposite direction. They pedal in a continuous loop on the two parallel causeways. How long does it take Todd to catch up with Kristie? (*Note:* She has a 24-mile head start.)

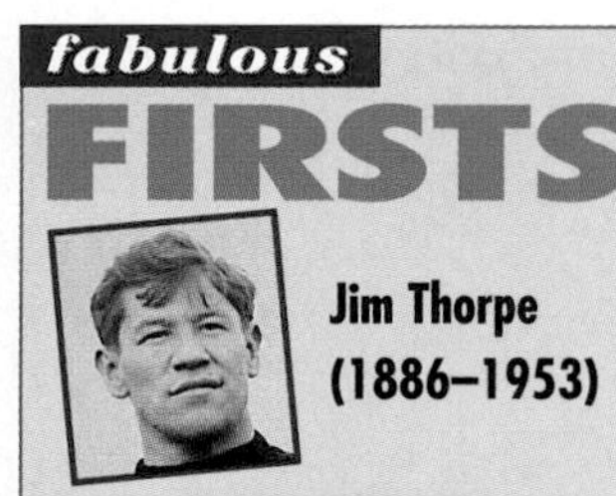

James Francis Thorpe was the first American-Indian sports hero, winner of Olympic Gold medals in 1912, a professional baseball and football player, and the first president of the American Professional Football Association, which later became the NFL. In 1950, he was selected by sportswriters and broadcasters as the greatest athlete of the first half of the 20th century.

47. Anthropology The formula $c = \frac{100w}{\ell}$ provides a measure called the cephalic index c. Anthropologists use this index to identify skulls by their ethnic characteristics. You calculate the cephalic index by using the width w of a person's head, ear to ear, and the length ℓ of the head from face to back.

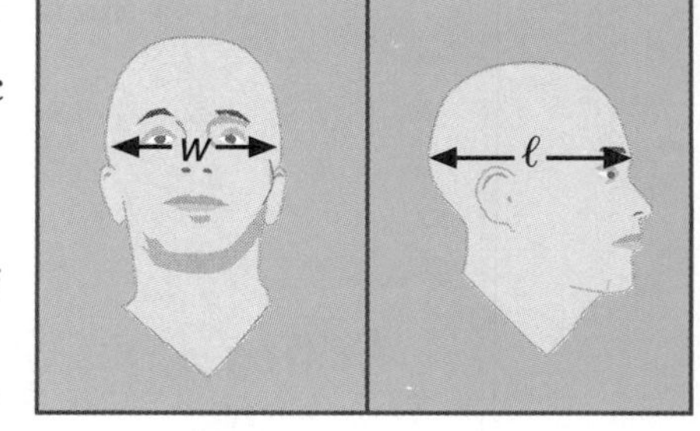

a. Solve the formula for w in terms of c and ℓ.

b. Solve the formula for ℓ in terms of c and w.

48. Psychology The formula $i = \frac{100m}{c}$ provides a measure i of intelligence, called the intelligence quotient, or I.Q. In the formula, m represents the person's mental age, and c represents the person's chronological age.

a. Solve the formula for m in terms of i and c.

b. Solve the formula for c in terms of i and m.

49. Baseball Chang has 32 hits in 128 times at bat. His current batting average is $\frac{32}{128} = 0.250$. How many consecutive hits must he get in his next x times at bat in order to get his average up to 0.300?

Mixed Review

50. Simplify $\dfrac{\frac{x^2 - 5x}{x^2 + x - 30}}{\frac{x^2 + 2x}{x^2 + 9x + 18}}$. (Lesson 12–7)

51. Simplify $\frac{a + 2}{b^2 + 4b + 4} \div \frac{4a + 8}{b + 4}$. (Lesson 12–3)

52. Find $(0.5a + 0.25b)^2$. (Lesson 9–8)

53. Sales The amounts in the picture at the right are the cash register totals of 20 customers on the Wednesday before Thanksgiving. Make a box-and-whisker plot of these data. (Lesson 7–7)

54. Statistics Refer to the data in Exercise 53. Find the range, median, upper quartile, lower quartile, and interquartile range of the data. Identify any outliers. (Lesson 5–7)

55. Write 12 pounds to 100 ounces as a fraction in simplest form. (Lesson 4–1)

56. Solve $3x = -15$. (Lesson 3–2)

57. Find $(-2)(3)(-3)$. (Lesson 2–6)

58. Complete: $3(2 + x) = 6 + \underline{\ ?\ }$. (Lesson 1–7)

CHAPTER 12 HIGHLIGHTS

VOCABULARY

After completing this chapter, you should be able to define each term, property, or phrase and give an example or two of each.

Algebra

excluded values (p. 660)

extraneous solutions (p. 697)

least common denominator (LCD) (p. 686)

least common multiple (LCM) (p. 685)

mixed expressions (p. 690)

rational equation (p. 696)

rational expression (p. 660)

Problem Solving

make an organized list (p. 685)

UNDERSTANDING AND USING THE VOCABULARY

State whether each sentence is *true* or *false*. If false, replace the underlined word or number to make a true sentence.

1. A <u>mixed</u> expression is an algebraic fraction whose numerator and denominator are polynomials.
2. The complex fraction $\frac{\frac{4}{5}}{\frac{2}{3}}$ can be simplfied as $\underline{\frac{6}{5}}$.
3. The equation $\frac{x}{x-1} + \frac{2x-3}{x-1} = 2$ has an extraneous solution of $\underline{1}$.
4. The mixed expression $6 - \frac{a-2}{a+3}$ can be rewritten as the rational expression $\underline{\frac{5a+16}{a+3}}$.
5. The least common multiple for $(x^2 - 144)$ and $(x + 12)$ is $\underline{x + 12}$.
6. The excluded values for $\frac{4x}{x^2 - x - 12}$ are $\underline{-3 \text{ and } 4}$.
7. The least common denominator is the <u>greatest common factor</u> of the denominators.

SKILLS AND CONCEPTS

OBJECTIVES AND EXAMPLES

Upon completing this chapter, you should be able to:

- simplify rational expressions (Lesson 12–1)

Simplify $\frac{x+y}{x^2+3xy+2y^2}$.

$$\frac{x+y}{x^2+3xy+2y^2} = \frac{\cancel{x+y}}{(\cancel{x+y})(x+2y)} = \frac{1}{x+2y}$$

REVIEW EXERCISES

Use these exercises to review and prepare for the chapter test.

Simplify each rational expression. State the excluded values of the variables.

8. $\frac{3x^2y}{12xy^3z}$
9. $\frac{z^2-3z}{z-3}$
10. $\frac{a^2-25}{a^2+3a-10}$
11. $\frac{3a^3}{3a^3+6a^2}$
12. $\frac{x^2+10x+21}{x^3+x^2-42x}$
13. $\frac{b^2-5b+6}{b^4-13b^2+36}$

- multiply rational expressions (Lesson 12–2)

$$\frac{1}{x^2+x-12} \cdot \frac{x-3}{x+5} = \frac{1}{(x+4)(\cancel{x-3})} \cdot \frac{\cancel{x-3}}{x+5} = \frac{1}{(x+4)(x+5)} = \frac{1}{x^2+9x+20}$$

Find each product. Assume that no denominator has a value of 0.

14. $\frac{7b^2}{9} \cdot \frac{6a^2}{b}$
15. $\frac{5x^2y}{8ab} \cdot \frac{12a^2b}{25x}$
16. $(3x+30) \cdot \frac{10}{x^2-100}$
17. $\frac{3a-6}{a^2-9} \cdot \frac{a+3}{a^2-2a}$
18. $\frac{x^2+x-12}{x+2} \cdot \frac{x+4}{x^2-x-6}$
19. $\frac{b^2+19b+84}{b-3} \cdot \frac{b^2-9}{b^2+15b+36}$

- divide rational expressions (Lesson 12–3)

$$\frac{y^2-16}{y^2-64} \div \frac{y+4}{y-8} = \frac{y^2-16}{y^2-64} \cdot \frac{y-8}{y+4} = \frac{(y-4)(\cancel{y+4})}{(\cancel{y-8})(y+8)} \cdot \frac{\cancel{y-8}}{\cancel{y+4}} = \frac{y-4}{y+8}$$

Find each quotient. Assume that no denominator has a value of 0.

20. $\frac{p^3}{2q} \div \frac{p^2}{4q}$
21. $\frac{y^2}{y+4} \div \frac{3y}{y^2-16}$
22. $\frac{3y-12}{y+4} \div (y^2-6y+8)$
23. $\frac{2m^2+7m-15}{m+5} \div \frac{9m^2-4}{3m+2}$

OBJECTIVES AND EXAMPLES

- divide a polynomial by a binomial. (Lesson 12–4)

$$\begin{array}{r} x^2 + x - 19 \\ x - 3\overline{)x^3 - 2x^2 - 22x + 21} \\ \underline{x^3 - 3x^2} \qquad\qquad \\ x^2 - 22x \qquad \\ \underline{x^2 - 3x} \qquad \\ -19x + 21 \\ \underline{-19x + 57} \\ -36 \end{array}$$

The quotient is $x^2 + x - 19 - \frac{36}{x - 3}$.

REVIEW EXERCISES

Find each quotient.

24. $(4a^2b^2c^2 - 8a^3b^2c + 6abc^2) \div (2ab^2)$
25. $(x^3 + 7x^2 + 10x - 6) \div (x + 3)$
26. $(x^3 - 7x + 6) \div (x - 2)$
27. $(x^4 + 3x^3 + 2x^2 - x + 6) \div (x - 2)$
28. $(48b^2 + 8b + 7) \div (12b - 1)$

- add and subtract rational expressions with like denominators. (Lesson 12–5)

$$\frac{m^2}{m+4} - \frac{16}{m+4} = \frac{m^2 - 16}{m+4}$$

$$= \frac{(m-4)\overset{1}{\cancel{(m+4)}}}{\underset{1}{\cancel{m+4}}}$$

$$= m - 4$$

Find each sum or difference. Express in simplest form.

29. $\frac{7a}{m^2} - \frac{5a}{m^2}$
30. $\frac{2x}{x-3} - \frac{6}{x-3}$
31. $\frac{m+4}{5} + \frac{m-1}{5}$
32. $\frac{-5}{2n-5} + \frac{2n}{2n-5}$
33. $\frac{a^2}{a-b} + \frac{-b^2}{a-b}$
34. $\frac{m^2}{m-n} - \frac{2mn - n^2}{m-n}$

- add and subtract rational expressions with unlike denominators (Lesson 12–6)

$$\frac{x}{x+3} - \frac{5}{x-2} = \frac{x}{x+3} \cdot \frac{x-2}{x-2} - \frac{5}{x-2} \cdot \frac{x+3}{x+3}$$

$$= \frac{x(x-2)}{(x+3)(x-2)} - \frac{5(x+3)}{(x+3)(x-2)}$$

$$= \frac{x^2 - 2x}{(x+3)(x-2)} - \frac{5x + 15}{(x+3)(x-2)}$$

$$= \frac{x^2 - 2x - 5x - 15}{(x+3)(x-2)}$$

$$= \frac{x^2 - 7x - 15}{x^2 + x - 6}$$

Find each sum or difference.

35. $\frac{7n}{3} - \frac{9n}{7}$
36. $\frac{7}{3a} - \frac{3}{6a^2}$
37. $\frac{2c}{3d^2} + \frac{3}{2cd}$
38. $\frac{2a}{2a+8} - \frac{4}{5a+20}$
39. $\frac{r^2 + 21r}{r^2 - 9} + \frac{3r}{r+3}$
40. $\frac{3a}{a-2} + \frac{5a}{a+1}$

OBJECTIVES AND EXAMPLES

- simplify mixed expressions and complex fractions. (Lesson 12–7)

$$\frac{y-\frac{40}{y-3}}{y+5} = \frac{\frac{y(y-3)}{(y-3)} - \frac{40}{y-3}}{y+5}$$

$$= \frac{\frac{y^2-3y-40}{y-3}}{y+5}$$

$$= \frac{\frac{(y-8)(y+5)}{y-3}}{y+5}$$

$$= \frac{(y-8)(y+5)}{y-3} \div (y+5)$$

$$= \frac{(y-8)\overset{1}{\cancel{(y+5)}}}{y-3} \cdot \frac{1}{\underset{1}{\cancel{y+5}}}$$

$$= \frac{y-8}{y-3}$$

REVIEW EXERCISES

Write each mixed expression as a rational expression.

41. $4 + \frac{m}{m-2}$

42. $2 - \frac{x+2}{x^2-4}$

Simplify.

43. $\frac{\frac{x^2}{y^3}}{\frac{3x}{9y^2}}$

44. $\frac{5+\frac{4}{a}}{\frac{a}{2}-\frac{3}{4}}$

45. $\frac{x-\frac{35}{x+2}}{x+\frac{42}{x+13}}$

46. $\frac{y+9-\frac{6}{y+4}}{y+4+\frac{2}{y+1}}$

- solve rational equations (Lesson 12–8)

$$\frac{3}{x} + \frac{1}{x-5} = \frac{1}{2x}$$

$$2x(x-5)\left(\frac{3}{x} + \frac{1}{x-5}\right) = \left(\frac{1}{2x}\right)2x(x-5)$$

$$6(x-5) + 2x = x - 5$$

$$6x - 30 + 2x = x - 5$$

$$7x = 25$$

$$x = \frac{25}{7}$$

Solve each equation.

47. $\frac{4x}{3} + \frac{7}{2} = \frac{7x}{12} - \frac{1}{4}$

48. $\frac{11}{2x} - \frac{2}{3x} = \frac{1}{6}$

49. $\frac{2}{3r} - \frac{3r}{r-2} = -3$

50. $\frac{x-2}{x} - \frac{x-3}{x-6} = \frac{1}{x}$

51. $-\frac{5}{m} = 19 - 4m$

52. $\frac{1}{h+1} + 2 = \frac{2h+3}{h-1}$

APPLICATIONS AND PROBLEM SOLVING

53. List Possibilities Wacky Wheels carries bicycles, tricycles, and wagons. They have an equal number of tricycles and wagons in stock. If there are 60 pedals and 180 wheels, how many bicycles, tricycles, and wagons are there in stock? (Lesson 12–6)

DENNIS THE MENACE

"MR. WILSON GAVE THEM TO ME! AND HE'S GOT LOTS MORE!"

54. Electronics Assume that $R_1 = 4$ ohms and $R_2 = 6$ ohms. What is the total resistance of the conductor if R_1 and R_2 are: (Lesson 12–8)

a. connected in series?

b. connected in parallel?

55. Finance Barrington High School is raising money to build a house for Habitat for Humanity by doing lawn work for friends and neighbors. Scott can rake a lawn and bag the leaves in 5 hours, while Kalyn can do it in 3 hours. If Scott and Kalyn work together, how long will it take them to rake a lawn and bag the leaves? (Lesson 12–8)

A practice test for Chapter 12 is provided on page 798.

ALTERNATIVE ASSESSMENT

COOPERATIVE LEARNING PROJECT

Fun Puzzle In this project, you will determine the mathematical equation that is used for a puzzle. Mara developed a puzzle in which the number that you start with ends up being the number that you end with. Edwin was intrigued by this and wanted to determine why it works and if it will work all the time.

Use the steps below to develop the algebraic equation for a puzzle. Then prove that it works by simplifying it. Will it work if the number chosen is zero? Will it work if the number chosen is a fraction? If not, find a counterexample. Will it always work?

1. Choose any whole number.
2. Multiply by three.
3. Add fifteen.
4. Multiply by four.
5. Subtract twice the chosen number.
6. Multiply by five.
7. Add ten times the chosen number.
8. Divide by six.
9. Subtract three times the chosen number.
10. Add thirteen.
11. Divide by seven.
12. Subtract nine.

Consider these ideas while accomplishing your task.

- Read the sequence and determine your variable.
- Develop the algebraic equation expressed by the sequence.
- Determine a plan for simplifying the algebraic equation.
- Simplify the equation.
- Determine when it works and when it doesn't work.

Write a report showing why it works and explaining if it works all the time.

THINKING CRITICALLY

- Write three different rational equations whose solution is the set of all real numbers except a.
- For all real numbers a, b, and x, tell whether each statement is *always true*, *sometimes true*, or *never true*. Give a reason.

$\frac{x}{x} = 1$

$\frac{ab^2}{b^2} = ab^3$

$\frac{x^2 + 6x - 5}{2x + 2} = \frac{x + 5}{2}$

PORTFOLIO

In the last chapter, you learned how to solve quadratic equations. Think about that process. Now choose a quadratic equation from your work in that lesson and solve it. Choose a rational equation from your work from this lesson and solve it right next to the quadratic equation that you solved. Refer to these two processes and compare and contrast solving quadratic equations and solving rational equations. Place this in your portfolio.

SELF EVALUATION

Applying a previous skill to help in developing a new skill is often a useful procedure. This can be refered to as building on a firm foundation. If a new idea is difficult to comprehend, stopping and relating it to a skill that has already been mastered can be helpful. The new skill is then "not so new" anymore and becomes more familiar.

Assess yourself. Do you build your skills? When developing a new idea, do you go back to something familiar and then build, or do you start from scratch? Give an example of when you could easily apply this method in learning a new skill in mathematics and in your daily life.

CUMULATIVE REVIEW

CHAPTERS 1–12

SECTION ONE: MULTIPLE CHOICE

There are eight multiple-choice questions in this section. After working each problem, write the letter of the correct answer on your paper.

1. **Geometry** The measure of the area of a square is 129 square inches. What is the perimeter of this square, rounded to the nearest hundredth?

   A. 11.36 in. **B.** 32.25 in.

   C. 45.43 in. **D.** 1040.06 in.

2. Express the relation shown as a set of ordered pairs.

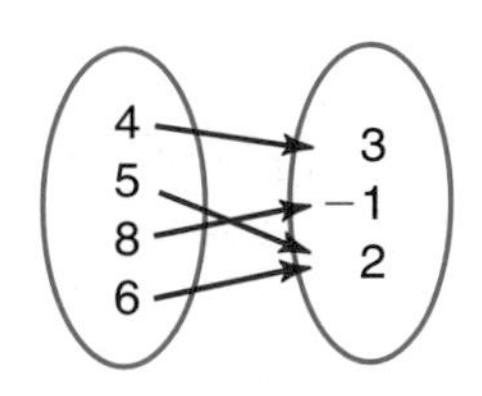

   A. $\{(3, 4), (-1, 8), (2, 5), (2, -6)\}$

   B. $\{(4, 3), (5, 2), (-1, 8), (-6, 2)\}$

   C. $\{4, 5, 8, -6, 3, -1, 2\}$

   D. $\{(4, 3), (5, 2), (8, -1), (-6, 2)\}$

3. **Geometry** Find the measure of the area of the rectangle shown below in simplest form.

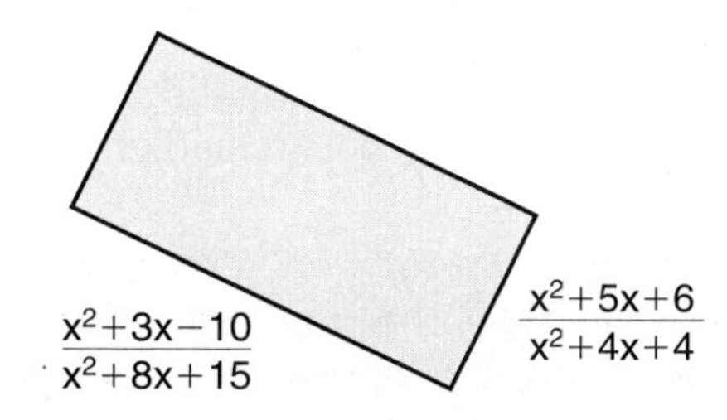

   A. $\frac{15}{32}x^2 - 1$ **B.** $\frac{x-2}{x+2}$

   C. $\frac{15x^2 - 32x - 60}{32x^2 + 92x + 60}$ **D.** $\frac{9x^2 - 42x + 49}{16x^2}$

4. **Probability** If a card is selected at random from a deck of 52 cards, what is the probability that it is not a face card?

   A. $\frac{3}{13}$ **B.** 3:10

   C. $\frac{13}{25}$ **D.** $\frac{10}{13}$

5. **Physics** Rafael tossed a rock off the edge of a 10-meter-high cliff with an initial velocity of 15 meters per second. To the nearest tenth of a second, determine when the rock will hit the ground by using the formula $H = -4.9t^2 + vt + h$.

   A. 3.6 s **B.** 3.1 s

   C. 4 s **D.** 3.9 s

6. Simplify $\dfrac{\frac{x^2 + 8x + 15}{x^2 + x - 6}}{\frac{x^2 + 2x - 15}{x^2 - 2x - 3}}$.

   A. $\frac{23}{30}$ **B.** $\frac{x^2 + 4x + 3}{x^2 + x - 6}$

   C. $\frac{x^2 + 10x + 25}{x^2 - x - 2}$ **D.** $\frac{x + 1}{x - 2}$

7. **Geometry** What is the value of x if the perimeter of a square is 60 cm and its area is $4x^2 - 28x + 49$ cm²?

   A. 4 only **B.** 11 only

   C. 4 and 11 **D.** −4 and 11

8. **Physics** The distance a force can move an object is $\frac{2a}{6a^2 - 17a - 3}$ yards. The distance a second force can move the same object is $\frac{a+2}{a^2 - 9}$ yards. How much farther did the object move when the second force was applied than when the first force was applied?

   A. $\frac{4a^2 + 7a + 2}{(a-3)(a+3)(6a+1)}$ **B.** $\frac{a+2}{5a^2 + 17a - 6}$

   C. $\frac{4a^2 + 17a - 2}{(a-3)(a+3)(6a-1)}$ **D.** $\frac{10a^2 + 11a + 1}{(3a-2)(2a+1)}$

SECTION TWO: SHORT ANSWER

This section contains nine questions for which you will provide short answers. Write your answer on your paper.

9. Evaluate
$42 \div 7 - 1 - 5 + 8 \cdot 2 + 14 \div 2 - 8$.

10. The rectangular penguin pond at the Bay Park Zoo is 12 meters long by 8 meters wide. The zoo wants to double the area of the pond by increasing the length and width by the same amount. By how much should the length and width be increased?

11. Determine the slope, y-intercept, and x-intercept of the graph of $2x - 3y = 13$.

12. Find the value of k if the remainder is 15 when $x^3 - 7x^2 + 4x + k$ is divided by $(x - 2)$.

13. The tens digit of a two-digit number exceeds twice its units digit by 1. If the digits are reversed, the number is 4 more than 3 times the sum of the digits. Find the number.

14. A long-distance cyclist pedaling at a steady rate travels 30 miles with the wind. He can travel only 18 miles against the wind in the same amount of time. If the rate of the wind is 3 miles per hour, what is the cyclist's rate without the wind?

15. Solve $-2 \leq 2x + 4 < 6$ and graph the solution set.

16. **Biology** The 2-inch long hummingbird flaps its wings about forty to fifty times each second. At this rate, how many times does it flap its wings in half of an hour? Express your answer in scientific notation.

17. **Construction** Muturi has 120 meters of fence to make a rectangular pen for his rabbits. If a shed is used as one side of the pen, what would be the maximum area of the pen?

SECTION THREE: OPEN-ENDED

This section contains three open-ended problems. Demonstrate your knowledge by giving a clear, concise solution to each problem. Your score on these problems will depend on how well you do the following.

- Explain your reasoning.
- Show your understanding of the mathematics in an organized manner.
- Use charts, graphs, and diagrams in your explanation.
- Show the solution in more than one way or relate it to other situations.
- Investigate beyond the requirements of the problem.

18. Mrs. Bloom bought some impatiens and petunias for her landscaping business for $111.25. How many flats of each did she buy if each flat of impatiens was $10.00, each flat of petunias was $8.75, and if two fewer flats of impatiens than petunias were bought?

19. Graph the quadratic function $y = -x^2 + 6x + 16$. Include the equation of the axis of symmetry, the coordinates of the vertex, and the roots of the related quadratic equation.

20. Solve the system of equations below by graphing. Explain how you determined the solution, and name at least three ordered pairs that satisfy the system.

$y \leq x + 3$

$2x - 2y < 8$

$2y + 3x > 4$

CHAPTER

13 Exploring Radical Expressions and Equations

Objectives

In this chapter, you will:

- use the Pythagorean theorem to solve problems,
- simplify radical expressions,
- solve problems involving radical equations,
- solve quadratic equations by completing the square, and
- solve problems by identifying subgoals.

Teens Talk to Parents

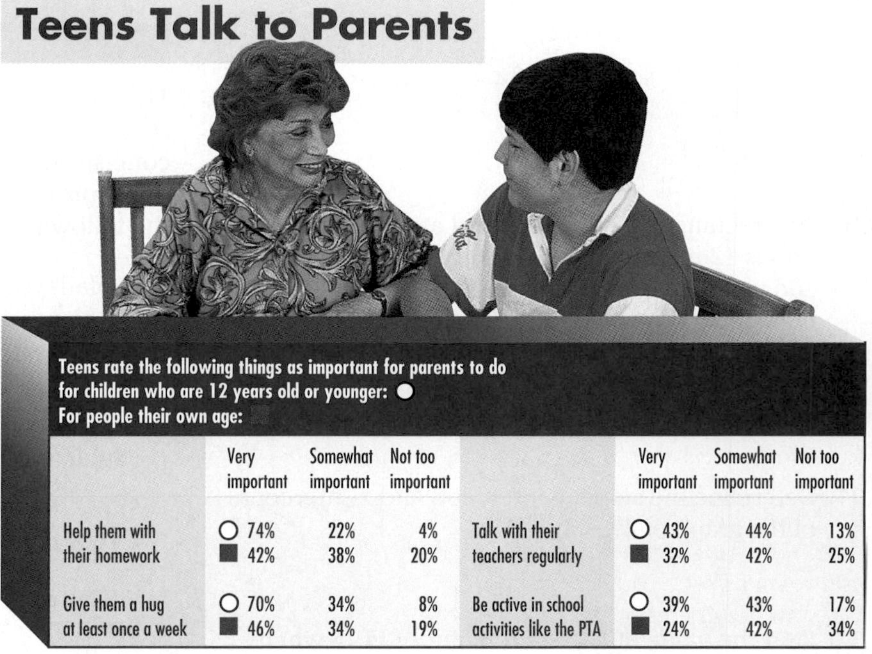

Teens rate the following things as important for parents to do for children who are 12 years old or younger: ○
For people their own age: ■

		Very important	Somewhat important	Not too important
Help them with their homework	○	74%	22%	4%
	■	42%	38%	20%
Give them a hug at least once a week	○	70%	34%	8%
	■	46%	34%	19%
Talk with their teachers regularly	○	43%	44%	13%
	■	32%	42%	25%
Be active in school activities like the PTA	○	39%	43%	17%
	■	24%	42%	34%

Source: *Oakland Press*, 1995

Almost everyone agrees that people who can communicate, who really listen to one another, have the healthiest relationships. This is certainly true when talking about teenagers and their parents. Kids can learn much from the experiences of their parents, and adults can benefit from the fresh ideas they share with their teens. What is the communication like in your home? What topics do you think are important to discuss with your parents?

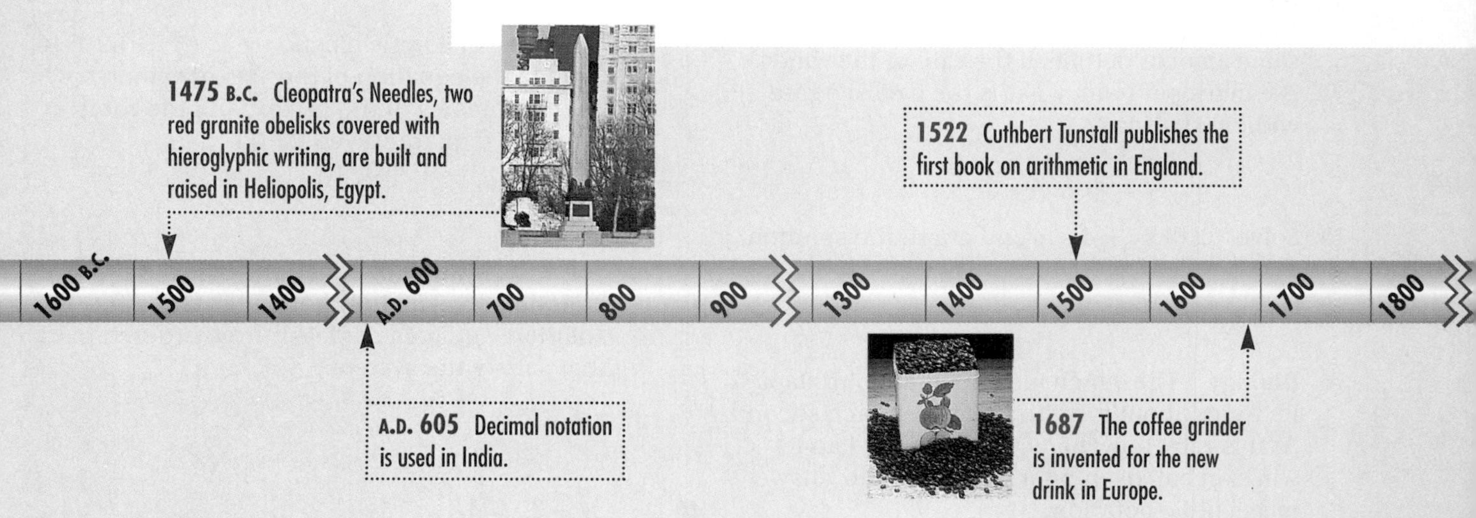

PEOPLE IN THE NEWS

At age 12, **Ryan Holladay** of Arlington, Virginia, is already a published author. His book *What Preteens Want Their Parents to Know* contains about 180 tips for parents on subjects from discipline and homework to love and respect. "It's tough growing up from being a little kid to a teenager," says Holladay. "I noticed lots of advice books written by adults, but none by kids. So I began writing one myself." Holladay surveyed children aged 9 to 12 for more ideas for his book, which took two years to complete. Since his book was published, Holladay has been on TV and radio shows to talk about it. His best advice to parents and children: "Be flexible."

Chapter Project

Form teams and conduct your own survey of what teens want from their parents or guardians. You can use the questions from the chart on page 710 or add your own. Keep track of the age, gender, and number of people questioned and their responses. See if you can make conclusions about the attitude of teens toward their parents' communication with them. Create charts or graphs using your data and report to the class on your findings.

1937 French artist Marie Laurenein creates her ethereal painting *The Dancers*.

1994 Kim Campbell becomes Canada's first female prime minister.

1890 | 1900 | 1910 | 1920 | 1930 | 1940 | 1950 | 1960 | 1970 | 1980 | 1990 | 2000

1963 Charles Moore takes his famous photograph for *Life* magazine, *Birmingham Riots,* during the civil rights struggle in Alabama.

1987 The African-American experience is portrayed in the photograph *Basket of Millet* by artist Elisabeth Sunday.

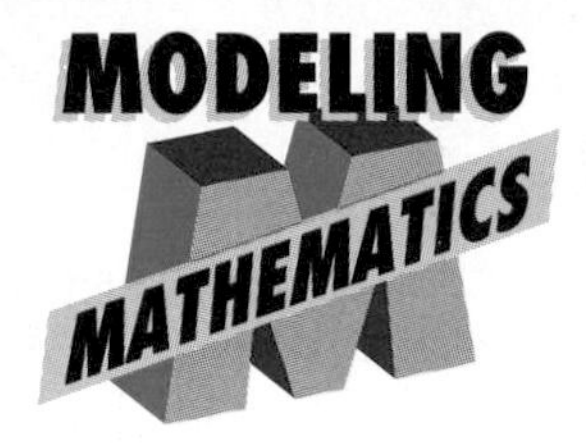

A Preview of Lesson 13–1

13–1A The Pythagorean Theorem

Materials:  geoboard  dot paper

In this activity, you will use the Pythagorean theorem to build squares on a geoboard or dot paper.

Activity Make a square with an area of 2 square units.

Step 1 Start with a right triangle like the one shown below.

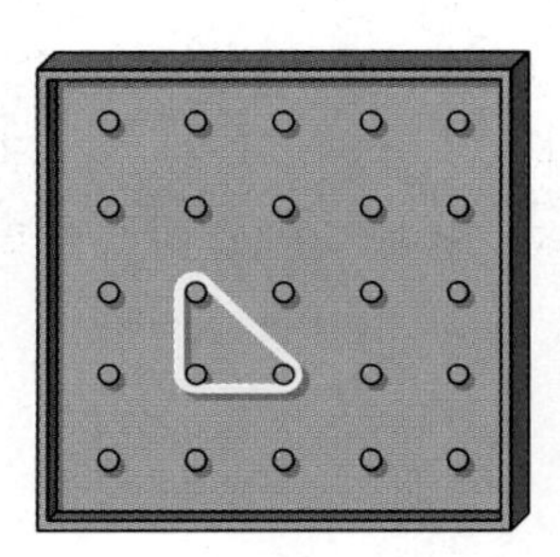

Step 2 Build squares on the two legs. Each square has an area of 1 square unit.

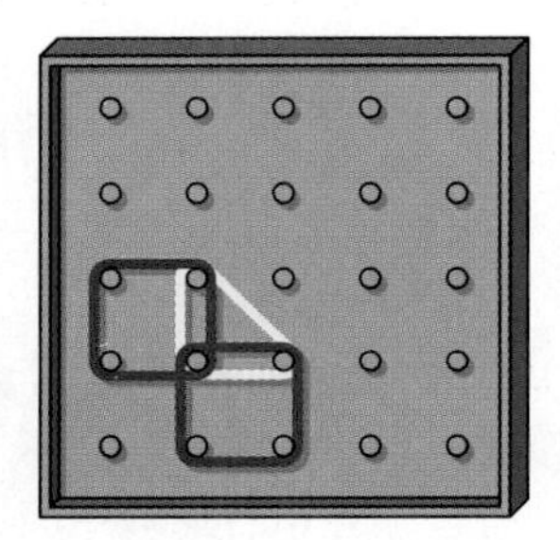

Step 3 Now build a square on the hypotenuse.

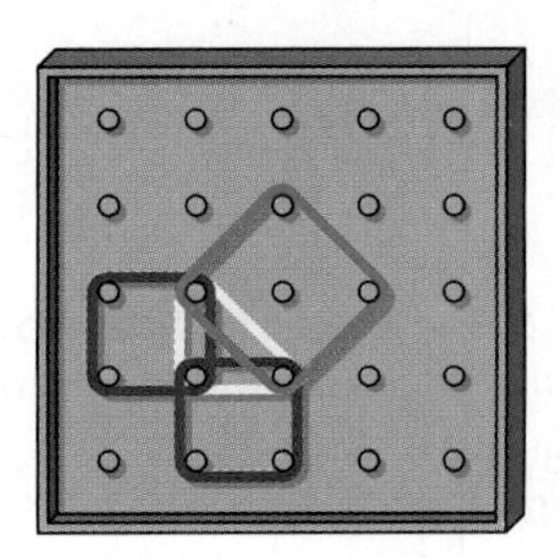

You can find the area of the square on the hypotenuse by using the Pythagorean theorem. Let c represent the measure of the hypotenuse and a and b represent the measures of the legs.

$c^2 = a^2 + b^2$

$= 1^2 + 1^2$ *Replace a with 1 and b with 1.*

$= 1 + 1$ or 2

The area of the square on the hypotenuse is 2 square units.

Model Build squares on each side of the triangles shown below using a geoboard or dot paper. Record the area of each square.

1.

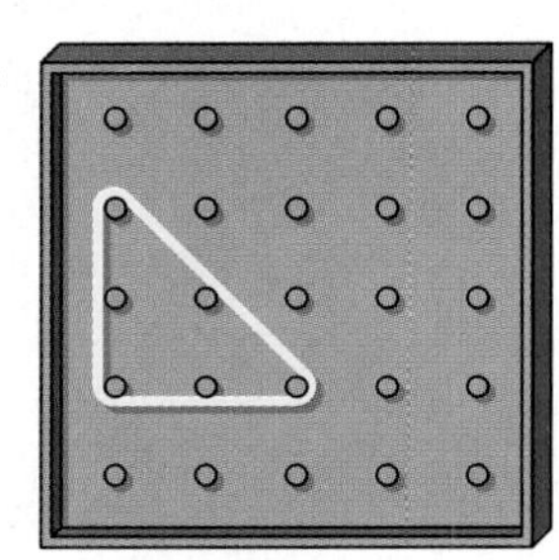

2.

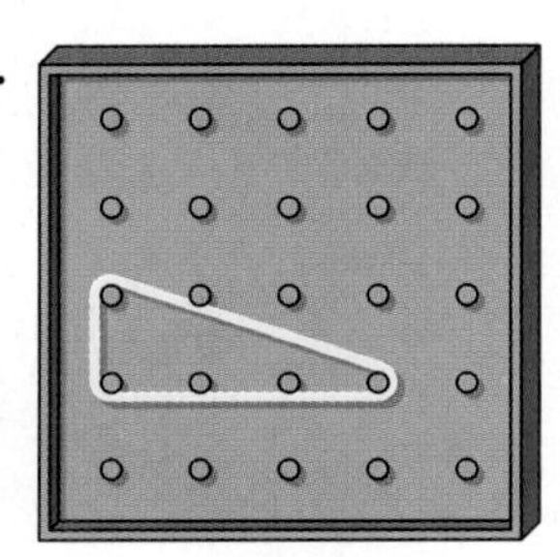

3.

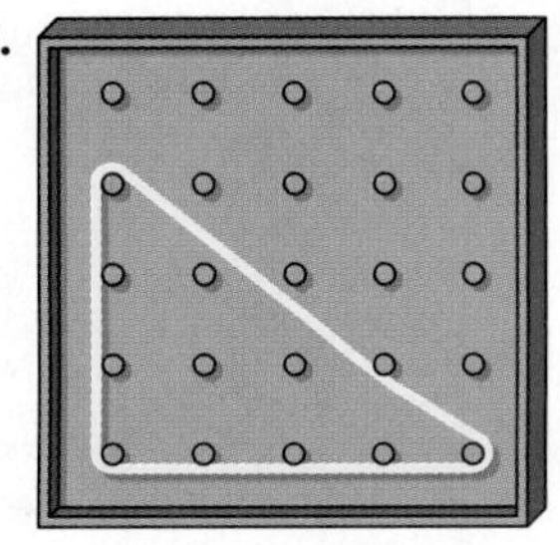

Draw Draw a square on dot paper having each area.

4. 4 square units **5.** 9 square units **6.** 8 square units

7. 13 square units **8.** 17 square units **9.** 32 square units

Write

10. Write a paragraph explaining how to find the total area of the shaded triangles in the drawing at the right.

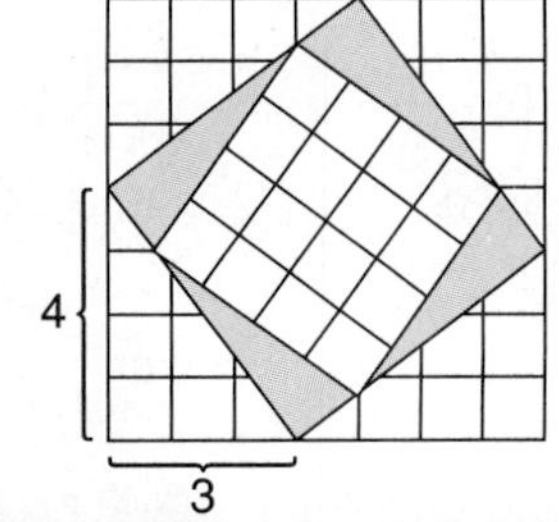

13-1 Integration: Geometry
The Pythagorean Theorem

What YOU'LL LEARN

- To use the Pythagorean theorem to solve problems.

Why IT'S IMPORTANT

You can use the Pythagorean theorem to solve problems involving sailing and travel.

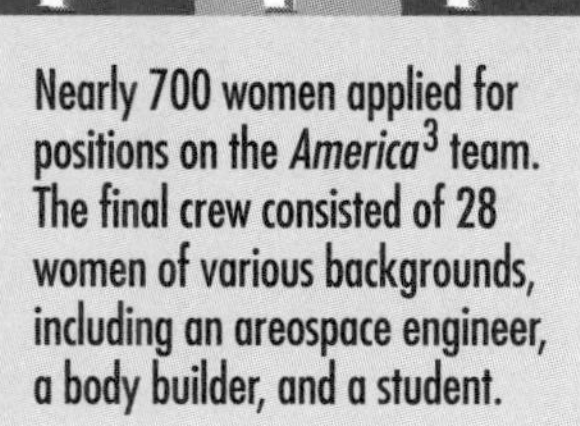

F Y I

Nearly 700 women applied for positions on the *America*[3] team. The final crew consisted of 28 women of various backgrounds, including an areospace engineer, a body builder, and a student.

Sailing

One of the most prestigious sailboat races in the world is the America's Cup. In 1995, for the first time in the 144-year history of the race, one of the sailboats, *America*^{3}, had an all-woman crew. Theirs was one of three U.S. boats vying for the most prized trophy in sailing.

A sailboat's *mast* and *boom* form a right angle. The sail itself, called a *mainsail*, is in the shape of a right triangle.

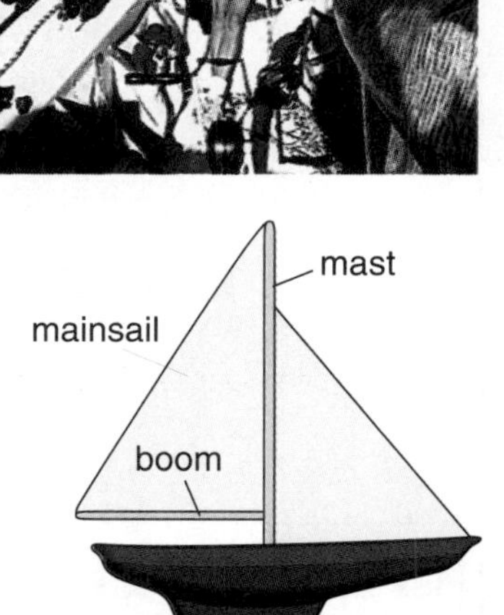

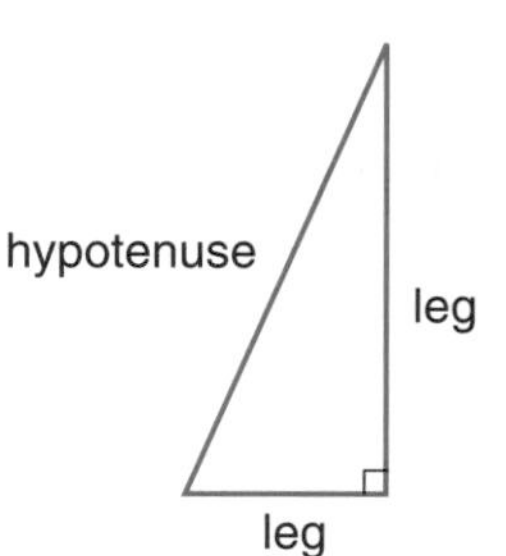

Recall that the side opposite the right angle in a right triangle is called the *hypotenuse*. This side is always the longest side of a right triangle. The other two sides are called the *legs* of the triangle.

To find the length of any side of a right triangle when the lengths of the other two are known, you can use a formula named for the Greek mathematician Pythagoras.

The Pythagorean Theorem	**If a and b are the measures of the legs of a right triangle and c is the measure of the hypotenuse, then $c^2 = a^2 + b^2$.**

You can use the Pythagorean theorem to find the length of the hypotenuse of a right triangle when the lengths of the legs are known.

Example 1 **Find the length of the hypotenuse of a right triangle if $a = 12$ and $b = 5$.**

LOOK BACK

Refer to Lesson 2-8 to review square roots.

$c^2 = a^2 + b^2$ *Pythagorean theorem*

$c^2 = 12^2 + 5^2$ *$a = 12$ and $b = 5$*

$c^2 = 144 + 25$

$c^2 = 169$

$c = \pm\sqrt{169}$

$c = \pm 13$ *Disregard -13. Why?*

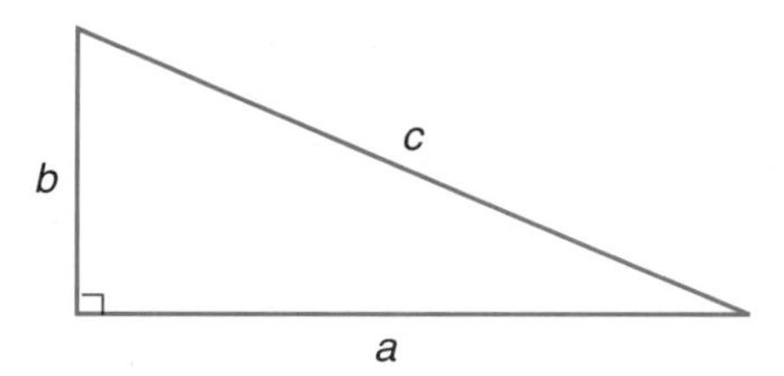

The length of the hypotenuse is 13 units.

Example 2 **Find the length of side a if $b = 9$ and $c = 21$. Round to the nearest hundredth.**

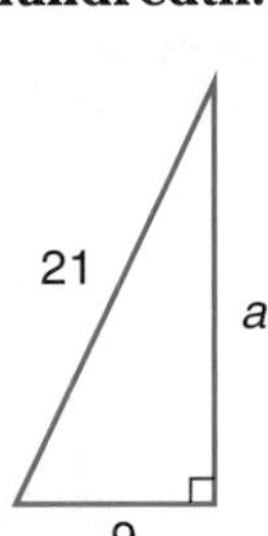

$c^2 = a^2 + b^2$ *Pythagorean theorem*

$21^2 = a^2 + 9^2$ *$b = 9$ and $c = 21$*

$441 = a^2 + 81$

$360 = a^2$

$\pm\sqrt{360} = a$ *Use a calculator to approximate $\sqrt{360}$ to the nearest hundredth.*

$18.97 \approx a$ *Only the positive value of a has meaning in this situation.*

The length of the leg, to the nearest hundredth, is 18.97 units.

The following corollary, based on the Pythagorean theorem, can be used to determine whether a triangle is a right triangle.

Corollary to the Pythagorean Theorem	**If c is the measure of the longest side of a triangle and $c^2 \neq a^2 + b^2$, then the triangle is not a right triangle.**

Example 3 **Determine whether the following side measures would form right triangles.**

a. 6, 8, 10

Since the measure of the longest side is 10, let $c = 10$, $a = 6$, and $b = 8$. Then determine whether $c^2 = a^2 + b^2$.

$10^2 \stackrel{?}{=} 6^2 + 8^2$

$100 \stackrel{?}{=} 36 + 64$

$100 = 100$ ✔

Since $c^2 = a^2 + b^2$, the triangle is a right triangle.

b. 7, 9, 12

Since the measure of the longest side is 12, let $c = 12$, $a = 7$, and $b = 9$. Then determine whether $c^2 = a^2 + b^2$.

$12^2 \stackrel{?}{=} 7^2 + 9^2$

$144 \stackrel{?}{=} 49 + 81$

$144 \neq 130$

Since $c^2 \neq a^2 + b^2$, the triangle is not a right triangle.

GLOBAL CONNECTIONS

One of the earliest inventions was the sail, and it was the most important means of powering boats for most of maritime history. Boats with sails appear in Egyptian rock carvings near the Nile, dating from 3300 B.C.

Example 4

APPLICATION
World Cultures

Agriculture was very important in ancient Aztec culture. Aztec farmers kept records of their farms including calculations of the dimensions and area. Because the terrain was so rough, very few of the farms were rectangular. Yet, by using measuring ropes that measured length with a unit called a *quahuitl* (about 2.5 meters), they were able to make accurate calculations. In the farm shown at the right, the farmer measured three sides of his farm. He had trouble measuring the fourth side because it was located in a dense forest. Find the measure of the fourth side.

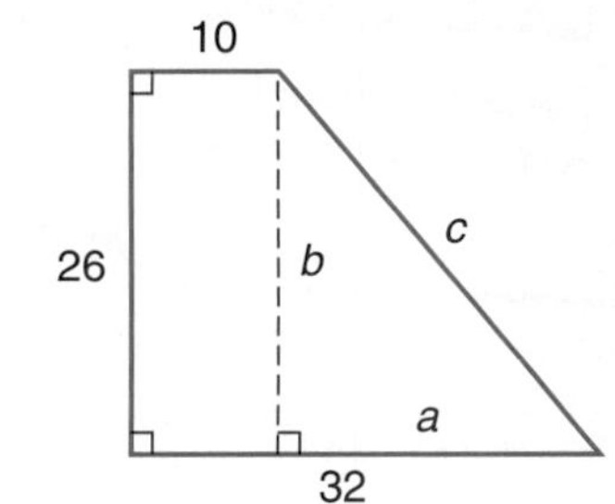

Explore Let c represent the length of the fourth side of the farm. Note that it is the hypotenuse of a right triangle.

Plan Use the Pythagorean theorem to find c. Let $a = 32 - 10$ or 22 and $b = 26$. Then solve the resulting equation.

Solve

$$c^2 = (22)^2 + (26)^2$$
$$c^2 = 484 + 676$$
$$c^2 = 1160$$
$$c \approx 34.06$$

The length of the forest side of the farm is approximately 34.06 quahuitls.

Examine Check the solution by substituting 34.06 for c in the Pythagorean theorem.

$$c^2 = a^2 + b^2$$
$$(34.06)^2 \stackrel{?}{=} (22)^2 + (26)^2$$
$$1160 = 1160 \checkmark$$

CHECK FOR UNDERSTANDING

Communicating Mathematics

Study the lesson. Then complete the following.

1. **Draw** a right triangle and label each side with a letter.
2. **Explain** how you can determine whether a triangle is a right triangle if you know the lengths of the three sides.
3. In 1955, Greece issued the stamp shown at the left to honor the 2500th anniversary of the Pythagorean School. Notice that there is a triangle bordered on each side by a checkerboard pattern.
   a. Count the number of squares along each side of the triangle.
   b. Use the Pythagorean theorem to show that it is a right triangle.
4. When taking the square root of a number, you can get a positive and a negative number. Why then, when using the Pythagorean theorem, is only the positive value used? Explain.

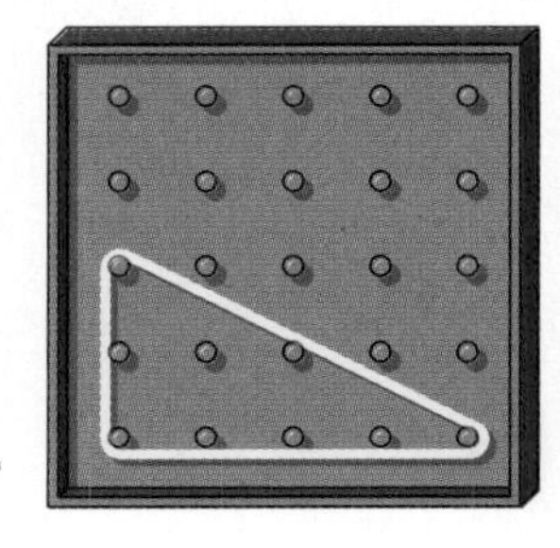

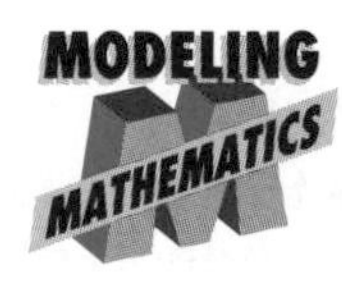

5. Use a geoboard or dot paper to build squares on each side of the triangle at the right. Record the areas of the squares.

Guided Practice

Solve each equation. Assume each variable represents a positive number.

6. $5^2 + 12^2 = c^2$
7. $a^2 + 24^2 = 25^2$
8. $16^2 + b^2 = 20^2$

Find the length of each missing side. Round to the nearest hundredth.

9.

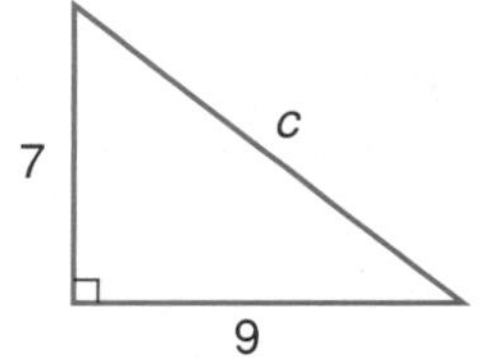

10.

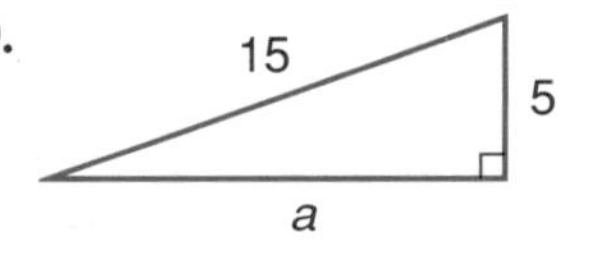

If c is the measure of the hypotenuse of a right triangle, find each missing measure. Round answers to the nearest hundredth, if necessary.

11. $a = 9, b = 12, c = ?$ **12.** $a = \sqrt{11}, c = 6, b = ?$

13. $b = \sqrt{30}, c = \sqrt{34}, a = ?$ **14.** $a = 7, b = 4, c = ?$

Determine whether the following side measures would form right triangles. Explain why or why not.

15. 12, 16, 20 **16.** 2, 8, 8

17. Baseball A baseball scout uses many different tests to determine whether or not to draft a particular player. One test for catchers is to see how quickly they can throw a ball from home plate to second base. On a baseball diamond, the distance from one base to the next is 90 feet. What is the distance from home plate to second base?

EXERCISES

Practice

Find the length of each missing side. Round to the nearest hundredth.

18.

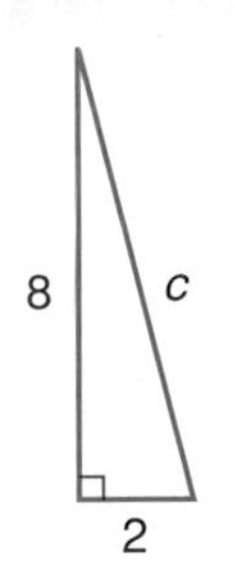

19.

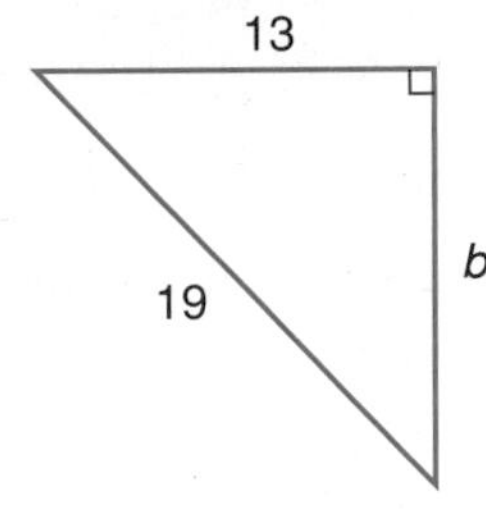

20.

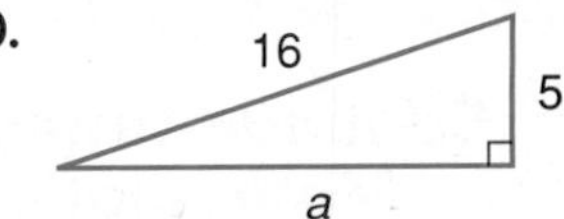

21.

22.

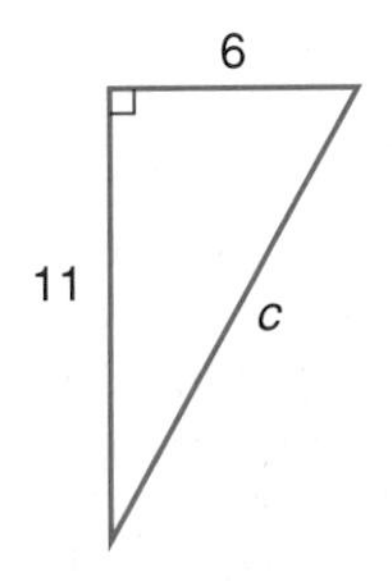

23.

If c is the measure of the hypotenuse of a right triangle, find each missing measure. Round answers to the nearest hundredth.

24. $a = 16, b = 30, c = ?$ **25.** $a = 11, c = 61, b = ?$

26. $b = 13, c = \sqrt{233}, a = ?$ **27.** $a = \sqrt{7}, b = \sqrt{9}, c = ?$

28. $a = 6, b = 3, c = ?$ **29.** $b = \sqrt{77}, c = 12, a = ?$

30. $b = 10, c = 11, a = ?$ **31.** $a = 4, b = \sqrt{11}, c = ?$

32. $a = 15, b = \sqrt{28}, c = ?$ **33.** $a = 12, c = 17, b = ?$

Determine whether the following side measures would form right triangles. Explain why or why not.

34. 6, 9, 12 **35.** 45, 60, 75 **36.** 30, 40, 50

37. 11, 12, 15 **38.** 16, $\sqrt{32}$, 20 **39.** 15, $\sqrt{31}$, 16

INTEGRATION
Geometry

Use an equation to solve each problem. Round answers to the nearest hundredth.

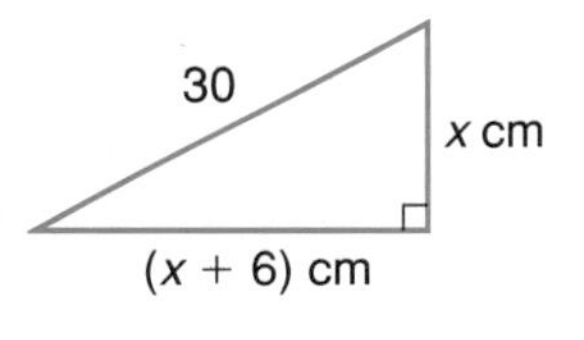

40. Find the length of the diagonal of a square if its area is 128 cm^2.
41. Find the length of the diagonal of a cube if each side of the cube is 5 inches long.
42. A right triangle has one leg that is 6 centimeters longer than the other. The hypotenuse is 30 centimeters long.

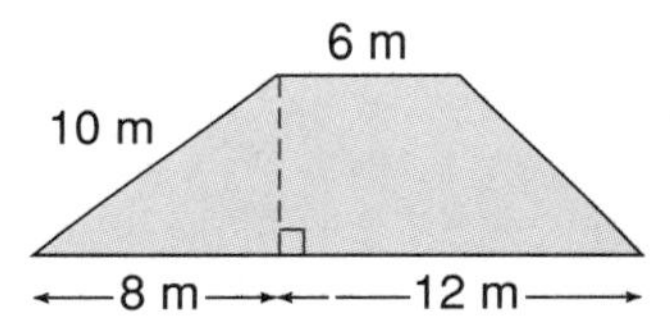

   a. Write an equation to find the length of the legs.

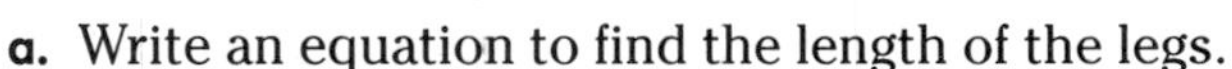

   b. Find the length of each leg of the triangle.
43. Look at the trapezoid at the right.
   a. Find the perimeter. (*Hint:* Drawing a second height will be helpful.)
   b. Find the area.

Programming

44. The graphing calculator program at the right calculates the hypotenuse of a right triangle and displays the right triangle being measured.

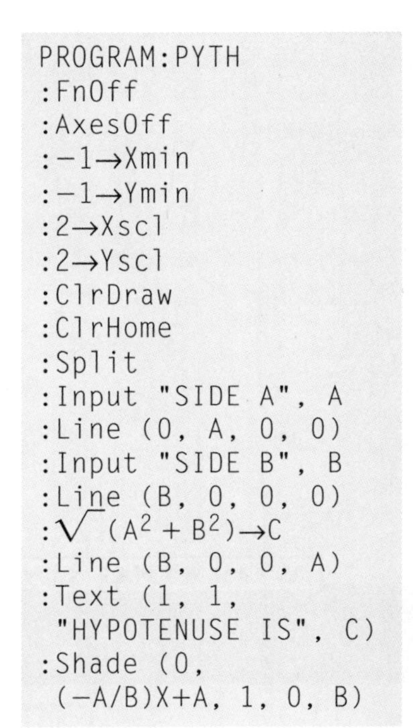

```
PROGRAM:PYTH
:FnOff
:AxesOff
:−1→Xmin
:−1→Ymin
:2→Xscl
:2→Yscl
:ClrDraw
:ClrHome
:Split
:Input "SIDE A", A
:Line (0, A, 0, 0)
:Input "SIDE B", B
:Line (B, 0, 0, 0)
:√(A²+B²)→C
:Line (B, 0, 0, A)
:Text (1, 1,
 "HYPOTENUSE IS", C)
:Shade (0,
 (−A/B)X+A, 1, 0, B)
```

Find the hypotenuse *c* of each right triangle below, given the measures of sides *a* and *b*. Round to the nearest hundredth.

a. $a = 2, b = 7$
b. $a = 9, b = 12$
c. $a = 13, b = 15$
d. $a = 6, b = 9$
e. $a = 12, b = 16$

Critical Thinking

45. The window in Julia's attic was a square with an area of 1 square foot. After she remodeled her home, the width and height of the new attic window were the same as the original, but its area was half that of the original window. If the new window is also a square, explain how this is possible. Include a drawing.

Applications and Problem Solving

46. **Sailing** Refer to the application at the beginning of the lesson. If the edge of the mainsail that is attached to the mast is 100 feet long and the edge of the mainsail that is attached to the boom is 60 feet long, what is the length of the longest edge of the mainsail?

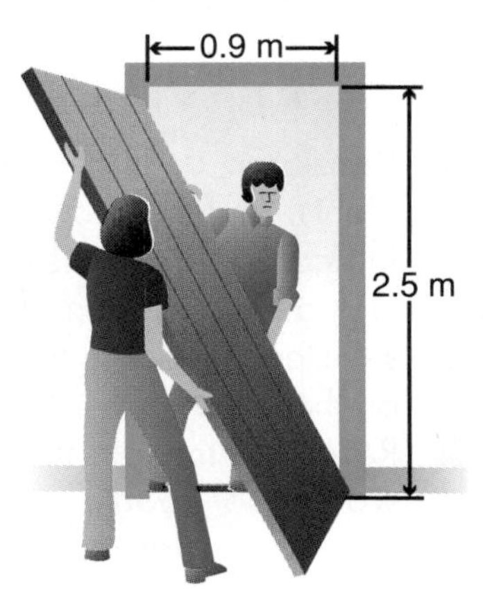

47. **Construction** The walls of the Downtown Recreation Center are being covered with paneling. The doorway into one room is 0.9 meters wide and 2.5 meters high. What is the length of the longest rectangular panel that can be taken through this doorway diagonally?

48. **Travel** The cruise ship M.S. Starward has a right triangle-shaped dance floor on the cabaret deck. The lengths of the two shortest sides of the dance floor are equal, and the longest side next to the stage is 36 feet long. How long is one side of the dance floor?

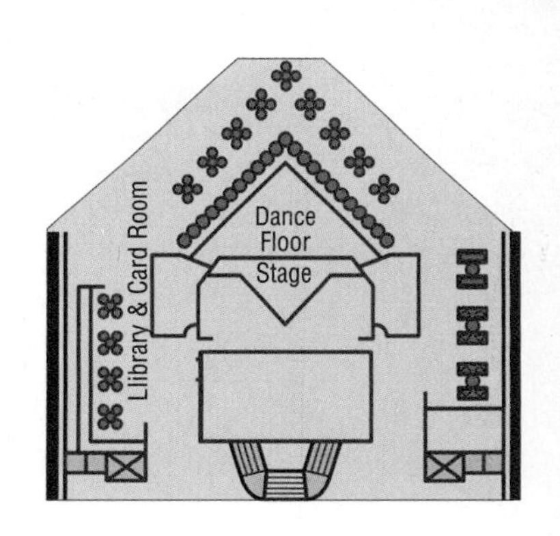

Mixed Review

49. Solve $\frac{a+2}{a} + \frac{a+5}{a} = 1$. (Lesson 12–8)

50. Find an equation of the axis of symmetry and the coordinates of the vertex of the graph of $y = x^2 - 6x + 8$. Then draw the graph. (Lesson 11–1)

51. Simplify $(4r^5)(3r^2)$. (Lesson 9–1)

52. **Number Theory** The sum of two numbers is 38. Their difference is 6. Find the numbers. (Lesson 8–3)

53. Solve $5c - 2 \geq c$. (Lesson 7–2)

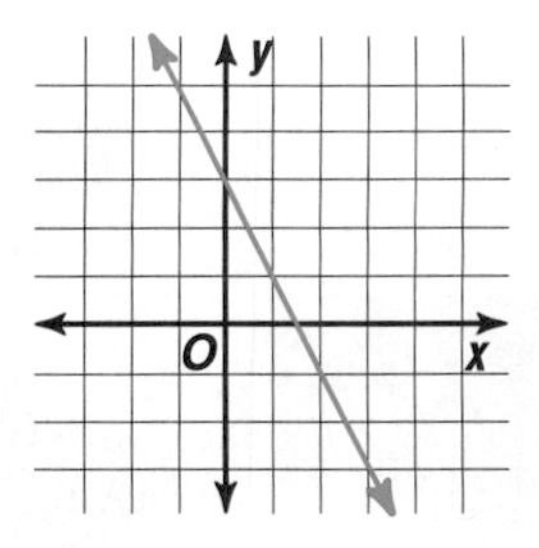

54. State the slope and y-intercept of the line graphed at the right. Then write an equation of the line in slope-intercept form. (Lesson 6–4)

55. Given $f(x) = 2x^2 - 5x + 8$, find $f(-3)$. (Lesson 5–5)

56. Solve $-\frac{5}{6}y = 15$. (Lesson 3–2)

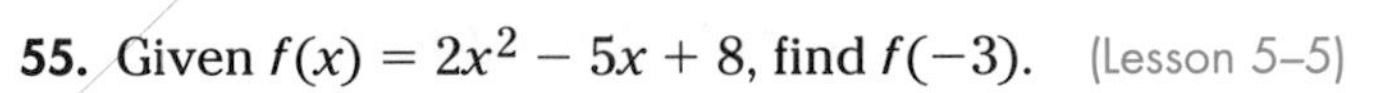

WORKING ON THE

In·ves·ti·ga·tion

Refer to the Investigation on pages 656–657.

A Growing Concern

1. Develop a detailed plan for planting the lawn, plants, and trees for the Sanchez family's yard. Indicate the dimensions of the lawn area as well as the location of the plants and trees you selected. Mark the length of each property line on the drawing.
2. Research the plants and trees that you feel would be good specimens for the areas in which you indicated that they would be planted. Explain why you chose those plants and trees. Think about all four seasons and which plants best suit each season. Also consider the amount of sunlight each area receives during the day.
3. Calculate the cost of materials, labor, and profit margin of planting these items and laying the sprinkler system. Justify your costs and make a case for the quantity and location of the plants, trees, and lawn area in your design.
4. Be sure to add the plants, trees, and sprinkler system to your overall design.

Add the results of your work to your Investigation Folder.

13–2 Simplifying Radical Expressions

What YOU'LL LEARN

- To simplify square roots, and
- to simplify radical expressions.

Why IT'S IMPORTANT

You can use radical expressions to solve problems involving physics and racing.

Physics

The period of a pendulum is the time in seconds that it takes the pendulum to make one complete swing back and forth. The formula for the period P of a pendulum is $P = 2\pi\sqrt{\frac{\ell}{32}}$, where ℓ is the length of the pendulum in feet. Suppose a clock makes one "tick" after each complete swing back and forth of a 2-foot-long pendulum. How many ticks would the clock make in one minute? *This problem will be solved in Example 4.*

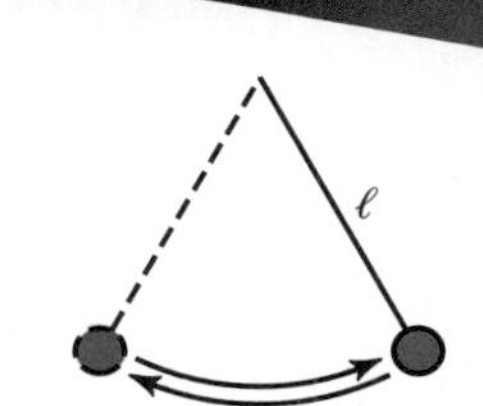

Can $2\pi\sqrt{\frac{\ell}{32}}$ be simplified? One rule used for simplifying radical expressions is that a radical expression is in *simplest form* if the **radicand,** the expression under the radical sign, contains no perfect square factors other than one. The following property can be used to simplify square roots.

Product Property of Square Roots	For any numbers a and b, where $a \geq 0$ and $b \geq 0$, $\sqrt{ab} = \sqrt{a} \cdot \sqrt{b}$.

The product property of square roots and prime factorization can be used to simplify radical expressions in which the radicand is not a perfect square.

Example 1 **Simplify.**

a. $\sqrt{18}$

$\sqrt{18} = \sqrt{3 \cdot 3 \cdot 2}$ *Prime factorization of 18*

$= \sqrt{3^2} \cdot \sqrt{2}$ *Product property of square roots*

$= 3\sqrt{2}$

b. $\sqrt{140}$

$140 = \sqrt{2 \cdot 2 \cdot 5 \cdot 7}$ *Prime factorization of 140*

$= \sqrt{2^2} \cdot \sqrt{5 \cdot 7}$ *Product property of square roots*

$= 2\sqrt{35}$

When finding the principal square root of an expression containing variables, be sure that the result is not negative. Consider the expression $\sqrt{x^2}$. Its simplest form is not x since, for example, $\sqrt{(-4)^2} \neq -4$. For radical expressions like $\sqrt{x^2}$, use absolute value to ensure nonnegative results.

$\sqrt{x^2} = |x|$ $\sqrt{x^3} = x\sqrt{x}$ $\sqrt{x^4} = x^2$ $\sqrt{x^5} = x^2\sqrt{x}$ $\sqrt{x^6} = |x^3|$

For $\sqrt{x^3}$, absolute value is not necessary. If x were negative, then x^3 would be negative, and $\sqrt{x^3}$ would not be defined as a real number. *Why is absolute value not necessary for $\sqrt{x^4}$?*

Example 2 **Simplify $\sqrt{72x^3y^4z^5}$.**

$$\sqrt{72x^3y^4z^5} = \sqrt{2^3 \cdot 3^2 \cdot x^3 \cdot y^4 \cdot z^5} \quad \textit{Prime factorization}$$
$$= \sqrt{2^2} \cdot \sqrt{2} \cdot \sqrt{3^2} \cdot \sqrt{x^2} \cdot \sqrt{x} \cdot \sqrt{y^4} \cdot \sqrt{z^4} \cdot \sqrt{z} \quad \textit{Product property}$$
$$= 2 \cdot 3 \cdot |x| \cdot y^2 \cdot z^2 \cdot \sqrt{2xz} \quad \textit{Simplify.}$$
$$= 6|x|y^2z^2\sqrt{2xz} \quad \textit{The absolute value of x ensures a nonegative result.}$$

The product property can also be used to multiply square roots.

Example 3 **Simplify $\sqrt{5} \cdot \sqrt{35}$.**

$$\sqrt{5} \cdot \sqrt{35} = \sqrt{5} \cdot \sqrt{5} \cdot \sqrt{7} \quad \textit{Product property of square roots}$$
$$= \sqrt{5^2 \cdot 7}$$
$$= 5\sqrt{7}$$

You can use a graphing calculator to explore and analyze radical expressions.

EXPLORATION — GRAPHING CALCULATORS

The formula $Y = 91.4 - (91.4 - T)[0.478 + 0.301(\sqrt{x} - 0.02x)]$ can be used to calculate the windchill factor. In this formula, Y represents the windchill, T represents the outside Fahrenheit temperature, and x represents the wind speed in miles per hour. When a meteorologist says that the temperature is 12 degrees, but it feels like 18 degrees below zero because of the wind, you can use the formula to find how fast the wind is blowing. Set the range at [0, 40] by 2 and [−50, 40] by 5.

Enter:

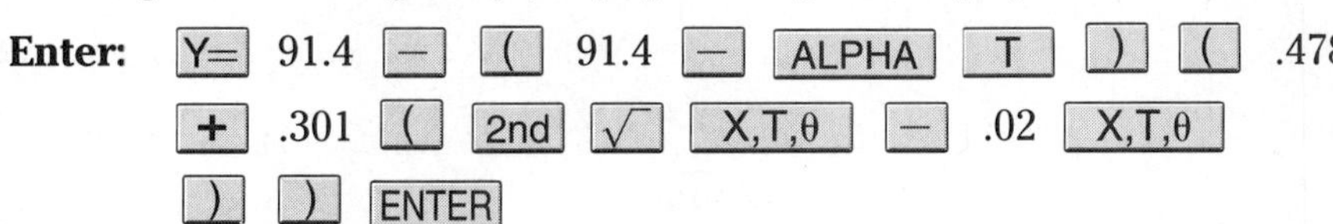

Store 12 into T by pressing 12 STO▶ ALPHA T ENTER. Graph the function and trace along the graph until Y = −18. At Y = −18, X = 10.2. So the wind is blowing at approximately 10 miles per hour.

Your Turn

a. Use the graph to find the wind speed if the temperature feels like −7°.

b. Enter the formula into Y1, Y2, and Y3 using different variables for T. Store three different temperatures for these variables and graph them simultaneously.

c. Analyze and compare the graphs.

CAREER CHOICES

A **meteorologist** studies the atmosphere, its physical characteristics, motions, processes, and effects on the environment.

A bachelor's degree in meteorology is required, but a graduate degree is needed for advancement.

For more information, contact:

American Meteorological Society
45 Beacon St.
Boston, MA 02108-3693

Example 4

Physics

Refer to the connection at the beginning of the lesson. How many ticks would the clock make in one minute? Use 3.14 for π and round to the nearest whole number.

Explore The clock makes one "tick" after each complete swing back and forth of its 2-foot-long pendulum. The formula for the period P of a pendulum is $P = 2\pi\sqrt{\frac{\ell}{32}}$.

Plan First find P, the number of seconds it takes for the pendulum to go back and forth one time. Then find $\frac{60}{P}$, the number of times the clock's pendulum swings back and forth in one minute.

Solve

$$P = 2\pi\sqrt{\frac{\ell}{32}}$$

$$\approx 2(3.14)\sqrt{\frac{2}{32}} \quad \textit{$\pi \approx 3.14$, $\ell = 2$}$$

$$\approx 6.28 \cdot \sqrt{\frac{1}{16}}$$

$$\approx 6.28 \cdot \frac{1}{4} \text{ or } 1.57$$

So it takes about 1.57 seconds for the pendulum to go back and forth once.

$$\frac{60}{1.57} \approx 38.22$$

Thus, the clock makes about 38 ticks per minute.

Examine Since 38.22×1.57 is about 40×1.5 or 60, the answer seems reasonable.

You can divide square roots and simplify radical expressions that involve division by using the quotient property of square roots.

Quotient Property of Square Roots	**For any numbers a and b, where $a \geq 0$ and $b > 0$,** $\sqrt{\frac{a}{b}} = \frac{\sqrt{a}}{\sqrt{b}}$.

A fraction containing radicals is in simplest form if no radicals are left in the denominator.

Example 5

a. Simplify $\frac{\sqrt{56}}{\sqrt{7}}$ and $\sqrt{\frac{34}{25}}$.

b. Compare the expressions using <, >, or =.

FYI

The world's largest pendulum clock is in Tokyo, Japan. It is part of a water-mill clock and is 73 feet 9.75 inches long.

a. $\frac{\sqrt{56}}{\sqrt{7}} = \sqrt{\frac{56}{7}}$ *Quotient property of square roots*

$= \sqrt{8}$

$= \sqrt{4} \cdot \sqrt{2}$

$= 2\sqrt{2}$

$\sqrt{\frac{34}{25}} = \frac{\sqrt{34}}{\sqrt{25}}$

$= \frac{\sqrt{34}}{5}$

b. You can compare these expressions by estimating their values and then using a scientific calculator to find approximations for each simplified expression.

Estimate: $2\sqrt{2} \rightarrow$ *Since 2 is a little more than 1, $2\sqrt{2}$ will be a little more than 2.*

$\frac{\sqrt{34}}{5} \rightarrow \frac{\sqrt{36}}{5} = \frac{6}{5}$ This will be a little more than 1. So, $2\sqrt{2} > \frac{\sqrt{34}}{5}$.

Verify by using a calculator.

Enter: 2 [X] 2 [$\sqrt{x}$] [=] 2.828427125

Enter: 34 [$\sqrt{x}$] [÷] 5 [=] 1.166190379

Since $2.8 > 1.2$, then $\frac{\sqrt{56}}{\sqrt{7}} > \sqrt{\frac{34}{25}}$.

PEANUTS®

PEANUTS reprinted by permission of United Feature Syndicate, Inc.

In the cartoon above, Woodstock simplified the radical expression by **rationalizing the denominator.** This method may be used to remove or eliminate radicals from the denominator of a fraction.

Example 6 Simplify.

a. $\frac{\sqrt{5}}{\sqrt{3}}$

$$\frac{\sqrt{5}}{\sqrt{3}} = \frac{\sqrt{5}}{\sqrt{3}} \cdot \frac{\sqrt{3}}{\sqrt{3}}$$ *Note that $\frac{\sqrt{3}}{\sqrt{3}} = 1$.*

$$= \frac{\sqrt{15}}{3}$$

b. $\frac{\sqrt{7}}{\sqrt{12}}$

$$\frac{\sqrt{7}}{\sqrt{12}} = \frac{\sqrt{7}}{\sqrt{2 \cdot 2 \cdot 3}}$$

$$= \frac{\sqrt{7}}{\sqrt{2 \cdot 2 \cdot 3}} \cdot \frac{\sqrt{3}}{\sqrt{3}}$$

$$= \frac{\sqrt{7}\sqrt{3}}{2 \cdot 3}$$

$$= \frac{\sqrt{21}}{6}$$

Binomials of the form $a\sqrt{b} + c\sqrt{d}$ and $a\sqrt{b} - c\sqrt{d}$ are called **conjugates** of each other. For example, $6 + \sqrt{2}$ and $6 - \sqrt{2}$ are conjugates. Conjugates are useful when simplifying radical expressions because their product is always a rational number with no radicals.

$$(6 + \sqrt{2})(6 - \sqrt{2}) = 6^2 - (\sqrt{2})^2$$
$$= 36 - 2$$
$$= 34$$

Use the pattern $(a - b)(a + b) = a^2 - b^2$ to simplify the product.

This is true because of the following.

$$(\sqrt{2})^2 = \sqrt{2} \cdot \sqrt{2}$$
$$= \sqrt{2 \cdot 2}$$
$$= \sqrt{2^2} \text{ or } 2$$

Conjugates are often used to rationalize the denominators of fractions containing square roots.

Example  **Simplify $\frac{4}{4 - \sqrt{3}}$.**

To rationalize the denominator, multiply both the numerator and denominator by $4 + \sqrt{3}$, which is the conjugate of $4 - \sqrt{3}$.

$$\frac{4}{4-\sqrt{3}} = \frac{4}{4-\sqrt{3}} \cdot \frac{4+\sqrt{3}}{4+\sqrt{3}}$$ *Notice that $\frac{4+\sqrt{3}}{4+\sqrt{3}} = 1$.*

$$= \frac{4(4) + 4\sqrt{3}}{4^2 - \left(\sqrt{3}\right)^2}$$ *Use the distributive property to multiply numerators. Use the pattern $(a - b)(a + b) = a^2 - b^2$ to multiply denominators.*

$$= \frac{16 + 4\sqrt{3}}{16 - 3}$$

$$= \frac{16 + 4\sqrt{3}}{13}$$

When simplifying radical expressions, check the following conditions to determine if the expression is in simplest form.

Simplest Radical Form	**A radical expression is in simplest form when the following three conditions have been met.** **1. No radicands have perfect square factors other than 1.** **2. No radicands contain fractions.** **3. No radicals appear in the denominator of a fraction.**

CHECK FOR UNDERSTANDING

Communicating Mathematics

Study the lesson. Then complete the following.

1. **Explain** why absolute values are sometimes needed when simplifying radical expressions containing variables.
2. **Describe** the steps you take to rationalize a denominator.
3. **You Decide** Niara showed the following equations to her friend Melanie and said, "I know that 6 can't equal 10, but all these steps make sense!" Melanie said, "One of the steps must be wrong." Who is correct? Can you explain the mistake?

$-60 = -60$	*Reflexive property of equality*
$36 - 96 = 100 - 160$	*Rewrite -60 as $36 - 96$ and $100 - 160$.*
$36 - 96 + 64 = 100 - 160 + 64$	*Add 64 to each side.*
$(6 - 8)^2 = (10 - 8)^2$	*Factor.*
$6 - 8 = 10 - 8$	*Take the square root of each side.*
$6 - 8 + 8 = 10 - 8 + 8$	*Add 8 to each side.*
$6 = 10$	*Simplify.*

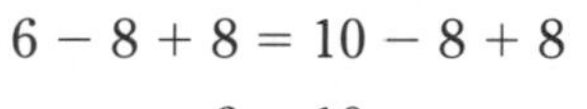

Math Journal

4. Refer to the cartoon on page 722. Obviously, Woodstock realized that rationalizing the denominator is important. What are other steps that you may have to use to simplify a radical expression?

Guided Practice

State the conjugate of each expression. Then multiply the expression by its conjugate.

5. $5 + \sqrt{2}$

6. $\sqrt{3} - \sqrt{7}$

State the fraction by which each expression should be multiplied to rationalize the denominator.

7. $\frac{4}{\sqrt{7}}$

8. $\frac{2\sqrt{5}}{4 - \sqrt{3}}$

Simplify. Leave in radical form and use absolute value symbols when necessary.

9. $\sqrt{18}$

10. $\frac{\sqrt{20}}{\sqrt{5}}$

11. $\sqrt{\frac{3}{7}}$

12. $\sqrt{\frac{2}{3}} \cdot \sqrt{\frac{5}{2}}$

13. $(\sqrt{2} + 4)(\sqrt{2} + 6)$

14. $(y - \sqrt{5})(y + \sqrt{5})$

15. $\frac{6}{3 - \sqrt{2}}$

16. $\sqrt{80a^2b^3}$

Compare each pair of expressions using <, >, or =.

17. $4\sqrt{3} \cdot \sqrt{3}, \sqrt{48} + \sqrt{8}$

18. $\sqrt{\frac{12}{7}}, \frac{\sqrt{18} \cdot \sqrt{2}}{\sqrt{7} \cdot \sqrt{3}}$

19. **Water Supply** There is a relationship between a city's capacity to supply water to its citizens and the city's size. Suppose a city has a population P (in thousands). Then the number of gallons per minute that are required to assure water adequacy is given by the expression $1020\sqrt{P}(1 - 0.01\sqrt{P})$. If a city has a population of 55,000 people, how many gallons per minute must the city's pumping stations be able to supply?

EXERCISES

Practice

Simplify. Leave in radical form and use absolute value symbols when necessary.

20. $\sqrt{75}$

21. $\sqrt{80}$

22. $\sqrt{280}$

23. $\sqrt{500}$

24. $\frac{\sqrt{7}}{\sqrt{3}}$

25. $\frac{\sqrt{5}}{\sqrt{10}}$

26. $\sqrt{\frac{2}{7}}$

27. $\sqrt{\frac{11}{32}}$

28. $5\sqrt{10} \cdot 3\sqrt{10}$

29. $7\sqrt{30} \cdot 2\sqrt{6}$

30. $\sqrt{\frac{3}{5}} \cdot \sqrt{\frac{7}{3}}$

31. $\sqrt{\frac{1}{6}} \cdot \sqrt{\frac{6}{11}}$

32. $\sqrt{40b^4}$

33. $\sqrt{54a^2b^2}$

34. $\sqrt{60m^2y^4}$

35. $\sqrt{147x^5y^7}$

36. $\sqrt{\frac{t}{8}}$

37. $\sqrt{\frac{27}{p^2}}$

38. $\sqrt{\frac{5n^5}{4m^5}}$

39. $\frac{\sqrt{9x^5y}}{\sqrt{12x^2y^6}}$

40. $(1 + 2\sqrt{5})^2$

41. $(y - \sqrt{7})^2$

42. $(\sqrt{m} + \sqrt{20})^2$

43. $\frac{14}{\sqrt{8} - \sqrt{5}}$

44. $\frac{9a}{6 + \sqrt{a}}$

45. $\frac{2\sqrt{5}}{-4 + \sqrt{8}}$

46. $\frac{3\sqrt{7}}{5\sqrt{3} + 3\sqrt{5}}$

47. $\frac{\sqrt{c} - \sqrt{d}}{\sqrt{c} + \sqrt{d}}$

48. $(\sqrt{2x} - \sqrt{6})(\sqrt{2x} + \sqrt{6})$

49. $(x - 4\sqrt{3})(x - \sqrt{3})$

Compare each pair of expressions using <, >, or =.

50. $\sqrt{\frac{8}{9}} \cdot \frac{2}{\sqrt{8}}, \frac{2}{\sqrt{51}} \cdot \sqrt{\frac{17}{3}}$

51. $\sqrt{10} \cdot \sqrt{30}, \frac{10}{\sqrt{5} + 9}$

52. $\frac{2}{\sqrt{6} - \sqrt{5}}, \frac{20}{6 + \sqrt{3}}$

53. $\frac{3\sqrt{2} - \sqrt{7}}{2\sqrt{3} - 5\sqrt{2}}, \frac{4\sqrt{5} - 3\sqrt{7}}{\sqrt{6}}$

Critical Thinking

54. Determine whether $\sqrt{a \cdot b} = \sqrt{a} \cdot \sqrt{b}$ is true for negative real numbers. Give examples to support your answer.

Applications and Problem Solving

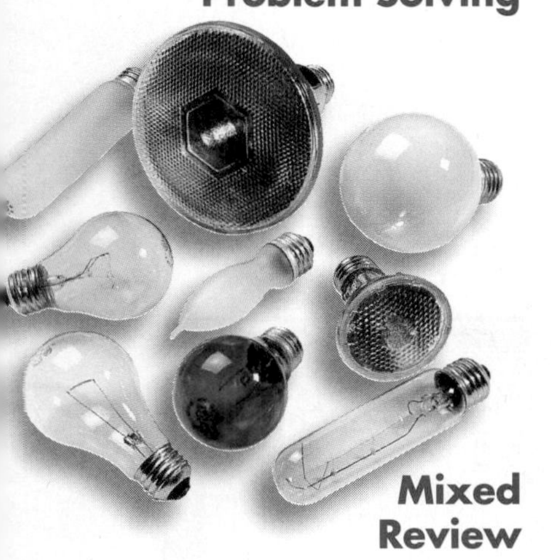

55. **Racing** In yacht racing from 1958 to 1987, 12-meter boats were not really 12 meters long, but the formula that governed their design contained numbers that equaled 12. In the expression $\frac{\sqrt{S} + L - F}{2.37}$, S is the area of the sails, L is the waterline length, and F is the distance from the deck of the boat to the waterline. The result must be less than or equal to 12 for a boat to be classified as a 12-meter boat. Determine if a boat for which $S = 158$ m^2, $L = 17.5$ m, and $F = 2$ m could be classified as a 12-meter boat.

56. **Electricity** The voltage V required for a circuit is given by $V = \sqrt{PR}$ where P is the power in watts and R is the resistance in ohms. Find the volts needed to light a 75-watt bulb with a resistance of 110 ohms.

Mixed Review

57. **Geometry** Find the length of the missing side of the triangle shown at the right. Round to the nearest hundredth. (Lesson 13–1)

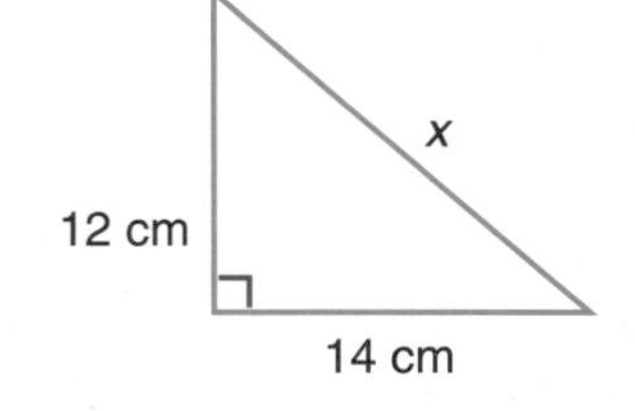

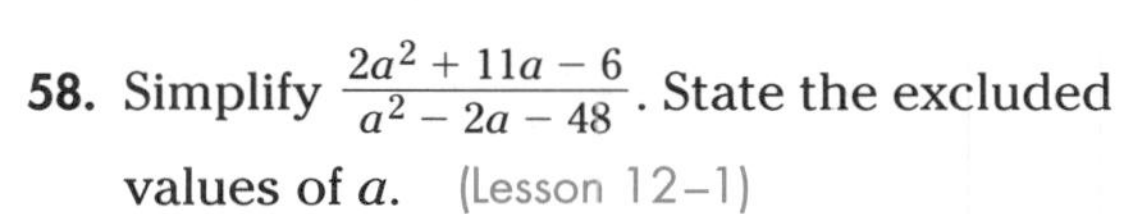

58. Simplify $\frac{2a^2 + 11a - 6}{a^2 - 2a - 48}$. State the excluded values of a. (Lesson 12–1)

59. Solve $3x^2 - 5x + 2 = 0$ by using the quadratic formula. (Lesson 11–3)

60. Factor $12a^2b^3 - 28ab^2c^2$. (Lesson 10–2)

61. **Forests** The largest forested areas in the world are located in northern Russia. They cover 2,700,000,000 acres, and they make up 25% of the world's forests. Express the number of acres in scientific notation. (Lesson 9–3)

62. Solve $4 - 2.3t < 17.8$. (Lesson 7–3)

63. **Geometry** Write an equation of the line that is perpendicular to the graph of $y = -5x + 2$ and passes through the point at (0, 6). (Lesson 6–6)

64. Express the relation shown in the graph at the right as a set of ordered pairs. Then state the domain and range of the relation. (Lesson 5–2)

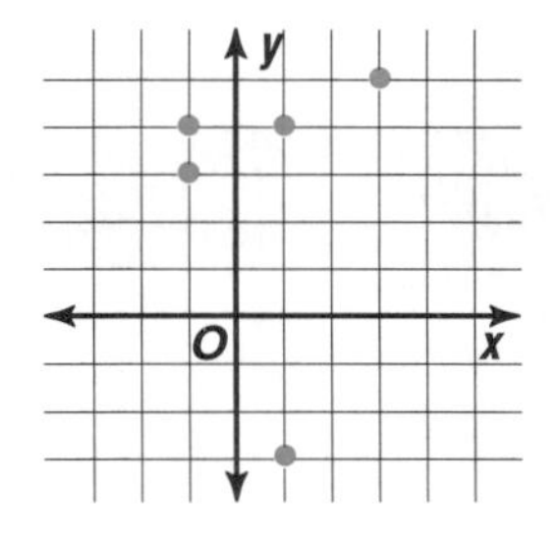

65. Solve $6(x + 3) = 3x$. (Lesson 3–5)

66. Write an algebraic expression for the verbal expression *one-fourth the square of a number.* (Lesson 1–1)

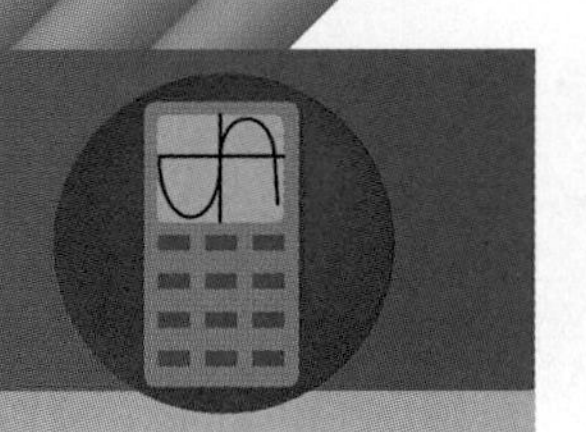

13–2B Graphing Technology
Simplifying Radical Expressions

An Extension of Lesson 13–2

The built-in square root feature of a graphing calculator allows us to simplify and approximate values of expressions containing radicals. In addition to obtaining approximate values for expressions, this feature is also useful for checking algebraic computations.

Example **Simplify each expression algebraically. Then check with a graphing calculator.**

a. $\sqrt{\frac{3}{5}}$

$$\sqrt{\frac{3}{5}} = \sqrt{\frac{3}{5}} \cdot \frac{\sqrt{5}}{\sqrt{5}}$$

$$= \frac{\sqrt{3 \cdot 5}}{5} \text{ or } \frac{\sqrt{15}}{5}$$

Verify with the calculator.

Enter: 2nd $\sqrt{\ }$ (3 ÷ 5)

ENTER .7745966692

2nd $\sqrt{\ }$ 15 ÷ 5

ENTER .7745966692

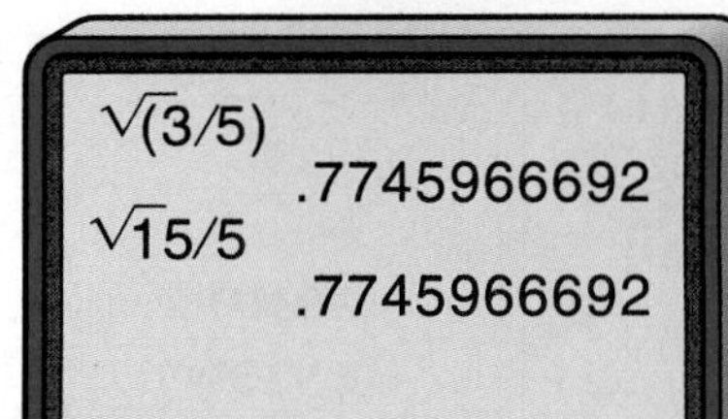

b. $\frac{1}{\sqrt{2} - 3}$

$$\frac{1}{\sqrt{2} - 3} = \frac{1}{\sqrt{2} - 3} \cdot \frac{\sqrt{2} + 3}{\sqrt{2} + 3}$$

$$= \frac{\sqrt{2} + 3}{\left(\sqrt{2}\right)^2 - 3^2} \text{ or } \frac{\sqrt{2} + 3}{-7}$$

Verify with the calculator.

Enter: 1 ÷ (2nd $\sqrt{\ }$ 2 −

3) ENTER −.6306019375

(2nd $\sqrt{\ }$ 2 + 3)

÷ (−) 7 ENTER

−.6306019375

1/(√2 −3)
−.6306019375
(√2+3)/−7
−.6306019375

EXERCISES

Simplify each expression. Then check with a graphing calculator. Round answers to the nearest hundredth.

1. $\sqrt{1372}$
2. $\sqrt{32} \cdot \sqrt{12}$
3. $\sqrt{2}(\sqrt{6} + 3)$
4. $\sqrt{\frac{5}{6}}$
5. $\frac{4}{\sqrt{7}}$
6. $\frac{2}{\sqrt{11} + 8}$
7. $\sqrt{12} + \sqrt{3}$
8. $\frac{4}{5}\sqrt{2} + \frac{3}{5}\sqrt{2}$
9. $\frac{\sqrt{3}}{2} - \frac{\sqrt{5}}{3} + \sqrt{18}$
10. $\sqrt{18} + \sqrt{108} + \sqrt{50}$
11. $\sqrt{\frac{2}{3}} + \frac{\sqrt{6}}{3} - 6\sqrt{6}$

13–3 Operations with Radical Expressions

What YOU'LL LEARN
- To simplify radical expressions involving addition, subtraction, and multiplication.

Why IT'S IMPORTANT

You can use radical expressions to solve problems involving travel and construction.

Largest Passenger Ships in the World (gross tonnage)

1. *Norway, 76,049*
2. *Majesty of the Seas, 73,937*
2. *Monarch of the Seas, 73,937*
4. *Sovereign of the Seas, 73,192*
5. *Sensation, 70,367*
5. *Ecstasy, 70,367*
5. *Fantasy, 70,367*

APPLICATION

Travel

The Norway cruise ship is the largest passenger ship in the Caribbean. Suppose the captain of the ship is on the star deck, which is 48 feet above the pool deck. The pool deck is 72 feet above the water. The captain sees the next island, but the passengers on the pool deck cannot. The equation $d = \sqrt{\frac{3h}{2}}$ represents the distance d in miles a person h feet high can see. So, $\sqrt{\frac{3(120)}{2}} - \sqrt{\frac{3(72)}{2}}$ describes how much farther the captain can see than the passengers. *You will find the value of this expression in Example 3.*

Radical expressions in which the radicands are alike can be added or subtracted in the same way that monomials are added or subtracted.

Monomials	**Radical Expressions**
$4x + 5x = (4 + 5)x$	$4\sqrt{5} + 5\sqrt{5} = (4 + 5)\sqrt{5}$
$= 9x$	$= 9\sqrt{5}$
$18y - 7y = (18 - 7)y$	$18\sqrt{2} - 7\sqrt{2} = (18 - 7)\sqrt{2}$
$= 11y$	$= 11\sqrt{2}$

Notice that the distributive property was used to simplify each radical expression.

Example 1 **Simplify each expression.**

a. $\mathbf{6\sqrt{7} + 5\sqrt{7} - 3\sqrt{7}}$

$$6\sqrt{7} + 5\sqrt{7} - 3\sqrt{7} = (6 + 5 - 3)\sqrt{7}$$
$$= 8\sqrt{7}$$

b. $\mathbf{5\sqrt{6} + 3\sqrt{7} + 4\sqrt{7} - 2\sqrt{6}}$

$$5\sqrt{6} + 3\sqrt{7} + 4\sqrt{7} - 2\sqrt{6} = 5\sqrt{6} - 2\sqrt{6} + 3\sqrt{7} + 4\sqrt{7}$$
$$= (5 - 2)\sqrt{6} + (3 + 4)\sqrt{7}$$
$$= 3\sqrt{6} + 7\sqrt{7}$$

In Example 1b, the expression $3\sqrt{6} + 7\sqrt{7}$ cannot be simplified further because the radicands are different. There are no common factors, and each radicand is in simplest form.

If the radicals in a radical expression are not in simplest form, simplify them first. Then use the distributive property wherever possible to further simplify the expression.

Example 2 **Simplify $4\sqrt{27} + 5\sqrt{12} + 8\sqrt{75}$. Then use a scientific calculator to verify your answer.**

$$\begin{aligned} 4\sqrt{27} + 5\sqrt{12} + 8\sqrt{75} &= 4\sqrt{3^2 \cdot 3} + 5\sqrt{2^2 \cdot 3} + 8\sqrt{5^2 \cdot 3} \\ &= 4(\sqrt{3^2} \cdot \sqrt{3}) + 5(\sqrt{2^2} \cdot \sqrt{3}) + 8(\sqrt{5^2} \cdot \sqrt{3}) \\ &= 4(3\sqrt{3}) + 5(2\sqrt{3}) + 8(5\sqrt{3}) \\ &= 12\sqrt{3} + 10\sqrt{3} + 40\sqrt{3} \\ &= 62\sqrt{3} \end{aligned}$$

The exact answer is $62\sqrt{3}$. Now, use a calculator to verify.

First, find a decimal approximation for the original expression.

Enter: 4 [×] 27 [√x] [+] 5 [×] 12 [√x] [+] 8 [×] 75 [√x] [=] 107.3871501

Next, find a decimal approximation for the simplified expression.

Enter: 62 [×] 3 [√x] [=] 107.3871501

Since the approximations are equal, the results have been verified.

Example 3

Travel

Refer to the application at the beginning of the lesson. How much farther is the captain able to see than the passengers on the pool deck?

$$\begin{aligned} d &= \sqrt{\frac{3(120)}{2}} - \sqrt{\frac{3(72)}{2}} \\ &= \sqrt{\frac{360}{2}} - \sqrt{\frac{216}{2}} \\ &= \sqrt{180} - \sqrt{108} \\ &= \sqrt{6^2 \cdot 5} - \sqrt{6^2 \cdot 3} \\ &= 6\sqrt{5} - 6\sqrt{3} \\ &\approx 3.02 \end{aligned}$$

The captain can see about 3 miles farther.

In the last lesson, you multiplied conjugates and expressions with like radicands. Multiplying two radical expressions with different radicands is similar to multiplying two binomials.

Example 4 **Simplify $(2\sqrt{3} - \sqrt{5})(\sqrt{10} + 4\sqrt{6})$.**

$(2\sqrt{3} - \sqrt{5})(\sqrt{10} + 4\sqrt{6})$

First terms, *Outer terms*, *Inner terms*, *Last terms*

$$= \underbrace{(2\sqrt{3})(\sqrt{10})}_{\text{First terms}} + \underbrace{(2\sqrt{3})(4\sqrt{6})}_{\text{Outer terms}} + \underbrace{(-\sqrt{5})(\sqrt{10})}_{\text{Inner terms}} + \underbrace{(-\sqrt{5})(4\sqrt{6})}_{\text{Last terms}}$$

$= 2\sqrt{30} + 8\sqrt{18} - \sqrt{50} - 4\sqrt{30}$ *Multiply.*

$= 2\sqrt{30} + 24\sqrt{2} - 5\sqrt{2} - 4\sqrt{30}$ *Simplify each term.*

$= -2\sqrt{30} + 19\sqrt{2}$ *Combine like terms.*

LOOK BACK

Refer to Lesson 9-7 to review multiplying polynomials.

CHECK FOR UNDERSTANDING

Communicating Mathematics

Study the lesson. Then complete the following.

1. **Write** three radical expressions that have the same radicand.
2. **Explain** why you should simplify each radical in a radical expression before adding or subtracting.
3. **Explain** why $\sqrt{x} + \sqrt{y} \neq \sqrt{x + y}$. Give an example using numbers.

4. Explain how you use the distributive property to simplify like radicands that are added or subtracted.

Guided Practice

Name the expressions in each group that will have the same radicand after each expression is written in simplest form.

5. $3\sqrt{5}, 5\sqrt{6}, 3\sqrt{20}$
6. $-5\sqrt{7}, 2\sqrt{28}, 6\sqrt{14}$
7. $\sqrt{24}, \sqrt{12}, \sqrt{18}, \sqrt{28}$
8. $9\sqrt{32}, 2\sqrt{50}, \sqrt{48}, 3\sqrt{200}$

Simplify.

9. $3\sqrt{6} + 10\sqrt{6}$
10. $2\sqrt{5} - 5\sqrt{2}$
11. $8\sqrt{7x} + 4\sqrt{7x}$

Simplify. Then use a calculator to verify your answer.

12. $8\sqrt{5} + 3\sqrt{5}$
13. $8\sqrt{3} - 2\sqrt{2} + 3\sqrt{2} + 5\sqrt{3}$
14. $2\sqrt{3} + \sqrt{12}$
15. $\sqrt{7} + \sqrt{\frac{1}{7}}$

Simplify.

16. $\sqrt{2}(\sqrt{18} + 4\sqrt{3})$
17. $(4 + \sqrt{5})(4 - \sqrt{5})$
18. **Geometry** Find the exact measures of the perimeter and area in simplest form for the rectangle at the right.

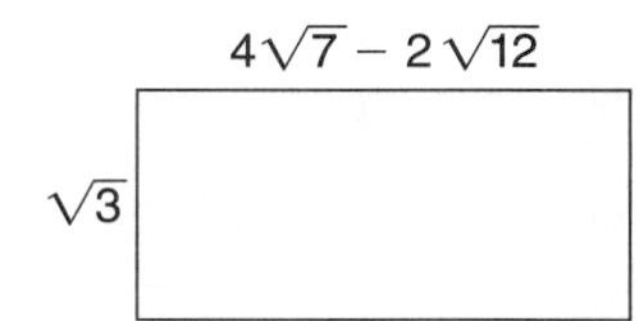

EXERCISES

Practice

Simplify.

19. $25\sqrt{13} + \sqrt{13}$
20. $7\sqrt{2} - 15\sqrt{2} + 8\sqrt{2}$
21. $2\sqrt{6} - 8\sqrt{3}$
22. $2\sqrt{11} - 6\sqrt{11} - 3\sqrt{11}$
23. $18\sqrt{2x} + 3\sqrt{2x}$
24. $3\sqrt{5m} - 5\sqrt{5m}$

Simplify. Then use a calculator to verify your answer.

25. $4\sqrt{3} + 7\sqrt{3} - 2\sqrt{3}$
26. $5\sqrt{5} + 3\sqrt{5} - 18\sqrt{5}$
27. $\sqrt{6} + 2\sqrt{2} + \sqrt{10}$
28. $4\sqrt{6} + \sqrt{7} - 6\sqrt{2} + 4\sqrt{7}$
29. $3\sqrt{7} - 2\sqrt{28}$
30. $2\sqrt{50} - 3\sqrt{32}$
31. $3\sqrt{27} + 5\sqrt{48}$
32. $2\sqrt{20} - 3\sqrt{24} - \sqrt{180}$
33. $\sqrt{80} + \sqrt{98} + \sqrt{128}$
34. $\sqrt{10} - \sqrt{\frac{2}{5}}$
35. $3\sqrt{3} - \sqrt{45} + 3\sqrt{\frac{1}{3}}$
36. $6\sqrt{\frac{7}{4}} + 3\sqrt{28} - 10\sqrt{\frac{1}{7}}$

Simplify.

37. $\sqrt{5}(2\sqrt{10} + 3\sqrt{2})$
38. $\sqrt{6}(\sqrt{3} + 5\sqrt{2})$
39. $(2\sqrt{10} + 3\sqrt{15})(3\sqrt{3} - 2\sqrt{2})$
40. $(\sqrt{5} - \sqrt{2})(\sqrt{14} + \sqrt{35})$
41. $(\sqrt{6} + \sqrt{8})(\sqrt{24} + \sqrt{2})$
42. $(5\sqrt{2} + 3\sqrt{5})(2\sqrt{10} - 3)$

Critical Thinking

43. Explain why the simplified form of $\sqrt{(x-5)^2}$ must have an absolute value sign, but $\sqrt{(x-5)^4}$ does not need one.

Applications and Problem Solving

44. **Construction** *Slip forming* is the fastest method of erecting tall concrete buildings. With this method, the 1815-foot CN Tower in Toronto, Canada, was built at an average speed of 20 feet per day. At the beginning of the week, construction workers were 530 feet above the ground. After one week of construction, they were 670 feet above the ground. How many more miles could they see from the top of the building at the end of the week than at the beginning? Write your answer in exact form and as an approximation to the nearest hundredth. (*Hint:* Use the formula from the application at the beginning of the lesson.)

45. **Construction** A wire is stretched from the top of a 12-foot pole to a stake in the ground and then to the base of the pole. If a total of 20 feet of wire is needed, how far is the stake from the pole? (*Hint:* In the figure, $a + b = 20$.)

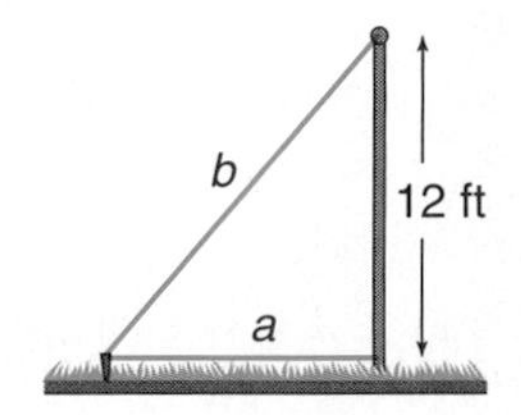

Mixed Review

46. Simplify $\frac{\sqrt{3}}{\sqrt{6}}$. (Lesson 13–2)

47. Find $\frac{x^2 - y^2}{3} \cdot \frac{9}{x + y}$. (Lesson 12–2)

48. Factor $3a^2 + 19a - 14$. (Lesson 10–3)

49. Find the degree of $16s^3t^2 + 3s^2t + 7s^6t$. (Lesson 9–4)

50. **Statistics** Tim's scores on the first four of five 50-point quizzes were 47, 45, 48, and 45. What score must he receive on the fifth quiz to have an average of at least 46 points for all the quizzes? (Lesson 7–3)

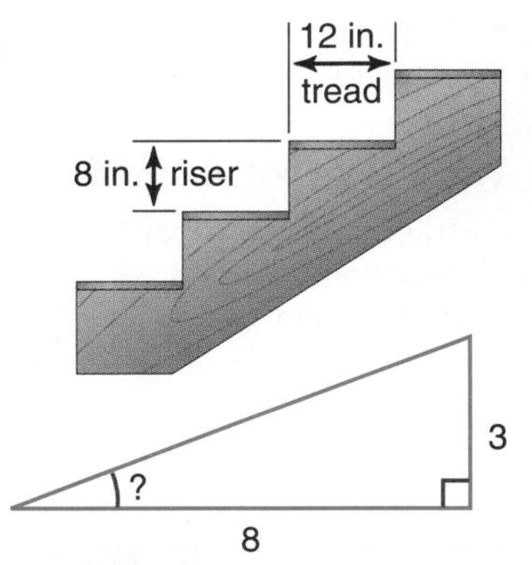

51. **Carpentry** When building a stairway, a carpenter considers the ratio of riser to tread. Write a ratio to describe the steepness of the stairs. (Lesson 6–1)

52. **Geometry** Find the measure of the marked acute angle to the nearest degree. (Lesson 4–3)

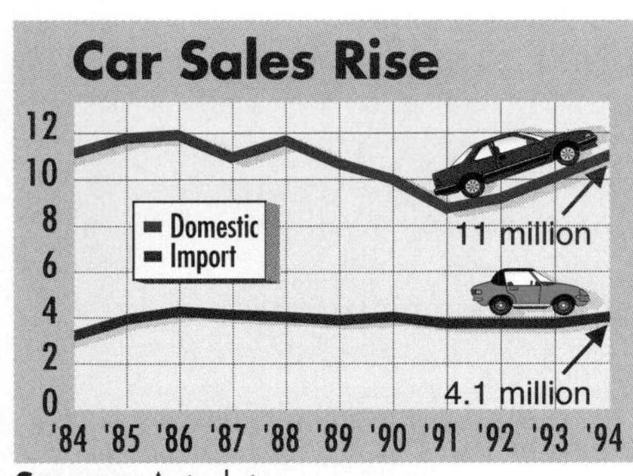

Source: Autodata

53. **Interpret Graphs** The graph at the left compares domestic and import brand auto sales from 1984 to 1994. (Lesson 1–9)
   a. During the ten-year period, when were domestic car sales the lowest? How many domestic cars were sold during that year?
   b. In 1990, how many more domestic cars were sold than imports?

SELF TEST

If c is the measure of the hypotenuse of a right triangle, find each missing measure. Round answers to the nearest hundredth. (Lesson 13–1)

1. $a = 21, b = 28, c = ?$
2. $a = \sqrt{41}, c = 8, b = ?$
3. $b = 28, c = 54, a = ?$

Simplify. Leave in radical form and use absolute value symbols when necessary. (Lesson 13–2)

4. $\sqrt{20}$
5. $2\sqrt{5} \cdot \sqrt{5}$
6. $\frac{\sqrt{42x^2}}{\sqrt{6y^3}}$

Simplify. (Lesson 13–3)

7. $8\sqrt{6} + 3\sqrt{6}$
8. $10\sqrt{17} + 9\sqrt{7} - 8\sqrt{17} + 6\sqrt{7}$
9. $(6 + \sqrt{3})(2\sqrt{5} - \sqrt{3})$

10. **Geometry** Find the perimeter and area of the figure at the right. Round answers to the nearest hundredth. (Lesson 13–1)

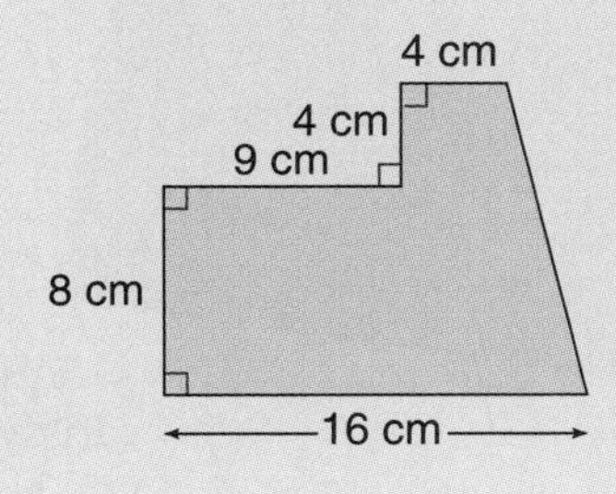

13-4 Radical Equations

Topographical map of the ocean surface

What YOU'LL LEARN

- To solve radical equations.

Why IT'S IMPORTANT

You can use radical equations to solve problems involving oceanography and recreation.

APPLICATION

Oceanography

The Tonga Trench in the Pacific Ocean is a potential source for a *tsunami* (su-nom'-ee), a large ocean wave generated by an undersea earthquake. The formula for a tsunami's speed s in meters per second is $s = 3.1\sqrt{d}$, where d is the depth of the ocean in meters.

Equations like $s = 3.1\sqrt{d}$ that contain radicals with variables in the radicand are called **radical equations**. To solve these equations, first isolate the radical on one side of the equation. Then square each side of the equation to eliminate the radical.

Find the depth of the Tonga Trench if a tsunami's speed is 322 meters per second.

$322 = 3.1\sqrt{d}$ *Replace s with 322.*

$\frac{322}{3.1} = \frac{3.1\sqrt{d}}{3.1}$ *Divide each side by 3.1.*

$\left(\frac{322}{3.1}\right)^2 = \left(\sqrt{d}\right)^2$ *Square each side of the equation.*

$\left(\frac{322}{3.1}\right)^2 = d$ Use a scientific calculator to simplify $\left(\frac{322}{3.1}\right)^2$.

Enter: [( ] 322 [÷] 3.1 [ )] [x^2] 10789.17794

The depth of the Tonga Trench is approximately 10,789 meters. *Check this result by substituting 10,789 for d into the original formula.*

FYI

A tsunami may begin as a 2-foot high wave. After traveling hundreds of miles across the ocean at speeds of 450 to 500 miles per hour, it could approach shallow coastal waters as a towering 50-foot wall of water, capable of destroying anything in its path.

Example 1 **Solve each equation.**

a. $\sqrt{x} + 4 = 7$

$\sqrt{x} + 4 = 7$

$\sqrt{x} + 4 - 4 = 7 - 4$ *Subtract 4 from each side.*

$\sqrt{x} = 3$ *Simplify.*

$\left(\sqrt{x}\right)^2 = 3^2$ *Square each side.*

$x = 9$ The solution is 9.

Check:

$\sqrt{x} + 4 = 7$

$\sqrt{9} + 4 \stackrel{?}{=} 7$ *x = 9*

$3 + 4 = 7$ ✔

b. $\sqrt{x + 3} + 5 = 9$

$\sqrt{x + 3} + 5 = 9$

$\sqrt{x + 3} + 5 - 5 = 9 - 5$ *Subtract 5 from each side.*

$\sqrt{x + 3} = 4$ *Simplify.*

$\left(\sqrt{x + 3}\right)^2 = 4^2$ *Square each side.*

$x + 3 = 16$

$x + 3 - 3 = 16 - 3$ *Subtract 3 from each side.*

$x = 13$ The solution is 13. *Check this result.*

Squaring each side of an equation does not necessarily produce results that satisfy the original equation. Therefore, you must check all solutions when you solve radical equations.

Example 2 **Solve $\sqrt{3x - 5} = x - 5$.**

$$\sqrt{3x - 5} = x - 5$$
$$(\sqrt{3x - 5})^2 = (x - 5)^2 \quad \textit{Square each side.}$$
$$3x - 5 = x^2 - 10x + 25 \quad \textit{Simplify.}$$
$$3x - 3x - 5 + 5 = x^2 - 10x + 25 - 3x + 5 \quad \textit{Add } -3x \textit{ and 5 to each side.}$$
$$0 = x^2 - 13x + 30 \quad \textit{Simplify.}$$
$$0 = (x - 10)(x - 3) \quad \textit{Factor.}$$
$$x - 10 = 0 \quad \text{or} \quad x - 3 = 0 \quad \textit{Use the zero product property.}$$
$$x = 10 \qquad x = 3$$

Check:

$$\sqrt{3x - 5} = x - 5 \qquad \sqrt{3x - 5} = x - 5$$
$$\sqrt{3(10) - 5} \stackrel{?}{=} 10 - 5 \qquad \sqrt{3(3) - 5} \stackrel{?}{=} 3 - 5$$
$$\sqrt{30 - 5} \stackrel{?}{=} 5 \qquad \sqrt{9 - 5} \stackrel{?}{=} -2$$
$$\sqrt{25} \stackrel{?}{=} 5 \qquad \sqrt{4} \stackrel{?}{=} -2$$
$$5 = 5 \checkmark \qquad 2 \neq -2$$

Since 3 does not satisfy the original equation, 10 is the only solution.

You can use the *Mathematics Exploration Toolkit* (*MET*) to solve equations involving square roots.

EXPLORATION GRAPHING SOFTWARE

The CALC commands below will be used.

ADD (add)	SUBTRACT (sub)	MULTIPLY (mult)
DIVIDE (div)	FACTOR (fac)	RAISETO (rai)
SIMPLIFY (simp)	STORE (sto)	SUBSTITUTE (subs)

To enter the square root symbol, type &.

Solve $\sqrt{x - 2} = x - 4$.

Enter:	**Result:**
$\&(x - 2) = x - 4$	$\sqrt{x - 2} = x - 4$
sto a	Saves the equation as a.
rai 2	$\left(\sqrt{x - 2}\right)^2 = (x - 4)^2$
simp	$x - 2 = x^2 - 8x + 16$
sub $x - 2$	$x - 2 - (x - 2) = x^2 - 8x + 16 - (x - 2)$
simp	$0 = x^2 - 9x + 18$
fac	$0 = (x - 6)(x - 3)$

By inspection, the solutions are $x = 6$ or $x = 3$. However, 3 does not satisfy the original equation. Therefore, 6 is the only solution.

Your Turn
Use CALC to solve each equation.

a. $3 + \sqrt{2x} = 7$

b. $\sqrt{x + 1} = x - 1$

c. $\sqrt{x} + 6 = 1$

d. $\sqrt{3x - 8} = 5$

e. $x + \sqrt{6 - x} = 4$

f. $\sqrt{3x - 9} = 2x + 6$

Example **3** **The geometric mean of a and b is x if $\frac{a}{x} = \frac{x}{b}$. Find two numbers that have a geometric mean of 8 given that one number is 12 more than the other.**

Number Theory

Explore Let n represent the lesser number.
Then $n + 12$ represents the greater number.

Plan Use the equation $\frac{a}{x} = \frac{x}{b}$. Replace each variable with the appropriate value.

$$\frac{n}{8} = \frac{8}{n + 12}$$ *The geometic mean of the numbers is 8.*

Solve

$$n^2 + 12n = 64$$

$$n^2 + 12n - 64 = 0$$

$$(n + 16)(n - 4) = 0 \quad \textit{Factor.}$$

$$n + 16 = 0 \quad \text{or} \quad n - 4 = 0 \quad \textit{Zero product property}$$

$$n = -16 \qquad n = 4$$

If $n = -16$, then $n + 12 = -4$. If $n = 4$, then $n + 12 = 16$.

Thus, the numbers are -16 and -4, or 4 and 16.

Examine $\frac{-16}{8} \stackrel{?}{=} \frac{8}{-4}$ or $\frac{4}{8} \stackrel{?}{=} \frac{8}{16}$

$64 = 64$ ✔ $\qquad$ $64 = 64$ ✔

CHECK FOR UNDERSTANDING

Communicating Mathematics

Study the lesson. Then complete the following.

1. **Explain** the first step you should do when solving a radical equation.
2. **Write** an expression for the geometric mean of 7 and y.
3. Refer to the application at the beginning of the lesson.
   a. Solve $s = 3.1\sqrt{d}$ for d.
   b. Use the equation you found in part a to find the depth of the ocean in meters if the speed of the tsunami is 400 meters per second.
4. **You Decide** Alberto says that if you have an equation that contains a radical, you can always get a real solution by squaring each side of the equation. Ellen disagrees. Who is correct? Explain.

5. a. **Assess Yourself** Explain in your own words the process or steps needed to solve a radical equation.
   b. Explain why it is important to check your answers when solving equations containing radicals.

Guided Practice

Square each side of the following equations.

6. $\sqrt{x} = 6$
7. $\sqrt{a + 3} = 2$
8. $13 = \sqrt{2y - 5}$

Solve each equation. Check your solution.

9. $\sqrt{m} = 4$
10. $\sqrt{b} = -3$
11. $-\sqrt{x} = -6$
12. $\sqrt{7x} = 7$
13. $\sqrt{-3a} = 6$
14. $\sqrt{y - 2} = 8$

15. **Engineering** It is possible to measure the speed of water using an L-shaped tube. You can find the speed V of the water in miles per hour by measuring the height h of the column of water above the surface in inches and by using the formula $V = \sqrt{2.5h}$. If you take the tube into a river and the height of the column is 6 inches, what is the speed of the water to the nearest tenth of a mile per hour?

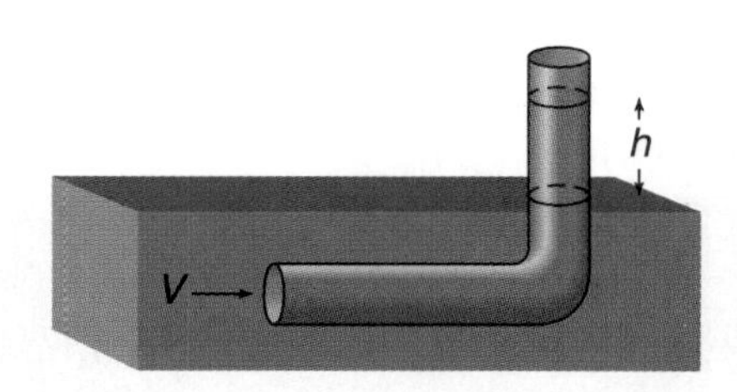

EXERCISES

Practice

Solve each equation. Check your solution.

16. $\sqrt{a} = 5\sqrt{2}$
17. $3\sqrt{7} = \sqrt{-x}$
18. $\sqrt{m} - 4 = 0$
19. $\sqrt{2d} + 1 = 0$
20. $10 - \sqrt{3y} = 1$
21. $3 + 5\sqrt{n} = 12$
22. $\sqrt{8s+1} = 5$
23. $\sqrt{4b+1} - 3 = 0$
24. $\sqrt{3r-5} + 7 = 3$
25. $\sqrt{\frac{w}{6}} = 2$
26. $\sqrt{\frac{4x}{5}} - 9 = 3$
27. $5\sqrt{\frac{4t}{3}} - 2 = 0$
28. $\sqrt{2x^2 - 121} = x$
29. $7\sqrt{3z^2 - 15} = 7$
30. $\sqrt{x+2} = x - 4$
31. $\sqrt{5x^2 - 7} = 2x$
32. $\sqrt{1-2m} = 1 + m$
33. $4 + \sqrt{b-2} = b$

Number Theory

34. The geometric mean of a certain number and 6 is 24. Find the number.
35. Find two numbers with a geometric mean of $\sqrt{30}$ given that one number is 7 more than the other.
36. Find two numbers with a geometric mean of 12 given that one number is 11 less than three times the other.

Solve each equation. Check your solution.

37. $\sqrt{x-12} = 6 - \sqrt{x}$
38. $\sqrt{x} + 4 = \sqrt{x+16}$
39. $\sqrt{x+7} = 7 + \sqrt{x}$

Solve each system of equations.

40. $2\sqrt{a} + 5\sqrt{b} = 6$
    $3\sqrt{a} - 5\sqrt{b} = 9$
41. $-3\sqrt{x} + 3\sqrt{y} = 1$
    $-4\sqrt{x} + 6\sqrt{y} = 3$
42. $s = 4t$
    $\sqrt{s} - 5\sqrt{t} = -6$

Critical Thinking

43. Solve for x if $x + 2 = x\sqrt{3}$.
44. Find two numbers such that the square root of their sum is 5 and the square root of their product is 12.

Applications and Problem Solving

45. **Recreation** The rangers at an aid station received a distress call from a group camping 60 miles east and 10 miles south of the station. A jeep sent to the campsite travels directly east for some number of miles and then turns and heads directly to the campsite. If the jeep traveled a total of 66 miles to get to the campsite, for how many miles did it travel due east?

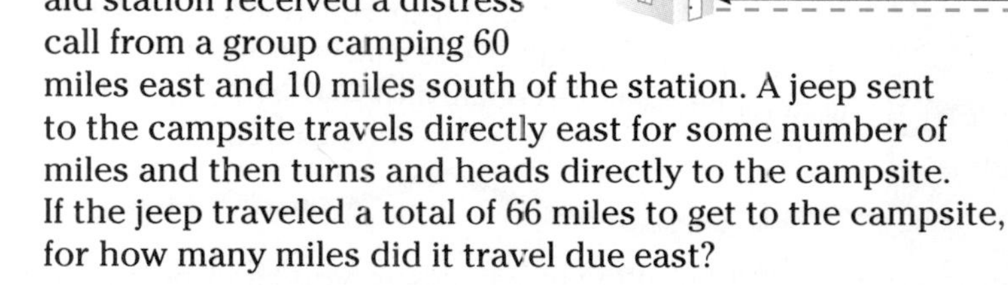

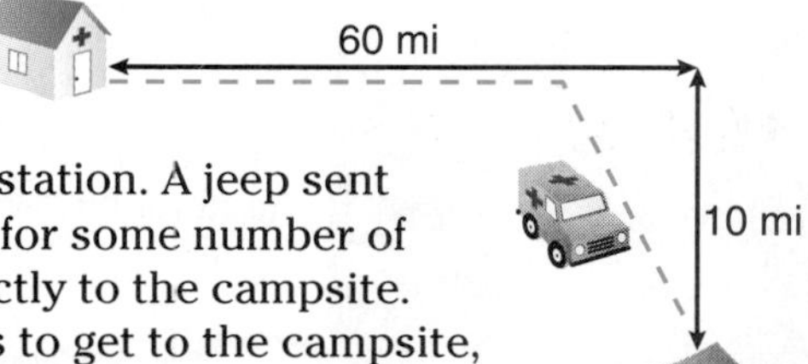

46. **Sound** The speed of sound near Earth's surface can be found with the equation $V = 20\sqrt{t + 273}$, where t is the surface temperature in degrees Celsius.
    a. Find the temperature if the speed of sound V is 356 meters per second.
    b. The speed of sound at Earth's surface is often given as 340 meters per second, but that's really only true at a certain temperature. On what temperature is the 340 m/s figure based?

47. **Travel** The speed s that a car is traveling in miles per hour, and the distance d in feet that it will skid when the brakes are applied, are related by the formula $s = \sqrt{30fd}$. In this formula, f is the coefficient of friction, which depends on the type and condition of the road. Sylvia Kwan told police she was traveling at about 30 miles per hour when she applied the brakes and skidded on a wet concrete road. The length of her skid marks was measured at 110 feet.

a. If $f = 0.4$ for a wet concrete road, should Ms. Kwan's car have skidded that far when she applied the brakes?

b. How fast was she traveling?

Mixed Review

48. Simplify $5\sqrt{6} - 11\sqrt{3} - 8\sqrt{6} + \sqrt{27}$. (Lesson 13–3)

49. Find $(6b^2 + 4b + 20) \div (b + 5)$. (Lesson 12–4)

50. Find the roots of $x^2 - 2x - 8 = 0$ by graphing its related function. (Lesson 11–2)

51. Find the GCF of $12x^2y^3$ and $42xy^4$. (Lesson 10–1)

52. Find $(6a - 2m) - (4a + 7m)$. (Lesson 9–5)

53. Use substitution to solve the system of equations. (Lesson 8–2)
$x + 4y = 16$
$3x + 6y = 18$

54. Write an equation in slope-intercept form of the line that passes through the points at $(6, -1)$ and $(3, 2)$. (Lesson 6–4)

55. **Probability** If the odds that an event will occur are 8:5, what is the probability that the event will occur? (Lesson 4–6)

56. Find $\frac{3}{7} + \left(-\frac{4}{9}\right)$. (Lesson 2–5)

Mathematics and SOCIETY

Nonlinear Math

The excerpt below appeared in an article in *Business Week* on September 5, 1994.

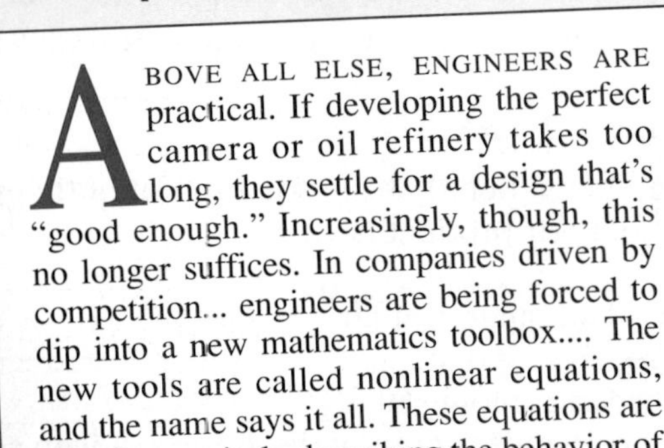

Above all else, engineers are practical. If developing the perfect camera or oil refinery takes too long, they settle for a design that's "good enough." Increasingly, though, this no longer suffices. In companies driven by competition... engineers are being forced to dip into a new mathematics toolbox.... The new tools are called nonlinear equations, and the name says it all. These equations are used for precisely describing the behavior of things with an unpredictable facet. That's nearly everything—from the workings of car engines to the actions of DNA molecules. Even baking a cake is nonlinear: turning up the oven's temperature twice as high won't bake the cake twice as fast. And with some industrial recipes, such as those for making drugs and plastics, a tiny change in ingredients or processing conditions can mean a huge difference in the finished product. Nonlinear math can help explain such lopsided effects. ■

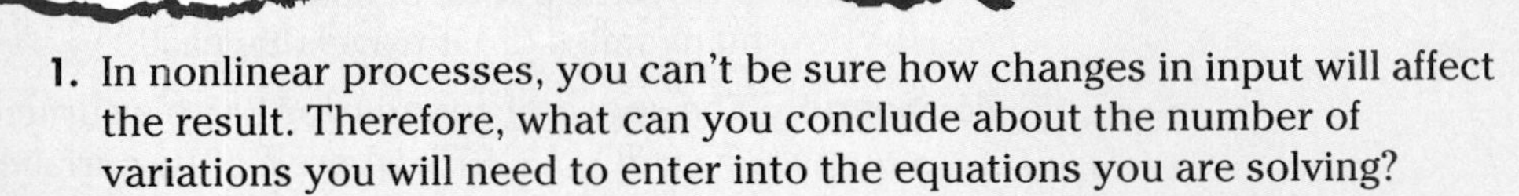

1. In nonlinear processes, you can't be sure how changes in input will affect the result. Therefore, what can you conclude about the number of variations you will need to enter into the equations you are solving?

2. Because of the huge number of variables that can be involved, solving nonlinear equations can require many millions of calculations. Why do you think the use of these equations has only recently begun to expand into many industries?

13–5 Integration: Geometry The Distance Formula

What YOU'LL LEARN

- To find the distance between two points in the coordinate plane.

Why IT'S IMPORTANT

You can use the distance formula to solve problems involving art and communication.

Geometry

In a coordinate plane, consider the points $A(-2, 6)$ and $B(5, 3)$. These two points do not lie on the same horizontal or vertical line. Therefore, you cannot find the distance between them by simply subtracting the x- or y-coordinates. A different method must be used.

Notice that a right triangle can be formed by drawing lines parallel to the axes through points at $(-2, 6)$ and $(5, 3)$. These lines intersect at $C(-2, 3)$. The measure of side b is the difference of the y-coordinates of the endpoints, $6 - 3$ or 3. The measure of side a is the difference of the x-coordinates of the endpoints, $5 - (-2)$ or 7.

Now the Pythagorean theorem can be used to find c, the distance between $A(-2, 6)$ and $B(5, 3)$.

$c^2 = a^2 + b^2$ *Pythagorean theorem*

$c^2 = 7^2 + 3^2$ *Replace a with 7 and b with 3.*

$c^2 = 49 + 9$

$c^2 = 58$

$c = \sqrt{58}$

$c \approx 7.62$

The distance between points A and B is approximately 7.62 units.

The method used for finding the distance between $A(-2, 6)$ and $B(5, 3)$ can also be used to find the distance between any two points in the coordinate plane. This method can be described by the following formula.

The Distance Formula

The distance d between any two points with coordinates (x_1, y_1) and (x_2, y_2) is given by the following formula.

$$d = \sqrt{(x_2 - x_1)^2 + (y_2 - y_1)^2}$$

Example  **Find the distance between the points with coordinates (3, 5) and (6, 4).**

$d = \sqrt{(x_2 - x_1)^2 + (y_2 - y_1)^2}$

$= \sqrt{(6 - 3)^2 + (4 - 5)^2}$ *$(x_1, y_1) = (3, 5)$ and $(x_2, y_2) = (6, 4)$*

$= \sqrt{3^2 + (-1)^2}$

$= \sqrt{9 + 1}$

$= \sqrt{10}$ or about 3.16 units

Example 2 **INTEGRATION Geometry**

Determine if triangle ABC with vertices $A(-3, 4)$, $B(5, 2)$, and $C(-1, -5)$ is an isosceles triangle.

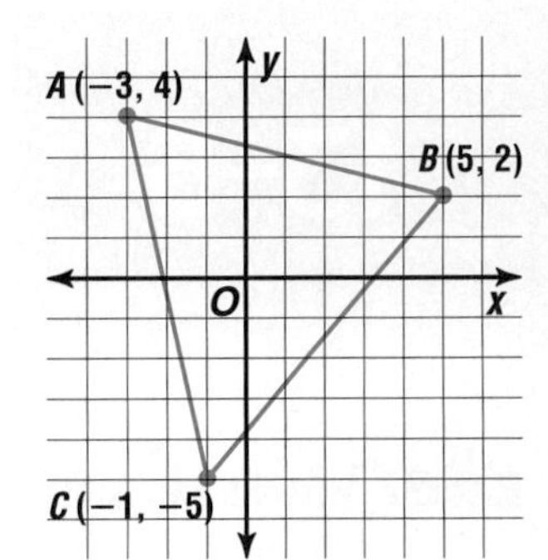

A triangle is isosceles if at least two sides are congruent. Find AB, BC, and AC.

$$AB = \sqrt{[5 - (-3)]^2 + (2 - 4)^2}$$
$$= \sqrt{8^2 + (-2)^2} \text{ or } \sqrt{68}$$
$$BC = \sqrt{(-1 - 5)^2 + (-5 - 2)^2}$$
$$= \sqrt{(-6)^2 + (-7)^2} \text{ or } \sqrt{85}$$
$$AC = \sqrt{[-1 - (-3)]^2 + (-5 - 4)^2}$$
$$= \sqrt{2^2 + (-9)^2} \text{ or } \sqrt{85}$$

Since $\overline{BC}$ and $\overline{AC}$ have the same length, $\sqrt{85}$, they are congruent. So, triangle ABC is an isosceles triangle.

Suppose you know the coordinates of a point, one coordinate of another point, and the distance between the two points. You can use the distance formula to find the missing coordinate.

Example 3 **Find the value of a if the distance between the points with coordinates $(-3, -2)$ and $(a, -5)$ is 5 units.**

$$d = \sqrt{(x_2 - x_1)^2 + (y_2 - y_1)^2}$$
$$5 = \sqrt{[a - (-3)]^2 + [-5 - (-2)]^2} \quad \text{Let } x_2 = a,\ x_1 = -3,\ y_2 = -5,\ y_1 = -2, \text{ and } d = 5.$$
$$5 = \sqrt{(a + 3)^2 + (-3)^2}$$
$$5 = \sqrt{a^2 + 6a + 9 + 9}$$
$$5 = \sqrt{a^2 + 6a + 18}$$
$$(5)^2 = \left(\sqrt{a^2 + 6a + 18}\right)^2 \quad \textit{Square each side.}$$
$$25 = a^2 + 6a + 18$$
$$0 = a^2 + 6a - 7$$
$$0 = (a + 7)(a - 1) \quad \textit{Factor.}$$
$$a + 7 = 0 \quad \text{or} \quad a - 1 = 0 \quad \textit{Zero product property}$$
$$a = -7 \qquad a = 1$$

The value of a is -7 or 1.

CHECK FOR UNDERSTANDING

Communicating Mathematics

Study the lesson. Then complete the following.

1. **Explain** why the value calculated under the radical sign in the distance formula will never be negative.
2. a. **Write** two ordered pairs and label them $A(x_1, y_1)$ and $B(x_2, y_2)$. Does it matter which ordered pair is first when using the distance formula? Explain.
   b. Find the distance between A and B.

3. a. **Explain** how you can find the distance between $X(12, 4)$ and $Y(3, 4)$ without using the distance formula.

   b. **Explain** how you can find the distance between $S(-2, 7)$ and $T(-2, -5)$ without using the distance formula.

4. Refer to Example 3. Check your answer by using the distance formula.

Guided Practice

Find the distance between each pair of points whose coordinates are given. Express answers in simplest radical form and as decimal approximations rounded to the nearest hundredth if necessary.

5. $(6, 8), (3, 4)$
6. $(3, 7), (-2, -5)$
7. $(2, 2), (5, -1)$
8. $(2, 7), (10, -4)$

Find the value of *a* if the points with the given coordinates are the indicated distance apart.

9. $(4, 7), (a, 3); d = 5$
10. $(5, a), (6, 1); d = \sqrt{10}$

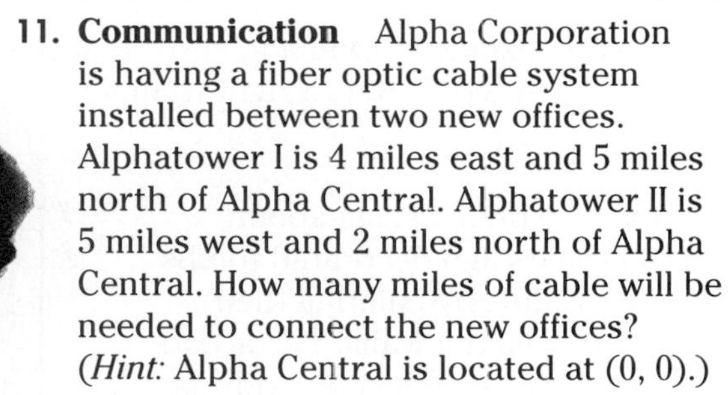

11. **Communication** Alpha Corporation is having a fiber optic cable system installed between two new offices. Alphatower I is 4 miles east and 5 miles north of Alpha Central. Alphatower II is 5 miles west and 2 miles north of Alpha Central. How many miles of cable will be needed to connect the new offices? (*Hint:* Alpha Central is located at (0, 0).)

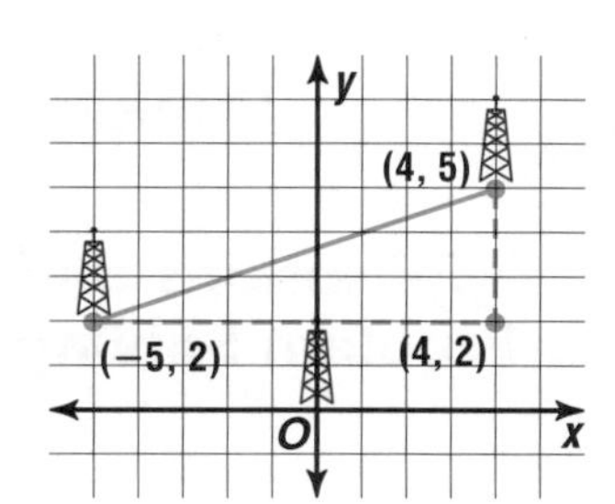

EXERCISES

Practice

Find the distance between each pair of points whose coordinates are given. Express answers in simplest radical form and as decimal approximations rounded to the nearest hundredth if necessary.

12. $(5, -1), (11, 7)$
13. $(-4, 2), (4, 17)$
14. $(-3, 8), (5, 4)$
15. $(-8, -4), (-3, -8)$
16. $(9, -2), (3, -6)$
17. $(4, 2), \left(6, -\frac{2}{3}\right)$
18. $\left(3, \frac{3}{7}\right), \left(4, -\frac{2}{7}\right)$
19. $\left(\frac{4}{5}, -1\right), \left(2, -\frac{1}{2}\right)$
20. $(4\sqrt{5}, 7), (6\sqrt{5}, 1)$
21. $(5\sqrt{2}, 8), (7\sqrt{2}, 10)$

Find the value of *a* if the points with the given coordinates are the indicated distance apart.

22. $(3, -1), (a, 7); d = 10$
23. $(-4, a), (4, 2); d = 17$
24. $(a, 5), (-7, 3); d = \sqrt{29}$
25. $(6, -3), (-3, a); d = \sqrt{130}$
26. $(10, a), (1, -6); d = \sqrt{145}$
27. $(20, -5), (a, 9); d = \sqrt{340}$

Geometry

Determine if the triangles with the following vertices are isosceles triangles.

28. $L(7, -4), M(-1, 2), N(5, -6)$
29. $T(1, -8), U(3, 5), V(-1, 7)$

30. Find the perimeter of square $QRST$ if two of the vertices are $Q(6, 7)$ and $R(-3, 4)$.

31. If the diagonals of a trapezoid have the same length, then the trapezoid is isosceles. Find the lengths of the diagonals of the trapezoid with vertices $A(-2, 2)$, $B(10, 6)$, $C(9, 8)$, and $D(0, 5)$ to determine if it is isosceles.

Programming

32. The program at the right calculates the distance between a pair of points whose coordinates are given.

Find the distance between each pair of points.

a. $A(6, -3)$, $B(12, 5)$
b. $M(-3, 5)$, $N(12, -2)$
c. $S(6.8, 9.9)$, $T(-5.9, 4.3)$

```
PROGRAM: DISTANCE
:ClrDraw
:Input "X1=", Q
:Input "Y1=", R
:Input "X2=", S
:Input "Y2=", T
:Line (Q, R, S, T)
:√((Q−S)² + (R−T)²)→D
:Text (5, 50, "DIST=", D)
```

Critical Thinking

33. Use the distance formula to show that the triangle with vertices at $(3, -2)$, $(-3, 7)$, and $(-9, 3)$ is a right triangle.

Applications and Problem Solving

34. **Art** Egyptian artists about 5000 years ago decorated tombs of pharaohs by painting their pictures on the walls. The artists used small sketches on grids as a reference. In Egypt, the main standard of length was the cubit. It was the length of a man's forearm from the elbow to the tip of the outstretched fingers. Use the grid at the right to find the length of a cubit to the nearest inch if each unit represents 3.3 inches.

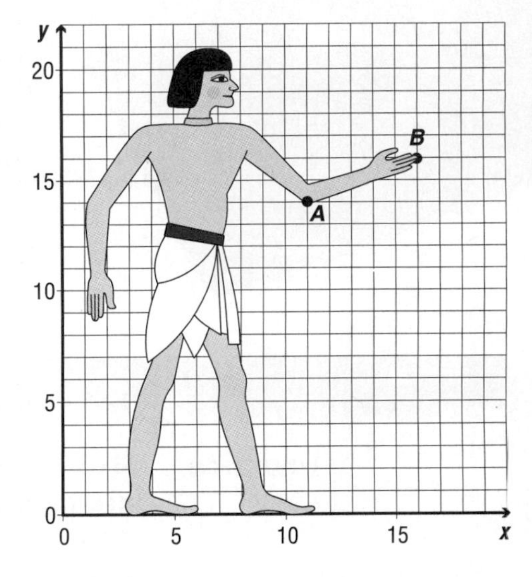

35. **Telecommunications** In order to set long distance rates, phone companies first superimpose an imaginary coordinate grid over the United States. Then the location of each exchange is represented by an ordered pair on the grid. The units of this grid are approximately equal to 0.316 miles. So, a distance of 3 units on the grid equals an actual distance of about 3(0.316) or 0.948 miles. Suppose the exchanges in two cities are at (132, 428) and (254, 105). Find the actual distance between these cities to the nearest mile.

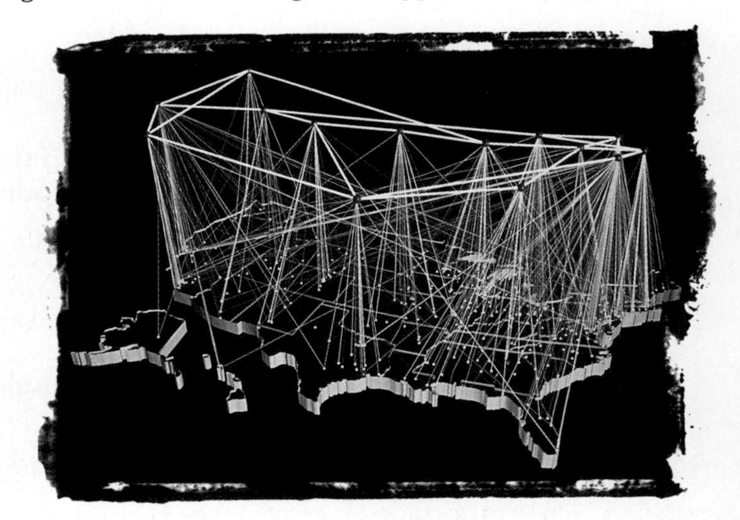

Granville T. Woods patented an improved telephone transmitter in 1884. This was the first of more than twenty patents that he received for his inventions.

Mixed Review

36. **Physics** The time t, in seconds, it takes an object to drop d feet is given by the formula $4t = \sqrt{d}$. Jessica and Lu-Chan each dropped a stone at the same time, but Jessica dropped hers from a spot higher than Lu-Chan's. Lu-Chan's stone hit the ground 1 second before Jessica's. If Jessica's stone dropped 112 feet farther than Lu-Chan's, how long did it take her stone to hit the ground? (Lesson 13–4)

37. Simplify $\sqrt{\frac{8}{9}}$. (Lesson 13–2)

38. Find $\frac{4p^3}{p-1} \div \frac{p^2}{p-1}$. (Lesson 12–3)

39. Graph the system of equations. Then determine whether the system has *one* solution, *no* solution, or *infinitely many* solutions. (Lesson 8–1)
$4x - y = 2$
$y - 4x = 4$

40. **Consumerism** Jackie wanted to buy a new coat that cost \$145. If she waited until the coat went on sale for 30% off the original price, how much money did Jackie save? (Lesson 4–5)

41. **Air Conditioning** The formula for determining the BTU (British Thermal Units) rating of the air conditioner necessary to cool a room is BTU = Area (sq ft) × Exposure Factor × Climate Factor. Use this formula to determine the BTU necessary to cool each of the rooms described in the following chart. (Lesson 2–6)

	Room Dimensions (feet)	Exposure Factor	Climate Factor
a.	22 by 16	North: 20	Buffalo: 1.05
b.	13 by 12	West: 25	Portland: 0.95
c.	17 by 14	East: 25	Topeka: 1.05
d.	26 by 18	South: 30	San Diego: 1.00
e.	23.5 by 15.3	North: 20	Tacoma: 0.95

WORKING ON THE

In·ves·ti·ga·tion

Refer to the Investigation on pages 656–657.

A Growing Concern

1 Review your scale drawing. Make sure you have included everything that you think the Sanchez family wanted or will want in the future. Make any changes that you feel need to be made now that the plan is complete.

2 The Sanchez family had asked for a 3-dimensional model of the design. On a piece of cardboard, draw the boundary lines of the Sanchez property. Use modeling clay, construction paper, paint, markers, and whatever else you need to create a 3-dimensional model of your design.

3 Use a flashlight to model the sun's movement during the day. Note the patterns and the length of time certain areas are shaded.

Add the results of your work to your Investigation Folder.

13–6A Completing the Square

A Preview of Lesson 13–6

Materials: algebra tiles equation mat

One way to solve a quadratic equation is by **completing the square.** To use this method, the quadratic expression on one side of the equation must be a perfect square. You can use algebra tiles as a model for completing the square.

Activity Use algebra tiles to complete the square for the equation $x^2 + 4x + 1 = 0$.

Step 1 Subtract 1 from each side of the equation.

$$x^2 + 4x + 1 - 1 = 0 - 1$$
$$x^2 + 4x = -1$$

Then model the equation $x^2 + 4x = -1$.

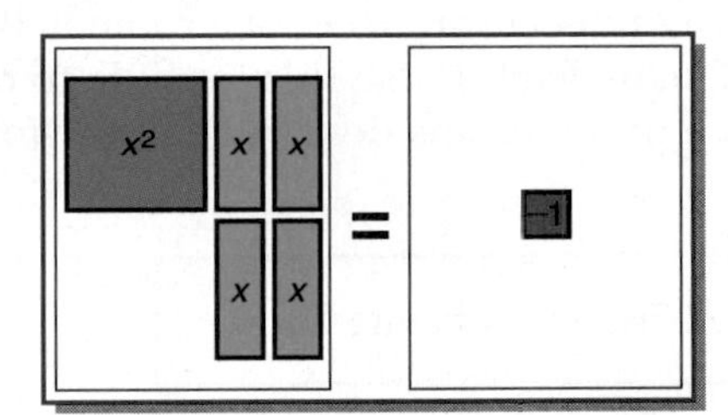

Step 2 Begin to arrange the x^2-tile and the x-tiles into a square.

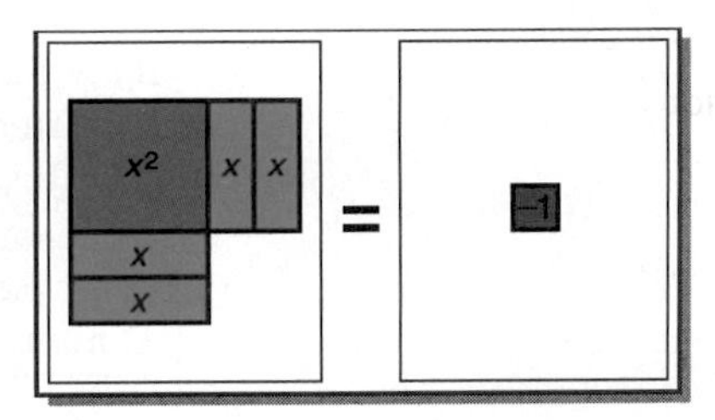

Step 3 In order to complete the square, you need to add 4 1-tiles to the left side of the mat. Since you are modeling an equation, add 4 1-tiles to the right side of the mat.

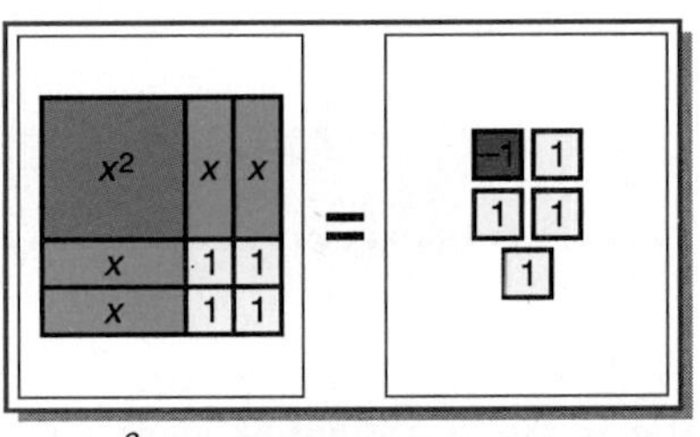

$x^2 + 4x + 4 = -1 + 4$

Step 4 Remove the zero pair on the right side of the mat. You have completed the square, and the equation is $x^2 + 4x + 4 = 3$ or $(x + 2)^2 = 3$.

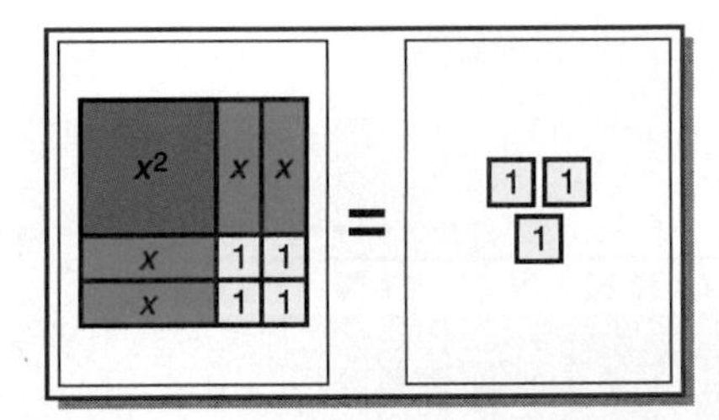

Model Use algebra tiles to complete the square for each equation.

1. $x^2 + 4x + 3 = 0$
2. $x^2 - 6x + 5 = 0$
3. $x^2 + 4x - 1 = 0$
4. $x^2 - 2x + 5 = 3$
5. $x^2 - 4x + 7 = 8$
6. $0 = x^2 + 8x - 3$

Draw

7. In the equations shown above, the coefficient of x was always an even number. Sometimes you have an equation like $x^2 + 3x - 1 = 0$ in which the coefficient of x is an odd number. Complete the square by making a drawing.

Write

8. Write a paragraph explaining how you could complete the square with models without first rewriting the equation. Include a drawing.

13-6 Solving Quadratic Equations by Completing the Square

What **YOU'LL LEARN**

- To solve quadratic equations by completing the square, and
- to solve problems by identifying subgoals.

Why **IT'S IMPORTANT**

You can solve quadratic equations to solve problems involving geography and construction.

Geography

The greatest flow of any river in the world is that of the Amazon, which discharges an average of 4.2 million cubic feet of water per second into the Atlantic Ocean. The rate at which water flows in a river varies depending on the distance from the shore.

Amazon River

Suppose the Castillos have a home on the bank of a river that is 40 yards wide. The rate of the river is given by the equation $y = -0.01x^2 + 0.4x$. Mr. and Mrs. Castillo do not want their children to wade in the water if the current is greater than 3 miles per hour. Find how many yards from shore the water flows at 3 miles per hour. *This problem will be solved in Example 4.*

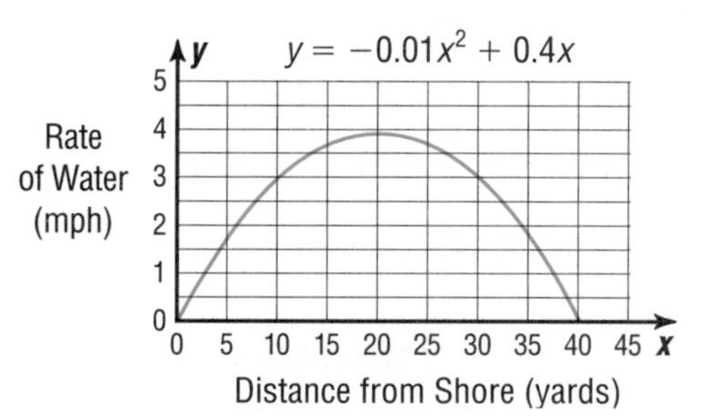

You can solve some quadratic equations by taking the square root of each side.

Example 1 **Solve $x^2 - 6x + 9 = 7$.**

$$x^2 - 6x + 9 = 7$$

$$(x - 3)^2 = 7 \quad \textit{$x^2 - 6x + 9$ is a perfect square trinomial.}$$

$$\sqrt{(x - 3)^2} = \sqrt{7} \quad \textit{Take the square root of each side.}$$

$$|x - 3| = \sqrt{7}$$

$$x - 3 = \pm\sqrt{7} \quad \textit{Why is this the case?}$$

$$x = 3 \pm \sqrt{7} \quad \textit{Add 3 to each side.}$$

The solution set is $\{3 + \sqrt{7}, 3 - \sqrt{7}\}$.

To use the method shown in Example 1, the quadratic expression on one side of the equation must be a perfect square. However, few quadratic expressions are perfect squares. To make any quadratic expression a perfect square, a method called **completing the square** may be used.

Consider the pattern for squaring a binomial such as $x + 5$.

$$(x + 5)^2 = x^2 + 2(5)x + 5^2$$

$$= x^2 + 10x + 25$$

$$\left(\frac{10}{2}\right)^2 \to 5^2 \quad \textit{Notice that one half of 10 is 5 and 5^2 is 25.}$$

To complete the square for a quadratic expression of the form $x^2 + bx$, you can follow the steps below.

Step 1 Find $\frac{1}{2}$ of b, the coefficient of x.

Step 2 Square the result of Step 1.

Step 3 Add the result of Step 2 to $x^2 + bx$, the original expression.

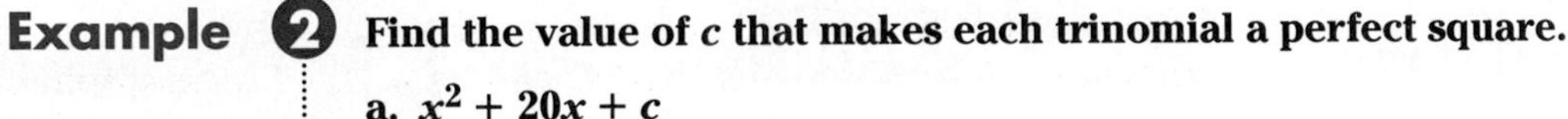

Example 2 **Find the value of c that makes each trinomial a perfect square.**

a. $x^2 + 20x + c$

Step 1 Find $\frac{1}{2}$ of 20. $\quad \frac{20}{2} = 10$

Step 2 Square the result of Step 1. $\quad 10^2 = 100$

Step 3 Add the result of Step 2 to $x^2 + 20x$. $\quad x^2 + 20x + 100$

Thus, $c = 100$. Notice that $x^2 + 20x + 100 = (x + 10)^2$.

b. $x^2 - 15x + c$

Step 1 Find $\frac{1}{2}$ of -15. $\quad \frac{-15}{2} = -7.5$

Step 2 Square the result of Step 1. $\quad (-7.5)^2 = 56.25$

Step 3 Add the result of Step 2 to $x^2 - 15x$. $\quad x^2 - 15x + 56.25$

Thus, $c = 56.25$. Notice that $x^2 - 15x + 56.25 = (x - 7.5)^2$.

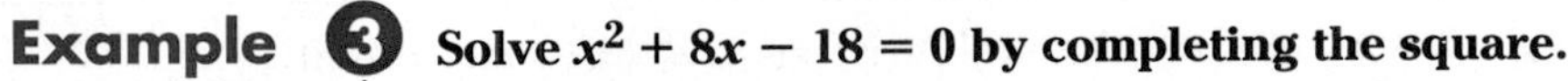

Example 3 **Solve $x^2 + 8x - 18 = 0$ by completing the square.**

$x^2 + 8x - 18 = 0$ *Notice that $x^2 + 8x - 18$ is not a perfect square.*

$x^2 + 8x = 18$ *Add 18 to each side. Then complete the square.*

$x^2 + 8x + 16 = 18 + 16$ *Since $\left(\frac{8}{2}\right)^2 = 16$, add 16 to each side.*

$(x + 4)^2 = 34$ *Factor $x^2 + 8x + 16$.*

$x + 4 = \pm\sqrt{34}$ *Take the square root of each side.*

$x = -4 \pm \sqrt{34}$ *Subtract 4 from each side.*

$x = -4 + \sqrt{34}$ or $x = -4 - \sqrt{34}$

The solution set is $\{-4 + \sqrt{34}, -4 - \sqrt{34}\}$. *Check this result.*

The method for solving quadratic equations cannot be used unless the coefficient of the first term is 1. To solve a quadratic equation in which the leading coefficient is not 1, divide each term by the coefficient.

Example 4 **Refer to the application at the beginning of the lesson. How far from shore will the rate of the current be 3 miles per hour?**

CONNECTION
Geography

$y = -0.01x^2 + 0.4x$

$3 = -0.01x^2 + 0.4x$ *Replace y with 3 since the rate is 3 mph.*

$-300 = x^2 - 40x$ *Divide each side by -0.01.*

$-300 + 400 = x^2 - 40x + 400$ *Complete the square. $\left(\frac{-40}{2}\right)^2 = 400$*

$100 = (x - 20)^2$ *Factor $x^2 - 40x + 400$.*

$\pm 10 = x - 20$ *Take the square root of each side.*

$20 \pm 10 = x$ *Add 20 to each side.*

The solutions are 20 + 10 or 30 and 20 − 10 or 10. Thus, the children should not be allowed to wade more than 10 yards from the shore; between 10 and 30 yards from the shore the water is flowing too fast.

Check this result with a graphing calculator by graphing the equation $y = -0.01x^2 + 0.4x$, and then estimating.

Sometimes finding the solution to a problem requires several steps. An important strategy for solving such problems is to **identify subgoals.** This strategy involves taking steps that will either produce part of the solution or make the problem easier to solve.

Example 5

PROBLEM SOLVING
Identify Subgoals

A square is extended in one direction by 14 centimeters. The resulting rectangle has an area of 51 square centimeters. What is the length of each side of the original square?

Finding an equation to represent this problem will be easier if you develop the equation in steps rather than trying to write one directly from the given information.

Step 1 First, let x be the length of each side of the original square. Then the area of the square is x^2 cm.

Step 2 The extension is 14 cm long and x cm wide, so its area is $14x$ cm^2.

Step 3 Add the measures of the areas and set them equal to 51.
$x^2 + 14x = 51$

Step 4 Complete the square to find the value of x.

$$x^2 + 14x = 51$$
$$x^2 + 14x + 49 = 51 + 49 \quad \textit{Since } \left(\tfrac{14}{2}\right)^2 = 49, \textit{ add 49 to each side.}$$
$$(x + 7)^2 = 100 \quad \textit{Factor } x^2 + 14x + 49.$$
$$x + 7 = \pm 10 \quad \textit{Find the square root of each side.}$$
$$x = -7 \pm 10$$
$$x = -7 + 10 \qquad x = -7 - 10$$
$$= 3 \qquad = -17$$

Since lengths cannot be negative, the length of each side of the original square is 3 centimeters. *Check this result.*

CHECK FOR UNDERSTANDING

Communicating Mathematics

Study the lesson. Then complete the following.

1. **Explain** which method for solving a quadratic equation always produces an exact solution, graphing or completing the square.
2. **Explain** the three steps used to complete the square for the expression $x^2 + bx$.
3. **Write** a quadratic equation that has no real solutions. After completing the square, explain how you could tell it had no real solutions.

4. Use algebra tiles to complete the square for the equation $x^2 + 6x + 2 = 0$.

Guided Practice

Find the value of c that makes each trinomial a perfect square.

5. $x^2 + 16x + c$
6. $a^2 - 7a + c$

Solve each equation by completing the square. Leave irrational roots in simplest radical form.

7. $x^2 + 4x + 3 = 0$
8. $d^2 - 8d + 7 = 0$
9. $a^2 - 4a = 21$
10. $4x^2 - 20x + 25 = 0$
11. $r^2 - 4r = 2$
12. $2t^2 + 3t - 20 = 0$

13. **Sports** The dimensions of a regulation high school basketball court are 50 feet by 84 feet. The builders of an indoor sports arena can afford to construct an arena of 5600 square feet. They want it to have a regulation basketball court and walkways the same width around the court. Find the dimensions of the walkway.

EXERCISES

Practice

Find the value of c that makes each trinomial a perfect square.

14. $x^2 - 6x + c$
15. $b^2 + 8b + c$
16. $m^2 - 5m + c$
17. $a^2 + 11a + c$
18. $9t^2 - 18t + c$
19. $\frac{1}{2}x^2 - 4x + c$

Solve each equation by completing the square. Leave irrational roots in simplest radical form.

20. $x^2 + 7x + 10 = -2$
21. $a^2 - 5a + 2 = -2$
22. $r^2 + 14r - 9 = 6$
23. $9b^2 - 42b + 49 = 0$
24. $x^2 - 24x + 9 = 0$
25. $t^2 + 4 = 6t$
26. $m^2 - 8m = 4$
27. $p^2 - 10p = 23$
28. $x^2 - \frac{7}{2}x + \frac{3}{2} = 0$
29. $5x^2 + 10x - 7 = 0$
30. $\frac{1}{2}d^2 - \frac{5}{4}d - 3 = 0$
31. $0.3t^2 + 0.1t = 0.2$
32. $b^2 + 0.25b = 0.5$
33. $3p^2 - 7p - 3 = 0$
34. $2r^2 - 5r + 8 = 7$

Find the value of c that makes each trinomial a perfect square.

35. $x^2 + cx + 81$
36. $4x^2 + cx + 225$
37. $cx^2 + 30x + 75$
38. $cx^2 - 18x + 36$

Solve each equation by completing the square. Leave irrational roots in simplest radical form.

39. $x^2 - 4x + c = 0$
40. $x^2 + bx + c = 0$
41. $x^2 + 4bx + b^2 = 0$

Critical Thinking

42. **Geometry** Consider the quadratic function $y = x^2 - 8x + 15$.
    a. Write the function in the form $y = (x - h)^2 + k$.
    b. Graph the function.
    c. What is the relationship of the point (h, k) to the graph?

Applications and Problem Solving

43. **Identify Subgoals** Two trains left the same station at the same time. One was traveling due north at a speed that was 10 mph faster than the other train, which was traveling due east. After one hour, the trains were 71 miles apart. How fast, to the nearest mile per hour, was each train traveling? (*Hint:* Use the Pythagorean theorem.)

44. **Construction** Arlando's Restaurant wants to add an outdoor café on the side of the restaurant. They are having a special water fountain shipped in that is 10 by 15 feet, and Arlando can afford to buy 1800 square feet of space next to his restaurant. He wishes to have a dining area around the fountain, of equal width all around.

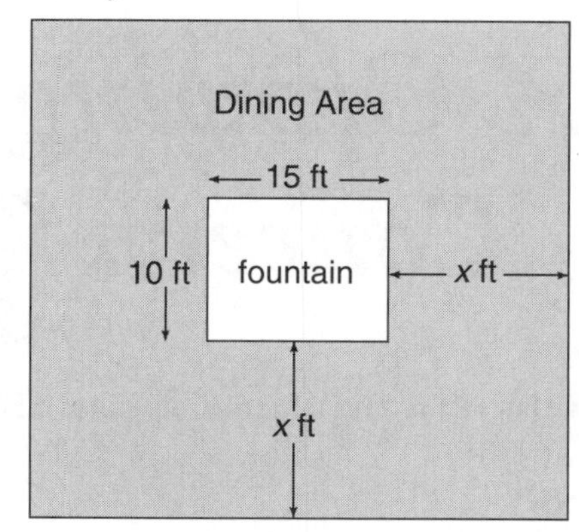

   a. Write an equation for x, the width of the dining area around the fountain and solve it.

   b. What should be the length and width of the piece of land Arlando buys for the café?

Mixed Review

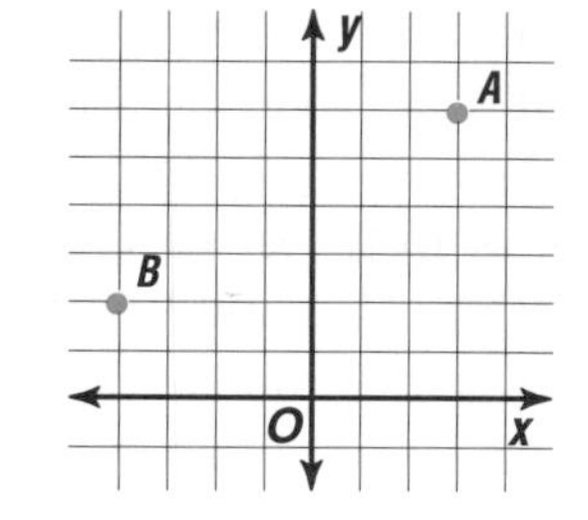

45. **Geometry** Find the distance between points A and B graphed at the right. Round your answer to the nearest hundredth. (Lesson 13–5)

46. **Geometry** The measures of the sides of a triangle are 5, 7, and 9. Determine whether this triangle is a right triangle. (Lesson 13–1)

47. **Astronomy** Earth, Jupiter, and Saturn revolve around the Sun about once every 1, 12, and 30 years, respectively. The last time Jupiter and Saturn appeared close to each other in Earth's night sky was in 1982. When will this happen again? (Lesson 12–6)

48. Find $(5y - 3)(y + 2)$. (Lesson 9–7)

49. Use elimination to solve the system of equations. (Lesson 8–4)

$5y - 4x = 2$

$2y + x = 6$

50. If the graph of $P(x, y)$ satisfies the given conditions, name the quadrant in which point P is located. (Lesson 5–1)

   a. $x > 0, y < 0$ b. $x < 0, y = 3$ c. $x = -1, y < 0$

51. **Food** The chart at the right compares the Calorie content of regular tacos and the light tacos at Taco Bell®. (Lesson 3–7)

   a. Find the mean number of Calories for the regular tacos.

   b. Find the mean number of Calories for the light tacos.

Item	Calories in Regular Tacos	Calories in Light Tacos
Taco	180	140
Soft Taco	220	180
Taco Supreme™	230	160
Soft Taco Supreme®	270	200
Chicken Soft Taco	223	180

52. Evaluate $|a| + |4b|$ if $a = -3$ and $b = -6$. (Lesson 2–3)

A Growing Concern

Refer to the Investigation on pages 656–657.

When landscape businesses prepare a proposal for a client, they prepare all the specifications for the bid and include a sketch of the proposal and a photo display of other work they have done for individuals or companies. They also include a list of references that they give to prospective clients. These clients can then call these people and see other work the company has done in order to verify their credentials. What else might be good to include in a sales presentation to a prospective client?

Analyze

You have made a scale drawing of your design and organized your computations in various ways. It is now time to analyze your design and verify your conclusions.

PORTFOLIO ASSESSMENT

You may want to keep your work on this Investigation in your portfolio.

1. Look over your scale drawing of the landscape design and verify the dimensions and placement of each item.
2. Refer to your data on the shade patterns in the yard. Review your knowledge of the plants and trees that you suggested and verify the amount of sun and/or shade that they will get or that they need. Make any necessary changes.
3. Organize a detailed financial bid for the Sanchez family's backyard design. The bid must include the cost of supplies, materials, labor costs, and a profit margin for each phase of the project.

Present

Select members of your class to represent the Sanchez family. Present your proposal, scale drawing, model, and written report as you would for the real Sanchez family.

4. Begin the presentation by stating the requirements that the Sanchez family had specified that they wanted or needed for their backyard design.
5. Explain the process you used to develop your plan. Justify your design and fully explain your bid costs.
6. Prepare a detailed plan for the Sanchez family. Submit in this plan one scale drawing, the model, a written description of the backyard design, and a detailed financial bid.
7. Make sure that each aspect of the design is justified in writing. Include any options that may negate the terms of the proposal, such as a change in the types of plants used.
8. Summarize your plan with a sales statement that details why your plan is superior.

CHAPTER 13 HIGHLIGHTS

VOCABULARY

After completing this chapter, you should be able to define each term, property, or phrase and give an example or two of each.

Algebra

completing the square (pp. 742, 743)
conjugate (p. 722)
distance formula (p. 737)
product property of square roots (p. 719)
quotient property of square roots (p. 721)
radical equations (p. 732)
radicand (p. 719)
rationalizing the denominator (p. 722)
simplest radical form (p. 723)

Geometry

Pythagorean theorem (p. 713)

Problem Solving

identify subgoals (p. 745)

UNDERSTANDING AND USING THE VOCABULARY

State whether each sentence is *true* or *false*. If false, replace the underlined word or number to make a true sentence.

1. The binomials $-3 + \sqrt{7}$ and $\underline{3 - \sqrt{7}}$ are conjugates.
2. In the expression $-4\sqrt{5}$, the radicand is $\underline{5}$.
3. The rational expression $\frac{1 + \sqrt{3}}{2 - \sqrt{5}}$ becomes $\underline{2 + 2\sqrt{3} + \sqrt{5} + \sqrt{15}}$ when the denominator is rationalized.
4. The value of c that makes the trinomial $t^2 - 3t + c$ a perfect square is $\underline{9}$.
5. The <u>longest</u> side of a right triangle is the hypotenuse.
6. The distance formula can be expressed using the equation $\underline{d^2 = (x_2 - x_1)^2 + (y_2 - y_1)^2}$.
7. After the first step in solving the rational equation $\sqrt{3x + 19} = x + 3$, you would have the equation $\underline{3x + 19 = x^2 + 9}$.
8. The two sides that form the right angle in a right triangle are called the <u>legs</u> of the triangle.
9. The expression $\underline{\frac{2x\sqrt{3x}}{\sqrt{6y}}}$ is in simplest radical form.
10. To verify whether a triangle with sides having lengths of 25, 20, and 15 is a right triangle, the Pythagorean Theorem would be used: $\underline{15^2 = 25^2 + 20^2}$.

SKILLS AND CONCEPTS

OBJECTIVES AND EXAMPLES

Upon completing this chapter, you should be able to:

- use the Pythagorean theorem to solve problems (Lesson 13–1)

Find the length of the missing side.

$c^2 = a^2 + b^2$

$25^2 = 15^2 + b^2$

$625 = 225 + b^2$

$400 = b^2$

$20 = b$

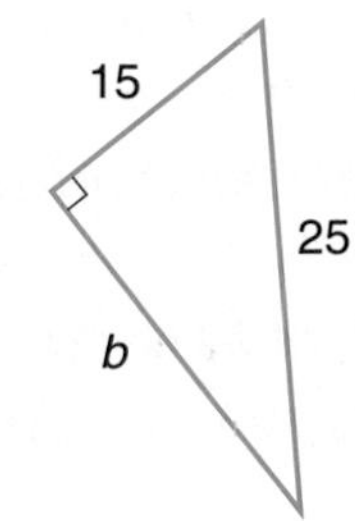

REVIEW EXERCISES

Use these exercises to review and prepare for the chapter test.

If *c* is the measure of the hypotenuse of a right triangle, find each missing measure. Round answers to the nearest hundredth.

11. $a = 30, b = 16, c = ?$

12. $a = 6, b = 10, c = ?$

13. $a = 10, c = 15, b = ?$

14. $b = 4, c = 56, a = ?$

15. $a = 18, c = 30, b = ?$

16. $a = 1.2, b = 1.6, c = ?$

Determine whether the following side measures would form right triangles. Explain why or why not.

17. 9, 16, 20

18. 20, 21, 29

19. 9, 40, 41

20. $18, \sqrt{24}, 30$

OBJECTIVES AND EXAMPLES

- simplify radical expressions (Lesson 13–2)

$$\sqrt{343x^2y^3} = \sqrt{7 \cdot 7^2 \cdot x^2 \cdot y \cdot y^2}$$
$$= \sqrt{7} \cdot \sqrt{7^2} \cdot \sqrt{x^2} \cdot \sqrt{y} \cdot \sqrt{y^2}$$
$$= 7|x|y\sqrt{7y}$$

$$\frac{3}{5-\sqrt{2}} = \frac{3}{5-\sqrt{2}} \cdot \frac{5+\sqrt{2}}{5+\sqrt{2}}$$
$$= \frac{3(5)+3\sqrt{2}}{5^2-(\sqrt{2})^2}$$
$$= \frac{15+3\sqrt{2}}{25-2}$$
$$= \frac{15+3\sqrt{2}}{23}$$

REVIEW EXERCISES

Simplify. Leave in radical form and use absolute value symbols when necessary.

21. $\sqrt{480}$

22. $\sqrt{\frac{60}{y^2}}$

23. $\sqrt{44a^2b^5}$

24. $\sqrt{96x^4}$

25. $(3 - 2\sqrt{12})^2$

26. $\frac{9}{3+\sqrt{2}}$

27. $\frac{2\sqrt{7}}{3\sqrt{5}+5\sqrt{3}}$

28. $\frac{\sqrt{3a^3b^4}}{\sqrt{8ab^{10}}}$

OBJECTIVES AND EXAMPLES

- simplify radical expressions involving addition, subtraction, and multiplication (Lesson 13–3)

$\sqrt{6} - \sqrt{54} + 3\sqrt{12} + 5\sqrt{3}$

$= \sqrt{6} - \sqrt{3^2 \cdot 6} + 3\sqrt{2^2 \cdot 3} + 5\sqrt{3}$

$= \sqrt{6} - (\sqrt{3^2} \cdot \sqrt{6}) + 3(\sqrt{2^2} \cdot \sqrt{3}) + 5\sqrt{3}$

$= \sqrt{6} - 3\sqrt{6} + 3(2\sqrt{3}) + 5\sqrt{3}$

$= \sqrt{6} - 3\sqrt{6} + 6\sqrt{3} + 5\sqrt{3}$

$= -2\sqrt{6} + 11\sqrt{3}$

REVIEW EXERCISES

Simplify. Then use a calculator to verify your answer.

29. $2\sqrt{6} - \sqrt{48}$

30. $2\sqrt{13} + 8\sqrt{15} - 3\sqrt{15} + 3\sqrt{13}$

31. $4\sqrt{27} + 6\sqrt{48}$

32. $5\sqrt{18} - 3\sqrt{112} - 3\sqrt{98}$

33. $\sqrt{8} + \sqrt{\frac{1}{8}}$

34. $4\sqrt{7k} - 7\sqrt{7k} + 2\sqrt{7k}$

- solve radical equations (Lesson 13–4)

Solve $\sqrt{5 - 4x} - 6 = 7$.

$\sqrt{5 - 4x} - 6 = 7$

$\sqrt{5 - 4x} = 13$

$(\sqrt{5 - 4x})^2 = (13)^2$

$5 - 4x = 169$

$-4x = 164$

$x = -41$

Solve each equation. Check your solution.

35. $\sqrt{3x} = 6$

36. $\sqrt{t} = 2\sqrt{6}$

37. $\sqrt{7x - 1} = 5$

38. $\sqrt{x + 4} = x - 8$

39. $\sqrt{r} = 3\sqrt{5}$

40. $\sqrt{3x - 14} + x = 6$

41. $\sqrt{\frac{4a}{3}} - 2 = 0$

42. $9 = \sqrt{\frac{5n}{4}} - 1$

43. $10 + 2\sqrt{b} = 0$

44. $\sqrt{a + 4} = 6$

- find the distance between two points in the coordinate plane (Lesson 13–5)

Find the distance between the pair of points with coordinates $(-5, 1)$ and $(1, 5)$.

$d = \sqrt{(x_2 - x_1)^2 + (y_2 - y_1)^2}$

$= \sqrt{(1 - (-5))^2 + (5 - 1)^2}$

$= \sqrt{6^2 + 4^2}$

$= \sqrt{36 + 16}$ or $\sqrt{52} \approx 7.21$

Find the distance between each pair of points whose coordinates are given.

45. $(9, -2), (1, 13)$

46. $(4, 2), (7, -9)$

47. $(4, -6), (-2, 7)$

48. $(2\sqrt{5}, 9), (4\sqrt{5}, 3)$

Find the value of a if the points with the given coordinates are the indicated distance apart.

49. $(-3, 2), (1, a)$; $d = 5$

50. $(5, -2), (a, -3)$; $d = \sqrt{170}$

51. $(1, 1), (4, a)$; $d = 5$

OBJECTIVES AND EXAMPLES

- solve quadratic equations by completing the square (Lesson 13–6)

Solve $y^2 + 6y + 2 = 0$ by completing the square.

$$y^2 + 6y + 2 = 0$$
$$y^2 + 6y = -2$$
$$y^2 + 6y + 9 = -2 + 9 \quad \text{Complete the square.}$$
$$(y + 3)^2 = 7 \quad \text{Since } \left(\frac{6}{2}\right)^2 = 9 \text{, add 9 to}$$
$$y + 3 = \pm\sqrt{7} \quad \text{each side}$$
$$y = -3 \pm \sqrt{7}$$

The solution set is $-3 + \sqrt{7}$ and $-3 - \sqrt{7}$.

REVIEW EXERCISES

Find the value of c that makes each trinomial a perfect square.

52. $y^2 - 12y + c$
53. $m^2 + 7m + c$
54. $b^2 + 18b + c$
55. $p^2 - \frac{2}{3}p + c$

Solve each equation by completing the square. Leave irrational roots in simplest radical form.

56. $x^2 - 16x + 32 = 0$
57. $m^2 - 7m = 5$
58. $4a^2 + 16a + 15 = 0$
59. $\frac{1}{2}y^2 + 2y - 1 = 0$
60. $n^2 - 3n + \frac{5}{4} = 0$

APPLICATIONS AND PROBLEM SOLVING

61. **Geometry** The sides of a triangle measure $4\sqrt{24}$ cm, $5\sqrt{6}$ cm, and $3\sqrt{54}$ cm. What is the perimeter of the triangle? (Lesson 13–3)

62. **Sight Distance** In the movie *Angels in the Outfield*, Roger decides to climb a tree in order to be able to see the Angels game better. The formula $V = 3.5\sqrt{h}$ relates height and distance, where h is your height in meters above the ground and V is the distance in kilometers that you can see. (Lesson 13–4)
    a. How far could Roger see if he had climbed 9 meters to the top of a tree?
    b. How high would someone have to be if she wanted to be able to see 56 kilometers?

63. **Geometry** What kind of triangle has vertices at (4, 2), (−3, 1), and (5, −4)? (Lesson 13–5)

64. **Number Theory** When 6 is subtracted from 10 times a number, the result is equal to 4 times the square of the number. Find the numbers. (Lesson 13–6)

65. **Nature** An 18-foot tall tree is broken by the wind. The top of the tree falls and touches the ground 12 feet from its base. How many feet from the base of the tree did the break occur? (Lesson 13–1)

66. **Physics** The time T (in seconds) required for a pendulum of length L (in feet) to make one complete swing back and forth is given by the formula $T = 2\pi\sqrt{\frac{L}{32}}$. How long does it take a pendulum 4 feet long to make one complete swing? (Lesson 13–2)

67. **Law Enforcement** Lina told the police officer that she was traveling at 55 mph when she applied the brakes and skidded. The skid marks at the scene were 240 feet long. Should Lina's car have skidded that far if it was traveling at 55 mph? Use the formula $s = \sqrt{15d}$. (Lesson 13–4)

A practice test for Chapter 13 is provided on page 799.

ALTERNATIVE ASSESSMENT

COOPERATIVE LEARNING PROJECT

Battleship In this project, you will determine the placement of ships for a game. A game that Ryan and Nicholas enjoy playing is a form of the game Battleship®. They determine the lengths of five ships that they will draw on their grids. Each time a ship intersects a grid coordinate, that is one of the points where the ship can be attacked and hit. A ship can be placed on the grid horizontally, vertically, or diagonally. When all of the grid coordinates for a ship are hit, the ship is sunk. The goal is to sink all of the other person's ships first.

For their first game, Ryan and Nicholas decided on $2\sqrt{5}$ units, 3 units, $\sqrt{13}$ units, $\sqrt{2}$ units, and 2 units for their ship lengths. Place these five ships on the grid below making sure that they have the above stated lengths. How many hits would another player have to get in order to sink all of the ships on your grid?

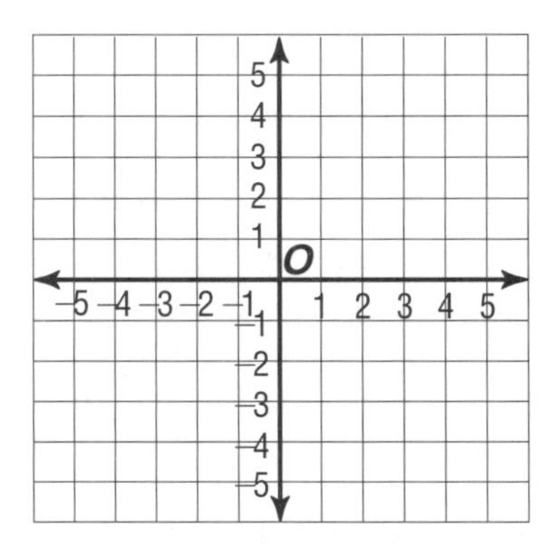

Follow these guidelines for your Battleship game.

- Determine whether each of the ships with the specified lengths have to be drawn horizontally, vertically, or diagonally.
- Devise a chart that will be helpful in organizing the data you will need; for example, length of ship, length of legs of triangle formed by diagonal ships, and number of grid coordinates for each ship.
- Create a grid that will incorporate the above specifications for the ships.
- Write a summary of the above game.
- Play the game with a fellow classmate.

THINKING CRITICALLY

- Is the sum of two irrational numbers always an irrational number? Explain.
- For any real number n, $\sqrt{n^2} = |n|$. What rule would be true for $\sqrt{n^t}$ if t were odd?

PORTFOLIO

Visualizing a concept or skill is a good way to help understand a concept. Using colored pencils is one way to visualize steps that need to be remembered or aid in "seeing" items that are easy to forget. When simplifying a rational expression, write the perfect square number or variables using colored pencils. This will serve to remind you what needs to be written outside the radical sign and what needs to stay inside the radical sign. Use one of the exercises in this chapter and use colored pencils to simplify it. Write how you chose what to put in color and how it aided you in the problem-solving process. Place this in your portfolio.

SELF EVALUATION

When solving a problem, one may look for alternative methods to use. The same method may not work in every situation. Being open-minded about a different process is a positive value. Comparing methods can also be beneficial.

Assess yourself. Are you a person that sticks to the same method for a solution or do you also look for other means? Do you compare methods or not? Think of two problems that you have had, one from your daily life and one from your mathematical experiences, where you used two different methods to solve each of them.

EXTRA PRACTICE

Pages 754–771 are contained in Volume One.

EXTRA PRACTICE

Lesson 7–1 Solve each inequality. Then check your solution.

1. $c + 9 \leq 3$
2. $d - (-3) < 13$
3. $z - 4 > 20$
4. $h - (-7) > -2$
5. $-11 > d - 4$
6. $2x > x - 3$
7. $2x - 3 \geq x$
8. $16 + w < -20$
9. $14p > 5 + 13p$
10. $-7 < 16 - z$
11. $-5 + 14b \leq -4 + 15b$
12. $2s - 6.5 \geq -11.4 + s$
13. $1.1v - 1 > 2.1v - 3$
14. $\frac{1}{2}t + \frac{1}{4} \geq \frac{3}{2}t - \frac{2}{3}$
15. $9x < 8x - 2$
16. $-2 + 9n \leq 10n$
17. $a - 2.3 \geq -7.8$
18. $5z - 6 > 4z$

Lesson 7–2 Solve each inequality. Then check your solution.

1. $7b \geq -49$
2. $-5j < -60$
3. $\frac{w}{3} > -12$
4. $\frac{p}{5} < 8$
5. $-8f < 48$
6. $\frac{t}{-4} \geq -10$
7. $\frac{128}{-g} < 4$
8. $-4.3x < -2.58$
9. $4c \geq -6$
10. $6 \leq 0.8n$
11. $\frac{2}{3}m \geq -22$
12. $-25 > \frac{a}{-6}$
13. $-15a < -28$
14. $-\frac{7}{9}x < 42$
15. $\frac{3y}{8} \leq 32$
16. $-7y \geq 91$
17. $0.8t > 0.96$
18. $\frac{4}{7}z \leq -\frac{2}{5}$

Lesson 7–3 Solve each inequality. Then check your solution.

1. $3y - 4 > -37$
2. $7s - 12 < 13$
3. $-5e + 9 > 24$
4. $-6v - 3 \geq -33$
5. $-2k + 12 < 30$
6. $-2x + 1 < 16 - x$
7. $15t - 4 > 11t - 16$
8. $13 - y \leq 29 + 2y$
9. $5q + 7 \leq 3(q + 1)$
10. $2(w + 4) \geq 7(w - 1)$
11. $-4t - 5 > 2t + 13$
12. $\frac{2t + 5}{3} < -9$
13. $\frac{z}{4} + 7 \geq -5$
14. $13r - 11 > 7r + 37$
15. $8c - (c - 5) > c + 17$
16. $-5(k + 4) \geq 3(k - 4)$
17. $9m + 7 < 2(4m - 1)$
18. $3(3y + 1) < 13y - 8$
19. $5x \leq 10(3x + 4)$
20. $3\left(a + \frac{2}{3}\right) \geq a - 1$
21. $0.7(n - 3) \leq n - 0.6(n + 5)$

Lesson 7–4 Solve each compound inequality. Then graph the solution set.

1. $2 + x < -5$ or $2 + x > 5$
2. $-4 + t > -5$ or $-4 + t < 7$
3. $3 \leq 2g + 7$ and $2g + 7 \leq 15$
4. $2v - 2 \leq 3v$ and $4v - 1 \geq 3v$
5. $3b - 4 \leq 7b + 12$ and $8b - 7 \leq 25$
6. $-9 < 2z + 7 < 10$
7. $5m - 8 \geq 10 - m$ or $5m + 11 < -9$
8. $12c - 4 \leq 5c + 10$ or $-4c - 1 \leq c + 24$
9. $2h - 2 \leq 3h \leq 4h - 1$
10. $3p + 6 < 8 - p$ and $5p + 8 \geq p + 6$
11. $2r + 8 > 16 - 2r$ and $7r + 21 < r - 9$
12. $4j + 3 < j + 22$ and $j - 3 < 2j - 15$
13. $2(q - 4) \leq 3(q + 2)$ or $q - 8 \leq 4 - q$
14. $\frac{1}{2}w + 5 \geq w + 2 \geq \frac{1}{2}w + 9$

Lesson 7-5

1. **Food** For breakfast at Paul's Place, you can select one item from each of the following categories for $1.99.

Meat	Potato	Bread	Beverage
ham sausage bacon	hash browns country potatoes	toast muffin bagel biscuit	juice coffee

   a. What is the probability that a customer will have ham for breakfast?

   b. What is the probability of selecting a biscuit and hash browns?

   c. What is the probability of having a bagel, country potatoes, and coffee?

2. **Music** In order to raise money for a trip to the opera, the music club has set up a lottery using two-digit numbers. The first digit will be a numeral from 1 to 4. The second digit will be a numeral from 3 to 8. The first digit in Trudy's lottery number is 2, but she can't remember the second digit. If only one two-digit lottery number is drawn, and that number has 2 as the first digit, what is the probability that Trudy will win?

3. **Law** A three-judge panel is being used to settle a dispute. Both sides in the dispute have decided that a majority decision will be upheld. If each judge will render a favorable decision based on the evidence presented two-thirds of the time, what is the probability that the correct side will win the dispute?

Lesson 7-6 Solve each open sentence. Then graph the solution set.

1. $|a-5| = -3$
2. $|g+6| > 8$
3. $|t-5| \leq 3$
4. $|a+5| \geq 0$
5. $|14-2z| = 16$
6. $|y-9| < 19$
7. $|2m-5| > 13$
8. $|14-w| \geq 20$
9. $|13-5y| = 8$
10. $|3p+5| \leq 23$
11. $|6b-12| \leq 36$
12. $|25-3x| < 5$
13. $|7+8x| > 39$
14. $|4c+5| \geq 25$
15. $|4-5s| > 46$

Lesson 7-7

1. **Travel** Speeds of the fastest train runs in the U.S. and Canada are given below in miles per hour. Make a box-and-whisker plot of the data.

93.5	82.5	89.3	83.8	81.8	86.8
90.8	84.9	95.0	83.1	83.2	88.2

2. **Basketball** The numbers below represent the 20 highest points-scored-per-game averages for a season in the NBA from 1947 to 1990. Make a box-and-whisker plot of this data.

35.0	33.5	32.5	37.9	31.2	38.4	34.5	34.7	32.9	44.8
31.7	37.1	36.5	34.0	50.4	32.3	33.6	34.8	33.1	35.6

3. **Baseball** The stem-and-leaf plot at the right shows the number of home runs hit by the home run leaders in the National League in 1990.

   a. Find the median, upper quartile, lower quartile, and interquartile range.

   b. Are there any outliers? If so, name them.

   c. Draw a box-and-whisker plot of the data.

Stem	Leaf
2	2 3 3 4 4 4 4 5 5 6 7 7 8
3	2 2 3 3 5 7
4	0

3|3 = 33

Lesson 7–8 Graph each inequality.

1. $y \leq -2$
2. $x < 4$
3. $x + y < -2$
4. $x + y > -4$
5. $y > 4x - 1$
6. $3x + y > 1$
7. $3y - 2x \leq 2$
8. $x < y$
9. $3x + y > 4$
10. $5x - y < 5$
11. $-4x + 3y \geq 12$
12. $-x + 3y \leq 9$
13. $y > -3x + 7$
14. $3x + 8y \leq 4$
15. $5x - 2y \geq 6$

Lesson 8–1 Graph each system of equations. Then determine whether the system has *one* solution, *no* solution, or *infinitely many* solutions. If the system has one solution, name it.

1. $y = 3x$; $4x + 2y = 30$
2. $x = -2y$; $3x + 5y = 21$
3. $y = x + 4$; $3x + 2y = 18$
4. $x + y = 6$; $x - y = 2$
5. $x + y = 6$; $3x + 3y = 3$
6. $y = -3x$; $4x + y = 2$
7. $x + y = 8$; $x - y = 2$
8. $\frac{1}{5}x - y = \frac{12}{5}$; $3x - 5y = 6$
9. $x + 2y = 0$; $y + 3 = -x$
10. $x + 2y = -9$; $x - y = 6$
11. $x + \frac{1}{2}y = 3$; $y = 3x - 4$
12. $\frac{2}{3}x + \frac{1}{2}y = 2$; $4x + 3y = 12$
13. $y = x - 4$; $x + \frac{1}{2}y = \frac{5}{2}$
14. $2x + y = 3$; $4x + 2y = 6$
15. $12x - y = -21$; $\frac{1}{2}x + \frac{2}{3}y = -3$

Lesson 8–2 Use substitution to solve each system of equations. If the system *does not* have exactly one solution, state whether it has *no* solution or *infinitely many* solutions.

1. $y = x$; $5x = 12y$
2. $y = 7 - x$; $2x - y = 8$
3. $x = 5 - y$; $3y = 3x + 1$
4. $3x + y = 6$; $y + 2 = x$
5. $x - 3y = 3$; $2x + 9y = 11$
6. $3x = -18 + 2y$; $x + 3y = 4$
7. $x + 2y = 10$; $-x + y = 2$
8. $2x = 3 - y$; $2y = 12 - x$
9. $6y - x = -36$; $y = -3x$
10. $\frac{3}{4}x + \frac{1}{3}y = 1$; $x - y = 10$
11. $x + 6y = 1$; $3x - 10y = 31$
12. $3x - 2y = 12$; $\frac{3}{2}x - y = 3$
13. $2x + 3y = 5$; $4x - 9y = 9$
14. $x = 4 - 8y$; $3x + 24y = 12$
15. $3x - 2y = -3$; $25x + 10y = 215$

Lesson 8-3 State whether addition, subtraction, or substitution would be most convenient to solve each system of equations. Then solve the system.

1. $x + y = 7$
   $x - y = 9$
2. $2x - y = 32$
   $2x + y = 60$
3. $-y + x = 6$
   $y + x = 5$
4. $s + 2t = 6$
   $3s - 2t = 2$
5. $x = y - 7$
   $2x - 5y = -2$
6. $3x + 5y = -16$
   $3x - 2y = -2$
7. $x - y = 3$
   $x + y = 3$
8. $x + y = 8$
   $2x - y = 6$
9. $2s - 3t = -4$
   $s = 7 - 3t$
10. $-6x + 16y = -8$
    $6x - 42 = 16y$
11. $3x + 0.2y = 7$
    $3x = 0.4y + 4$
12. $9x + 2y = 26$
    $1.5x - 2y = 13$
13. $\frac{2}{3}x - \frac{1}{2}y = 14$
    $\frac{5}{6}x - \frac{1}{2}y = 18$
14. $4x - \frac{1}{3}y = 8$
    $5x + \frac{1}{3}y = 6$
15. $2x - y = 3$
    $\frac{2}{3}x - y = -1$

Lesson 8-4 Use elimination to solve each system of equations.

1. $x + 8y = 3$
   $4x - 2y = 7$
2. $4x - y = 4$
   $x + 2y = 3$
3. $3y - 8x = 9$
   $y - x = 2$
4. $x + 4y = 30$
   $2x - y = -6$
5. $3x - 2y = 0$
   $4x + 4y = 5$
6. $9x - 3y = 5$
   $x + y = 1$
7. $-3x + 2y = 10$
   $-2x - y = -5$
8. $2x + 5y = 13$
   $4x - 3y = -13$
9. $5x + 3y = 4$
   $-4x + 5y = -18$
10. $2x - 7y = 9$
    $-3x + 4y = 6$
11. $2x - 6y = -16$
    $5x + 7y = -18$
12. $6x - 3y = -9$
    $-8x + 2y = 4$
13. $\frac{1}{3}x - y = -1$
    $\frac{1}{5}x - \frac{2}{5}y = -1$
14. $3x - 5y = 8$
    $4x - 7y = 10$
15. $x - 0.5y = 1$
    $0.4x + y = -2$

Lesson 8-5 Solve each system of inequalities by graphing.

1. $x > 3$
   $y < 6$
2. $y > 2$
   $y > -x + 2$
3. $x \leq 2$
   $y - 3 \geq 5$
4. $x + y \leq -1$
   $2x + y \leq 2$
5. $y \geq 2x + 2$
   $y \geq -x - 1$
6. $y \leq x + 3$
   $y \geq x + 2$
7. $x + 3y \geq 4$
   $2x - y < 5$
8. $y - x > 1$
   $y + 2x \leq 10$
9. $5x - 2y > 15$
   $2x - 3y < 6$
10. $4x + 3y > 4$
    $2x - y < 0$
11. $4x + 5y \geq 20$
    $y \geq x + 1$
12. $-4x + 10y \leq 5$
    $-2x + 5y < -1$

EXTRA PRACTICE

Lesson 9-1 Simplify.

1. $a^5(a)(a^7)$
2. $(r^3t^4)(r^4t^4)$
3. $(x^3y^4)(xy^3)$
4. $(bc^3)(b^4c^3)$
5. $(-3mn^2)(5m^3n^2)$
6. $[(3^3)^2]^2$
7. $(3s^3t^2)(-4s^3t^2)$
8. $x^3(x^4y^3)$
9. $(1.1g^2h^4)^3$
10. $-\frac{3}{4}a(a^2b^3c^4)$
11. $\left(\frac{1}{2}w^3\right)^2(w^4)^2$
12. $\left(\frac{2}{3}y^3\right)(3y^2)^3$
13. $[(-2^3)^3]^2$
14. $(10s^3t)(-2s^2t^2)^3$
15. $(-0.2u^3w^4)^3$

Lesson 9-2 Simplify. Assume no denominator is equal to zero.

1. $\frac{b^6c^5}{b^3c^2}$
2. $\frac{(-a)^4b^8}{a^4b^7}$
3. $\frac{(-x)^3y^3}{x^3y^6}$
4. $\frac{12ab^5}{4a^4b^3}$
5. $\frac{24x^5}{-8x^2}$
6. $\frac{-9h^2k^4}{18h^5j^3k^4}$
7. $\frac{a^0}{2a^{-3}}$
8. $\frac{9a^2b^7c^3}{2a^5b^4c}$
9. $\frac{-15xy^5z^7}{-10x^4y^6z^4}$
10. $\frac{(u^{-3}v^3)^2}{(u^3v)^{-3}}$
11. $\frac{(-r)s^5}{r^{-3}s^{-4}}$
12. $\frac{28a^{-4}b^0}{14a^3b^{-1}}$
13. $\frac{(j^2k^3l)^4}{(jk^4)^{-1}}$
14. $\left(\frac{-2x^4y}{4y^2}\right)^0$
15. $\frac{3m^7n^2p^4}{9m^2np^3}$

Lesson 9-3 Express each number in scientific notation.

1. 6500
2. 953.56
3. 0.697
4. 843.5
5. 568,000
6. 0.0000269
7. 0.121212
8. 543×10^4
9. 739.9×10^{-5}
10. 6480×10^{-2}
11. 0.366×10^{-7}
12. 167×10^3

Evaluate. Express each result in scientific and standard notation.

13. $(2 \times 10^5)(3 \times 10^{-8})$
14. $\frac{4.8 \times 10^3}{1.6 \times 10^1}$
15. $(4 \times 10^2)(1.5 \times 10^6)$
16. $\frac{8.1 \times 10^2}{2.7 \times 10^{-3}}$
17. $\frac{7.8 \times 10^{-5}}{1.3 \times 10^{-7}}$
18. $(2.2 \times 10^{-2})(3.2 \times 10^5)$
19. $(3.1 \times 10^4)(4.2 \times 10^{-3})$
20. $(78 \times 10^6)(0.01 \times 10^3)$
21. $\frac{2.31 \times 10^{-2}}{3.3 \times 10^{-3}}$

Lesson 9–4 State whether each expression is a polynomial. If the expression is a polynomial, identify it as a *monomial,* a *binomial,* or a *trinomial* and find the degree of the polynomial.

1. $5x^2y + 3xy + 7$
2. 0
3. $\frac{5}{k} - k^2y$
4. $3a^2x - 5a$
5. $a + \frac{5}{c}$
6. $14abcd - 6d^3$
7. $\frac{a^3}{3}$
8. $-4h^3$
9. $x^2 - \frac{x}{2} + \frac{1}{3}$

Arrange the terms of each polynomial so that the powers of x are in descending order.

10. $5x^2 - 3x^3 + 7 + 2x$
11. $-6x + x^5 + 4x^3 - 20$
12. $5b + b^3x^2 + \frac{2}{3}bx$
13. $21p^2x + 3px^3 + p^4$
14. $3ax^2 - 6a^2x^3 + 7a^3 - 8x$
15. $\frac{1}{3}s^2x^3 + 4x^4 - \frac{2}{5}s^4x^2 + \frac{1}{4}x$

Lesson 9–5 Find each sum or difference.

1.
$$\begin{array}{r} -7t^2 + 4ts - 6s^2 \\ (+)\ -5t^2 - 12ts + 3s^2 \\ \hline \end{array}$$

2.
$$\begin{array}{r} 6a^2 - 7ab - 4b^2 \\ (-)\ 2a^2 + 5ab + 6b^2 \\ \hline \end{array}$$

3.
$$\begin{array}{lllll} & 4a^2 & - 10b^2 & + 7c^2 & \\ & -5a^2 & & + 2c^2 & + 2b \\ (+) & & 7b^2 & - 7c^2 & + 7a \\ \hline \end{array}$$

4.
$$\begin{array}{r} z^2 + 6z - 8 \\ (-)\ 4z^2 - 7z - 5 \\ \hline \end{array}$$

5. $(4d + 3e - 8f) - (-3d + 10e - 5f + 6)$
6. $(7g + 8h - 9) + (-g - 3h - 6k)$
7. $(9x^2 - 11xy - 3y^2) - (x^2 - 16xy + 12y^2)$
8. $(-3m + 9mn - 5n) + (14m - 5mn - 2n)$
9. $(4x^2 - 8y^2 - 3z^2) - (7x^2 - 14z^2 - 12)$
10. $(17z^4 - 5z^2 + 3z) - (4z^4 + 2z^3 + 3z)$
11. $(6 - 7y + 3y^2) + (3 - 5y - 2y^2) + (-12 - 8y + y^2)$
12. $(-3x^2 + 2x - 5) + (2x - 6) + (5x^2 + 3) + (-9x^2 - 7x + 4)$

Lesson 9–6 Find each product.

1. $-3(8x + 5)$
2. $3b(5b + 8)$
3. $1.1a(2a + 7)$
4. $\frac{1}{2}x(8x - 6)$
5. $7xy(5x^2 - y^2)$
6. $5y(y^2 - 3y + 6)$
7. $-ab(3b^2 + 4ab - 6a^2)$
8. $4m^2(9m^2n + mn - 5n^2)$
9. $4st^2(-4s^2t^3 + 7s^5 - 3st^3)$
10. $-\frac{1}{3}x(9x^2 + x - 5)$
11. $-2mn(8m^2 - 3mn + n^2)$
12. $-\frac{3}{4}ab^2\left(\frac{1}{3}b^2 - \frac{4}{9}b + 1\right)$

Solve.

13. $-3(2a - 12) + 48 = 3a - 3$
14. $-6(12 - 2w) = 7(-2 - 3w)$
15. $a(a - 6) + 2a = 3 + a(a - 2)$
16. $11(a - 3) + 5 = 2a + 44$
17. $q(2q + 3) + 20 = 2q(q - 3)$
18. $w(w + 12) = w(w + 14) + 12$
19. $x(x + 8) - x(x + 3) - 23 = 3x + 11$
20. $y(y - 12) + y(y + 2) + 25 = 2y(y + 5) - 15$
21. $x(x - 3) + 4x - 3 = 8x + 4 + x(3 + x)$
22. $c(c - 3) + 4(c - 2) = 12 - 2(4 + c) - c(1 - c)$

Lesson 9–7 **Find each product.**

1. $(d + 2)(d + 3)$
2. $(z + 7)(z - 4)$
3. $(m - 8)(m - 5)$
4. $(2x - 5)(x + 6)$
5. $(7a - 4)(2a - 5)$
6. $(4x + y)(2x - 3y)$
7. $(7v + 3)(v + 4)$
8. $(7s - 8)(3s - 2)$
9. $(4g + 3h)(2g - 5h)$
10. $(4a + 3)(2a - 1)$
11. $(7y - 1)(2y - 3)$
12. $(2x + 3y)(5x + 2y)$
13. $(12r - 4s)(5r + 8s)$
14. $(x - 2)(x^2 + 2x + 4)$
15. $(3x + 5)(2x^2 - 5x + 11)$
16. $(4s + 5)(3s^2 + 8s - 9)$
17. $(3a + 5)(-8a^2 + 2a + 3)$
18. $(5x - 2)(-5x^2 + 2x + 7)$
19. $(x^2 - 7x + 4)(2x^2 - 3x - 6)$
20. $(a^2 + 2a + 5)(a^2 - 3a - 7)$
21. $(5x^4 - 2x^2 + 1)(x^2 - 5x + 3)$

Lesson 9–8 **Find each product.**

1. $(t + 7)^2$
2. $(w - 12)(w + 12)$
3. $(q - 4h)^2$
4. $(10x + 11y)(10x - 11y)$
5. $(4e + 3)^2$
6. $(2b - 4d)(2b + 4d)$
7. $(a + 2b)^2$
8. $(4x + y)^2$
9. $(6m + 2n)^2$
10. $(5c - 2d)^2$
11. $(5b - 6)(5b + 6)$
12. $(1 + x)^2$
13. $(4x - 9y)^2$
14. $(8a - 2b)(8a + 2b)$
15. $\left(\frac{1}{2}a + b\right)^2$
16. $(5a - 12b)^2$
17. $(a - 3b)^2$
18. $(7a^2 + b)(7a^2 - b)$
19. $(x + 2)(x - 2)(2x + 5)$
20. $(4x - 1)(4x + 1)(x - 4)$
21. $(x - 3)(x + 3)(x - 4)(x + 4)$

Lesson 10–1 **Find the factors of each number.**

1. 17
2. 21
3. 81
4. 24
5. 18
6. 22

State whether each number is *prime* or *composite*. If the number is composite, find its prime factorization.

7. 39
8. 89
9. 72
10. 41
11. 57
12. 60

Factor each expression completely. Do not use exponents.

13. -64
14. -26
15. -240
16. -231
17. $44rs^2t^3$
18. $756(mn)^2$

Find the GCF of the given monomials.

19. 16, 60
20. 15, 50
21. -80, 45
22. 29, -58
23. 305, 55
24. 252, 126
25. 128, 245
26. $7y^2$, $14y^2$
27. $4xy$, $-6x$
28. $35t^2$, $7t$
29. $16pq^2$, $12p^2q$
30. 5, 15, 10
31. $12mn$, $10mn$, $15mn$
32. 14, 12, 20
33. $26jk^4$, $16jk^3$, $8j^2$

Lesson 10-2 Complete. In exercises with two blanks, both blanks represent the same expression.

1. $6x + 3y = 3(\underline{?} + y)$
2. $8x^2 - 4x = 4x(2x - \underline{?})$
3. $12a^2b + 6a = 6a(\underline{?} + 1)$
4. $14r^2t - 42t = 14t(\underline{?} - 3)$
5. $24x^2 + 12y^2 = 12(\underline{?} + y^2)$
6. $12xy + 12x^2 = \underline{?}\,(y + x)$
7. $(bx + by) + (3ax + 3ay) = b(\underline{?}) + 3a(\underline{?})$
8. $(10x^2 - 6xy) + (15x - 9y) = 2x(\underline{?}) + 3(\underline{?})$
9. $(6x^3 + 6x) + (7x^2y + 7y) = 6x(\underline{?}) + 7y(\underline{?})$

Factor each polynomial.

10. $10a^2 + 40a$
11. $15wx - 35wx^2$
12. $27a^2b + 9b^3$
13. $11x + 44x^2y$
14. $16y^2 + 8y$
15. $14mn^2 + 2mn$
16. $25a^2b^2 + 30ab^3$
17. $2m^3n^2 - 16m^2n^3 + 8mn$
18. $2ax + 6xc + ba + 3bc$
19. $6mx - 4m + 3rx - 2r$
20. $3ax - 6bx + 8b - 4a$
21. $a^2 - 2ab + a - 2b$
22. $8ac - 2ad + 4bc - bd$
23. $2e^2g + 2fg + 4e^2h + 4fh$

Lesson 10-3 Complete.

1. $p^2 + 9p - 10 = (p + \underline{?})(p - 1)$
2. $y^2 - 2y - 35 = (y + 5)(y - \underline{?})$
3. $4a^2 + 4a - 63 = (2a - 7)(2a \underline{?} 9)$
4. $4r^2 - 25r + 6 = (r - 6)(\underline{?} - 1)$
5. $b^2 + 12b + 35 = (b + 5)(b + \underline{?})$
6. $3x^2 - 7x - 6 = (3x + 2)(x \underline{?} 7)$
7. $3a^2 - 2a - 21 = (a \underline{?} 3)(3a + 7)$
8. $4y^2 + 11y + 6 = (\underline{?} + 3)(y + 2)$
9. $2z^2 - 11z + 15 = (\underline{?} - 5)(z - 3)$
10. $6n^2 + 7n - 3 = (2n + \underline{?})(3n - 1)$

Factor each trinomial, if possible. If the trinomial cannot be factored using integers, write *prime*.

11. $5x^2 - 17x + 14$
12. $a^2 - 9a - 36$
13. $x^2 + 2x - 15$
14. $n^2 - 8n + 15$
15. $b^2 + 22b + 21$
16. $c^2 + 2c - 3$
17. $x^2 - 5x - 24$
18. $2n^2 - 11n + 7$
19. $8m^2 - 10m + 3$
20. $z^2 + 15z + 36$
21. $s^2 - 13st - 30t^2$
22. $6y^2 + 2y - 2$
23. $2r^2 + 3r - 14$
24. $5x - 6 + x^2$
25. $x^2 - 4xy - 5y^2$
26. $5r^2 - 3r + 15$
27. $18v^2 + 42v + 12$
28. $4k^2 + 2k - 12$

Lesson 10-4 Factor each polynomial, if possible. If the polynomial cannot be factored, write *prime*.

1. $x^2 - 9$
2. $a^2 - 64$
3. $t^2 - 49$
4. $4x^2 - 9y^2$
5. $1 - 9z^2$
6. $16a^2 - 9b^2$
7. $8x^2 - 12y^2$
8. $a^2 - 4b^2$
9. $x^2 - y^2$
10. $75r^2 - 48$
11. $x^2 - 36y^2$
12. $3a^2 - 16$
13. $12t^2 - 75$
14. $9x^2 - 100y^2$
15. $49 - a^2b^2$
16. $12a^2 - 48$
17. $169 - 16t^2$
18. $8r^2 - 4$
19. $-45m^2 + 5$
20. $9x^4 - 16y^2$
21. $36b^2 - 64$
22. $5g^2 - 20h^2$
23. $\frac{1}{4}n^2 - 16$
24. $\frac{1}{4}t^2 - \frac{4}{9}p^2$
25. $(r - t)^2 + t^2$
26. $12x^3 - 27xy^2$
27. $0.01n^2 - 1.69r^2$
28. $0.04m^2 - 0.09n^2$
29. $(x - y)^2 - y^2$
30. $162m^4 - 32n^8$

Lesson 10-5 Determine whether each trinomial is a perfect square trinomial. If so, factor it.

1. $x^2 + 12x + 36$
2. $n^2 - 13n + 36$
3. $a^2 + 4a + 4$
4. $b^2 - 14b + 49$
5. $x^2 + 20x - 100$
6. $y^2 - 10y + 100$
7. $9b^2 - 6b + 1$
8. $4x^2 + 4x + 1$
9. $2n^2 + 17n + 21$
10. $9x^2 - 10x + 4$
11. $9y^2 + 8y - 16$
12. $4a^2 - 20a + 25$

Factor each polynomial, if possible. If the polynomial cannot be factored, write *prime*.

13. $n^2 - 8n + 16$
14. $4k^2 - 4k + 1$
15. $x^2 + 16x + 64$
16. $t^2 - 4t + 1$
17. $x^2 + 22x + 121$
18. $s^2 + 30s + 225$
19. $1 - 10z + 25z^2$
20. $9p^2 - 56p + 49$
21. $9n^2 - 36nm + 36m^2$
22. $16a^2 + 81 - 72a$
23. $9x^2 + 12xy + 4y^2$
24. $m^2 + 16mn + 64n^2$
25. $8t^4 + 56t^3 + 98t^2$
26. $4p^2 + 12pr + 9r^2$
27. $16m^4 - 72m^2n^2 + 81n^4$

Lesson 10-6 Solve each equation. Check your solutions.

1. $y(y - 12) = 0$
2. $2x(5x - 10) = 0$
3. $7a(a + 6) = 0$
4. $(b - 3)(b - 5) = 0$
5. $(p - 5)(p + 5) = 0$
6. $(4t + 4)(2t + 6) = 0$
7. $(3x - 5)^2 = 0$
8. $x^2 - 6x = 0$
9. $n^2 + 36n = 0$
10. $2x^2 + 4x = 0$
11. $2x^2 = x^2 - 8x$
12. $7y - 1 = -3y^2 + y - 1$
13. $\frac{1}{2}y^2 - \frac{1}{4}y = 0$
14. $\frac{5}{6}x^2 - \frac{1}{3}x = \frac{1}{3}x$
15. $\frac{2}{3}x = \frac{1}{3}x^2$
16. $\frac{3}{4}a^2 + \frac{7}{8}a = a$
17. $n^2 - 3n = 0$
18. $3x^2 - \frac{3}{4}x = 0$
19. $8a^2 = -4a$
20. $(2y + 8)(3y + 24) = 0$
21. $(4x - 7)(3x + 5) = 0$

Lesson 11-1 Write the equation of the axis of symmetry and find the coordinates of the vertex of the graph of each equation. State if the vertex is a maximum or minimum. Then graph the equation.

1. $y = x^2 + 6x + 8$
2. $y = -x^2 + 3x$
3. $y = -x^2 + 7$
4. $y = x^2 + x + 3$
5. $y = -x^2 + 4x + 5$
6. $y = 3x^2 + 6x + 16$
7. $y = -x^2 + 2x - 3$
8. $y = 3x^2 + 24x + 80$
9. $y = x^2 - 4x - 4$
10. $y = 5x^2 - 20x + 37$
11. $y = 3x^2 + 6x + 3$
12. $y = 2x^2 + 12x$
13. $y = x^2 - 6x + 5$
14. $y = \frac{1}{2}x^2 + 3x + \frac{9}{2}$
15. $y = \frac{1}{4}x^2 - 4x + \frac{15}{4}$
16. $y = 4x^2 - 1$
17. $y = -2x^2 - 2x + 4$
18. $y = 6x^2 - 12x - 4$
19. $y = x^2 - 1$
20. $y = -x^2 + x + 1$
21. $y = -5x^2 - 3x + 2$
22. $y = x^2 - x - 6$
23. $y = 2x^2 + 5x - 2$
24. $y = -3x^2 - 18x - 15$

Lesson 11–2 State the real roots of each quadratic equation whose related function is graphed below.

1. 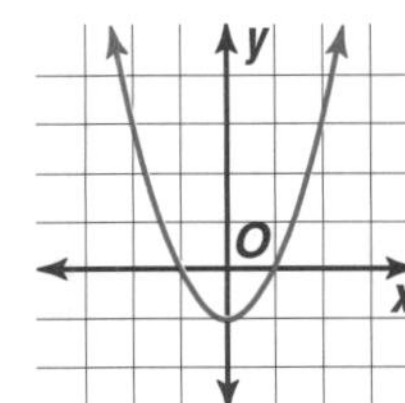
2. 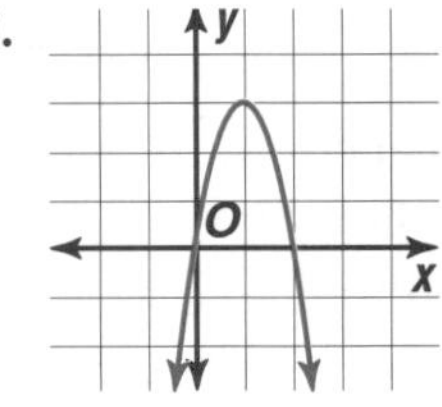
3. 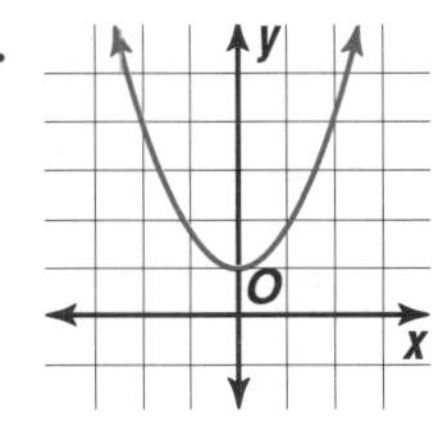
4. 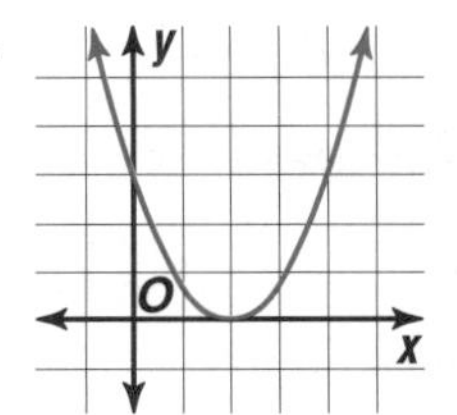

Solve each equation by graphing. If exact roots cannot be found, state the consecutive integers between which the roots lie.

5. $x^2 + 2x - 3 = 0$
6. $-x^2 + 6x - 5 = 0$
7. $-a^2 - 2a + 3 = 0$
8. $2r^2 - 8r + 5 = 0$
9. $-3x^2 + 6x - 9 = 0$
10. $c^2 + c = 0$
11. $3t^2 + 2 = 0$
12. $-b^2 + 5b + 2 = 0$
13. $3x^2 + 7x = 1$
14. $x^2 + 5x - 24 = 0$
15. $8 - k^2 = 0$
16. $x^2 - 7x = 18$
17. $a^2 + 12a + 36 = 0$
18. $64 - x^2 = 0$
19. $-4x^2 + 2x = -1$

Lesson 11–3 Solve each equation by using the quadratic formula. Approximate irrational roots to the nearest hundredth.

1. $x^2 - 8x - 4 = 0$
2. $x^2 + 7x + 6 = 0$
3. $x^2 + 5x - 6 = 0$
4. $y^2 - 7y - 8 = 0$
5. $m^2 - 2m = 35$
6. $4n^2 - 20n = 0$
7. $m^2 + 4m + 2 = 0$
8. $2t^2 - t - 15 = 0$
9. $5t^2 = 125$
10. $t^2 + 16 = 0$
11. $-4x^2 + 8x = -3$
12. $3k^2 + 2 = -8k$
13. $8t^2 + 10t + 3 = 0$
14. $3x^2 - \frac{5}{4}x - \frac{1}{2} = 0$
15. $-5b^2 + 3b - 1 = 0$
16. $s^2 + 8s + 7 = 0$
17. $d^2 - 14d + 24 = 0$
18. $3k^2 + 11k = 4$
19. $n^2 - 3n + 1 = 0$
20. $2z^2 + 5z - 1 = 0$
21. $3h^2 = 27$
22. $3f^2 + 2f = 6$
23. $2x^2 = 0.7x + 0.3$
24. $3w^2 - 8w + 2 = 0$
25. $2r^2 - r - 3 = 0$
26. $x^2 - 9x = 5$
27. $6t^2 - 4t - 9 = 0$

Lesson 11–4 Use a calculator to determine the approximate value of each expression to the nearest hundredth.

1. $3^{1.6}$
2. $10^{-0.2}$
3. $\left(\frac{1}{3}\right)^{-1.4}$
4. $\left(\frac{2}{3}\right)^{5.1}$
5. $40(2^{-0.5})$
6. $10(2^{-1.6})$
7. $0.3(4^{0.8})$
8. $30(0.75^{-3.6})$
9. $5^{1.75}$

Graph each function. State the *y*-intercept.

10. $y = 3^x + 1$
11. $y = 2^x - 5$
12. $y = 2^{x+3}$
13. $y = 3^{x+1}$
14. $y = \left(\frac{1}{4}\right)^x$
15. $y = 5\left(\frac{2}{5}\right)^x$
16. $y = 3 \cdot 2^x$
17. $y = 4 \cdot 5^x$
18. $y = 6^x$
19. $y = 3^x$
20. $y = \left(\frac{1}{8}\right)^x$
21. $y = \left(\frac{3}{4}\right)^x$

Solve each equation.

22. $6^{3x-4} = 6^x$
23. $3^4 = 3^{2x+2}$
24. $4^x = 4^{5x+8}$
25. $2^x = 4^{x+1}$
26. $5^{4x} = 5^{-4}$
27. $2^{x+3} = 2^{-5}$

EXTRA PRACTICE

Lesson 11-5 Determine whether each exponential equation represents growth or decay.

1. $y = 3.89(1.05)^x$
2. $y = 476(0.35)^x$
3. $y = 19{,}520(0.98)^x$
4. $y = 16(1.0432)^x$
5. $y = 1.01(1.099)^x$
6. $y = 84(0.03)^x$

7. **Education** Marco withdrew all of the $2500 in his savings account to pay the tuition for his first semester at college. The account had earned 12% interest compounded monthly, and no withdrawals or additional deposits were made.
   a. If Marco's original deposit was $1250, how long ago did he open the account?
   b. If Marco's original deposit was $1500, how long ago did he open the account?
8. **Finance** Erin saved $500 of the money she earned working at the Dairy Dream last summer. She deposited the money in a certificate of deposit that earns 8.75% interest compounded monthly. If she rolls over the CD at the same rate each year, when will Erin's CD have a balance of $800?
9. **Demographics** In 1994, the metropolitan area of Pensacola, Florida, had a population of 371,000. The growth rate from 1990 to 1994 was 7.7%.
   a. Write an exponential equation for the area's growth.
   b. Compute the estimated population for Pensacola in the year 2000.

Lesson 12-1 Simplify each rational expression. State the excluded values of the variables.

1. $\frac{13a}{39a^2}$
2. $\frac{38x^2}{42xy}$
3. $\frac{14y^2z}{49yz^3}$
4. $\frac{p+5}{2(p+5)}$
5. $\frac{79a^2b}{158a^3bc}$
6. $\frac{a+b}{a^2-b^2}$
7. $\frac{y+4}{(y-4)(y+4)}$
8. $\frac{c^2-4}{(c+2)^2}$
9. $\frac{a^2-a}{a-1}$
10. $\frac{(w-4)(w+4)}{(w-2)(w-4)}$
11. $\frac{m^2-2m}{m-2}$
12. $\frac{x^2+4}{x^4-16}$
13. $\frac{r^3-r^2}{r-1}$
14. $\frac{3m^3}{6m^2-3m}$
15. $\frac{4t^2-8}{4t-4}$
16. $\frac{6y^3-12y^2}{12y^2-18}$
17. $\frac{x-3}{x^2+x-12}$
18. $\frac{5x^2+10x+5}{3x^2+6x+3}$

Lesson 12-2 Find each product. Assume that no denominator has a value of 0.

1. $\frac{a^2b}{b^2c} \cdot \frac{c}{d}$
2. $\frac{6a^2n}{8n^2} \cdot \frac{12n}{9a}$
3. $\frac{2a^2d}{3bc} \cdot \frac{9b^2c}{16ad^2}$
4. $\frac{10n^3}{6x^3} \cdot \frac{12n^2x^4}{25n^2x^2}$
5. $\left(\frac{2a}{b}\right)^2 \cdot \frac{5c}{6a}$
6. $\frac{6m^3n}{10a^2} \cdot \frac{4a^2m}{9n^3}$
7. $\frac{5n-5}{3} \cdot \frac{9}{n-1}$
8. $\frac{a^2}{a-b} \cdot \frac{3a-3b}{a}$
9. $\frac{2a+4b}{5} \cdot \frac{25}{6a+8b}$
10. $\frac{4t}{4t+40} \cdot \frac{3t+30}{2t}$
11. $\frac{3k+9}{k} \cdot \frac{k^2}{k^2-9}$
12. $\frac{7xy^3}{11z^2} \cdot \frac{44z^3}{21x^2y}$
13. $\frac{3}{x-y} \cdot \frac{(x-y)^2}{6}$
14. $\frac{x+5}{3x} \cdot \frac{12x^2}{x^2+7x+10}$
15. $\frac{a^2-b^2}{4} \cdot \frac{16}{a+b}$
16. $\frac{4a+8}{a^2-25} \cdot \frac{a-5}{5a+10}$
17. $\frac{r^2}{r-s} \cdot \frac{r^2-s^2}{s^2}$
18. $\frac{a^2-b^2}{a-b} \cdot \frac{7}{a+b}$

Lesson 12-3 Find each quotient. Assume that no denominator has a value of 0.

1. $\frac{5m^2n}{12a^2} \div \frac{30m^4}{18an}$
2. $\frac{25g^7h}{28t^3} \div \frac{5g^5h^2}{42s^2t^3}$
3. $\frac{6a+3b}{36} \div \frac{3a+2b}{45}$
4. $\frac{x^2y}{18z} \div \frac{2yz}{3x^2}$
5. $\frac{p^2}{14qr^3} \div \frac{2r^2p}{7q}$
6. $\frac{5e-f}{5e+f} \div (25e^2 - f^2)$
7. $\frac{t^2-2t-15}{t-5} \div \frac{t+3}{t+5}$
8. $\frac{5x+10}{x+2} \div (x+2)$
9. $\frac{3d}{2d^2-3d} \div \frac{9}{2d-3}$
10. $\frac{3v^2-27}{15v} \div \frac{v+3}{v^2}$
11. $\frac{3g^2+15g}{4} \div \frac{g+5}{g^2}$
12. $\frac{b^2-9}{4b} \div (b-3)$
13. $\frac{p^2}{y^2-4} \div \frac{p}{2-y}$
14. $\frac{k^2-81}{k^2-36} \div \frac{k-9}{k+6}$
15. $\frac{2a^3}{a+1} \div \frac{a^2}{a+1}$
16. $\frac{x^2-16}{16-x^2} \div \frac{7}{x}$
17. $\frac{y}{5} \div \frac{y^2-25}{5-y}$
18. $\frac{3m}{m+1} \div (m-2)$

Lesson 12-4 Find each quotient.

1. $(2x^2 - 11x - 20) \div (2x + 3)$
2. $(a^2 + 7a + 12) \div (a + 3)$
3. $(m^2 + 9m + 20) \div (m + 5)$
4. $(x^2 - 2x - 35) \div (x - 7)$
5. $(c^2 + 12c + 36) \div (c + 9)$
6. $(y^2 - 2y - 30) \div (y + 7)$
7. $(3t^2 - 14t - 24) \div (3t + 4)$
8. $(2r^2 - 3r - 35) \div (2r + 7)$
9. $\frac{12n^2+36n+15}{6n+3}$
10. $\frac{10x^2+29x+21}{5x+7}$
11. $\frac{4t^3+17t^2-1}{4t+1}$
12. $\frac{2a^3+9a^2+5a-12}{a+3}$
13. $\frac{4m^3+5m-21}{2m-3}$
14. $\frac{6t^3+5t^2+12}{2t+3}$
15. $\frac{27c^2-24c+8}{9c-2}$
16. $\frac{3b^3+8b^2+b-7}{b+2}$
17. $\frac{t^3-19t+9}{t-4}$
18. $\frac{9d^3+5d-8}{3d-2}$

Lesson 12-5 Find each sum or difference. Express in simplest form.

1. $\frac{4}{z} + \frac{3}{z}$
2. $\frac{a}{12} + \frac{2a}{12}$
3. $\frac{5}{2t} + \frac{-7}{2t}$
4. $\frac{y}{2} + \frac{y}{2}$
5. $\frac{b}{x} + \frac{2}{x}$
6. $\frac{5x}{24} - \frac{3x}{24}$
7. $\frac{7p}{p} - \frac{8p}{p}$
8. $\frac{8k}{5m} - \frac{3k}{5m}$
9. $\frac{y}{2} + \frac{y-6}{2}$
10. $\frac{a+2}{6} - \frac{a+3}{6}$
11. $\frac{8}{m-2} - \frac{6}{m-2}$
12. $\frac{x}{x+1} + \frac{1}{x+1}$
13. $\frac{2n}{2n-5} + \frac{5}{5-2n}$
14. $\frac{y}{b+6} - \frac{2y}{b+6}$
15. $\frac{x-y}{2-y} + \frac{x+y}{y-2}$
16. $\frac{r^2}{r-s} + \frac{s^2}{r-s}$
17. $\frac{12n}{3n+2} + \frac{8}{3n+2}$
18. $\frac{6x}{x+y} + \frac{6y}{x+y}$

EXTRA PRACTICE

Lesson 12-6 Find each sum or difference.

1. $\frac{s}{3} + \frac{2s}{7}$
2. $\frac{5}{2a} + \frac{-3}{6a}$
3. $\frac{2n}{5} - \frac{3m}{4}$
4. $\frac{6}{5x} + \frac{7}{10x^2}$
5. $\frac{3z}{7w^2} - \frac{2z}{w}$
6. $\frac{s}{t^2} - \frac{r}{3t}$
7. $\frac{5}{xy} + \frac{6}{yz}$
8. $\frac{2}{t} + \frac{t+3}{s}$
9. $\frac{a}{a-b} + \frac{b}{2b+3a}$
10. $\frac{a}{a^2-4} - \frac{4}{a+2}$
11. $\frac{4a}{2a+6} + \frac{3}{a+3}$
12. $\frac{m}{1(m-n)} - \frac{5}{m}$
13. $\frac{-3}{a-5} + \frac{-6}{a^2-5a}$
14. $\frac{3t+2}{3t-6} - \frac{t+2}{t^2-4}$
15. $\frac{y+5}{y-5} + \frac{2y}{y^2-25}$
16. $\frac{-18}{y^2-9} + \frac{7}{3-y}$
17. $\frac{c}{c^2-4c} - \frac{5c}{c-4}$
18. $\frac{t+10}{t^2-100} + \frac{1}{t-10}$

Lesson 12-7 Write each mixed expression as a rational expression.

1. $4 + \frac{2}{x}$
2. $8 + \frac{5}{3t}$
3. $3b + \frac{b+1}{2b}$
4. $2n + \frac{4+n}{n}$
5. $a^2 + \frac{2}{a-2}$
6. $3r^2 + \frac{4}{2r+1}$

Simplify.

7. $\frac{3\frac{1}{2}}{4\frac{3}{4}}$
8. $\frac{\frac{x^2}{y}}{\frac{y}{x^3}}$
9. $\frac{\frac{t^4}{u}}{\frac{t^3}{u^2}}$
10. $\frac{\frac{x^3}{y^2}}{\frac{x+y}{x-y}}$
11. $\frac{\frac{y}{3}+\frac{5}{6}}{2+\frac{5}{y}}$
12. $\frac{\frac{1}{x}+\frac{1}{y}}{\frac{1}{y}-\frac{1}{x}}$
13. $\frac{t-2}{\frac{t^2-4}{t^2+5t+6}}$
14. $\frac{\frac{y^2-1}{y^2+3y-4}}{y+1}$

Lesson 12-8 Solve each equation.

1. $\frac{k}{6} + \frac{2k}{3} = -\frac{5}{2}$
2. $\frac{3x}{5} + \frac{3}{2} = \frac{7x}{10}$
3. $\frac{18}{b} = \frac{3}{b} + 3$
4. $\frac{3}{5x} + \frac{7}{2x} = 1$
5. $\frac{2a-3}{6} = \frac{2a}{3} + \frac{1}{2}$
6. $\frac{x+1}{x} + \frac{x+4}{x} = 6$
7. $\frac{2b-3}{7} - \frac{b}{2} = \frac{b+3}{14}$
8. $\frac{2y}{y-4} - \frac{3}{5} = 3$
9. $\frac{2t}{t+3} + \frac{3}{t} = 2$
10. $\frac{5x}{x+1} + \frac{1}{x} = 5$
11. $\frac{r-1}{r+1} - \frac{2r}{r-1} = -1$
12. $\frac{m}{m+1} + \frac{5}{m-1} = 1$
13. $\frac{5}{5-p} - \frac{p^2}{5-p} = -2$
14. $\frac{14}{b-6} = \frac{1}{2} + \frac{6}{b-8}$
15. $\frac{r}{3r+6} - \frac{r}{5r+10} = \frac{2}{5}$
16. $\frac{4x}{2x+3} - \frac{2x}{2x-3} = 1$
17. $\frac{2a-3}{a-3} - 2 = \frac{12}{a+2}$
18. $\frac{z+3}{z-1} + \frac{z+1}{z-3} = 2$

Lesson 13-1 **If c is the measure of the hypotenuse of a right triangle, find each missing measure. Round answers to the nearest hundredth.**

1. $b = 20, c = 29, a = ?$
2. $a = 7, b = 24, c = ?$
3. $a = 2, b = 6, c = ?$
4. $b = 10, c = \sqrt{200}, a = ?$
5. $a = 3, c = 3\sqrt{2}, b = ?$
6. $a = 6, c = 14, b = ?$
7. $a = \sqrt{11}, c = \sqrt{47}, b = ?$
8. $a = \sqrt{13}, b = 6, c = ?$
9. $a = \sqrt{6}, b = 3, c = ?$
10. $b = \sqrt{75}, c = 10, a = ?$
11. $b = 9, c = \sqrt{130}, a = ?$
12. $a = 9, c = 15, b = ?$
13. $b = 5, c = 11, a = ?$
14. $a = \sqrt{33}, b = 4, c = ?$

Determine whether the following side measures would form right triangles.

15. 14, 48, 50
16. 20, 30, 40
17. 21, 72, 75
18. 5, 12, $\sqrt{119}$
19. 15, 39, 36
20. $\sqrt{5}$, 12, 13
21. 10, 12, $\sqrt{22}$
22. 2, 3, 4
23. $\sqrt{7}$, 8, $\sqrt{71}$

Lesson 13-2 **Simplify. Leave in radical form and use absolute value symbols when necessary.**

1. $\sqrt{50}$
2. $\sqrt{20}$
3. $\sqrt{162}$
4. $\sqrt{700}$
5. $\frac{\sqrt{3}}{\sqrt{5}}$
6. $\frac{\sqrt{72}}{\sqrt{6}}$
7. $\sqrt{\frac{8}{7}}$
8. $\sqrt{\frac{7}{32}}$
9. $\sqrt{10} \cdot \sqrt{20}$
10. $\sqrt{7} \cdot \sqrt{3}$
11. $6\sqrt{2} \cdot \sqrt{3}$
12. $5\sqrt{6} \cdot 2\sqrt{3}$
13. $\sqrt{4x^4y^3}$
14. $\sqrt{200m^2y^3}$
15. $\sqrt{12ts^3}$
16. $\sqrt{175a^4b^6}$
17. $\sqrt{\frac{54}{g^2}}$
18. $\sqrt{99x^3y^7}$
19. $\sqrt{\frac{32c^5}{9d^2}}$
20. $\sqrt{\frac{27p^4}{3p^2}}$
21. $\frac{1}{3 + \sqrt{5}}$
22. $\frac{2}{\sqrt{3} - 5}$
23. $\frac{\sqrt{3}}{\sqrt{3} - 5}$
24. $\frac{\sqrt{6}}{7 - 2\sqrt{3}}$
25. $(\sqrt{p} + \sqrt{10})^2$
26. $(2\sqrt{5} + \sqrt{7})(2\sqrt{5} - \sqrt{7})$
27. $(t - 2\sqrt{3})(t - \sqrt{3})$

Lesson 13-3 **Simplify.**

1. $3\sqrt{11} + 6\sqrt{11} - 2\sqrt{11}$
2. $6\sqrt{13} + 7\sqrt{13}$
3. $2\sqrt{12} + 5\sqrt{3}$
4. $9\sqrt{7} - 4\sqrt{2} + 3\sqrt{2} + 5\sqrt{7}$
5. $3\sqrt{5} - 5\sqrt{3}$
6. $4\sqrt{8} - 3\sqrt{5}$
7. $2\sqrt{27} - 4\sqrt{12}$
8. $8\sqrt{32} + 4\sqrt{50}$
9. $\sqrt{45} + 6\sqrt{20}$
10. $2\sqrt{63} - 6\sqrt{28} + 8\sqrt{45}$
11. $14\sqrt{3t} + 8\sqrt{3t}$
12. $7\sqrt{6x} - 12\sqrt{6x}$
13. $5\sqrt{7} - 3\sqrt{28}$
14. $7\sqrt{8} - \sqrt{18}$
15. $7\sqrt{98} + 5\sqrt{32} - 2\sqrt{75}$
16. $4\sqrt{6} + 3\sqrt{2} - 2\sqrt{5}$
17. $-3\sqrt{20} + 2\sqrt{45} - \sqrt{7}$
18. $4\sqrt{75} + 6\sqrt{27}$
19. $10\sqrt{\frac{1}{5}} - \sqrt{45} - 12\sqrt{\frac{5}{9}}$
20. $\sqrt{15} - \sqrt{\frac{3}{5}}$
21. $3\sqrt{\frac{1}{3}} - 9\sqrt{\frac{1}{12}} + \sqrt{243}$

Lesson 13-4 Solve each equation. Check your solution.

1. $\sqrt{5x} = 5$
2. $4\sqrt{7} = \sqrt{-m}$
3. $\sqrt{t} - 5 = 0$
4. $\sqrt{3b} + 2 = 0$
5. $\sqrt{x - 3} = 6$
6. $5 - \sqrt{3x} = 1$
7. $2 + 3\sqrt{y} = 13$
8. $\sqrt{3g} = 6$
9. $\sqrt{a} - 2 = 0$
10. $\sqrt{2j} - 4 = 8$
11. $5 + \sqrt{x} = 9$
12. $\sqrt{5y + 4} = 7$
13. $7 + \sqrt{5c} = 9$
14. $2\sqrt{5t} = 10$
15. $\sqrt{44} = 2\sqrt{p}$
16. $4\sqrt{x - 5} = 15$
17. $4 - \sqrt{x - 3} = 9$
18. $\sqrt{10x^2 - 5} = 3x$
19. $\sqrt{2a^2 - 144} = a$
20. $\sqrt{3y + 1} = y - 3$
21. $\sqrt{2x^2 - 12} = x$
22. $\sqrt{b^2 + 16} + 2b = 5b$
23. $\sqrt{m + 2} + m = 4$
24. $\sqrt{3 - 2c} + 3 = 2c$

Lesson 13-5 Find the distance between each pair of points whose coordinates are given. Express answers in simplest radical form and as decimal approximations rounded to the nearest hundredth.

1. $(4, 2), (-2, 10)$
2. $(-5, 1), (7, 6)$
3. $(4, -2), (1, 2)$
4. $(-2, 4), (4, -2)$
5. $(3, 1), (-2, -1)$
6. $(-2, 4), (7, -8)$
7. $(-5, 0), (-9, 6)$
8. $(5, -1), (5, 13)$
9. $(2, -3), (10, 8)$
10. $(-7, 5), (2, -7)$
11. $(-6, -2), (-5, 4)$
12. $(8, -10), (3, 2)$
13. $(4, -3), (7, -9)$
14. $(6, 3), (9, 7)$
15. $(10, 0), (9, 7)$
16. $(2, -1), (-3, 3)$
17. $(-5, 4), (3, -2)$
18. $(0, -9), (0, 7)$
19. $(-1, 7), (8, 4)$
20. $(-9, 2), (3, -3)$
21. $(3\sqrt{2}, 7), (5\sqrt{2}, 9)$
22. $(6, 3), (10, 0)$
23. $(3, 6), (5, -5)$
24. $(-4, 2), (5, 4)$

Lesson 13-6 Find the value of c that makes each trinomial a perfect square.

1. $a^2 + 6a + c$
2. $x^2 + 10x + c$
3. $t^2 + 12t + c$
4. $y^2 - 9y + c$
5. $p^2 - 14p + c$
6. $b^2 + 5b + c$

Solve each equation by completing the square. Leave irrational roots in simplest radical form.

7. $x^2 - 4x = 5$
8. $t^2 + 12t - 45 = 0$
9. $b^2 + 4b - 12 = 0$
10. $a^2 - 8a - 84 = 0$
11. $c^2 + 6 = -5c$
12. $t^2 - 7t = -10$
13. $p^2 - 8p + 5 = 0$
14. $a^2 + 4a + 2 = 0$
15. $2y^2 + 7y - 4 = 0$
16. $t^2 + 3t = 40$
17. $x^2 + 8x - 9 = 0$
18. $y^2 + 5y - 84 = 0$
19. $x^2 + 2x - 6 = 0$
20. $t^2 + 12t + 32 = 0$
21. $2x - 3x^2 = -8$
22. $2y^2 - y - 9 = 0$
23. $2z^2 - 5z - 4 = 0$
24. $4t^2 - 6t - \frac{1}{2} = 0$

Pages 787–792 are contained in Volume One.

CHAPTER 7 TEST

Solve each inequality. Then check your solution.

1. $-12 \leq d + 7$
2. $7x < 6x - 11$
3. $z - 1 \geq 2z - 3$
4. $5 - 4b > -23$
5. $-\frac{2}{3}r \leq \frac{7}{12}$
6. $8y + 3 < 13y - 9$
7. $8(1 - 2z) \leq 25 + z$
8. $0.3(m + 4) > 0.5(m - 4)$
9. $\frac{2n - 3}{-7} \leq 5$
10. $y + \frac{5}{8} > \frac{11}{24}$

Solve each compound inequality. Then graph the solution set.

11. $x + 1 > -2$ and $3x < 6$
12. $2n + 1 \geq 15$ or $2n + 1 \leq -1$
13. $8 + 3t > 2$ and $-12 > 11t - 1$
14. $|2x - 1| < 5$
15. $|5 - 3b| \geq 1$
16. $|3 - 5y| < 8$

Define a variable, write an inequality, and solve each problem. Then check your solution.

17. Twice a number subtracted from 12 is no less than the number increased by 27.
18. Seven less than twice a number is between 71 and 83.
19. The product of two integers is no less than 30. One of the integers is 6. What is the other integer?
20. The average of four consecutive odd integers is less than 20. What are the greatest integers that satisfy this condition?

Graph each inequality.

21. $y \geq 5x + 1$
22. $x - 2y > 8$
23. $3x - 2y < 6$

Solve.

24. **Business** Two men and three women are each waiting for a job interview. There is only enough time to interview two people before lunch. Two people are chosen at random.
    a. What is the probability that both people are women?
    b. What is the probability that at least one person is a woman?
    c. Which is more likely, one of the people is a woman and the other is a man, or both people are either men or women?

25. **Fire Safety** The city council of McBride is investigating the efficiency of the fire department. The time taken by the fire department to respond to a fire alarm was surveyed. It was found that the response times in minutes for 17 alarms were as follows.

    1, 3, 2, 2, 1, 9, 4, 6, 1, 10, 1, 4, 5, 10, 1, 3, 6

    a. Draw a box-and-whisker plot of the data.
    b. Between what two values of the data is the middle 50% of the data?

CHAPTER 8 TEST

Graph each system of equations. Then determine whether the system has *one* solution, *no* solution, or *infinitely many* solutions. If the system has one solution, name it.

1. $y = x + 2$
   $y = 2x + 7$
2. $x + 2y = 11$
   $x = 14 - 2y$
3. $2x + 5y = 16$
   $5x - 2y = 11$
4. $3x + y = 5$
   $2y - 10 = -6x$
5. $y + 2x = -1$
   $y - 4 = -2x$
6. $2x + y = -4$
   $5x + 3y = -6$

Use substitution or elimination to solve each system of equations.

7. $y = 7 - x$
   $x - y = -3$
8. $x = 2y - 7$
   $y - 3x = -9$
9. $x + y = 8$
   $x - y = 2$
10. $3x - y = 11$
    $x + 2y = -36$
11. $3x + y = 10$
    $3x - 2y = 16$
12. $5x - 3y = 12$
    $-2x + 3y = -3$
13. $2x + 5y = 12$
    $x - 6y = -11$
14. $x + y = 6$
    $3x - 3y = 13$
15. $3x + \frac{1}{3}y = 10$
    $2x - \frac{5}{3}y = 35$
16. $8x - 6y = 14$
    $6x - 9y = 15$
17. $5x - y = 1$
    $y = -3x + 1$
18. $7x + 3y = 13$
    $3x - 2y = -1$

Solve each system of inequalities by graphing.

19. $y \leq 3$
    $y > -x + 2$
20. $x \leq 2y$
    $2x + 3y \leq 7$
21. $x > y + 1$
    $2x + y \geq -4$

Solve.

22. **Number Theory** The units digit of a two-digit number exceeds twice the tens digit by 1. Find the number if the sum of its digits is 10.
23. **Geometry** The difference between the length and width of a rectangle is 7 cm. Find the dimensions of the rectangle if its perimeter is 50 cm.
24. **Finance** Last year, Jodi invested $10,000, part at 6% annual interest and the rest at 8% annual interest. If she received $760 in interest at the end of the year, how much did she invest at each rate?
25. **Organize Data** Joey sold 30 peaches from his fruit stand for a total of $7.50. He sold small ones for 20 cents each and large ones for 35 cents each. How many of each kind did he sell?

CHAPTER 9 TEST

Simplify. Assume that no denominator is equal to zero.

1. $(a^2b^4)(a^3b^5)$

2. $(-12abc)(4a^2b^4)$

3. $\left(\frac{3}{5}m\right)^2$

4. $(-3a)^4(a^5b)^2$

5. $(-5a^2)(-6b^3)^2$

6. $(5a)^2b + 7a^2b$

7. $\frac{y^{11}}{y^6}$

8. $\frac{mn^4}{m^3n^2}$

9. $\frac{9a^2bc^2}{63a^4bc}$

10. $\frac{48a^2bc^5}{(3ab^3c^2)^2}$

11. $\frac{14ab^{-3}}{21a^2b^{-5}}$

12. $\frac{(10a^2bc^4)^{-2}}{(5^{-1}a^{-1}b^{-5})^2}$

Express each number in scientific notation.

13. 46,300

14. 0.003892

15. 284×10^3

16. 0.0031×10^4

Evaluate. Express each result in scientific notation.

17. $(3 \times 10^3)(2 \times 10^4)$

18. $\frac{2.5 \times 10^3}{5 \times 10^{-3}}$

19. $\frac{14.72 \times 10^{-4}}{3.2 \times 10^{-3}}$

20. $(15 \times 10^{-7})(3.1 \times 10^4)$

21. Find the degree of $5ya^3 - 7 - y^2a^2 + 2y^3a$ and arrange the terms so that the powers of y are in descending order.

Find each sum or difference.

22.
$$\begin{array}{r} 5ax^2 + 3a^2x - 7a^3 \\ \underline{(+)\ 2ax^2 - 8a^2x + 4} \end{array}$$

23.
$$\begin{array}{r} x^3 - 3x^2y + 4xy^2 + y^3 \\ \underline{(-)\ 7x^3 + x^2y - 9xy^2 + y^3} \end{array}$$

24. $(n^2 - 5n + 4) - (5n^2 + 3n - 1)$

25. $(ab^3 - 4a^2b^2 + ab - 7) + (-2ab^3 + 4ab^2 + 3ab + 2)$

Simplify.

26. $(h - 5)^2$

27. $(2x - 5)(7x + 3)$

28. $(4x - y)(4x + y)$

29. $(2a^2b + b^2)^2$

30. $3x^2y^3(2x - xy^2)$

31. $(4m + 3n)(2m - 5n)$

32. $x^2(x - 8) - 3x(x^2 - 7x + 3) + 5(x^3 - 6x^2)$

33. $(x - 6)(x^2 - 4x + 5)$

CHAPTER 10 TEST

Find the GCF of the given monomials.

1. 48, 64
2. $18a^2b$, $28a^3b^2$
3. $6x^2y^3$, $12x^2y^2z$, $15x^2y$

Factor each polynomial, if possible. If the polynomial cannot be factored using integers, write *prime*.

4. $25y^2 - 49w^2$
5. $t^2 - 16t + 64$
6. $x^2 + 14x + 24$
7. $28m^2 + 18m$
8. $a^2 - 11ab + 18b^2$
9. $12x^2 + 23x - 24$
10. $2h^2 - 3h - 18$
11. $6x^3 + 15x^2 - 9x$
12. $4my - 20m + 3py - 15p$
13. $x^3 - 4x^2 - 9x + 36$
14. $36a^2b^3 - 45ab^4$
15. $36m^2 + 60mn + 25n^2$
16. $\frac{1}{4}a^2 - \frac{4}{9}$
17. $64p^2 - 63p + 16$
18. $15a^2b + 5a^2 - 10a$
19. $6y^2 - 5y - 6$
20. $4s^2 - 100t^2$
21. $2d^2 + d - 1$
22. $3g^2 + g + 1$
23. $2xz + 2yz - x - y$

Solve each equation. Check your solutions.

24. $(4x - 3)(3x + 2) = 0$
25. $18s^2 + 72s = 0$
26. $4x^2 = 36$
27. $t^2 + 25 = 10t$
28. $a^2 - 9a - 52 = 0$
29. $x^3 - 5x^2 - 66x = 0$
30. $2x^2 = 9x + 5$
31. $3b^2 + 6 = 11b$

Solve.

32. **Geometry** A rectangle is 4 inches wide by 7 inches long. When the length and width are increased by the same amount, the area is increased by 26 square inches. What are the dimensions of the new rectangle?

33. **Construction** A rectangular lawn is 24 feet wide by 32 feet long. A sidewalk will be built along the inside edges of all four sides. The remaining lawn will have an area of 425 square feet. How wide will the walk be?

Write the equation of the axis of symmetry and find the coordinates of the vertex of the graph of each equation. State if the vertex is a maximum or minimum. Then graph the equation.

1. $y = x^2 - 4x + 13$
2. $y = -3x^2 - 6x + 4$
3. $y = 2x^2 + 3$
4. $y = -1(x - 2)^2 + 1$

Solve each equation by graphing. If exact roots cannot be found, state the consecutive integers between which the roots lie.

5. $x^2 - 2x + 2 = 0$
6. $x^2 + 6x = -7$
7. $x^2 + 24x + 144 = 0$
8. $2x^2 - 8x = 42$

Solve each equation.

9. $x^2 + 7x + 6 = 0$
10. $2x^2 - 5x - 12 = 0$
11. $6n^2 + 7n = 20$
12. $3k^2 + 2k = 5$
13. $y^2 - \frac{3y}{5} + \frac{2}{25} = 0$
14. $-3x^2 + 5 = 14x$
15. $4^{x-2} = 16^{2x+5}$
16. $1000^x = 10{,}000^{6x+4}$
17. $5^{x^2} = 5^{15-2x}$
18. $\left(\frac{1}{2}\right)^{x-2} = 4^{5x}$

Graph each function. State the *y*-intercept.

19. $y = \left(\frac{1}{2}\right)^x$
20. $y = 4 \cdot 2^x$
21. $y = \left(\frac{1}{3}\right)^x - 3$

Solve.

22. **Automobile** Adina Ley needs to replace her car. If she leases a car, she will pay $410 a month for 2 years and then has the option to buy the car for $14, 458. The price of the car now is $17,369. If the car depreciates at 16% per year, how will the depreciated price compare with the buyout price of the lease?

23. **Geometry** The area of a certain square is one-half the area of the rectangle formed if the length of one side of the square is increased by 2 cm and the length of an adjacent side is increased by 3 cm. What are the dimensions of the square?

24. **Number Theory** Find two integers whose sum is 21 and whose product is 90.

25. **Investment** After 6 years, a certain investment is worth $8479. If the money was invested at 9% interest compounded semiannually, find the original amount that was invested.

CHAPTER 12 TEST

CHAPTER TEST

Simplify each rational expression. State the excluded values of the variables.

1. $\frac{5 - 2m}{6m - 15}$
2. $\frac{3 + x}{2x^2 + 5x - 3}$
3. $\frac{4c^2 + 12c + 9}{2c^2 - 11c - 21}$

Simplify each expression.

4. $\frac{1 - \frac{9}{t}}{1 - \frac{81}{t^2}}$
5. $\frac{\frac{5}{6} + \frac{u}{t}}{\frac{2u}{t} - 3}$
6. $\frac{x + 4 + \frac{5}{x - 2}}{x + 6 + \frac{15}{x - 2}}$

Perform the indicated operations.

7. $\frac{2x}{x - 7} - \frac{14}{x - 7}$
8. $\frac{n + 3}{2n - 8} \cdot \frac{6n - 24}{2n + 1}$
9. $(10m^2 + 9m - 36) \div (2m - 3)$
10. $\frac{x^2 + 4x - 32}{x + 5} \cdot \frac{x - 3}{x^2 - 7x + 12}$
11. $\frac{z^2 + 2z - 15}{z^2 + 9z + 20} \div (z - 3)$
12. $\frac{4x^2 + 11x + 6}{x^2 - x - 6} \div \frac{x^2 + 8x + 16}{x^2 + x - 12}$
13. $(10z^4 + 5z^3 - z^2) \div 5z^3$
14. $\frac{y}{7y + 14} + \frac{6}{3y + 6}$
15. $\frac{x + 5}{x + 2} + 6$
16. $\frac{x^2 - 1}{x + 1} - \frac{x^2 + 1}{x - 1}$
17. $\frac{-3}{a - 5} + \frac{15}{a^2 - 5a}$
18. $\frac{8}{m^2} \cdot \left(\frac{m^2}{2c}\right)^2$

Solve each equation.

19. $\frac{2}{3t} + \frac{1}{2} = \frac{3}{4t}$
20. $\frac{2e}{e - 4} - 2 = \frac{4}{e + 5}$
21. $\frac{4}{h - 4} = \frac{3h}{h + 3}$

Solve each formula for the variable indicated.

22. $F = G\left(\frac{Mm}{d^2}\right)$, for G
23. $\frac{1}{R_T} = \frac{1}{R_1} + \frac{1}{R_2}$, for R_2

Solve.

24. **Keyboarding** Willie can type a 200 word essay in 6 hours. Myra can type the same essay in $4\frac{1}{2}$ hours. If they work together, how long will it take them to type the essay?

25. **Electronics** Three appliances are connected in parallel: a lamp of resistance 120 ohms, a toaster of resistance 20 ohms, and an iron of resistance 12 ohms. Find the total resistance.

CHAPTER 13 TEST

Simplify. Leave in radical form and use absolute value symbols when necessary.

1. $\sqrt{480}$
2. $\sqrt{72} \cdot \sqrt{48}$
3. $\sqrt{54x^4y}$
4. $\sqrt{\frac{32}{25}}$
5. $\sqrt{\frac{3x^2}{4n^3}}$
6. $\sqrt{6} + \sqrt{\frac{2}{3}}$
7. $\left(x + \sqrt{3}\right)^2$
8. $\frac{7}{7 + \sqrt{5}}$
9. $3\sqrt{50} - 2\sqrt{8}$
10. $\sqrt{\frac{10}{3}} \cdot \sqrt{\frac{4}{30}}$
11. $2\sqrt{27} + \sqrt{63} - 4\sqrt{3}$
12. $\left(1 - \sqrt{3}\right)\left(3 + \sqrt{2}\right)$

Find the distance between each pair of points whose coordinates are given. Express answers in simplest radical form.

13. $(4, 7), (4, -2)$
14. $(-9, 2), \left(\frac{2}{3}, \frac{1}{2}\right)$
15. $(-1, 1), (1, -5)$

Find the length of each missing side. Round to the nearest hundredth.

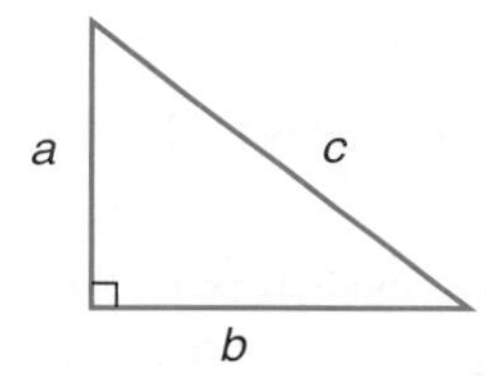

16. $a = 8, b = 10, c = ?$
17. $a = 12, c = 20, b = ?$
18. $a = 6\sqrt{2}, c = 12, b = ?$
19. $b = 13, c = 17, a = ?$

Solve each equation. Check your solution.

20. $\sqrt{4x + 1} = 5$
21. $\sqrt{4x - 3} = 6 - x$
22. $y^2 - 5 = -8y$
23. $2x^2 - 10x - 3 = 0$

Solve.

24. **Geometry** Find the measures of the perimeter and area, in simplest form, for the rectangle shown at the right.

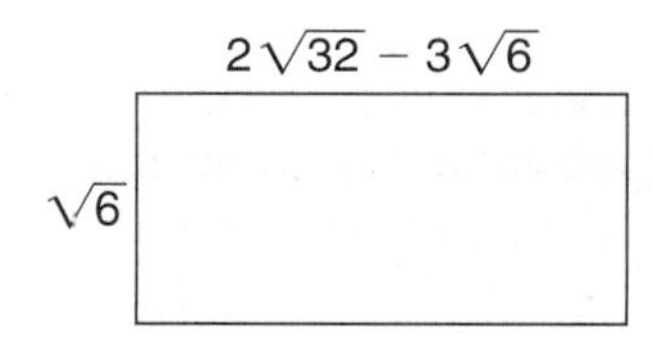

25. **Sports** A hiker leaves her camp in the morning. How far is she from camp after walking 9 miles due west and then 12 miles due north?

GLOSSARY

This glossary contains terms from Volumes One and Two. Pages 1–447 are contained in Volume One. Pages 382–753 are contained in Volume Two.

A

absolute value (85) The absolute value of a number is its distance from zero on a number line.

acute triangle (165) In an acute triangle, all of the angles measure less than 90 degrees.

adding integers (86)

1. To add integers with the *same* sign, add their absolute values. Give the result the same sign as the integers.
2. To add integers with *different* signs, subtract the lesser absolute value from the greater absolute value. Give the result the same sign as the integer with the greater absolute value.

addition property for inequality (385) For all numbers a, b, and c, the following are true:
1. if $a > b$, then $a + c > b + c$.
2. if $a < b$, then $a + c < b + c$.

addition property of equality (144) For any numbers a, b, and c, if $a = b$, then $a + c = b + c$.

additive identity (37) For any number a, $a + 0 = 0 + a = a$.

additive inverse property (87) For any number a, $a + (-a) = 0$.

algebraic expression (6) An expression consisting of one or more numbers and variables along with one or more arithmetic operations.

angle of depression (208) An angle of depression is formed by a horizontal line and a line of sight below it.

angle of elevation (208) An angle of elevation is formed by a horizontal line and a line of sight above it.

associative property (51) For any numbers a, b, and c, $(a + b) + c = a + (b + c)$ and $(ab)c = a(bc)$.

axes (254) Two perpendicular number lines that are used to locate points on a coordinate plane.

axis of symmetry (612) The equation of the axis of symmetry for the graph of $y = ax^2 + bx + c$, where $a \neq 0$, is $x = -\frac{b}{2a}$.

B

back-to-back stem-and-leaf plot (27) A back-to-back stem-and-leaf plot is used to compare two sets of data. The same stem is used for the leaves of both plots.

base **1.** (7) In an expression of the form x^n, the base is x. **2.** (215) The number that is divided into the percentage in the percent proportion.

best-fit line (341) A line drawn on a scatter plot that passes close to most of the data points.

binomial (514) The sum of two monomials.

boundary (437) A boundary of an inequality is a line that separates the coordinate plane into half-planes.

box-and-whisker plot (427) A type of diagram or graph that shows the quartiles and extreme values of data.

C

coefficient (47) The numerical factor in a term.

commutative property (51) For any numbers a and b, $a + b = b + a$ and $ab = ba$.

comparison property (94) For any two numbers a and b, exactly one of the following sentences is true.

$$a < b \qquad a = b \qquad a > b$$

complementary angles (163) Two angles are complementary if the sum of their measures is 90 degrees.

complete graph (278) A complete graph shows the origin, the points at which the graph crosses the x- and y-axes, and other important characteristics of the graph.

completeness property (121) Each real number corresponds to exactly one point on the number line. Each point on the number line corresponds to exactly one real number.

completeness property for points in the plane (256)

1. Exactly one point in the plane is named by a given ordered pair of numbers.

2. Exactly one ordered pair of numbers names a given point in the plane.

completing the square (742, 743) To add a constant term to a binomial of the form $x^2 + bx$ so that the resulting trinomial is a perfect square.

complex fraction (114) If a fraction has one or more fractions in the numerator or denominator, it is called a complex fraction.

composite numbers (558) A whole number, greater than 1, that is not prime.

compound event (414) A compound event consists of two or more simple events.

compound inequality (405) Two inequalities connected by *and* or *or*.

congruent angles (164) Angles that have the same measure.

conjugates (722) Two binomials of the form $a\sqrt{b} + c\sqrt{d}$ and $a\sqrt{b} - c\sqrt{d}$.

consecutive integers (158) Consecutive integers are integers in counting order.

consistent (455) A system of equations is said to be consistent when it has at least one ordered pair that satisfies both equations.

constant of variation (239) The number k in equations of the form $y = kx$ and $xy = k$.

constants (496) Monomials that are real numbers.

coordinate (73) The number that corresponds to a point on a number line.

coordinate plane (254) The plane containing the x- and y-axes.

corresponding angles (201) Matching angles in similar triangles, which have equal measures.

corresponding sides (201) The sides opposite the corresponding angles in similar triangles.

cosine (206) In a right triangle with acute angle A, the cosine of angle A = $\frac{\text{measure of leg adjacent to angle } A}{\text{measure of hypotenuse}}$.

cross products (94) When two fractions are compared, the cross products are the products of the terms on the diagonals.

D

data (25) Numerical information.

defining the variable (127) Choosing a variable to represent one of the unspecified numbers in a problem.

degree **1.** (515) The degree of a monomial is the sum of the exponents of its variables.
2. (516) The degree of a polynomial is the degree of the term of the greatest degree.

density property (96) Between every pair of distinct rational numbers, there are infinitely many rational numbers.

dependent (456) A system of equations that has an infinite number of solutions.

dependent variable (58) The variable in a function whose value is determined by the independent variable.

difference of squares (545) Two perfect squares separated by a subtraction sign, $a^2 - b^2 = (a + b)(a - b)$.

dimensional analysis (174) The process of carrying units throughout a computation.

direct variation (239) A direct variation is described by an equation of the form $y = kx$, where $k \neq 0$.

discrete mathematics (88) A branch of mathematics that deals with finite or discontinuous quantities.

discriminant (632) In the quadratic formula, the expression $b^2 - 4ac$.

distance formula (737) The distance d between any two points with coordinates (x_1, y_1) and (x_2, y_2) is given by the following formula.

$$d = \sqrt{(x_2 - x_1)^2 + (y_2 - y_1)^2}$$

distributive property (46) For any numbers a, b, and c:

1. $a(b + c) = ab + ac$ and $(b + c)a = ba + ca$.

2. $a(b - c) = ab - ac$ and $(b - c)a = ba - ca$.

dividing rational numbers (112) The quotient of two rational numbers having the same sign is positive. The quotient of two rational numbers having different signs is negative.

division property for inequality (393) For all numbers a, b, and c, the following are true:

1. If c is positive and $a < b$, then $\frac{a}{c} < \frac{b}{c}$, and if c is positive and $a > b$, then $\frac{a}{c} > \frac{b}{c}$.

2. If c is negative and $a < b$, then $\frac{a}{c} > \frac{b}{c}$, and if c is negative and $a > b$, then $\frac{a}{c} < \frac{b}{c}$.

division property of equality (151) For any numbers a, b, and c, with $c \neq 0$, if $a = b$, then $\frac{a}{c} = \frac{b}{c}$.

domain (263) The set of all first coordinates from the ordered pairs in a relation.

draw a diagram (406) A problem-solving strategy that is often used as an organizational tool.

element (33) A member of a set.

elimination (469) The elimination method of solving a system of equations is a method that uses addition or subtraction to eliminate one of the variables to solve for the other variable.

equally likely (229) Outcomes that have an equal chance of occurring.

equation (33) A mathematical sentence that contains an equals sign, =.

equation in two variables (271) An equation in two variables contains two unknown values.

equilateral triangle (164) A triangle in which all the sides have the same length and all the angles have the same measure.

equivalent equation (144) Equations that have the same solution.

equivalent expressions (47) Expressions that denote the same number.

evaluate (8) To find the value of an expression when the values of the variables are known.

excluded value (660) A value is excluded from the domain of a variable because if that value were substituted for the variable, the result would have a denominator of zero.

exponent (7) In an expression of the form x^n, the exponent is n.

exponential function (634, 635) A function that can be described by an equation of the form $y = a^x$, where $a > 0$ and $a \neq 1$.

extraneous solutions (697) Solutions derived from an equation that are not solutions of the original equation.

extreme values (427) The least value and the greatest value in a set of data.

extremes (196) *See* proportion.

F

factored form (559) A monomial is written in factored form when it is expressed as the product of prime numbers and variables where no variable has an exponent greater than 1.

factoring (564) To express a polynomial as the product of monomials and polynomials.

factoring by grouping (567) A method of factoring polynomials with four or more terms.

factors (6) In a multiplication expression, the quantities being multiplied are called factors.

family of graphs (354) A family of graphs includes graphs and equations of graphs that have at least one characteristic in common.

FOIL method (537) To multiply two binomials, find the sum of the products of

F the first terms,
O the outside terms,
I the inside terms, and
L the last terms.

formula (128) An equation that states a rule for the relationship between certain quantities.

function **1.** (56) A relationship between input and output in which the output depends on the input. **2.** (287) A relation in which each element of the domain is paired with exactly one element of the range.

functional notation (289) In functional notation, the equation $y = x + 5$ is written as $f(x) = x + 5$.

G

general equation for exponential decay (644) The general equation for exponential decay is represented by the formula $A = C(1 - r)^t$.

general equation for exponential growth (643) The general equation for exponential growth is represented by the formula $A = C(1 + r)^t$.

graph (73, 255) To draw, or plot, the points named by certain numbers or ordered pairs on a number line or coordinate plane, respectively.

greatest common factor (GCF) (559) The greatest common factor of two or more integers is the greatest number that is a factor of all the integers.

guess and check (574) A problem-solving strategy in which several values or combinations of values are tried in order to find a solution to a problem.

H

half-life (638) The half-life of an element is defined as the time it takes for one-half a quantity of a radioactive element to decay.

half-plane (437) The region of a graph on one side of a boundary is called a half-plane.

horizontal axis (56) The horizontal line in a graph that represents the independent variable.

hypotenuse (206) The side of a right triangle opposite the right angle.

I

identify subgoals (745) A problem-solving strategy that uses a series of small steps, or subgoals.

identity (170) An equation that is true for every value of the variable.

inconsistent (456) A system of equations is said to be inconsistent when it has no ordered pair that satisfies both equations.

independent (456) A system of equations is said to be independent if the system has exactly one solution.

independent variable (58) The variable in a function whose value is subject to choice is the independent variable. The independent variable affects the value of the dependent variable.

inequality (33) A mathematical sentence having the symbols $<$, $\leq$, $>$, or $\geq$.

integers (73) The set of numbers represented as $\{..., -3, -2, -1, 0, 1, 2, 3, ...\}$.

interquartile range (304, 306) The difference between the upper quartile and the lower quartile of a set of data. It represents the middle half, or 50%, of the data in the set.

intersection (406) The intersection of two sets A and B is the set of elements common to both A and B.

inverse of a relation (264) The inverse of any relation is obtained by switching the coordinates in each ordered pair.

inverse variation (241) An inverse variation is described by an equation of the form $xy = k$, where $k \neq 0$.

irrational numbers (120) A number that cannot be expressed in the form $\frac{a}{b}$, where a and b are integers and $b \neq 0$.

isosceles triangle (164) In an isosceles triangle, at least two angles have the same measure and at least two sides have the same length.

L

least common denominator (LCD) (686) The least common denominator is the least common multiple of the denominators of two or more fractions.

least common multiple (LCM) (685) The least common multiple of two or more integers is the least positive integer that is divisible by each of the integers.

legs (206) The sides of a right triangle that are not the hypotenuse.

like terms (47) Terms that contain the same variables, with corresponding variables with the same power.

linear equation (280) An equation whose graph is a line.

linear function (278) An equation whose graph is a nonvertical line.

line plot (78) Numerical data displayed on a number line.

look for a pattern (13, 497, 637) A problem-solving strategy often involving the use of tables to organize information so that a pattern may be determined.

lower quartile (306) The lower quartile divides the lower half of a set of data into two equal parts.

M

make an organized list (685) A problem-solving strategy that uses an organized list to arrange and evaluate data in order to determine a solution.

mapping (263) A mapping pairs one element in the domain with one element in the range.

matrix (88) A matrix is a rectangular arrangement of elements in rows and columns.

maximum (611) The highest point on the graph of a curve, such as a the vertex of parabola that opens downward.

mean (178) The mean of a set of data is the sum of the numbers in the set divided by the number of numbers in the set.

means (196) *See* proportion.

measures of central tendency (178) Numbers known as measures of central tendency are often used to describe sets of data because they represent a centralized, or middle, value.

measures of variation (306) Measures of variation are used to describe the distribution of data.

median (178) The median is the middle number of a set of data when the numbers are arranged in numerical order.

midpoint (369) A point that is halfway between the endpoints of a segment.

minimum (611) The lowest point on the graph of a curve, such as a the vertex of parabola that opens upward.

mixed expression (690) An algebraic expression that contains a monomial and a rational expression.

mode (178) The mode of a set of data is the number that occurs most often in the set.

monomial (496) A monomial is a number, a variable, or a product of a number and one or more variables.

multi-step equations (157) Multi-step equations are equations that need more than one operation to solve them.

multiplicative identity (38) For any number a, $a \cdot 1 = 1 \cdot a = a$.

multiplicative inverse (38) For every nonzero number $\frac{a}{b}$, where $a, b \neq 0$, there is exactly one number $\frac{b}{a}$ such that $a \cdot b = 1$.

multiplication property for inequality (393) For all numbers a, b, and c, the following are true.

1. If c is positive and $a < b$, then $ac < bc$, $c \neq 0$, and if c is positive and $a > b$, then $ac > bc$, $c \neq 0$.
2. If c is negative and $a < b$, then $ac > bc$, $c \neq 0$, and if c is negative and $a > b$, then $ac < bc$, $c \neq 0$.

multiplicative property of −1 (107) The product of any number and −1 is its additive inverse.

$$-1(a) = -a \text{ and } a(-1) = -a$$

multiplicative property of equality (150) For any numbers a, b, and c, if $a = b$, then $a \cdot c = b \cdot c$.

multiplicative property of zero (38) For any number a, $a \cdot 0 = 0 \cdot a = 0$.

N

negative correlation (340) There is a negative correlation between x and y if the values are related in opposite ways.

negative exponent (503) For any nonzero number a and any integer n, $a^{-n} = \frac{1}{a^n}$.

negative number (73) Any number that is less than zero.

number line (72) A line with equal distances marked off to represent numbers.

number theory (158) The study of numbers and the relationships between them.

O

obtuse triangle (165) An obtuse triangle has one angle with measure greater than 90 degrees.

odds (229) The odds of an event occurring is the ratio of the number of ways the event can occur (successes) to the number of ways the event cannot occur (failures).

open sentences (32) Mathematical statements with one or more variables, or unknown numbers.

opposites (87) The opposite of a number is its additive inverse.

order of operations (19)

1. Simplify the expressions inside grouping symbols, such as parentheses, brackets, and braces, and as indicated by fraction bars.
2. Evaluate all powers.
3. Do all multiplications and divisions from left to right.
4. Do all additions and subtractions from left to right.

ordered pair (57) Pairs of numbers used to locate points in the coordinate plane.

organize data (464) Organizing data is useful before solving a problem. Some ways to organize data are to use tables, charts, different types of graphs, or diagrams.

origin (254) The point of intersection of the two axes in the coordinate plane.

outcomes (413) Outcomes are all possible combinations of a counting problem.

outlier (307) In a set of data, a value that is much greater or much less than the rest of the data can be called a outlier.

parabola (610, 611) The general shape of the graph of a quadratic function.

parallel lines (362) Lines in the plane that never intersect. Nonvertical parallel lines have the same slope.

parallelogram (362) A quadrilateral in which opposite sides are parallel.

parent graph (354) The simplest of the graphs in a family of graphs

percent (215) A percent is a ratio that compares a number to 100.

percentage (215) The number that is divided by the base in a percent proportion.

percent of decrease (222) The ratio of an amount of decrease to the previous amount, expressed as a percent.

percent of increase (222) The ratio of an amount of increase to the previous amount, expressed as a percent.

percent proportion (215) $\frac{\text{Percentage}}{\text{Base}} = \frac{r}{100}$

perfect square (119) A rational number whose square root is a rational number.

perfect square trinomial (587) A trinomial which, when factored, has the form $(a + b)^2 = (a + b)(a + b)$ or $(a - b)^2 = (a - b)(a - b)$.

perpendicular lines (362) Lines that meet to form right angles.

point-slope form (333) For any point (x_1, y_1) on a nonvertical line having slope m, the point-slope form of a linear equation is as follows:

$$y - y_1 = m(x - x_1).$$

polynomial (513, 514) A polynomial is a monomial or a sum of monomials.

positive correlation (340) There is a positive correlation between x and y if the values are related in the same way.

power (7) An expression of the form x^n is known as a power.

power of a monomial (498) For any numbers a and b, and any integers m, n, and p,

$$(a^m b^n)^p = a^{mp} b^{np}.$$

power of a power (498) For any number a, and all integers m and n, $(a^m)^n = a^{mn}$.

power of a product (498) For all numbers a and b, and any integer m, $(ab)^m = a^m b^m$.

prime factorization (558) A whole number expressed as a product of factors that are all prime numbers.

prime number (558) A prime number is a whole number, greater than 1, whose only factors are 1 and itself.

prime polynomial (577) A polynomial that cannot be written as a product of two polynomials with integral coefficients is called a prime polynomial.

principal square root (119) The nonnegative square root of an expression.

probability (228) The ratio that tells how likely it is that an event will take place.

$$P(\text{event}) = \frac{\text{number of favorable outcomes}}{\text{total number of possible outcomes}}$$

problem-solving plan (126)

1. Explore the problem.
2. Plan the solution.
3. Solve the problem.
4. Examine the solution.

product (6) The result of multiplication.

product of powers (497) For any number a, and all integers m and n, $a^m \cdot a^n = a^{m+n}$.

product property of square roots (719) For any number a and b, where $a \geq 0$ and $b \geq 0$, $\sqrt{ab} = \sqrt{a} \cdot \sqrt{b}$.

proportion (195) In a proportion, the product of the extremes is equal to the product of the means. If $\frac{a}{b} = \frac{c}{d}$, then $ad = bc$.

Pythagorean theorem (713) If a and b are the measures of the legs of a right triangle and c is the measure of the hypotenuse, then $c^2 = a^2 + b^2$.

Pythagorean triple (583) Three whole numbers a, b, and c such that $a^2 + b^2 = c^2$.

quadrant (254) One of the four regions into which the x- and y-axes separate the coordinate plane.

quadratic equation (620) A quadratic equation is one in which the value of the related quadratic function is 0.

quadratic formula (628) The roots of a quadratic equation in the form $ax^2 + bx + c = 0$, where $a \neq 0$, are given by the formula $x = \frac{-b \pm \sqrt{b^2 - 4ac}}{2a}$.

quadratic function (610, 611) A quadratic function is a function that can be described by an equation of the form $y = ax^2 + bx + c$, where $a \neq 0$.

quartiles (304, 306) In a set of data, the quartiles are values that divide the data into four equal parts.

quotient of powers (501) For all integers m and n and any nonzero number a, $\frac{a^m}{a^n} = a^{m-n}$.

quotient property of square roots (721) For any numbers a and b, where $a > 0$ and $b > 0$, $\sqrt{\frac{a}{b}} = \frac{\sqrt{a}}{\sqrt{b}}$.

radical equations (732) Equations that contain radicals with variables in the radicand.

radical sign (119) The symbol $\sqrt{\ }$, indicating the principal or nonnegative root of an expression.

radicand (719) The radicand is the expression under the radical sign.

random (229) When an outcome is chosen without any preference, the outcome occurs at random.

range **1.** (263) The set of all second coordinates from the ordered pairs in the relation. **2.** (306) The difference between the greatest and the least values of a set of data.

rate **1.** (197) The ratio of two measurements having different units of measure. **2.** (215) In the percent proportion, the rate is the fraction with a denominator of 100.

ratio (195) A ratio is a comparison of two numbers by division.

rational equation (696) A rational equation is an equation that contains rational expressions.

rational expression (660) A rational expression is an algebraic fraction whose numerator and denominator are polynomials.

rational numbers (93) A rational number is a number that can be expressed in the form $\frac{a}{b}$, where a and b are integers and $b \neq 0$.

rationalizing the denominator (722) Rationalizing the denominator of a radical expression is a method used to remove or eliminate the radicals from the denominator of a fraction.

real numbers (121) The set of rational numbers and the set of irrational numbers together form the set of real numbers.

reciprocal (38) The multiplicative inverse of a number.

reflexive property of equality (39) For any number a, $a = a$.

regression line (342) The most accurate best-fit line for a set of data, and can be determined with a graphing calculator or computer.

relation (260, 263) A relation is a set of ordered pairs.

replacement set (33) A set of numbers from which replacements for a variable may be chosen.

right triangle (165) A right triangle has one angle with a measure of 90 degrees.

rise (325) The vertical change in a line.

roots (620) The solutions of a quadratic equation.

run (325) The horizontal change in a line.

scalar multiplication (108) In scalar multiplication, each element of a matrix is multiplied by a constant.

scale (197) A ratio called a scale is used when making a model to represent something that is too large or too small to be conveniently drawn at actual size.

scatter plot (339) In a scatter plot, the two sets of data are plotted as ordered pairs in the coordinate plane.

scientific notation (506) A number is expressed in scientific notation when it is in the form $a \times 10^n$, where $1 \leq a < 10$ and n is an integer.

set (33) A collection of objects or numbers.

set-builder notation (385) A notation used to describe the members of a set. For example, $\{y \mid y < 17\}$ represents the set of all numbers y such that y is less than 17.

sequence (13) A set of numbers in a specific order.

similar triangles (201) If two triangles are similar, the measures of their corresponding sides are proportional, and the measures of their corresponding angles are equal.

simple events (414) A single event in a probability problem.

simple interest (217) The amount paid or earned for the use of money. The formula $I = prt$ is used to solve simple interest problems.

simplest form (47) An expression is in simplest form when it is replaced by an equivalent expression having no like terms and no parentheses.

simplest radical form (723) A radical expression is in simplest radical form when the following three conditions have been met.

1. No radicands have perfect square factors other than one.
2. No radicands contain fractions.
3. No radicals appear in the denominator of a fraction.

sine (206) In a right triangle with acute angle A, sine of angle $A = \frac{\text{measure of leg opposite angle } A}{\text{measure of the hypotenuse}}$.

slope (324, 325) The ratio of the rise to the run as you move from one point to another along a line.

slope-intercept form (347) An equation of the form $y = mx + b$, where m is the slope and b is the y-intercept of a given line.

solution (32) A replacement for the variable in an open sentence that results in a true sentence.

solution of an equation in two variables (271) If a true statement results when the numbers in an ordered pair are substituted into an equation in two variables, then the ordered pair is a solution of the equation.

solution set (33) The set of all replacements for the variable in an open sentence that result in a true sentence.

solve an equation (145) To solve an equation means to isolate the variable having a coefficient of 1 on one side of the equation.

solving an open sentence (32) Finding a replacement for the variable that results in a true sentence.

solving a triangle (208) Finding the measures of all sides and angles of a right triangle.

square of a difference (544) If a and b are any numbers, $(a - b)^2 = (a - b)(a - b) = a^2 - 2ab + b^2$.

square of a sum (543) If a and b are any numbers, $(a + b)^2 = (a + b)(a + b) = a^2 + 2ab + b^2$.

square root (118, 119) One of two identical factors of a number.

standard form (333) The standard form of a linear equation is $Ax + By = C$, where A, B, and C are integers, $A \geq 0$, and A and B are not both zero.

statistics (25) A branch of mathematics concerned with methods of collecting, organizing, and interpreting data.

stem-and-leaf plot (26) In a stem-and-leaf plot, each piece of data is separated into two numbers that are used to form a stem and a leaf. The data are organized into two columns. The column on the left contains the stem and the column on the right contains the leaves.

substitution (462) The substitution method of solving a system of equations is a method that uses substitution of one equation into the other equation to solve for the other variable.

substitution property of equality (39) If $a = b$, then a may be replaced by b in any expression.

subtracting integers (87) To subtract a number, add its additive inverse. For any numbers a and b, $a - b = a + (-b)$.

subtraction property for inequality (385) For all numbers a, b, and c, the following are true:

1. if $a > b$, then $a - c > b - c$.
2. if $a < b$, then $a - c < b - c$.

subtraction property of equality (146) For any numbers a, b, and c, if $a = b$, then $a - c = b - c$.

supplementary angles (162) Two angles are supplementary if the sum of their measures is 180 degrees.

symmetric property of equality (39) For any numbers a and b, if $a = b$, then $b = a$.

symmetry (612) Symmetrical figures are those in which the figure can be folded and each half matches the other exactly.

system of equations (455) A set of equations with the same variables.

system of inequalities (482) A set of inequalities with the same variables.

T

tangent (206) In a right triangle, the tangent of angle $A = \frac{\text{measure of leg opposite angle } A}{\text{measure of leg adjacent to angle } A}$.

term **1.** (13) A number in a sequence. **2.** (47) A number, a variable, or a product or quotient of numbers and variables.

transitive property of equality (39) For any numbers a, b, and c, if $a = b$ and $b = c$, then $a = c$.

tree diagram (413) A tree diagram is a diagram used to show the total number of possible outcomes.

triangle (163) A triangle is a polygon with three sides and three angles.

trigonometric ratios (206)
$\sin A = \frac{a}{c}$ $\cos A = \frac{b}{c}$ $\tan A = \frac{a}{b}$

trinomials (514) A trinomial is the sum of three monomials.

U

uniform motion (235) When an object moves at a constant speed, or rate, it is said to be in uniform motion.

union (408) The union of two sets A and B is the set of elements contained in both A or B.

unique factorization theorem (558) The prime factorization of every number is unique except for the order in which the factors are written.

unit cost (95) The cost of one unit of something.

upper quartile (306) The upper quartile divides the upper half of a set of data into two equal parts.

use a model (339) A problem-solving strategy that uses models, or simulations, of mathematical situations that are difficult to solve directly.

use a table (255) A problem-solving strategy that uses tables to organize and solve problems.

V

variable (6) Variables are symbols that are used to represent unspecified numbers.

Venn diagrams (73) Venn diagrams are diagrams that use circles or ovals inside a rectangle to show relationships of sets.

vertex (610, 611) The maximum or minimum point of a parabola.

vertical axis (56) The vertical line in a graph that represents the dependent variable.

vertical line test (289) If any vertical line passes through no more than one point of the graph of a relation, then the relation is a function.

W

weighted average (233) The weighted average M of a set of data is the sum of the product of each number in the set and its weight divided by the sum of all the weights.

whiskers (428) The whiskers of a box-and-whisker plot are the segments that are drawn from the lower quartile to the least value and from the upper quartile to the greatest value.

whole numbers (72) The set of whole numbers is represented by $\{0, 1, 2, 3, ...\}$.

work backward (156) A problem-solving strategy that uses inverse operations to determine an original value.

X

***x*-axis** (254) The horizontal number line.

***x*-coordinate** (254) The first number in an ordered pair.

***x*-intercept** (346) The coordinate at which a graph intersects the x-axis.

Y

***y*-axis** (254) The vertical number line.

***y*-coordinate** (254) The second number in an ordered pair.

***y*-intercept** (346) The coordinate at which a graph intersects the y-axis.

Z

zero exponent (502) For any nonzero number a, $a^0 = 1$.

zero product property (594) For all numbers a and b, if $ab = 0$, then $a = 0$, $b = 0$, or both a and b equal 0.

zeros (620) The zeros of a function are the roots, or x-intercepts, of the function.

SPANISH GLOSSARY

This glossary contains terms from Volumes One and Two. Pages 1–447 are contained in Volume One. Pages 382–753 are contained in Volume Two.

absolute value/valor absoluto (85) El valor absoluto de un número equivale al número de unidades que dicho número dista de cero en la recta numérica.

acute triangle/triángulo agudo (165) En un triángulo agudo, todos los ángulos miden menos de 90°.

adding integers/suma de enteros (86)

1. Para sumar enteros del *mismo* signo, suma los valores absolutos de los números. Da al resultado el mismo signo de los números.
2. Para sumar enteros de *distinto* signo, resta el valor menor del valor mayor. Da al resultado el mismo signo que el número con el mayor valor absoluto.

addition property of equality/propiedad de adición de la igualdad (144) Para cualquiera de los números a, b y c, si $a = b$, entonces $a + c = b + c$.

addition property of inequality/propiedad de adición de la desigualdad (385) Para todos los números a, b y c:

1. si $a > b$, entonces $a + c > b + c$;
2. si $a < b$, entonces $a + c < b + c$.

additive identity/identidad aditiva (37) Para cualquier número a, $a + 0 = 0 + a = a$.

additive inverse property/propiedad del inverso de la adición (87) Para cualquier número a, $a + (-a) = 0$.

algebraic expression/expresión algebraica (6) Una expresión que consiste de uno o más números y variables, además de una o más operaciones aritméticas.

angle of depression/ángulo de depresión (208) El ángulo de depresión se forma por una línea horizontal y una línea visual por debajo de la misma.

angle of elevation/ángulo de elevación (208) Un ángulo de elevación se forma por una línea horizontal y una línea visual por encima de la misma.

associative property/propiedad asociativa (51) Para cualquiera de los números a, b, y c, $(a + b) + c = a + (b + c)$ y $(ab)c = a(bc)$.

axes/ejes (254) Dos rectas numéricas perpendiculares que se usan para ubicar puntos en un plano de coordenadas.

axis of symmetry/eje de simetría (612) La ecuación para el eje de simetría de la gráfica $y = ax^2 + bx + c$, en la cual $a \neq 0$ y $x = -\frac{b}{2a}$.

back-to-back stem-and-leaf plot/diagrama de tallo y hojas consecutivo (27) Un diagrama de tallo y hojas consecutivo se usa para comparar dos conjuntos de datos. El mismo tallo se usa para las hojas de ambos diagramas.

base/base **1.** (7) En una expresión de la forma x^n, la base es x. **2.** (215) El número que se divide entre el porcentaje en una proporción de porcentaje.

best-fit line/línea de mejor encaje (341) Es una línea que se dibuja en un diagrama de dispersión y que pasa cerca de la mayoría de los puntos de datos.

binomial/binomio (514) La suma de dos monomios.

boundary/frontera (437) La frontera de una desigualdad es una línea que separa el plano de coordenadas en dos mitades de planos.

box-and-whisker plot/diagrama de caja y patillas (427) Un tipo de diagrama o gráfica que muestra los valores cuartílicos y los extremos de los datos.

coefficient/coeficiente (47) El factor numérico de un término.

commutative property/propiedad conmutativa (51) Para cualquiera de los números a y b, $a + b = b + a$ y $ab = ba$.

comparison property/propiedad de comparación (94) Para cualquier par de números a y b, exactamente una de las siguientes operaciones es válida.

$$a < b \qquad a = b \qquad a > b$$

complementary angles/ángulos complementarios (163) Dos ángulos son complementarios si la suma de sus medidas es 90 grados.

complete graph/gráfica completa (278) Una gráfica completa muestra el origen, los puntos en donde la gráfica cruza el eje de coordenadas x y el eje de coordenadas y, además de otras características importantes en la gráfica.

completeness property/propiedad del completo (121) Cada número real corresponde exactamente a un punto en la recta numérica. Cada punto en la recta numérica corresponde exactamente a un número real.

completeness property for points in the plane/propiedad del completo de los puntos en el plano (256)

1. Exactamente un punto en el plano es nombrado por un par ordenado de números dado.

2. Exactamente un par ordenado de números nombra un punto dado en el plano.

completing the square/completar el cuadrado (742, 743) Sumar un término constante a un binomio de la forma $x^2 = bx$, de modo que el trinomio resultante sea un cuadrado perfecto.

complex fraction/fracción compleja (114) Si una fracción tiene uno o más fracciones en el numerador o denominador, entonces se llama una fracción compleja.

composite number/número compuesto (558) Un número entero, mayor de 1, que no es un número primo.

compound event/evento compuesto (414) Un evento compuesto consiste de dos o más eventos simples.

compound inequality/desigualdad compuesta (405) Dos desigualdades conectadas por *y* u *o*.

congruent angles/ángulos congruentes (164) Ángulos que tienen la misma medida.

conjugates/conjugados (722) Dos binomios de la forma $a\sqrt{b} + c\sqrt{d}$ y $a\sqrt{b} - c\sqrt{d}$.

consecutive integers/números enteros consecutivos (158) Son los números enteros en el orden de contar.

consistent/consistente (455) Se dice que un sistema de ecuaciones es consistente cuando tiene por lo menos un par ordenado que satisface ambas ecuaciones.

constant of variation/constante de variación (239) El número k en ecuaciones de la forma $y = kx$ y $xy = k$.

constants/constantes (469) Los monomios que son números reales.

coordinate/coordenada (73) El número que corresponde a un punto en una recta numérica.

coordinate plane/plano de coordenadas (254) El plano que contiene el eje de coordenadas x y el eje de coordenadas y.

corresponding angles/ángulos correspondientes (201) Ángulos que encajan en triángulos semejantes y que tienen medidas iguales.

corresponding sides/lados correspondientes (201) Los lados opuestos a los ángulos correspondientes en triángulo semejantes.

cosine/coseno (206) En un triángulo rectángulo con ángulo agudo A, el coseno del

$$\text{ángulo } A = \frac{\text{medida del cateto adyacente al ángulo } A}{\text{medida de la hipotenusa}}.$$

cross products/productos cruzados (94) Cuando se comparan dos fracciones, los productos cruzados son los productos de los términos en diagonales.

D

data/datos (25) La información numérica.

defining the variable/definir la variable (127) El escoger una variable para representar uno de los números no especificados en un problema.

degree/grado **1.** (515) El grado de un monomio es la suma de los exponentes de sus variables. **2.** (516) El grado de un polinomio es el grado del término con el grado más alto.

density property/propiedad de la densidad (96) Entre cada par de números racionales distintos, existe una infinidad de números racionales.

dependent/dependiente (456) Un sistema de ecuaciones que tiene un número infinito de soluciones.

dependent variable/variable dependiente (58) La variable en una función cuyo valor lo determina la variable independiente.

difference of squares/diferencia de cuadrados (545) Dos cuadrados perfectos separados por un signo de sustracción, $a^2 - b^2 = (a + b)(a - b)$.

dimensional analysis/análisis dimensional (174) El proceso de llevar unidades a lo largo de un cómputo.

direct variation/variación directa (239) Una función lineal descrita por una ecuación de la forma $y = kx$, en que $k \neq 0$.

discrete mathematics/matemáticas de números discretos (88) Una rama de las matemáticas que estudia los conjuntos de números finitos o interrumpidos.

discriminant/discriminante (632) En la fórmula cuadrática, la expresión $b^2 - 4ac$ se denomina la discriminante.

distance formula/fórmula de distancia (737) La distancia d entre cualquier par de puntos con coordenadas (x_1, y_1) y (x_2, y_2) es dada por la siguiente fórmula.

$$d = \sqrt{(x_2 - x_1)^2 + (y_2 - y_1)^2}.$$

distributive property/propiedad distributiva (46) Para cualquiera de los números a, b y c:

1. $a(b + c) = ab + ac$ y $(b + c)a = ba + ca$.

2. $a(b - c) = ab - ac$ y $(b - c)a = ba - ca$.

dividing rational numbers/división de números racionales (112) El cociente de dos números racionales que tienen el mismo signo es positivo. El cociente de dos números racionales que tienen diferente signo es negativo.

division property of equality/propiedad de división de la igualdad (151) Para cualquiera de los números a, b y c, en que $c \neq 0$, si $a = b$, entonces $\frac{a}{c} = \frac{b}{c}$.

division property of inequality/propiedad de división de la desigualdad (393) Para cualquiera de los números reales a, b y c:

1. si c es positivo y $a < b$, entonces $\frac{a}{c} < \frac{b}{c}$, y

   si c es positivo y $a > b$, entonces $\frac{a}{c} > \frac{b}{c}$.

2. si c es negativo y $a < b$, entonces $\frac{a}{c} > \frac{b}{c}$, y

   si c es negativo y $a > b$, entonces $\frac{a}{c} < \frac{b}{c}$.

domain/dominio (263) El conjunto de todas las primeras coordenadas de los pares ordenados de una relación.

draw a diagram/trazar un diagrama (406) Una estrategia para resolver problemas que a menudo se usa como una herramienta organizadora.

E

element/elemento (33) Un miembro de un conjunto.

elimination/eliminación (469) El método de eliminación para resolver un sistema de ecuaciones es un método que usa adición o sustracción para eliminar una de las variables y así despejar la otra variable.

equally likely/igualmente verosímil (229) Respuestas que tienen igual posibilidad de ocurrir.

equation/ecuación (33) Un enunciado matemático que contiene un signo de igualdad, =.

equation in two variables/ecuación en dos variables (271) Una ecuación en dos variables contiene dos valores desconocidos.

equilateral triangle/triángulo equilátero (164) Un triángulo en el cual todos los lados y todos los ángulos tienen la misma medida.

equivalent equation/ecuación equivalente (144) Ecuaciones que tienen la misma solución.

equivalente expressions/expresiones equivalentes (47) Expresiones que denotan el mismo número.

evaluate/evaluar (8) El método de hallar el valor de una expresión cuando se conocen los valores de las variables.

excluded value/valor excluido (660) Se excluye un valor del dominio de una variable porque si se sustituyera ese valor por la variable, el resultado tendría un cero en el denominador.

exponent/exponente (7) En una expresión de la forma x^n, el exponente es n.

exponential function/función exponencial (634, 635) Una función que se puede describir por una ecuación de la forma $y = a^x$, en que $a > 0$ y $a \neq 1$.

extraneous solutions/soluciones extrañas (697) Soluciones obtenidas de una ecuación y las cuales no son soluciones aceptables para la ecuación original.

extreme values/valores extremos (427) Los valores extremos son el menor valor y el mayor valor en un conjunto de datos.

extremes/extremos (196) *Ver* proporción.

factored form/forma factorial (559) Un monomio está escrito en forma factorial cuando está expresado como el producto de números primos y variables y las variables no tienen un exponente mayor de 1.

factoring/factorizar (564) Expresar un polinomio como el producto de monomios o polinomios.

factoring by grouping/factorizando por grupos (567) Un método de factorización de polinomios de cuatro o más términos.

factors/factores (6) En una expresión de multiplicación, los factores son las cantidades que se multiplican.

family of graphs/familia de gráficas (354) Una familia de gráficas incluye gráficas y ecuaciones de gráficas que tienen por lo menos una característica en común.

FOIL method/método FOIL (537) Para multiplicar dos binomios, halla la suma de los productos de los primeros términos, los términos de afuera, los términos de adentro y los últimos términos.

formula/fórmula (128) Una ecuación que enuncia una regla para la relación entre ciertas cantidades.

function/función **1.** (56) Una relación entre los datos de entrada y de salida, en que los datos de salida dependen de los datos de entrada. **2.** (287) Una relación en que cada elemento del dominio se aparea exactamente con un elemento de la amplitud.

functional notation/notación funcional (289) En notación funcional, la ecuación $y = x + 5$ se escribe $f(x) = x + 5$.

general equation for exponential decay/ ecuación general para la disminución exponencial (644) La ecuación general para la disminución exponencial está representada por $A = C(1 - r)^t$.

general equation for exponential growth/ ecuación general para el crecimiento exponencial (643) La ecuación general para el crecimiento exponencial está representada por $A = C(1 + r)^t$.

graph/gráfica (73, 255) Consiste en dibujar, o trazar sobre una recta numérica o plano de coordenadas, los puntos nombrados por ciertos números o pares ordenados.

greatest common factor (GCF)/máximo común divisor (MCD) (559) El máximo común divisor de dos o más números enteros es el número mayor que es un factor de todos los números.

guess and check/conjetura y cotejo (574) Una estrategia para resolver problemas en la cual se prueban varios valores o combinaciones de valores para hallar una solución al problema.

H

half-life/media vida (638) La media vida de un elemento radioactivo es el tiempo que tarda en desintegrarse la mitad del elemento.

half-plane/medio plano (437) La región de un plano en un lado de una recta en el plano.

horizontal axis/eje horizontal (56) La recta horizontal en una gráfica que representa la variable independiente.

hypotenuse/hipotenusa (206) El lado de un triángulo rectángulo opuesto al ángulo recto.

I

identify subgoals/identificación de submetas (745) Una estrategia para resolver problemas que utiliza una serie de pasos secundarios o submetas.

identity/identidad (170) Una ecuación que es cierta para cualquier valor de la variable.

inconsistent/inconsistente (456) Se dice que un sistema de ecuaciones es inconsistente cuando no tiene pares ordenados que satisfacen ambas ecuaciones.

independent/independiente (456) Se dice que un sistema de ecuaciones es independiente si el sistema tiene exactamente una solución.

independent variable/variable independiente (58) La variable en una función cuyo valor está sujeto a elección se dice que es una variable independiente. La variable independiente determina el valor de la variable dependiente.

inequality/desigualdad (33) Un enunciado matemático que contiene los símbolos $<$, $\leq$, $>$, o $\geq$.

integers/enteros (73) El conjunto de enteros representados por $\{\ldots, -3, -2, -1, 0, 1, 2, 3, \ldots\}$.

interquartile range/amplitud intercuartílica (304, 306) La diferencia entre el cuartil superior y el inferior de un conjunto de datos. Representa la mitad inferior, ó 50%, de los datos en un conjunto.

intersection/intersección (406) Para dos conjuntos A y B, es el conjunto de elementos comunes a ambos A y B.

inverse of a relation/inverso de una relación (264) El inverso de cualquier relación se obtiene intercambiando las coordenadas en cada par ordenado.

inverse variation/variación inversa (241) Una variación inversa se describe por una ecuación de la forma $xy = k$, en que $k \neq 0$.

irrational number/números irracional (120) Un número que no se puede expresar en la forma $\frac{a}{b}$, en que a y b son números enteros y $b \neq 0$.

isosceles triangle/triángulo isósceles (164) En un triángulo isósceles, por lo menos dos ángulos tienen la misma medida y por lo menos dos lados tienen la misma longitud.

L

least common denominator (LCD)/mínimo común denominador (MCD) (686) El mínimo común múltiplo de los denominadores de dos o más fracciones.

least common multiple (LCM)/mínimo común múltiplo (MCM) (685) Para dos o más enteros, el MCM es el menor número entero positivo divisible entre cada uno de los enteros.

legs/catetos (206) Los lados de un triángulo rectángulo que forman el ángulo recto.

like terms/términos semejantes (47) Términos que contienen las mismas variables, con variables correspondientes que tienen la misma potencia.

line plot/esquema lineal (78) Datos numéricos desplegados sobre una recta numérica.

linear equation/ecuación lineal (280) Una ecuación que puede escribirse de la forma $Ax + By = C$, en que A y B no son ambos iguales a cero. La gráfica de una ecuación lineal es una recta.

linear function/función lineal (278) Ecuación cuya gráfica es una recta no vertical.

look for a pattern/busca un patrón (13, 497, 637) Una estrategia para resolver problemas que a menudo involucra el uso de tablas para organizar la información de manera que se pueda determinar un patrón.

lower quartile/cuartil inferior (306) El cuartil inferior divide la mitad inferior de un conjunto de datos en dos partes iguales.

M

make an organized list/haz una lista organizada (685) Una estrategia para resolver problemas que utiliza una lista organizada para arreglar y evaluar los datos y determinar una solución.

mapping/relación (263) Una relación aparea un elemento en el dominio con un elemento en la amplitud.

matrix/matriz (88) Una matriz es un arreglo rectangular de elementos en hileras y columnas.

maximum/máximo (611) El punto más alto en la gráfica de una curva, tal como el vértice de la parábola que se abre hacia abajo.

mean/media (178) La media de un conjunto de datos es la suma de los números en el conjunto dividida entre el número de números en el conjunto.

means/media proporcional (196) *Ver* proporción.

measures of central tendency/medidas de tendencia central (178) Los números que se usan a menudo para describir conjuntos de datos porque estos representan un valor centralizado o en el medio.

measures of variation/medidas de variación (306) Números usados para describir la amplitud o distribución de los datos.

median/mediana (178) La mediana es el número en el centro de un conjunto de datos cuando los números se organizan en orden numérico.

midpoint/punto medio (369) El punto medio de un segmento es el punto equidistante entre los extremos del segmento.

minimum/mínimo (611) El punto más bajo en la gráfica de una curva, tal como el vértice de una parábola que se abre hacia arriba.

mixed expression/expresión mixta (690) Una expresión algebraica que contiene un monomio y una expresión racional.

mode/modal (178) El número que ocurre con más frecuencia en un conjunto.

monomial/monomio (496) Un número, una variable o el producto de un número y una o más variables.

multi-step equations/ecuaciones múltiples (157) Ecuaciones que requieren más de una operación para resolverlas.

multiplicative identity/identidad de multiplicación (38) Para cualquier número a, $a \cdot 1 = 1 \cdot a = a$.

multiplicative inverse/inverso multiplicativo (38) Para cualquier número no cero a, hay exactamente un número $\frac{1}{a}$, tal que $a \cdot \frac{1}{a} = \frac{1}{a} \cdot a = 1$.

multiplicative property of −1/propiedad multiplicativa de −1 (107) El producto de cualquier número y −1 es el inverso aditivo del número.

$$-1(a) = -a \text{ y } a(-1) = -a.$$

multiplicative property of equality/propiedad multiplicativa de la igualdad (150) Para cualquiera de los números a, b y c, si $a = b$, entonces $a \cdot c = b \cdot c$.

multiplication property of inequality/propiedad de multiplicación de la desigualdad (393) Para todos los números a, b y c, lo siguiente es cierto:

1. Si c es positivo y $a < b$, entonces $ac < bc$, y si c es positivo y $a > b$, entonces $ac > bc$.
2. Si c es negativo y $a < b$, entonces $ac > bc$, y si c es negativo y $a > b$, entonces $ac < bc$.

multiplicative property of zero/propiedad multiplicativa de cero (38) Para cualquier número a, $a \cdot 0 = 0 \cdot a = 0$.

N

negative correlation/correlación negativa (340) Existe una correlación negativa entre x y y si los valores están relacionados de maneras opuestas.

negative exponent/exponente negativo (503) Para cualquiera de los números no cero a y cualquier número entero n, $a^{-n} = \frac{1}{a^n}$

negative number/número negativo (73) Cualquier número que es menos de cero.

number line/recta numérica (72) Una recta con marcas equidistantes que se usa para representar números.

number theory/teoría de números (158) El estudio de los números y de las relaciones entre los mismos.

O

obtuse triangle/triángulo obtuso (165) Un triángulo obtuso tiene un ángulo cuya medida es mayor de 90 grados.

odds/posibilidades (229) Las posibilidades de que un evento ocurra son la proporción del número de formas en que el evento puede ocurrir (éxitos) comparada con el número de formas en que puede no ocurrir (fracasos).

open sentences/enunciados abiertos (32) Enunciado matemático que contiene una o más variables, o incógnitas.

opposites/opuestos (87) El opuesto de un número es su inverso aditivo.

order of operations/orden de operaciones (19)

1. Simplifica las expresiones dentro de los símbolos de agrupación, tales como paréntesis, corchetes y como lo indiquen las barras de fracción.
2. Evalúa todas las potencias.
3. Realiza todas las multiplicaciones y las divisiones de izquierda a derecha.
4. Realiza todas las sumas y las restas de izquierda a derecha.

ordered pair/par ordenado (57) Pares de números que se usan para ubicar puntos en el plano de coordenadas.

organize data/organizar datos (464) Una estrategia útil antes de resolver un problema. Algunas formas de organizar los datos son el uso de tablas, esquemas, diferentes tipos de gráficas o diagramas.

origin/origen (254) El punto de intersección de los dos ejes en el plano de coordenadas.

outcomes/resultados (413) Todas las maneras en que puede ocurrir un evento.

outlier/valor atípico (307) Cualquier elemento en un conjunto de datos que es por lo menos 1.5 veces el valor de la amplitud intercuartílica mayor que el cuartil superior, o menor que el cuartil inferior.

P

parabola/parábola (610, 611) La forma general de la gráfica de una función cuadrática.

parallel lines/rectas paralelas (362) Rectas en el plano que nunca se intersecan. Las rectas paralelas no verticales tienen la misma pendiente.

parallelogram/paralelogramo (362) Un cuadrilátero cuyos lados opuestos son paralelos.

parent graph/gráfica principal (354) La gráfica más simple en una familia de gráficas.

percent/por ciento (215) Un por ciento es una proporción que compara un número con 100.

percentage/porcentaje (215) El número que se divide entre la base en un por ciento de proporción.

percent of decrease/porcentaje de disminución (222) La proporción de una cantidad de disminución comparada con una cantidad previa, expresada en forma de por ciento.

percent of increase/porcentaje de aumento (222) La proporción de una cantidad de aumento comparada con una cantidad previa, expresada en forma de por ciento.

percent proportion/proporción de porcentaje (215)

$$\frac{\text{Percentaje}}{\text{Base}} = \frac{r}{100}$$

perfect square/cuadrado perfecto (119) Un número racional cuya raíz cuadrada es un número racional.

perfect square trinomial/cuadrado perfecto trinómico (587) Un trinomio que al factorizarse tiene la forma $(a + b)^2 = (a + b)(a + b)$ o $(a - b)^2 = (a - b)(a - b)$.

perpendicular lines/rectas perpendiculares (362) Rectas que se encuentran para formar ángulos rectos.

point-slope form/forma punto–pendiente (333) Para cualquier punto (x_1, y_1) sobre una recta no vertical cuya pendiente es m, la forma punto–pendiente de una ecuación lineal es la siguiente:

$$y - y_1 = m(x - x_1).$$

polynomial/polinomio (513, 514) Un polinomio es un monomio o la suma de monomios.

positive correlation/correlación positiva (340) Existe una correlación positiva entre x y y si los valores están relacionados de la misma forma.

power/potencia (7) Una expresión de la forma x^n se conoce como una potencia.

power of a monomial/potencia de un monomio (498) Para cualquiera de los números a y b, y para todos los números enteros m, n y p, $(a^m b^n)^p = a^{mp} b^{np}$.

power of a power/potencia de una potencia (498) Para cualquier número a y todos los números enteros m y n, $(a^m)^n = a^{mn}$.

power of a product/potencia de un producto (498) Para todos los números a y b y cualquier número entero m, $(ab)^m = a^m b^m$.

prime factorization/factorización prima (558) Un número entero expresado como un producto de factores que son todos números primos.

prime number/número primo (558) Un número primo es un número entero mayor que 1 cuyos únicos factores son 1 y el número mismo.

prime polynomial/polinomio primo (577) Un polinomio que no se puede escribir como el producto de dos polinomios con coeficientes integrales se llama un polinomio primo.

principal square root/raíz cuadrada principal (119) La raíz cuadrada no negativa de una expresión.

probability/probabilidad (228) Una razón que expresa la posibilidad de que algún evento suceda.

$$P(\text{evento}) = \frac{\text{número de resultados favorables}}{\text{número de resultados posibles}}$$

problem-solving plan/plan para solucionar problemas (126)

1. Explorar el problema.
2. Planificar la solución.
3. Resolver el problema.
4. Examinar la solución.

product/producto (6) El resultado de la multiplicación.

product of powers/producto de potencias (497) Para cualquiera de los números a y todos los números enteros m y n, $a^m \cdot a^n = a^{m+n}$.

product property of square roots/propiedad del producto de raíces cuadradas (719) Para cualquiera de los números a y b, en que $a \geq 0$ y $b \geq 0$, $\sqrt{ab} = \sqrt{a} \cdot \sqrt{b}$.

proportion/proporción (195) En una proporción, el producto de los extremos es igual al producto de las medias. Si $\frac{a}{b} = \frac{c}{d}$, entonces $ad = bc$.

Pythagorean theorem/teorema de Pitágoras (713) Si a y b son las medidas de los catetos de un triángulo rectángulo y c es la medida de la hipotenusa, entonces, $c^2 = a^2 + b^2$.

Pythagorean triple/triplete de Pitágoras (583) Tres números enteros a, b, y c, tales que $a^2 + b^2 = c^2$.

Q

quadrant/cuadrante (254) Una de las cuatro regiones en que el eje x y el eje y separan el plano de coordenadas.

quadratic equation/ecuación cuadrática (620) Una ecuación cuadrática es una en que el valor de la función cuadrática relacionada es 0.

quadratic formula/fórmula cuadrática (628) Las raíces de una ecuación cuadrática en la forma $ax^2 + bx + c = 0$, en la cual $a \neq 0$, son dadas por la fórmula

$$x = \frac{-b \pm \sqrt{b^2 - 4ac}}{2a}.$$

quadratic function/función cuadrática (610, 611) Función que se puede describir con una equación de la forma $ax^2 + bx + c = 0$, en la cual $a \neq 0$.

quartiles/cuartiles (304, 306) En un conjunto de datos, los cuartiles son valores que dividen los datos en cuatro partes iguales.

quotient of powers/cociente de potencias (501) Para todos los números enteros m y n y todo número no cero a, $\frac{a^m}{a^n} = a^{m-n}$.

quotient property of square roots/propiedad del cociente de raíces cuadradas (721) Para cualquiera de los números a y b, en que $a > 0$ y $b > 0$, $\sqrt{\frac{a}{b}} = \frac{\sqrt{a}}{\sqrt{b}}$.

radical equations/ecuaciones radicales (732) Ecuaciones que contienen radicales con variables en el radicando.

radical sign/signo radical (119) El símbolo $\sqrt{\ }$ que se usa para indicar la raíz cuadrada principal no negativa de una expresión.

radicand/radicando (719) El radicando es la expresión debajo del signo radical.

random/al azar (229) Cuando se escoge un resultado sin ninguna preferencia, el resultado ocurre al azar.

range/amplitud **1.** (263) El conjunto de todas las segundas coordenadas de los pares ordenados de una relación. **2.** (306) La diferencia entre los valores mayor y menor en un conjunto de datos.

rate/razón (197) La proporción de dos medidas que se dan en diferentes unidades de medida.

rate/tasa (215) En una proporción de por ciento, la tasa es la fracción con 100 como denominador.

ratio/razón (195) Una razón es una comparación de dos números mediante división.

rational equation/ecuación racional (696) Una ecuación racional es una ecuación que contiene expresiones racionales.

rational expression/expresión racional (660) Una expresión racional es una fracción algebraica cuyo numerador y cuyo denominador son polinomios.

rational number/número racional (93) Un número racional es un número que se puede expresar en la forma $\frac{a}{b}$, en que a y b son números enteros y $b \neq 0$.

rationalizing the denominator/racionalizando el denominador (722) Un proceso que se usa para quitar o eliminar radicales del denominador de una fracción, en una expresión radical.

real numbers/números reales (121) El conjunto de números irracionales junto con los números racionales.

reciprocal/recíproco (38) El inverso multiplicativo de un número.

reflexive property of equality/propiedad reflexiva de la igualdad (39) Para cualquier número a, $a = a$.

regression line/línea de regresión (342) La línea más exacta de mejor ajuste para un conjunto de datos. Se puede determinar con una calculadora de gráficar o una computadora.

relation/relación (260, 263) Una relación es un conjunto de pares ordenados.

replacement set/conjunto de substitución (33) Un conjunto de números de los cuales se pueden escoger números para reemplazar una variable.

right triangle/triángulo rectángulo (165) Un triángulo que tiene un ángulo con una medida de 90 grados.

rise/altura (325) El cambio vertical en una recta.

roots/raíces (620) Las soluciones para una ecuación cuadrática.

run/carrera (325) El cambio horizontal en una recta.

scalar multiplication/multiplicación escalar (108) En multiplicación escalar, cada elemento de una matriz se multiplica por una constante.

scale/escala (197) Una razón llamada una escala se usa en la construcción de un modelo para representar algo que es muy grande o muy pequeño para ser dibujado a su tamaño real.

scatter plot/diagrama de dispersión (339) Gráfica que muestra dos conjuntos de datos trazados como puntos (pares ordenados) en el plano de coordenadas.

scientific notation/notación científica (506) Un número está expresado en notación científica cuando está en la forma de $a \times 10^n$, en que $1 \leq a < 10$ y n es un número entero.

set/conjunto (33) Una colección de objetos o números.

set-builder notation/notación de construcción de conjuntos (385) Una notación que se usa para describir los miembros de un conjunto. Por ejemplo, $\{y \mid y < 17\}$ representa el conjunto de todos los números y de modo que y es menor que 17.

sequence/sucesión (13) Un conjunto de números en un orden específico.

similar triangles/triángulos semejantes (201) Si dos triángulos son semejantes, las medidas de sus lados correspondientes son proporcionales y las medidas de sus ángulos correspondientes son iguales.

simple event/evento simple (414) Evento sencillo en un problema de probabilidad.

simple interest/interés simple (217) La cantidad pagada o ganada por el uso de una cantidad de dinero. La fórmula $I = prt$ se usa para resolver problemas de interés simple.

simplest form/forma reducida (47) Una expresión está en su forma reducida cuando ha sido reemplazada por una expresión similar que no tiene términos semejantes ni paréntesis.

simplest radical form/forma radical reducida (723) Una expresión radical se encuentra en forma reducida cuando se satisfacen las siguientes condiciones:

1. Ningún radicando tiene factores cuadrados perfectos además de uno.
2. Ningún radicando contiene fracciones.
3. No aparece ningún radical en el denominador de una fracción.

sine/seno (206) En un triángulo rectángulo con ángulo agudo A, el seno del ángulo $A = \frac{\text{medida del cateto opuesto al ángulo } A}{\text{medida de la hipotenusa}}$.

slope/pendiente (324, 325) El cambio vertical (altura) al cambio horizontal (carrera) a medida que te mueves de un punto a otro a lo largo de la recta.

slope-intercept form/forma pendiente-intersección (347) Una ecuación de la forma $y = mx + b$, en que m es la pendiente y b es la intersección en y de una recta dada.

solution/solución (32) Una sustitución por una variable en una ecuación que resulta en una ecuación válida.

solution of an equation in two variables/ solución de una ecuación de dos variables (271) Si al sustituir los números en un par ordenado se satisface una ecuación de dos variables, entonces el par ordenado es una solución de la ecuación.

solution set/conjunto de solución (33) El conjunto de todos los sustitutos para una variable en un enunciado abierto que satisfacen el enunciado.

solve an equation/resuelve una ecuación (145) Resolver una ecuación quiere decir aislar la variable cuyo coeficiente es 1, en un lado de la ecuación.

solving an open sentence/resolviendo un enunciado abierto (32) Hallar un sustituto que satisface la variable.

solving a triangle/resolviendo un triángulo (208) El proceso de hallar las medidas de todos los lados y ángulos de un triángulo rectángulo.

square of a difference/cuadrado de una diferencia (544) Si a y b son cualquier par de números, $(a - b)^2 = (a - b)(a - b) = a^2 - 2ab + b^2$.

square of a sum/cuadrado de una suma (543) Si a y b son cualquier par de números, $(a + b)^2 = (a + b)(a + b) = a^2 + 2ab + b^2$.

square root/raíz cuadrada (118, 119) Uno de los dos factores idénticos de un número.

standard form/forma estándar (333) La forma estándar de una ecuación lineal es $Ax + By = C$, en la cual A, B y C son números enteros, $A \geq 0$ y A y B no son ceros ambos.

statistics/estadística (25) Una rama de las matemáticas que tiene que ver con los métodos de recolección, organización e interpretación de datos.

stem-and-leaf plot/gráfica de tallo y hojas (26) En una gráfica de tallo y hojas, cada dato se separa en dos números que se usan para formar un tallo y las hojas. Los datos se organizan en dos columnas. La columna de la izquierda contiene el tallo y la columna de la derecha las hojas.

substitution/sustitución (462) El método de sustitución para resolver un sistema de ecuaciones es un método que sustituye una ecuación en la otra ecuación para despejar la otra variable.

substitution property of equality/propiedad de sustitución de la igualdad (39) Si $a = b$, entonces, a se puede reemplazar por b en cualquier expresión.

subtracting integers/sustracción de enteros (87) Para restar un entero, suma su inverso aditivo. Para cualquiera de los enteros a y b, $a - b = a + (-b)$.

subtraction property of equality/propiedad de sustracción de la igualdad (146) Para cualquiera de los números a, b y c, si $a = b$, entonces $a - c = b - c$.

subtraction property of inequality/propiedad de sustracción de la desigualdad (385) Para todos los números a, b y c los siguientes son ciertos:

1. si $a > b$, entonces $a - c > b - c$;
2. si $a < b$, entonces $a - c < b - c$.

supplementary angles/ángulos suplementarios (162) Dos ángulos son suplementarios si la suma de sus medidas es 180 grados.

symmetric property of equality/propiedad simétrica de la igualdad (39) Para cualquiera de los números a y b, si $a = b$, entonces $b = a$.

symmetry/simetría (612) Las figuras simétricas son aquellas en que la figura se puede doblar y cada mitad es exactamente igual a la otra.

system of equations/sistema de ecuaciones (455) Un conjunto de ecuaciones con las mismas variables.

system of inequalities/sistema de desigualdades (482) Un conjunto de desigualdades con las mismas variables.

tangent/tangente (206) En un triángulo rectángulo con ángulo agudo A, la tangente del ángulo $A = \frac{\text{medida del cateto opuesto al ángulo } A}{\text{medida del cateto adyacente al ángulo } A}$.

term/término 1. (13) Un número en una sucesión. 2. (47) Un número, una variable o un producto o cociente de números y variables.

transitive property of equality/propiedad transitiva de la igualdad (39) Para cualquiera de los números a, b y c, si $a = b$ y $b = c$, entonces $a = c$.

tree diagram/diagrama de árbol (413) Un diagrama de árbol es un diagrama que se usa para mostrar el número total de posibles resultados.

triangle/triángulo (163) Un triángulo es un polígono con tres lados y tres ángulos.

trigonometric ratios/razones trigonométricas (206) Para el triángulo rectángulo ABC con ángulo agudo A, el seno $A = \frac{a}{c}$, coseno $A = \frac{b}{c}$, tangente $A = \frac{a}{b}$.

trinomial/trinomio (514) La suma de tres monomios.

U

uniform motion/movimiento uniforme (235) Cuando un objeto se mueve a una velocidad, o a un ritmo constante, se dice que se mueve en movimiento uniforme.

union/unión (408) Para dos conjuntos A y B, el conjunto de los elementos contenidos en ambos A o B o A y B.

unique factorization theorem/teorema de la factorización única (558) La factorización prima de cada número es única excepto por el orden en que se escriben los factores.

unit cost/costo unitario (95) El costo de una unidad de un artículo.

upper quartile/cuartil superior (306) El cuartil superior divide la parte superior de un conjunto de datos en dos partes iguales.

use a model/usa un modelo (339) Una estrategia para resolver problemas que usa modelos, o simulaciones, de situaciones matemáticas difíciles de resolver directamente.

use a table/usa una tabla (255) Una estrategia para resolver problemas que usa tablas para organizar y resolver problemas.

variable/variable (6) Las variables son símbolos que se usan para representar números desconocidos.

Venn diagrams/diagramas de Venn (73) Los diagramas de Venn son diagramas que usan círculos u óvalos dentro de un rectángulo para mostrar relaciones de conjuntos.

vertex/vértice (610, 611) El punto máximo o mínimo de una parábola.

vertical axis/eje vertical (56) La recta vertical en una gráfica que representa la variable dependiente.

vertical line test/prueba de recta vertical (289) Si cualquier recta vertical pasa por un solo punto de la gráfica de una relación, entonces la relación es una función.

weighted average/promedio ponderado (233) El promedio ponderado M de un conjunto de datos es la suma del producto de cada número en el conjunto y su peso divididos entre la suma de todos los pesos.

whiskers/patillas (428) Las patillas de un diagrama de caja y patillas son los segmentos que se dibujan desde el cuartil inferior hasta el mínimo valor y desde el cuartil superior hasta el máximo valor.

whole numbers/números enteros (72) El conjunto de números enteros se representa por {0,1,2,3,...}.

work backward/trabaja al revés (156) Una estrategia para resolver problemas que usa operaciones inversas para determinar un valor inicial.

x*-axis/eje *x (254) La recta numérica horizontal en el plano de coordenadas.

x*-coordinate/coordenada *x (254) El primer número en un par ordenado.

x*-intercept/intersección con el eje *x (346) La coordenada x de un punto donde una gráfica interseca el eje x.

y*-axis/eje *y (254) La recta numérica vertical en el plano de coordenadas.

y*-coordinate/coordenada *y (254) El segundo número en un par ordenado.

y*-intercept/intersección con el eje *y (346) La coordenada y de un punto donde una gráfica interseca el eje y.

Z

zero exponent/exponente cero (502) Para cualquier número no cero a, $a^0 = 1$.

zero product property/propiedad del producto cero (594) Para todos los números a y b, si $ab = 0$, entonces $a = 0$, $b = 0$, o ambos a y b son iguales a cero.

zeros/ceros (620) Las raíces, o intersecciones con el eje x de la gráfica de una función.

CHAPTER B SELECTED ANSWERS

Page B3 Pretest
1. C **3.** A **5.** D **7.** B **9.** C **11.** A **13.** C **15.** C **17.** A **19.** B **21.** A **23.** A **25.** B

Page B4 Review Lesson 1–1
1, 3. Sample answers are given. **1.** one more than z **3.** fifty-seven decreased by three times q **5.** $6n$ **7.** $n \div 4$ **9.** $7n - 25$ **11.** $74 - 2n^3$ **13.** $27a^3$ **15.** 27 **17.** 10,000

Page B5 Review Lesson 1–2
1.

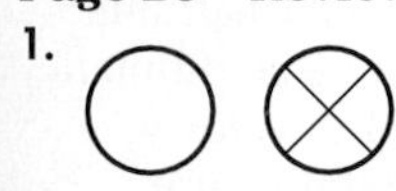

3. $w - 25, w - 20$ **5.** 4:46 P.M.

Page B6 Review Lesson 1–3
1. 12 **3.** 39 **5.** 28 **7.** 10 **9a.** $15(d + j) + 9(d + j)$ **9b.** \$921

Page B7 Review Lesson 1–4
1a. 800 feet **1b.** 500 feet **1c.** 600 feet

Page B8 Review Lesson 1–5
1. false **3.** true **5.** 22 **7.** $2\frac{3}{11}$

Page B9 Review Lesson 1–6
1. 1 **3.** 15 **5.** multiplication property of zero **7.** multiplicative identity **9.** additive identity **11.** substitution
13. $70 \div 10 + 3(6 - 3 \cdot 2) - 6 \div 2$
$= 70 \div 10 + 3(6 - 6) - 6 \div 2$ *substitution (=)*
$= 70 \div 10 + 3(0) - 6 \div 2$ *substitution (=)*
$= 70 \div 10 + 0 - 6 \div 2$ *mult. prop. of 0*
$= 7 + 0 - 3$ *substitution (=)*
$= 7 - 3$ *add. ident.*
$= 4$ *substitution (=)*
15a. $0.27 + 6(0.11)$
15b. $0.27 + 6(0.11)$
$= 0.27 + 0.66$ *substitution (=)*
$= 0.93$ *substitution (=)*

Page B10 Review Lesson 1–7
1. 12, 7, 3, 4; $12r^3, 3r^3$ **3.** 162 **5.** $35w + 21$ **7.** $7c$ **9.** $17t + 7y$ **11.** $2w - 21v$ **13a.** $12(23 + 9.95)$ **13b.** \$395.40

Page B11 Review Lesson 1–8
1. $17a + 6b$ **3.** $15q + 13r + 15$ **5.** $13st + 9s^2t$ **7.** $2q + 9r$ **9.** $\frac{7}{5} + \frac{3}{10}b$ **11.** $4.8s + 4.7t$ **13.** commutative property (+) **15.** multiplicative identity **17.** No, the order of these steps cannot be reversed.

Page B12 Review Lesson 1–9
1. a

3.
No. of pins
Time

Page B13 Chapter 1 Test
1. $3x + 7$ **3.** 100, 50 **5.** 108 **7.** 70, 74, 76, 77, 81, 82, 89, 93, 93, 95, 98 **9.** 1.89 **11.** distributive property **13.** substitution **15.** commutative property (+) **17.** $6m + 12n$ **19.**

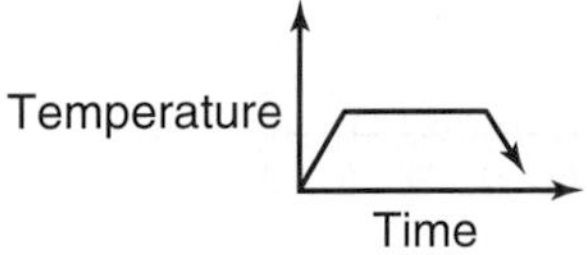

Page B14 Review Lesson 2–1
1. 2 **3.** 8 **5.** −11
7.

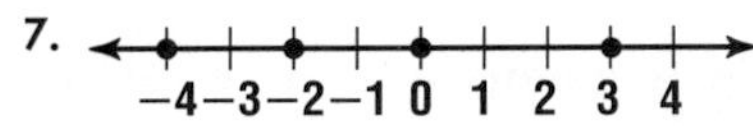

9.
−6 −5 −4 −3 −2 −1 0 1 2

11.
−5 −4 −3 −2 −1 0 1 2 3

13. 2 **15.** 0 **17.** −8

Page B15 Review Lesson 2–2
1a.
2 million 4 million 6 million 8 million

1b. 4 counties **1c.** Los Angeles County has three times as many residents as the three smallest counties in the table.

Page B16 Review Lesson 2–3
1. −20 **3.** 101 **5.** 66 **7.** $-9a$ **9.** $29x$ **11.** −25 **13.** 11

Page B17 Review Lesson 2–4
1. $<$ **3.** $=$ **5.** $-\frac{9}{7}, -1, \frac{1}{8}$ **7.** 20-ounce box

Page B18 Review Lesson 2–5
1. $-\frac{11}{13}$ **3.** 0.043 **5.** 2.42 **7.** 0.28 **9.** −2.75 **11.** 1.8 **13.** $\frac{3}{8}$

Page B19 Review Lesson 2–6
1. 54 **3.** −15 **5.** −132 **7.** 4 **9.** $-9s - 4st$ **11.** $28x - 24.4y$ **13a.** 906 mg **13b.** No, 906 mg is less than 1200 mg. **13c.** Yes, 204 + 2(52) + 906 = 1214, which is greater than 1200.

Page B20 Review Lesson 2–7
1. $-6q$ **3.** 5 **5.** $-26y$ **7.** $4w$ **9.** $-6d + 4c$ **11.** $11x - 7$

13a. $1\frac{1}{8}$ cups **13b.** $1\frac{3}{4}$ tablespoons

Page B21 Review Lesson 2–8

1. 8 **3.** $-\frac{9}{7}$ **5.** $\pm\frac{12}{5}$ **7.** ± 7.68 **9.** Q **11.** 62.6 miles

Page B22 Review Lesson 2–9

1a. $\frac{1}{3}$ **1b.** $\frac{1}{2}$ **3.** $A = s^2$ **5.** $xy > 7(x + y)$ **7.** Let x = the number of paperbacks, then $\frac{1}{3}x + 10$ = the number of hardbacks; $x + \frac{1}{3}x + 10 = 90$.

Page B23 Chapter 2 Test

1.

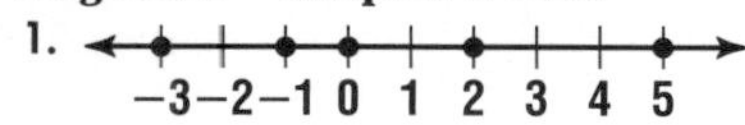

3.

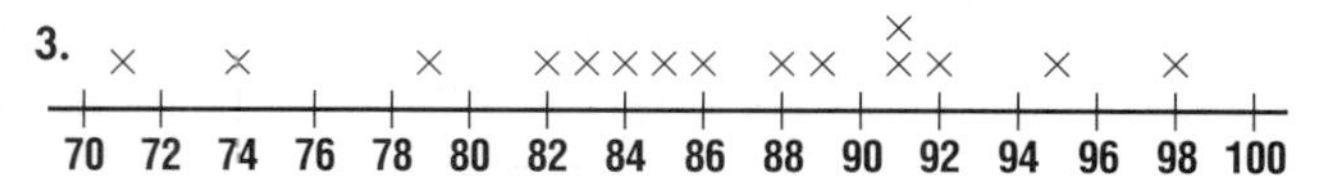

5. −58 **7.** $-\frac{1}{16}$ **9.** 5 **11.** −14 **13.** $-7q$ **15.** $-\frac{8}{3}$
17. $2(a + b) = 46$

Page B24 Review Lesson 3–1

1. −27 **3.** −25 **5.** 16 **7.** −27 **9.** −5.8 **11.** $\frac{5}{6}$
13a. $\ell = \frac{1}{8} + (d + t)$ **13b.** $\frac{3}{16}$ inch

Page B25 Review Lesson 3–2

1. 12 **3.** −7 **5.** $\frac{1}{14}$ **7.** $7n = 63$; 9 **9.** $-3n = -93$; 31
11. 112 **13.** 11.25

Page B26 Review Lesson 3–3

1. 13 **3.** 7 **5.** 133 **7.** 25 **9.** 8 **11.** $n + (n + 2) = 76$; 37, 39

Page B27 Review Lesson 3–4

1. 19°; 109° **3.** 65°; 155° **5.** $(76 - y)°$; $(166 - y)°$ **7.** 10°
9. $(180 - 3x)°$ **11.** 23°, 157°

Page B28 Review Lesson 3–5

1. −6 **3.** 5 **5.** 12 **7.** 88 **9.** −14 **11.** identity
13. 10 years

Page B29 Review Lesson 3–6

1. $y = \frac{6 - z}{2}$ **3.** $y = 6 - b$ **5.** $y = \frac{9n}{3 + m}, m \neq -3$
7. $y = \frac{t - s}{4}$ **9.** $y = \frac{5u + 3v}{8}$ **11.** \$1600

Page B30 Review Lesson 3–7

1. 90.6; 90; none **3.** 14; 14; 10 and 18 **5.** $\frac{83}{120}; \frac{17}{24}; \frac{3}{4}$
7. 63; 62 and 77 **9a.** 5; 5; 5 **9b.** They all represent the data equally well because they are all the same value.

Page B31 Chapter 3 Test

1. −44 **3.** $\frac{17}{15}$ **5.** −6 **7.** 5 **9.** 14 **11.** −11.7 **13.** 28°
15. $d = 21 - 17y$ **17.** $d = \frac{-12 - 7w}{3}$ **19.** \$70

Page B32 Review Lesson 4–1

1. 8 **3.** 3 **5.** 5 **7.** $\frac{13}{5}$ **9.** 1900 miles

Page B33 Review Lesson 4–2

1. $\triangle MNO$ and $\triangle TUS$ **3.** $\overline{MN}$ and $\overline{TU}$; $\overline{MO}$ and $\overline{TS}$; $\overline{NO}$ and $\overline{US}$

Page B34 Review Lesson 4–3

1. A: 0.441, 0.897, 0.492; B: 0.897, 0.441, 2.033 **3.** 0.1736
5. 0.8480 **7.** 28° **9.** 60°

Page B35 Review Lesson 4–4

1. 56.25% **3.** 36 **5.** 125 **7.** 60% **9a.** 338 people **9b.** 41 people **9c.** 341 people

Page B36 Review Lesson 4–5

1. \$9.60 **3.** \$243.60 **5.** \$103.58 **7.** 6.6%

Page B37 Review Lesson 4–6

1. $\frac{1}{3}$ **3.** $\frac{3}{10}$ **5.** 4:1

Page B38 Review Lesson 4–7

1. 45 grams **3.** 40 pounds

Page B39 Review Lesson 4–8

1. 40 **3.** $\frac{3}{64}$ **5.** 25 **7.** $\frac{5611}{500}$ **9.** $83\frac{3}{5}$ pounds

Page B40 Chapter 4 Test

1. 7 **3.** 4 **5.** 34° **7.** 42% **9.** 928 **11.** 15% **13.** 8:7
15. 11 P.M. **17.** $42\frac{6}{7}$

Page B41 Review Lesson 5–1

1–6.

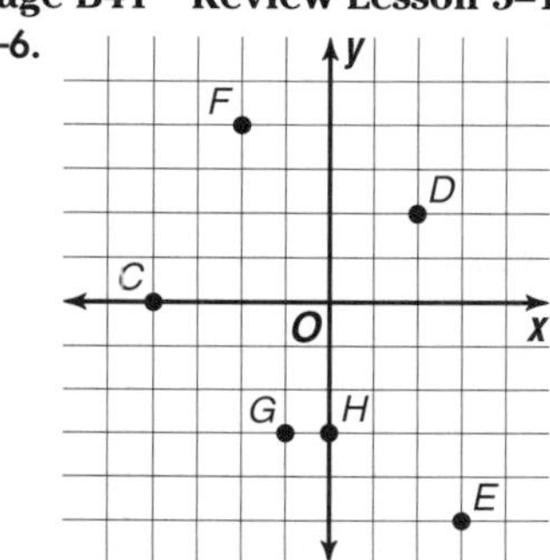

7. (1, 2); I **9.** (−2, 3); II **11.** (0, 1); none

Page B42 Review Lesson 5–2

1. D = {−3, −2, 4, 5}; R = {0, 4, 5} **3.** $D = \left\{-\frac{7}{9}, \frac{1}{3}, 2\frac{1}{5}\right\}$; $R = \left\{-\frac{1}{4}, \frac{3}{8}, 4\right\}$ **5.** {(−1, 6), (3, 0), (2, 4), (8, 0)}; D = {−1, 2, 3, 8}; R = {0, 4, 6}; I = {(6, −1), (0, 3), (4, 2), (0, 8)}

Page B43 Review Lesson 5–3

1. b, c, d **3.** {(−2, −5), (−1, −4), (0, −3), (1, −2), (2, −1)} **5.** $\left\{\left(-2, -3\frac{2}{3}\right), (-1, -3), \left(0, -2\frac{1}{3}\right), \left(1, -1\frac{2}{3}\right), (2, -1)\right\}$

7. $\left\{\left(-2, 3\frac{6}{7}\right), \left(-1, 3\frac{3}{7}\right), (0, 3), \left(1, 2\frac{4}{7}\right), \left(2, 2\frac{1}{7}\right)\right\}$

9a. $y = x + 1$ **9b.**

x	y
3	4
4	5
5	6
6	7

Page B44 Review Lesson 5–4

1. yes; $5x - 3y = 6$ **3.** yes; $y = 27$

5.

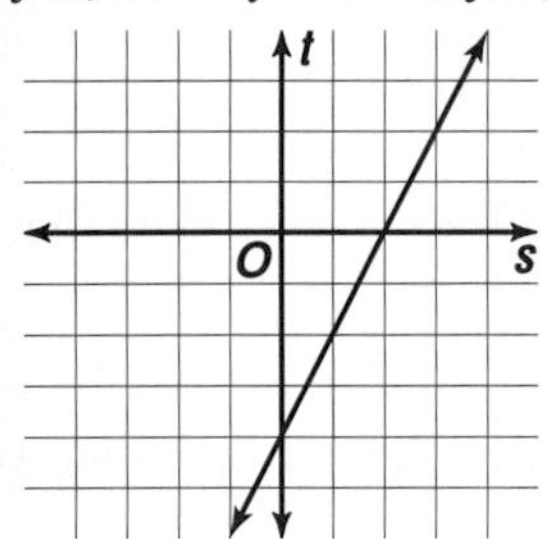

7.

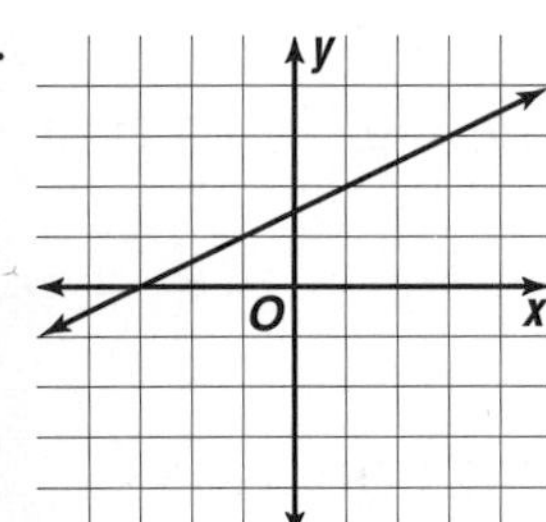

9.

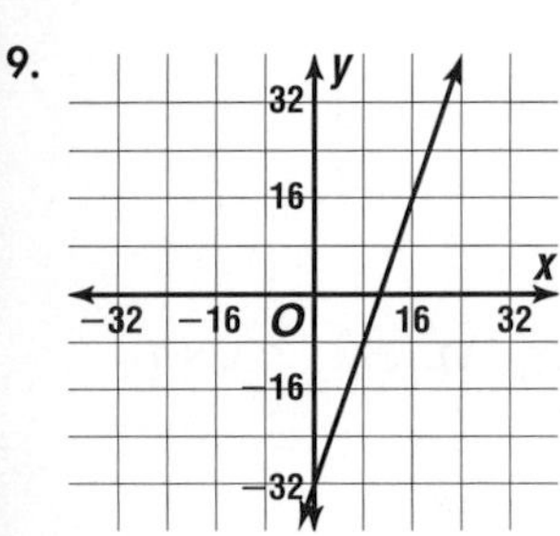

Page B45 Review Lesson 5–5

1. no **3.** no **5.** no **7.** yes **9.** yes **11.** −14 **13.** 7 **15.** −2

Page B46 Review Lesson 5–6

1. $b = \frac{1}{4}a$ **3.** $y = 2x + 5$ **5a.** $f(x) = 0.02x$ **5b.** \$340

Page B47 Review Lesson 5–7

1. 29; 86.5; 90.5; 82; 8.5 **3.** 16; 28; 34; 24.5; 9.5 **5.** 10; 5; 6.5; 2.5; 4 **7.** 26; 63; 77; 60; 17 **9.** 12,000,000; 7,500,000; 12,500,000; 5,000,000; 7,500,000

Page B48 Chapter 5 Test

1. $\{(3, 12), (0, 5), (-4, -8)\}$

3.

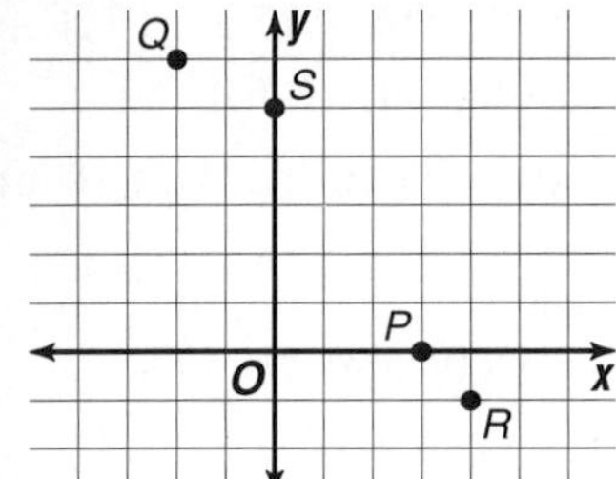

5.

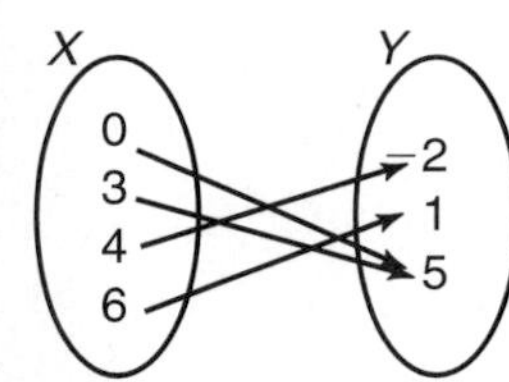

7. $\{(-2, -5), (0, -4), (2, -3), (4, -2)\}$

9.

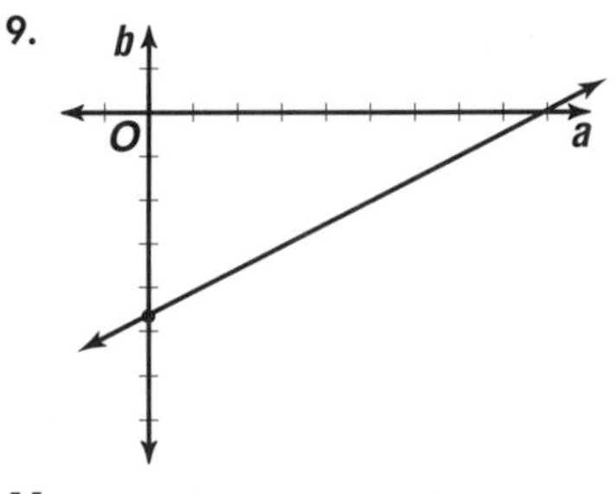

11.

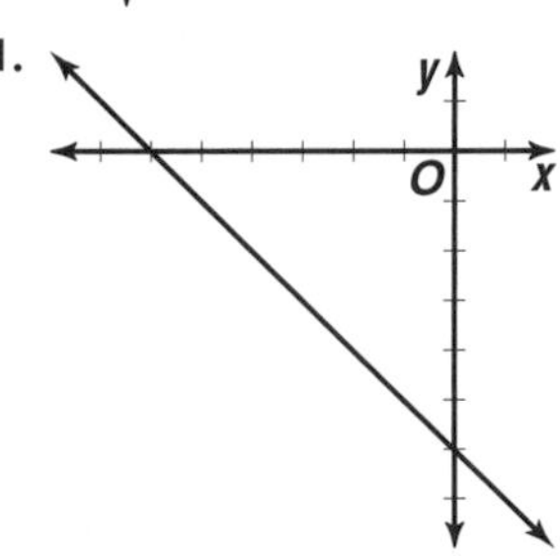

13. 33 **15.** $n = 3m - 2$ **17.** 8.1; 11.8

Page B49 Review Lesson 6–1

1. −2 **3.** $\frac{1}{6}$ **5.** 9 **7.** 5 **9.** −3 **11.** 3 **13.** 692 feet

Page B50 Review Lesson 6–2

1. $7x - 10y = -25$ **3.** $3x + y = 13$ **5.** $x - 6y = 3$ **7.** $y - 5 = \frac{3}{10}(x - 7)$ **9.** $y - 4 = 6(x - 4)$ **11.** $y - 6 = -1(x - 7)$ **13a.** $y - 2 = \frac{5}{7}(x - 3)$ **13b.** no

Page B51 Review Lesson 6–3

1a.

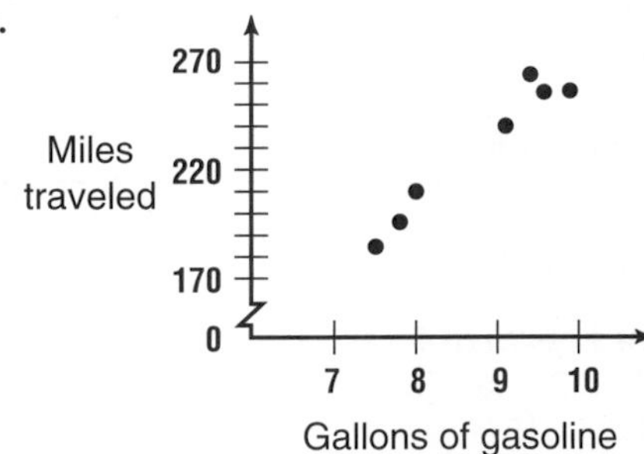

1b. gallons of gasoline and the miles traveled **1c.** Yes, there is a strong positive correlation.

3a.

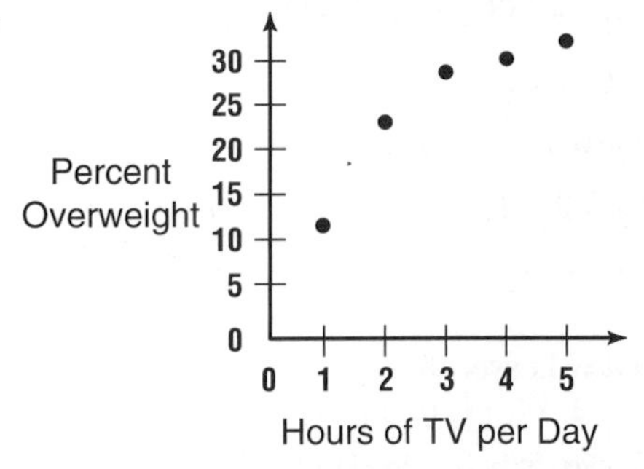

3b. $y = 4.9x + 10.5$ **3c.** The more television a child watches, the greater the likelihood that he or she will be overweight.

Page B52 Review Lesson 6–4

1. 4; 2 **3.** $\frac{9}{4}$; $-\frac{9}{2}$ **5.** $y = 5$; $y = 5$ **7.** $y = -9x + 8$; $9x + y = 8$ **9.** $-\frac{1}{7}$, $\frac{22}{7}$; $y = -\frac{1}{7}x + \frac{22}{7}$ **11.** 4; −4; $y = 4x - 4$ **13a.** $y = 2.50x + 4.95$ **13b.** Switch to the unlimited usage plan.

Page B53 Review Lesson 6–5

1.

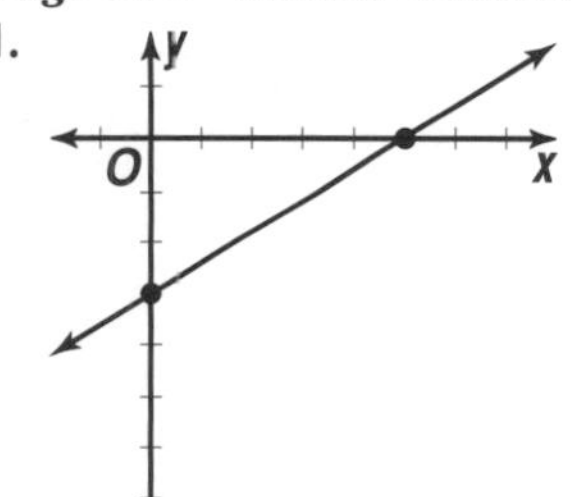

3.

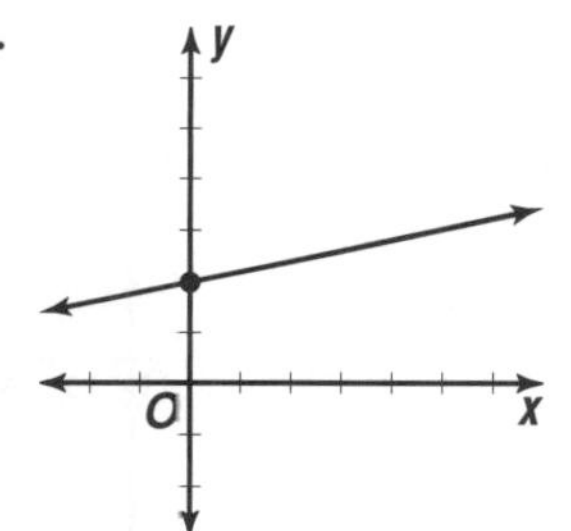

5a.

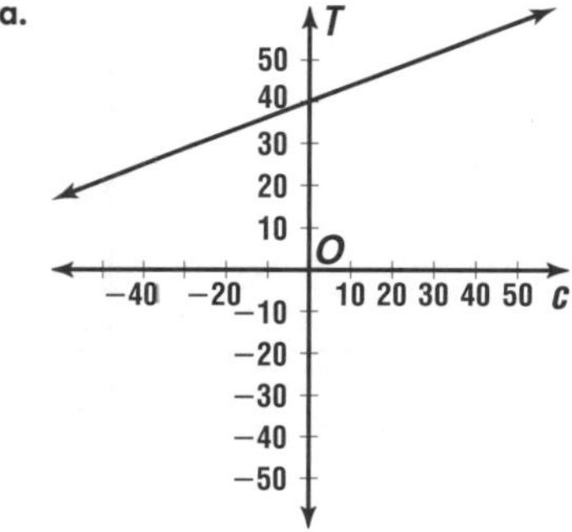

5b. 73°

Page B54 Review Lesson 6–6

1. $y = 4x + 2$ **3.** $y = \frac{4}{9}x + \frac{10}{3}$ **5.** $y = -\frac{1}{5}x + \frac{4}{5}$ **7.** $y = 0.73x + 8.3$

Page B55 Review Lesson 6–7

1. (5, 3) **3.** $\left(\frac{9}{2}, 0\right)$ **5.** $\left(4, -\frac{7}{2}\right)$ **7.** (6, 8) **9.** $(3a, 3b)$ **11.** $M(7, -31)$ **13.** $N(49, -4)$ **15.** $M(-21, -7)$

Page B56 Chapter 6 Test

1. $-\frac{1}{5}$ **3.** 8 **5.** $y = 3$

7.

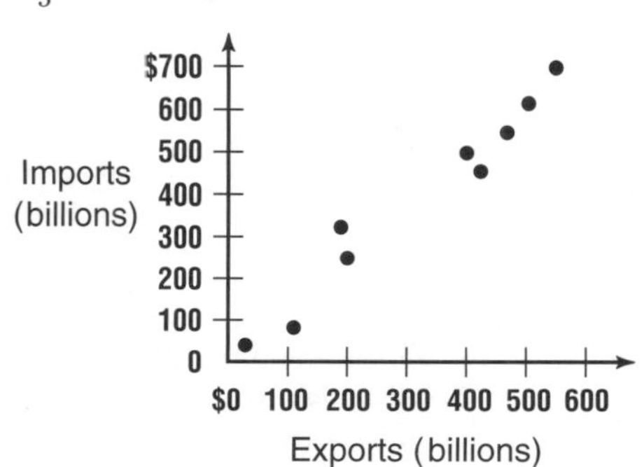

9. $\frac{5}{9}$; $-\frac{5}{3}$ **11.** $-\frac{3}{14}$; $\frac{20}{7}$

13.

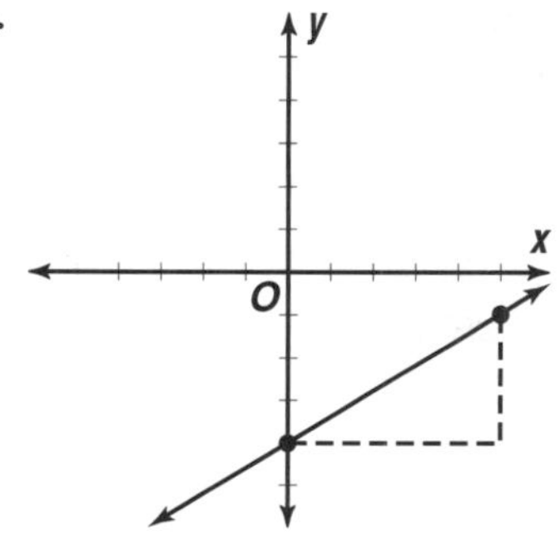

15. $y = -\frac{1}{5}x + 9$ **17.** $\left(5, -\frac{3}{2}\right)$ **19.** $M(-21, 15)$

Page B57 Posttest

1. C **3.** A **5.** D **7.** A **9.** D **11.** A **13.** D **15.** B **17.** B **19.** C **21.** C **23.** C

CHAPTER 7 SOLVING LINEAR EQUATIONS

Pages 388–390 Lesson 7–1

7. c **9.** d **11.** $\{x \mid x > -5\}$ **13.** $\{y \mid y < -5\}$ **15.** $x - 17 < -13$, $\{x \mid x < 4\}$

17. $\{a \mid a < 18\}$

13 14 15 16 17 18 19 20 21

19. $\{x \mid x \leq 1\}$

−4 −3 −2 −1 0 1 2 3 4

21. $\left\{x \mid x > \frac{11}{3}\right\}$

1 2 3 $\frac{11}{3}$ 5 6 7 8 9

23. $\{x \mid x > 2\}$

−1 0 1 2 3 4 5 6 7

25. $\left\{x \mid x < \frac{3}{8}\right\}$ **27.** $\{x \mid x \leq 15\}$ **29.** $\{x \mid x < 0.98\}$

31. $\{r \mid r < 10\}$ **33.** $x - (-4) \geq 9$, $\{x \mid x \geq 5\}$ **35.** $3x < 2x + 8$, $\{x \mid x < 8\}$ **37.** $20 + x < 53$, $\{x \mid x < 33\}$ **39.** $2x > x - 6$, $\{x \mid x > -6\}$ **41.** 12 **43.** −2 **45a.** no **45b.** yes **45c.** yes **45d.** yes **47.** The value of x falls between −2.4 and 3.6. **49a.** $x \leq \$12.88$ **49b.** Sample answer: There may be sales tax on his purchases. **51.** $y = -3x + 3$ **53.** 42, 131, 145, 159, 28 **55.** 12 **57.** <

Pages 366–368 Lesson 6-6

7. $\frac{2}{3}, -\frac{3}{2}$ **9.** perpendicular **11.** $y = \frac{5}{6}x - \frac{21}{2}$
13. $y = x - 9$ **15.** $y = \frac{9}{5}x$ **17.** perpendicular
19. parallel **21.** perpendicular **23.** perpendicular
25. $y = x + 1$ **27.** $y = \frac{8}{7}x + \frac{2}{7}$ **29.** $y = 2.5x - 5$
31. $y = \frac{2}{3}x + 4$ **33.** $y = -\frac{9}{2}x + 14$ **35.** $y = 3x - 19$
37. $y = -\frac{1}{5}x - 1$ **39.** $y = -\frac{7}{2}x + 11$ **41.** $y = -3$
43. $y = \frac{5}{4}x$ **45.** $y = \frac{1}{3}x - 6$ **47.** No, because the slope of $\overline{AC}$ is $\frac{6}{7}$ and the slope of $\overline{BC}$ is $-\frac{2}{3}$. These slopes are not negative reciprocals of each other, so the lines are not perpendicular and the figure is not a rhombus.
49a. $y = \frac{1}{2}x + \frac{7}{2}$, $y = -2x + 11$ **49b.** right or 90° angle
51.

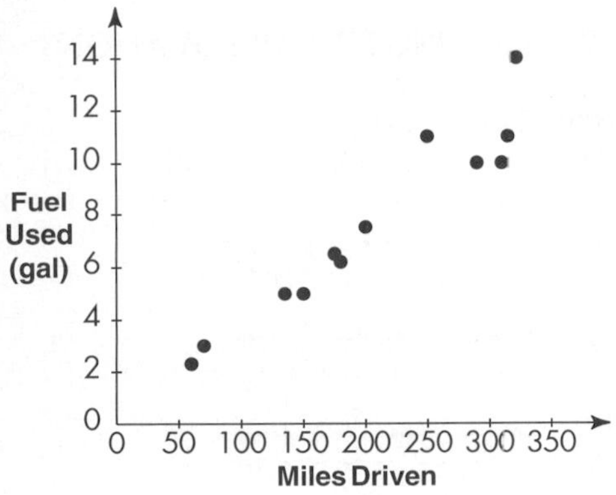

53. \$242.80 **55a.** $\frac{5}{12}$; $\frac{7}{12}$ **55b.** 16-karat gold
57. multiplicative identity

Pages 372–374 Lesson 6-7

5. (1.5, 6) **7.** (−6, 1) **9.** (−11, 7) **11.** (8, 9.8)
13. (12.5, 6) **15.** (1, 5) **17.** (3, 2) **19.** $\left(\frac{1}{2}, 1\right)$
21. (0.8, 2.7) **23.** (4x, 9y) **25.** (−7, 0) **27.** (21, −6)
29. (9, 10) **31.** $\left(\frac{5}{6}, \frac{1}{3}\right)$ **33.** $B(2.3, 6.8)$ **35.** $P(6.65, -1.85)$
37. (−1, 5) **39.** $\left(1, \frac{5}{2}\right)$ **41a.** $N(6, 3)$, $M(10, 3)$
41b. parallel, $MN = \frac{1}{2}AB$ **43a.** $P(-4, 1)$, $Q(10, -1)$, $R(2, 9)$ **43b.** 62 square units; Sample answer: The area of the smaller triangle is $\frac{1}{2}bh$. Since the base of the larger triangle is twice that of the smaller one and the height is also twice the length of the small one, the area of the larger is $\frac{1}{2}(2b)(2h)$, or $2bh$. This is 4 times the area of the small one. **45.** $y = -\frac{7}{9}x - \frac{8}{3}$ **47.** yes; $9x - 6y = 7$
49. −20 **51.** $3.1x + 1.54$

Page 375 Chapter 6 Highlights

1. parallel **3.** midpoint **5.** perpendicular
7. slope-intercept **9.** slope

Pages 376–378 Chapter 6 Study Guide and Assessment

11. $-\frac{1}{3}$ **13.** $\frac{2}{5}$ **15.** $\frac{25}{3}$ **17.** $y + 3 = -2(x - 4)$
19. $y - 7 = 0$ **21.** $y - 3 = \frac{3}{5}x$ **23.** $y - 1 = -\frac{6}{7}(x - 4)$

25. $3x - y = 18$ **27.** $3x - 4y = 22$ **29.** $y = 5$ **31.** $x = -2$
33a. Yes; it is positive **33b.** Sample answer: 35 stories
33c. $y = 0.03x + 21$

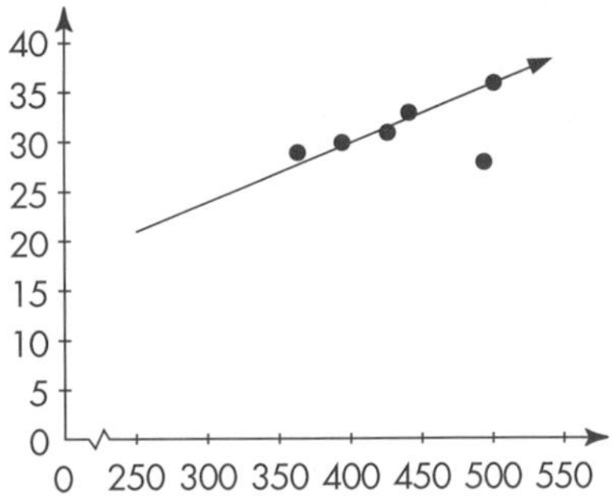

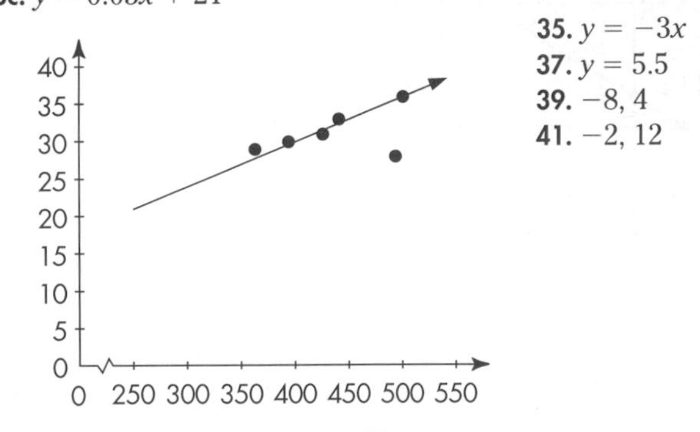

35. $y = -3x$
37. $y = 5.5$
39. −8, 4
41. −2, 12

43.

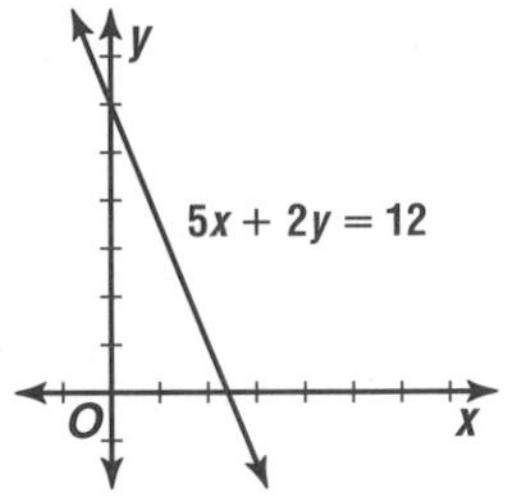

45.

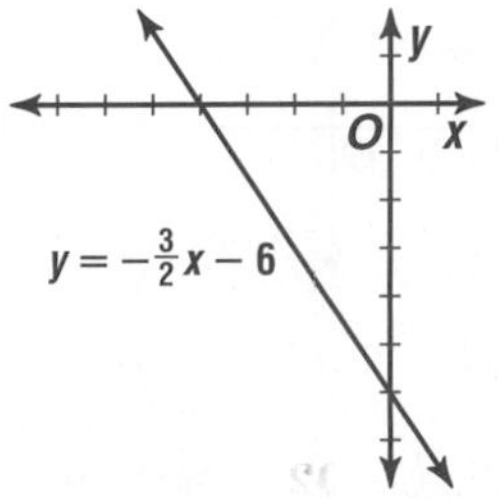

47.

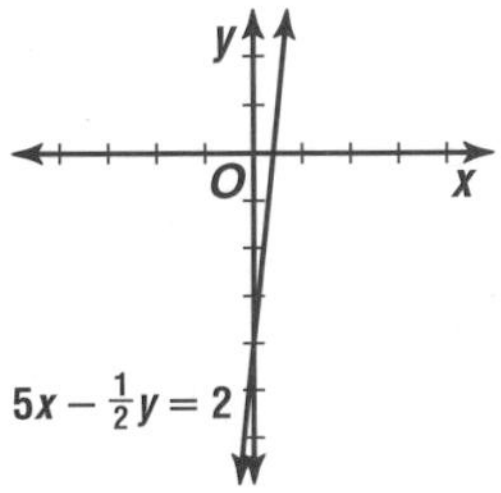

49.

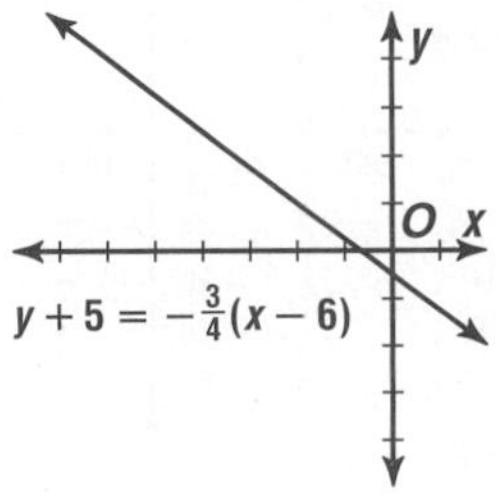

51. $y = -\frac{7}{2}x - 14$ **53.** $y = -\frac{3}{8}x + \frac{13}{2}$ **55.** $y = 5x - 15$
57. $\left(1, -\frac{5}{2}\right)$ **59.** $\left(5, \frac{11}{2}\right)$ **61.** $\left(\frac{7}{2}, -\frac{3}{2}\right)$ **63.** (13, 11)
65. (−5, 20) **67.** $d = 45t - 10$

CHAPTER 7 SOLVING LINEAR EQUATIONS

Pages 388–390 Lesson 7-1

7. c **9.** d **11.** $\{x \mid x > -5\}$ **13.** $\{y \mid y < -5\}$
15. $x - 17 < -13$, $\{x \mid x < 4\}$
17. $\{a \mid a < 18\}$

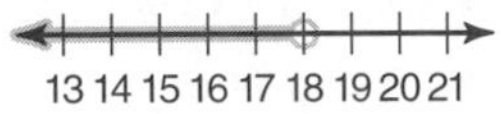

19. $\{x \mid x \le 1\}$

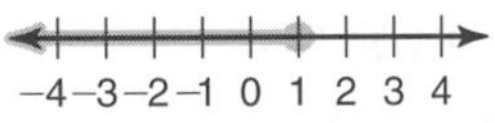

21. $\left\{x \mid x > \frac{11}{3}\right\}$

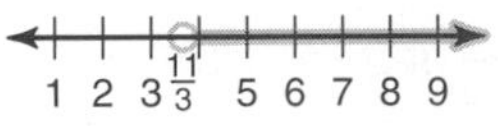

23. $\{x \mid x > 2\}$

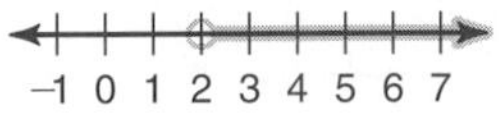

25. $\left\{x \mid x < \frac{3}{8}\right\}$ **27.** $\{x \mid x \le 15\}$ **29.** $\{x \mid x < 0.98\}$
31. $\{r \mid r < 10\}$ **33.** $x - (-4) \ge 9$, $\{x \mid x \ge 5\}$ **35.** $3x < 2x + 8$, $\{x \mid x < 8\}$ **37.** $20 + x < 53$, $\{x \mid x < 33\}$ **39.** $2x > x - 6$, $\{x \mid x > -6\}$ **41.** 12 **43.** −2 **45a.** no **45b.** yes **45c.** yes **45d.** yes **47.** The value of x falls between −2.4 and 3.6. **49a.** $x \le \$12.88$ **49b.** Sample answer: There may be sales tax on his purchases. **51.** $y = -3x + 3$
53. 42, 131, 145, 159, 28 **55.** 12 **57.** <

Page 391 Lesson 7–2A

1. Sample answer: The variable remains on the left, but the inequality symbol is reversed. **3.** When the coefficient of x is positive, you can solve the inequality like an equation and retain the same inequality symbol. If the coefficient of x is negative, you can solve like an equation, but the symbol must be reversed.

Pages 396–398 Lesson 7–2

9. multiply by $-\frac{1}{6}$ or divide by -6; yes; $\{y \mid y \le 4\}$ **11.** multiply by 4; no; $\{x \mid x < -20\}$ **13.** $\{x \mid x < 30\}$ **15.** $\{t \mid t \le -30\}$ **17.** $\frac{1}{5}x \le 4.025$; $\{x \mid x \le 20.125\}$ **19.** $s \ge 12$ **21.** $\{b \mid b > -12\}$ **23.** $\{x \mid x \ge -44\}$ **25.** $\{r \mid r < -6\}$ **27.** $\{t \mid t < 169\}$ **29.** $\{g \mid g \ge 7.5\}$ **31.** $\{x \mid x \ge -0.7\}$ **33.** $\left\{r \mid r < -\frac{1}{20}\right\}$ **35.** $\{x \mid x < -27\}$ **37.** $\{m \mid m \ge -24\}$ **39.** $36 \ge \frac{1}{2}x$; $\{x \mid x \le 72\}$ **41.** $\frac{3}{4}x \le -24$; $\{x \mid x \le -32\}$ **43.** $-8x \le 144$; -18 or greater **45.** $y < 7.14$ meters **47.** $\ge$ **49.** $<$ **51.** up to 416 miles **53.** at least 5883 signatures **55.** $(-1, 1)$ **57.** -2 **59.** \$155.64 **61.** 65 yd by 120 yd

Pages 402–404 Lesson 7–3

7. c **9.** $\{x \mid x > 2\}$ **11.** $\{d \mid d > -125\}$ **13.** $\{2, 3\}$ **15a.** $x + (x + 2) > 75$ **15b.** $x > 36.5$ **15c.** Sample answer: 38 and 40. **17.** $\{-10, -9, \ldots, 2, 3\}$ **19.** $\{-10, -9, \ldots, -5, -4\}$ **21.** $\{t \mid t > 3\}$ **23.** $\{w \mid w \le 15\}$ **25.** $\{n \mid n > -9\}$ **27.** $\{m \mid m < 15\}$ **29.** $\{x \mid x < -15\}$ **31.** $\left\{p \mid p \le \frac{14}{3}\right\}$ **33.** $\{x \mid x > -10\}$ **35.** $\{k \mid k \le -1\}$ **37.** $\{y \mid y < -1\}$ **39.** $3(x + 7) > 5x - 13$; $\{x \mid x < 17\}$ **41.** $2x + 2 \le 18$ for $x > 0$; 7 and 9; 5 and 7; 3 and 5; 1 and 3 **43.** no solution $\{\varnothing\}$ **45a.** $x \le -8$ **45b.** $x > 8$ **45c.** $x > 2$ **45d.** $x \le -1$ **47.** $x + 0.04x + 0.15(x + 0.04x) \le \$50$, $x \le \$41.80$ **49.** at least \$571,428.57 **51a.** at most 2.9 weeks **51b.** no change **51c.** at most 4.1 weeks **53.** $\{y \mid y > 10\}$ **55.** $3x + 2y = 14$ **57.** $\{-5, -3, -2, 4, 16\}$ **59.** 25.1; 23.5; no mode

Pages 409–412 Lesson 7–4

7. $0 \le x \le 9$

9. $-3 < x \le 1$ **11.** The solution is the empty set. There are no numbers greater than 5 but less than -3.

13. $\{h \mid h \le -7$ or $h \ge 1\}$

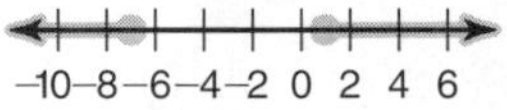

15. $\{w \mid 1 > w \ge -5\}$

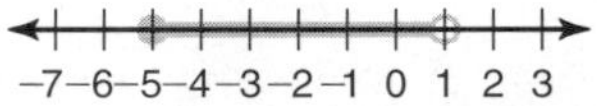

17. Drawings will vary; 16 pieces. **19.** $\varnothing$

21. **23.**

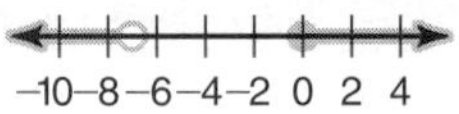

25. $-4 \le x \le 5$ **27.** $x \le -2$ or $x > 1$

29. $\{x \mid -1 < x < 5\}$

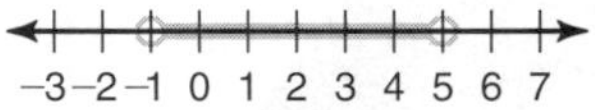

31. $\{x \mid x < -2$ or $x > 3\}$

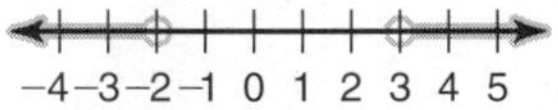

33. $\{c \mid c < 7\}$

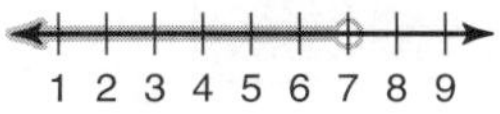

35. $\varnothing$

37. $\{x \mid x$ is a real number.$\}$

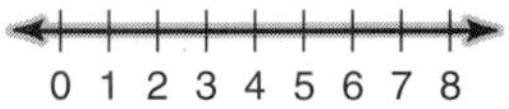

39. $\{y \mid y > 3$ and $y \ne 6\}$

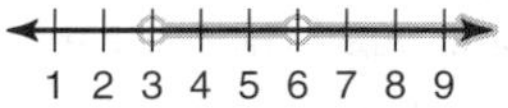

41. $\{x \mid x$ is a real number.$\}$

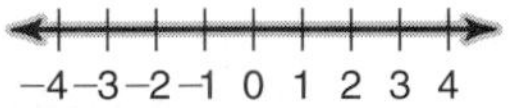

43. $\{w \mid w < 4\}$

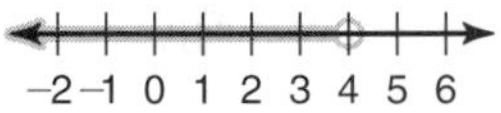

45. Sample answer: $x > 5$ and $x < -4$ **47.** $n + 2 \le 6$ or $n + 2 \ge 10$; $\{n \mid n \le 4$ or $n \ge 8\}$ **49.** $31 \le 6n - 5 \le 37$; $\{n \mid 6 \le n \le 7\}$

51. $\{m \mid -4 < m < 1\}$

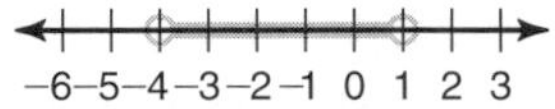

53a. $\{x \mid x < -7$ or $x > 1\}$ **53b.** $\{x \mid -5 \le x < 1\}$ **55.** $-4 \le x \le -1.5$ or $x \ge 2$ **57.** $4.4 < x < 6.7$ **59.** $\left\{m \mid m \ge \frac{44}{3}\right\}$ **61.** -5

63.

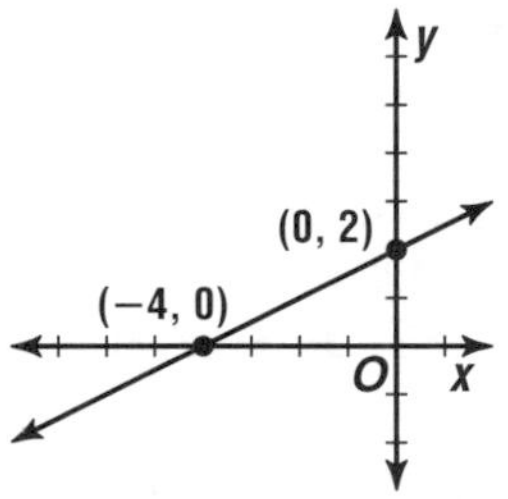

(0, 2)
(−4, 0)

65. a little more than half a mile **67.** $18px - 15bg$

Page 412 Self Test

1. $\{y \mid y \ge -17\}$ **3.** $\{n \mid n < 4\}$ **5.** $\{g \mid g < -5\}$ **7.** c **9.** more than 17 points

Pages 415–419 Lesson 7–5

7a. outcomes from tree diagram: burger, soup, lemonade; burger, soup, soft drink; burger, salad, lemonade; burger, salad, soft drink; burger, french fries, lemonade; burger, french fries, soft drink; sandwich, soup, lemonade; sandwich, soup, soft drink; sandwich, salad, lemonade; sandwich, salad, soft drink; sandwich, french fries, lemonade; sandwich, french fries, soft drink; taco, soup, lemonade; taco, soup, soft drink; taco, salad, lemonade; taco, salad, soft drink; taco, french fries, lemonade; taco, french fries, soft drink; pizza, soup, lemonade; pizza, soup, soft drink; pizza, salad, lemonade; pizza, salad, soft drink; pizza, french fries, lemonade; pizza, french fries, soft drink **7b.** $\frac{1}{3}$ or $0.\overline{3}$ **7c.** $\frac{1}{12}$ or $0.08\overline{3}$ **7d.** $\frac{1}{24}$ or $0.041\overline{6}$ **9a.** 15 **9b.** $\frac{1}{5}$ or 0.2 **11.** $\frac{1}{3}$ or $0.\overline{3}$ **13a.** R3-G5, R3-R10, R3-B10, R3-G1, R3-Y14, B3-G5, B3-R10, B3-B10, B3-G1, B3-Y14, R5-G5, R5-R10, R5-B10, R5-G1, R5-Y14, R14-G5, R14-R10, R14-B10, R14-G1, R14-Y14, Y10-G5, Y10-R10, Y10-B10, Y10-G1, Y10-Y14 **13b.** $\frac{3}{25}$ or 0.12 **13c.** $\frac{2}{25}$ or 0.08 **13d.** 0 **13e.** $\frac{14}{25}$ or 0.56 **15a.** about 5.6% **15b.** about 26.3% **17.** 32% **19.** between 83 and 99, inclusive **21.** 3; -9 **23a.** 50° **23b.** 130° **23c.** yes

Pages 423–426 Lesson 7–6

5. c **7.** c **9.** d

11. $\{m \mid m \le -5$ or $m \ge 5\}$

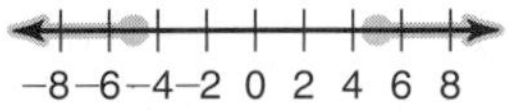

13. $\{r \mid -9 < r < 3\}$

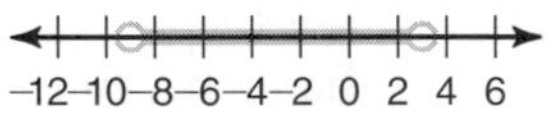

15. $|x| = 2$

17. $\{-2, 6\}$

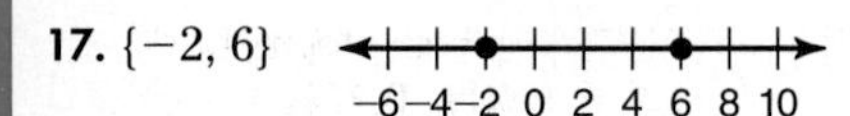

19. $\varnothing$

21. $\{y \mid 1 \leq y \leq 3\}$

−2 −1 0 1 2 3 4 5 6

23. $\varnothing$

25. $\left\{e \mid \frac{5}{3} < e < 3\right\}$

1 2 3 4

27. $\{y \mid y$ is a real number.$\}$

−4 −3 −2 −1 0 1 2 3 4

29. $\{w \mid 0 \leq w \leq 18\}$

0 2 4 6 8 10 12 14 16 18

31. $\{-2, 3\}$

−4 −3 −2 −1 0 1 2 3 4

33. $\left\{x \mid x \leq -\frac{8}{3} \text{ or } x \geq 4\right\}$

−4 −3 −2 −1 0 1 2 3 4 5 6

35. $|p - 1| \leq 0.01$ **37.** $|t - 50| > 50$
39. $|x + 1| = 3$ **41.** $|x - 1| \leq 1$ **43.** $|x - 8| \geq 3$
45. $\{-2, -1, 0, 1, 2\}$ **47.** $2a + 1$ **49.** $a \neq 0$; never
51. $\frac{8}{13}$ or 0.61 **53.** no; $52 \leq s \leq 66$ **55.** $\$16{,}500 \leq p \leq \$18{,}000$ **57a.** Outcomes from tree diagram: BBBB, BBBG, BBGB, BBGG, BGBB, BGBG, BGGB, BGGG, GBBB, GBBG, GBGB, GBGG, GGBB, GGBG, GGGB, GGGG
57b. $\frac{1}{16}$ or 0.0625, regardless of gender **57c.** $\frac{3}{8}$ or 0.375
59. $\{x \mid x \geq -1\}$ **61.** $\{k \mid k \geq -15\}$

63.

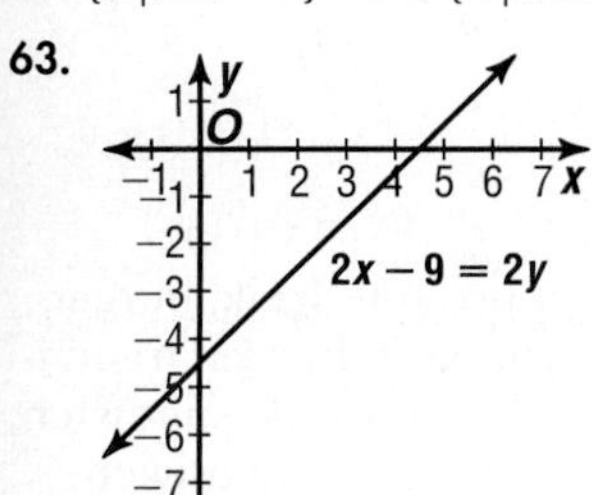

65. $m = -6 - \frac{n}{2}$

Pages 430–432 Lesson 7-7

7a. A; 25, 65, 30, 60, 40; B; 20, 70, 40, 60, 45 **7b.** B **7c.** A
7d. B **9a.** Q2 = 6.5, Q3 = 16, Q1 = 5, IQR = 11 **9b.** no

9c.

11a.

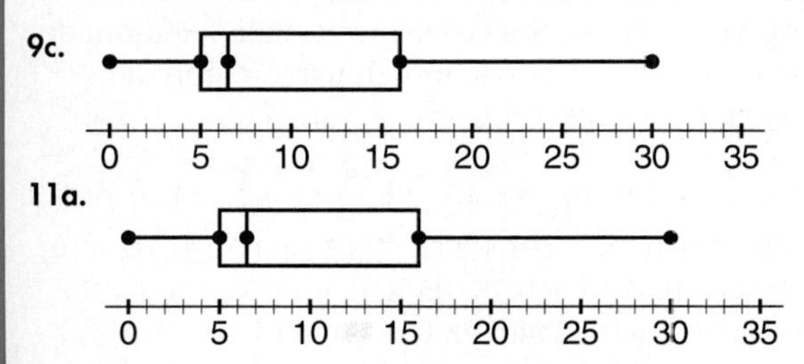

11b. clustered with lots of outliers **11c.** There are four western states that have more American Indian people than other states. **11d.** It is greater than the median.

13a.

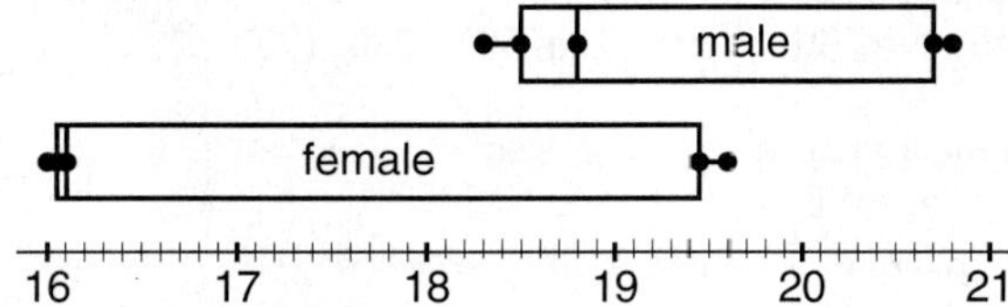

13b. 1990 **15.** Sample answer: Class A appears to be a more difficult class than B because the students don't do as well. **17.** $\{m \mid m > 1\}$

19.

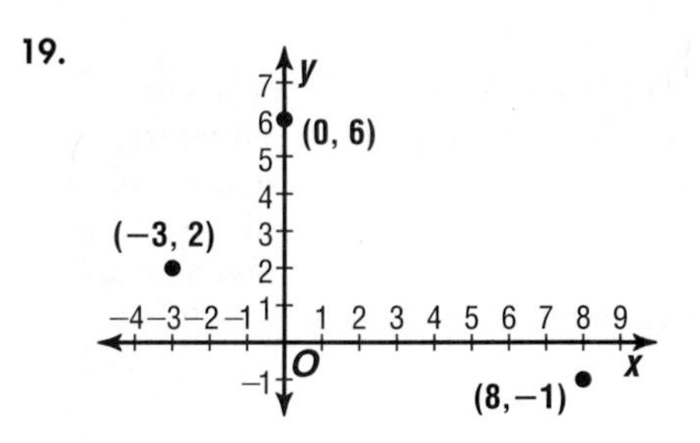

Page 434 Lesson 7-7B

1. women identifying objects with their left hands
3. The left hand data are more clustered. **5.** males

Page 435 Lesson 7-8A

1.

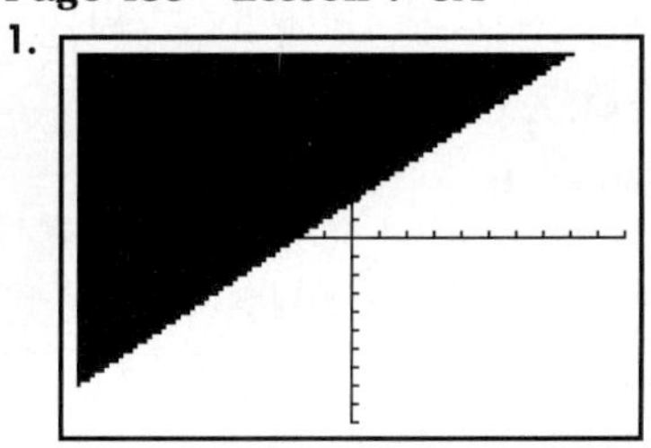

3.

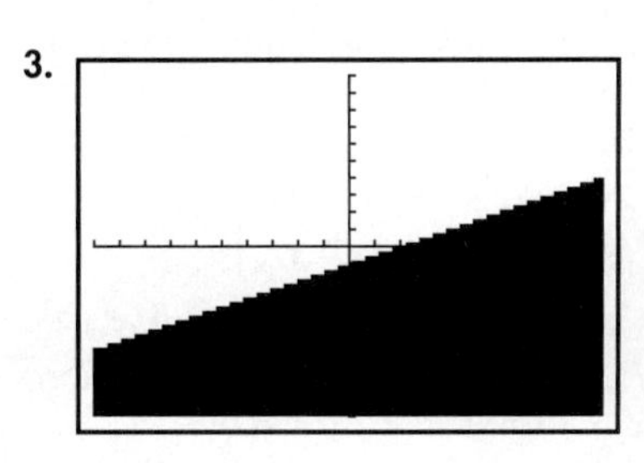

5.

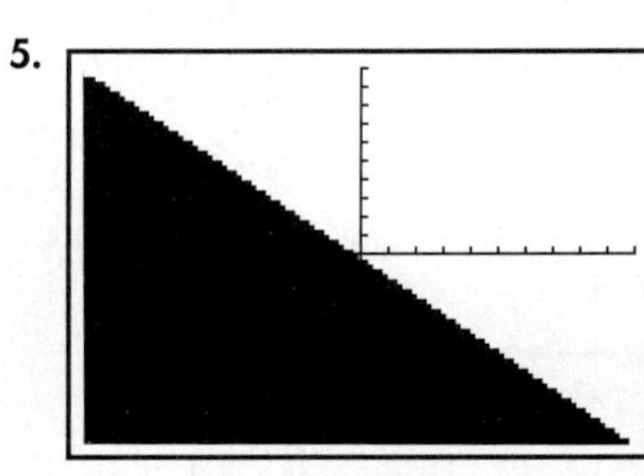

7.

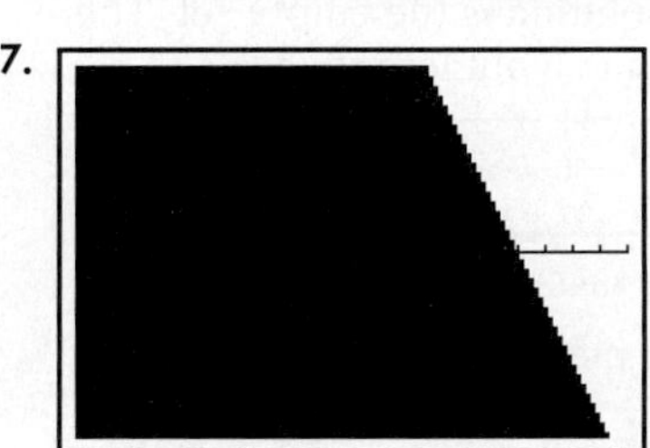

9.

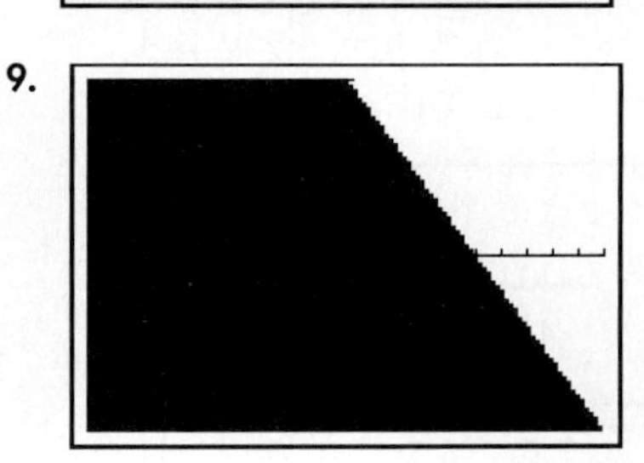

Page 439–441 Lesson 7-8

5. a **7.** b **9.** a, c; no

11. 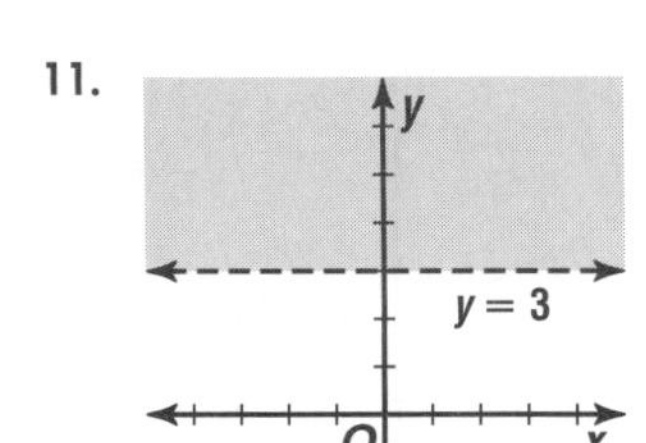

13. 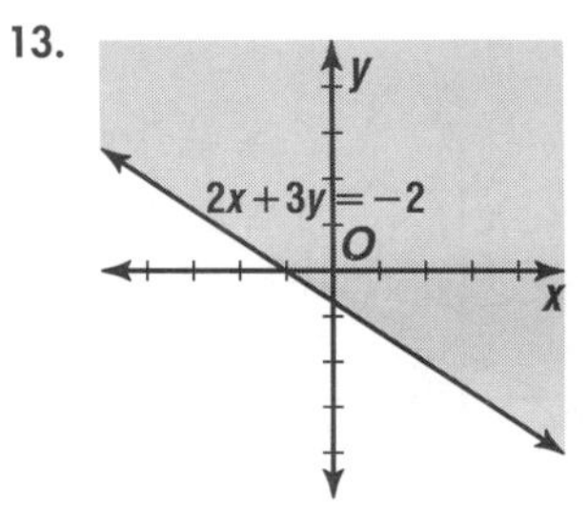

15. 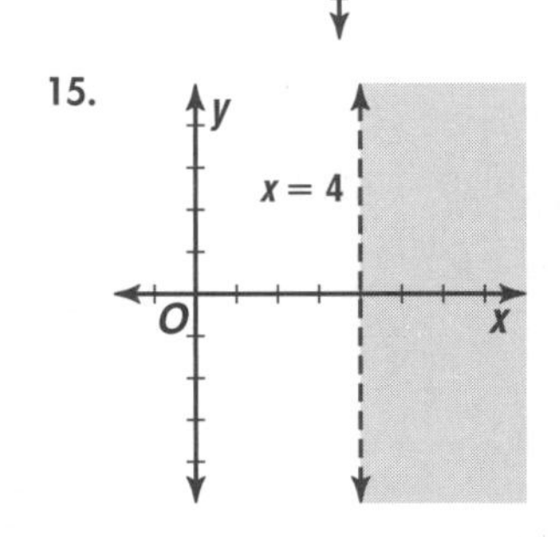

17. 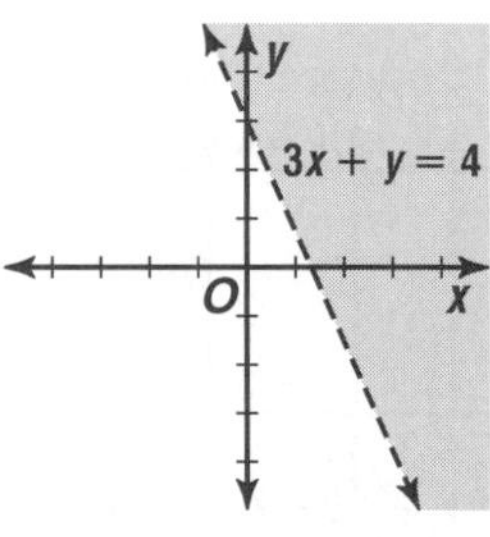

19. {(1, 1), (1, 2)} **21.** ∅

23. 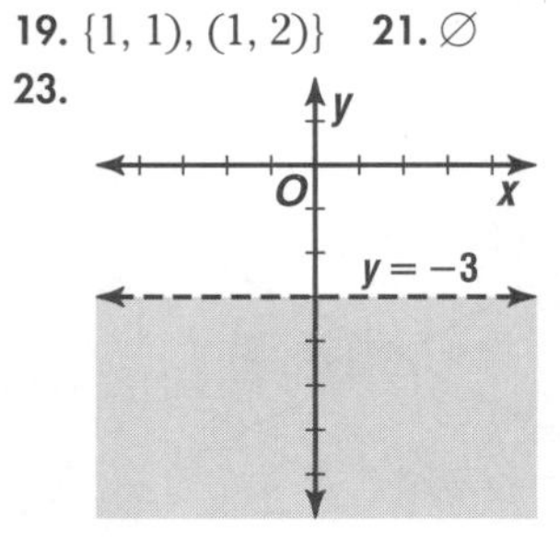

25. 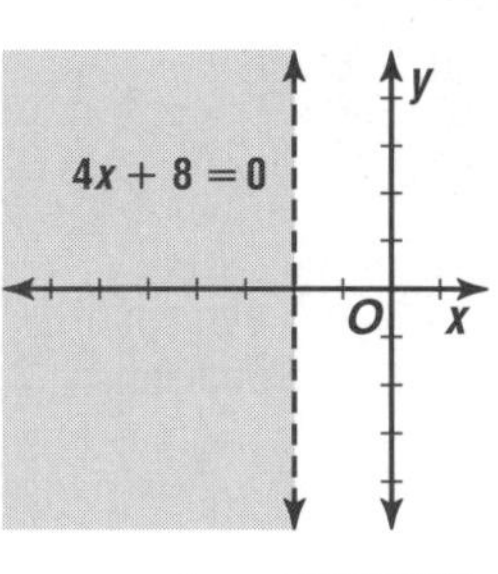

27. 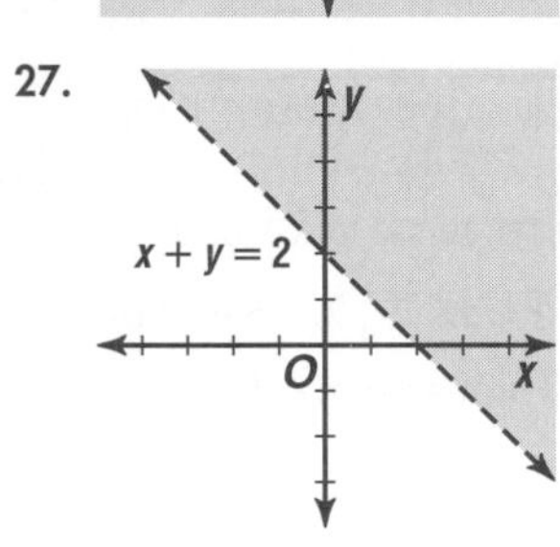

29. 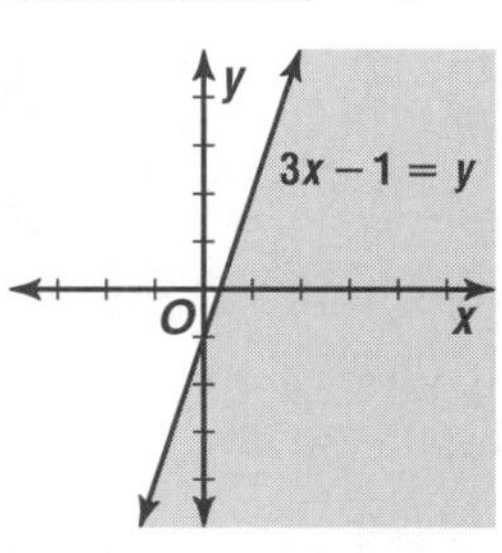

31. 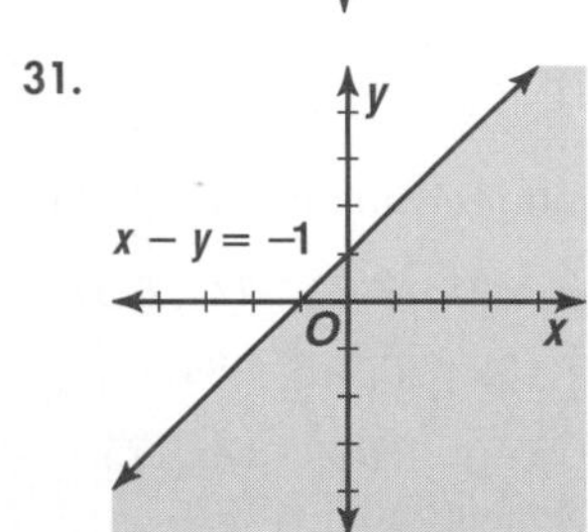

33. 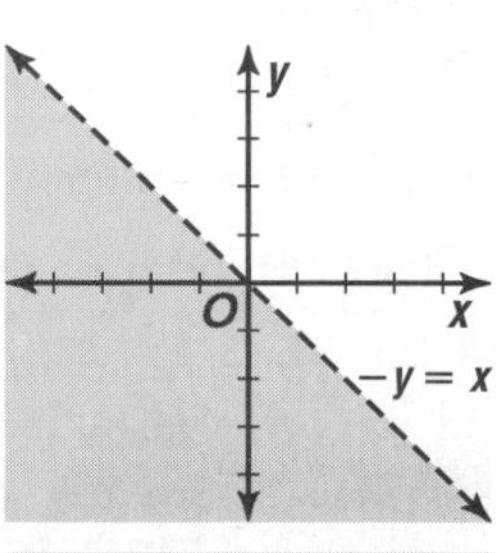

35. 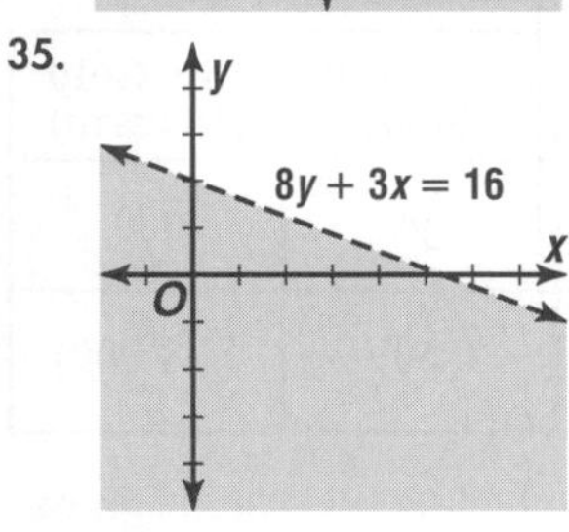

37. 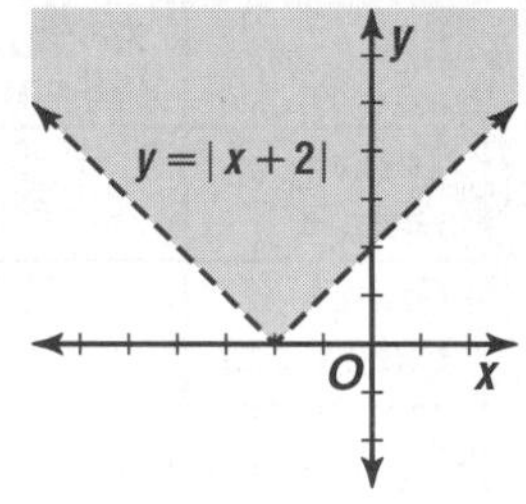

39. 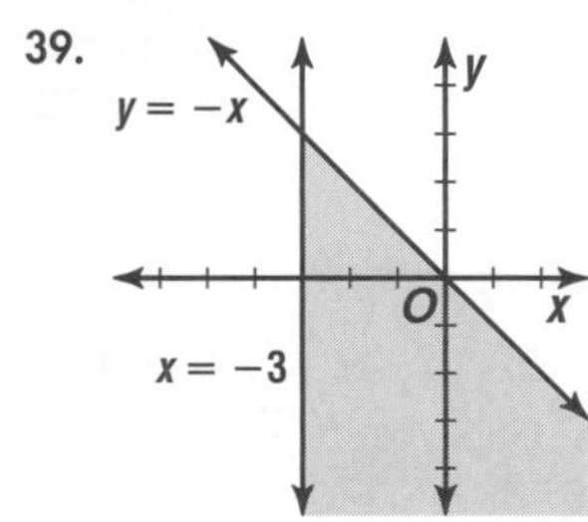

41. 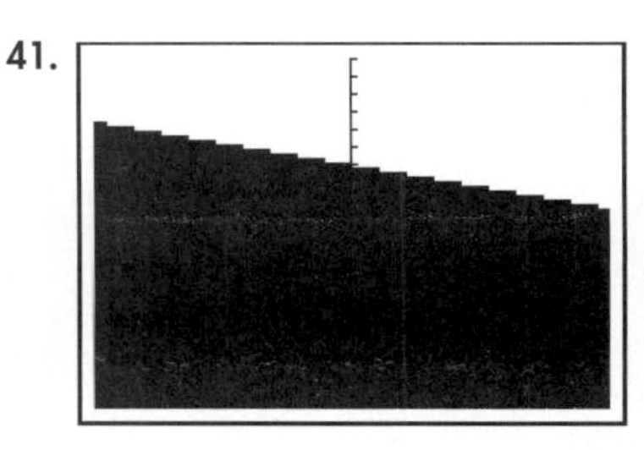

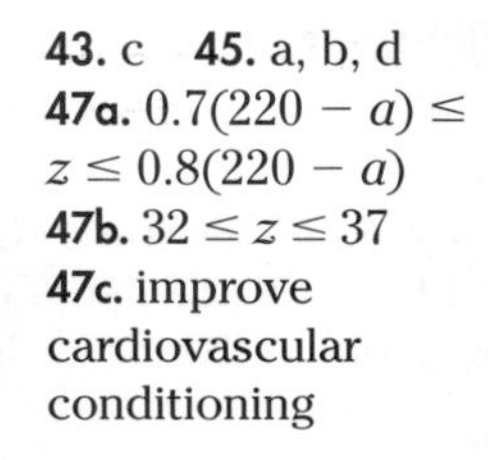
43. c **45.** a, b, d
47a. $0.7(220 - a) \leq z \leq 0.8(220 - a)$
47b. $32 \leq z \leq 37$
47c. improve cardiovascular conditioning

49. 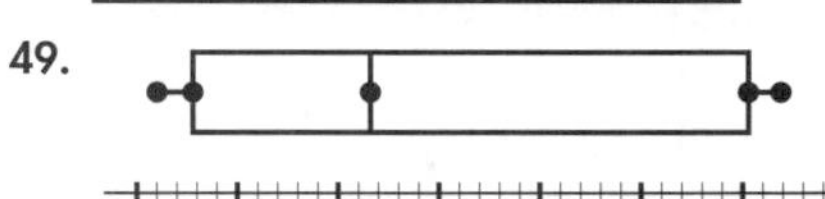

51. $y = 4x - 2$

53. 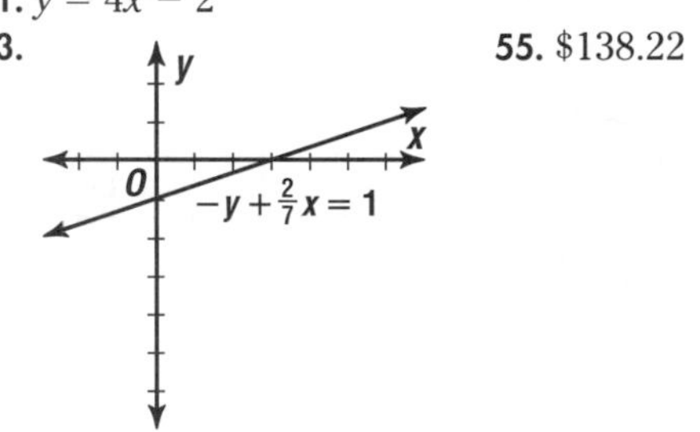

55. $138.22

Page 443 Chapter 7 Highlights

1. h **3.** i **5.** c, f **7.** d **9.** g **11.** a

Pages 444–446 Chapter 7 Study Guide and Assessment

13. $\{n \mid n > -35\}$ **15.** $\{p \mid p \leq -18\}$ **17.** $3n > 4n - 8$, $\{n \mid n < 8\}$ **19.** $\{w \mid w \leq -15\}$ **21.** $\{x \mid x > 32\}$
23. $-\frac{3}{4}n \leq 30$, $\{n \mid n \geq -40\}$ **25.** $\{-5, -4, \ldots 0, 1\}$

27. $\left\{y \mid y \leq -\frac{9}{2}\right\}$ **29.** $\left\{x \mid x > -\frac{5}{2}\right\}$ **31.** $\{z \mid z \leq 20\}$

33. $\{a \mid a$ is a real number$\}$
−4 −3 −2 −1 0 1 2 3 4

35. $\{b \mid b \leq 5\}$
−1 0 1 2 3 4 5 6 7

37a. $\frac{2}{7}$ **37b.** $\frac{1}{2}$ **37c.** $\frac{2}{7}$ **37d.** 0 **37e.** $\frac{3}{14}$ **37f.** $\frac{1}{14}$

39. $\{y \mid y > -5$ or $y < -5\}$
−9 −8 −7 −6 −5 −4 −3 −2 −1

41. $\{k \mid 3 \geq k \geq -4\}$
−5 −4 −3 −2 −1 0 1 2 3 4 5

43. $\left\{y \mid y \geq \frac{21}{5} \text{ or } y \leq 1\right\}$
−2 −1 0 1 2 3 4 5 6

45. 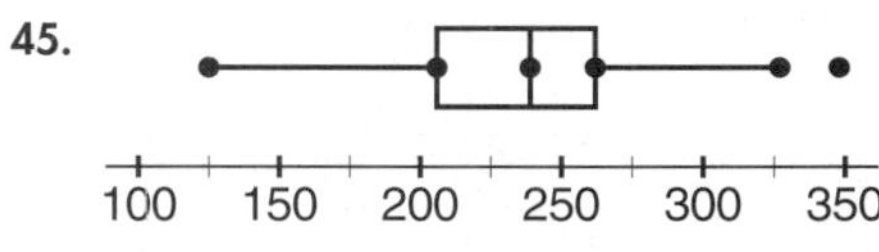

47. {(2, −1), (−1, 1)} **49.** {(5, 10), (3, 6)}

51. 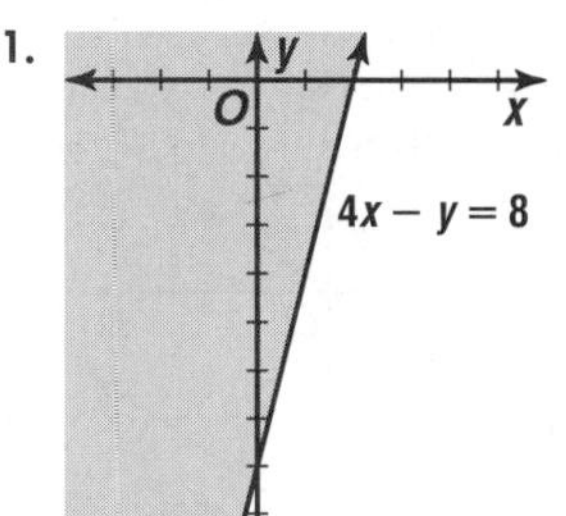

53. 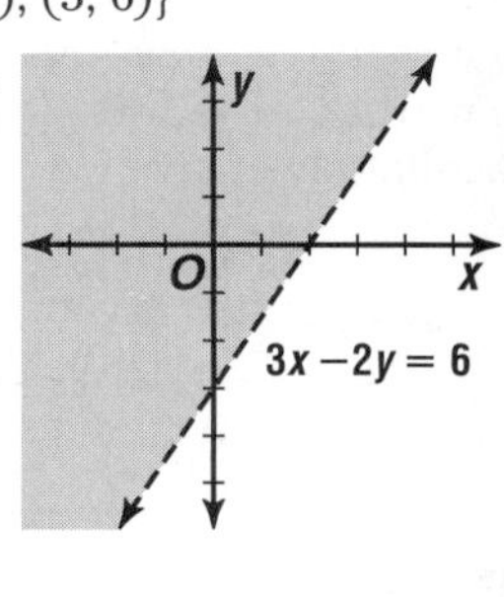

55. 17 to 20 books

CHAPTER 8 SYSTEMS OF LINEAR EQUATIONS AND INEQUALITIES

Page 453 Lesson 8–1A

1. (1, 8) **3.** (2.86, 4.57) **5.** (2.28, 3.08) **7.** (−2.9, 5.6) **9.** (1.14, −3.29)

Pages 458–461 Lesson 8-1

9. no solution **11.** one; (−6, 2) **13.** yes

15. (3, 5)

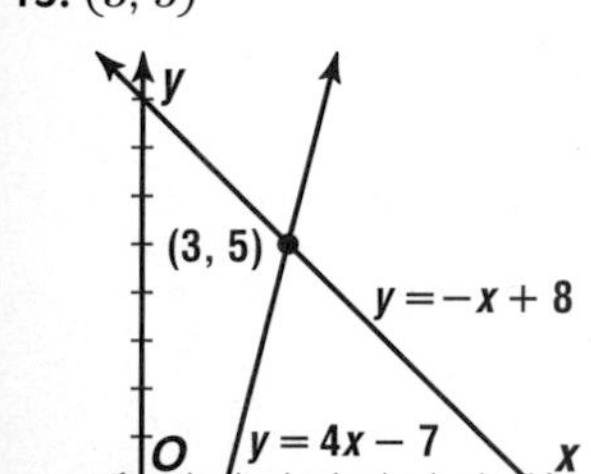

17. (2, −6)

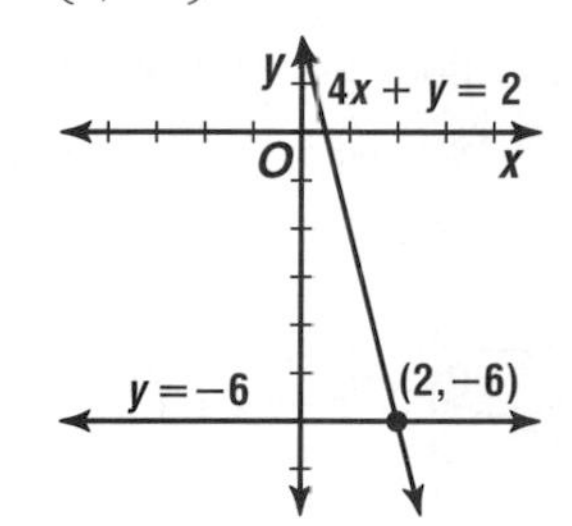

19. (−6, 8)

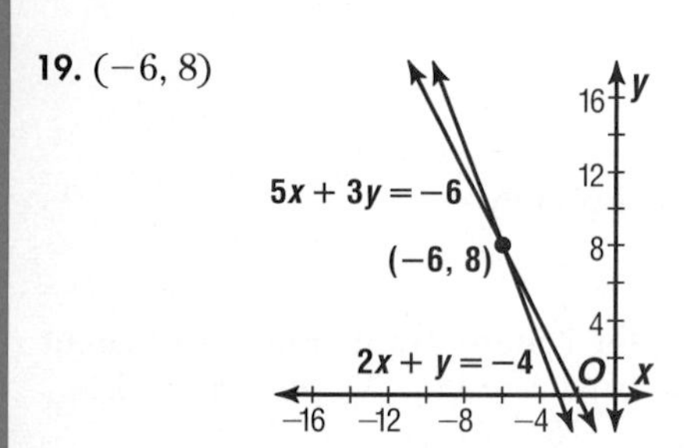

21. one, (3, −1) **23.** no solution **25.** one, (3, 3)

27. (2, −2)

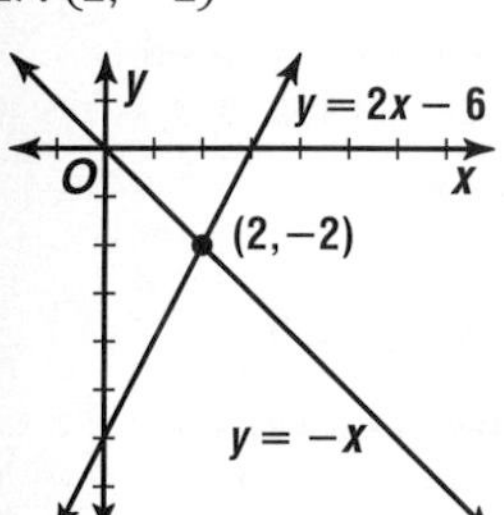

29. (−1, 3)

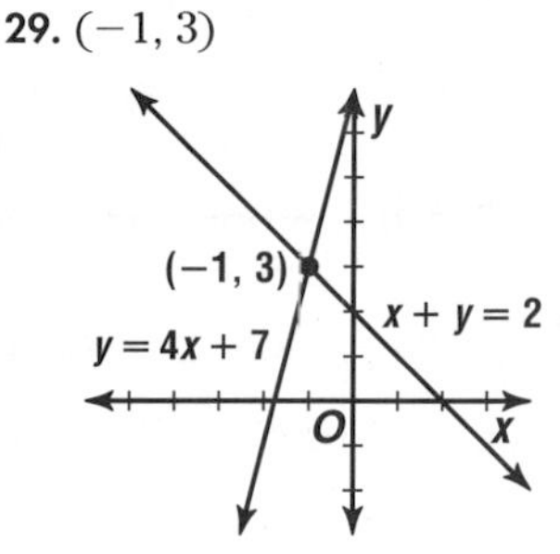

31. (−2, 4)

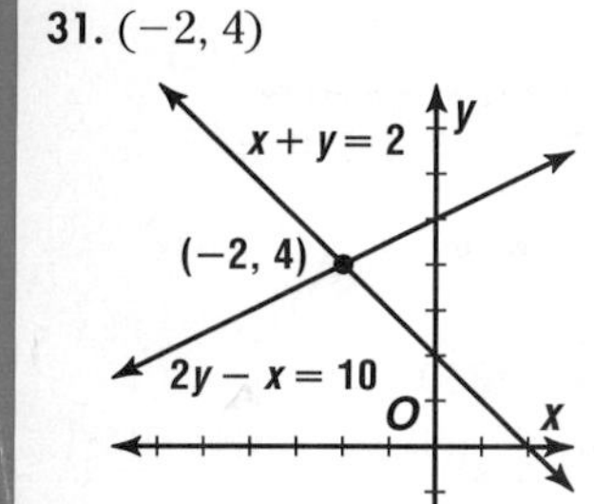

33. (2, 0)

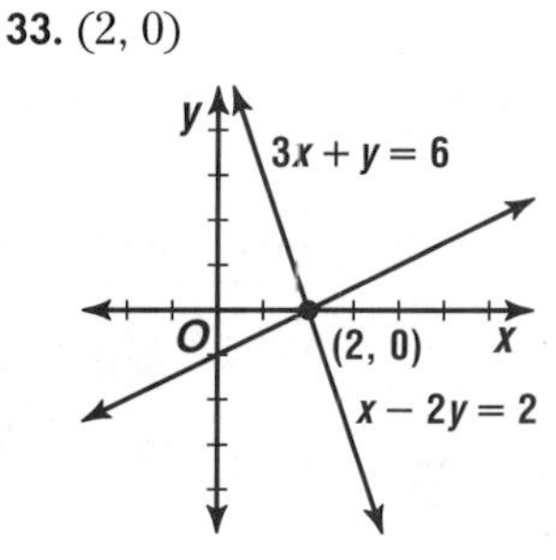

35. infinitely many

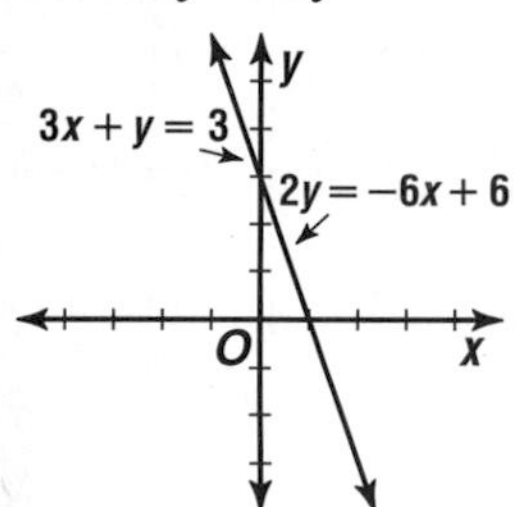

37. no solution

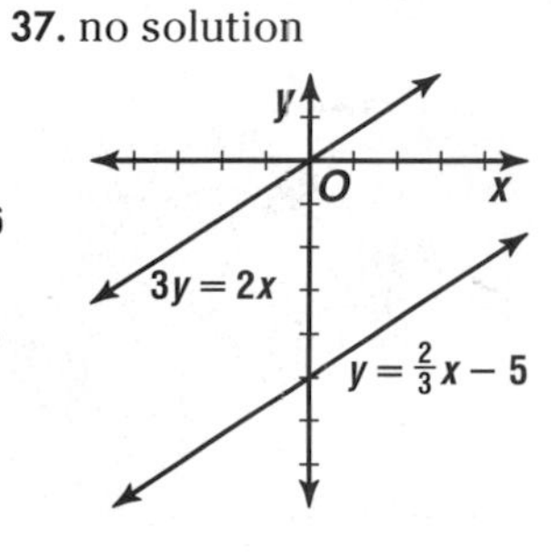

39. (8, 6)

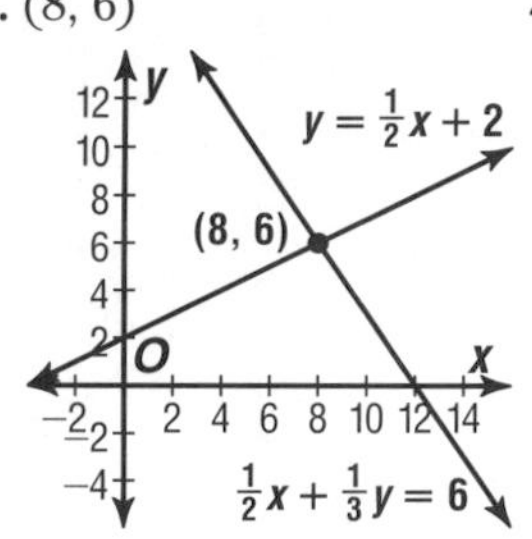

41. infinitely many

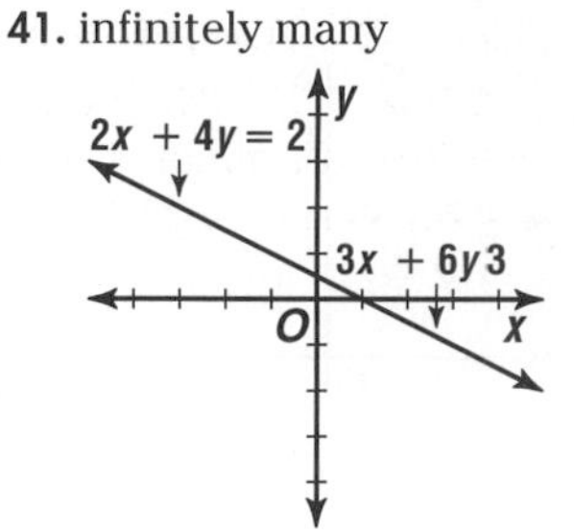

43. 15 square units

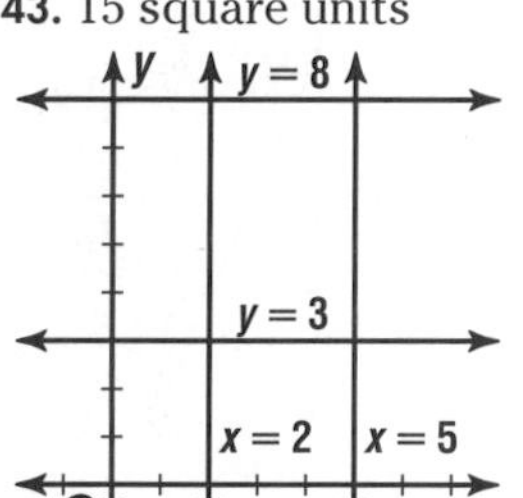

45. (1.71, −2.57)

47. (−0.25, −3.25)

49. $A = -3, B = 2$

51. A.D. 376

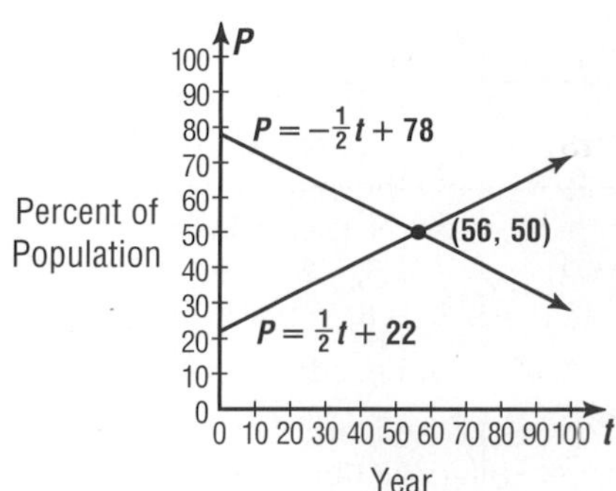

53. −1.5, −13.5 **55.** $y = -2x + 2$ **57.** $7\frac{1}{2}\%$ **59a.** how many three- and four-bedroom homes will be built **59b.** $100 - h$ **59c.** 80 homes

Pages 466–468 Lesson 8–2

7. $x = 8 - 4y; y = 2 - \frac{1}{4}x$ **9.** $x = -\frac{0.75}{0.8}y - 7.5; y = -\frac{0.8}{0.75}x - 8$

11. $\left(3, \frac{3}{2}\right)$ **13.** no solution **15.** infinitely many **17.** (3, 1)

19. (−4, 4) **21.** (4, −1) **23.** (2, 0) **25.** (9, 1) **27.** (2, 5)

29. (4, 2) **31.** (5, 2) **33.** $\left(\frac{8}{3}, \frac{13}{3}\right)$ **35.** (36, −6, −84)

37. (14, 27, −6) **39a.** $y = 1000 + 5x, y = 13x$ **39b.** 125 tickets **41a.** 26.5 years **41b.** 33.8 seconds

43a.

	75% Gold (18-carat)	50% Gold (12-carat)	58% Gold (14-carat)
Total Grams	x	y	300
Grams of Pure Gold	$0.75x$	$0.50y$	0.58(300)

43b. $x + y = 300;\ 0.75x + 0.50y = 0.58(300)$ **43c.** 96 grams of 18-carat gold, 204 grams of 12-carat gold **45.** 37 shares **47.** {(−1, −7), (4, 8), (7, 17), (13, 35)} **49.** −3 **51.** $m - 12$

Pages 472–474 Lesson 8–3

5. addition, (1, 0) **7.** subtraction, $\left(-\frac{5}{2}, -2\right)$

9. substitution, (1, 4) **11.** 8, 48 **13.** (1, −4), (1.29, −4.05) **15.** +; (6, 2) **17.** −; (4, −1) **19.** −; (4, −7) **21.** +; (5, 1) **23.** sub; infinitely many **25.** −; (−2, 3) **27.** −; $\left(\frac{1}{2}, 1\right)$

29. +; (10, −15) **31.** +; (1.75, 2.5) **33.** 11, 53 **35.** 5, 8 **37.** (2, 3, 7) **39.** (14, 27, −6) **41.** Ling, 1.45 hours or 1 hour, 27 minutes; José, 1.15 hours, or 1 hour, 9 minutes **43.** 320 gal of 25% and 180 gal of 50% **45.** −3 **47.** $-\frac{7}{6}$ **49.** substitution (=)

Page 474 Self Test

1. (1, −2)

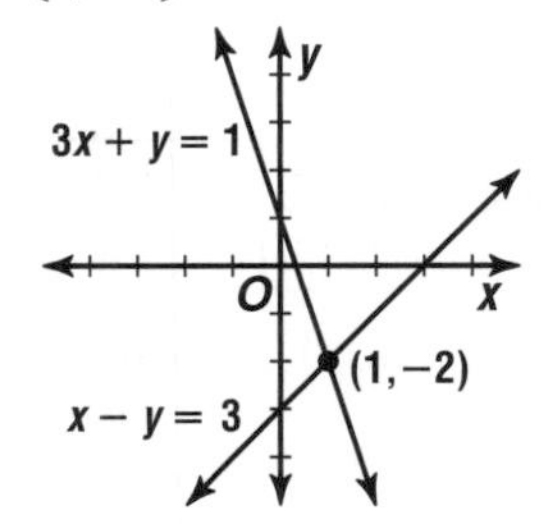

3. infinitely many

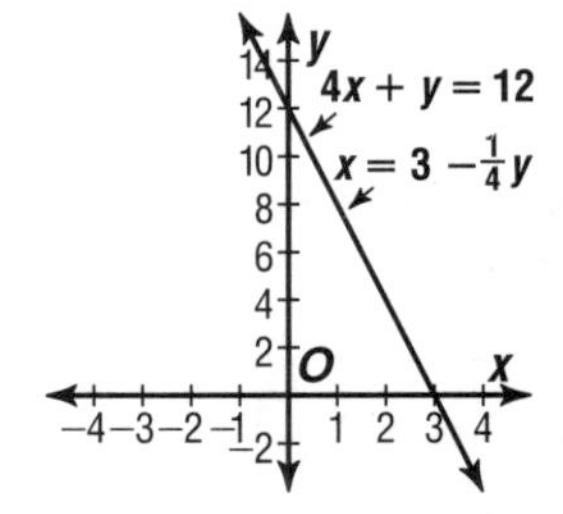

5. (−9, −7) **7.** (10, 15) **9.** (4, −2)

Pages 478–481 Lesson 8–4

5. (−1, 1); Multiply the first equation by −3, then add. **7.** (−9, −13); Multiply the second equation by 5, then add. **9.** (−1, −2); Multiply the first equation by 5, multiply the second equation by −8, then add. **11.** b; (2, 0) **13.** c; (4, 1) **15.** (2, 1) **17.** (5, −2) **19.** (2, −5) **21.** (−4, −7) **23.** (−1, −2) **25.** (4, −6) **27.** (13, −2) **29.** (10, 12) **31.** 6, 9 **33.** elimination, addition; $\left(2, \frac{1}{8}\right)$ **35.** substitution or elimination, multiplication; infinitely many **37.** elimination, subtraction; (24, 4) **39.** (11, 12) **41.** $\left(\frac{1}{3}, \frac{1}{6}\right)$ **43.** (−2, 7), (2, 2), (7, 5) **45.** 6 2-seat tables, 11 4-seat tables **47.** (3, −4) **49.** $\frac{1}{3}$ **51.** 165 yd **53.** 2 cups

Pages 485–486 Lesson 8–5

7.

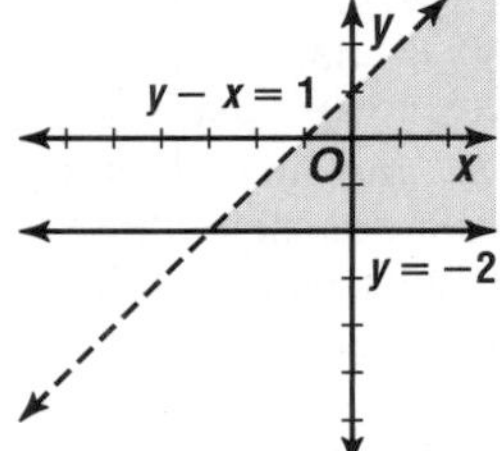

9.

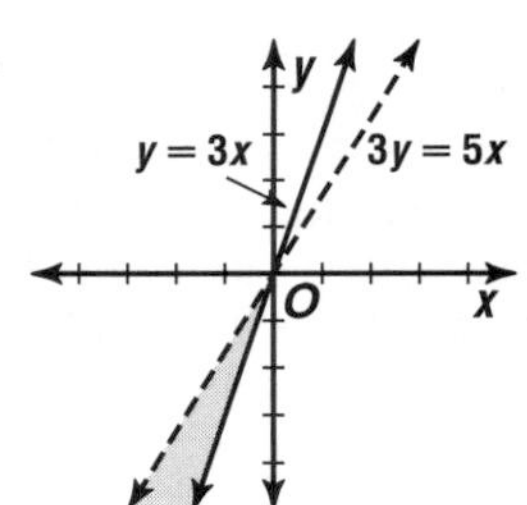

11.

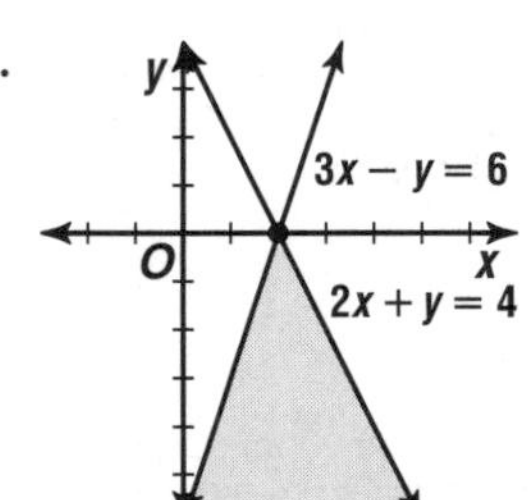

13. $y > x$, $y \leq x + 4$

15.

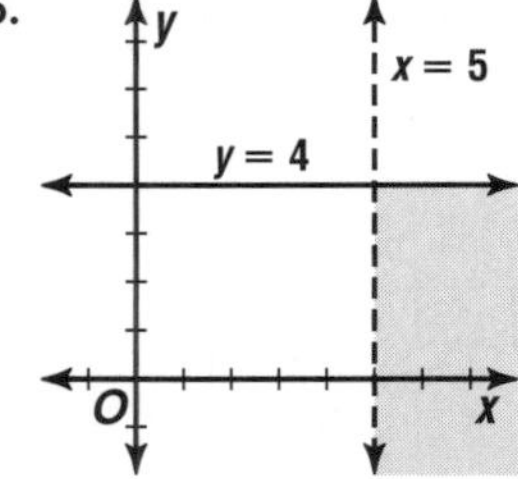

17.

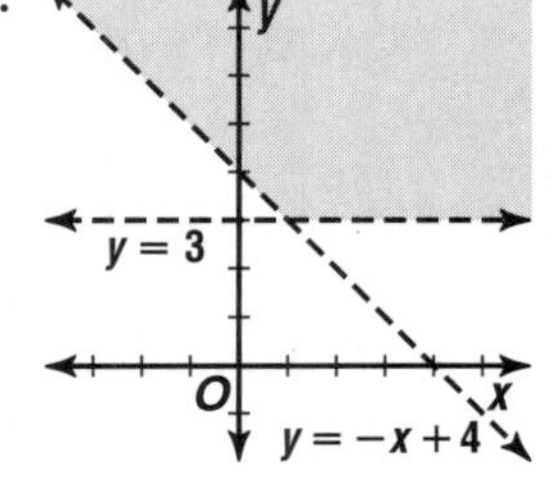

19.

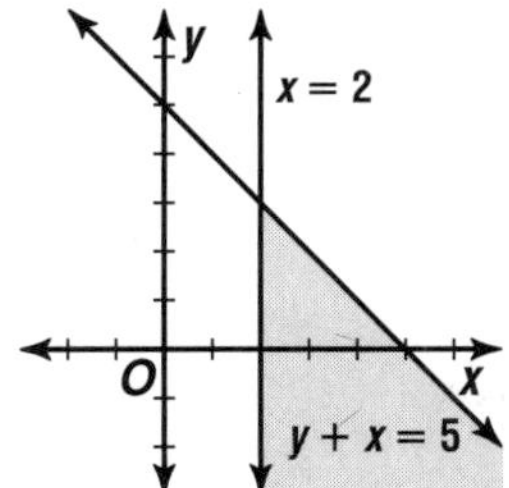

21.

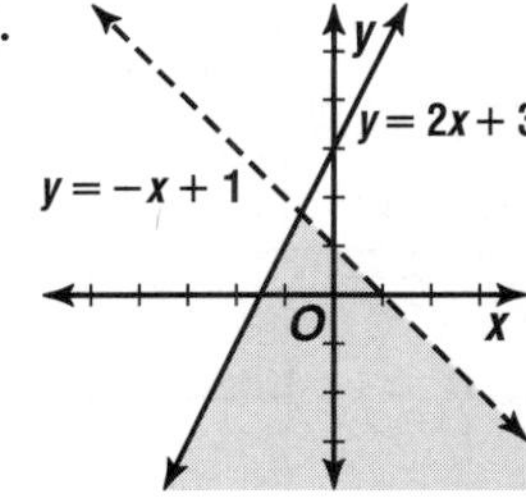

23.

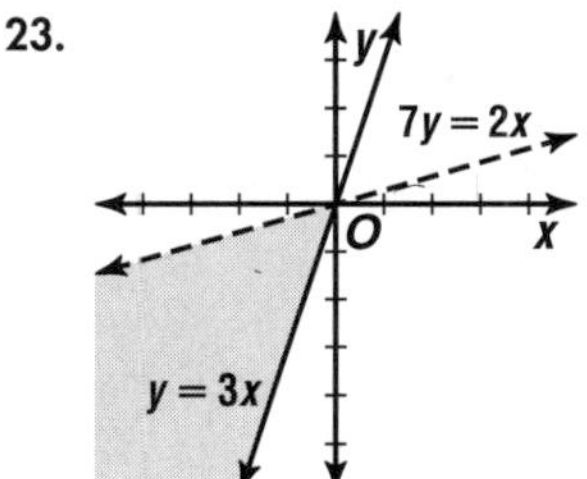

25.

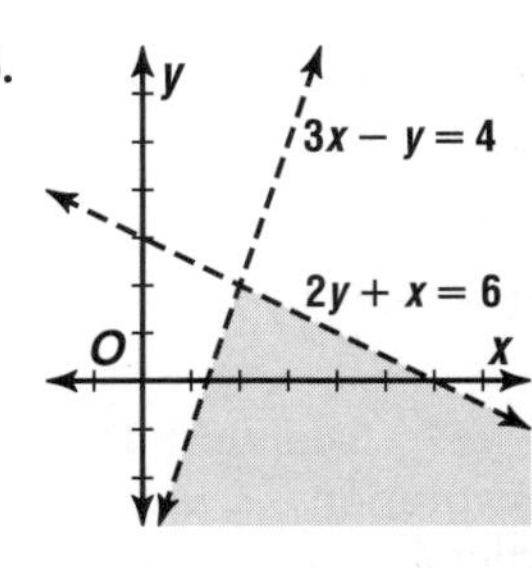

27.

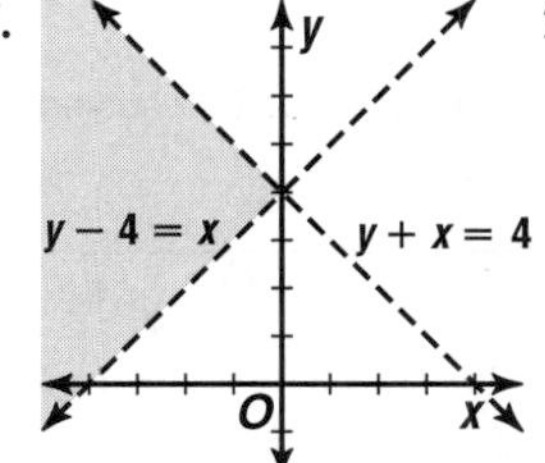

29.

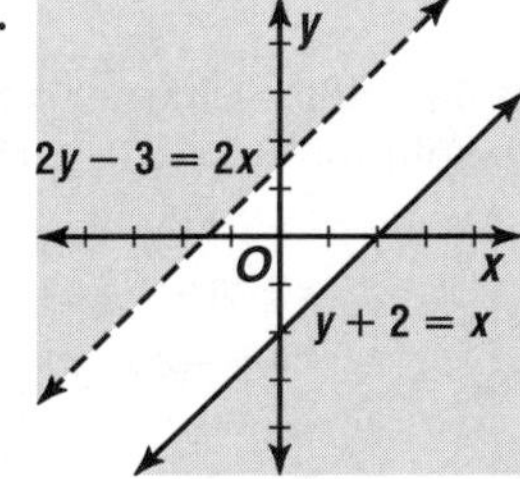

31.

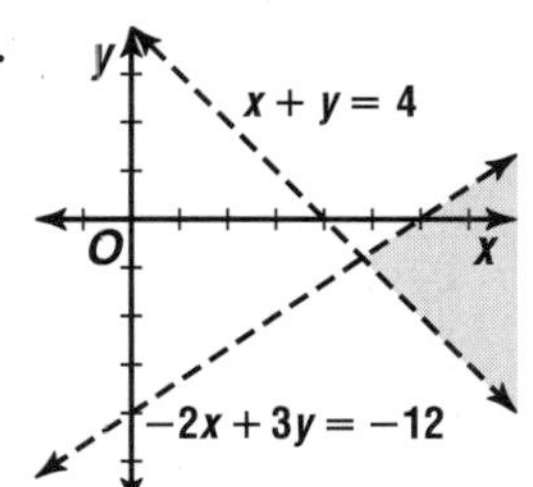

33. $y > -1$, $x \geq -2$ **35.** $y \leq x$, $y > x - 3$ **37.** $x \geq 0$, $y \geq 0$, $x + 2y \leq 6$

39.

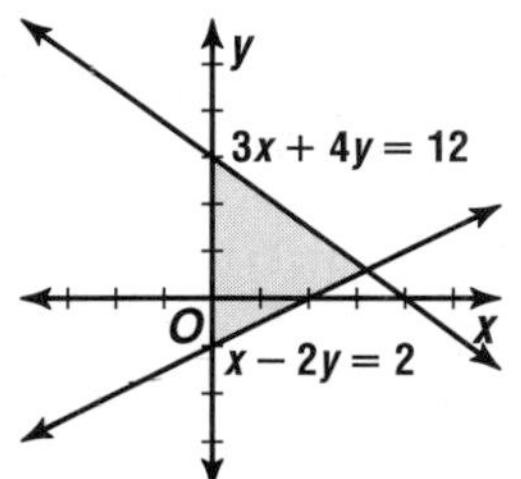

41.

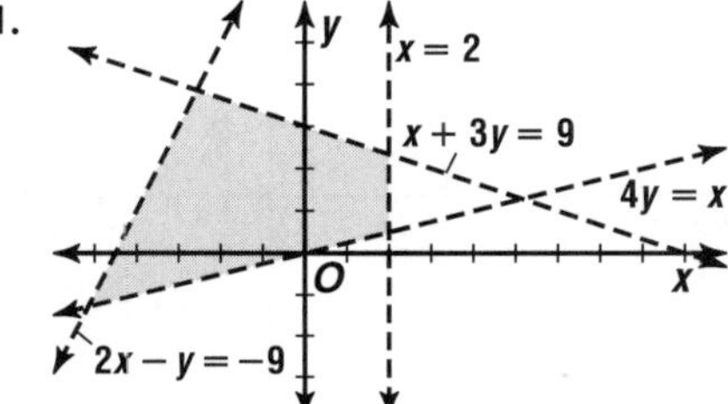

43.

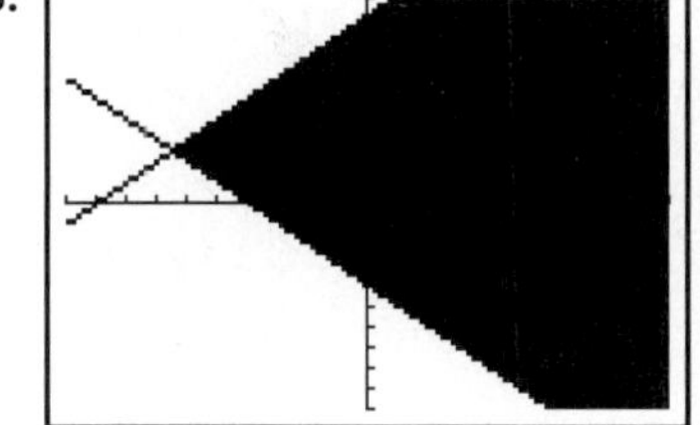

45.

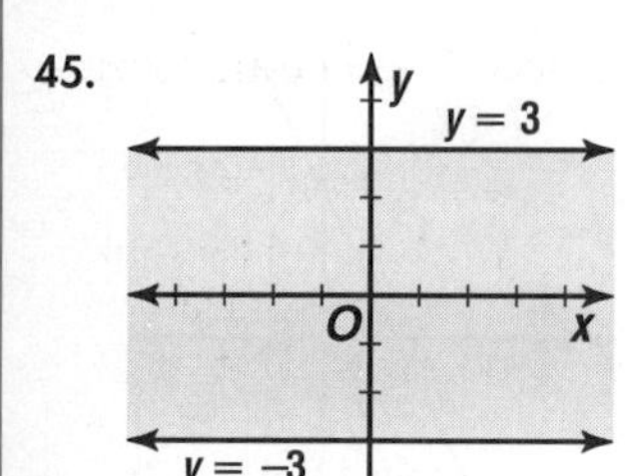

47. Sample answer: walk, 15 min, jog, 15 min; walk, 10 min, jog, 20 min; walk, 5 min, jog, 25 min. **49.** 4 \$5 bills, 8 \$20 bills **51.** $y = -\frac{1}{2}x + \frac{9}{2}$ **53.** 10 **55.** Let y = the number of yards gained in both games; $y = 134 + (134 - 17)$

Page 487 Chapter 8 Highlights

1. substitution **3.** inconsistent **5.** elimination **7.** infinitely many **9.** no **11.** second

Pages 488–490 Chapter 8 Study Guide and Assessment

13. no solution

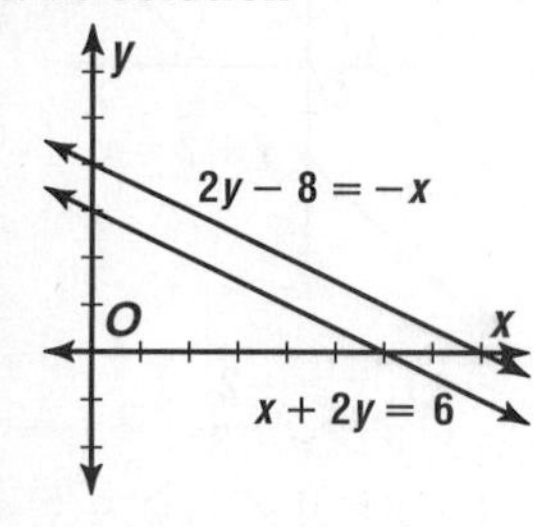

15. $(-2, -7)$

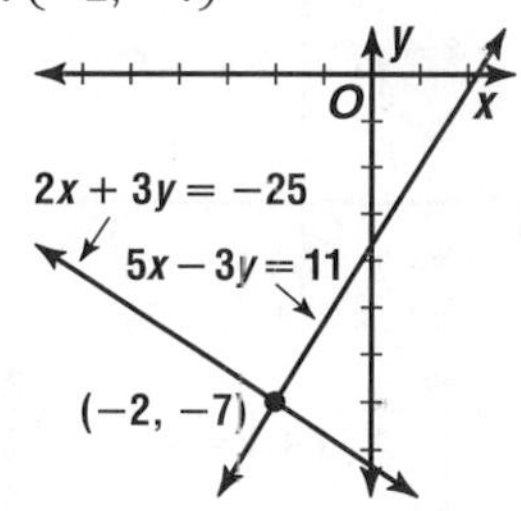

17. no solution

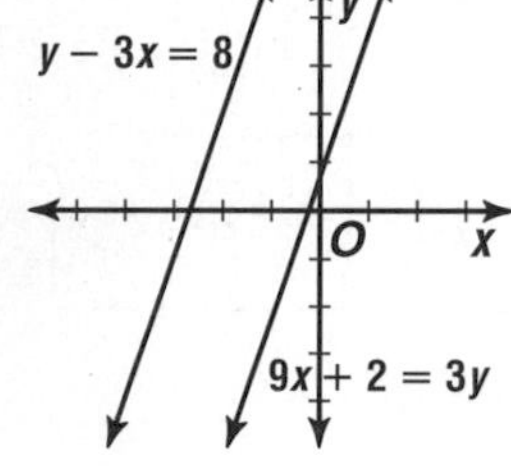

19. one, $(2, -2)$

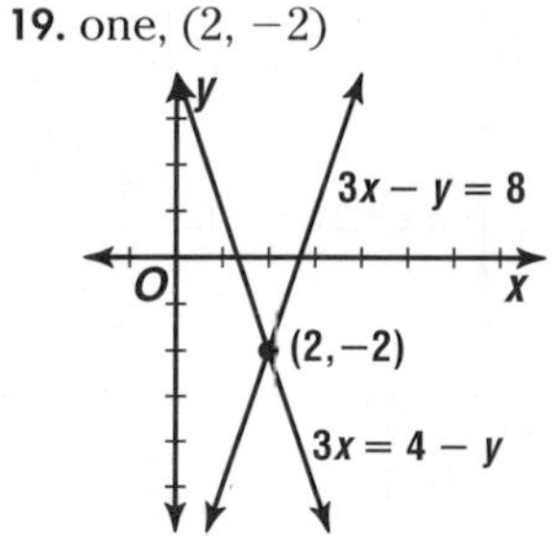

21. $(0, 2)$ **23.** $\left(\frac{1}{2}, \frac{1}{2}\right)$ **25.** $(2, -1)$ **27.** $(5, 1)$ **29.** $(8, -2)$ **31.** $(-9, -7)$ **33.** $\left(\frac{3}{5}, 3\right)$ **35.** $(2, -1)$ **37.** $\left(\frac{7}{9}, 0\right)$ **39.** $(0, 0)$ **41.** $(13, -2)$

43.

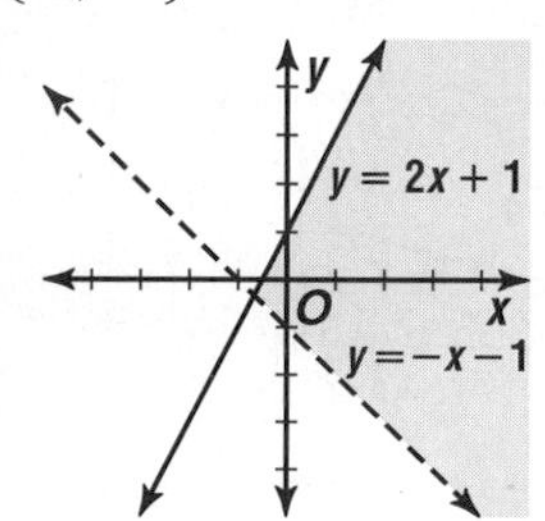

45.

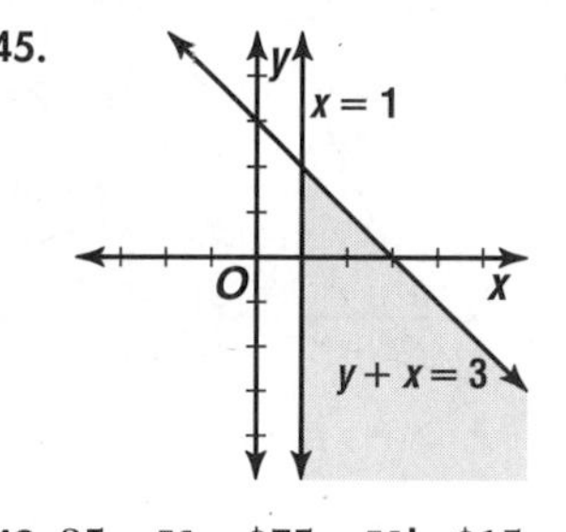

47.

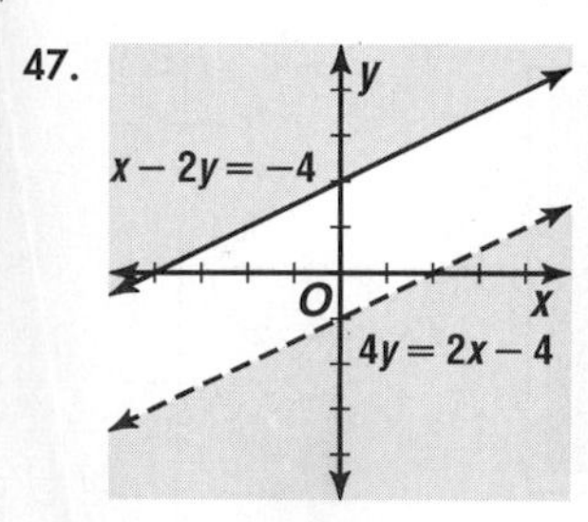

49. 35 **51a.** \$75 **51b.** \$15

CHAPTER 9 EXPLORING POLYNOMIALS

Pages 499–500 Lesson 9–1

5. no **7.** no **9.** a^{12} **11.** 3^{16} or 43,046,721 **13.** $9a^2y^6$ **15.** $15a^4b^3$ **17.** m^4n^3 **19.** 2^{12} or 4096 **21.** $a^{12}x^8$ **23.** $6x^4y^4z^4$ **25.** $a^2b^2c^2$ **27.** $\frac{4}{25}d^2$ **29.** $0.09x^6y^4$ **31.** $90y^{10}$ **33.** $4a^3b^5$ **35.** $-520x^9$ **37.** -2^4 equals $-(2)(2)(2)(2)$ or -16 and $(-2)^4$ equals $(-2)(-2)(-2)(-2)$ or 16. **39.** 301 parts

41. one; $(-6, 3)$

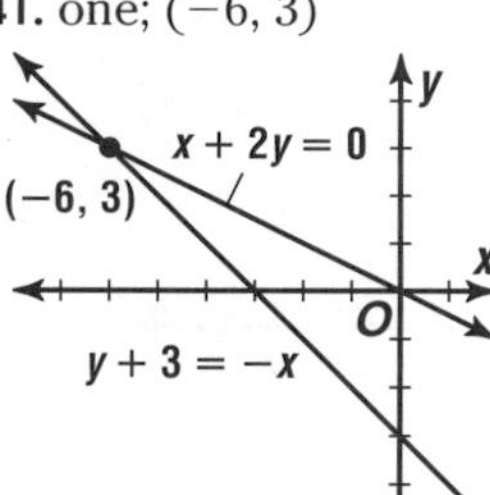

43. $n = 2m + 1$

m	−3	−2	−1	0	1
n	−5	−3	−1	1	3

45. 136° **47.** $1.11y + 0.06$

Pages 504–505 Lesson 9–2

7. $\frac{1}{121}$ **9.** 36 **11.** $\frac{6}{r^4}$ **13.** $\frac{\pi}{4}$ **15.** a^2 **17.** m^6 **19.** $\frac{m^3}{3}$ **21.** $\frac{c^3}{b^8}$ **23.** b^8 **25.** $-s^6$ **27.** $-\frac{4b^3}{c^3}$ **29.** $\frac{1}{64a^6}$ **31.** $\frac{s^3}{r^3}$ **33.** $\frac{a}{4c^2}$ **35.** m^{3+n} **37.** 3^{4x-6} **39.** $\frac{1}{q^{18}}$ **41.** \$1257.14 **43.** $98a^5b^4$ **45.** $|x + 1| < 3$ **47.** $(3, -5)$ **49.** $\frac{52}{41}$

Pages 509–512 Lesson 9–3

5. 43,400,000; 1.515×10^3 **7.** 507,000,000; 4.4419×10^4 **9.** 4,551,400,000; 7.14×10^2 **11.** 1.672×10^{-21} mg **13.** 4×10^{-6} in. **15.** 6.2×10^{-7}; 0.00000062 **17.** 6×10^7; 60,000,000 **19.** 9.5×10^{-3} **21.** 8.76×10^{10} **23.** 3.1272×10^8 **25.** 9.0909×10^{-2} **27.** 7.86×10^4 **29.** 7×10^{-10} **31.** 9.9×10^{-6} **33.** 6×10^{-3}; 0.006 **35.** 8.992×10^{-7}; 0.0000008992 **37.** 4×10^{-2}; 0.04 **39.** 6.5×10^{-6}; 0.0000065 **41.** 6.6×10^{-6}; 0.0000066 **43.** 1.2×10^{-4}; 0.00012 **45.** 2.4336×10^{-1} **47.** 2.8×10^5 **49a.** Sample answer: overflow **49b.** Multiply 3.7 and 5.6 and multiply 10^{112} and 10^{10}. Then write the product in scientific notation. **49c.** 2.072×10^{123} **51.** 1,000,000,000,001 **53.** 6.75×10^{18} molecules **55a.** about 90.1 kg

57.

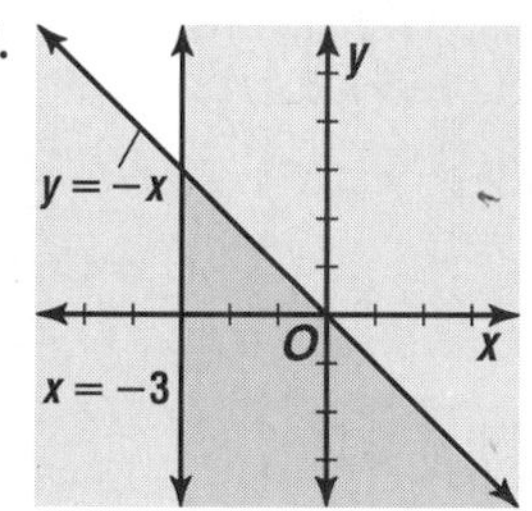

59. 25% **61.** $y = x + 5$ **63.** yes **65.** 27° **67.** −163

Page 513 Lesson 9–4A

1.

3.

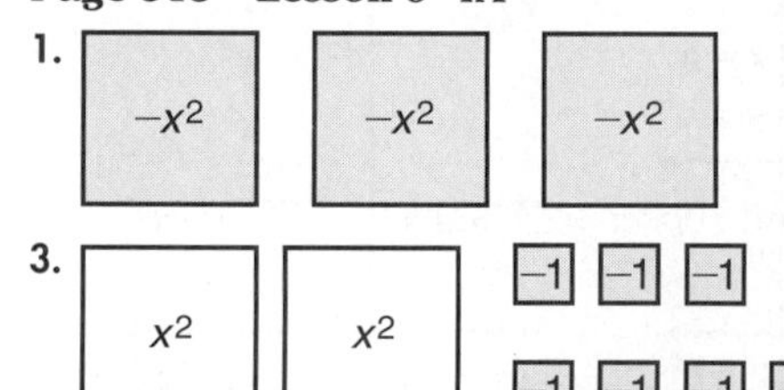

5. $x^2 - 3x + 2$ **7.** $2x^2 - x - 5$ **9.** x^2, x, and 1 represent the areas of the tiles.

Pages 517–519 Lesson 9–4

5. yes; trinomial **7.** yes; binomial **9.** 0 **11.** $x^8 - 12x^6 + 5x^3 - 11x$ **13a.** $\ell w - \pi r^2 - s^2$ **13b.** about 91.43 square units **15.** yes; trinomial **17.** yes; binomial **19.** yes; trinomial **21.** 5 **23.** 3 **25.** 9 **27.** 4 **29.** $x^5 + 3x^3 + 5$ **31.** $-x^7 + abx^2 - bcx + 34$ **33.** $1 + x^2 + x^3 + x^5$ **35.** $7a^3x + \frac{2}{3}x^2 - 8a^3x^3 + \frac{1}{5}x^5$ **37.** $2ab + \pi b^2$; about 353.10 square units **39.** $ab - 4x^2$; 116 square units **41b.** $8a^4 + 9a^3 + 4a^2 + 3a^1 + 5a^0$ **43.** about 153 eggs **45.** 4.235×10^4 **47.** 7, 8, 9 **49.** $\left\{(-3, 8), \left(1, \frac{8}{3}\right), (3, 0), (9, -8)\right\}$ **51.** \$38.75

Page 521 Lesson 9–5A

1. $-x^2 + 6$ **3.** $-2x^2 - 6x$ **5.** $3x^2 - x + 4$

7. true

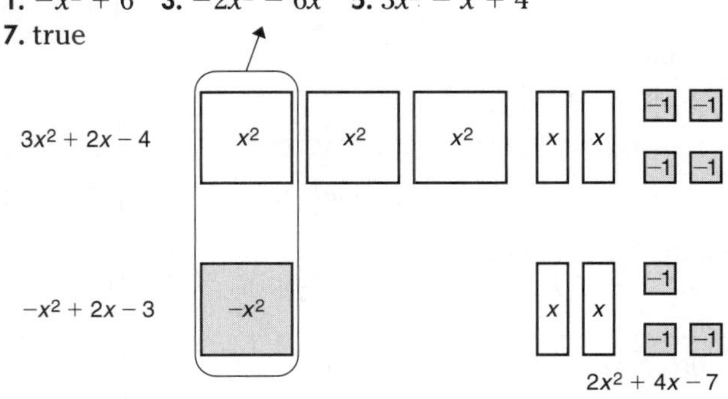

9. Method from Activity 2:

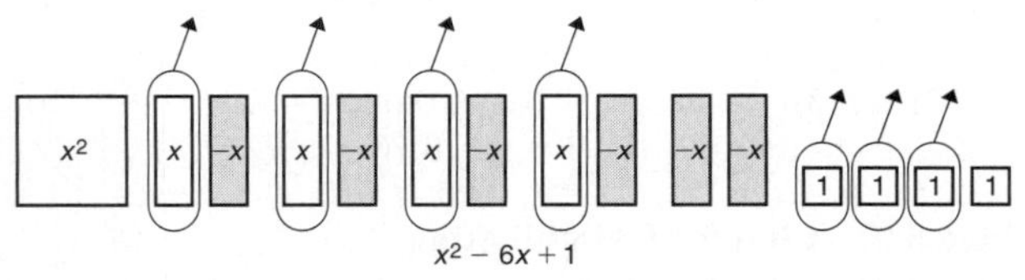

You need to add zero-pairs so that you can remove 4 green x-tiles.

Method from Activity 4:

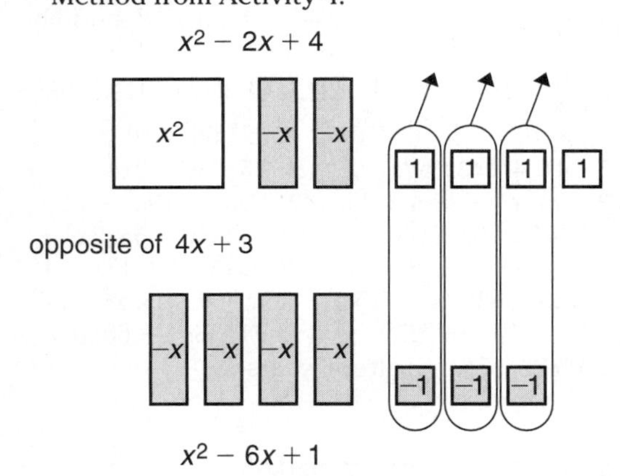

You remove all zero-pairs to find the difference in simplest form.

Pages 524–527 Lesson 9–5

7. $6a^2 - 3$ **9.** $4x^2 + 3y^2 - 8y - 7x$ **11.** $-8y^2$ and $3y^2$; $2x$ and $4x$ **13.** $3p^3q$ and $10p^3q$; $-2p$ and $-p$ **15.** $6m^2n^2 + 8mn - 28$ **17.** $-4y^2 + 5y + 3$ **19.** $7p^3 - 3p^2 - 2p - 7$ **21.** $10x^2 + 13xy$ **23.** $4a^3 + 2a^2b - b^2 + b^3$ **25.** $-4a + 6b - 5c$ **27.** $3a - 11m$ **29.** $-2n^2 + 7n + 5$ **31.** $13x - 2y$ **33.** $-y^3 + 3y + 3$ **35.** $4z^3 - 2z^2 + z$ **37.** $2x + 3y$ **39.** $353 - 18x$ **41.** $-2n^2 - n + 4$ **43.** $719x^2$ cubic stories **45.** $-3x - 2x^3 + 4x^5$

47. (3, 1)

y
(3, 1)
O
x
2x – 5y = 1
5x – 3y = 12

49. Sample answer: Let h = the number of hours for the repair and let c = the total charge; $c = 34h + 15$.

51. $-\frac{50}{3}$

Page 527 Self Test

1. $-6n^5y^7$ **3.** $-12ab^4$ **5.** 5.67×10^6 **7.** about 1.53×10^4 seconds or 4.25 hours **9.** $5x^2 - 4x - 14$

Page 528 Lesson 9–6A

1. $x^2 + 2x$ **3.** $2x^2 + 2x$ **5.** $2x^2 + x$

7. true

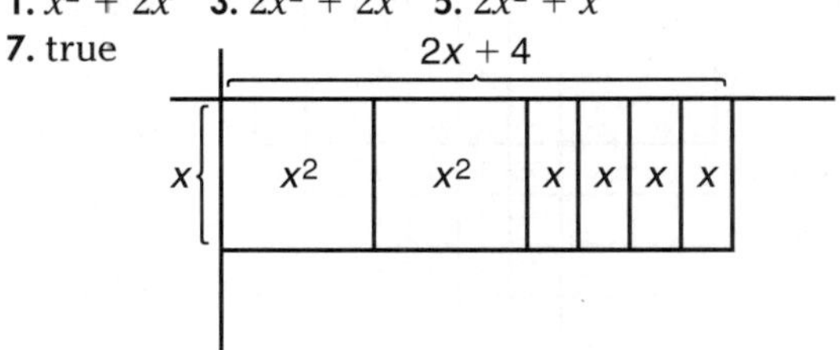

9a. $3x$ and $x + 15$

9b. $(3x^2 + 45x)$ square feet

x + 15
3x
x²
x

Pages 531–533 Lesson 9–6

5. $-63b^4c - 7b$ **7.** $10y^2 - 26y$ **9.** $3w^2 - 2w$ **11.** $\frac{103}{19}$ **13a.** $a^2 + a$ **13b.** $a^2 + 2a$ **15.** $\frac{1}{3}x^2 - 9x$ **17.** $-20m^5 - 8m^4$ **19.** $30m^5 - 40m^4n + 60m^3n^3$ **21.** $-28d^3 + 16d^2 - 12d$ **23.** $-32r^2s^2 - 56r^2s + 112rs^3$ **25.** $\frac{36}{5}x^3y + x^3 - 24x^2y$ **27.** $36t^2 - 42$ **29.** $61y^3 - 16y^2 + 167y - 18$ **31.** $53a^3 - 57a^2 + 7a$ **33.** $-\frac{77}{8}$ **35.** 0 **37.** $\frac{23}{24}$ **39.** 2 **41.** $15p^2 + 32p$ **43.** Sample answer: $1(8a^2b + 18ab)$, $a(8ab + 18b)$, $b(8a^2 + 18a)$, $2(4a^2b + 9ab)$, $(2a)(4ab + 9b)$, $(2b)(4a^2 + 9a)$, $(ab)(8a + 18)$, $(2ab)(4a + 9)$ **45.** $1.50t + 1.25mt$

47a.

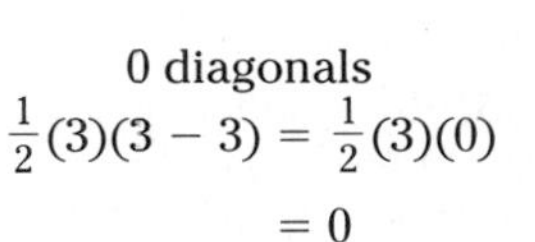

0 diagonals

$\frac{1}{2}(3)(3 - 3) = \frac{1}{2}(3)(0)$

$= 0$

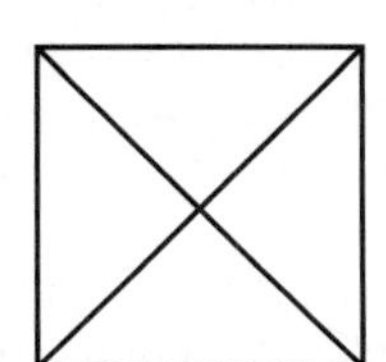

2 diagonals

$\frac{1}{2}(4)(4 - 3) = \frac{1}{2}(4)(1)$

$= 2$

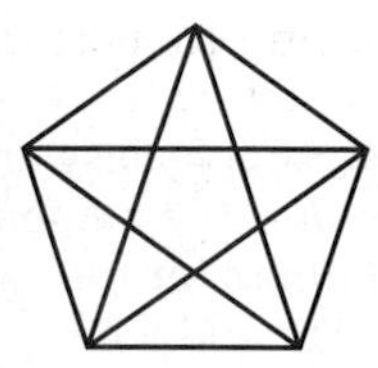

5 diagonals

$\frac{1}{2}(5)(5-3) = \frac{1}{2}(5)(2)$

$= 5$

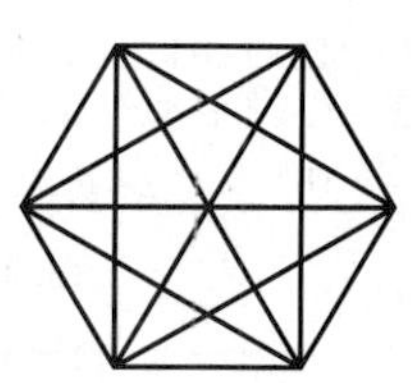

9 diagonals

$\frac{1}{2}(6)(6-3) = \frac{1}{2}(6)(3)$

$= 9$

47b. $\frac{1}{2}n^2 - \frac{3}{2}n$ **47c.** 90 diagonals **49.** 75 gal of 50%, 25 gal of 30% **51.** $a^2 - 1$ **53.** 11 days **55.** 25.7

Page 535 Lesson 9–7A

1. $x^2 + 3x + 2$ **3.** $x^2 - 6x + 8$ **5.** $2x^2 - 2$

7. false

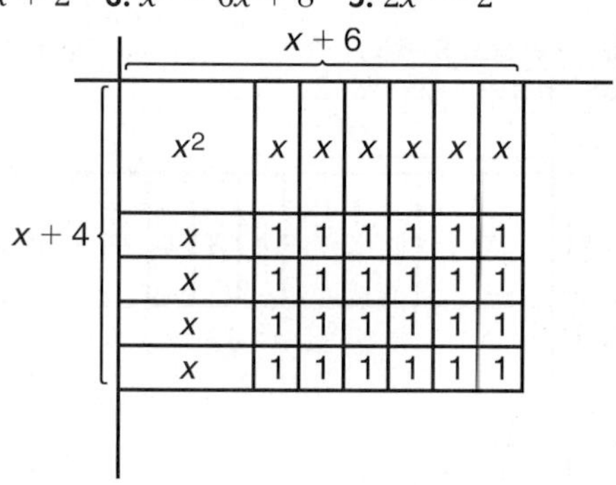

9. false

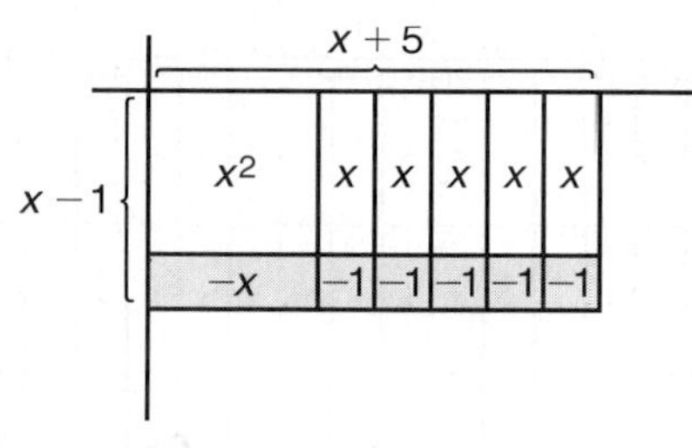

11. By the distributive property, $(x + 3)(x + 2) = x(x + 2) + 3(x + 2)$. The top row represents $x(x + 2)$ or $x^2 + 2x$. The bottom row represents $3(x + 2)$ or $3x + 6$.

Pages 539–541 Lesson 9–7

5. $d^2 + 10d + 16$ **7.** $y^2 - 4y - 21$ **9.** $2x^2 + 9x - 5$ **11.** $10a^2 + 11ab - 6b^2$ **13a.** $a^3 + 3a^2 + 2a$ **13c.** The result is the same as the product in part b. **15.** $c^2 - 10c + 21$ **17.** $w^2 - 6w - 27$ **19.** $10b^2 - b - 3$ **21.** $169x^2 - 9$ **23.** $0.15v^2 - 2.9v - 14$ **25.** $\frac{2}{3}a^2 + \frac{1}{18}ab - \frac{1}{3}b^2$ **27.** $0.63p^2 + 3.9pq + 6q^2$ **29.** $6x^3 + 11x^2 - 68x + 55$ **31.** $9x^3 - 45x^2 + 62x - 16$ **33.** $20d^4 - 9d^3 + 73d^2 - 39d + 99$ **35.** $10x^4 + 3x^3 + 51x^2 - 16x - 48$ **37.** $a^4 - a^3 - 8a^2 - 29a - 35$ **39.** $63y^3 - 57y^2 - 36y$ **41a.** 28 cm, 20 cm, 16 cm **41b.** 8960 cm^3 **41c.** 8960 **41d.** They are the same measure. **43.** $-8x^4 - 6x^3 + 24x^2 - 12x + 80$ **45.** $-3x^3 - 5x^2 - 6x + 24$ **47a.** Sample answer: $x - 2$, $x + 3$ **47b.** Sample answer: $x^2 + x - 6$ **47c.** larger, 2 sq ft **49.** $\{y \mid y < -6\}$ **51.** $y = 2x + 1$ **53.** $a = 4, y = 9$ **55.** 15°C

Pages 546–547 Lesson 9–8

7. $m^2 - 6mn + 9n^2$ **9.** $m^4 + 8m^2n + 16n^2$ **11.** $25 - 10x + x^2$ **13.** $x^2 + 8xy + 16y^2$ **15.** $9b^2 - 6ab + a^2$ **17.** $81p^2 - 4q^2$ **19.** $25b^2 - 120ab + 144a^2$ **21.** $x^6 + 2x^3a^2 + a^4$ **23.** $64x^4 - 9y^2$ **25.** $1.21g^2 + 2.2gh^5 + h^{10}$ **27.** $\frac{16}{9}x^4 - y^2$ **29.** $9x^3 - 45x^2 - x + 5$ **31.** $a^3 + 9a^2b + 27ab^2 + 27b^3$ **33.** $x^2 + y^2 + z^2 + 2xy + 2yz + 2xz$

	x	y	z
x	x^2	xy	xz
y	xy	y^2	yz
z	xz	yz	z^2

35a. $2\pi s + 7\pi$ square meters **35b.** about 28.27 square meters **35c.** about 40.84 square meters **37.** $6t^2 - 3t - 3$

39a.

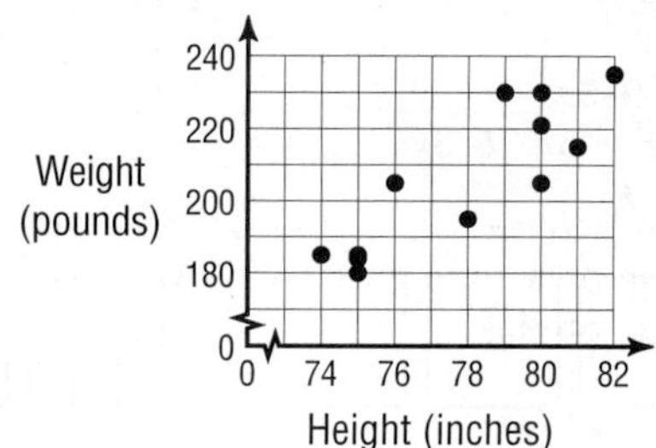

39b. the taller the player, the greater the weight

41.

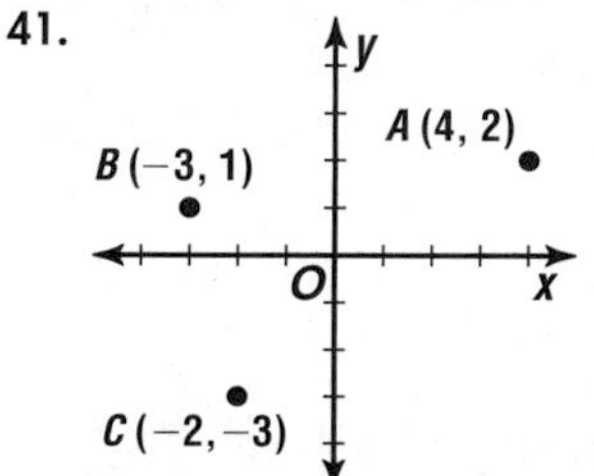

43. 5

Page 549 Chapter 9 Highlights

1. e **3.** h **5.** i **7.** f **9.** a

Pages 550–552 Chapter 9 Study Guide and Assessment

11. y^7 **13.** $20a^5x^5$ **15.** $576x^5y^2$ **17.** $-\frac{1}{2}m^4n^8$ **19.** y^4 **21.** $3b^3$ **23.** $\frac{a^4}{2b}$ **25.** 2.4×10^5 **27.** 4.88×10^9 **29.** 7.96×10^5 **31.** 6×10^{11} **33.** 6×10^{-1} **35.** 1.68×10^{-1} **37.** 2 **39.** 5 **41.** 4 **43.** $3x^4 + x^2 - x - 5$ **45.** $-3x^3 + x^2 - 5x + 5$ **47.** $16m^2n^2 - 2mn + 11$ **49.** $21m^4 - 10m - 1$ **51.** $12a^3b - 28ab^3$ **53.** $8x^5y - 12x^4y^3 + 4x^2y^5$ **55.** $2x^2 - 17xy^2 + 10x + 10y^2$ **57.** $r^2 + 4r - 12$ **59.** $4x^2 + 13x - 12$ **61.** $18x^2 - 0.125$ **63.** $2x^3 + 15x^2 - 11x - 9$ **65.** $x^2 - 36$ **67.** $16x^2 + 56x + 49$ **69.** $25x^2 - 9y^2$ **71.** $36a^2 - 60ab + 25b^2$ **73.** $305.26 **75.** no; after 6 years

CHAPTER 10 USING FACTORING

Pages 561–563 Lesson 10–1

5. 1, 2, 4 **7.** prime **9.** $-1 \cdot 2 \cdot 3 \cdot 5$ **11.** 4 **13.** $6d$ **15.** $4gh$ **17.** 40 in. **19.** 1, 67 **21.** 1, 2, 4, 5, 8, 10, 16, 20, 40, 80 **23.** 1, 5, 10, 19, 25, 38, 50, 95, 190, 950 **25.** composite; $3^2 \cdot 7$ **27.** prime **29.** composite; $2^2 \cdot 5 \cdot 7 \cdot 11$ **31.** $-1 \cdot 3 \cdot 3 \cdot 13$ **33.** $2 \cdot 2 \cdot b \cdot b \cdot b \cdot d \cdot d$ **35.** $-1 \cdot 2 \cdot 7 \cdot 7 \cdot a \cdot a \cdot b$ **37.** 9 **39.** 1 **41.** 19 **43.** $7pq$ **45.** $15r^2t^2$ **47.** 12 **49.** 6 **51.** $8m^2n$ **53.** $-12x^3yz^2$ **55.** $3m^2n$ **57.** 29 cm by 47 cm **59.** 3, 5; 5, 7; 11, 13; 17, 19; 29, 31; 41, 43; 59, 61; 71, 73

61a. $2b^3 \times 1 \times 1$, $2b^2 \times b \times 1$, $2b \times b \times b$, $b^3 \times 2 \times 1$, $b^2 \times 2b \times 1$, and $b^2 \times 2 \times b$ **61c.** $4b^3 + 1$ or 865, $2b^3 + 2b^2 + 1$ or 505, $5b^2$ or 180, $3b^3 + 2$ or 650, $2b^3 + b^2 + b$ or 474, and $b^3 + 2b^2 + 2b$ or 300, respectively **61d.** Though the volume remains constant, the surface areas vary greatly. **63.** 1500 squares of sod **65.** $3b$
67.

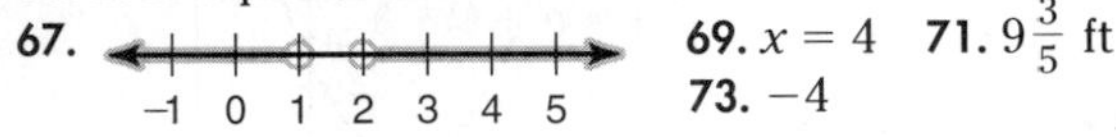

69. $x = 4$ **71.** $9\frac{3}{5}$ ft **73.** -4

Page 564 Lesson 10–2A

1. $3(x + 3)$ **3.** $x(3x + 4)$
5. no

x	x	
1	1	1

7. yes

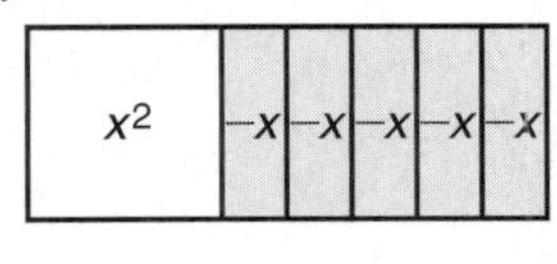

9. Binomials can be factored if they can be represented by a rectangle. Examples: $4x + 4$ can be factored and $4x + 3$ cannot be factored.

Pages 569–571 Lesson 10–2

7. 3 **9.** 1 **11.** $4xy^2$ **13.** $(a + b)(x + y)$ **15.** $(x - y)(3a + 4b)$ **17.** $3t$ **19.** $x(29y - 3)$ **21.** $3c^2d(1 - 2d)$ **23.** $(r + k)(x + 2y)$ **25a.** $g = \frac{1}{2}n(n - 1)$ **25b.** 91 games **25c.** 63 games **27.** $8rs$ **29.** $2y - 5$ **31.** $5k - 7p$ **33.** $2xz(7 - 9z)$ **35.** $a(17 - 41ab)$ **37.** $(m + x)(2y + 7)$ **39.** $3xy(x^2 - 3y + 12)$ **41.** $(2x^2 - 5y^2)(x - y)$ **43.** $(2x - 5y)(2a - 7b)$ **45.** $7abc(4abc + 3ac - 2)$ **47.** $2(2m + r)(3x - 2)$ **49.** $(7x + 3t - 4)(a + b)$ **51.** $8a - 4b + 8c + 16d + ab + 64$ **53.** $4r^2(4 - \pi)$ **55.** $(4z + 3m)$ cm by $(z - 6)$ cm **57.** Sample answer: $(3a + 2b)(ab + 6)$, $(3a + ab)(2b + 6)$, $(2b + ab)(3a + 6)$ **59.** $(2s - 3)(s + 8)$ **61.** A and C sharp **63.** 9 **65.** $\{z \mid z \geq -1.654\}$ **67.** $\left\{(-3, 10), \left(0, \frac{11}{2}\right), (1, 4), \left(2, \frac{5}{2}\right), (5, -2)\right\}$ **69.** $y = -\frac{4}{3}x + \frac{7}{3}$ **71.** 7

Page 573 Lesson 10–3A

1. $(x + 1)(x + 5)$ **3.** $(x + 3)(x + 4)$ **5.** $(x - 1)(x - 2)$
7. $(x - 1)(x + 5)$
9. yes

x^2	x	x	x	x	x
x	1	1	1	1	1
x	1	1	1	1	1

11. no

x^2	x	x	x
x	–1	–1	
x	–1	–1	

13. Trinomials can be factored if they can be represented by a rectangle. Sample answers: $x^2 + 4x + 4$ can be factored and $x^2 + 6x + 4$ cannot be factored.

Pages 578–580 Lesson 10–3

5. 3, 8 **7.** 8, 5 **9.** $-2, -6$ **11.** – **13.** $(t + 3)(t + 4)$ **15.** $2(y + 2)(y - 3)$ **17.** prime **19.** 9, -9, 15, -15 **21.** $(3x^2 + 2x)$ m^2 **23.** – **25.** 5 **27.** 5 **29.** $(m - 4)(m - 10)$ **31.** prime **33.** $(2x + 7)(x - 3)$ **35.** $(2x + 3)(x - 4)$ **37.** $(2n - 7)(2n + 5)$ **39.** $(2 + 3m)(5 + 2m)$ **41.** prime **43.** $2x(3x + 8)(2x - 5)$ **45.** $2a^2b(5a - 7b)(2a - 3b)$ **47.** 7, -7, 11, -11 **49.** 1, -1, 11, -11, 19, -19, 41, -41 **51.** 6, 4 **53.** r cm, $(5r + 6)$ cm, $(3r - 7)$ cm **55.** no; $(2x + 3)(x - 1)$ **57.** no; $(x - 3)(x - 3)$ **59.** 27 ft^3 **61.** $(3x - 10)$ shares **63.** $(5, -2)$ **65.** $y = -\frac{2}{3}x + \frac{14}{3}$ **67.** D = {0, 1, 2}; R = {2, −2, 4} **69.** 90° **71.** $3x + 4y$

Page 580 Self Test

1. $10n^2$ **3.** $6xy(3y - 4x)$ **5.** $(2q + 3)(q - 6)$ **7.** $(3y - 5)(y - 1)$ **9.** 41,312,432 or 23,421,314

Pages 584–586 Lesson 10–4

7. yes **9.** no **11.** d **13.** a **15.** $(1 - 4g)(1 + 4g)$ **17.** $5(2m - 3n)(2m + 3n)$ **19.** $(x - y)(x + y)(x^2 + y^2)$ **21.** 5 **23.** $(2 - v)(2 + v)$ **25.** $(10d - 1)(10d + 1)$ **27.** $2(z - 7)(z + 7)$ **29.** prime **31.** $17(1 - 2k)(1 + 2k)$ **33.** prime **35.** prime **37.** $(ax - 0.8y)(ax + 0.8y)$ **39.** $\frac{1}{2}(3a - 7b)(3a + 7b)$ **41.** $(a + b - c - d)(a + b + c + d)$ **43.** $(x^2 - 2y)(x^2 + 2y)(x^4 + 4y^2)$ **45.** $(a^2 + 5b^2)(a - 2b)(a + 2b)$ **47.** 624 **49.** $(2a - b)$ in., $(2a + b)$ in. **51.** $(x - 2)$ ft, $(x + 4)$ ft **53.** $(a - 5b)$ in., $(a + 5b)$ in., $(5a + 3b)$ in. **55a.** square **55b.** 25 cm^2 **57.** 9, 12, 15 **59.** \$3.16, \$1.50, \$1.25, \$1.20 **61.** (4, 16)
63. **65.**

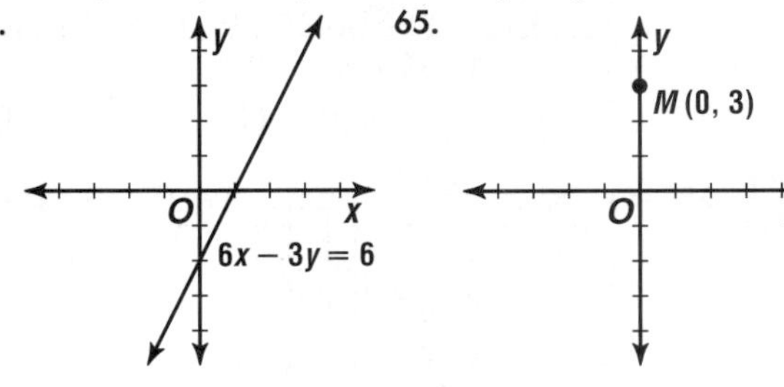

67. 5.14, 3.6, 0.6

Pages 591–593 Lesson 10–5

7. 8a **9.** 6c **11.** yes; $(2n - 7)^2$ **13.** yes; $(4b - 7c)^2$ **15.** $4(a - 3b)(a + 3b)$ **17.** $2(5g + 2)^2$ **19.** $(2a - 3b)(2a + 3b)(5x - y)$ **21.** yes; $(r - 4)^2$ **23.** yes; $(7p - 2)^2$ **25.** no **27.** yes; $(2m + n)^2$ **29.** no **31.** no **33.** yes; $\left(\frac{1}{2}a + 3\right)^2$ **35.** $a(45a - 32b)$ **37.** $(v - 15)^2$ **39.** prime **41.** $3(y - 7)(y + 7)$ **43.** $2(3a - 4)^2$ **45.** $(y^2 + z^2)(x - 1)(x + 1)$ **47.** $(a^2 + 2)(4a + 3b^2)$ **49.** $0.7(p - 3q)(p - 2q)$ **51.** $(g^2 - 3h)(g + 3)^2$ **53.** -110, 110 **55.** 9 **57.** $(6y + 26)$ cm **59.** $(8x^2 - 22x + 14)$ cm^2 **61a.** $a \geq b$ **61b.** $a \leq b$ **61c.** $a = b$ **63a.** \$1166.40 **63b.** $p(1 + r)^2$ **63c.** \$1144.90 **65.** $50.6t^2 + 21t - 102$ **67.** 20 hours
69a.

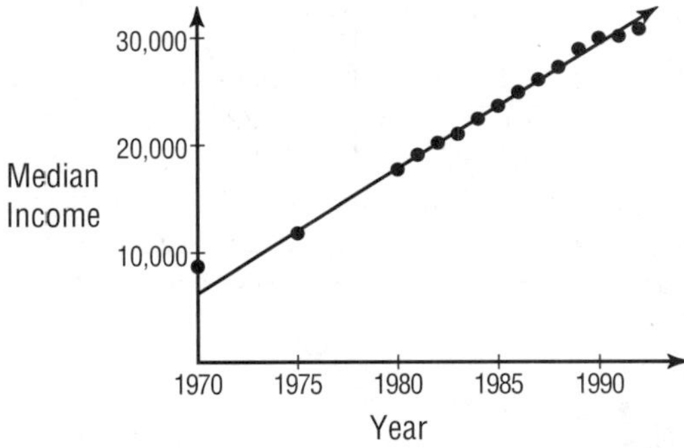

69b. Sample answer: Yes; $I = 1200y + 6000$, where I is the median income and y is the number of years since 1970. **71.** 1:3 **73.** 152 ft

Pages 598–600 Lesson 10-6

5. $\{0, -5\}$ **7.** $\left\{0, \frac{5}{3}\right\}$ **9.** $\{5\}$ **11.** 6 **13.** $\{0, 24\}$ **15.** $\left\{\frac{3}{2}, \frac{8}{3}\right\}$ **17.** $\{-9, -4\}$ **19.** $\{-8, 8\}$ **21.** $\{0, 4\}$ **23.** $\left\{-\frac{1}{3}, -\frac{5}{2}\right\}$ **25.** $\{12, -4\}$ **27.** $\{-9, 0, 9\}$ **29.** $\{-5, 7\}$ **31.** -14 and -12 or 12 and 14 **33.** 5 cm **35a.** $\{-6, 1\}$; $\{-6, 1\}$ **35b.** They are equivalent; they have the same solution. **37a.** about 3.35 s **37b.** His ideas about falling objects differed from what most people thought to be true. He believed that Earth is a moving planet and that the sun and planets do not revolve around Earth. **39.** about 90,180 ft or 17 mi **41.** 0.5 km **43.** $40q^2 + rq - 6r^2$ **45.** 14; -4 **47.** 4.355 minutes or about 4 minutes 21 seconds

Page 601 Chapter 10 Highlights

1. false; composite **3.** false; sample answer: 64 **5.** false; $2^4 \cdot 3$ **7.** true **9.** false; zero product property

Pages 602–604 Chapter 10 Study Guide and Assessment

11. composite, $2^2 \cdot 7$ **13.** composite, $2 \cdot 3 \cdot 5^2$ **15.** prime **17.** 5 **19.** $4ab$ **21.** $5n$ **23.** $7x^2$ **25.** $13(x + 2y)$ **27.** $6ab(4ab - 3)$ **29.** $12pq(3pq - 1)$ **31.** $(a - 4c)(a + b)$ **33.** $(4k - p^2)(4k^2 - 7p)$ **35.** $(8m - 3n)(3a + 5b)$ **37.** $(y + 3)(y + 4)$ **39.** prime **41.** $(r - 4)(2r + 5)$ **43.** $(b - 4)(b + 4)$ **45.** $(4a - 9b^2)(4a + 9b^2)$ **47.** prime **49.** $(a + 9)^2$ **51.** $(2 - 7r)^2$ **53.** $6b(b - 2g)^2$ **55.** $(5x - 12)^2$ **57.** $\left\{\frac{2}{3}, -\frac{7}{4}\right\}$ **59.** $\{0, -17\}$ **61.** $\{-5, -8\}$ **63.** $\left\{-\frac{2}{5}\right\}$ **65.** 384 **67.**

		3
×	5	4
1	6	2

69. 9, 11; -11, -9

CHAPTER 11 EXPLORING QUADRATIC AND EXPONENTIAL FUNCTIONS

Page 610 Lesson 11–1A

1. $(-8, -5)$

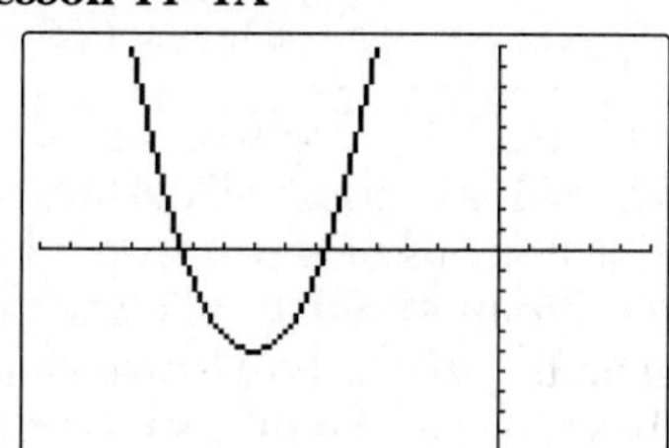

3. $(5, 0)$

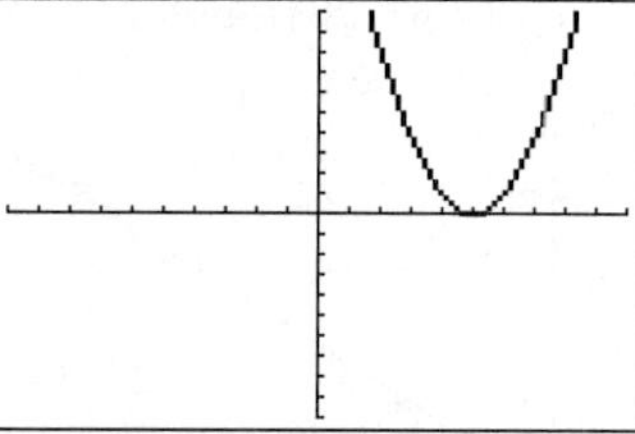

5. $(10, 14)$

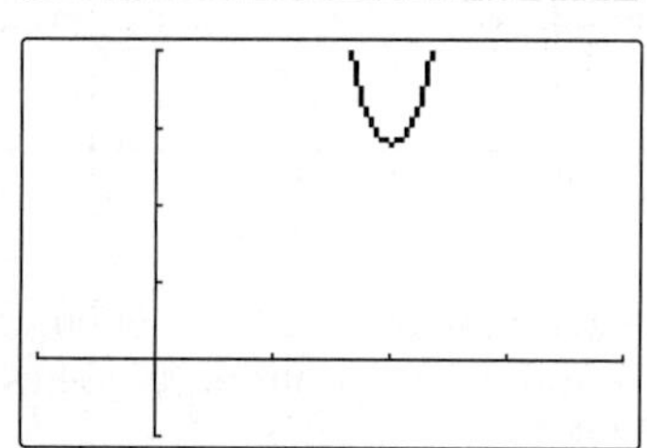

Pages 615–617 Lesson 11-1

7. $x = 0$, $(0, 2)$, min.

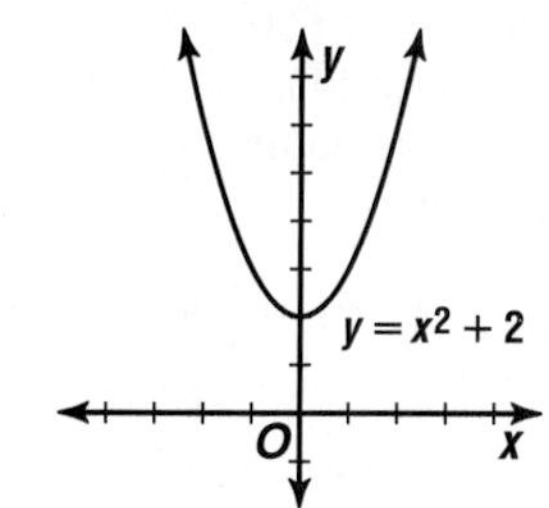

9. $x = -2$, $(-2, -13)$, min.

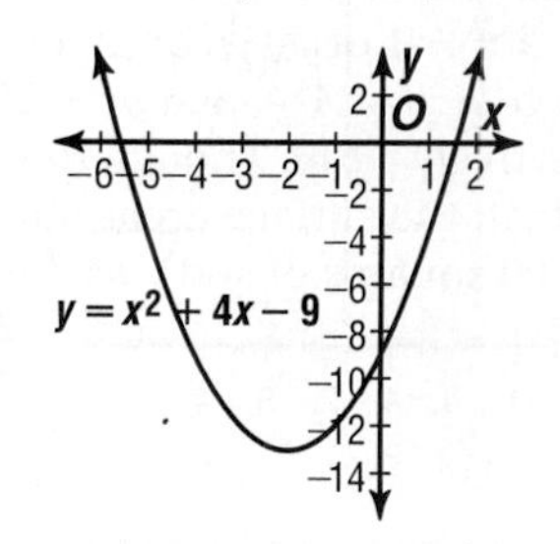

11. $x = 2.5$, $(2.5, 12.25)$, max. **13.** c

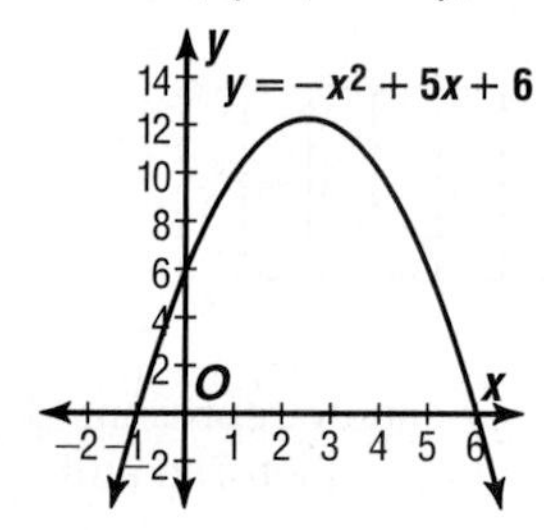

15. $x = 0$, $(0, 0)$, min.

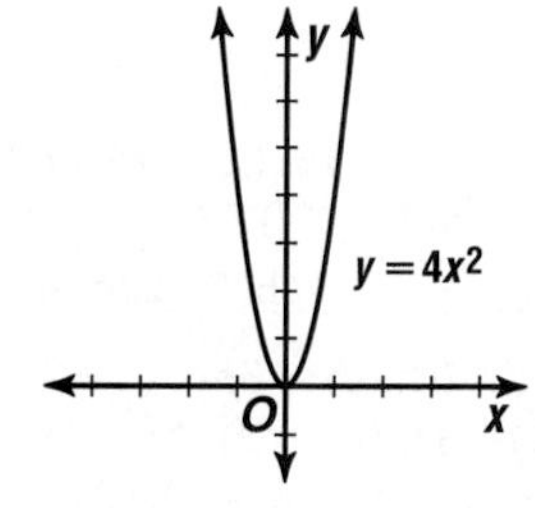

17. $x = -1$, $(-1, 17)$, min.

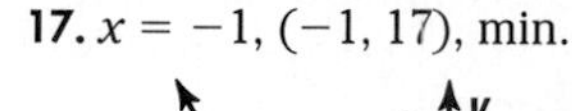
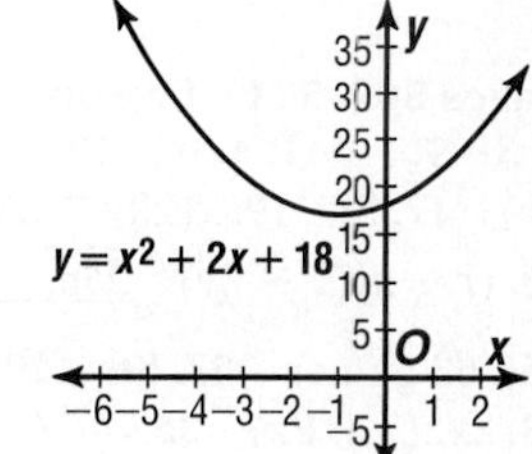

19. $x = 0$, $(0, -5)$, min.

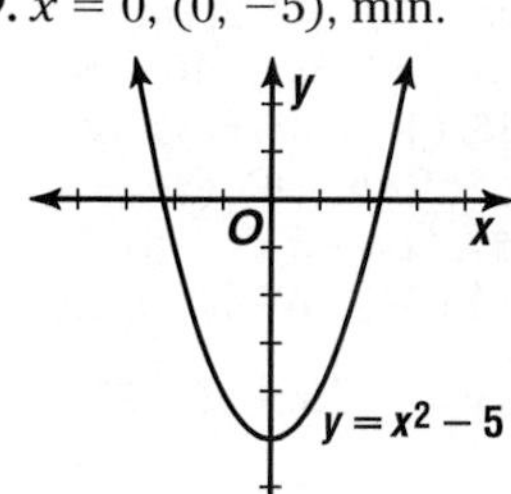

21. $x = -3$, $(-3, -29)$, min.

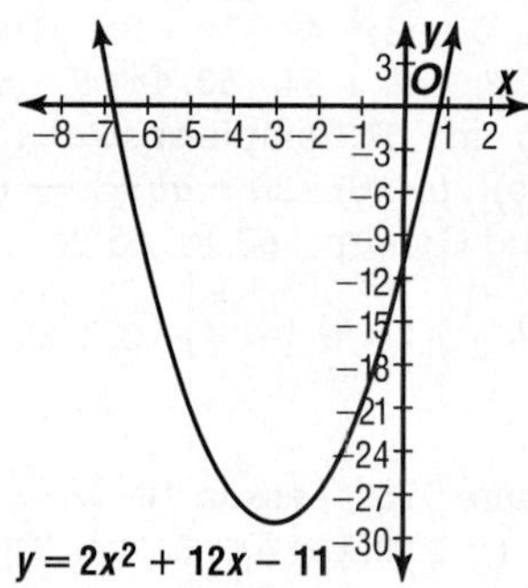

23. $x = 0$, $(0, -25)$, min.

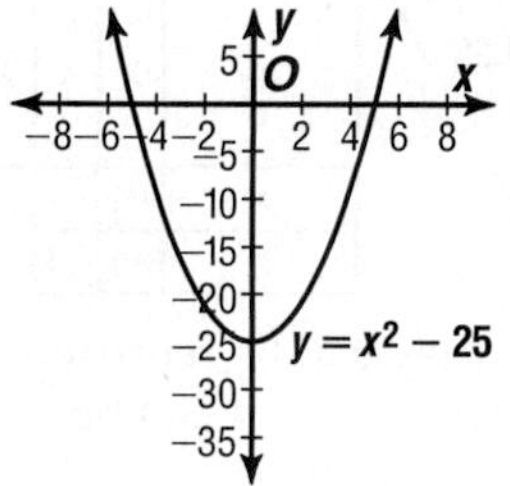
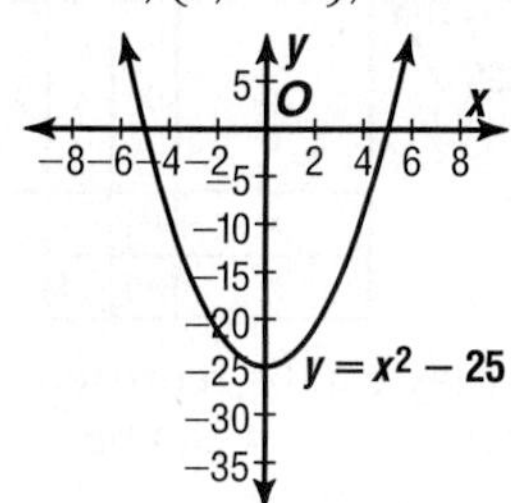

25. $x = -1$, $(-1, 7)$, max.

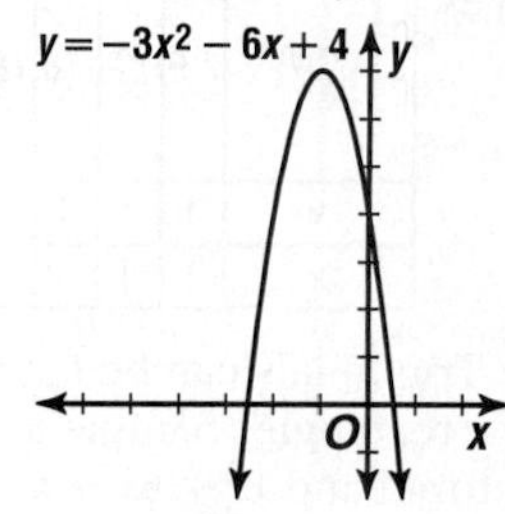

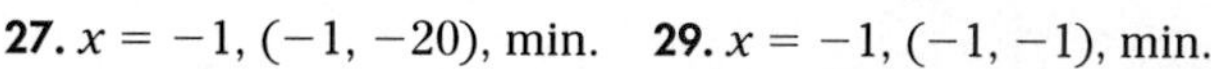

27. $x = -1$, $(-1, -20)$, min.

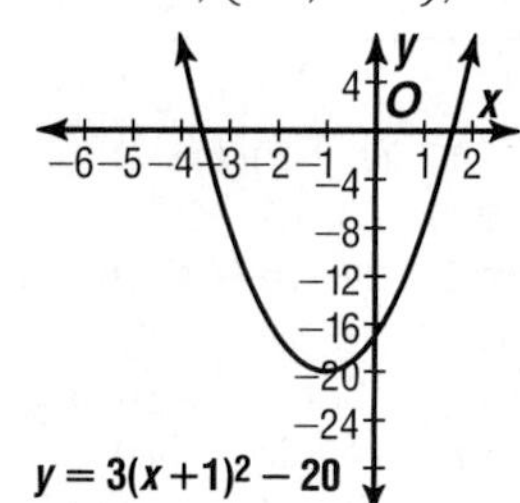

29. $x = -1$, $(-1, -1)$, min.

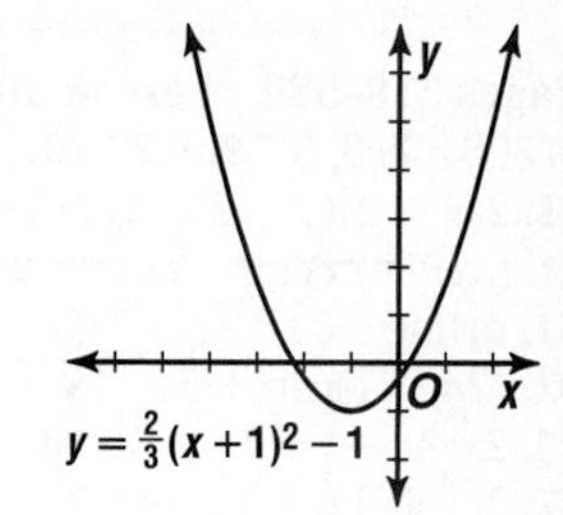

31. *c* **33.**

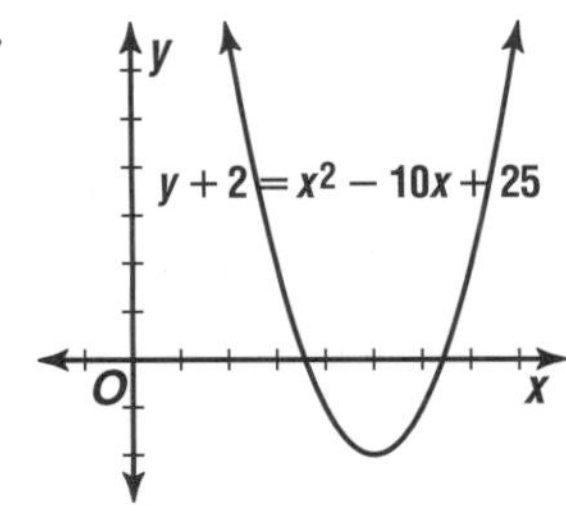

35.

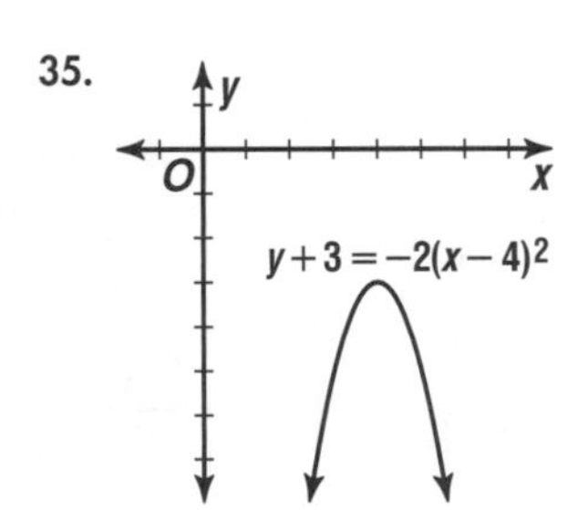

37. $x = -1$ **39.** $x = 2$

41. $(-1.10, 125.8)$

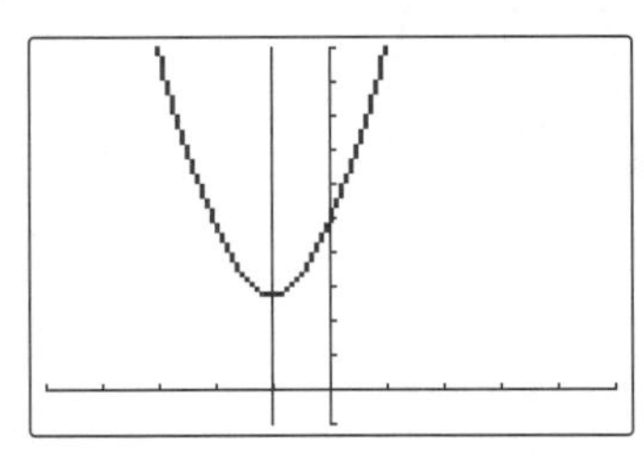

43. $(-0.15, 90.14)$

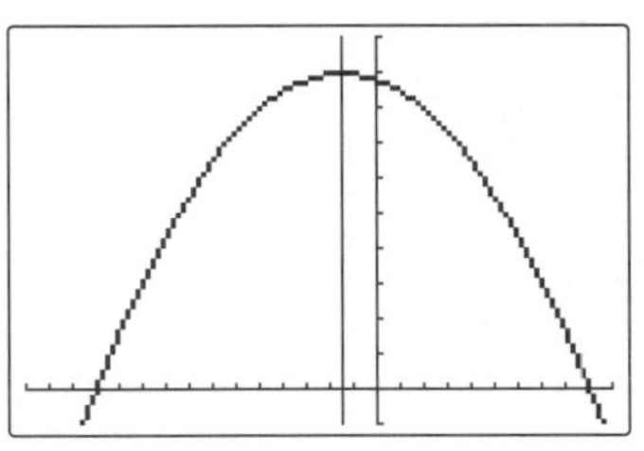

45a.

Year	t	$U(t)$
1970	0	329.96
1975	5	370.31
1980	10	559.16
1985	15	896.51
1990	20	1382.36
1993	23	1745.15

45b. D: $0 \le t \le 23$; R: $326 < U(t) < 1746$

45c.

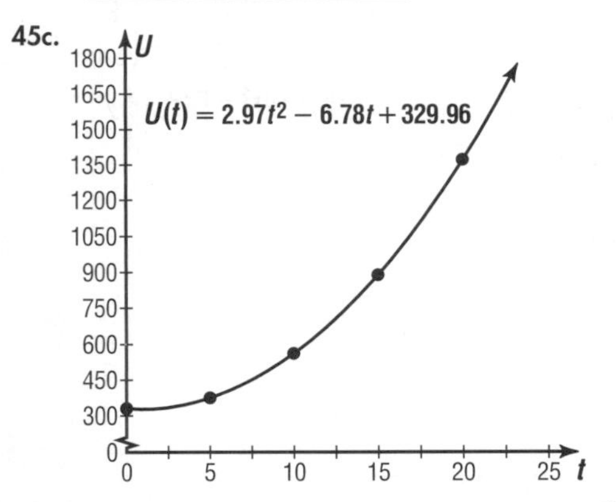

47. 0.15 km **49a.** 6790 km **49b.** 2.2792×10^8 km

51a.

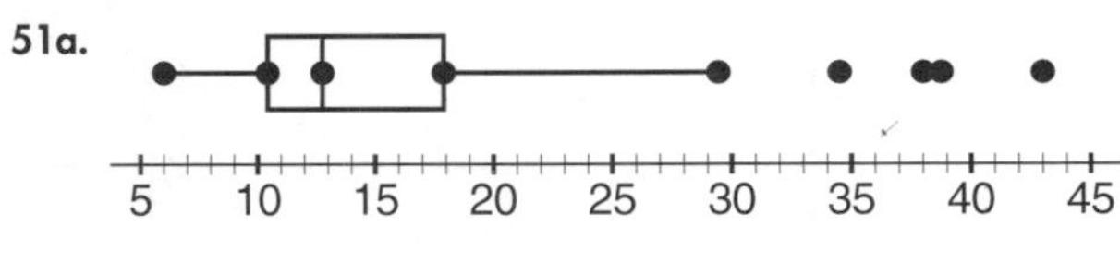

51b. 34.5, 38, 38.7, 43 **53.** -5

Page 619 Lesson 11–1B

1. All the graphs open downward from the origin. $y = -2x^2$ is narrower than $y = -x^2$ and $y = -5x^2$ is the narrowest.

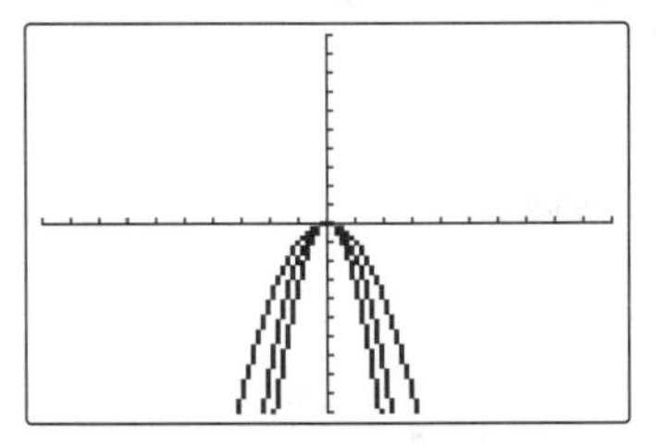

3. All open downward, have the same shape, and have vertices along the x-axis. However, each vertex is different.

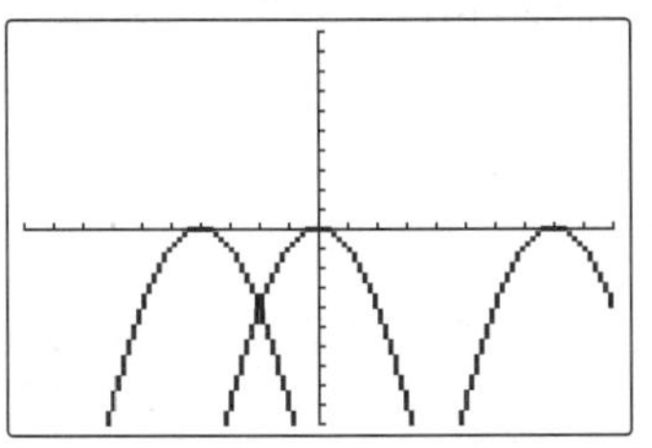

5. It will open upward, have a vertex at the origin, and be narrower than $y = x^2$.

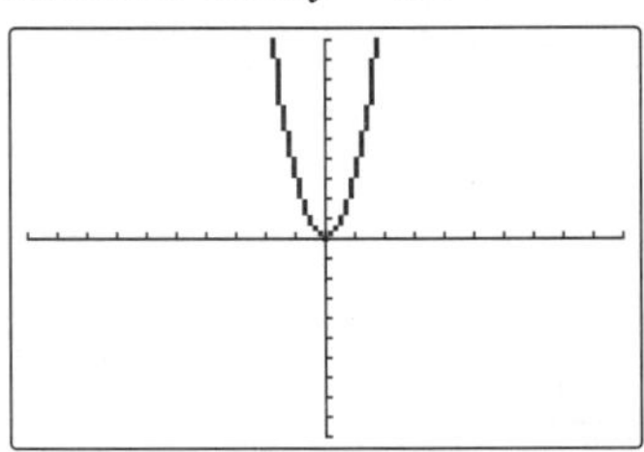

7. It will have vertex at the origin, open downward, and be wider than $y = x^2$.

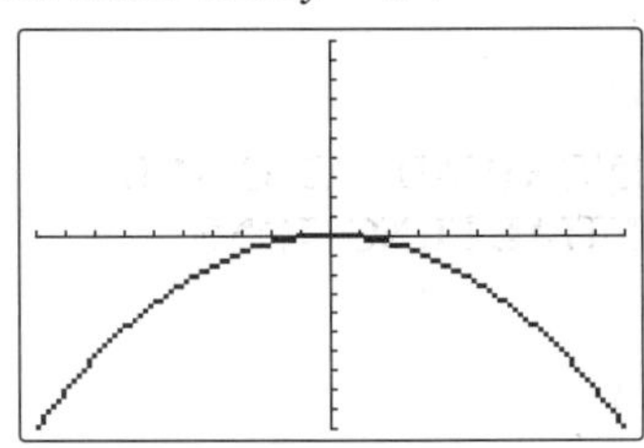

9a. If $|a| > 1$, the graph is narrower than the graph of $y = x^2$. If $0 < |a| < 1$, the graph is wider than the graph of $y = x^2$. If $a < 0$, it opens downward; if $a > 0$, it opens upward. **9b.** The graph has the same shape as $y = x^2$, but is shifted a units (up if $a > 0$, down if $a < 0$). **9c.** The graph has the same shape as $y = x^2$, but is shifted a units (left if $a > 0$, right if $a < 0$). **9d.** The graph has the same shape as $y = x^2$ but is shifted a units left or right and b units up or down as prescribed in 9b and 9c.

Pages 624–627 Lesson 11–2

7. 2 real roots **9.** $-1, 2$

11. $-3, 8$

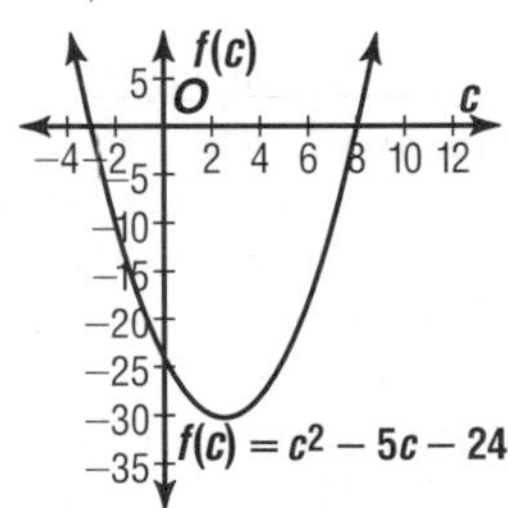

13. $-2 < w < -1, 4 < w < 5$

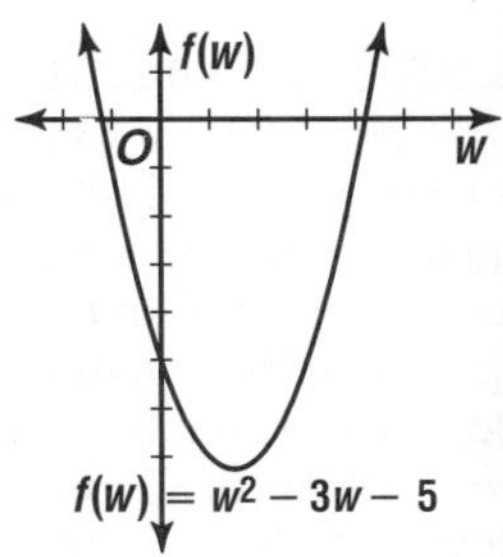

15. 5

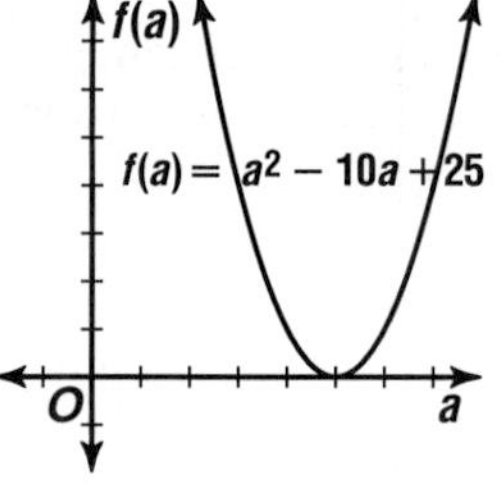

17. −2, −6 **19.** 2 **21.** −4, 4 **23.** −3 **25.** ∅ **27.** −4, 1 **29.** 4 **31.** $8 < a < 9$, $-1 < a < 0$ **33.** 3, $-1 < n < 0$

35.

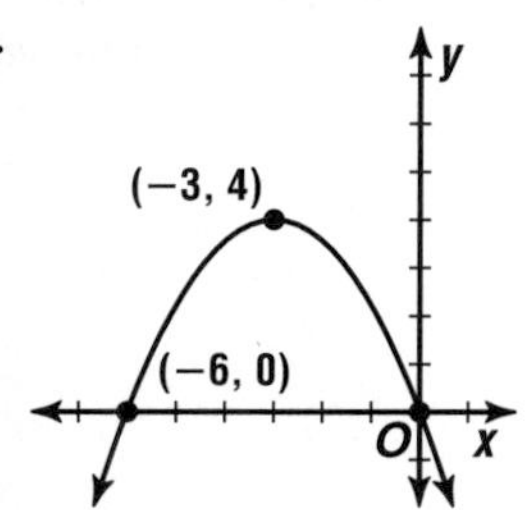

37.

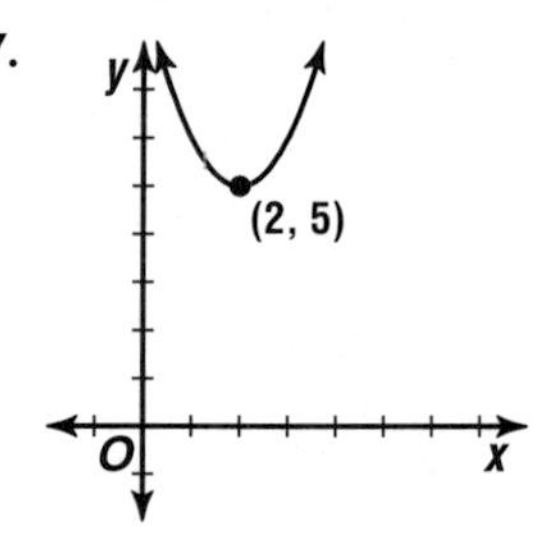

39. −4, −8 or 4, 8 **41.** no solution **43.** −2, −6 **45.** no *y*-intercepts **47.** no real roots **49.** 0.29, 1.71 **51.** −0.79, 2.54 **53.** −2, $k > -2$, $k < -2$

55a.

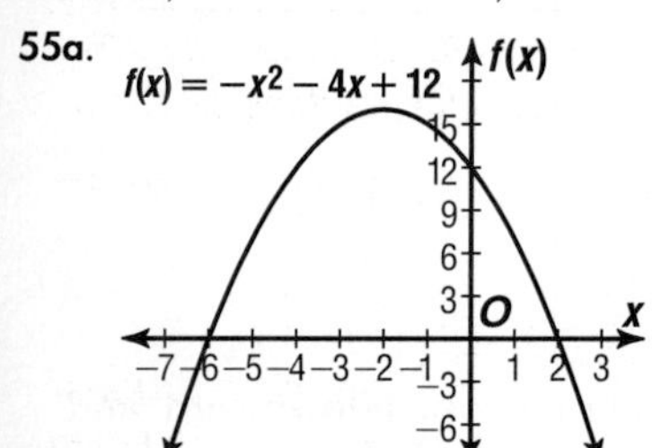

55b. 8 feet **55c.** 16 feet **55d.** $85\frac{1}{3}$ ft² **55e.** $297

57. $x = 0$; (0, 4)

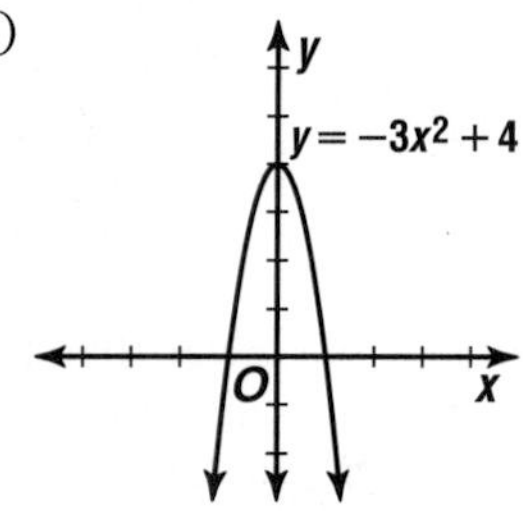

59. $(4x - 3)$ m by $(2x - 1)$ m **61.** 6 **63.** $-4 < x < \frac{4}{3}$ **65.** $24.75

Pages 631–633 Lesson 11–3

5. 1, 3, −18; 3, −6 **7.** 4, −2, 15; no real roots **9.** 1.83, −2.33 **11.** −1, −6 **13.** about 29.4 ft/s **15.** −8.61, −1.39 **17.** $-\frac{4}{5}$, 1 **19.** $-\frac{5}{3}$, 4 **21.** −2.91, 2.41 **23.** 7 **25.** $-\frac{3}{2}$, $-\frac{5}{3}$ **27.** −6, 6 **29.** $-\frac{1}{3}$, 1 **31.** −3, $-\frac{1}{2}$ **33.** 0.60, −0.25 **35.** 0.5, 2 **37.** −0.2, 1.4 **39.** Sample answer: 4, 20, 23; $4x^2 + 20x + 23 = 0$ **41a.** 10, 1 **41b.** none **41c.** −0.5 **41d.** −1, 0.7142857143 **43a.** Discriminant is not a perfect square. **43b.** Discriminant is a perfect square but the expression is not an integer **43c.** Discriminant is a perfect square and the equation can be factored.
43d. Discriminant is 0 and the equation is a perfect square.
45a. Y, 10 mg; C, 8.75 mg **45b.** $0 = a^2 - 11a + 12$; 1.2 yr, 9.8 yr **47.** {0, −9} **49.** 24, 18 **51.** yes **53.** −4°F

Page 633 Self Test

1. $x = \frac{1}{2}$, $\left(\frac{1}{2}, -\frac{25}{4}\right)$

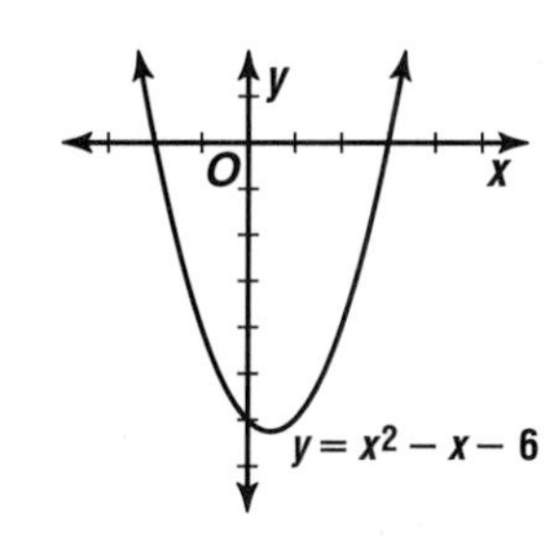

3. $x = 0$, (0, 7)

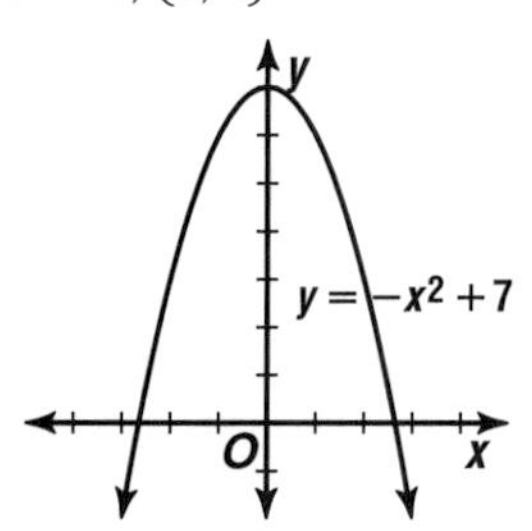

5. −9, 9

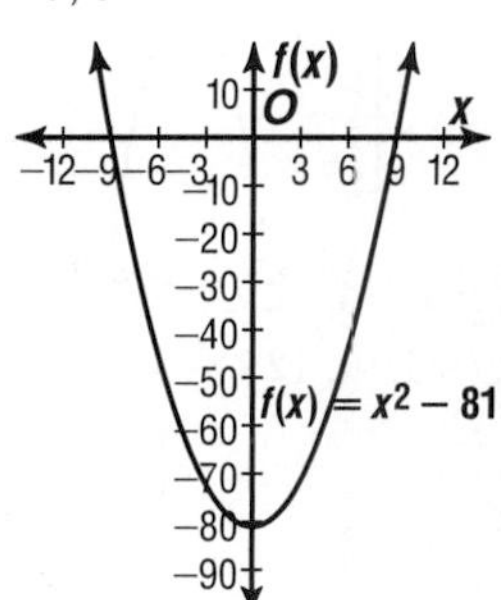

7. ∅

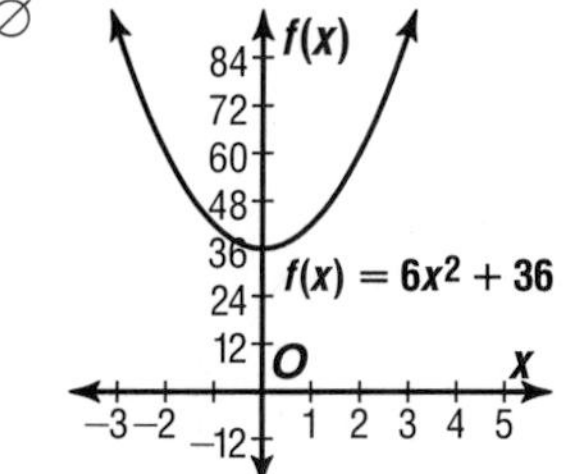

9. ∅

Page 634 Lesson 11–4A

1.

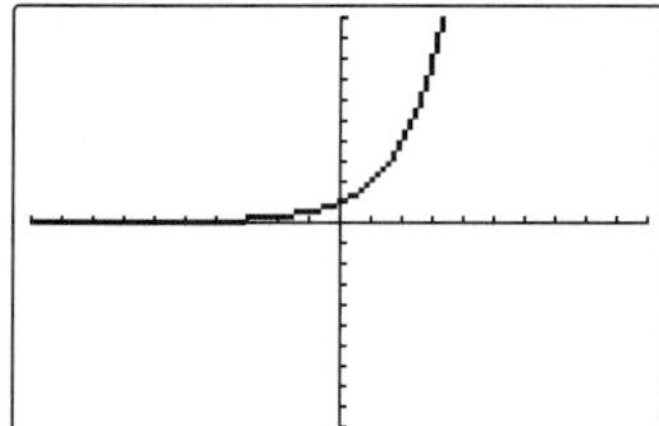

3.

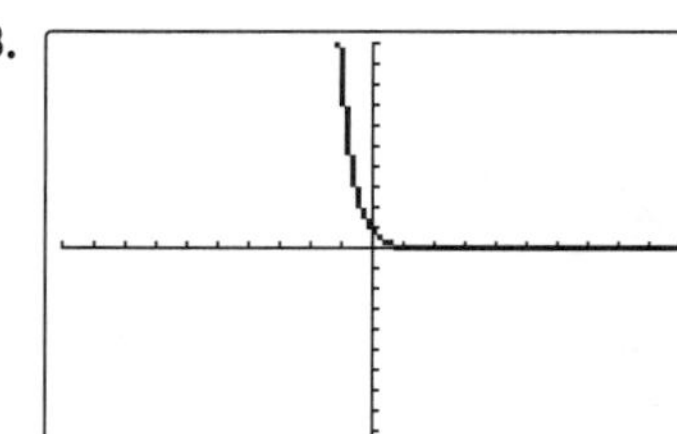

5.

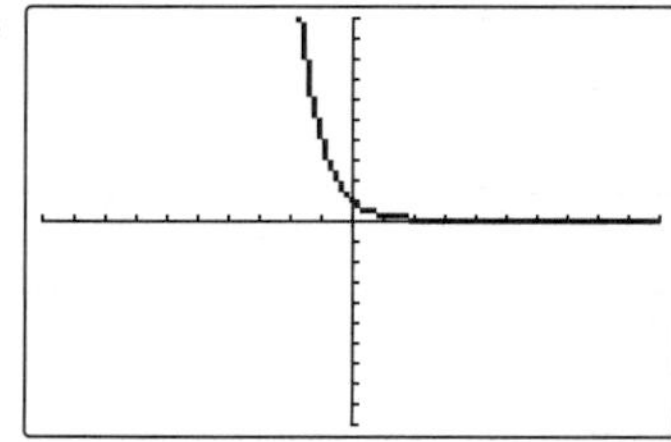

7.

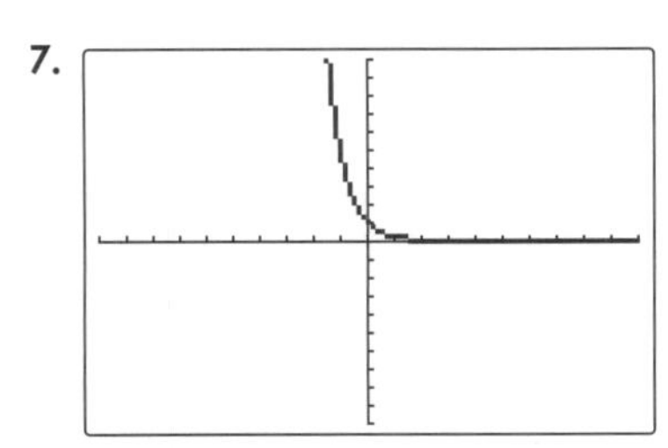

9.

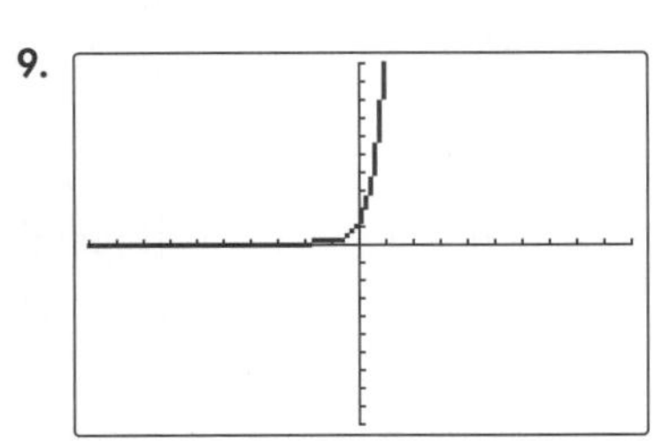

Pages 640–642 Lesson 11–4

7. 5.20 **9.** 12.51
11. 7

13. −2 **15.** −2 **17.** 10.56 **19.** 1.63 **21.** 25.86 **23.** 0.86
25. 5

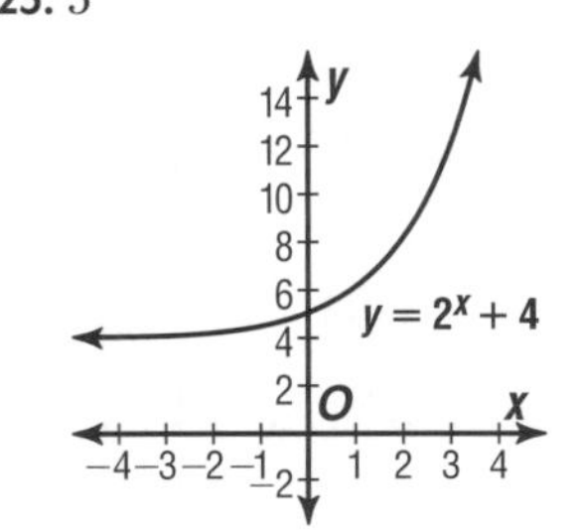

27. 3

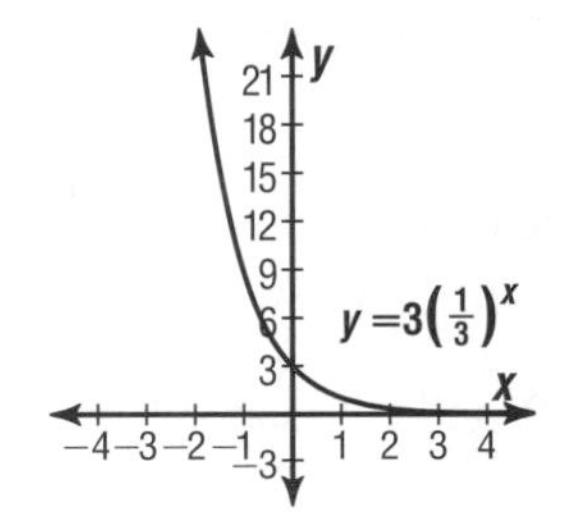

29. 1

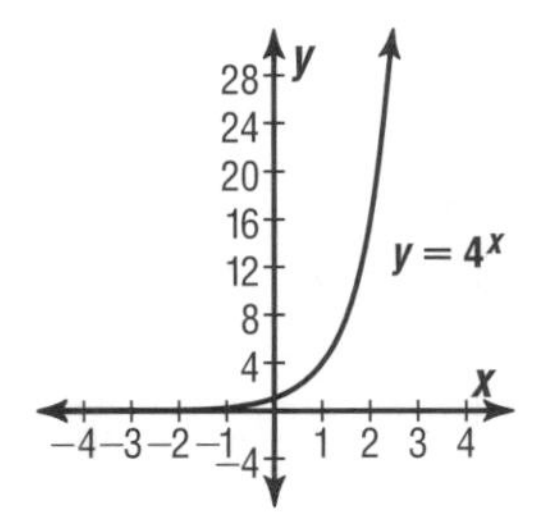

31. no, linear **33.** no, no pattern **35.** −7 **37.** −3
39. 3 **41.** 3, −1
43. All have the same shape but different y-intercepts.

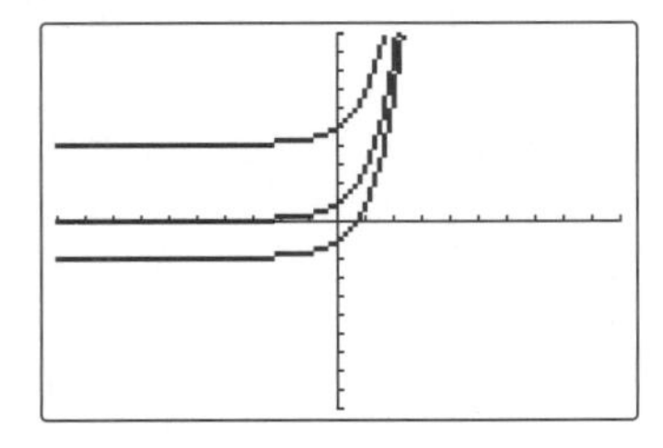

45. All have the same shape but are positioned at different places along the x-axis.

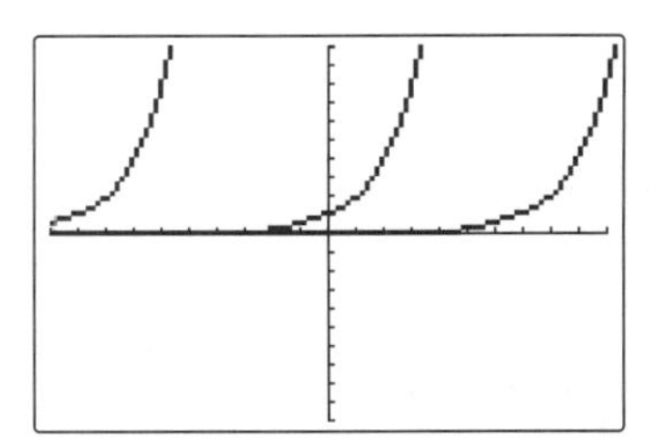

47a. 2.51 **47b.** You cannot write 10 as a power or 2.5.
49. 1600 bacteria **51.** $-3, -\frac{1}{2}$ **53.** $\frac{1}{5}ab(4a - 3b - 1)$
55. 184

57a–b.

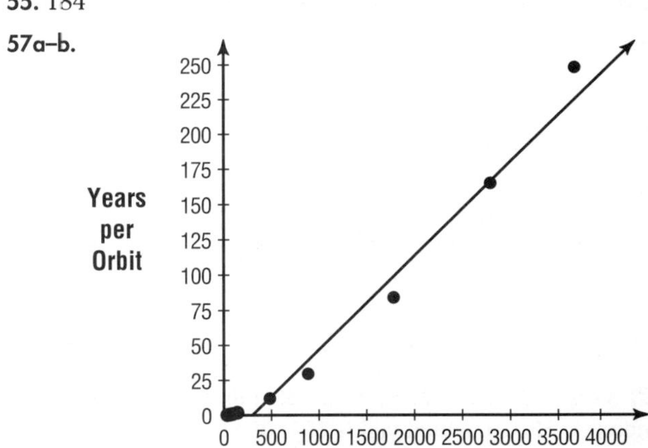

Sample answer: $y = 0.07x - 12$ **57c.** 275 years **59.** 60

Pages 646–649 Lesson 11–5

3a. 1 **3b.** 2 **3c.** 4 **3d.** 365 **5.** decay **7.** about \$70,000,000,000 **9.** \$661.44 **11.** \$10,962.19 **13a.** Each equation represents growth and t is the number of years since 1994. $y = 58.7(1.025)^t$, $y = 919.9(1.019)^t$, $y = 35.6(1.021)^t$, $y = 2.9(1.013)^t$ **13b.** 68.1 million, 1029.9 million, 40.3 million, 3.1 million
13c.

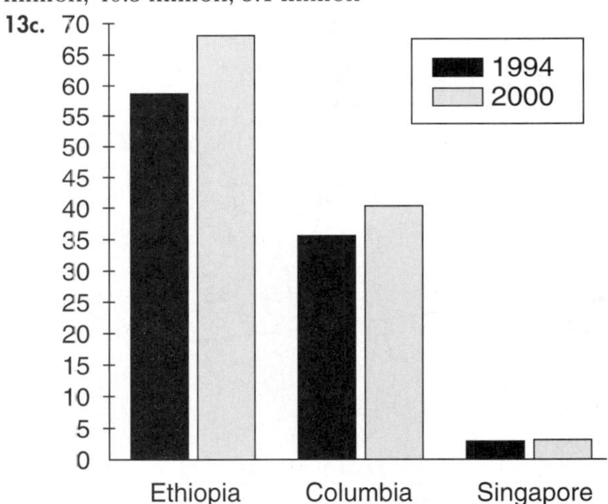

15a. $y = 231.5(1.0963)^t$, t = years since 1950, growth
15b. 57.6 trillion **17a.** \$6931.53 **17b.** \$6954.58
17c. \$23.05 **17d.** daily **21.** $a > 1$, growth; $0 < a < 1$, decay; x represents time **23.** $-\frac{1}{2}$ **25.** $-\frac{r^{10}t^2}{3}$ **27.** $y = -2.5x + 2$

Page 651 Chapter 11 Highlights

1. d **3.** i **5.** c **7.** b **9.** f

Pages 652–654 Chapter 11 Study Guide and Assessment

11. $x = 0$; $(0, 4)$ **13.** $x = -1$; $(-1, -20)$ **15.** $x = -1$; $(-1, -1)$

17.

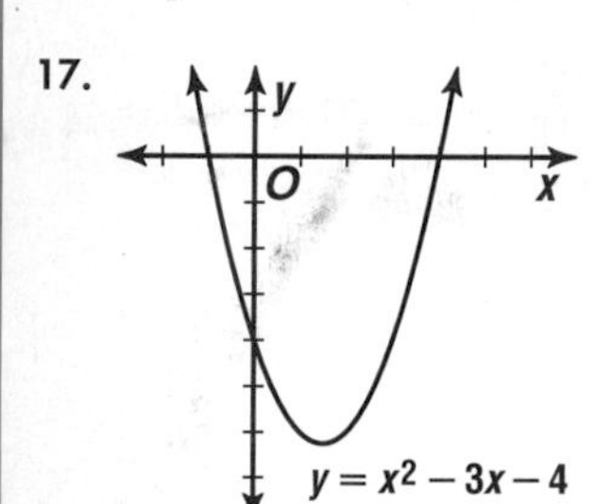

19.

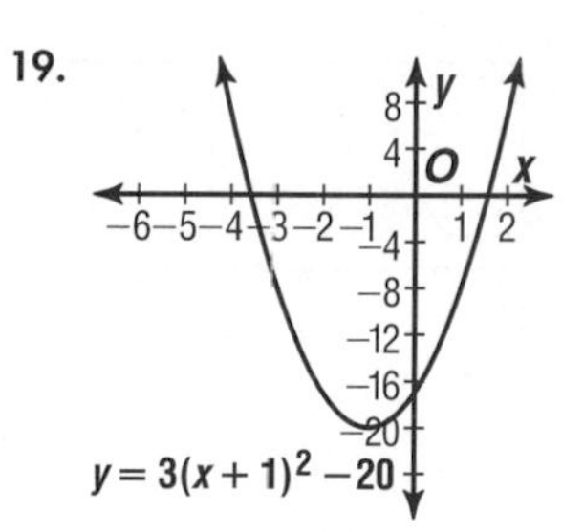

21. −3, 4 **23.** $-5 < x < -4,\ 0 < x < 1$ **25.** 3, 7 **27.** 10, −2 **29.** $\frac{3}{2}, -\frac{5}{2}$ **31.** 1, $\frac{4}{9}$ **33.** 2, 3 **35.** 0.47, −0.71

37. 7

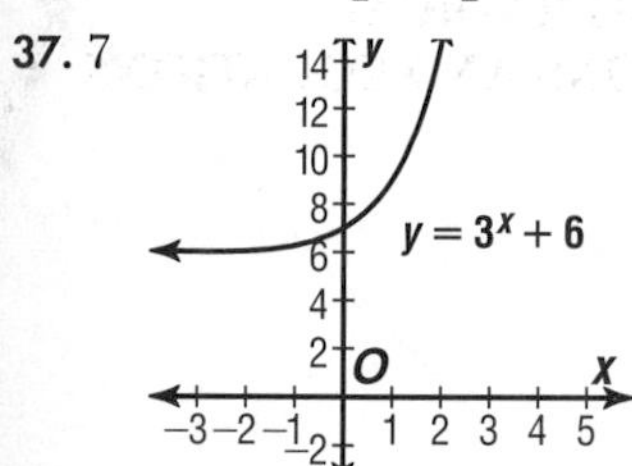

39. 1

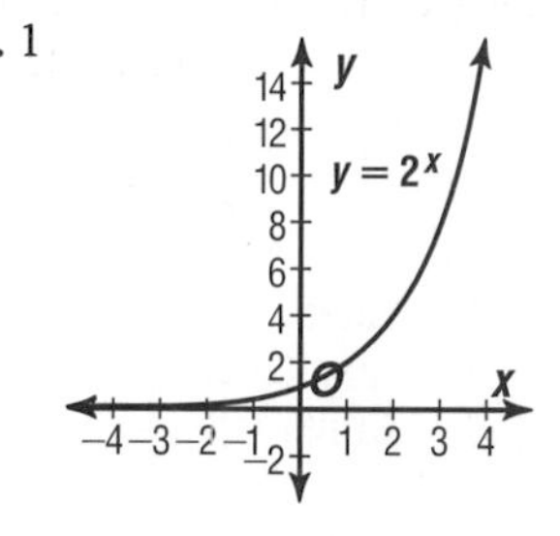

41. −3 **43.** −6 **45.** −3, $\frac{3}{2}$ **47.** \$12,067.68 **49.** \$24,688.37 **51.** 1.3 seconds and 4.7 seconds **53.** between 4 meters and 5 meters **55.** CD, which yields \$541.50 vs. savings at \$530.92

CHAPTER 12 EXPLORING RATIONAL EXPRESSIONS AND EQUATIONS

Pages 663–665 Lesson 12–1

7. $7a^2b$; $\frac{-b^2}{3a^3}$; $a \neq 0, b \neq 0$ **9.** b; $\frac{3}{b+5}$; $b \neq 0, b \neq -5$ **11.** $m - 3$; $\frac{1}{m+3}$; $m \neq \pm 3$ **13a.** 5 **13b.** 750 lb **15.** $\frac{5y}{2z}$; $y \neq 0, z \neq 0$ **17.** $\frac{4}{5x}$; $x \neq 0, y \neq 0$ **19.** y; $y \neq -\frac{1}{3}$ **21.** $y + 7x$; $y \neq 7x$ **23.** $\frac{1}{a-4}$; $a \neq 4, -3$ **25.** $\frac{1}{x+4}$; $x \neq -4$ **27.** $\frac{a+6}{a+4}$; $a \neq -4, 2$ **29.** $\frac{b+1}{b-9}$; $b \neq 9, 4$ **31.** $\frac{4}{5}$; $x \neq -1$ **33.** $-\frac{5}{3}$ **35.** not possible, $y \neq -2$ **37a.** 192 **37b.** 3840 lb **39a.** 9.5°/mile **39b.** $-\frac{95}{10} = -\frac{19}{2}$

39c.

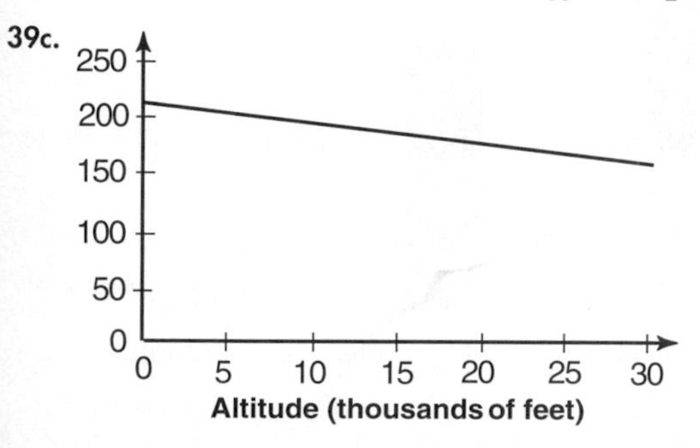

41. ± 11

43.

45.

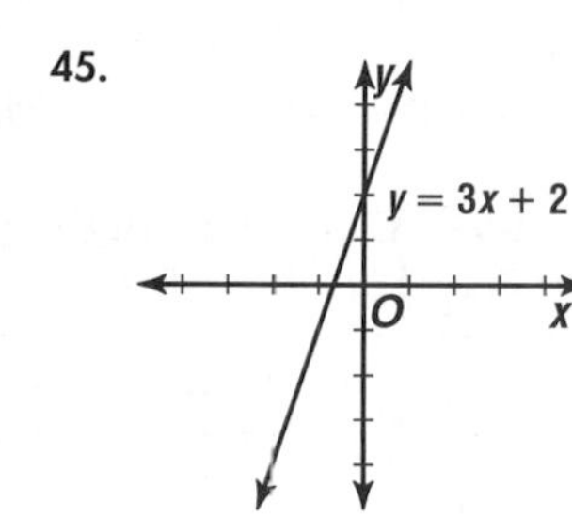

Page 666 Lesson 12–1B

1. $\frac{x-5}{x+5}$; −5 **3.** $\frac{1}{2x}$; 0, 4.5 **5.** $\frac{x-1}{x-8}$; 8 **7.** −1; 3 **9.** $-\frac{x+5}{x+6}$; −6, 5

Pages 669–670 Lesson 12–2

5. $\frac{9m}{2xy}$ **7.** $\frac{4}{x+3}$ **9.** $\frac{4}{3}$ **11a.** 91.44 cm/yd **11b.** changing centimeters to yards **13.** $\frac{3a}{2}$ **15.** $\frac{2}{x}$ **17.** $8y$ **19.** $\frac{4m}{3(m+5)}$ **21.** $\frac{2m}{a+9}$ **23.** 2 **25.** $3(x - y)$ **27.** 36 **29.** $\frac{4x}{3x-5}$ **31.** $\frac{4}{5(x+5)}$ **33.** 21.8 mph; converts ft/s to mph **35.** 10,800 Calories **37.** Sample answer: $\frac{3(x+2)}{x+7} \cdot \frac{2(x-3)}{x-4}$; $\frac{6}{x+7} \cdot \frac{x^2-x-1}{x-4}$ **39a.** $\frac{1 \text{ franc}}{0.1981 \text{ dollars}}$; $\frac{1 \text{ dollar}}{5.05 \text{ francs}}$ **39b.** 12,500 pesos $\cdot \frac{0.1597 \text{ dollars}}{1 \text{ peso}} \cdot \frac{1 \text{ franc}}{0.1981 \text{ dollars}}$; about 10,077 francs **39c.** Convert to American dollars and then to Hong Kong's dollar. **41.** 729 **43.** 4 **45.** 0

Pages 672–674 Lesson 12–3

5. $\frac{5}{x}$ **7.** $\frac{y+3}{x^2-9}$ **9.** $\frac{1}{x^2+2x+5}$ **11.** $\frac{(m+2)^2}{4}$ **13.** $\frac{x+4}{x+6}$ **15.** 0.5 ft³ **17.** $\frac{a}{a+11}$ **19.** $\frac{b^2m}{c^2}$ **21.** $\frac{m}{m+5}$ **23.** $\frac{4}{z+3}$ **25.** $\frac{2(x+5)}{x+1}$ **27.** $x + 3$ **29.** $x + 5$ **31.** $(x + 5)(x + 3)$ **33.** 88 ft/s; sample answer: changes 60 mph to ft/s **35.** 27.5 yd³; changes ft³ to yd³ **37.** $\frac{x-y}{4}$ **39a.** 5500 miles **39b.** 41.3 trains **39c.** 10,618.2 miles; 27.3 trains **41.** $\frac{8}{m-5}$ **43.** $(4a - 3b^2)^2$ **45.** $x < -2$ **47.** $7 + 2 = 2 + 7$

Pages 678–680 Lesson 12–4

5. $3b^2 - 5$ **7.** $t - 1$ **9.** $2m + 3 + \frac{-3}{m+2}$ **11.** $(2x + 3)$ m **13.** $\frac{b}{3} + 3 - \frac{7}{3b}$ **15.** $\frac{m}{7} + 1 - \frac{4}{m}$ **17.** $a^2 + 2a + 12 + \frac{3}{a-2}$ **19.** $m - 3 - \frac{2}{m+7}$ **21.** $2x + 6 + \frac{1}{x-7}$ **23.** $3m^2 + 4k^2 - 2$ **25.** $a + 7 - \frac{1}{a+3}$ **27.** $3x - 2 + \frac{21}{2x+3}$ **29.** $t + 4 - \frac{5}{4t+1}$ **31.** $8x^2 + \frac{32x}{7} - 9$ **33a.** $x^2 + 4x + 4$ **33b.** $5t^2 - 3t - 2$ **33c.** $2a^2 + 3a - 4$ **35.** −15 **37.** \$480,000 **39.** $-\frac{x}{7}$ **41.** $3(x - 7)(x + 5)$ **43.** (1, 3)

45.

Page 680 Self Test

1. $\frac{25x}{36y}$ **3.** $\frac{(x+4)(4x^2+2x-3)}{(x+3)(x+2)(x-1)}$ **5.** $\frac{x-5}{x+4}$ **7.** $2x - 5$ **9.** $4(3x - 1)$ or $(12x - 4)$ units

Pages 683–684 Lesson 12–5

5. $\frac{10}{x}$ **7.** $\frac{10}{a+2}$ **9.** 3 **11.** m **13.** $-\frac{a}{6}$ **15.** 2 **17.** $\frac{7}{x+7}$ **19.** 1 **21.** 1 **23.** $\frac{4x}{x+2}$ **25.** $\frac{2y-6}{y+3}$ **27.** 2 **29.** $\frac{25x-93}{2x+5}$ **31.** $\frac{24x+26y}{7x-2y}$ **33.** b **35.** 442 days **37.** $\pm\frac{6}{5}$ **39.** −3, 4 **41.** $\frac{1}{216}$

Pages 687–689 Lesson 12-6

3. x^2 **5.** $90m^2b^2$ **7.** $(x-3)(x+3)$ **9.** $\frac{7+9m}{15m^2}$ **11.** $\frac{-20}{3(x+2)}$ **13.** $\frac{22-7y}{6y^2}$ **15.** 168 members **17.** $\frac{-5x}{63}$ **19.** $\frac{7yz+3}{xyz}$ **21.** $\frac{2s-3t}{s^2t^2}$ **23.** $\frac{7z+8}{(z+5)(z-4)}$ **25.** $\frac{k^2+k-10}{(k+5)(k+3)}$ **27.** $\frac{-17r-32}{(3r-2)(r-5)}$ **29.** $\frac{10+m}{2(2m-3)}$ **31.** $\frac{3w-4}{3(5w+2)}$ **33.** $\frac{-5n-4}{(n+3)(n+1)}$ **35.** $\frac{-17t+27}{(t+3)(10t-9)}$ **37.** $\frac{v^2+11v-2}{(v+4)(v-1)}$

39a. 360, 3, 120, 360 **39b.** 540, 6, 90, 540 **39c.** The GCF times the LCM of two numbers is equal to the product of the two numbers. **39d.** Divide the GCF into the product of the two numbers to find the LCM. **41.** 15 months later or July 20, 1997 **43.** 2 **45.** $\frac{3 \pm \sqrt{41}}{4} \approx 2.35$ or -0.85 **47.** $-15a^6b^3$ **49.** 4

Pages 693–695 Lesson 12-7

3. $\frac{8x+3}{x}$ **5.** $\frac{6m^2+m+1}{2m}$ **7.** $\frac{9}{5}$ **9.** $\frac{y}{2}$ **11.** $\frac{a-b}{3}$ **13.** $\frac{3x+15}{x+3}$ **15.** $\frac{6x-1}{2x+1}$ **17.** $\frac{5r^2+r-48}{r^2-9}$ **19.** $\frac{4}{3}$ **21.** ab **23.** $\frac{6}{x}$ **25.** $\frac{a-5}{a-2}$ **27.** $\frac{1}{x+3}$ **29.** $\frac{m+6}{m+2}$ **31.** $\frac{y-3}{y+2}$ **33.** $\frac{t^2+2t+2}{-t^2}$ **35.** $\frac{32b^2}{35}$ **37a.** $x \neq 0, x \neq -3$ **37b.** $x \neq 0, x \neq -3$ **37c.** yes **39a.** $\frac{fs}{s-v}$ **39b.** 413.5 **39c.** 2 **39d.** 6 **41.** $\frac{2x+3}{x+3}$ **43.** $(2x+1)(2x-1)$ **45.** $(-3, 4)$ **47.** $w = \frac{P-2\ell}{2}$

Pages 700–702 Lesson 12-8

5. -12 **7.** $-\frac{1}{2}$ **9.** 0 **11a.** $\frac{1}{5}$ **11b.** $\frac{3}{5}$ **11c.** $\frac{x}{5}$ **13.** 3.429 ohms **15.** 8 ohms, 4 ohms **17.** $\frac{50}{3}$ **19.** $\frac{5}{4}$ **21.** 0 **23.** 2 or $-\frac{1}{3}$ **25.** 4 or $-\frac{1}{3}$ **27.** -13 **29.** $\frac{3}{5}$ **31.** 3 **33.** 6 **35.** 4 ohms **37.** $R_1 = \frac{R_2R_T}{R_2 - R_T}$ **39.** $n = \frac{IR}{E - Ir}$ **41.** 10 **43.** 96 ohms **45a.** 2.5 hours **45b.** 2.19 hours **45c.** after 923 miles **47a.** $w = \frac{c\ell}{100}$ **47b.** $\ell = \frac{100w}{c}$ **49.** 10 **51.** $\frac{b+4}{4(b+2)^2}$

53.

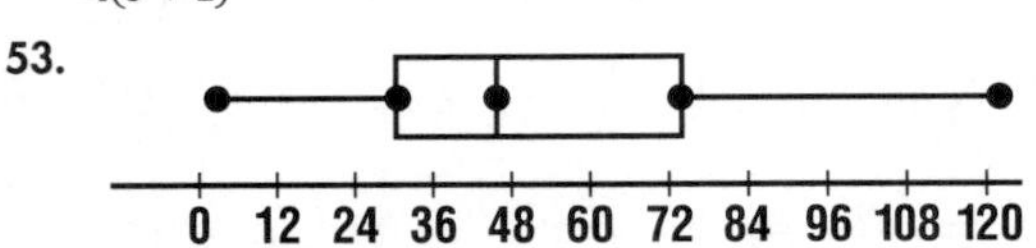

55. $\frac{48}{25}$ **57.** 36

Page 703 Chapter 12 Highlights

1. false, rational **3.** true **5.** false, $x^2 - 144$ **7.** false, least common multiple

Pages 704–706 Chapter 12 Study Guide and Assessment

9. $z, z \neq 3$ **11.** $\frac{a}{a+2}, a \neq 0, -2$ **13.** $\frac{1}{(b+3)(b+2)}$, $b \neq \pm 2, \pm 3$ **15.** $\frac{3axy}{10}$ **17.** $\frac{3}{a^2-3a}$ **19.** $b + 7$ **21.** $\frac{y^2-4y}{3}$ **23.** $\frac{2m-3}{3m-2}$ **25.** $x^2 + 4x - 2$ **27.** $x^3 + 5x^2 + 12x + 23 + \frac{52}{x-2}$ **29.** $\frac{2a}{m^2}$ **31.** $\frac{2m+3}{5}$ **33.** $a + b$ **35.** $\frac{22n}{21}$ **37.** $\frac{4c^2+9d}{6cd^2}$ **39.** $\frac{4r}{r-3}$ **41.** $\frac{5m-8}{m-2}$ **43.** $\frac{3x}{y}$ **45.** $\frac{x^2+8x-65}{x^2+8x+12}$ **47.** -5 **49.** $-\frac{1}{4}$ **51.** 5, $-\frac{1}{4}$ **53.** 6 bicycles, 24 tricycles, and 24 wagons **55.** $1\frac{7}{8}$ hours

CHAPTER 13 EXPLORING RADICAL EXPRESSIONS AND EQUATIONS

Page 712 Lesson 13-1A

1. $4 + 4 = 8$ **3.** $16 + 9 = 25$

5.

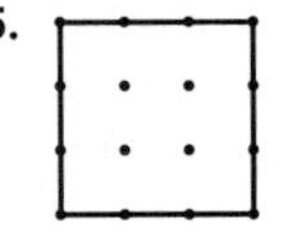

7. **9.**

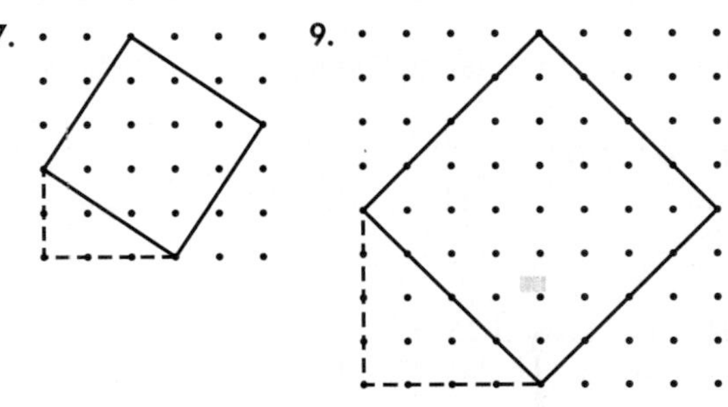

Pages 715–718 Lesson 13-1

7. 7 **9.** 11.40 **11.** 15 **13.** 2 **15.** yes **17.** 127.28 ft **19.** 13.86 **21.** 13.08 **23.** 14.70 **25.** 60 **27.** 4 **29.** $\sqrt{67} \approx 8.19$ **31.** $\sqrt{27} \approx 5.20$ **33.** $\sqrt{145} \approx 12.04$ **35.** yes **37.** no; $11^2 + 12^2 \neq 15^2$ **39.** yes **41.** $\sqrt{75}$ in. or about 8.66 in. **43a.** 44.49 m **43b.** 78 m^2

45.

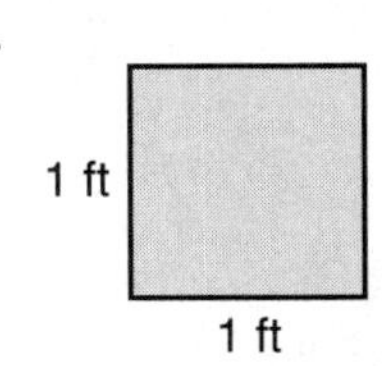

Area = 1 ft^2 or 144 in^2

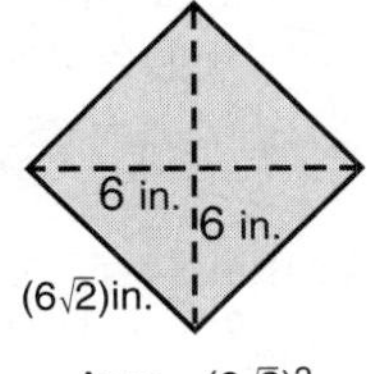

Area = $(6\sqrt{2})^2$ or 72 in^2

47. about 2.66 m **49.** -7 **51.** $12r^7$ **53.** $\left\{c \mid c \geq \frac{1}{2}\right\}$ **55.** 41

Pages 724–725 Lesson 13-2

5. $5 - \sqrt{2}$; 23 **7.** $\frac{\sqrt{7}}{\sqrt{7}}$ **9.** $3\sqrt{2}$ **11.** $\frac{\sqrt{21}}{7}$ **13.** $10\sqrt{2} + 26$ **15.** $\frac{18+6\sqrt{2}}{7}$ **17.** > **19.** about 7003.5 gal/min **21.** $4\sqrt{5}$ **23.** $10\sqrt{5}$ **25.** $\frac{\sqrt{2}}{2}$ **27.** $\frac{\sqrt{22}}{8}$ **29.** $84\sqrt{5}$ **31.** $\frac{\sqrt{11}}{11}$ **33.** $3|ab|\sqrt{6}$ **35.** $7x^2y^3\sqrt{3xy}$ **37.** $\frac{3\sqrt{3}}{|p|}$ **39.** $\frac{x\sqrt{3xy}}{2y^3}$ **41.** $y^2 - 2y\sqrt{7} + 7$ **43.** $\frac{28\sqrt{2}+14\sqrt{5}}{3}$ **45.** $\frac{-2\sqrt{5}-\sqrt{10}}{2}$ **47.** $\frac{c - 2\sqrt{cd} + d}{c - d}$ **49.** $x^2 - 5x\sqrt{3} + 12$ **51.** > **53.** < **55.** Yes; the result is about 11.84. **57.** 18.44 cm **59.** 1, $\frac{2}{3}$ **61.** 2.7×10^9 acres

63. $y = \frac{1}{5}x + 6$ **65.** -6

Page 726 Lesson 13–2B

1. $14\sqrt{7}$ or 37.04 **3.** $2\sqrt{3} + 3\sqrt{2}$ or 7.71 **5.** $\frac{4\sqrt{7}}{7}$ or 1.51 **7.** $3\sqrt{3}$ or 5.20 **9.** $\frac{\sqrt{3}}{2} - \frac{\sqrt{5}}{3} + 3\sqrt{2}$ or 4.36 **11.** $\frac{-16\sqrt{6}}{3}$ or -13.06

Pages 729–731 Lesson 13–3

5. $3\sqrt{5}, 3\sqrt{20}$ **7.** none **9.** $13\sqrt{6}$ **11.** $12\sqrt{7x}$ **13.** $13\sqrt{3} + \sqrt{2}$; 23.93 **15.** $\frac{8}{7}\sqrt{7}$; 3.02 **17.** 11 **19.** $26\sqrt{13}$ **21.** in simplest form **23.** $21\sqrt{2x}$ **25.** $9\sqrt{3}$; 15.59 **27.** $\sqrt{6} + 2\sqrt{2} + \sqrt{10}$; 8.44 **29.** $-\sqrt{7}$; -2.65 **31.** $29\sqrt{3}$; 50.23 **33.** $4\sqrt{5} + 15\sqrt{2}$; 30.16 **35.** $4\sqrt{3} - 3\sqrt{5}$; 0.22 **37.** $10\sqrt{2} + 3\sqrt{10}$ **39.** $19\sqrt{5}$ **41.** $10\sqrt{3} + 16$ **43.** Both $(x - 5)^2$ and $(x - 5)^4$ must be nonnegative, but $x - 5$ may be negative. **45.** $6\frac{2}{5}$ ft **47.** $3x - 3y$ **49.** 7 **51.** $\frac{2}{3}$ **53a.** 1991; about 9 million **53b.** 6 million

Page 731 Self Test

1. 35 **3.** 46.17 **5.** 10 **7.** $11\sqrt{6}$ **9.** $12\sqrt{5} - 6\sqrt{3} + 2\sqrt{15} - 3$

Pages 734–736 Lesson 13–4

7. $a + 3 = 4$ **9.** 16 **11.** 36 **13.** -12 **15.** 3.9 mph **17.** -63 **19.** no real solution **21.** $\frac{81}{25}$ **23.** 2 **25.** 24 **27.** $\frac{3}{25}$ **29.** $\pm\frac{4}{3}\sqrt{3}$ **31.** $\sqrt{7}$ **33.** 6 **35.** $-3, -10$ or $3, 10$ **37.** 16 **39.** no real solution **41.** $\left(\frac{1}{4}, \frac{25}{36}\right)$ **43.** $\sqrt{3} + 1$ **45.** $54\frac{2}{3}$ mi **47a.** No, at 30 mph, her car should have skidded 75 feet after the breaks were applied and not 110 feet. **47b.** about 36 miles per hour **49.** $6b - 26 + \frac{150}{b + 5}$ **51.** $6xy^3$ **53.** $(-4, 5)$ **55.** $\frac{8}{13}$

Pages 739–741 Lesson 13–5

5. 5 **7.** $3\sqrt{2}$ or 4.24 **9.** 7 or 1 **11.** about 9.49 mi **13.** 17 **15.** $\sqrt{41}$ or 6.40 **17.** $\frac{10}{3}$ or 3.33 **19.** $\frac{13}{10}$ or 1.30 **21.** $2\sqrt{3}$ or 3.46 **23.** 17 or -13 **25.** -10 or 4 **27.** 8 or 32 **29.** no **31.** $\sqrt{157} \neq \sqrt{101}$; Trapezoid is not isosceles. **33.** The distance between $(3, -2)$ and $(-3, 7)$ is $3\sqrt{13}$ units. The distance between $(-3, 7)$ and $(-9, 3)$ is $2\sqrt{13}$ units. The distance between $(3, -2)$ and $(-9, 3)$ is 13 units. Since $\left(3\sqrt{13}\right)^2 + \left(2\sqrt{13}\right)^2 = 13^2$, the triangle is a right triangle. **35.** 109 miles **37.** $\frac{2\sqrt{2}}{3}$

39. no solution

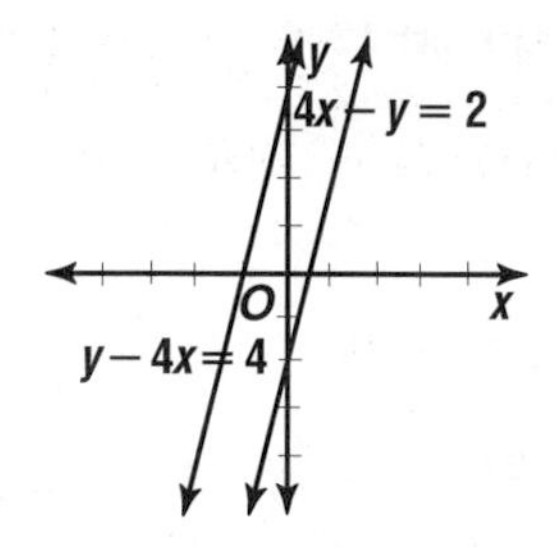

41a. 7392 BTU **41b.** 3705 BTU **41c.** 6247.5 BTU **41d.** 14,040 BTU **41e.** 6831.45 BTU

Page 742 Lesson 13–6A

1. $(x + 2)^2 = 1$ **3.** $(x + 2)^2 = 5$ **5.** $(x - 2)^2 = 5$ **7.** $(x + 1.5)^2 = 3.25$

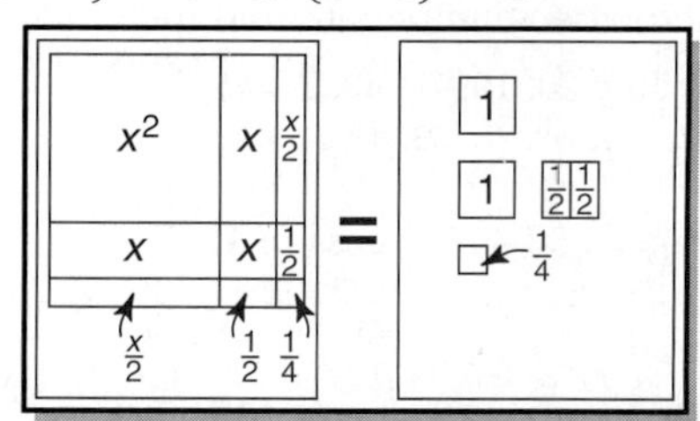

Pages 746–747 Lesson 13–6

5. 64 **7.** $-1, -3$ **9.** $7, -3$ **11.** $2 \pm \sqrt{6}$ **13.** 4.9 ft **15.** 16 **17.** $\frac{121}{4}$ **19.** 8 **21.** 4, 1 **23.** $\frac{7}{3}$ **25.** $3 \pm \sqrt{5}$ **27.** $5 \pm 4\sqrt{3}$ **29.** $\frac{-5 \pm 2\sqrt{15}}{5}$ **31.** $\frac{2}{3}, -1$ **33.** $\frac{7 \pm \sqrt{85}}{6}$ **35.** $18, -18$ **37.** 3 **39.** $2 \pm \sqrt{4 - c}$ **41.** $b(-2 \pm \sqrt{3})$ **43.** 45 mph, 55 mph **45.** 8.06 **47.** 2042 **49.** $(2, 2)$ **51a.** 224.6 **51b.** 172

Page 749 Chapter 13 Highlights

1. false, $-3 - \sqrt{7}$ **3.** false, $-2 - 2\sqrt{3} - \sqrt{5} - \sqrt{15}$ **5.** true **7.** false, $3x + 19 = x^2 + 6x + 9$ **9.** false, $\frac{x\sqrt{2xy}}{y}$

Pages 750–752 Chapter 13 Study Guide and Assessment

11. 34 **13.** $5\sqrt{5} \approx 11.18$ **15.** 24 **17.** no; $9^2 + 16^2 \neq 20^2$ **19.** yes **21.** $4\sqrt{30}$ **23.** $2|a|b^2\sqrt{11b}$ **25.** $57 - 24\sqrt{3}$ **27.** $\frac{3\sqrt{35} - 5\sqrt{21}}{-15}$ **29.** $2\sqrt{6} - 4\sqrt{3}$ **31.** $36\sqrt{3}$ **33.** $\frac{9\sqrt{2}}{4}$ **35.** 12 **37.** $\frac{26}{7}$ **39.** 45 **41.** 3 **43.** no solution **45.** 17 **47.** $\sqrt{205} \approx 14.32$ **49.** 5 or -1 **51.** 5 or -3 **53.** $\frac{49}{4}$ **55.** $\frac{1}{9}$ **57.** $\frac{7 \pm \sqrt{69}}{2}$ **59.** $-2 \pm \sqrt{6}$ **61.** $22\sqrt{6} \approx 53.9$ **63.** scalene triangle **65.** 5 ft **67.** No, it should skid about 201.7 ft.

PHOTO CREDITS

Cover (t) FPG International, (b) Nawrocki Stock Photo, (background) SuperStock; **iii** (t) courtesy Home of the Hamburger, Inc., (b) Aaron Haupt; **viii** (l) Terry Eiler, (r) Paul L. Ruben; **ix** (t) Geoff Butler, (b) Skip Comer; **x** (t) Stephen Webster, (bl) Aaron Haupt (br) Mark Peterson/Tony Stone Images; **xi** (t) Morton & White Photographic, (b) Tom Tietz; **xii** Stan Fellerman; **xiii** (t) Tom McCarthy/The Stock Market, (c) Geoff Butler, (b) Charles H. Kuhn Collection, The Ohio State University Cartoon, Graphic & Photographic Arts Research Library/Reprinted with special permission of King Features Syndicate; **xiv** (t) Clark James Mishler/The Stock Market, (b) Julian Baum/Science Photo Library/Photo Researchers; **xv** (t) Jonathan Daniel/Allsport USA, (c) The Image Bank/Steve Niedorf, (b) The Image Bank/Jody Dole; **xvi** Tate Gallery, London/Art Resource, NY; **xvii** (t) Morton & White Photographic, (b) NCSA, University of IL/Science Photo Library/ Photo Researchers; **382** (t) Bob Daemmrich/Tony Stone Images, (b) Ron Sheridan/Ancient Art & Architecture Collection; **383** (tl) Phil Matt, (tr) Ken Edward/Science Source/Photo Researchers, (cl) The Bettmann Archive, (cr) courtesy University of TX at Brownsville, (b) Brock May/Photo Researchers; **384, 386** Aaron Haupt; **390** Doug Martin, (background) KS Studio; **392** The Gordon Hart Collection, Bluffton IN; **394** Life Images; **395** Kenji Kerins; **397** Doug Martin; **398** (l) Duomo/Paul Sutton, (r) Mark Lawrence/The Stock Market; **399** Aaron Haupt; **404** (t) Kenji Kerins, (b) Doug Martin; **405** (t) Alain Evrard/Photo Researchers, (c) Nikolas/Konstantinou/Tony Stone Images, (b) Norbert Wu/Tony Stone Images; **407** Aaron Haupt; **411** Pal Hermansen/Tony Stone Images; **413** (t) Suzanne Murphy-Larronde/FPG International, (b) Bill Ross/Westlight; **415** KS Studio, (background) Duomo/William Sallaz; **416** Travelpix/ FPG International; **417** (t) KS Studio, (b) Matt Meadows; **418** (t) Eclipse Studios, (b) Skip Comer; **419** Aaron Haupt; **420** (t) The Image Bank/Joseph Drivas, (b) The Image Bank/Michael Salas; **422** Aaron Haupt; **423** Scott Cunningham; **425** (t) David R. Frazier/ Photo Researchers, (b) MAK-1; **429** Tom Tracy/FPG International; **430** Gary Buss/FPG International; **431** Lawrence Migdale; **432** courtesy The White House; **433** KS Studio; **436** Lori Adamski Peek/Tony Stone Images; **438** Court Mast/FPG International; **440** Kenji Kerins; **442** (tl) Aaron Haupt, (tr) Jane Shemilt/Science Photo Library/Photo Researchers, (b) Photo Researchers; **449** Steve Lissau; **450** (t) David Madison, (bl) Ronald Sheridan/ Ancient Art & Architecture Collection, (br) The Bettmann Archive; **451** (tl) Tom Lynn/Sports Illustrated, (tr) AP Photo/Elise Amendola/Wide World Photos, (c) Reuters/Bettmann, (b) The Bettmann Archive; **455** Mark Lenniham/AP/Wide World Photos; **460** (t) Dinodia, (b) Dinodia/Ravi Shekhar; **461** Matt Meadows; **462** Clark James Mishler/ The Stock Market; **465** (t) Bob Daemmrich/Tony Stone Images, (b) Aaron Haupt; **467** (l) Aaron Haupt, (r) PEANUTS reprinted by permission of United Feature Syndicate, Inc.; **468** (t) The Image Bank/Ronald Johnson, (b) Brian Brake/Photo Researchers; **469** (l) Charles H. Kuhn Collection, The Ohio State University Cartoon, Graphic and Photographic Arts Research Library/Reprinted with Special Permission of King Features Syndicate, (r) Archive Photos; **471** Doug Martin; **473** M. Gerber/LGI Photo Agency; **475** (r) J-L Charmet/Science Photo Library/Photo Researchers; **477** (t) Bob Daemmrich, (b) AP/Wide World Photos; **481** (t) Aaron Haupt, (b) Tim Courlas; **482** Aaron Haupt; **483** James Blank/FPG International; **485** The Image Bank/Nino Mascardi; **486** Elaine Shay; **490** Tracy I. Borland; **491** Aaron Haupt; **494** (t) Reprinted with special permission of King Features Syndicate, (c, b) The Bettmann Archive; **495** (tr) Gary Buss/FPG International, (tl) Richard Lee/Detroit Free Press, (cr) Culver Pictures, (b) The Bettmann Archive; **496** (t) People Weekly © 1995 Keith Philpott, (b) AP/Wide World Photos; **500** (t) Will Crocker/The Image Bank, (b) Doug Martin; **501** Aaron Haupt; **505** (t) Julie Marcotte/Westlight, (t, inset) Aaron Haupt, (b) Kenji Kerins; **506** (t) Aaron Haupt, (b) Mason Morfit/FPG International; **507** David R. Frazier Photolibrary; **508** (t) Julian Baum/Science Photo Library/Photo Researchers, (b) Archive Photos/PA News; **509** (tl) Photo Researchers, (tr) Centers For Disease Control, (b) UPI/Bettmann; **510** (l) Demmie Todd/ Shooting Star, (r) Adam Hart-Davis/Science Photo Library/Photo Researchers; **511** (t) Prof. Aaron Pollack/Science Photo Library/ Photo Researchers, (b) Doug Martin; **512** (t) The Image Bank/Co Rentmeester, (b) The Image Bank/Garry Gay; **514** (t) Tate Gallery, London/Art Resource, N.Y., (bl) The Metropolitan Museum of Art, Gift of Adele R. Levy, 1958/ Malcolm Varon, (br) Erich Lessing/Art Resource, NY; **515** David R. Frazier/Photo Researchers; **518** Gene Frazier; **519** Scott Cunningham; **522** (t) Aaron Haupt, (b) Doug Martin; **526** Bill Ross/Westlight; **527** (l) Richard Hutchings/Photo Researchers, (r) Aaron Haupt; **529** (t) Aaron Haupt, (b) The Image Bank/Jeff Spielman; **531** (t) Robert Tringali/Sportschrome East/West, (b) Mark Gibson/The Stock Market; **533** (t) Eric A. Wessman/Viesti Associates Inc., (b) Charlotte Raymond/Photo Researchers; **536** (t) Jon Blumb, (b) Mark Burnett/Photo Researchers; **540** Steve Elmore/The Stock Market; **541** Aaron Haupt; **542** (t) Larry Lefever from Grant Heilman, (b) Aaron Haupt; **547** Aaron Haupt; **553** (l) Geoff Butler, (r) Morton & White Photographic; **554** Steve Lissau; **555** (l) The Image Bank/Ira Block, (r) David R. Frazier Photolibrary; **556** (t) Culver Pictures, (bl br) The Bettmann Archive; **557** (tl) Everett Collection, (tr) Morton & White Photographic, (cr) Universal/ Shooting Star, (b) UPI/Bettmann; **558** Morton & White Photographic; **559** National Archives; **560** (t) Archive Photos, (b) Morton & White Photographic; **561–562** Morton & White Photographic; **563** (t) Morton & White Photographic, (b) Jeff Shallit; **565** (t) The Image Bank/Barros & Barros, (b) KS Studio; **566** Kenji Kerins; **569** (t) Larry Hamill, (c) Otto Greule/Allsport USA, (b) Duomo/Ben Van Hook; **570** Kenji Kerins; **579** The Image Bank/Derek Berwin; **583** (t) The Bettmann Archive, (b) Ron Sheridan/Ancient Art & Architecture Collection; **586** Brent Turner/BLT Productions; **593** The Image Bank/L.D.Gordon; **594** Piffret/Photo Researchers; **596** The Image Bank/Janeart LTD.; **597** (t) Joe Towers/The Stock Market, (b) Roy Marsch/The Stock Market; **598** The Image Bank/ Steve Niedorf; **599** (l) Chuck Doyle, (r) Archive Photos; **600** Kenji Kerins; **608** (t) Tim Courlas, (c) North Wind Picture Archive, (b) Corbis-Bettmann; **609** (t) Kathleen Cabble/St. Petersburg Times, (c) Archive Photos (bl) UPI/Corbis-Bettmann, (br) Archive Photos/Frank Edwards; **611** Arteaga Photo Ltd.; **613** (l) American Photo Library, (r) Ed & Chris Kumler; **615** Morton & White Photographic; **616** (t) Mark Burnett, (b) Morton & White Photographic; **620** Aaron Haupt; **626** David R. Frazier Photolibrary; **628** (t) Library of Congress, (b) Aaron Rezny/The Stock Market; **631** The Image Bank/Harald Sund; **633** (t) The Image Bank/Marc Grimberg, (b) Morton & White Photographic; **635** Morton & White Photographic; **636** David M. Dennis; **641** Howard Sochurek/The Stock Market; **642** (t) The Image Bank/Jody Dole, (b) courtesy Bob Paz/Caltech; **645** (t) The Image Bank/Michael Melford, (c) James N. Westwater, (b) The Image Bank/Gary Cralle; **646** Culver Pictures; **647** Elaine Shay; **648** (t) The Image Bank/ Herbert Schaible, (b) The Image Bank/Hans Wol; **650** (t) Morton & White Photographic, (bl) John Phelan/The Stock Market, (br) John Maher/The Stock Market; **654** (l) Doug Martin, (r) The Image Bank/Co Rentmeester; **655** (l) Mark Burnett, (c) G. Petrov/Washington Stock Photo, (r) M. Winn/Washington Stock Photo; **657** Tony Freeman/ PhotoEdit; **658** (t) Lou Jacobs Jr./Grant Heilman Photography Inc., (bl) Ron Sheridan/Ancient Art & Architecture Collection, (br) David York/Medichrome/The Stock Shop; **659** (t) courtesy Joe Maktima, (c,background) M. Miller/H. Armstrong Roberts, (c) Lawrence Migdale, (b) Grace Welty; **660** (t) Morton & White Photographic, (b) MAK-1; **662, 663** Morton & White Photographic; **664** (t) Ron Watts/Westlight, (b) Doug Martin; **665** (t) Ned Gillette/The Stock Market, (b) Lynn Campbell; **667** Benjamin Rondel/The Stock Market; **669** (t) KS Studio, (b) Morton & White Photographic; **670** (t) Morton & White Photographic, (b) Bob Mullenix; **671** (l) Stephen Dunn/Allsport USA, (r) Brett Palmer/The Stock Market; **672** (l) Mark Gibson/The Stock Market, (r) Larry Hamill; **674** Aaron Haupt; **675, 677** KS Studio; **680** (t) Ward's Natural Science, (bl br) Aaron Haupt; **681** The Bettmann Archive; **684** Dave Merrill & Co./National Geographic Society; **685** (br) Caltech & Carnegie Institution, (others) NASA; **687** Roy Morsch/The Stock Market; **688** (t) UPI/Bettmann, (b) Morton & White Photographic; **689** (t) The Image Bank/GK & Vikki Hart, (b) Morton & White Photographic; **690** Morton & White Photographic; **694** The Image Bank/Andrew Unangst; **696** Kenji Kerins; **697** (l) Patrick W. Stoll/Westlight, (r) Morton & White Photographic; **698** Doug Martin; **701** (l) Morton & White Photographic, (r) The Image Bank/Guido Alberto Rossi; **702** Archive Photos; **706** "Dennis The Menace" used by permission of Hank Ketcham and North American Syndicate; **707** Morton & White Photographic; **710** (t) Glencoe photo, (c) Ron Sheridan/ Ancient Art & Architecture Collection, (b) Matt Meadows; **711** (tl) Annie Griffiths Belt, (tr) Doug Martin, (cl) The Bettmann Archive, (cr) Archive Photos/Reuters/Peter Jones, (b) Lucille Day/Black Star; **713** Stock Newport/Daniel Forster; **715** (t) Ulrike Welsch/Photo Researchers, (b) Glencoe photo; **717** David R. Frazier; **718** Archive Photos/David Lees;

719 W. Warren/Westlight; **720** Doug Martin; **721** Gerard Photography; **722** PEANUTS reprinted by permission of United Features Syndicate, Inc.; **723** Bob Daemmrich; **724** (l) The Image Bank/Flip Chalfant, (r) Glencoe Photo; **725** KS Studio; **727** (t) Tom Bean/The Stock Market, (b) D. Logan/H. Armstrong Roberts; **728** The Image Bank/William Kennedy; **730** K. Scholz/H. Armstrong Roberts; **732** (t) The Image Bank/Kaz Mori, (c) DLR, ESA/Science Photo Library/Photo Researchers, (b) Greg Davis/The Stock Market; **736** (t) The Image Bank/Joe Azzara, (b) KS Studio; **739** Ken Davies/Masterfile; **740** (l) Glencoe photo, (r) NCSA, University of IL/Science Photo Library/Photo Researchers; **743** (t) David R. Frazier Photolibrary, (c) Dallas & John Heaton/Westlight, (b) Wong How Man/Westlight; **746** Doug Martin; **747** Sylvain Grandadam/Photo Researchers; **748** David R. Frazier Photolibrary; **752** The Walt Disney Co./Shooting Star; **753** Aaron Haupt.

APPLICATIONS & CONNECTIONS INDEX

This index contains terms from Volumes One and Two. Pages 1–447 are contained in Volume One. Pages 382–753 are contained in Volume Two.

C

E

F

I

INDEX

This index contains terms from Volumes One and Two. Pages 1–447 are contained in Volume One. Pages 382–753 are contained in Volume Two.

A

F

G

H

I

L

M

N

O

Q

R

S

T